FUNDAMENTAL AND PHYSICAL CONSTANTS

quantity	symbol	English	SI
Charge			
electron	e		-1.6022×10^{-19} C
proton	p		$+1.6021 \times 10^{-19}$ [illegible]
Density			
air [STP]		0.0805 lbm/ft^3	1.29 [illegible]
air [70°F, (20°C), 1 atm]		0.0749 lbm/ft^3	1.20 k[illegible]
earth [mean]		345 lbm/ft^3	5520 kg[illegible]
mercury		849 lbm/ft^3	1.360 × [illegible]
sea water		64.0 lbm/ft^3	1025 kg/[illegible]
water [mean]		62.4 lbm/ft^3	1000 kg/m[illegible]
Distance [mean]			
earth radius		2.09×10^{7} ft	6.370×10^{6} m
earth-moon separation		1.26×10^{9} ft	3.84×10^{8} m
earth-sun separation		4.89×10^{11} ft	1.49×10^{11} m
moon radius		5.71×10^{6} ft	1.74×10^{6} m
sun radius		2.28×10^{9} ft	6.96×10^{8} m
first Bohr radius	a_0	1.736×10^{-10} ft	5.292×10^{-11} m
Gravitational Acceleration			
earth [mean]	g	32.174 (32.2) ft/sec^2	9.8067 (9.81) m/s^2
moon [mean]		5.47 ft/sec^2	1.67 m/s^2
Mass			
atomic mass unit	u	3.66×10^{-27} lbm	1.6606×10^{-27} kg
earth		4.11×10^{23} slugs	6.00×10^{24} kg
earth [customary U.S.]		1.32×10^{25} lbm	n.a.
electron [rest]	m_e	2.008×10^{-30} lbm	9.109×10^{-31} kg
moon		1.623×10^{23} lbm	7.36×10^{22} kg
neutron [rest]	m_n	3.693×10^{-27} lbm	1.675×10^{-27} kg
proton [rest]	m_p	3.688×10^{-27} lbm	1.673×10^{-27} kg
sun		4.387×10^{30} lbm	1.99×10^{30} kg
Pressure, atmospheric		14.696 (14.7) lbf/in^2	1.0133×10^{5} Pa
Temperature, standard		32°F (492°R)	0°C (273K)
Velocity			
earth escape		3.67×10^{4} ft/sec	1.12×10^{4} m/s
light [vacuum]	c	9.84×10^{8} ft/sec	2.9979 (3.00) $\times 10^{8}$ m/s
sound [air, STP]	a	1090 ft/sec	331 m/s
[air, 70°F (20°C)]		1130 ft/sec	344 m/s
Volume, molal ideal gas [STP]		359 ft^3/lbmol	22.41 m^3/kmol
Fundamental Constants			
Avogadro's number	N_A		6.022×10^{23} mol^{-1}
Bohr magneton	μ_B		9.2732×10^{-24} J/T
Boltzmann constant	k	5.65×10^{-24} ft-lbf/°R	1.3807×10^{-23} J/K
Faraday constant	F		96 487 C/mol
gravitational constant	g_c	32.174 lbm-ft/lbf-sec^2	
gravitational constant	G	3.44×10^{-8} ft^4/lbf-sec^4	6.672×10^{-11} N·m^2/kg^2
nuclear magneton	μ_N		5.050×10^{-27} J/T
permeability of a vacuum	μ_0		1.2566×10^{-6} N/A^2 (H/m)
permittivity of a vacuum	ϵ_0		8.854×10^{-12} C^2/N·m^2 (F/m)
Planck's constant	h		6.6256×10^{-34} J·s
Rydberg constant	R_∞		1.097×10^{7} m^{-1}
specific gas constant, air	R	53.3 ft-lbf/lbm-°R	287 J/kg·K
Stefan-Boltzmann constant		1.71×10^{-9} BTU/ft^2-hr-°R^4	5.670×10^{-8} W/m^2·K^4
triple point, water		32.02°F, 0.0888 psia	0.01109°C, 0.6123 kPa
universal gas constant	R^*	1545 ft-lbf/lbmol-°R	8314 J/kmol·K
	R^*	1.986 BTU/lbmol-°R	

MATHEMATICAL CONSTANTS

symbol	name	value (rounded)
π	Archimedes number (pi)	3.14159 26536
e	base of natural logs	2.71828 18285
C or τ	Euler constant	0.57721 56649
G	Catalan constant	1.64493 69032

ABOUT THE  SYMBOL

The *EXAM USE* symbol you see on the cover is Professional Publications' way of indicating that this book meets the seven criteria listed below. Professional Publications puts this symbol only on its reference manuals with these characteristics, and does so voluntarily.

1. **This book is primarily concept- and theory-oriented.** Less than 10 percent of the space in this book contains solved problems. The remaining space contains explanations, theory, and data presented at the college level or higher.

2. **This book does not use old examination problems as source material.** Solved example problems in this book are used to illustrate isolated concepts, not to illustrate typical exam problems. The format of examples and practice problems in this book does not imitate that of examination problems.

3. **This book safeguards examination security.** This book contains a highly visible statement that copying examination problems is not permitted. This book discourages problem copying and other subversive activities. All significant blank space is blocked out to discourage you from using that space to copy examination problems.

4. **This book cannot be used to find similar problems.** Solved examples are not indexed, listed, or categorized for fast access. Subjects and explanations are organized logically by theory, not by problem type. Solved examples are spread throughout each chapter, not consolidated for rapid reference.

5. **This book is ethical in authorship.** All authors and contributors are listed, and no pseudonyms are used. Information about examination subjects was obtained from public and published sources. No subversive methods were used.

6. **This book is ethical in presentation.** Primary sources of nonoriginal data and other material are acknowledged. No references are made to actual exam problems. Acceptability to, approval of, or affiliation with the National Council of Examiners for Engineering and Surveying is not claimed. No inflated claims or guarantees about passing the exam related to this book are made.

7. **This is a bona fide book.** It contains standard book elements, such as a title page identifying the author and publisher, a copyright, and an index. This book is permanently bound in softcover, hardcover, or wire format.

Each state's engineering registration board establishes its own rules. The *EXAM USE* symbol does not guarantee you will be able to use this book in an examination. Books from Professional Publications and other publishers without the *EXAM USE* symbol might also be permitted in an examination. Contact your state's engineering licensing board for more information.

IMPORTANT NAMES, DATES AND ADDRESSES

This book belongs to

____

____

____

phone ______________________________

examination date: ____________ hours: ________

examination location: ______________________________

____

____

____

tape your cancelled check here

phone number of your registration board: ______________________________

address of your registration board: ______________________________

____

____

names of contacts at your registration board: ______________________________

____

____

tape your proof of mailing here

date you sent your application: ______________________________

registered/certified mail receipt number: ____________________

date confirmation was received: ______________________________

names of examination proctors: ______________________________

____

____

tape dime here

booklet number: ____________ (A.M.) ____________ (P.M.)

tape dime here

problems you disagreed with on the examination

problem no. reason

____

____

____

CHEMICAL ENGINEERING REFERENCE MANUAL

Fourth Edition

Randall N. Robinson, P.E.

PROFESSIONAL PUBLICATIONS, INC.
Belmont, CA 94002

In the ENGINEERING REFERENCE MANUAL SERIES

Engineer-In-Training Reference Manual
Engineering Fundamentals Quick Reference Cards
Engineer-In-Training Sample Examinations
Mini-Exams for the E-I-T Exam
1001 Solved Engineering Fundamentals Problems
E-I-T Review: A Study Guide
Diagnostic F.E. Exam for the Macintosh
Fundamentals of Engineering Video Series: Thermodynamics
Civil Engineering Reference Manual
Civil Engineering Quick Reference Cards
Civil Engineering Sample Examination
Civil Engineering Review Course on Cassettes
101 Solved Civil Engineering Problems
Seismic Design of Building Structures
Seismic Design Fast
Timber Design for the Civil P.E. Exam
Fundamentals of Reinforced Masonry Design
246 Solved Structural Engineering Problems
Mechanical Engineering Reference Manual
Mechanical Engineering Quick Reference Cards
Mechanical Engineering Sample Examination
101 Solved Mechanical Engineering Problems
Mechanical Engineering Review Course on Cassettes
Consolidated Gas Dynamics Tables
Fire and Explosion Protection Systems
Electrical Engineering Reference Manual
Electrical Engineering Quick Reference Cards
Electrical Engineering Sample Examination
Chemical Engineering Reference Manual
Chemical Engineering Quick Reference Cards
Chemical Engineering Practice Exam Set
Land Surveyor Reference Manual
1001 Solved Surveying Fundamentals Problems
Land Surveyor-In-Training Sample Examination
Engineering Economic Analysis
Engineering Law, Design Liability, and Professional Ethics
Engineering Unit Conversions

In the ENGINEERING CAREER ADVANCEMENT SERIES

How to Become a Professional Engineer
The Expert Witness Handbook—A Guide for Engineers
Getting Started as a Consulting Engineer
Intellectual Property Protection—A Guide for Engineers
E-I-T/P.E. Course Coordinator's Handbook
Becoming a Professional Engineer
Engineering Your Start-Up
High-Technology Degree Alternatives
Metric in Minutes

CHEMICAL ENGINEERING REFERENCE MANUAL
Fourth Edition

Printed in the United States of America

ISBN: 0-932276-75-X

Professional Publications, Inc.
1250 Fifth Avenue, Belmont, CA 94002
(415) 593-9119

Current printing of this edition: 6

Revised and reprinted in 1994.

TABLE OF CONTENTS

4 THERMODYNAMICS

5 FLUID STATICS AND DYNAMICS

6 HEAT TRANSFER: CONDUCTION AND RADIATION

7 HEAT TRANSFER: CONVECTION AND EQUIPMENT

8 VAPOR-LIQUID PROCESSES

PREFACE

The fourth edition has been extensively revised and reformatted. With new material, including new tables and figures for most of the chapters, this edition reflects the latest NCEE Professional Engineering examination format.

Many examples and practice problems have been included to illustrate the problem types that will be encountered in the exam. By using this book for home study or as a text in a review course, you will be well prepared for the NCEE Chemical Engineering examination for a Professional Engineering license.

Recently, as few as 15% of all applicants to the Chemical Engineering portion of the NCEE examination have passed; whereas ten years ago, the Chemical Engineering passing rate was 65%. This downward trend is attributed to the move toward more practical rather than theoretical problems on the exam.

The standards and methods of grading the exam problems have moved away from emphasis on the numerical answers and toward emphasis on assumptions and methods of solution. I have told all of my students to be sure that they have listed every assumption and reference at the beginning of each solution. In this manner, the criteria for solutions predetermined by the examiners will be met.

Randall N. Robinson, P.E.
San Jose, CA

PROFESSIONAL PUBLICATIONS, INC. ● Belmont, CA

ACKNOWLEDGMENTS

page 2-9
Figure 2.4, "Mean Specific Heats of Gases," reprinted with permission from *Chemical Process Principles*, Part One, 2nd ed., O.A. Hougen, K.M. Watson, and R.A. Ragatz, John Wiley & Sons, 1954

page 2-10
Table 2.1, "Combustion Data for Gases," reprinted with permission from *Steam: Its Generation and Use*, 39th ed., Babcock & Wilcox, 1978

page 4-25
Appendix A, "Nelson-Obert Generalized Compressibility Chart: Low Pressure Range," reprinted with permission from *Standards of Tubular Exchanger Manufacturers Association*, 6th ed., 1978

page 4-26
Appendix B, "Nelson-Obert Generalized Compressibility Chart: Intermediate Pressure Range," reprinted with permission from *Standards of Tubular Exchanger Manufacturers Association*, 6th ed., 1978

page 4-27
Appendix C, "Nelson-Obert Generalized Compressibility Chart: High Pressure Range," reprinted with permission from *Standards of Tubular Exchanger Manufacturers Association*, 6th ed., 1978

pages 4-28 to 4-30
Appendix D, "Saturated Steam: Temperature," reprinted with permission from *ASME Steam Tables*, 5th ed., by the American Society of Mechanical Engineers, 1983

page 4-31
Appendix E, "Saturated Steam: Pressure," reprinted with permission from *ASME Steam Tables*, 5th ed., by the American Society of Mechanical Engineers, 1983

pages 4-32 to 4-34
Appendix F, "Superheated Steam," reprinted with permission from *ASME Steam Tables*, 5th ed., by the American Society of Mechanical Engineers, 1983

pages 4-35 to 4-41
Appendix G, "Air Table," reprinted with permission from *Gas Tables: Thermodynamic Properties of Air*, J.H. Keenan, F.G. Keyes, and J. Chao, John Wiley & Sons, Inc., 1980

pages 4-42 to 4-49
Appendix H, "Heats and Free Energies of Formation," reprinted with permission from *Chemical Engineers Handbook*, 5th ed., J. Perry, McGraw-Hill, Inc., 1973

pages 4-50 to 4-52
Appendix I, "Heats of Combustion," data taken from *Selected Values of Properties of Hydrocarbons*, National Bureau of Standards Circular C461, November, 1947, reprinted through courtesy of the American Petroleum Institute

page 4-53
Appendix J, "Specific Heats of Miscellaneous Materials," reprinted with permission from *Chemical Engineers Handbook*, 5th ed., J. Perry, McGraw-Hill, Inc., 1973

page 4-54
Appendix K, "Mollier Chart," reprinted with permission from *Steam: Its Generation and Use*, 39th ed., Babcock & Wilcox, 1978

page 5-9
Figure 5.8, "Moody Friction Factor Chart," reprinted with permission from *Friction Factors for Pipe Flow*, L.F. Moody, ASME Transactions, Vol. 66, No. 8, American Society of Mechanical Engineers, 1944

page 5-22
Tables 5.9 and 5.10, "Properties of Water at Atmospheric Pressure," and "Properties of Air at Atmospheric Pressure," data derived with permission from *ASCE Manual of Engineering Practices*, No. 25, American Society of Civil Engineers, 1942

page 5-23
Table 5.11, "Equivalent Length of Straight Pipe for Various Fittings," reprinted with permission from *De*

Laval Engineering Handbook, 3rd ed., Hans Gartmann, McGraw-Hill, Inc., 1970

page 5-26
Appendix C, "Properties of Liquids," reprinted with permission from *Cameron Hydraulic Data*, 15th ed., Ingersoll Rand Co., 1977

pages 5-27 to 5-30
Appendix D, "Dimensions of Welded and Seamless Steel Pipe," extracted from *American Standard Wrought Steel and Wrought Iron Pipe* (ASA B36, 10–1959), American Society of Mechanical Engineers

page 5-31
Appendix E, "Dimensions of Copper Water Tubing," reprinted with permission from *Heating and Air Conditioning*, B.H. Jennings, International Textbook Publishers, 1956

page 5-32
Appendix F, "Dimensions of Brass and Copper Tubing," reprinted with permission from *Heating and Air Conditioning*, B.H. Jennings, International Textbook Publishers, 1956

page 5-33
Appendix G, "Specific Gravity of Hydrocarbons," reprinted with permission from *Cameron Hydraulic Data*, 15th ed., Ingersoll Rand Co., 1977

page 5-35
Appendix I, "Net Expansion Factor, Y for Compressible Flow Through Nozzles and Orifices," data extracted from *Fluid Meters, Their Theory and Application*, 4th ed., 1937, and *Orifice Meters with Supercritical Flow*, R.G. Cunningham, reprinted with permission from the American Society of Mechanical Engineers.

page 6-2
Figures 6.1 and 6.2 "Thermal Conductivities of Some Good Conductors," and "Thermal Conductivities of Some Poor Conductors," reprinted with permission from *Process Heat Transfer*, D.Q. Kern, McGraw-Hill, Inc., 1950

page 6-11
Figure 6.7, "Transient Cooling Charts for Solid Objects," reprinted with permission from *Chemical Engineers Handbook*, 5th ed., J. Perry, McGraw-Hill, Inc., 1973

page 7-11
Table 7.7, "Characteristics of Tubing," reprinted with permission from *Standards of Tubular Exchanger Manufacturers Association*, 6th ed., 1978

page 7-13
Tables 7.8 and 7.9, "Typical Overall Heat Transfer Coefficients in Tubular Heat Exchangers," and "Typical Overall Heat Transfer Coefficients in Refinery Service," reprinted with permission from *Chemical Engineers Handbook*, 5th ed., J. Perry, McGraw-Hill, Inc., 1973

pages 7-14 to 7-21
Figure 7.8, "Correction Factors for Multiple-Pass Heat Exchangers," reprinted with permission from *Standards of Tubular Exchanger Manufacturers Association*, 6th ed., 1978

page 7-22
Figure 7.9, "Tube-Side Heat Transfer Curve," reprinted with permission from *Process Heat Transfer*, D.Q. Kern, McGraw-Hill, Inc., 1950

page 7-23
Figure 7.10, "Tube-Side Water-Heat Transfer Curve," reprinted with permission from *Process Heat Transfer*, D.Q. Kern, McGraw-Hill, Inc., 1950

page 7-25
Figure 7.13, "Colburn Plot for Multiple and Finned Tube Exchangers," reprinted with permission from *Process Heat Transfer*, D.Q. Kern, McGraw-Hill, Inc., 1950

page 7-27
Figure 7.15, "Condensing Coefficients," reprinted with permission from *Process Heat Transfer*, D.Q. Kern, McGraw-Hill, Inc., 1950

page 7-28
Figure 7.16, "Logarithmic Mean Temperature Differences," reprinted with permission from *Standards of Tubular Exchanger Manufacturers Association*, 6th ed., 1978

page 8-3
Figure 8.1, "Vapor Pressure of Normal Paraffin Hydrocarbons and Water (Cox Chart)," reprinted with permission from *Volumetric and Phase Behavior of Hydrocarbons*, Bruce Sage and William Lacey, Gulf Publishing Co.

page 8-5
Figure 8.2,"Vapor-Liquid Equilibrium Constants for Light Hydrocarbons (Low Temperature)," from *American Institute of Chemical Engineers Symposium Series, Vol. 47*, No. 8, C.L. DePriester, reproduced by permission of the American Institute of Chemical Engineers, 1951–1953

page 8-6
Figure 8.3, "Vapor-Liquid Equilibrium Constants for Light Hydrocarbons (Low Temperature)," from *American Institute of Chemical Engineers Symposium Series, Vol. 47*, No. 8, C.L. DePriester, reproduced by permission of the American Institute of Chemical Engineers, 1951–1953

page 8-6
Table 8.2, "Force Constants $(\frac{\varepsilon}{k})$ and Collision Diameters (σ) for Various Gases," data derived with permission from *Transport Phenomena*, Bird, Stewart, & Lightfoot, John Wiley & Sons, Inc.

page 8-9
Figure 8.5, "Plot of Ω as a Function of $\frac{kT}{\varepsilon}$," data derived with permission from *Transport Phenomena*, Bird, Stewart,& Lightfoot, John Wiley & Sons, Inc.

page 8-12
Figure 8.9, "Number of Units in an Absorption Column for Constant $\frac{mG_M}{L_M}$ (Colburn Chart)," reprinted with permission from *Chemical Engineers Handbook*, 5th ed., J. Perry, McGraw-Hill, Inc., 1973

page 8-17
Figure 8.13, "Kremser-Brown Absorption Factors," reprinted with permission from *Mass Transfer Operations*, 2nd ed., R.E. Treybal, McGraw-Hill, Inc.

page 9-15
Figure 9.11, "Psychrometric Chart," Catalog No. 794-002 (AC467) reproduced by permission from the Carrier Corporation, 1982

page 11-7
Table 11.3, "Integrated Forms of the Constant Volume, Irreversible Batch Reactor," data derived from *Chemical Reaction Engineering: An Introduction to the Design of Chemical Reactors*, Octave Levenspiel, John Wiley & Sons, Inc., 1962

page 11-8
Table 11.4, "Integrated Forms of the Variable Volume, Irreversible Batch Reactor," data derived from *Chemical Reaction Engineering: An Introduction to the Design of Chemical Reactors*, Octave Levenspiel, John Wiley & Sons, Inc., 1962

page 11-9
Table 11.5, "Integrated Forms of the Constant Volume, Reversible Batch Reactor," data derived from *Chemical Reaction Engineering: An Introduction to the Design of Chemical Reactors*, Octave Levenspiel, John Wiley & Sons, Inc., 1962

page 11-17
Table 11.7, "Performance Equations of the Constant Volume, Irreversible CSTR," data derived from *Chemical Reaction Engineering: An Introduction to the Design of Chemical Reactors*, Octave Levenspiel, John Wiley & Sons, Inc., 1962

page 11-18
Table 11.8, "Performance Equations of the Variable Volume, Irreversible CSTR," data derived from *Chemical Reaction Engineering: An Introduction to the Design of Chemical Reactors*, Octave Levenspiel, John Wiley & Sons, Inc., 1962

page 11-19
Tables 11.9 and 11.10, "Integrated Forms of the Constant Volume, Irreversible PFR," and "Integrated Forms of the Variable Volume, Irreversible PFR," data derived from *Chemical Reaction Engineering: An Introduction to the Design of Chemical Reactors*, Octave Levenspiel, John Wiley & Sons, Inc., 1962

OUTLINE OF SUBJECTS FOR SELF-STUDY

Week	Subject	Date to be Started	Date to be Completed	Check When Completed
1	Mathematics	________	________	☐
2	Stoichiometry, Heat and Material Balances	________	________	☐
3	Engineering Economic Analysis	________	________	☐
4	Thermodynamics	________	________	☐
5	Fluid Statics and Dynamics	________	________	☐
6	Heat Transfer: Conduction and Radiation	________	________	☐
7	Heat Transfer: Convection and Equipment	________	________	☐
8	Vapor-Liquid Processes	________	________	☐
9	Distillation, Evaporation, and Humidification	________	________	☐
10	Liquid-Liquid and Solid-Liquid Processes	________	________	☐
11	Kinetics	________	________	☐

INTRODUCTION

Purpose of Registration

As an engineer, you may choose to obtain your professional engineering license. This is done through procedures which have been established by the state in which you reside. These procedures are designed to protect the public by preventing unqualified individuals from legally practicing as engineers.

There are many reasons for wanting to become a professional engineer. Some of the reasons include:

- You may wish to become an independent consultant. By law, consulting engineers must be registered.
- Your company might require a professional engineering license as a requirement for employment or advancement.
- Your state might require registration as a professional engineer if you use the title *engineer*.
- Your state might require registration as a professional engineer before you can testify in court as an "expert" in an engineering matter.

The Registration Process

The registration procedure is similar in most states. You probably will take two 8-hour written examinations. The first examination is the *Engineer-in-Training* examination, also known as the *Intern Engineer* exam and the *Fundamentals of Engineering* exam. The initials E-I-T, I.E., and F.E. are also used. The second examination is the *Professional Engineer* (P.E.) exam, which differs from the E-I-T exam in format and content.

If you have significant experience in engineering, you may be allowed to skip the E-I-T examination. However, actual details of registration requirements, experience requirements, minimum education levels, fees, and examination schedules vary from state to state. You should contact your state's Board of Registration for Professional Engineers for details of registration within your state.

Reciprocity Among States

All states use the NCEES P.E. examination.[1] If you pass the P.E. examination in one state, your certificate will probably be honored by other states which have used the NCEES examination. It will not be necessary to retake the P.E. examination.

The simultaneous administration of identical examinations in multiple states has led to the term *Uniform Examination*. However, each state is free to choose its own minimum passing score or to add special questions to the NCEES examination. Therefore, this Uniform Examination does not automatically ensure reciprocity among states.

Of course, you may apply for and receive a professional engineering license from another state. However, a license from one state will not permit you to practice engineering in another state. You must have a professional engineering license in each state in which you work.

Applying for the Examination

Each state charges different fees, requires different qualifications, and uses different forms. Therefore, it will be necessary for you to request an application and an information packet from the state in which you reside or in which you plan to take the exam. It generally is sufficient to phone for this information. Telephone numbers for all of the U.S. state boards of registration are given below.

[1] The National Council of Examiners for Engineering and Surveying (NCEES) in Clemson, South Carolina produces, distributes, and grades the national examinations. It does not distribute applications to take the P.E. examination.

Phone Numbers of State Boards of Registration

Alabama	(205) 261-5568
Alaska	(907) 465-2540
Arizona	(602) 255-4053
Arkansas	(501) 371-2517
California	(916) 920-7466
Colorado	(303) 866-2396
Connecticut	(203) 566-3386
Delaware	(302) 656-7311
District of Columbia	(202) 727-7454
Florida	(904) 488-9912
Georgia	(404) 656-3926
Guam	(671) 646-1079
Hawaii	(808) 548-4100
Idaho	(208) 334-3860
Illinois	(217) 782-8556
Indiana	(317) 232-2980
Iowa	(515) 281-5602
Kansas	(913) 296-3053
Kentucky	(502) 564-2680
Louisiana	(504) 568-8450
Maine	(207) 289-3236
Maryland	(301) 333-6322
Massachusetts	(617) 727-3055
Michigan	(517) 335-1669
Minnesota	(612) 296-2388
Mississippi	(601) 359-6160
Missouri	(314) 751-2334
Montana	(406) 444-4285
Nebraska	(402) 471-2021
Nevada	(702) 789-0231
New Hampshire	(603) 271-2219
New Jersey	(201) 648-2660
New Mexico	(505) 827-7316
New York	(518) 474-3846
North Carolina	(919) 781-9499
North Dakota	(701) 258-0786
Ohio	(614) 466-8948
Oklahoma	(405) 521-2874
Oregon	(503) 378-4180
Pennsylvania	(717) 783-7049
Puerto Rico	(809) 722-2121
Rhode Island	(401) 277-2565
South Carolina	(803) 734-9166
South Dakota	(605) 394-2510
Tennessee	(615) 741-3221
Texas	(512) 440-7723
Utah	(801) 530-6628
Vermont	(802) 828-2363
Virgin Islands	(809) 774-3130
Virginia	(804) 367-8512
Washington	(206) 753-6966
West Virginia	(304) 348-3554
Wisconsin	(608) 266-1397
Wyoming	(307) 777-6155

Examination Format

The NCEES Professional Engineering examination in Chemical Engineering consists of two 4-hour sessions separated by a 1-hour lunch period. Both the morning and the afternoon sessions contain ten problems. Most states do not have required problems.

Each examinee is given an exam booklet which contains problems for civil, mechanical, electrical, and chemical engineers. Some states, such as California, will only allow you to work problems from the chemical part of the booklet. Other states, such as New York, will allow you to work problems from the entire booklet. Read the examination instructions carefully on this point.

Based on previous examinations, a typical exam would contain problems from the following areas:

Subjects	Estimated Number of Problems
Thermodynamics	2
Separations	2
Mass Transfer	6
Heat Transfer	4
Kinetics	3
Fluids	2
Economics	1

Since the examination structure is not rigid, it is not possible to give the exact number of problems that will appear in each subject area. Only economic analysis can be considered a permanent part of the examination.

The examination is open book. Usually, all forms of solution aids are allowed in the examination, including monographs, specialty slide rules, and pre-programmed and programmable calculators. Since their use says little about the depth of your knowledge, such aids should be used only to check your work.

Most states do not limit the number and types of books you can bring into the exam.[2] Loose-leaf papers, including Post-it Notes™ and writing tablets, are usually forbidden, although you may be able to bring in loose reference papers in a three-ring binder. References used in the afternoon session need not be the same as for the morning session.

Any battery-powered, silent calculator may be used. There are usually no restrictions on programmable or preprogrammed calculators. Printers cannot be used.

You will not be permitted to share books, calculators, or any other items with other examinees.

---

[2] Check with your state to see if review books can be brought into the examination. Most states do not have any restrictions. Some states ban only collections of solved problems, such as Schaum's Outline Series. A few prohibit all review books.

You will receive the results of your examination by mail. Allow 12-14 weeks for notification. Your score may or may not be revealed to you, depending on your state's procedure.

Examination Dates

The NCEES examinations are administered on the same weekend in all states. Each state decides independently whether to offer the examination on Thursday, Friday, or Saturday of the examination period. Upcoming examination dates are given in the table below:

Year	Spring Exam	Fall Exam
1990	April 19-21	October 25-27
1991	April 11-13	October 24-26
1992	April 9-11	October 29-31
1993	April 15-17	October 28-30
1994	April 14-16	October 27-29
1995	April 6-8	October 26-28

Objectively Scored Problems

Objectively scored problems were added to the April 1988 professional engineering exam. Such problems appear in multiple-choice, true/false, and data-selection formats.

The single remaining economics problems on the exam is one of the problems that has been converted to objective scoring. In addition, NCEES has specifically targeted the following subjects for objective scoring:

- fluids
- heat transfer
- mass transfer
- thermodynamics

Objectively scored problems appear in both the morning and afternoon parts of the examination. In all, 50% (5 problems) of each session is objectively scored.

Grading the Examination

Full credit is achieved by correctly working four problems in the morning and four problems in the afternoon. You may not claim credit for more than a total of eight worked problems or for more than four worked problems per session. All solutions are recorded in official solution booklets.

At its August 13, 1983, board meeting, NCEES adopted a new method of grading its professional engineering exams. The minutes of that meeting include the following paragraph.

> "The following is the method for determining the recommended passing standard for an applicant for the P.E. examination, effective April 1984. The Principles and Practice of Engineering Examination's recommended passing standard will be established as a minimum raw score of 48 based on eight questions."

NCEES has decided to name the new grading method the *Criterion-Referenced Method* to distinguish it from the old *Norm-Referenced Method.* The criteria are the specific elements you must include in your solution to receive credit for your solution. The criteria are determined in advance, prior to the administration of the examination.

Getting ten points on a problem requires solving it correctly to completion and making no mathematical errors in the solution. For each mathematical error, a point or two would be lost from the problem's score.

This grading method has been validated by correlating passing rates from previous examinations graded by both methods. (Previous examinations, however, used the old grading method.) It is not expected that passing percentages will change markedly. Only examinees with marginal scores will be affected.

Each state is still free to specify its own cut-off score and passing requirements. NCEES's recommended method or score does not have to be used (although most states do).

Preparing for the Exam

You should develop an examination strategy early in the preparation process. This strategy will depend on your background. One of the following two general strategies is recommended:

- A broad approach has been successful for examinees who have recently completed academic studies. Their strategy has been to review the fundamentals of a broad range of undergraduate chemical engineering subjects. The examination includes enough fundamental problems to give merit to this strategy.
- Working engineers who have been away from classroom work for a long time have found it much better to concentrate on the subjects in which they have had extensive professional experience. By studying the list of examination subjects, they have been able to choose those areas which will give them a good probability of finding enough problems that they can solve.

Do not make the mistake of studying only a few subjects in hopes of finding enough problems to work. The more subjects you are familiar with, the better your chances will be of passing the examination. More important than strategy are fast recall and stamina. You must be able to quickly recall solution procedures, formulas, and important data; and this sharpness must be maintained for eight hours.

In order to develop this recall and stamina, you should work the sample problems at the end of each chapter and compare your answers to the solutions provided. This will enable you to become familiar with problem types and solution methods. You will not have time in the exam to derive solution methods; you must know them instinctively.

It is imperative that you develop and adhere to a review outline and schedule. If you are not taking a classroom review course where order of preparation is determined by the lectures, you should use the accompanying *Outline of Subjects for Self-Study* to schedule your preparation.

It is unnecessary to take a large quantity of books to the examination. This book, a dictionary, and one or two other references of your choice should be sufficient. The examination is very fast-paced. You will not have time to look up solution procedures, data, or equations with which you are not familiar. Although the examination is open-book, there is insufficient time to use books with which you are not thoroughly familiar.

To minimize time spent searching for often-used formulas and data, you should prepare a one-page summary of all important formulas and information in each subject area. You can then use these summaries during the examination instead of searching for the correct page in your book.

What to do Before the Exam

Engineers who have taken the P.E. exam in previous years have made the suggestions listed below. These suggestions will make your examination experience as comfortable and successful as possible.

- Keep a copy of your examination application. Send the original application by certified mail and request a receipt of delivery. Tape your delivery receipt on the first page of this book.
- Visit the exam site the day before your examination. This is especially important if you are not familiar with the area. Find the examination room, the parking area, and the rest rooms.
- Plan on arriving at least 30 minutes before the examination starts. This will assure you a convenient parking place and adequate time for site, room, and seating changes.
- If you live a considerable distance from the examination site, consider getting a hotel room in which to spend the night before.
- Take off the day before the examination to relax. Don't cram the last night. Rather, get a good night's sleep.
- Be prepared to find that the examination room is not ready at the designated time. Take an interesting novel or magazine to read in the interim and at lunch.
- If you make arrangements for baby sitters or transportation, allow for a delayed completion.
- Prepare your examination kit the day before. Here is a checklist of items to take with you to the examination.

[ ] copy of your application

[ ] proof of delivery receipt

[ ] letter admitting you to the exam

[ ] photographic identification

[ ] reference books

[ ] course notes in a three-ring binder

[ ] calculator and a spare

[ ] spare calculator batteries or battery pack

[ ] battery charger and 20′ extension cord

[ ] chair cushions (a large, thick bath mat works well)

[ ] earplugs

[ ] desk expander – if you are taking the exam in theater chairs with tiny, fold-up writing surfaces, you should take a long, wide board to place across the arm rests

[ ] a cardboard box to use as a bookcase to set on your desk

[ ] twist-to-advance pencils with extra leads

[ ] large eraser

[ ] snacks such as raisins, nuts, or trail mix

[ ] beverage in a thermos

[ ] a light lunch

[ ] a collection of graph paper

[ ] scissors, stapler, and staple remover

[ ] construction paper for stopping drafts and sunlight

[ ] transparent and masking tapes

[ ] sunglasses

[ ] prescription glasses, if you wear them

[ ] aspirin

[ ] travel pack of Kleenex

[ ] Webster's Dictionary

[ ] dictionary of scientific terms

[ ] $10.00 in change for parking meters, candy machines, and pay toilets

[ ] a light comfortable sweater

[ ] raincoat, boots, gloves, hat, and umbrella if there is any chance of inclement weather

[ ] local street maps

[ ] note to the parking patrol for your windshield

[ ] pad of scratch paper with holes for three-ring binder

[ ] straightedge, ruler, compass, protractor, and French curves

Exam Strategies and Techniques

Previous examinees have reported that the following strategies and techniques have helped them considerably.

- Read through all of the problems before starting your first solution. In order to save you from rereading and re-evaluating each problem later in the day, you should classify each problem at the beginning of the 4-hour session. The following categories are suggested:

  · problems you can do easily

  · problems you can do with effort

  · problems for which you can get partial credit

  · problems you cannot do

- Do all of the problems in order of increasing difficulty. All problems on the examination are worth 10 points, so there is nothing to be gained by attempting the difficult or long problems if easier or shorter problems are available.

- Follow these guidelines when solving a problem:

  · Do not rewrite the problem statement.

  · Do not unnecessarily redraw any figures.

  · Use pencil only.

  · Be neat. (Print all text. Use a straightedge or template where possible.)

  · Draw a box around each answer.

  · Label each answer with a symbol.

  · Give the units.

  · List your sources whenever you use obscure solution methods or data.

  · Make complicated problems easily visible by drawing a diagram simple and large enough to include the basic data.

  · Use a schematic showing the streams entering and leaving, if possible.

  · All reaction equations should be written out and balanced.

  · When heat balances are required, the standard heats of reaction should be included.

  · Identify conversion factors.

  · Write on one side of the page only.

  · Use one page per problem, no matter how short the solution is.

  · Go through all calculations a second time and check for mathematical errors, or solve by an alternative method.

- Record the details of any problem which you think is impossible to solve with the information given. Your ability to point out an error may later give you the margin needed to pass.

Notice to Examinees

Do not copy, memorize, or distribute problems from the Principles and Practice of Engineering (P&P) Examination. These acts are considered to be exam subversion.

The P&P examination is copyrighted by the National Council of Examiners for Engineering and Surveying. Copying and reproducing P&P exam problems for commercial purposes is a violation of federal copyright law. Reporting examination problems to other examinees invalidates the examination process and threatens the health and welfare of the public.

MATH

1 MATHEMATICS

1 INTRODUCTION

It is usually unnecessary for chemical engineers to work complex mathematical problems by hand. Most mathematics problems are limited to algebra, simple geometry or trigonometry, and simple differentiations or integrations in calculus. This chapter is designed to serve as a reference for formulas and techniques required to solve most common chemical engineering problems.

Symbols Used in this Book

The symbols, letters, and Greek characters used to represent variables in each chapter are defined in the nomenclature sections of each chapter. Some of the symbols which are used as operators in this book are listed in table 1.1. The Greek alphabet is found in table 1.2.

Table 1.1
Symbols for Operators Used in this Book

symbol	name	typical use
Σ	sigma	series addition
π	pi	series multiplication
Δ	delta	change in quantity
$\overline{x}$	over bar	average value
$\dot{Q}$	over dot	per unit time
!	factorial	
$\vert x \vert$	absolute value	
$\approx$	approximately equal to	
$\propto$	proportional to	
∞	infinity	
log	base 10 logarithm	
ln	base e (natural) logarithm	
EE	scientific notation (power of 10)	
exp	exponential (power of e)	

Table 1.2
The Greek Alphabet

A	α	alpha	N	ν	nu
B	β	beta	Ξ	ξ	xi
Γ	γ	gamma	O	o	omicron
Δ	δ	delta	Π	π	pi
E	ϵ	epsilon	P	ρ	rho
Z	ς	zeta	Σ	σ	sigma
H	η	eta	T	τ	tau
Θ	θ	theta	Υ	υ	upsilon
I	ι	iota	Φ	ϕ	phi
K	κ	kappa	X	χ	chi
Λ	λ	lambda	Ψ	ψ	psi
M	μ	mu	Ω	ω	omega

2 MENSURATION

Nomenclature

A	total surface area
d	distance
D	diameter
h	height
p	perimeter
r	radius
s	side (edge) length, or arc length
V	volume
θ	vertex angle, in radians
ϕ	central angle, in radians

Circle

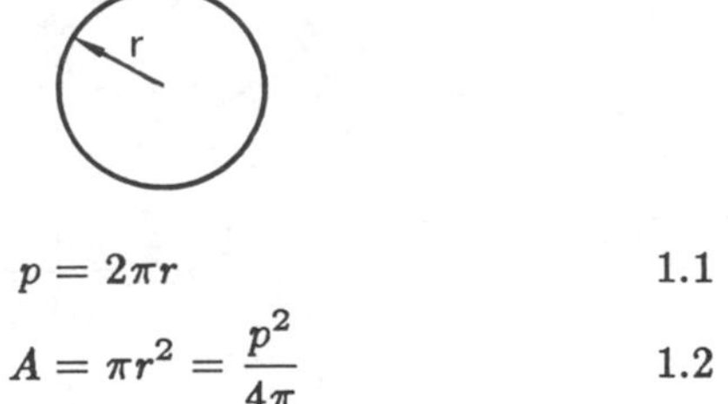

$$p = 2\pi r \tag{1.1}$$

$$A = \pi r^2 = \frac{p^2}{4\pi} \tag{1.2}$$

MATH

Circular Segment

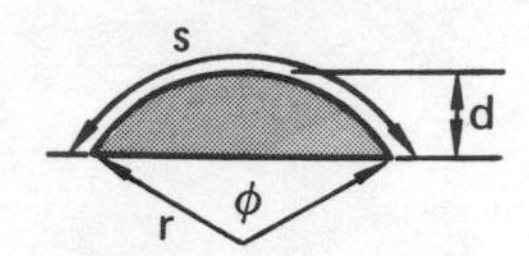

$$A = \frac{1}{2}r^2(\phi - \sin\phi) \quad 1.3$$

$$\phi = \frac{s}{r} = 2\left(\arccos\frac{r-d}{r}\right) \quad 1.4$$

Triangle

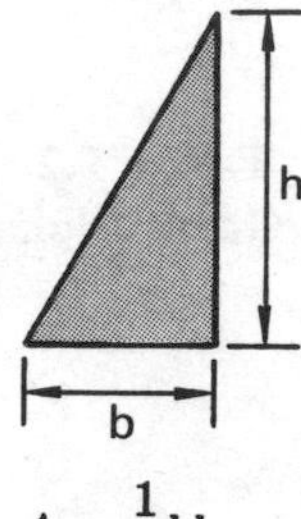

$$A = \frac{1}{2}bh \quad 1.5$$

Parabola

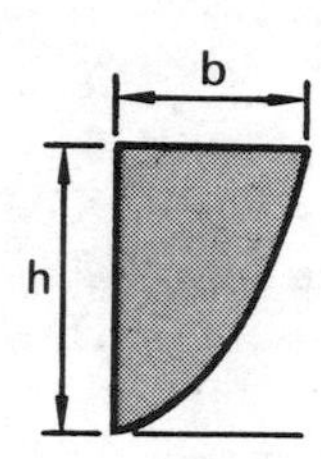

$$A = \frac{2bh}{3} \quad 1.6$$

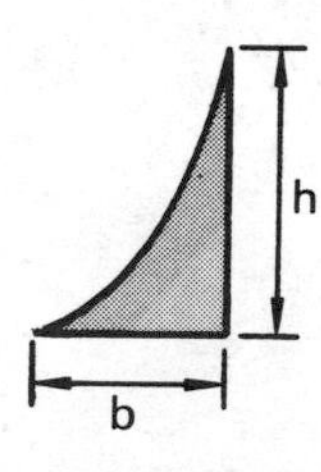

$$A = \frac{1}{3}bh \quad 1.7$$

Circular Sector

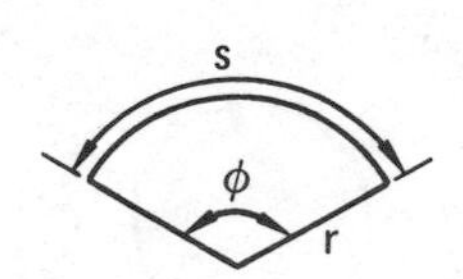

$$A = \frac{1}{2}\phi r^2 = \frac{1}{2}sr \quad 1.8$$

$$\phi = \frac{s}{r} \quad 1.9$$

Ellipse

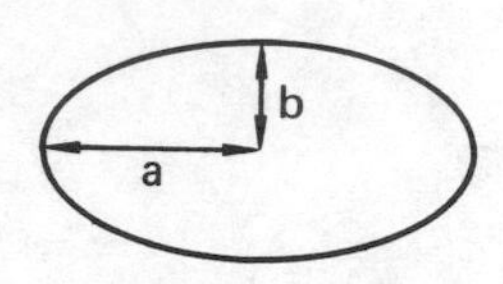

$$A = \pi ab \quad 1.10$$

$$p = 2\pi\sqrt{\frac{1}{2}(a^2 + b^2)} \quad 1.11$$

Trapezoid

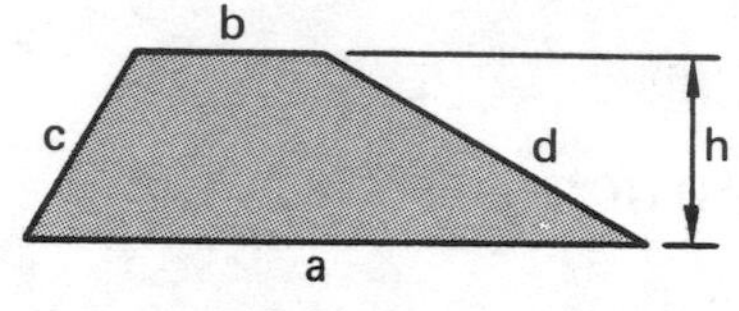

$$p = a + b + c + d \quad 1.12$$

$$A = \frac{1}{2}h(a + b) \quad 1.13$$

The trapezoid is *isosceles* if $c = d$.

Parallelogram

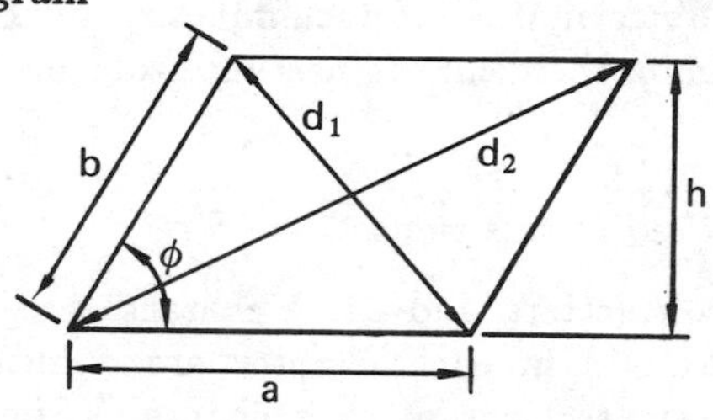

$$p = 2(a + b) \quad 1.14$$

$$d_1 = \sqrt{a^2 + b^2 - 2ab(\cos\phi)} \quad 1.15$$

$$d_2 = \sqrt{a^2 + b^2 + 2ab(\cos\phi)} \quad 1.16$$

$$d_1^2 + d_2^2 = 2(a^2 + b^2) \quad 1.17$$

$$A = ah = ab(\sin\phi) \quad 1.18$$

If $a = b$, the parallelogram is a *rhombus*.

Regular Polygon (n equal sides)

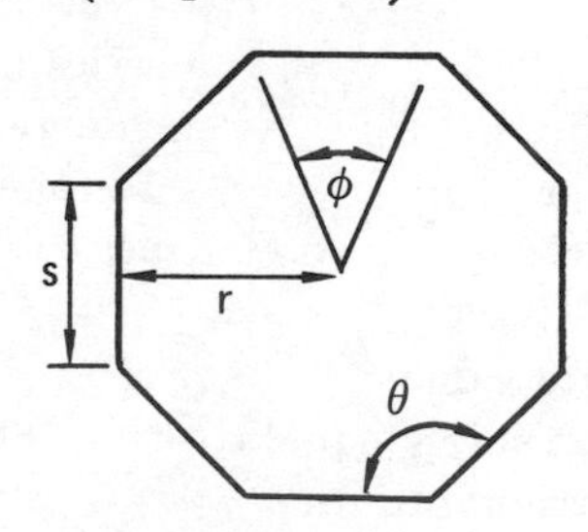

$$\phi = \frac{2\pi}{n} \quad 1.19$$

$$\theta = \frac{\pi(n-2)}{n} \quad 1.20$$

$$p = ns \quad 1.21$$

$$s = 2r\left[\tan\left(\frac{\phi}{2}\right)\right] \qquad 1.22$$

$$A = \frac{1}{2}nsr \qquad 1.23$$

Table 1.3
Polygons

number of sides	name of polygon
3	triangle
4	rectangle
5	pentagon
6	hexagon
7	heptagon
8	octagon
9	nonagon
10	decagon

Table 1.4
Polyhedrons

number of faces	form of faces	total surface area	volume
4	equilateral triangle	$1.7321\ s^2$	$0.1179\ s^3$
6	square	$6.0000\ s^2$	$1.0000\ s^3$
8	equilateral triangle	$3.4641\ s^2$	$0.4714\ s^3$
12	regular pentagon	$20.6457\ s^2$	$7.6631\ s^3$
20	equilateral triangle	$8.6603\ s^2$	$2.1817\ s^3$

Sphere

$$V = \frac{4\pi r^3}{3} \qquad 1.24$$

$$A = 4\pi r^2 \qquad 1.25$$

Right Circular Cone

$$V = \frac{\pi r^2 h}{3} \qquad 1.26$$

$$A = \pi r\sqrt{r^2 + h^2} \qquad 1.27$$

(does not include base area)

Right Circular Cylinder

$$V = \pi h r^2 \qquad 1.28$$

$$A = 2\pi r h \qquad 1.29$$

(does not include end area)

Paraboloid of Revolution

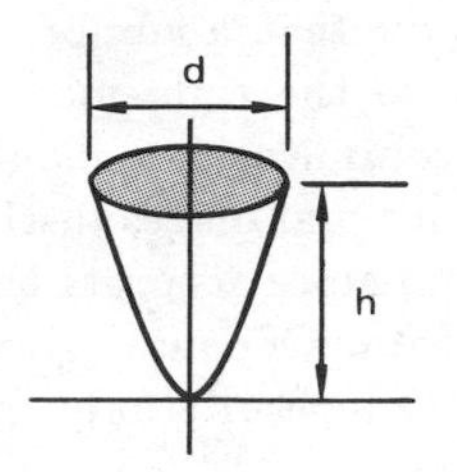

$$V = \frac{\pi h d^2}{8} \qquad 1.30$$

Regular Polyhedron

$$r = \frac{3V}{A} \qquad 1.31$$

Example 1.1

What is the hydraulic radius of a 6″ pipe filled to a depth of 2″?

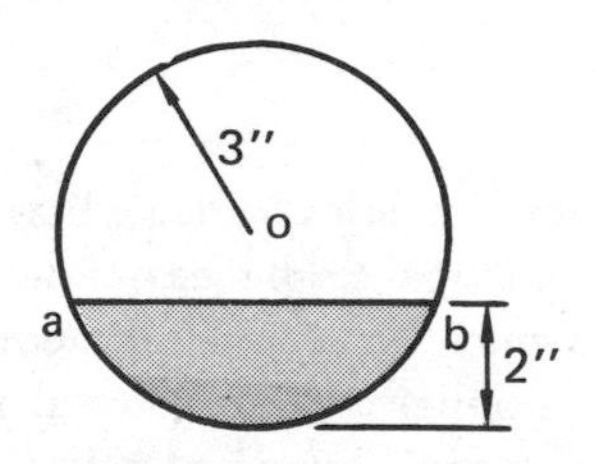

The hydraulic radius is defined as

$$r_h = \frac{\text{area in flow}}{\text{length of wetted perimeter}} = \frac{A}{s}$$

Points o, a, and b may be used to find the central angle of the circular segment.

$$\frac{1}{2}(\text{angle } aob) = \arccos\left(\frac{1}{3}\right) = 70.53°$$

$$\phi = 141.06° = 2.46 \text{ radians}$$

Then,

$$A = \frac{1}{2}(3)^2(2.46 - 0.63) = 8.235 \text{ in}^2$$

$$s = (3)(2.46) = 7.38 \text{ in}$$

$$r_h = \frac{8.235}{7.38} = 1.12 \text{ in}$$

MATH

3 SIGNIFICANT DIGITS

The significant digits in a number include the left-most, non-zero digits to the right-most digit written. Final answers from computations should be rounded off to the number of decimal places justified by the data. The answer can be no more accurate than the least accurate number in the data. Of course, rounding should be done on final calculation results only. It should not be done on interim results.

Table 1.5
Significant Digits

number as written	number of significant digits	implied range
341	3	340.5 to 341.5
34.1	3	34.05 to 34.15
0.00341	3	0.003405 to 0.003415
3410.	4	3409.5 to 3410.5
341 EE7	3	340.5 EE7 to 341.5 EE7
3.41 EE-2	3	3.405 EE-2 to 3.415 EE-2

4 ALGEBRA

Algebra provides the rules which allow complex mathematical relationships to be expanded or condensed. Algebraic laws may be applied to complex numbers, variables, and numbers. The general rules for changing the form of a mathematical relationship are given here:

Commutative Law for Addition

$$a + b = b + a \qquad 1.32$$

Commutative Law for Multiplication

$$ab = ba \qquad 1.33$$

Associative Law for Addition

$$a + (b + c) = (a + b) + c \qquad 1.34$$

Associative Law for Multiplication

$$a(bc) = (ab)c \qquad 1.35$$

Distributive Law

$$a(b + c) = ab + ac \qquad 1.36$$

A. POLYNOMIAL EQUATIONS

Standard Forms

$$(a + b)(a - b) = a^2 - b^2 \qquad 1.37$$

$$(a \pm b)^2 = a^2 \pm 2ab + b^2 \qquad 1.38$$

$$(a \pm b)^3 = a^3 \pm 3a^2b + 3ab^2 \pm b^3 \qquad 1.39$$

$$(a^3 \pm b^3) = (a \pm b)(a^2 \mp ab + b^2) \qquad 1.40$$

$$(a^n + b^n) = (a + b)(a^{n-1} - a^{n-2}b + \dots + b^{n-1}) \text{(for } n \text{ odd)} \qquad 1.41$$

$$(a^n - b^n) = (a - b)(a^{n-1} + a^{n-2}b + \dots + b^{n-1}) \qquad 1.42$$

General Forms and Reversions

$$y = P_n(x) = 1 + ax + bx^2 + cx^3 + \dots + sx^n \qquad 1.43$$

Equation 1.43 can be reverted to x as a function of y if it converges, by letting

$$Y = y - 1$$

$$x = Q_n(Y) = AY + BY^2 + CY^3 + \dots + SY^n \qquad 1.44$$

Then,

$$aA = 1 \qquad 1.45$$

$$a^3B = -b \qquad 1.46$$

$$a^5C = -ca + 2b^2 \qquad 1.47$$

$$a^7D = -da^2 + 5abc - 5b^3 \qquad 1.48$$

$$a^9E = -ea^3 + 6a^2bd + 3a^2c^2 - 21ab^2c + 14b^4 \qquad 1.49$$

Example 1.2

If $y = P_3(x) = 2x + x^2 + \frac{x^3}{6}$, what is $Q_3(y)$?

$$x = Ay + By^2 + Cy^3$$

$$2A = 1;\ A = \frac{1}{2}$$

$$8B = -1;\ B = \frac{-1}{8}$$

$$32C = \frac{-1}{3} + 2;\ C = \frac{5}{96}$$

$$x = \frac{y}{2} - \frac{y^2}{8} + \frac{5y^3}{96}$$

Operations with Polynomials

Consider three polynomials defined as follows:

$$p = P_n(x) = 1 + a_1x + a_2x^2 + a_3x^3 + \ldots \quad 1.50$$

$$q = Q_n(x) = 1 + b_1x + b_2x^2 + b_3x^3 + \ldots \quad 1.51$$

$$r = R_n(x) = 1 + c_1x + c_2x^2 + c_3x^3 + \ldots \quad 1.52$$

Table 1.6 tabulates the values of the coefficients of r for the indicated operations performed on polynomials p and q.

Quadratic Equations

Given a quadratic equation $ax^2 + bx + c = 0$, the roots x_1^* and x_2^* may be found from

$$x_1^*,\ x_2^* = \frac{-b \pm \sqrt{b^2 - 4ac}}{2a} \quad 1.53$$

$$x_1^* + x_2^* = \frac{-b}{a} \quad 1.54$$

$$x_1^* x_2^* = \frac{c}{a} \quad 1.55$$

Cubic Equations

Cubic and higher order equations occur infrequently in most engineering problems. However, they usually are difficult to factor when they do occur. Trial and error solutions are usually unsatisfactory except for finding the general region in which a root occurs. Graphical means can be used to obtain only a fair approximation of the root.

Numerical analysis techniques must be used if extreme accuracy is needed. The more efficient numerical analysis techniques are too complicated to present here. However, the bisection method illustrated in example 1.3 usually can provide the required accuracy with only a few simple iterations.

The bisection method starts out with two values of the independent variable, L_0 and R_0, which straddle a root. Since the function has a value of zero at a root, $f(L_0)$ and $f(R_0)$ will have opposite signs. The following algorithm describes the remainder of the bisection method.

Let n be the iteration number. Then, for $n = 0, 1, 2 \ldots$ perform the following steps until sufficient accuracy is attained.

step 1: Set $m = \frac{1}{2}(L_n + R_n)$

step 2: Calculate $f(m)$.

step 3: If $f(L_n)f(m) < 0$, set $L_{n+1} = L_n$ and $R_{n+1} = m$, otherwise, set $L_{n+1} = m$ and $R_{n+1} = R_n$.

$f(x)$ has at least one root in the interval (L_{n+1}, R_{n+1}).

Table 1.6
Polynomial Operations

operation	c_1	c_2	c_3	c_4
$r = \frac{1}{p}$	$-a_1$	$a_1^2 - a_2$	$2a_1a_2 - a_3 - a_1^3$	$2a_1a_3 - 3a_1^2a_2 - a_4 + a_2^2 + a_1^4$
$r = \frac{1}{p^2}$	$-2a_1$	$3a_1^2 - 2a_2$	$6a_1a_2 - 2a_3 - 4a_1^3$	$6a_1a_3 + 3a_2^2 - 2a_4 - 12a_1^2a_2 + 5a_1^4$
$r = \sqrt{p}$	$\frac{1}{2}a_1$	$\frac{1}{2}a_2 - \frac{1}{8}a_1^2$	$\frac{1}{2}a_3 - \frac{1}{4}a_1a_2 + \frac{1}{16}a_1^3$	$\frac{1}{2}a_4 - \frac{1}{4}a_1a_3 - \frac{1}{8}a_2^2 + \frac{3}{16}a_1^2a_2 - \frac{5}{128}a_1^4$
$r = \frac{1}{\sqrt{p}}$	$-\frac{1}{2}a_1$	$\frac{3}{8}a_1^2 - \frac{1}{2}a_2$	$\frac{3}{4}a_1a_2 - \frac{1}{2}a_3 - \frac{5}{16}a_1^3$	$\frac{3}{4}a_1a_3 + \frac{3}{8}a_2^2 - \frac{1}{2}a_4 - \frac{15}{16}a_1^2a_2 + \frac{35}{128}a_1^4$
$r = p^n$	na_1	$\frac{1}{2}(n-1)c_1a_1 + na_2$	$c_1a_2(n-1)$ $+\frac{1}{6}c_1a_1^2(n-1)(n-2)$ $+na_3$	$na_4 + c_1a_3(n-1) + \frac{1}{2}n(n-1)a_2^2$ $+\frac{1}{2}(n-1)(n-2)c_1a_1a_2$ $+\frac{1}{24}(n-1)(n-2)(n-3)c_1a_1^3$
$r = pq$	$a_1 + b_1$	$b_2 + a_1b_1 + a_2$	$b_3 + a_1b_2 + a_2b_1 + a_3$	$b_4 + a_1b_3 + a_2b_2 + a_3b_1 + a_4$
$r = \frac{p}{q}$	$a_1 - b_1$	$a_2 - (b_1c_1 + b_2)$	$a_3 - (b_1c_2 + b_2c_1 + b_3)$	$a_4 - (b_1c_3 + b_2c_2 + b_3c_1 + b_4)$
$r = e^{(p-1)}$	a_1	$a_2 + \frac{1}{2}a_1^2$	$a_3 + a_1a_2 + \frac{1}{6}a_1^2$	$a_4 + a_1a_3 + \frac{1}{2}a_2^2 + \frac{1}{2}a_2a_1^2 + \frac{1}{24}a_1^4$
$r = 1 + \ln p$	a_1	$a_2 - \frac{1}{2}a_1c_1$	$a_3 - \frac{1}{3}(a_2c_1 + 2a_1c_2)$	$a_4 - \frac{1}{4}(a_3c_1 + 2a_2c_2 + 3a_1c_2)$

The estimated value of that root, x^*, is

$$x^* \approx \frac{1}{2}(L_{n+1} + R_{n+1}) \qquad 1.56$$

The maximum error is $\frac{1}{2}(R_{n+1} - L_{n+1})$. The iterations continue until the maximum error is reasonable for the accuracy of the problem.

Example 1.3

Use the bisection method to find the roots of

$$f(x) = x^3 - 2x - 7$$

The first step is to find L_0 and R_0, which are the values of x which straddle a root and have opposite signs. A table can be made and values of $f(x)$ calculated for random values of x.

x	−2	−1	0	+1	+2	+3
$f(x)$	−11	−6	−7	−8	−3	+14

Since $f(x)$ changes sign between $x = 2$ and $x = 3$,

$$L_0 = 2 \text{ and } R_0 = 3$$

Iteration 0:

$$m = \frac{1}{2}(2 + 3) = 2.5$$

$$f(2.5) = (2.5)^3 - 2(2.5) - 7 = 3.625$$

Since $f(2.5)$ is positive, a root must exist in the interval (2, 2.5). Therefore,

$$L_1 = 2 \text{ and } R_1 = 2.5$$

At this point, the best estimate of the root is

$$x^* \approx \frac{1}{2}(2 + 2.5) = 2.25$$

The maximum error is $\frac{1}{2}(2.5 - 2) = 0.25$.

Iteration 1:

$$m = \frac{1}{2}(2 + 2.5) = 2.25$$

$$f(2.25) = -0.1094$$

Since $f(m)$ is negative, a root must exist in the interval (2.25, 2.5). Therefore,

$$L_2 = 2.25 \text{ and } R_2 = 2.5$$

The best estimate of the root is

$$x^* \approx \frac{1}{2}(2.25 + 2.5) = 2.375$$

The maximum error is $\frac{1}{2}(2.5 - 2.25) = 0.125$.

This procedure continues until the maximum error is acceptable. Of course, this method does not automatically find any other roots that may exist on the real number line.

Finding Roots to General Expressions

There is no specific technique that will work with all general expressions for which roots are needed. If graphical means are not used, some combination of factoring and algebraic simplification must be used. However, multiplying each side of an equation by a power of a variable may introduce extraneous roots. Such an extraneous root will not satisfy the original equation, even though it was derived correctly according to the rules of algebra.

Although it is always a good idea to check your work, this step is particularly necessary whenever you have squared an expression or multiplied it by a variable.

Example 1.4

Find the value of x which will satisfy the following expression:

$$\sqrt{x - 2} = \sqrt{x} + 2$$

First, square both sides.

$$x - 2 = x + 4\sqrt{x} + 4$$

Next, subtract x from both sides and combine constants.

$$4\sqrt{x} = -6$$

Solving for x yields $x^* = \frac{9}{4}$. However, $\frac{9}{4}$ does not satisfy the original expression since it is an extraneous root.

B. SIMULTANEOUS LINEAR EQUATIONS

Given n independent equations and n unknowns, the n values which simultaneously solve all n equations can be found by the methods illustrated in example 1.5.

Example 1.5

Solve

$$2x + 3y = 12 \qquad \text{(a)}$$
$$3x + 4y = 8 \qquad \text{(b)}$$

Method 1 Substitution

step 1: From equation (a), solve for $x = 6 - 1.5y$

step 2: Substitute $(6 - 1.5y)$ into equation (b) wherever x appears. $3(6 - 1.5y) + 4y = 8$, or $y^* = 20$

step 3: Solve for x^* from either equation.

$$x^* = 6 - 1.5(20) = -24$$

step 4: Check that (–24, 20) solves both original equations.

Method 2 Reduction

step 1: Multiply each equation by a number chosen to make the coefficient of one of the variables the same in each equation.

$$3 \times \text{ equation (a)} : 6x + 9y = 36 \ (c)$$
$$2 \times \text{ equation (b)} : 6x + 8y = 16 \ (d)$$

step 2: Subtract one equation from the other. Solve for one of the variables.

$$(c) - (d): \ y^* = 20$$

step 3: Solve for the remaining variable.

step 4: Check that the calculated values of (x^*, y^*) solve both original equations.

Method 3 Cramer's Rule

This method is best for three or more simultaneous equations. (The calculation of determinants is covered later in this chapter.)

step 1: Find x^* and y^* which satisfy

$$a_1x + b_1y = c_1$$
$$a_2x + b_2y = c_2$$

step 2: Calculate the determinants,

$$D_1 = \begin{vmatrix} a_1 & b_1 \\ a_2 & b_2 \end{vmatrix}$$

$$D_2 = \begin{vmatrix} c_1 & b_1 \\ c_2 & b_2 \end{vmatrix}$$

$$D_3 = \begin{vmatrix} a_1 & c_1 \\ a_2 & c_2 \end{vmatrix}$$

Then, if $D_1 \neq 0$, the unique numbers satisfying the two simultaneous equations are

$$x^* = \frac{D_2}{D_1}$$
$$y^* = \frac{D_3}{D_1}$$

If D_1 (the determinant of the coefficients matrix) is zero, the system of simultaneous equations may still have a solution. However, Cramer's rule cannot be used to find that solution. If the system is homogeneous (i.e., has the general form $Ax = 0$), then a non-zero solution exists if and only if D_1 is zero.

Example 1.6

Solve the following system of simultaneous equations:

$$2x + 3y - 4z = 1$$
$$3x - y - 2z = 4$$
$$4x - 7y - 6z = -7$$

step 1: Calculate the determinants:

$$D_1 = \begin{vmatrix} 2 & 3 & -4 \\ 3 & -1 & -2 \\ 4 & -7 & -6 \end{vmatrix} = 82$$

$$D_2 = \begin{vmatrix} 1 & 3 & -4 \\ 4 & -1 & -2 \\ -7 & -7 & -6 \end{vmatrix} = 246$$

$$D_3 = \begin{vmatrix} 2 & 1 & -4 \\ 3 & 4 & -2 \\ 4 & -7 & -6 \end{vmatrix} = 82$$

$$D_4 = \begin{vmatrix} 2 & 3 & 1 \\ 3 & -1 & 4 \\ 4 & -7 & -7 \end{vmatrix} = 164$$

step 2: Then,

$$x^* = \frac{D_2}{D_1} = 3$$
$$y^* = \frac{D_3}{D_1} = 1$$
$$z^* = \frac{D_4}{D_1} = 2$$

C. SIMULTANEOUS QUADRATIC EQUATIONS

Although simultaneous non-linear equations are best solved graphically, a specialized method exists for simultaneous quadratic equations. This method is known as *eliminating the constant term.*

step 1: Isolate the constant terms of both equations on the right-hand side of the equalities.

step 2: Multiply both sides of one equation by a number chosen to make the constant terms of both equations the same.

step 3: Subtract one equation from the other to obtain a difference equation.

step 4: Factor the difference equation into terms.

step 5: Solve for one of the variables from one of the factor terms.

step 6: Substitute the formula for the variable into one of the original equations and complete the solution.

step 7: Check the solution.

Example 1.7

Solve for the simultaneous values of x and y.

step 1:

$$2x^2 - 3xy + y^2 = 15$$
$$x^2 - 2xy + y^2 = 9$$

steps 2 & 3:

$$\begin{array}{r} 6x^2 - 9xy + 3y^2 = 45 \\ \underline{-(5x^2 - 10xy + 5y^2) = 45} \\ x^2 + xy - 2y^2 = 0 \end{array}$$

steps 4 & 5: $x^2+xy-2y^2$ factors into $(x+2y)(x-y)$, from which we obtain $x = -2y$.

step 6: Substituting $x = -2y$ into $(2x^2 - 3xy + y^2 = 15)$ gives $y^* = \pm 1$, from which $x^* = \pm 2$ can be derived by further substitution.

D. EXPONENTIATION

(x is any variable or constant)

$$x^m x^n = x^{(n+m)} \qquad 1.57$$
$$\frac{x^m}{x^n} = x^{(m-n)} \qquad 1.58$$
$$(x^n)^m = x^{(mn)} \qquad 1.59$$
$$a^{\frac{m}{n}} = \sqrt[n]{a^m} \qquad 1.60$$
$$\left(\frac{a}{b}\right)^n = \frac{a^n}{b^n} \qquad 1.61$$
$$\sqrt[n]{x} = (x)^{\frac{1}{n}} \qquad 1.62$$
$$x^{-n} = \frac{1}{x^n} \qquad 1.63$$
$$x^0 = 1 \qquad 1.64$$

E. LOGARITHMS

Logarithms are exponents. That is, the exponent x in the expression $b^x = n$ is the logarithm of n to the base b. Therefore, $(\log_b n) = x$ is equivalent to $(b^x = n)$.

The base for common logs is 10. Usually, *log* will be written when common logs are desired, although $\log_{10}$ appears occasionally. The base for *natural (naperien) logs* is 2.718..., a number which is given the symbol e. When natural logs are desired, usually *ln* will be written, although $\log_e$ is also used.

Most logarithms will contain an integer part (the *characteristic*) and a fractional part (the *mantissa*). The logarithm of any number less than one is negative. If the number is greater than one, its logarithm is positive. Although the logarithm may be negative, the mantissa is always positive.

For common logarithms of numbers greater than one, the characteristics will be positive and equal to one less than the number of digits in front of the decimal. If the number is less than one, the characteristic will be negative and equal to one more than the number of zeros immediately following the decimal point.

Example 1.8

What is $\log_{10}$ (0.05)?

Since the number is less than one and there is one leading zero, the characteristic is −2. From the logarithm tables, the mantissa of 5.0 is .699. Two ways of combining the mantissa and characteristic are possible.

Method 1: $\bar{2}.699$

Method 2: 8.699 − 10

If the logarithm is to be used in a calculation, it must be converted to operational form: $-2+0.699 = -1.301$. Notice that -1.301 is not the same as $\bar{1}.301$.

F. LOGARITHM IDENTITIES

$$x^a = \text{antilog}[a \log(x)] \quad 1.65$$

$$\log(x^a) = a \log(x) \quad 1.66$$

$$\log(xy) = \log(x) + \log(y) \quad 1.67$$

$$\log\left(\frac{x}{y}\right) = \log(x) - \log(y) \quad 1.68$$

$$ln(x) = \frac{\log_{10} x}{\log_{10} e} \approx 2.3(\log_{10} x) \quad 1.69$$

$$\log_b(b) = 1 \quad 1.70$$

$$\log(1) = 0 \quad 1.71$$

$$\log_b(b^n) = n \quad 1.72$$

Example 1.9

The surviving fraction, x, of a radioactive isotope is given by

$$x = e^{-0.005t}$$

For what value of t will the surviving fraction be 7%?

$$0.07 = e^{-0.005t}$$

Taking the natural log of both sides,

$$ln(0.07) = ln(e^{-0.005t})$$

$$-2.66 = -0.005t$$

$$t = 532$$

G. PARTIAL FRACTIONS

Given some rational fraction $H(x) = \frac{P(x)}{Q(x)}$ where $P(x)$ and $Q(x)$ are polynomials, the polynomials and constants A_i and $Y_i(x)$ are needed such that

$$H(x) = \sum_i \frac{A_i}{Y_i(x)} \quad 1.73$$

case 1: $Q(x)$ factors into n different linear terms. That is,

$$Q(x) = (x - a_1)(x - a_2)\ldots(x - a_n) \quad 1.74$$

Then,

$$H(x) = \sum_{i=1}^{n} \frac{A_i}{x - a_i} \quad 1.75$$

case 2: $Q(x)$ factors into n identical linear terms. That is,

$$Q(x) = (x - a)(x - a)\ldots(x - a) \quad 1.76$$

Then,

$$H(x) = \sum_{i=1}^{n} \frac{A_i}{(x - a)^i} \quad 1.77$$

case 3: $Q(x)$ factors into n different quadratic terms, $(x^2 + p_i x + q_i)$. Then,

$$H(x) = \sum_{i=1}^{n} \frac{A_i x + B_i}{x^2 + p_i x + q_i} \quad 1.78$$

case 4: $Q(x)$ factors into n idential quadratic terms, $(x^2 + px + q)$. Then,

$$H(x) = \sum_{i=1}^{n} \frac{A_i x + B_i}{(x^2 + px + q)^i} \quad 1.79$$

case 5: $Q(x)$ factors into any combination of the above. The solution is illustrated by example 1.10.

Example 1.10

Resolve

$$H(x) = \frac{x^2 + 2x + 3}{x^4 + x^3 + 2x^2}$$

into partial fractions.

Here, $Q(x) = x^4 + x^3 + 2x^2$, which factors into $x^2(x^2 + x + 2)$. This is a combination of cases *2* and *3*. We set

$$H(x) = \frac{A_1}{x} + \frac{A_2}{x^2} + \frac{A_3 + A_4 x}{x^2 + x + 2}$$

Cross-multiplying to obtain a common denominator yields

$$\frac{(A_1+A_4)x^3+(A_1+A_2+A_3)x^2+(2A_1+A_2)x+2A_2}{x^4 + x^3 + 2x^2}$$

Since the original numerator is known, the following simultaneous equations result:

$$\begin{aligned} A_1 + A_4 &= 0 \\ A_1 + A_2 + A_3 &= 1 \\ 2A_1 + A_2 &= 2 \\ 2A_2 &= 3 \end{aligned}$$

The solutions are: $A_1^* = 0.25$; $A_2^* = 1.5$; $A_3^* = -0.75$; $A_4^* = -0.25$. So,

$$H(x) = \frac{1}{4x} + \frac{3}{2x^2} - \frac{x+3}{4(x^2+x+2)}$$

H. LINEAR AND MATRIX ALGEBRA

A matrix is a rectangular collection of variables or scalars contained within a set of square or round brackets. In the discussion that follows, matrix A will be assumed to have m rows and n columns. There are several classifications of matrices:

If $n = m$, the matrix is *square.*

A *diagonal* matrix is a square matrix with all zero values except for the a_{ij} values, for all $i = j$.

An *identity* matrix is a diagonal matrix with all non-zero entries equal to one. (This usually is designated as I.)

A *scalar* matrix is a square diagonal matrix with all non-zero entries equal to some constant.

A *triangular* matrix has zeros in all positions above or below the diagonal. This is not the same as an *echelon* matrix since the diagonal entries are non-zero.

Matrices are used to simplify the presentation and solution of linear equations (hence the name 'linear algebra'). For example the system of equations in example 1.5 can be written in matrix form as

$$\begin{pmatrix} 2 & 3 \\ 3 & 4 \end{pmatrix} \begin{pmatrix} x \\ y \end{pmatrix} = \begin{pmatrix} 12 \\ 8 \end{pmatrix} \qquad 1.80$$

The above expression implies that there is a set of algebraic operations that can be performed with matrices. The important algebraic operations are listed here, along with their extensions to linear algebra.

- *Equality of Matrices:* For two matrices to be equal they must have the same number of rows and columns. Corresponding entries must all be the same.
- *Inequality of Matrices:* There are no 'less-than' or 'greater-than' relationships in linear algebra.
- *Addition and Subtraction of Matrices:* Addition (or subtraction) of two matrices can be accomplished by adding (or subtracting) the corresponding entries of two matrices which have the same shape.
- *Multiplication of Matrices:* Multiplication can be done only if the left-hand matrix has the same number of columns as the right-hand matrix has rows. Multiplication is accomplished by multiplying the elements in each right-hand matrix column, adding the products, and then placing the sum at the intersection point of the involved row and column. This is illustrated by example 1.11
- *Division of Matrices:* Division can be accomplished only by multiplying by the inverse of the denominator matrix.

Example 1.11

$$\begin{pmatrix} 1 & 4 & 3 \\ & & \\ 5 & 2 & 6 \end{pmatrix} \begin{pmatrix} 7 & 12 \\ 11 & 8 \\ 9 & 10 \end{pmatrix} = C$$

$$\begin{aligned} [(1)(7) + (4)(11) + (3)(9)] &= 78 \\ [(1)(12) + (4)(8) + (3)(10)] &= 74 \\ [(5)(7) + (2)(11) + (6)(9)] &= 111 \\ [(5)(12) + (2)(8) + (6)(10)] &= 136 \end{aligned}$$

$$C = \begin{pmatrix} 78 & 74 \\ 111 & 136 \end{pmatrix}$$

Other operations which can be performed on a matrix are described and illustrated below.

The *transpose* is an $(n \times m)$ matrix formed from the original $(m \times n)$ matrix by taking its ith row and making it the ith column. The diagonal is unchanged in this operation. The transpose of a matrix A is indicated as A^t.

Example 1.12

Find the transpose of

$$A = \begin{pmatrix} 1 & 6 & 9 \\ 2 & 3 & 4 \\ 7 & 1 & 5 \end{pmatrix}$$

The transpose of A is

$$A^t = \begin{pmatrix} 1 & 2 & 7 \\ 6 & 3 & 1 \\ 9 & 4 & 5 \end{pmatrix}$$

The *determinant*, D, is a scalar calculated from a square matrix. The determinant of a matrix is indicated by enclosing the matrix in vertical lines.

For a (2×2) matrix,

$$A = \begin{pmatrix} a & b \\ c & d \end{pmatrix} \qquad 1.81$$

$$D = \begin{vmatrix} a & b \\ c & d \end{vmatrix} = ad - bc \qquad 1.82$$

For a (3×3) matrix,

$$A = \begin{pmatrix} a & b & c \\ d & e & f \\ g & h & i \end{pmatrix} \qquad 1.83$$

$$D = a\begin{vmatrix} e & f \\ h & i \end{vmatrix} - d\begin{vmatrix} b & c \\ h & i \end{vmatrix} + g\begin{vmatrix} b & c \\ e & f \end{vmatrix} \qquad 1.84$$

There are several rules governing the calculation of determinants:

- If A has a row or column of zeros, the determinant is zero.
- If A has two identical rows or columns, the determinant is zero.
- If A is triangular, the determinant is equal to the product of the diagonal entries.
- If B is obtained from A by multiplying a row or column by a scalar k, then $D_B = k(D_A)$.
- If B is obtained from A by switching two rows or columns, then $D_B = -D_A$.
- If B is obtained from A by adding a multiple of a row or column to another, then $D_B = D_A$.

Example 1.13

Find the determinant of

$$\begin{pmatrix} 2 & 3 & -4 \\ 3 & -1 & -2 \\ 4 & -7 & -6 \end{pmatrix}$$

$$\begin{aligned} D &= 2\begin{vmatrix} -1 & -2 \\ -7 & -6 \end{vmatrix} - 3\begin{vmatrix} 3 & -4 \\ -7 & -6 \end{vmatrix} + 4\begin{vmatrix} 3 & -4 \\ -1 & -2 \end{vmatrix} \\ &= 2(6 - 14) - 3(-18 - 28) + 4(-6 - 4) \\ &= 82 \end{aligned}$$

The *cofactor* of an entry in a matrix is the determinant of the matrix formed by omitting the entry's row and column in the original matrix. The sign of the cofactor is determined from the following positional matrices.

For a (2×2) matrix,

$$\begin{pmatrix} + & - \\ - & + \end{pmatrix} \qquad 1.85$$

For a (3×3) matrix,

$$\begin{pmatrix} + & - & + \\ - & + & - \\ + & - & + \end{pmatrix} \qquad 1.86$$

Example 1.14

What is the cofactor of the (-3) in the following matrix?

$$\begin{pmatrix} 2 & 9 & 1 \\ -3 & 4 & 0 \\ 7 & 5 & 9 \end{pmatrix}$$

The resulting matrix is

$$\begin{pmatrix} 9 & 1 \\ 5 & 9 \end{pmatrix}$$

with determinant 76. The cofactor is -76.

The *classical adjoint* is a matrix formed from the transposed cofactor matrix with the conventional sign arrangement. The resulting matrix is represented as A_{adj}.

Example 1.15

Find the classical adjoint of

$$\begin{pmatrix} 2 & 3 & -4 \\ 0 & -4 & 2 \\ 1 & -1 & 5 \end{pmatrix}$$

The matrix of cofactors (considering the sign convention) is

$$\begin{pmatrix} -18 & 2 & 4 \\ -11 & 14 & 5 \\ -10 & -4 & -8 \end{pmatrix}$$

The transposed cofactor matrix is

$$A_{adj} = \begin{pmatrix} -18 & -11 & -10 \\ 2 & 14 & -4 \\ 4 & 5 & -8 \end{pmatrix}$$

The inverse, A^{-1}, of A is a matrix such that $(A)(A^{-1}) = I$. (I is a square matrix with ones along to the left-to-right diagonal and zeros elsewhere.)

For the (2×2) matrix

$$\begin{pmatrix} a & b \\ c & d \end{pmatrix} \qquad 1.87$$

the inverse is

$$\frac{1}{D}\begin{pmatrix} d & -b \\ -c & a \end{pmatrix} \qquad 1.88$$

For larger matrices, the inverse is best calculated by dividing every entry in the classical adjoint by the determinant of the original matrix.

Example 1.16

Find the the inverse of

$$\begin{pmatrix} 4 & 5 \\ 2 & 3 \end{pmatrix}$$

The determinant is 2. The inverse is

$$\frac{1}{2}\begin{pmatrix} 3 & -5 \\ -2 & 4 \end{pmatrix} = \begin{pmatrix} \frac{3}{2} & -\frac{5}{2} \\ -1 & 2 \end{pmatrix}$$

5 TRIGONOMETRY

A. DEGREES AND RADIANS

360 degrees = one complete circle = 2π radians

90 degrees = right angle = $\frac{1}{2}\pi$ radians

one radian = 57.3 degrees

one degree = 0.0175 radians

multiply degrees by $\left(\frac{\pi}{180}\right)$ to obtain radians

multiply radians by $\left(\frac{180}{\pi}\right)$ to obtain degrees

B. RIGHT TRIANGLES

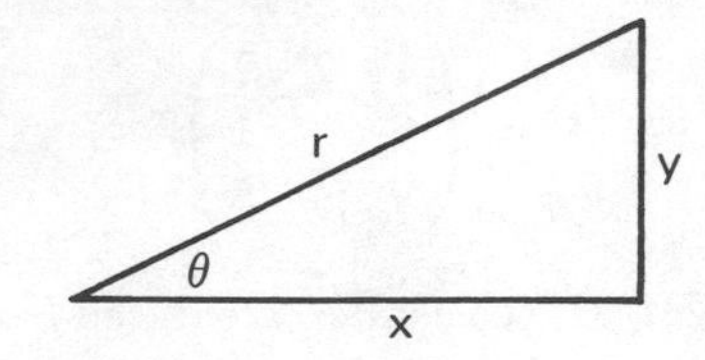

Figure 1.1 Right Triangle

Pythagorean Theorem

$$x^2 + y^2 = r^2 \qquad 1.89$$

Trigonometric Functions

$$\sin\theta = \frac{y}{r} \qquad 1.90$$

$$\cos\theta = \frac{x}{r} \qquad 1.91$$

$$\tan\theta = \frac{y}{x} \qquad 1.92$$

$$\cot\theta = \frac{x}{y} \qquad 1.93$$

$$\csc\theta = \frac{r}{y} \qquad 1.94$$

$$\sec\theta = \frac{r}{x} \qquad 1.95$$

Relationship of the Trigonometric Functions to the Unit Circle

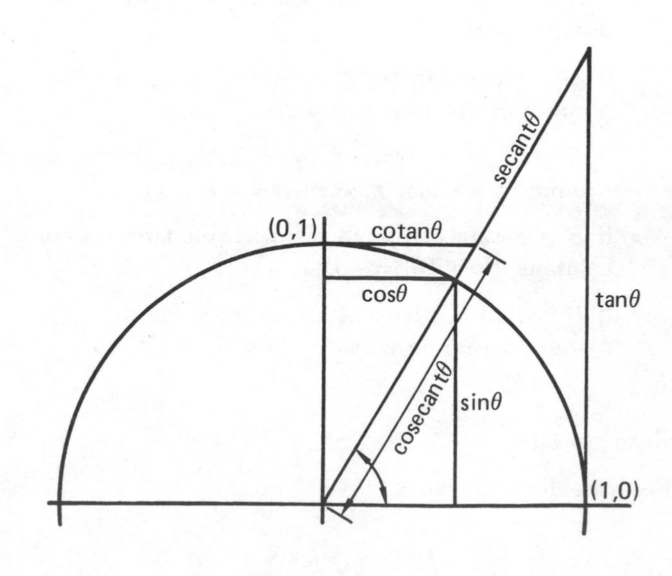

Figure 1.2 The Unit Circle

Table 1.7
Signs of the Trigonometric Functions

quadrants	
II	I
III	IV

function	I	II	III	IV
sin	+	+	−	−
cos	+	−	−	+
tan	+	−	+	−

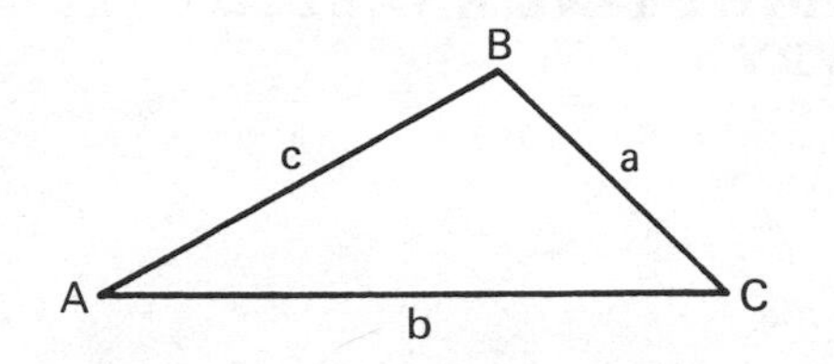

Figure 1.3 A General Triangle

Table 1.8
Functions of the Related Angles

$f(\theta)$	$-\theta$	$90-\theta$	$90+\theta$	$180-\theta$	$180+\theta$
sin	$-\sin\theta$	$\cos\theta$	$\cos\theta$	$\sin\theta$	$-\sin\theta$
cos	$\cos\theta$	$\sin\theta$	$-\sin\theta$	$-\cos\theta$	$-\cos\theta$
tan	$-\tan\theta$	$\cot\theta$	$-\cot\theta$	$-\tan\theta$	$\tan\theta$

Trigonometric Identities

$$\sin^2\theta + \cos^2\theta = 1 \qquad 1.96$$

$$1 + \tan^2\theta = \sec^2\theta \qquad 1.97$$

$$1 + \cot^2\theta = \csc^2\theta \qquad 1.98$$

$$\sin\ 2\theta = 2(\sin\theta)(\cos\theta) \qquad 1.99$$

$$\cos\ 2\theta = \cos^2\theta - \sin^2\theta = 1 - 2\ \sin^2\theta \qquad 1.100$$

$$\sin\theta = 2\left[\sin\left(\frac{\theta}{2}\right)\cos\left(\frac{\theta}{2}\right)\right] \qquad 1.101$$

$$\sin\left(\frac{\theta}{2}\right) = \pm\sqrt{\frac{1}{2}(1-\cos\theta)} \qquad 1.102$$

Two-Angle Formulas

$$\sin(\theta+\phi) = [\sin\theta][\cos\phi] + [\cos\theta][\sin\phi] \qquad 1.103$$

$$\sin(\theta-\phi) = [\sin\theta][\cos\phi] - [\cos\theta][\sin\phi] \qquad 1.104$$

$$\cos(\theta+\phi) = [\cos\theta][\cos\phi] - [\sin\theta][\sin\phi] \qquad 1.105$$

$$\cos(\theta-\phi) = [\cos\theta][\cos\phi] + [\sin\theta][\sin\phi] \qquad 1.106$$

C. GENERAL TRIANGLES

(*Law of Sines*)
$$\frac{\sin\ A}{a} = \frac{\sin\ B}{b} = \frac{\sin\ C}{c} \qquad 1.107$$

(*Law of Cosines*)
$$a^2 = b^2 + c^2 - 2bc(\cos\ A) \qquad 1.108$$

$$\text{Area} = \frac{1}{2}ab(\sin\ C) \qquad 1.109$$

D. HYPERBOLIC FUNCTIONS

Hyperbolic functions are specific equations containing the terms e^x and e^{-x}. These combinations of e^x and e^{-x} appear regularly in certain types of problems. In order to simplify the mathematical equations in which they appear, these hyperbolic functions are given special names and symbols.

$$\sinh\ x = \frac{e^x - e^{-x}}{2} \qquad 1.110$$

$$\cosh\ x = \frac{e^x + e^{-x}}{2} \qquad 1.111$$

$$\tanh\ x = \frac{e^x - e^{-x}}{e^x + e^{-x}} = \frac{\sinh\ x}{\cosh\ x} \qquad 1.112$$

$$\coth\ x = \frac{e^x + e^{-x}}{e^x - e^{-x}} = \frac{\cosh\ x}{\sinh\ x} \qquad 1.113$$

$$\text{sech}\ x = \frac{2}{e^x + e^{-x}} = \frac{1}{\cosh\ x} \qquad 1.114$$

$$\text{csch}\ x = \frac{2}{e^x - e^{-x}} = \frac{1}{\sinh\ x} \qquad 1.115$$

The hyperbolic identities are somewhat different from the standard trigonometric identities. Several of the most common identities are presented below.

$$\cosh^2 x - \sinh^2 x = 1 \qquad 1.116$$

$$1 - \tanh^2 x = \text{sech}^2 x \qquad 1.117$$

$$1 - \coth^2 x = -\text{csch}^2 x \qquad 1.118$$

$$\cosh\ x + \sinh\ x = e^x \qquad 1.119$$

$$\cosh\ x - \sinh\ x = e^{-x} \qquad 1.120$$

$$\sinh(x+y) = [\sinh x][\cosh y] + [\cosh x][\sinh y] \qquad 1.121$$

$$\cosh(x+y) = [\cosh x][\cosh y] + [\sinh x][\sinh y] \qquad 1.122$$

6 STRAIGHT LINE ANALYTIC GEOMETRY

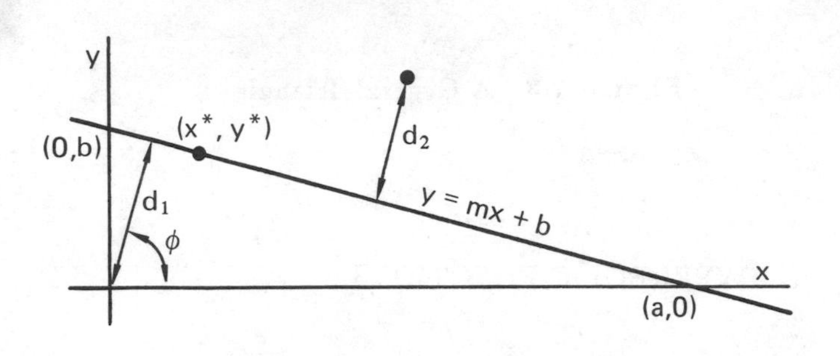

Figure 1.4 A Straight Line

A. EQUATIONS OF A STRAIGHT LINE

General Form

$$Ax + By + C = 0 \quad 1.123$$

Slope Form

$$y = mx + b \quad 1.124$$

Point-slope Form

$$(y - y^*) = m(x - x^*) \quad 1.125$$

$(x^*,\ y^*)$ is any point on the line.

Intercept Form

$$\frac{x}{a} + \frac{y}{b} = 1 \quad 1.126$$

Two-point Form

$$\frac{y - y_1^*}{x - x_1^*} = \frac{y_2^* - y_1^*}{x_2^* - x_1^*} \quad 1.127$$

Normal Form

$$x(\cos\phi) + y(\sin\phi) - d_1 = 0 \quad 1.128$$

Polar Form

$$r(\sin\theta) = d_1 \quad 1.129$$

B. POINTS, LINES, AND DISTANCES

The distance d_2 between a point and a line is

$$d_2 = \frac{|Ax_1 + By_1 + C|}{\sqrt{A^2 + B^2}} \quad 1.130$$

The distance between two points is

$$d = \sqrt{(x_2 - x_1)^2 + (y_2 - y_1)^2} \quad 1.131$$

Parallel Lines

$$\frac{A_1}{A_2} = \frac{B_1}{B_2} \quad 1.132$$

$$m_1 = m_2 \quad 1.133$$

Perpendicular Lines

$$A_1A_2 = -B_1B_2 \quad 1.134$$

$$m_1 = \frac{-1}{m_2} \quad 1.135$$

Point of Intersection of Two Lines

$$x_1 = \frac{B_2C_1 - B_1C_2}{A_2B_1 - A_1B_2} \quad 1.136$$

$$y_1 = \frac{A_1C_2 - A_2C_1}{A_2B_1 - A_1B_2} \quad 1.137$$

Smaller Angle Between Two Intersecting Lines

$$\tan\theta = \frac{A_1B_2 - A_2B_1}{A_1A_2 + B_1B_2} = \frac{m_2 - m_1}{1 + m_1m_2} \quad 1.138$$

$$\theta = |\arctan(m_1) - \arctan(m_2)| \quad 1.139$$

Example 1.17

What is the angle between the lines?

$$y_1 = -0.577x + 2$$
$$y_2 = +0.577x - 5$$

method 1:

$$\arctan\left[\frac{m_2 - m_1}{1 + m_1 m_2}\right] =$$
$$\arctan\left[\frac{0.577 - (-0.577)}{1 + (0.577)(-0.577)}\right] = 60°$$

method 2: Write both equations in general form.

$$-0.577x - y_1 + 2 = 0$$
$$0.577x - y_2 - 5 = 0$$
$$\arctan\left[\frac{A_1 B_2 - A_2 B_1}{A_1 A_2 + B_1 B_2}\right] =$$
$$\arctan\left[\frac{(-0.577)(-1) - (0.577)(-1)}{(-0.577)(0.577) + (-1)(-1)}\right] = 60°$$

method 3:

$$\theta = |\arctan(-0.577) - \arctan(0.577)|$$
$$= |-30° - 30°| = 60°$$

C. LINEAR AND CURVILINEAR REGRESSION

If it is necessary to draw a straight line through n data points (x_1, y_1), (x_2, y_2),...(x_n, y_n), the following method based on the theory of least squares can be used:

step 1: Calculate the following quantities.

$$\sum x_i,\ \sum x_i^2,\ (\sum x_i)^2,\ \overline{x} = \left(\frac{\sum x_i}{n}\right),\ \sum x_i y_i$$
$$\sum y_i,\ \sum y_i^2,\ (\sum y_1)^2,\ \overline{y} = \left(\frac{\sum y_i}{n}\right)$$

step 2: Calculate the slope of the line $y = mx + b$.

$$m = \frac{n\sum(x_i y_i) - (\sum x_i)(\sum y_i)}{n\sum x_i^2 - (\sum x_i)^2} \qquad 1.140$$

step 3: Calculate the y intercept.

$$b = \overline{y} - m\overline{x} \qquad 1.141$$

step 4: To determine the goodness of fit, calculate the *correlation coefficient.*

$$r = \frac{n\sum(x_i y_i) - (\sum x_i)(\sum y_i)}{\sqrt{|n\sum x_i^2 - (\sum x_i)^2||n\sum y_i^2 - (\sum y_i)^2|}} \qquad 1.142$$

If m is positive, r will be positive. If m is negative, r will be negative. As a general rule, if the absolute value of r exceeds 0.85, the fit is good. Otherwise, the fit is poor. r equals 1.0 if the fit is a perfect straight line.

Example 1.18

An experiment is performed in which the dependent variable (y) is measured against the independent variable (x). The results are as follows:

x	y
1.2	0.602
4.7	5.107
8.3	6.984
20.9	10.031

What is the least squares straight line equation which represents this data?

step 1:

$$\sum x_i = 35.1$$
$$\sum y_i = 22.72$$
$$\sum x_i^2 = 529.23$$
$$\sum y_i^2 = 175.84$$
$$(\sum x_i)^2 = 1232.01$$
$$(\sum y_i)^2 = 516.19$$
$$\overline{x} = 8.775$$
$$\overline{y} = 5.681$$
$$\sum x_i y_i = 292.34$$
$$n = 4$$

step 2:

$$m = \frac{(4)(292.34) - (35.1)(22.72)}{(4)(529.23) - (35.1)^2} = 0.42$$

step 3:

$$b = 5.681 - (0.42)(8.775) = 2.0$$

step 4: From equation 1.142, $r = 0.91$.

A low value of r does not eliminate the possibility of a non-linear relationship existing between x and y. It is possible that the data describes a parabolic, logarithmic, or other non-linear relationship. (Usually this will be apparent if the data are graphed.) It may be necessary to convert one or both variables to new variables by taking squares, square roots, cubes, or logs, to name a few of the possibilities.

The apparent shape of the line through the data will give a clue to the type of variable tranformation that is required. The following curves may be used as guides to some of the simpler variable transformations.

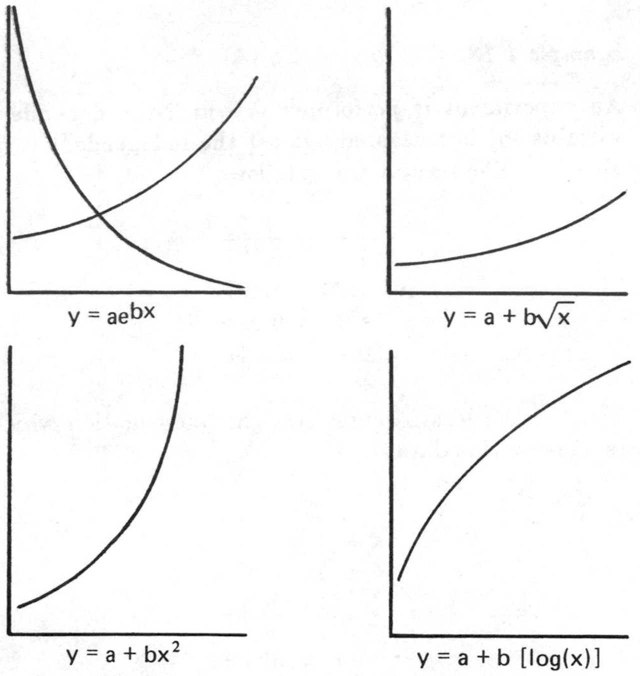

Figure 1.5 Non-Linear Data Plots

Example 1.19

Repeat example 1.18 assuming that the relationship between the variables is non-linear.

The first step is to graph the data. Since the graph has the appearance of the fourth case, it can be assumed that the relationship between the variables has the form of $y = a + b[\log(x)]$. Therefore, the variable change $z = \log(x)$ is made, resulting in the following set of data:

z	y
0.0792	0.602
0.672	5.107
0.919	6.984
1.32	10.031

If the regression analysis is performed on this set of data, the resulting equation and correlation coefficient are

$$y = -0.036 + 7.65z$$

$$r = 0.999$$

This a very good fit. The relationship between the variable x and y is approximately

$$y = -0.036 + 7.65[\log(x)]$$

Figure 1.6 illustrates several common problems encountered in trying to fit and evaluate curves from experimental data. Figure 1.6(a) shows a graph of clustered data with several extreme points. There will be moderate correlation due to the weighting of the extreme points, although there is little actual correlation at low values of the variables. The extreme data should be excluded or the range should be extended by obtaining more data.

Figure 1.6(b) shows that good correlation exists in general, but extreme points are missed, and the overall correlation is moderate. If the results within the small linear range can be used, the extreme points should be excluded. Otherwise, additional data points are needed, and curvilinear relationships should be investigated.

Figure 1.6(c) illustrates the problem of drawing conclusions of cause and effect. There may be a predictable relationship between variables, but that does not imply a cause and effect relationship. In the case shown, both variables are functions of a third variable, the city population. But, there is no direct relationship between the plotted variables.

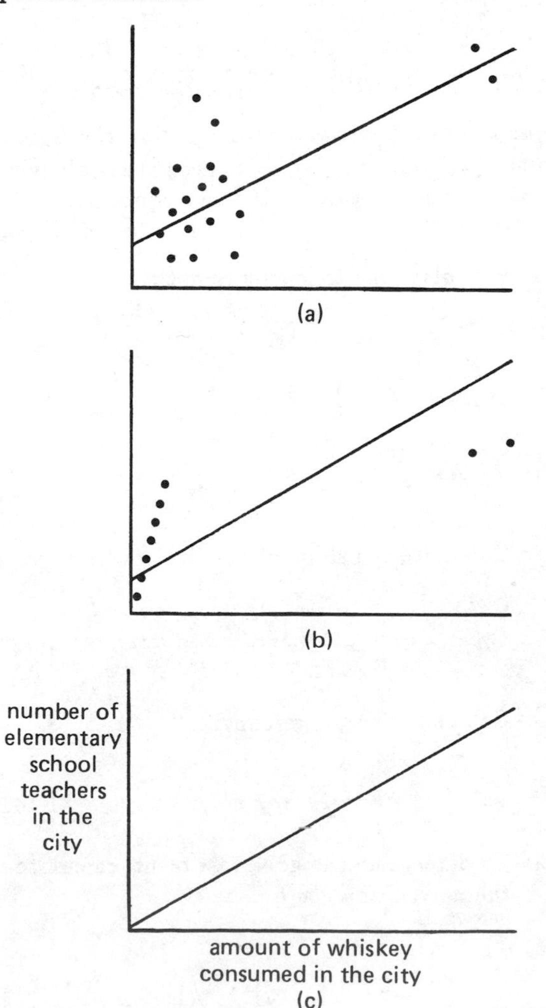

Figure 1.6 Common Regression Difficulties

D. THREE POINT METHOD

Three pairs of (x, y) data fully determine the parameters of a second degree equation characterized by three constants. The most common applications are parabola, hyperbola, and some exponential models with three constants. Special forms of *Hoerl's equations* can also be fitted. These equations can be fitted exactly through the three pairs of data by simultaneous solution of the equation model. For example, most heat capacity data can be made to fit the model $C_p = a + bT + \frac{c}{T^2}$. Another example, the *Antoine equation* for vapor pressure, is a three-constant hyperbola.

When solving the model equations, the difference and ratios of two data points occur frequently. To simplify the models, the following notation is used

$$e = x_2^2 - x_1^2 \qquad 1.143$$

$$f = x_3^2 - x_2^2 \qquad 1.144$$

$$m = x_2 - x_1 \qquad 1.145$$

$$n = x_3 - x_2 \qquad 1.146$$

$$o = y_2 - y_1 \qquad 1.147$$

$$p = y_3 - y_2 \qquad 1.148$$

$$q = x_2^z - x_1^z \qquad 1.149$$

$$r = x_3^z - x_2^z \qquad 1.150$$

$$s = \frac{x_2}{x_1} \qquad 1.151$$

$$t = \frac{x_3}{x_2} \qquad 1.152$$

$$u = \frac{y_2}{y_1} \qquad 1.153$$

$$v = \frac{y_3}{y_2} \qquad 1.154$$

The procedures for each model should be followed exactly in the order presented for the calculation of the constants. Each constant calculated is used in successive equations, until all of the constants are determined.

Parabola

$$y = a + bx + cx^2 \qquad 1.155$$

The constants are calculated directly:

$$c = \frac{no - mp}{ne - mf} \qquad 1.156$$

$$b = \frac{o - ce}{m} \qquad 1.157$$

$$a = y_1 - bx_1 - cx_1^2 \qquad 1.158$$

Hyperbola

$$y = a + \frac{b}{c + x} \qquad 1.59$$

The constants are calculated directly:

$$c = \frac{x_1 no - x_3 mp}{mp - no} \qquad 1.160$$

$$b = \frac{o(c + x_1)(c + x_2)}{-m} \qquad 1.161$$

$$a = y_1 - \frac{b}{c + x_1} \qquad 1.162$$

Hyperbola

$$y = c\left(\frac{a - x}{b + x}\right) \qquad 1.163$$

The constants are calculated directly:

$$b = \frac{mx_3y_3 + nx_1y_1 - (m + n)x_2y_2}{no - mp} \qquad 1.164$$

$$c = \frac{bo + x_2y_2 - x_1y_1}{-m} \qquad 1.165$$

$$a = \frac{by_1 + (c + y_1)x_1}{c} \qquad 1.166$$

Exponential

$$y = ae^{bx} + c \qquad 1.167$$

Finding the value of c requires trial and error solution of the following equation.

$$\frac{m}{n} = \frac{ln\left(\frac{y_2 - c}{y_1 - c}\right)}{ln\left(\frac{y_3 - c}{y_2 - c}\right)} \qquad 1.168$$

Once c is known,

$$b = \frac{ln\left(\frac{y_2 - c}{y_1 - c}\right)}{m} = \frac{ln\left(\frac{y_3 - c}{y_2 - c}\right)}{n} \qquad 1.169$$

$$a = \frac{y_1 - c}{e^{bx_1}} \qquad 1.170$$

Exponential

$$y = a + bx^c \qquad 1.171$$

Finding the value of a requires iteration of the following equation:

$$ln\ (y_1 - a)ln\left(\frac{1}{t}\right) + ln(y_2 - a)ln(st) + ln(y_3 - a)ln\left(\frac{1}{s}\right) = 0 \qquad 1.172$$

Once a is known,

$$c = \frac{ln\left(\frac{y_2 - a}{y_3 - a}\right)}{ln\left(\frac{1}{t}\right)} \qquad 1.173$$

$$b = \frac{y_2 - a}{x_2^c} \qquad 1.174$$

Hoerl's

$$y = ax^b e^{cx} \qquad 1.175$$

The coefficients are calculated directly:

$$c = \frac{ln\ u\ ln\ t - ln\ v\ ln\ s}{m\ ln\ t - n\ ln\ s} \qquad 1.176$$

$$b = \frac{ln\ u - cm}{ln\ s} \qquad 1.177$$

$$a = \frac{y_3}{x_3^b e^{cx_3}} \qquad 1.178$$

Heat Capacity Curve #1

$$y = a + bx + \frac{c}{x^2} \qquad 1.179$$

The constants are calculated directly:

$$b = \frac{fo - pes^2t^2}{fm - nes^2t^2} \qquad 1.180$$

$$c = \frac{(bm - o)x_1^4}{es^2} = \frac{(bn - p)x_2^4 t^2}{f} \qquad 1.181$$

$$a = y_1 - bx_1 - \frac{c}{x_1^2} \qquad 1.182$$

Heat Capacity Curve #2

$$y = a + bx + \frac{c}{x} \qquad 1.183$$

The constants are calculated directly:

$$b = \frac{no - pmst}{nm - nmst} \qquad 1.184$$

$$c = x_2^2\frac{(bm - o)}{ms} = \frac{x_2^2 t(bn - p)}{n} \qquad 1.185$$

$$a = y_1 - bx_1 - \frac{c}{x_1} \qquad 1.186$$

Log Relationship

$$y = a\ ln\ (b + x) + c \qquad 1.187$$

Iterate the first equation to find b.

$$\frac{o}{p} = \frac{ln\left(\frac{b + x_2}{b + x_1}\right)}{ln\left(\frac{b + x_3}{b + x_2}\right)} \qquad 1.188$$

$$a = \frac{o}{ln\left(\frac{b + x_2}{b + x_1}\right)} = \frac{p}{ln\left(\frac{b + x_3}{b + x_2}\right)} \qquad 1.189$$

$$c = y_3 - a\ ln\ (b + x_3) \qquad 1.190$$

S Curve

$$y = \frac{1}{a + be^{-cx}} \quad 1.191$$

Iterate first equation to find a.

$$\frac{m}{n} = \frac{ln\left(\dfrac{\frac{1}{y_2} - a}{\frac{1}{y_3} - a}\right)}{ln\left(\dfrac{\frac{1}{y_3} - a}{\frac{1}{y_2} - a}\right)} \quad 1.192$$

$$c = \frac{ln\left(\dfrac{\frac{1}{y_3} - a}{\frac{1}{y_1} - a}\right)}{-m} \quad 1.193$$

$$b = \left(\frac{1}{y_3} - a\right) e^{cx_3} \quad 1.194$$

Parabolic

$$y = a + bx + cx^z \text{ (where } z \neq 0, 1, \text{ or } 2) \quad 1.195$$

A fourth set of (x, y) pairs is required. The notation is

$$x_4^2 - x_3^2 = g \quad 1.196$$

$$x_4 - x_3 = j \quad 1.197$$

$$y_4 - y_3 = k \quad 1.198$$

$$x_4^z - x_3^z = l \quad 1.199$$

The value of z is found by iterating the equation

$$\frac{mp - no}{mr - nq} - \frac{nk - jp}{nl - jr} = 0 \quad 1.200$$

The constants c, b, and a are then calculated directly:

$$c = \frac{np - no}{mr - nq} \quad 1.201$$

$$b = \frac{o - cq}{m} \quad 1.202$$

$$a = y_1 - bx_1 - cx^z \quad 1.203$$

Cubic

$$a + bx + cx^2 + dx^3 \quad 1.204$$

$$d = \frac{(fj - gn)(fo - ep) - (en - fm)(gp - fk)}{(fj - gn)(fq - er) - (en - fm)(gr - fl)} \quad 1.205$$

$$b = \frac{f(o - dq) - e(p - dr)}{en - mf} = \frac{g(p - dr) - f(k - dl)}{(fj - gn)} \quad 1.206$$

$$c = \frac{o - bm - dq}{e} = \frac{p - bn - dr}{f} = \frac{k - bi - dl}{g} \quad 1.207$$

$$a = y_1 - bx_1 - cx_1^2 - dx_1^3 \quad 1.208$$

7 CONIC SECTIONS

A. CIRCLE

The center-radius form of a circle with radius r and center at (h, k) is

$$(x - h)^2 + (y - k)^2 = r^2 \quad 1.209$$

The x-intercept is found by letting $y = 0$ and solving for x. The y-intercept is found similarly.

The general form is

$$x^2 + y^2 + Dx + Ey + F = 0 \quad 1.210$$

This can be converted to the center-radius form.

$$\left(x + \frac{D}{2}\right)^2 + \left(y + \frac{E}{2}\right)^2 = \frac{1}{4}(D^2 + E^2 - 4F) \quad 1.211$$

If the right-hand side is greater than zero, the equation is that of a circle with center at $(-\frac{1}{2}D, -\frac{1}{2}E)$ and radius given by the square root of the right-hand side. If the right-hand side is zero, the equation is that of a point. If the right-hand side is negative, the plot is imaginary.

B. PARABOLA

A parabola is formed by a locus of points equidistant from point F and a line called the *directrix*.

$$(y - k)^2 = 4p(x - h) \quad 1.212$$

Equation 1.212 represents a parabola with *vertex* at (h, k), focus at $(p + h, k)$, and directrix equation $x = h - p$. The parabola points to the left if $p > 0$ and points to the right if $p < 0$.

$$(x-h)^2 = 4p(y-k) \qquad 1.213$$

Equation 1.213 represents a parabola with vertex at (h, k), focus at $(h, p+k)$, and directrix equation $y = k-p$. The parabola points down if $p > 0$ and points up if $p < 0$.

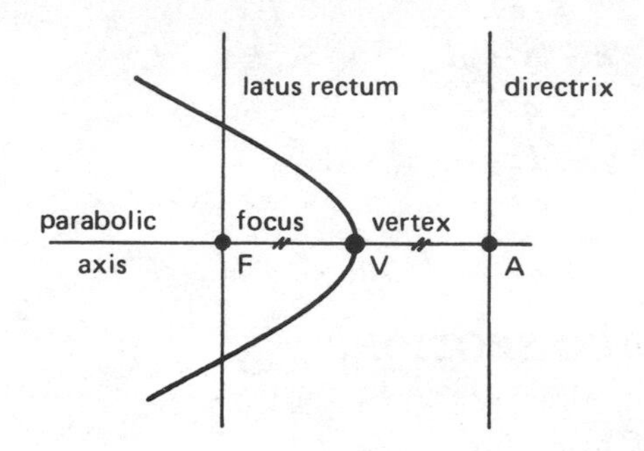

Figure 1.7 A Parabola

An alternate form of the vertically-oriented parabola is

$$y = Ax^2 + Bx + C \qquad 1.214$$

This parabola has a vertex at

$$\left(\frac{-B}{2A}, C - \frac{B^2}{4A}\right) \qquad 1.215$$

and points down if $A > 0$ and points up if $A < 0$.

C. ELLIPSE

An ellipse is formed from a locus of points such that the sum of distances from the two foci is constant. The distance between the two foci is $2c$. The sum of those distances is

$$F_1P + PF_2 = 2a \qquad 1.216$$

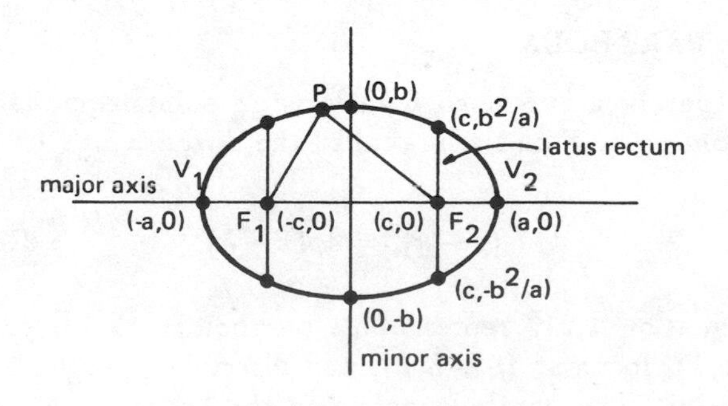

Figure 1.8 An Ellipse

The eccentricity of an ellipse is less than 1, and is equal to

$$e = \frac{\sqrt{a^2 - b^2}}{a} \qquad 1.217$$

For an ellipse centered at the origin:

$$\left(\frac{x}{a}\right)^2 + \left(\frac{y}{b}\right)^2 = 1 \qquad 1.218$$

$$b^2 = a^2 - c^2 \qquad 1.219$$

If $a > b$, the ellipse is wider than it is tall. If $a < b$, it is taller than it is wide.

For an ellipse centered at (h, k),

$$\frac{(x-h)^2}{a^2} + \frac{(y-k)^2}{b^2} = 1 \qquad 1.220$$

The general form of an ellipse is

$$Ax^2 + Cy^2 + Dx + Ey + F = 0 \qquad 1.221$$

If $A \neq C$ and both have the same sign, the general form can be written as

$$A\left(x + \frac{D}{2A}\right)^2 + C\left(y + \frac{E}{2C}\right)^2 = M \qquad 1.222$$

$$M = \frac{D^2}{4A} + \frac{E^2}{4C} - F \qquad 1.223$$

If $M = 0$, the graph is a single point at

$$\left(\frac{-D}{2A}, \frac{-E}{2C}\right) \qquad 1.224$$

If $M < 0$, the graph is the null set.

If $M > 0$, then the ellipse is centered

$$\left(-\frac{D}{2A}, -\frac{E}{2C}\right) \qquad 1.225$$

and the equation can be rewritten

$$\frac{\left(x + \frac{D}{2A}\right)^2}{\frac{M}{A}} + \frac{\left(y + \frac{E}{2C}\right)^2}{\frac{M}{C}} = 1 \qquad 1.226$$

D. HYPERBOLA

A hyperbola is a locus of points such that $F_1P - PF_2 = 2a$. The distance between the foci is $2c$, and $a < c$.

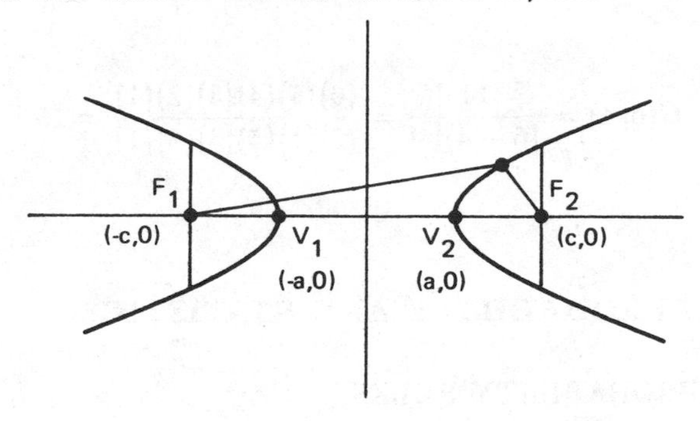

Figure 1.9 A Hyperbola

For hyperbola centered at the origin with foci on the x-axis,

$$\left(\frac{x}{a}\right)^2 - \left(\frac{y}{b}\right)^2 = 1 \text{ with } b^2 = c^2 - a^2 \qquad 1.227$$

If the foci are on the y-axis,

$$\left(\frac{y}{a}\right)^2 - \left(\frac{x}{b}\right)^2 = 1 \qquad 1.228$$

The coordinates and length of the *latus recta* are the same as for the ellipse. The hyperbola is asymptotic to the lines

$$y = \pm\left(\frac{b}{a}\right)x \qquad 1.229$$

The asymptotes need not be perpendicular, but if they are, the hyperbola is known as a *rectangular hyperbola.*

If the asymptotes are the x and y axes, the equation of the hyperbola is

$$xy = \pm a^2 \qquad 1.230$$

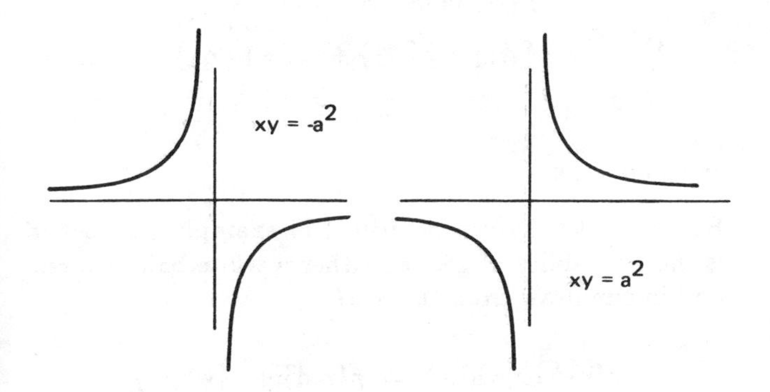

Figure 1.10 Rectangular Hyperbolas

In general, for a hyperbola with transverse axis parallel to the x-axis and center at (h, k),

$$\frac{(x-h)^2}{a^2} - \frac{(y-k)^2}{b^2} = 1 \qquad 1.231$$

The general form of the hyperbolic equation is

$$Ax^2 + Cy^2 + Dx + Ey + F = 0 \qquad 1.232$$

If $AC < 0$, the equation can be rewritten as

$$A\left(x + \frac{D}{2A}\right)^2 + C\left(y + \frac{E}{2C}\right)^2 = M \qquad 1.233$$

where

$$M = \frac{D^2}{4A} + \frac{E^2}{4C} - F \qquad 1.234$$

If $M = 0$, the graph is two intersecting lines.

If $M \neq 0$, the graph is a hyperbola with center at

$$\left(-\frac{D}{2A}, -\frac{E}{2C}\right) \qquad 1.235$$

The transerse axis is horizontal if (M/A) is positive. It is vertical if (M/C) is positive.

8 SPHERES

The equation of a sphere whose center is at the point (h, k, l) and whose radius is r is

$$(x-h)^2 + (y-k)^2 + (z-l)^2 = r^2 \qquad 1.236$$

If the sphere is centered at the origin, the equation is

$$x^2 + y^2 + z^2 = r^2 \qquad 1.237$$

9 PERMUTATIONS AND COMBINATIONS

Suppose you have n objects, and you wish to work with a subset of r of them. An order-conscious arrangement of n objects taken r at a time is known as *permutation.* The permutation is said to be order-conscious because the arrangement of two objects (say A and B) as AB is different from the arrangement BA. There are a number of ways of taking n objects r at a time. The total number of possible permutations is

$$P(n,r) = \frac{n!}{(n-r)!} \quad 1.238$$

Example 1.20

A shelf only has room for three vases. If four different vases are available, how many ways can the shelf be arranged?

$$P(4,3) = \frac{4!}{(4-3)!} = \frac{(4)(3)(2)(1)}{(1)} = 24$$

The special cases of n objects taken n at a time are illustrated by the following examples.

Example 1.21

How many ways can seven resistors be connected end-to-end in a single unit?

$$P(7,7) = \frac{7!}{(7-7)!} = \frac{7!}{0!} = 7! = 5040$$

Example 1.22

Five people are to sit at a round table with five chairs. How many ways can these five people be arranged so that they all have different companions?

This is known as a *ring permutation*. Since the starting point of the arrangement around the circle does not affect the number of permutations, the answer is

$$(5-1)! = 4! = 24$$

An arrangement of n objects taken r at a time is known as a *combination* if the arrangement is not order-conscious. The total number of possible combinations is

$$C(n,r) = \frac{n!}{(n-r)!r!} \quad 1.239$$

Example 1.23

How many possible ways can six people fit into a four-seat boat?

$$C(6,4) = \frac{6!}{(6-4)!4!} = \frac{(6)(5)(4)(3)(2)(1)}{(2)(1)(4)(3)(2)(1)} = 15$$

10 PROBABILITY AND STATISTICS

A. PROBABILITY RULES

The following rules are applied to sample space A and B

$$A = [A_1, A_2, A_3, \ldots, A_n]$$
$$B = [B_1, B_2, B_3, \ldots, B_n] \quad 1.240$$

where the A_i and B_i are independent.

Rule 1:

$$p(\phi) = \text{ probability of an impossible event } = 0 \quad 1.241$$

Example 1.24

An urn contains five white balls, two red balls, and three green balls. What is the probability of drawing a blue ball from the urn?

$$p\{\text{blue ball}\} = p\{\phi\} = 0$$

Rule 2:

$$p\{A_1 \text{ or } A_2 \text{ or} \ldots \text{ or } A_n\}$$
$$= p\{A_1\} + p\{A_2\} + \ldots + p\{A_n\} \quad 1.242$$

Example 1.25

Returning to the urn described in example 1.24, what is the probablity of getting either a white ball or a red ball in one draw from the urn?

$$p\{\text{red or white}\} = p\{\text{red}\} + p\{\text{white}\}$$
$$= 0.2 + 0.5 = 0.7$$

Rule 3:

$$p\{A_i \text{ and } B_i \text{ and} \ldots Z_i\} = p\{A_i\}p\{B_i\}\ldots p\{Z_i\} \qquad 1.243$$

Example 1.26

Given two identical urns (as described in example 1.24), what is the probability of getting a red ball from the first urn and a green ball from the second urn, given one draw from each urn?

$$p\{\text{red and green}\} = p\{\text{red}\}p\{\text{green}\} = (0.2)(0.3) = .06$$

Rule 4:

$$p\{\text{not } A\} = \text{probability of event } A \text{ not occuring} = 1 - p\{A\}$$

Example 1.27

Given the urn of example 1.24, what is the probability of not getting a red ball from the urn in one draw?

$$p\{\text{not red}\} = 1 - p\{\text{red}\} = 1 - 0.2 = 0.8$$

Rule 5:

$$p\{A_i \text{ or } B_i\} = p\{A_i\} + p\{B_i\} - p\{A_i\}p\{B_i\} \qquad 1.245$$

Example 1.28

Given one urn as described in example 1.24 and a second urn containing eight red balls and two black balls, what is the probability of drawing either a white ball from the first urn or a red ball from the second urn, given one draw each?

$$p\{\text{red or white}\} = p\{\text{white}\} + p\{\text{red}\} - p\{\text{white}\}p\{\text{red}\} = 0.5 + 0.8 - (0.5)(0.8) = 0.9$$

Rule 6:

$p\{\frac{A}{B}\}$ = probability that A will occur given that B has already occurred, where the two events are dependent.

$$= \frac{p\{A \text{ and } B\}}{p\{B\}} \qquad 1.246$$

The above equation is known as *Bayes Theorem.*

B. PROBABILITY DENSITY FUNCTIONS

Probability density functions are mathematical functions giving the probabilities of numerical events. A *numerical event* is any occurrence that can be described by an integer or real number. For example, obtaining heads in a coin toss is not a numerical event. However, a concrete sample having a compressive strength less than 5000 psi is a numerical event.

Discrete density functions give the probability that the event x will occur. That is,

$$f(x) = \text{probability of a process having a value of } x \qquad 2.247$$

Important discrete functions are the binomial and Poisson distributions.

Binomial

n is the number of trials
x is the number of successes
p is the probability of success in a single trial
q is the probability of failure, $1 - p$
$\binom{n}{x}$ is the binomial coefficient $= \frac{n!}{(n-x)!x!}$
$x! = x(x-1)(x-2)\ldots(2)(1)$

Then, the probability of obtaining x successes in n trials is

$$f\{x\} = \binom{n}{x} p^x q^{(n-x)} \qquad 1.248$$

The mean of the binomial distribution is np. The variance of the distribution is npq.

Example 1.29

In a large quantity of items, 5% are defective. If seven items are sampled, what is the probability that exactly three will be defective?

$$f\{3\} = \binom{7}{3}(0.05)^3(0.95)^4 = 0.0036$$

Poisson

Suppose an event occurs, on the average, λ times per period. The probability that the event will occur x times per period is

$$f\{x\} = \frac{e^{-\lambda}\lambda^x}{x!} \qquad 1.249$$

λ is both the distribution mean and the variance. λ must be a number greater than zero.

Example 1.30

The number of customers arriving in some period is distributed as Poisson with a mean of eight. What is the probability that six customers will arrive in any given period?

$$f\{6\} = \frac{e^{-8}8^6}{6!} = 0.122$$

Continuous probability density functions are used to find the cumulative distribution functions, $F(x)$. Cumulative distribution functions give the probability of event x or less occurring.

$$x = \text{ any value, not necessarily an integer} \qquad 1.250$$

$$f\{x\} = \frac{dF\{x\}}{dx} \qquad 1.251$$

$$F\{x\} = \text{ probability of } x \text{ or less occurring} \qquad 1.252$$

Exponential

$$f\{x\} = u(e^{-ux}) \qquad 1.253$$

$$F\{x\} = 1 - e^{-ux} \qquad 1.254$$

The mean of the exponential distribution is $\frac{1}{u}$. The variance is $\left(\frac{1}{u}\right)^2$.

Example 1.31

The reliability of a unit is exponentially distributed with mean time to failure (MTBF) of 1000 hours. What is the probability that the unit will be operational at $t = 1200$ hours?

The reliability of an item is (1 − probability of failing before time t). Therefore,

$$R\{t\} = 1 - F\{t\} = 1 - (1 - e^{-ux}) = e^{-ux}$$

$$u = \frac{1}{\text{MTBF}} = \frac{1}{1000} = 0.001$$

$$R\{1200\} = e^{-(0.001)(1200)} = 0.3$$

Normal

Although $f\{x\}$ can be expressed mathematically for the normal distribution, tables are used to evaluate $F\{x\}$ since $f\{x\}$ cannot be easily integrated. Since the x axis of the normal distribution will seldom correspond to actual sample variables, the sample values are converted into standard values. Given the mean, u, and the standard deviation, σ, the standard normal variable is

$$z = \frac{\text{sample value } - u}{\sigma} \qquad 1.255$$

Then, the probability of a sample exceeding the given sample value is equal to the area in the tail past point z.

Example 1.32

Given a population that is normally distributed with mean of 66 and standard deviation of five, what percent of the population exceeds 72?

$$z = \frac{72 - 66}{5} = 1.2$$

Then, from table 1.9,

$$p\{\text{exceeding } 72\} = 0.5 - 0.3849 = 0.1151 \text{ or } 11.5\%$$

C. STATISTICAL ANALYSIS OF EXPERIMENTAL DATA

Experiments can take on many forms. An experiment might consist of measuring the weight of one cubic foot of concrete. Or, an experiment might consist of measuring the speed of a car on a roadway. Generally, such experiments are performed more than once to increase the precision and accuracy of the results.

Table 1.9
Standard Normal Variables

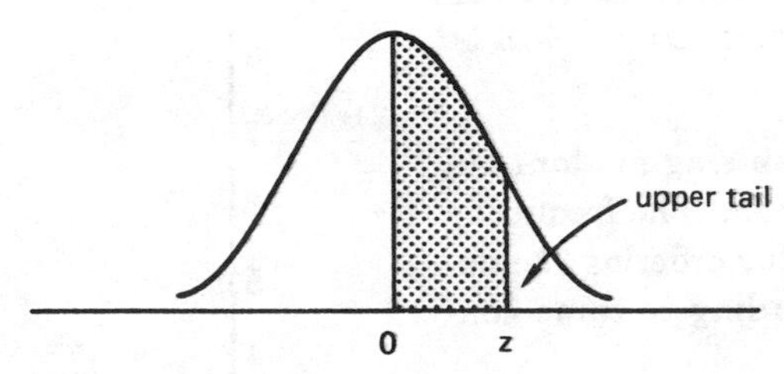

z	0	1	2	3	4	5	6	7	8	9
0.0	.0000	.0040	.0080	.0120	.0160	.0199	.0239	.0279	.0319	.0359
0.1	.0398	.0438	.0478	.0517	.0557	.0596	.0636	.0675	.0714	.0754
0.2	.0793	.0832	.0871	.0910	.0948	.0987	.1026	.1064	.1103	.1141
0.3	.1179	.1217	.1255	.1293	.1331	.1368	.1406	.1443	.1480	.1517
0.4	.1554	.1591	.1628	.1664	.1700	.1736	.1772	.1808	.1844	.1879
0.5	.1915	.1950	.1985	.2019	.2054	.2088	.2123	.2157	.2190	.2224
0.6	.2258	.2291	.2324	.2357	.2389	.2422	.2454	.2486	.2518	.2549
0.7	.2580	.2612	.2642	.2673	.2704	.2734	.2764	.2794	.2823	.2852
0.8	.2881	.2910	.2939	.2967	.2996	.3023	.3051	.3078	.3106	.3133
0.9	.3159	.3186	.3212	.3238	.3264	.3289	.3315	.3340	.3365	.3389
1.0	.3413	.3438	.3461	.3485	.3508	.3531	.3554	.3577	.3599	.3621
1.1	.3643	.3665	.3686	.3708	.3729	.3749	.3770	.3790	.3810	.3830
1.2	.3849	.3869	.3888	.3907	.3925	.3944	.3962	.3980	.3997	.4015
1.3	.4032	.4049	.4066	.4082	.4099	.4115	.4131	.4147	.4162	.4177
1.4	.4192	.4207	.4222	.4236	.4251	.4265	.4279	.4292	.4306	.4319
1.5	.4332	.4345	.4357	.4370	.4382	.4394	.4406	.4418	.4429	.4441
1.6	.4452	.4463	.4474	.4484	.4495	.4505	.4515	.4525	.4535	.4545
1.7	.4554	.4564	.4573	.4582	.4591	.4599	.4608	.4616	.4625	.4633
1.8	.4641	.4649	.4656	.4664	.4671	.4678	.4686	.4693	.4699	.4706
1.9	.4713	.4719	.4726	.4732	.4738	.4744	.4750	.4756	.4761	.4767
2.0	.4772	.4778	.4783	.4788	.4793	.4798	.4803	.4808	.4812	.4817
2.1	.4821	.4826	.4830	.4834	.4838	.4842	.4846	.4850	.4854	.4857
2.2	.4861	.4864	.4868	.4871	.4875	.4878	.4881	.4884	.4887	.4890
2.3	.4893	.4896	.4898	.4901	.4904	.4906	.4909	.4911	.4913	.4916
2.4	.4918	.4920	.4922	.4925	.4927	.4929	.4931	.4932	.4934	.4936
2.5	.4938	.4940	.4941	.4943	.4945	.4946	.4948	.4949	.4951	.4952
2.6	.4953	.4955	.4956	.4957	.4959	.4960	.4961	.4962	.4963	.4964
2.7	.4965	.4966	.4967	.4968	.4969	.4970	.4971	.4972	.4973	.4974
2.8	.4974	.4975	.4976	.4977	.4977	.4978	.4979	.4979	.4980	.4981
2.9	.4981	.4982	.4982	.4983	.4984	.4984	.4985	.4985	.4986	.4986
3.0	.4987	.4987	.4987	.4988	.4988	.4989	.4989	.4989	.4990	.4990
3.1	.4990	.4991	.4991	.4991	.4992	.4992	.4992	.4992	.4993	.4993
3.2	.4993	.4993	.4994	.4994	.4994	.4994	.4994	.4995	.4995	.4995
3.3	.4995	.4995	.4995	.4996	.4996	.4996	.4996	.4996	.4996	.4997
3.4	.4997	.4997	.4997	.4997	.4997	.4997	.4997	.4997	.4997	.4998
3.5	.4998	.4998	.4998	.4998	.4998	.4998	.4998	.4998	.4998	.4998
3.6	.4998	.4998	.4999	.4999	.4999	.4999	.4999	.4999	.4999	.4999
3.7	.4999	.4999	.4999	.4999	.4999	.4999	.4999	.4999	.4999	.4999
3.8	.4999	.4999	.4999	.4999	.4999	.4999	.4999	.4999	.4999	.4999
3.9	.5000	.5000	.5000	.5000	.5000	.5000	.5000	.5000	.5000	.5000

Of course, the intrinsic variability of the process being measured will cause the observations to vary, and we would not expect the experiment to yield the same result each time it was performed. Eventually, a collection of experimental outcomes (observations) will be available for analysis.

One fundamental technique for organizing random observations is the *frequency distribution*. The frequency distribution is a systematic method for ordering the observations from small to large, according to some convenient numerical characteristic.

Example 1.33

The number of cars that travel through an intersection between 12 noon and 1 p.m. is measured for 30 consecutive working days. The results of the 30 observations are:

79,66,72,70,68,66,68,76,73,71,74,70,71,69,67,
74,70,68,69,64,75,70,68,69,64,69,62,63,63,61

What is the frequency distribution using an interval of two cars per hour?

cars per hour	frequency of occurrence
60-61	1
62-63	3
64-65	2
66-67	3
68-69	8
70-71	6
72-73	2
74-75	3
76-77	1
78-79	1

In example 1.33, two cars per hour is known as the *step interval*. The step interval should be chosen so that the data is presented in a meaningful manner. If there are too many intervals, many of them will have zero frequencies. If there are too few intervals, the frequency distribution will have little value. Generally, 10 to 15 intervals are used.

Once the frequency distribution is complete, it can be represented graphically as a histogram. The procedure in drawing a histogram is to mark off the interval limits on a number line and then draw bars with lengths that are proportional to the frequencies in the intervals. If it is necessary to show the continuous nature of the data, a frequency polygon can be drawn.

Example 1.34

Draw the frequency histogram and frequency polygon for the data given in example 1.33.

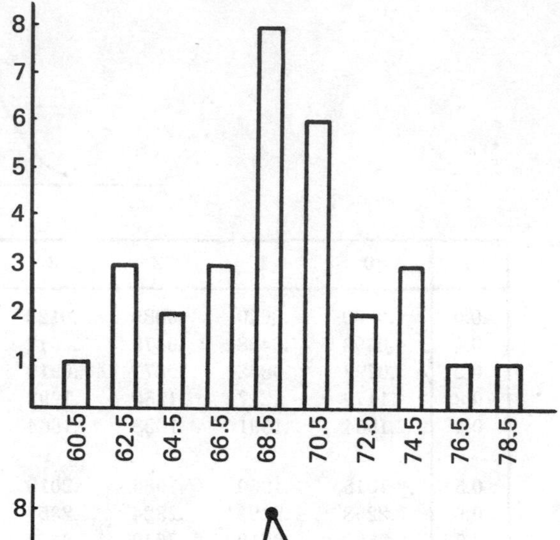

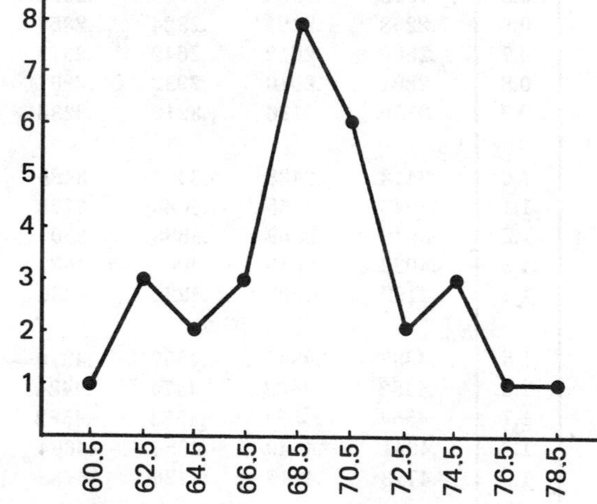

If it necessary to know the number or percentage of observations that occur up to and including some value, the cumulative frequency table can be formed. This procedure is illustrated in the following example.

Example 1.35

From the cumulative frequency distribution and graph for the data given in example 1.33.

cars per hour	frequency	cumulative frequency	cumulative percent
60-61	1	1	3
62-63	3	4	13
64-65	2	6	20
66-67	3	9	30
68-69	8	17	57
70-71	6	23	77
72-73	2	25	83
74-75	3	28	93
76-77	1	29	97
78-79	1	30	100

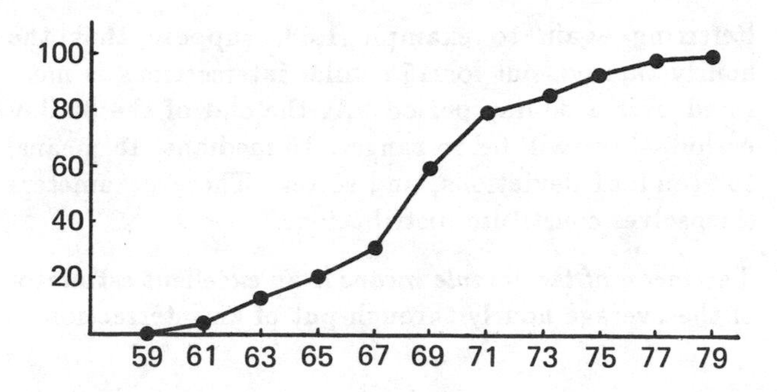

It is often unnecessary to present the experimental data in its entirety, either in tabular or graphical form. In such cases, the data and distribution can be represented by various parameters. One type of parameter is a measure of *central tendency*. Mode, median, and mean are measures of central tendency. The other type of parameter is a measure of dispersion. Standard deviation and variance are measures of dispersion.

The *mode* is the observed value which occurs most frequently. The mode may vary greatly between series of observations. Therefore, its main use is as a quick measure of the central value since no computation is required to find it. Beyond this, the usefulness of the mode is limited.

The *median* is the point in the distribution which divides the total observations into two parts containing equal numbers of observations. It is not influenced by the extremity of scores on either side of the distribution. The median is found by counting up (from either end of the frequency distribution) until half of the observations have been accounted for. The procedure is more difficult if the median falls within an interval, as illustrated in example 1.36.

Similar in concept to the median are *percentile ranks, quartiles*, and *deciles*. The median could also have been called the *50th percentile observation*. Similarly, the 80th percentile would be the number of cars per hour for which the cumulative frequency was 80%. The quartile and decile points on the distribution divide the observations or distributions into segments of 25% and 10%, respectively.

The *arithmetic mean* is the arithmetic average of the observations. The *mean* may be found without ordering the data (which was necessary to find the mode and median). The mean can be found from the following formula:

$$x = \left(\frac{1}{n}\right)(x_1 + x_2 + \ldots + x_n) = \frac{\sum x_i}{n} \qquad 1.256$$

The *geometric mean* is used occasionally when it is necessary to average ratios. The geometric mean is calculated as

$$\text{geometric mean} = \sqrt[n]{x_1 x_2 x_3 \ldots x_n} \qquad 1.257$$

The *harmonic mean* is defined as

$$\text{harmonic mean} = \frac{n}{\frac{1}{x_1} + \frac{1}{x_2} + \ldots + \frac{1}{x_n}} \qquad 1.258$$

The *root-mean-squared value* (rms) of a series of observations is defined as

$$x_{rms} = \sqrt{\frac{\sum x_i^2}{n}} \qquad 1.259$$

Example 1.36

Find the mode, median, and arithmetic mean of the distribution represented by the data given in example 1.33.

The mode is the interval 68-69, since this interval has the highest frequency. If 68.5 is taken as the interval center, then 68.5 would be the mode.

Since there are 30 observations, the median is the value which separates the observations into two groups of 15. From example 1.35, the median occurs somewhere within the 68-69 interval. Up through interval 66-67, there are nine observations, so six more are needed to make 15. Interval 68-69 has eight observations, so the median is found to be $\left(\frac{6}{8}\right)$ or $\left(\frac{3}{4}\right)$ of the way through the interval. Since the real limits of the interval are 67.5 and 69.5, the median is located at

$$67.5 + \frac{3}{4}(69.5 - 67.5) = 69$$

The mean can be found from the raw data or from the grouped data using the interval center as the assumed observation value. Using the raw data,

$$\overline{x} = \frac{\sum x}{n} = \frac{2069}{30} = 68.97$$

The simplest statistical parameter which describes the variation in observed data is the *range*. The range is found by subtracting the smallest value from the largest. Since the range is influenced by extreme (low probability) observations, its use as a measure of variability is limited.

The *standard deviation* is a better estimate of variability because it considers every observation. The standard deviation can be found from

$$\sigma = \sqrt{\frac{\sum (x_i - \overline{x})^2}{n}} = \sqrt{\frac{\sum x_i^2}{n} - (\overline{x})^2} \qquad 1.260$$

MATH

The above formula assumes that n is a large number, such as above 50. Theoretically, n is the size of the entire population. If a small sample (less that 50) is used to calculate the standard deviation of the distribution, the formulas are changed. The *sample standard deviation* is

$$s = \sqrt{\frac{\sum(x_i - \bar{x})^2}{n-1}} = \sqrt{\frac{\sum x_i^2 - \frac{(\sum x_i)^2}{n}}{n-1}} \qquad 1.261$$

The difference is small when n is large, but care must be taken in reading the problem. If the *standard deviation of the sample* is requested, calculate σ. If an estimate of the *population standard deviation* or *sample standard deviation* is requested, calculate s. (Note that the standard deviation of the sample is not the same as the sample standard deviation.)

The *relative dispersion* is defined as a measure of dispersion divided by a measure of central tendency. The *coefficient of variation* is a relative dispersion calculated from the standard deviation and the mean. That is,

$$\text{coefficient of variation} = \frac{s}{\bar{x}} \qquad 1.262$$

Skewness is a measure of frequency distribution's lack of symmetry. It is calculated as

$$\text{skewness} = \frac{\bar{x} - \text{mode}}{s}$$

$$\approx \frac{3(\bar{x} - \text{median})}{s} \qquad 1.263$$

Example 1.37

Calculate the range, standard deviation of the sample, and population variance from the data given in example 1.33.

$$\sum x = 2069$$

$$\left(\sum x\right)^2 = 4{,}280{,}761$$

$$\sum x^2 = 143{,}225$$

$$n = 30 \quad \bar{x} = 68.97$$

$$\sigma = \sqrt{\frac{143{,}225}{30} - (68.97)^2} = 4.16$$

$$s = \sqrt{\frac{143{,}224 - \frac{(4{,}280{,}761)}{30}}{29}} = 4.29$$

$$s^2 = 18.4 \text{ (sample variance)}$$

$$\sigma^2 = 17.3 \text{ (population variance)}$$

$$R = 79 - 61 = 18$$

Referring again to example 1.33, suppose that the hourly through-put for 15 similar intersections is measured over a 30-day period. At the end of the 30-day period, there will be 15 ranges, 15 medians, 15 means, 15 standard deviations, and so on. These parameters themselves constitute distributions.

The *mean of the sample means* is an excellent estimator of the average hourly through-put of an intersection μ.

$$\mu = \left(\frac{1}{15}\right)\sum \bar{x} \qquad 1.264$$

The standard deviation of the sample means is known as the *standard error of the mean* to distinguish it from the standard deviation of the raw data. The standard error is written as $\sigma_{\bar{x}}$.

The standard error is not a good estimator of the population standard deviation σ'.

In general, if k sets of n observations each are used to estimate the population mean (μ) and the population standard deviation (σ'), then,

$$\mu \approx \left(\frac{1}{k}\right)\sum \bar{x} \qquad 1.265$$

$$\sigma' \approx \sqrt{k}\sigma_{\bar{x}} \qquad 1.266$$

11 BASIC HYPOTHESIS TESTING

Suppose a distribution is $\sim N(\mu, \sigma'^2)$.[1] If samples of size n are taken k times, the values of the sample means $\bar{x}$ will form a distribution themselves. These means also will be distributed normally with the form

$$\sim N\left(\mu, \frac{\sigma'^2}{n}\right) \qquad 1.267$$

That is, the mean of the sample means will be identical to the original population, but the variance and standard deviation will be much smaller. This is known as the *central limit theorem*.

Thus, the probability that $\bar{x}$ exceeds some value, say x^*, is

$$p\left\{z > \frac{x^* - \mu}{\frac{\sigma'}{\sqrt{n}}}\right\} \qquad 1.268$$

[1] This is the standard method of saying the distribution is normally distributed with mean μ and a variance of σ'^2.

This can be solved as an *exceedance problem* (see example 1.32), or a hypothesis test can be performed. A *hypothesis test* has the following characteristics:

- a sample is taken in an experiment
- a parameter (usually $\bar{x}$) is measured
- it is desired to know if the sample could have come from a population $\sim N(\mu, \sigma'^2)$

There are many types of hypothesis tests, depending on the type of population (i.e., whether or not normal), the parameter being tested (i.e., central tendency or dispersion), and the size of the sample.

If the sample size is not much greater than 30, if the native population is assumed to be normal, and if μ and σ' are known, the following procedure can be used.

step 1: Assume random sampling from a normal population.

step 2: Choose the desired confidence level, C. Usually a 95% confidence level result is said to be ***significant***. 99% test results are said to be ***highly significant***.

step 3: Decide on a 1-tail or 2-tail test. If the question is worded as, "Has the population mean changed?" or, "Are the populations the same?", then a ***2-tail test*** is needed. If the question is, "Has the mean increased?" or, "... decreased?", then a ***1-tail test*** is needed.

step 4: From the normal table find the value z' for a table entry equal to

$$\frac{1-C}{\#\text{tails in the test}} \qquad 1.269$$

step 5: Calculate

$$z = \frac{|\bar{x}-\mu|}{\dfrac{\sigma'}{\sqrt{n}}} \qquad 1.270$$

If $z \geq z'$, then the distributions are not the same.

Example 1.38

When operating properly, a chemical plant has a product output which is normally distributed with mean 880 tons/day and standard deviation of 21 tons. The output is measured on 50 consecutive days, and the mean output is 871 tons/day. Is the plant operating correctly?

step 1: Assume a random sampling from the normal distribution.

step 2: Choose $C = 0.95$ for significant results.

step 3: Wanting to know if the plant is operating correctly is the same as asking, "Has anything changed?" There is no mention of *direction* (i.e., the question was not, "Has the output decreased?"). Therefore, choose a 2-tail test.

step 4: $1 - C = 0.05$. The 0.05 outside lower limit in table 1.9 is $z' = 1.96$. (This corresponds to an area under the curve of $0.5 - 0.025 = 0.475$.)

step 5:

$$z = \frac{|871-880|}{\dfrac{21}{\sqrt{50}}} = 3.03$$

Since $3.03 > 1.96$, the distributions are not the same. There is a 95% chance that the plant is not operating correctly.

12 DIFFERENTIAL CALCULUS

A. TERMINOLOGY

Given y, a function of x, the first derivative with respect to x may be written as

$$Dy, y', \text{ or } \frac{dy}{dx} \qquad 1.271$$

The first derivative corresponds to the slope of the line described by the function y. The second derivative may be written as

$$D^2y, \; y'', \; \text{or } \frac{d^2y}{dx^2} \qquad 1.272$$

B. BASIC OPERATIONS

In the formulas that follow, f and g are functions of x. D is the derivative operator, and a is a constant.

$$D(a) = 0 \qquad 1.273$$

$$D(af) = aD(f) \qquad 1.274$$

$$D(f+g) = D(f) + D(g) \qquad 1.275$$

$$D(f-g) = D(f) - D(g) \qquad 1.276$$

$$D(f \cdot g) = fD(g) + gD(f) \qquad 1.277$$

$$D\left(\frac{f}{g}\right) = \frac{gD(f) - fD(g)}{g^2} \qquad 1.278$$

$$D(x^n) = nx^{n-1} \qquad 1.279$$

$$D(f^n) = nf^{n-1}D(f) \qquad 1.280$$

$$D(f(g)) = \frac{df(g)}{dg}D(g) \qquad 1.281$$

$$D(\ln x) = \frac{1}{x} \qquad 1.282$$

$$D(e^{ax}) = ae^{ax} \qquad 1.283$$

Example 1.39

A function is given as $f(x) = x^3 - 2x$. What is the slope of the line at $x = 3$?

$$y' = 3x^2 - 2$$

$$y'(3) = 27 - 2 = 25$$

Note the meaning of $y'(3)$ is $\frac{dy}{dx}$ at $x = 3$.

C. TRANSCENDENTAL FUNCTIONS

$$D(\sin x) = \cos x \qquad 1.284$$

$$D(\cos x) = -\sin x \qquad 1.285$$

$$D(\tan x) = \sec^2 x \qquad 1.286$$

$$D(\cot x) = -\csc^2 x \qquad 1.287$$

$$D(\sec x) = (\sec x)(\tan x) \qquad 1.288$$

$$D(\csc x) = (-\csc x)(\cot x) \qquad 1.289$$

$$D(\arcsin x) = \frac{1}{\sqrt{1-x^2}} \qquad 1.290$$

$$D(\arctan x) = \frac{1}{1+x^2} \qquad 1.291$$

$$D(\text{arcsec } x) = \frac{1}{x\sqrt{x^2-1}} \qquad 1.292$$

$$D(\arccos x) = -D(\arcsin x) \qquad 1.293$$

$$D(\text{arccot } x) = -D(\arctan x) \qquad 1.294$$

$$D(\text{arccsc } x) = -D(\text{arcsec } x) \qquad 1.295$$

D. VARIATIONS ON DIFFERENTIATION

Partial Differentiation

If the function has two or more independent variables, a partial derivative is found by considering all extraneous variables as constants. The geometric interpretation of the partial derivative $(\partial z/\partial x)$ is the slope of a line tangent to the 3-dimensional surface in a plane of constant y, and parallel to the x axis. Similarly, the interpretation of $(\partial z/\partial y)$ is the slope of a line tangent to the surface in a plane of constant x, and parallel to the y axis.

Example 1.40

A surface has the equation $x^2 + y^2 + z^2 = 9$. What is the slope of a line tangent to (1,2,2) and parallel to the x axis?

$$z = \sqrt{9 - x^2 - y^2}$$

$$\frac{\partial z}{\partial x} = \frac{-x}{\sqrt{9 - x^2 - y^2}}$$

At the point (1,2,2),

$$\frac{\partial z}{\partial x} = \frac{-1}{2}$$

Implicit Differentiation

If a relationship between n variables cannot be manipulated to yield an explicit function of $(n-1)$ independent variables, the relationship implicitly defines the nth remaining variable. The derivative of the implicit variable taken with respect to any other variable is found by a process known as *implicit differentiation.*

If $f(x,y) = 0$ is a function, the implicit derivative is

$$\frac{dy}{dx} = \frac{-\partial f}{\partial x} \Big/ \frac{\partial f}{\partial y} \qquad 1.296$$

If $f(x,y,z) = 0$ is a function, the implicit derivatives are

$$\frac{\partial z}{\partial x} = \frac{-\partial f}{\partial x} \Big/ \frac{\partial f}{\partial z} \qquad 1.297$$

$$\frac{\partial z}{\partial y} = \frac{-\partial f}{\partial y} \Big/ \frac{\partial f}{\partial z} \qquad 1.298$$

Example 1.41

If $f = x^2 + xy + y^3$, what is $\frac{dy}{dx}$?

Since this function cannot be written as an explicit function of x, implicit differentiation is required.

$$\frac{\partial f}{\partial x} = 2x + y$$

$$\frac{\partial f}{\partial y} = x + 3y^2$$

$$\frac{dy}{dx} = \frac{-(2x+y)}{x+3y^2}$$

Example 1.42

Solve example 1.40 using implicit differentiation.

$$f = x^2 + y^2 + z^2 - 9$$
$$\frac{\partial f}{\partial x} = 2x$$
$$\frac{\partial f}{\partial z} = 2z$$
$$\frac{\partial z}{\partial x} = \frac{-2x}{2z} = \frac{-x}{z}$$

and at (1,2,2),

$$\frac{\partial z}{\partial x} = \frac{-1}{2}$$

Extrema and Optimization

Derivatives can be used to locate local *maxima, minima,* and *points of inflection.* No distinction is made between local and global extrema. The end points of the interval always should be checked against the local extrema located by the method below. The following rules define the extreme points.

$f'(x) = 0$ at any extrema
$f''(x) = 0$ at an inflection point
$f''(x)$ is negative at a maximum
$f''(x)$ is positive at a minimum

Example 1.43

Find the global extreme points of function $f(x) = x^3 + x^2 - x + 1$ on the interval [-2, +2].

$$f'(x) = 3x^2 + 2x - 1$$
$$f'(x) = 0 \text{ at } x = \frac{1}{3} \text{ and } x = -1$$
$$f''(x) = 6x + 2$$
$$f(-1) = 2$$
$$f''(-1) = -4$$

So, $x = -1$ is a maximum.

$$f\left(\frac{1}{3}\right) = \frac{22}{27}$$
$$f''\left(\frac{1}{3}\right) = +4$$

So, $x = \frac{1}{3}$ is a minimum.

Checking the end points,

$$f(-2) = -1$$
$$f(+2) = +11$$

Therefore, the absolute extreme points are the end points.

13 INTEGRAL CALCULUS

A. FUNDAMENTAL THEOREM

The *fundamental theorem of calculus* is

$$\int_{x_1}^{x_2} f'(x) = f(x_2) - f(x_1) \qquad 1.299$$

B. INTEGRATION BY PARTS

If f and g are functions, then

$$\int f\,dg = fg - \int g\,df \qquad 1.300$$

Example 1.44

Evaluate the following integral: $\int xe^x dx$

Use integration by parts.

Let $f = x$. Then, $df = dx$.

Let $dg = e^x dx$. Then, $g = \int e^x dx = e^x$.

Therefore,

$$\int xe^x dx = xe^x - \int e^x dx + c$$
$$= xe^x - e^x + c$$

C. INDEFINITE INTEGRALS ("...+ C" omitted)

$$\int dx = x \qquad 1.301$$
$$\int au\,dx = a\int u\,dx \qquad 1.302$$
$$\int (u + v)dx = \int u\,dx + \int v\,dx \qquad 1.303$$
$$\int x^m dx = \frac{x^{(m+1)}}{m+1} \quad m \neq -1 \qquad 1.304$$

MATH

$$\int \frac{dx}{x} = ln|x| \quad 1.305$$

$$\int e^{ax}dx = \frac{1}{a}e^{ax} \quad 1.306$$

$$\int xe^{ax}dx = \frac{1}{a^2}e^{ax}(ax-1) \quad 1.307$$

$$\int \cosh x\ dx = \sinh x \quad 1.308$$

$$\int \sinh x\ dx = \cosh x \quad 1.309$$

$$\int \sin x\ dx = -\cos x \quad 1.310$$

$$\int \cos x\ dx = \sin x \quad 1.311$$

$$\int \tan x\ dx = ln|\sec x| \quad 1.312$$

$$\int \cot x\ dx = ln|\sin x| \quad 1.313$$

$$\int \sec x\ dx = ln|\sec x + \tan x| \quad 1.314$$

$$\int \csc x\ dx = ln|\csc x - \cot x| \quad 1.315$$

$$\int \frac{dx}{1+x^2} = \arctan x \quad 1.316$$

$$\int \frac{dx}{\sqrt{1-x^2}} = \arcsin x \quad 1.317$$

$$\int \frac{dx}{x\sqrt{x^2-1}} = \mathrm{arcsec}\, x \quad 1.318$$

D. USES OF INTEGRALS

Finding Areas

The area bounded by $x = a$, $x = b$, $f_1(x)$ above, and $f_2(x)$ below is given by

$$A = \int_b^a [f_1(x) - f_2(x)]dx \quad 1.319$$

Surfaces of Revolution

The surface area obtained by rotating $f(x)$ about the x axis is

$$A_s = 2\pi \int_a^b f(x)\sqrt{1+[f'(x)]^2}dx \quad 1.320$$

Rotation of a Function

The volume of a function rotated about the x axis is

$$V = \pi \int_a^b (f(x))^2 dx \quad 1.321$$

The volume of a function rotated about the y axis is

$$V = 2\pi \int_b^a x f(x) dx \quad 1.322$$

Length of a Curve

The length of a curve given by $f(x)$ is

$$L = \int_a^b \sqrt{1+(f'(x))^2}dx \quad 1.323$$

Example 1.45

For the shaded area shown, find (a) the area, and (b) the volume enclosed by the curve rotated about the x axis.

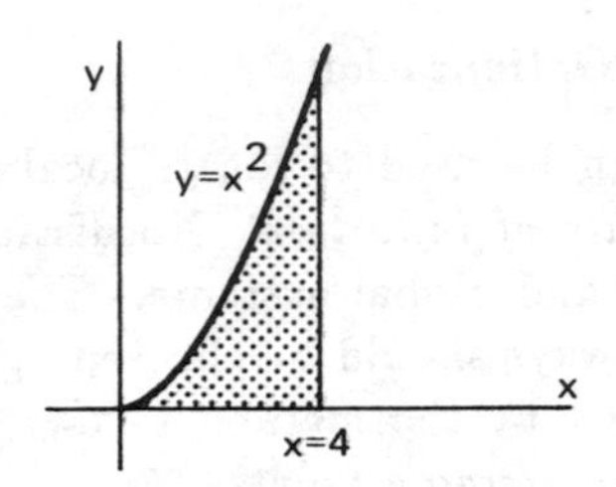

(a)

$$f_2(x) = 0 \qquad f_1(x) = x^2$$

$$A = \int_0^4 x^2 dx = \left[\frac{x^3}{3}\right]_0^4 = 21.33$$

(b)

$$V = \pi \int_0^4 (x^2)^2 dx = \pi \left[\frac{x^5}{5}\right]_0^4 = 204.8\pi$$

14 DIFFERENTIAL EQUATIONS

A differential equation is a mathematical expression containing a dependent variable and one or more of that variable's derivatives. First order differential equations contain only the first derivative of the dependent variable. Second order equations contain the second derivative.

The differential equation is said to be *linear* if all terms containing the dependent variable are multiplied only by real scalars. The equation is said to be *homogeneous* if there are no terms which do not contain the dependent variable or one of its derivatives.

Most differential equations are difficult to solve. However, there are several forms which are fairly simple. These are presented here.

A. FIRST ORDER LINEAR

The first order linear differential equation has the general form given by equation 1.324. $p(t)$ and $g(t)$ may be constants or any function of t.

$$y' + p(t)y = g(t) \qquad 1.324$$

The solution depends on an *integrating factor* defined as

$$u = \exp[\int p(t)dt] \qquad 1.325$$

The solution to the first order linear differential equation is

$$y = \frac{1}{u}\left[\int ug(t)dt + c\right] \qquad 1.326$$

Example 1.46

Find a solution to the differential equation

$$y' - y = 2te^{2t} \quad y(0) = 1$$

This meets the definition of a first order linear equation with

$$p(t) = -1 \text{ and } g(t) = 2te^{2t}$$

The integrating constant is

$$u = \exp\left[\int -1dt\right] = e^{-t}$$

Then, y is

$$\begin{aligned} y &= \left(\frac{1}{e^{-t}}\right)\left[\int e^{-t}2te^{2t}dt + c\right] \\ &= e^t\left[\int 2te^t dt + c\right] \\ &= e^t[2te^t - 2e^t + c] \end{aligned}$$

But, $y(0) = 1$, so

$$\begin{aligned} c &= +3 \\ y &= e^t[2e^t(t-1) + 3] \end{aligned}$$

B. SECOND ORDER HOMOGENEOUS WITH CONSTANT COEFFICIENTS

This type of differential equation has the following general form

$$c_1y'' + c_2y' + c_3y = 0 \qquad 1.327$$

The solution can be found by first solving the characteristic quadratic equation for its roots k_1^* and k_2^*. This characteristic equation is derived directly from the differential equation

$$c_1k^2 + c_2k + c_3 \qquad 1.328$$

The form of the solution depends on the values of k_1^* and k_2^*. If $k_1^* \neq k_2^*$ and both are real, then

$$y = a_1(e^{k_1^*t}) + a_2(e^{k_2^*t}) \qquad 1.329$$

If $k_1^* = k_2^*$, then

$$y = a_1(e^{k_1^*t}) + a_2t(e^{k_2^*t}) \qquad 1.330$$

If $k^* = (r \pm iu)$, then

$$y = a_1(e^{rt})\cos(ut) + a_2(e^{rt})\sin(ut) \qquad 1.331$$

In all three cases, a_1 and a_2 must be found from the given initial conditions.

Example 1.47

Solve the following differential equation for x.

$$y'' + 6y' + 9y = 0 \quad y(0) = 0,\ y'(0) = 1$$

The characteristic equation is

$$k^2 + 6k + 9 = 0$$

This has roots of $k_1^* = k_2^* = -3$; therefore, the solution has the form

$$y(t) = a_1e^{-3t} + a_2te^{-3t}$$

But, $y(0) = 0$,

$$\begin{aligned} 0 &= a_1(e^0) + a_2(0)(e^0) \\ 0 &= a_1(1) + 0 \\ 0 &= a_1 \end{aligned}$$

Also, $y'(0) = 1$. The derivative of $y(t)$ is

$$\begin{aligned} y'(t) &= -3a_2te^{-3t} + a_2e^{-3t} \\ 1 &= -3a_2(0)(e^0) + a_2e^0 \\ 1 &= 0 + a_2 \\ 1 &= a_2 \end{aligned}$$

The final solution is

$$y = te^{-3t}$$

15 LAPLACE TRANSFORMS

Traditional methods of solving non-homogeneous differential equations are very difficult. The *Laplace transformation* can be used to reduce the solution of many complex differential equations to simple algebra.

Every mathematical function can be converted into a Laplace function by use of the following transformation definition.

$$\mathcal{L}[f(t)] = \int_0^\infty e^{-st} f(t)dt \qquad 1.332$$

The variable s is equivalent to the derivative operator. However, it may be thought of as a simple variable.

Example 1.48

Let $f(t)$ be the unit step. That is, $f(t) = 0$ for $t < 0$ and $f(t) = 1$ for $t \geq 0$.

Then, the Laplace transform of $f(t) = 1$ is

$$\begin{aligned} \mathcal{L}[f(t)] &= \int_0^\infty e^{-st}(1)dt = \left.\frac{-e^{-st}}{s}\right|_0^\infty \\ &= 0 - \left(\frac{-1}{s}\right) = \frac{1}{s} \end{aligned}$$

Example 1.49

What is the Laplace transformation of $f(t) = e^{at}$?

$$\begin{aligned} \mathcal{L}[e^{at}] &= \int_0^\infty e^{-st}e^{at}dt \\ &= \int_0^\infty e^{-(s-a)^t}dt \\ &= \left.\frac{-e^{-(s-a)t}}{s-a}\right|_0^\infty \\ &= \frac{1}{s-a} \end{aligned}$$

Generally, it is unnecessary to actually obtain a function's Laplace transform by use of equation 1.332. Tables of the transforms are readily available. A small collection of the most frequently required transforms is given at the end of this chapter.

The Laplace transform method can be used with any linear differential equation with constant coefficients. Assuming the dependent variable is y, the basic procedure is as follows:

step 1: Put the differential equation in standard form.

step 2: Use superposition and take the Laplace transform of each term.

step 3: Use the following relationships to expand terms:

$$\mathcal{L}(y'') = s^2\mathcal{L}(y) - sy(0) - y'(0) \qquad 1.333$$

$$\mathcal{L}(y') = s\mathcal{L}(y) - y(0) \qquad 1.334$$

step 4: Solve for $\mathcal{L}(y)$. Simplify the resulting expression using partial fractions.

step 5: Find y by applying the inverse transform.

This method reduces the solutions of differential equations to simple algebra. However, a complete set of transforms is required.

Working with Laplace transforms is simplified by the following two theorems:

Linearity theorem: If c is constant, then

$$\mathcal{L}[cf(t)] = c\mathcal{L}[f(t)] \qquad 1.335$$

Superposition theorem: If $f(t)$ and $g(t)$ are different functions, then

$$\mathcal{L}[f(t) \pm g(t)] = \mathcal{L}[f(t)] \pm \mathcal{L}[g(t)] \qquad 1.336$$

Example 1.50

Suppose the following differential equation results from the analysis of a mechanical system.

$$y'' + 2y' + 2y = \cos(t)$$

$$y(0) = 1, y'(0) = 0$$

y is the dependent variable. Start by taking the Laplace transform of both sides

$$\mathcal{L}(y'') + 2\mathcal{L}(y') + 2\mathcal{L}(y) = \mathcal{L}(\cos(t))$$

$$s^2\mathcal{L}(y) - sy(0) - y'(0) + 2s\mathcal{L}(y) - 2y(0) + 2\mathcal{L}(y) = \mathcal{L}\cos(t)$$

But, $y(0) = 1$ and $y'(0) = 0$. Also, the Laplace transform of $\cos(t)$ can be found from the appendix at the end of this chapter.

$$s^2\mathcal{L}(y) - s + 2s\mathcal{L}(y) - 2 + 2\mathcal{L}(y) = \frac{s}{s^2+1}$$
$$\mathcal{L}(y)[s^2 + 2s + 2] - s - 2 = \frac{s}{s^2+1}$$

$$\mathcal{L}(y) = \frac{s^3 + 2s^2 + 2s + 2}{(s^2+1)(s^2+2s+2)}$$

This is now expanded by partial fractions:

$$\begin{aligned}
&\frac{s^3 + 2s^2 + 2s + 2}{(s^2+1)(s^2+2s+2)} \\
&= \frac{A_1 s + B_1}{s^2+1} + \frac{A_2 s + B_2}{s^2+2s+2} \\
&= [s^3(A_1 + A_2) + s^2(2A_1 + B_1 + B_2) \\
&\quad + s(2A_1 + 2B_1 + A_2) + 2B_1 + B_2] \\
&\quad \div [(s^2+1)(s^2+2s+2)]
\end{aligned}$$

The following simultaneous equations result:

$$\begin{aligned}
A_1 + A_2 &= 1 \\
2A_1 + B_1 + B_2 &= 2 \\
2A_1 + A_2 + 2B_1 &= 2 \\
2B_1 + B_2 &= 2
\end{aligned}$$

These equations have the solutions

$$A_1^* = \frac{1}{5} \quad A_2^* = \frac{4}{5}$$
$$B_1^* = \frac{2}{5} \quad B_2^* = \frac{6}{5}$$

Therefore, y can be found by taking the following inverse transform:

$$y = \mathcal{L}^{-1}\left[\frac{\frac{s}{5} + \frac{2}{5}}{s^2+1} + \frac{\frac{4s}{5} + \frac{6}{5}}{s^2+2s+2}\right]$$

The solution is

$$y = \frac{1}{5}\cos(t) + \frac{2}{5}\sin(t) + \frac{4}{5}e^{-t}\cos(t) + \frac{2}{5}e^{-t}\sin(t)$$

16 APPLICATIONS OF DIFFERENTIAL EQUATIONS

A. FLUID MIXTURE PROBLEMS

The typical fluid mixing problem involves a tank containing some fluid. There may be an initial solute in the liquid, or the liquid may be pure. Liquid and solute are added at known rates. A drain usually removes some of the liquid which is assumed to be thoroughly mixed. The problem is to find the weight or concentration of solute in the tank at some time t. The following symbols are used.

$Q_1(t)$	liquid inflow rate at time t
$I(t)$	liquid inflow rate from all sources at time t
k	a constant
$Q_2(t)$	liquid outflow rate due to all drains at time t
$S_1(t)$	solute inflow rate at time t (this may have to be calculated from the incoming concentration and $Q_1(t)$)
$S_2(t)$	solute outflow rate at time t $S_2(t) = \frac{Q_2(t)W(t)}{V(t)}$
V_o	original volume of liquid in the tank at time $= 0$
$V(t)$	volume of liquid in the tank at time $= t$ (equal to $V_o + \int Q_1(t)dt - \int Q_2(t)dt$)
W_o	initial weight of solute in tank at $t = 0$
$W(t)$	weight of solute in tank at time $= t$

case 1: Constant liquid in tank volume ($V(t) = V_o$)

$$S_1(t) - S_2(t) = \frac{d}{dt}(W(t)) \qquad 1.337$$

The differential equation is

$$W'(t) + \frac{Q_2(t)W(t)}{V_o} = S_1(t) \qquad 1.338$$

This is a first order linear equation because Q_2, V_o, and S_1 are constants.

case 2: Changing volume

$$\begin{aligned}
W'(t) &= S_1(t) - S_2(t) && 1.339 \\
&= S_1(t) - \frac{Q_2(t)W(t)}{V(t)} && 1.340
\end{aligned}$$

The differential equation is

$$W'(t) + \frac{Q_2(t)W(t)}{V(t)} = S_1(t) \qquad 1.341$$

Example 1.51

A tank contains 100 gallons of pure water at the beginning of an experiment. 1 gpm of pure water flows into the tank, as does 1 gpm of water containing $\frac{1}{4}$ pound of salt per gallon. A perfectly mixed solution drains from the tank at the rate of 2 gpm. How much salt is in the tank eight minutes after the experiment has begun?

Choose $W(t)$ as the variable giving the weight of salt in the tank at time t. $\frac{1}{4}$ pound of salt enters the tank per minute. What goes out depends on the concentration in the tank. Specifically, the leaving salt is

$$\begin{aligned}\text{salt leaving} &= (2\text{ gpm})(\#\text{ lbs salt per gallon})\\ &= (2\text{ gpm})\left(\frac{\#\text{ lbs salt total}}{100}\right)\\ &= 0.02W(t)\end{aligned}$$

The difference between the inflow and the outflow is given by equation 1.337.

$$W'(t) = \frac{1}{4} - 0.02\,W(t)$$

This is a first order linear differential equation. It can be solved using the integrating factor (equation 1.326), simple constant coefficient methods, or Laplace transforms. The solution is

$$W(t) = 12.5 - 12.5e^{-0.02t}$$

At $t = 8$, $W(t) = 1.85$ pounds.

B. DECAY PROBLEMS

A given quantity is known to decrease at a rate proportional to the amount present. The original amount is known, and the amount at some time t is desired.

k	a negative proportionality constant
A_o	original amount present
$A(t)$	amount present at time t
t	time
$t_{\frac{1}{2}}$	half-life

The differential equation is

$$A'(t) = kA(t) \qquad 1.342$$

The solution is

$$A(t) = A_o e^{kt} \qquad 1.343$$

If A^* is known for some time t^*, k can be found from

$$k = \left(\frac{1}{t^*}\right) ln \left(\frac{A^*}{A_0}\right) \qquad 1.344$$

k can also be found from the half-life

$$k = \frac{-0.693}{t_{\frac{1}{2}}} \qquad 1.345$$

C. SURFACE TEMPERATURE

k	a constant
t	time
T	absolute temperature of the surface
T_o	ambient temperature

Assuming the surface temperature changes at a rate proportional to the difference in surface and ambient temperatures, the differential equation is

$$\frac{dT}{dt} = k(T - T_o) \qquad 1.346$$

Equation 1.346 is known as *Newton's Law of Cooling*.

D. SURFACE EVAPORATION

A	exposed surface area
k	proportionality constant
r	radius
s	side length
t	time
V	object volume

The equation is

$$\frac{dV}{dt} = -kA \qquad 1.347$$

For a spherical drop, this reduces to

$$\frac{dr}{dt} = -k \qquad 1.348$$

For a cube, this reduces to

$$\frac{ds}{dt} = -2k \qquad 1.349$$

Appendix A
Laplace Transforms

f(t)	$\mathcal{L}$[f(t)]
Unit impulse at t=0	1
Unit impulse at t=c	e^{-cs}
Unit step at t=0	(1/s)
Unit step at t=c	$\frac{e^{-cs}}{s}$
t	$\frac{1}{s^2}$
$\frac{t^{n-1}}{(n-1)!}$	$\frac{1}{s^n}$
sinAt	$\frac{A}{s^2+A^2}$
At − sinAt	$\frac{A^3}{s^2(s^2+A^2)}$
sinh(At)	$\frac{A}{s^2-A^2}$
t sinAt	$\frac{2As}{(s^2+A^2)^2}$
cosAt	$\frac{s}{s^2+A^2}$
1 − cosAt	$\frac{A^2}{s(s^2+A^2)}$
cosh(At)	$\frac{s}{s^2-A^2}$
t cosAt	$\frac{s^2-A^2}{(s^2+A^2)^2}$
t^n (n is a positive integer)	$\frac{n!}{s^{(n+1)}}$
e^{At}	$\frac{1}{s-A}$
$e^{At}\sin Bt$	$\frac{B}{(s-A)^2+B^2}$
$e^{At}\cos Bt$	$\frac{s-A}{(s-A)^2+B^2}$
$e^{At}t^n$ (n is positive integer)	$\frac{n!}{(s-A)^{n+1}}$
$1 - e^{-At}$	$\frac{A}{s(s+A)}$
$e^{-At} + At - 1$	$\frac{A^2}{s^2(s+A)}$
$\frac{e^{-At} - e^{-Bt}}{B-A}$	$\frac{1}{(s+A)(s+B)}$
$\frac{(C-A)e^{-At} - (C-B)e^{-Bt}}{B-A}$	$\frac{s + C}{(s+A)(s+B)}$
$\frac{1}{AB} + \frac{Be^{-At} - Ae^{-Bt}}{AB(A-B)}$	$\frac{1}{s(s+A)(s+B)}$

Appendix B
Conversion Factors

To Convert	Into	Multiply by
Acres	hectares	0.4047
Acres	square feet	43,560.0
Acres	square miles	1.562 EE − 3
Ampere hours	coulombs	3,600.0
Angstrom units	inches	3.937 EE − 9
Angstrom units	microns	1 EE − 4
Astronomical units	kilometers	1.495 EE8
Atmospheres	cms of mercury	76.0
BTU's	horsepower-hrs	3.931 EE − 4
BTU's	kilowatt-hrs	2.928 EE − 4
BTU/hr	watts	0.2931
Bushels	cubic inches	2,150.4
Calories, gram (mean)	BTU (mean)	3.9685 EE − 3
Centares	square meters	1.0
Centimeters	kilometers	1 EE − 5
Centimeters	meters	1 EE − 2
Centimeters	millimeters	10.0
Centimeters	feet	3.281 EE − 2
Centimeters	inches	0.3937
Chains	inches	792.0
Coulombs	faradays	1.036 EE − 5
Cubic centimeters	cubic inches	0.06102
Cubic centimeters	pints (U.S. liq.)	2.113 EE − 3
Cubic feet	cubic meters	0.02832
Cubic feet/min.	pounds water/min.	62.43
Cubic feet/sec.	gallons/min.	448.831
Cubits	inches	18.0
Days	seconds	86,400.0
Degrees (angle)	radians	1.745 EE − 2
Degrees/sec.	revolutions/min.	0.1667
Dynes	grams	1.020 EE − 3
Dynes	joules/meter (newtons)	1 EE − 5
Ells	inches	45.0
Ergs	BTU's	9.480 EE − 11
Ergs	foot-pounds	7.3670 EE − 8
Ergs	kilowatt-hours	2.778 EE − 14
Faradays/sec.	amperes (absolute)	96,500
Fathoms	feet	6.0
Feet	centimeters	30.48
Feet	meters	0.3048
Feet	miles (nautical)	1.645 EE − 4
Feet	miles (statute)	1.894 EE − 4
Feet/min.	centimeters/sec.	0.5080
Feet/sec.	knots	0.5921
Feet/sec.	miles/hour	0.6818
Foot-pounds	BTU's	1.286 EE − 3
Foot-pounds	kilowatt-hours	3.766 EE − 7
Furlongs	miles (U.S.)	0.125
Furlongs	feet	660.0
Gallons	liters	3.785
Gallons of water	pounds of water	8.3453
Gallons/min.	cubic feet/hour	8.0208
Grams	ounces (avoirdupois)	3.527 EE − 2
Grams	ounces (troy)	3.215 EE − 2
Grams	pounds	2.205 EE−3
Hectares	acres	2.471
Hectares	square feet	1.076 EE5
Horsepower	BTU/min.	42.42
Horsepower	kilowatts	0.7457
Horsepower	watts	745.7
Hours	days	4.167 EE − 2
Hours	weeks	5.952 EE − 3
Inches	centimeters	2.540
Inches	miles	1.578 EE − 5
Joules	BTU's	9.480 EE − 4
Joules	ergs	1 EE7
Kilograms	pounds	2.205
Kilometers	feet	3,281.0
Kilometers	meters	1,000.0
Kilometers	miles	0.6214
Kilometers/hr.	knots	0.5396
Kilowatts	horsepower	1.341
Kilowatt-hours	BTU'S	3,413.0
Knots	feet/hour	6,080.0
Knots	nautical miles/hr.	1.0
Knots	statute miles/hr.	1.151
Light years	miles	5.9 EE12
Links (surveyor's)	inches	7.92
Liters	cubic centimeters	1,000.0
Liters	cubic inches	61.02
Liters	gallons (U.S. liq.)	0.2642
Liters	milliliters	1,000.0
Liters	pints (U.S. liq.)	2.113
Meters	centimeters	100.0
Meters	feet	3.281
Meters	kilometers	1 EE − 3
Meters	miles (nautical)	5.396 EE − 4
Meters	miles (statute)	6.214 EE − 4
Meters	millimeters	1,000.0
Microns	meters	1 EE − 6
Miles (nautical)	feet	6,080.27
Miles (statute)	feet	5,280.0
Miles (nautical)	kilometers	1.853
Miles (statute)	kilometers	1.609
Miles (nautical)	miles (statute)	1.1516
Miles (statute)	miles (nautical)	0.8684
Miles/hour	feet/min.	88.0
Milligrams/liter	parts/million	1.0
Milliliters	liters	1 EE − 3
Millimeters	inches	3.937 EE − 2
Newtons	dynes	1 EE5
Ohms (international)	ohms (absolute)	1.0005
Ounces	grams	28.349527
Ounces	pounds	6.25 EE − 2
Ounces (troy)	ounces (avoirdupois)	1.09714
Parsecs	miles	19 EE12
Parsecs	kilometers	3.084 EE13
Pints (liq.)	cubic centimeters	473.2
Pints (liq.)	cubic inches	28.87
Pints (liq.)	gallons	0.125
Pints (liq.)	quarts (liq.)	0.5
Pounds	kilograms	0.4536
Pounds	ounces	16.0
Pounds	ounces (troy)	14.5833
Pounds	pounds (troy)	1.21528
Quarts (dry)	cubic inches	67.20

Appendix B
Conversion Factors

To Convert	Into	Multiply by
Quarts (liq.)	cubic inches	57.75
Quarts (liq.)	gallons	0.25
Quarts (liq.)	liters	0.9463
Radians	degrees	57.30
Radians	minutes	3,438.0
Revolutions	degrees	360.0
Revolutions/min.	degrees/sec.	6.0
Rods	meters	5.029
Rods	feet	16.5
Rods (surveyor's measure)	yards	5.5
Seconds	minutes	1.667 EE − 2
Slugs	pounds	32.17
Tons (long)	kilograms	1,016.0
Tons (short)	kilograms	907.1848
Tons (long)	pounds	2,240.0
Tons (short)	pounds	2,000.0
Tons (long)	tons (short)	1.120
Tons (short)	tons (long)	0.89287
Volt (absolute)	statvolts	3.336 EE − 3
Watts	BTU/hour	3.4129
Watts	horsepower	1.341 EE − 3
Yards	meters	0.9144
Yards	miles (nautical)	4.934 EE − 4
Yards	miles (statute)	5.682 EE − 4

Appendix C
Computational Values of Fundamental Constants

Constant	SI	English
charge on electron	−1.602 EE−19 C	
charge on proton	+1.602 EE−19 C	
atomic mass unit	1.66 EE−27 kg	
electron rest mass	9.11 EE−31 kg	
proton rest mass	1.673 EE−27 kg	
neutron rest mass	1.675 EE−27 kg	
earth weight		1.32 EE25 lb
earth mass	6.00 EE24 kg	4.11 EE23 slug
mean earth radius	6.37 EE3 km	2.09 EE7 ft
mean earth density	5.52 EE3 kg/m³	345 lbm/ft³
earth escape velocity	1.12 EE4 m/s	3.67 EE4 ft/sec
distance from sun	1.49 EE11 m	4.89 EE11 ft
Boltzmann constant	1.381 EE−23 J/°K	5.65 EE−24 $\frac{\text{ft-lbf}}{°\text{R}}$
permeability of a vacuum	1.257 EE−6 H/m	
permittivity of a vacuum	8.854 EE−12 F/m	
Planck constant	6.626 EE−34 J·s	
Avogadro's number	6.022 EE23 molecules/gmole	2.73 EE26 $\frac{\text{molecules}}{\text{pmole}}$
Faraday's constant	9.648 EE4 C/gmole	
Stefan-Boltzmann constant	5.670 EE−8 W/m²−K⁴	1.71 EE−9 $\frac{\text{BTU}}{\text{ft}^2\text{-hr-}°\text{R}^4}$
gravitational constant (G)	6.672 EE−11 m³/s²−kg	3.44 EE−8 $\frac{\text{ft}^4}{\text{lbf-sec}^4}$
universal gas constant	8.314 J/°K−gmole	1545 $\frac{\text{ft-lbf}}{°\text{R-pmole}}$
speed of light	3.00 EE8 m/s	9.84 EE8 ft/sec
speed of sound, air, STP	3.31 EE2 m/s	1.09 EE3 ft/sec
speed of sound, air, 70°F, one atmosphere	3.44 EE2 m/s	1.13 EE3 ft/sec
standard atmosphere	1.013 EE5 N/m²	14.7 psia
standard temperature	0°C	32°F
molar ideal gas volume (STP)	22.4138 EE−3 m³/gmole	359 ft³/pmole
standard water density	1 EE3 kg/m³	62.4 lbm/ft³
air density, STP	1.29 kg/m³	8.05 EE−2 lbm/ft³
air density, 70°F, 1 atm	1.20 kg/m³	7.49 EE−2 lbm/ft³
mercury density	1.360 EE4 kg/m³	8.49 EE2 lbm/ft³
gravity on moon	1.67 m/s²	5.47 ft/sec²
gravity on earth	9.81 m/s²	32.17 ft/sec²

Appendix D
Periodic Table of the Elements

The number of electrons in filled shells is shown in the column at the extreme left; the remaining electrons for each element are shown immediately below the symbol for each element. Atomic numbers are enclosed in brackets. Atomic weights (rounded, based on Carbon-12) are shown above the symbols. Atomic weight values in parentheses are those of the isotopes of longest half-life for certain radioactive elements whose atomic weights cannot be precisely quoted without knowledge of origin of the element.

METALS — TRANSITION METALS — NON-METALS

periods	I A	II A	III B	IV B	V B	VI B	VII B	VIII	VIII	VIII	I B	II B	III A	IV A	V A	VI A	VII A	O
1 0	1.0079 H[1] 1																1.0079 H[1] 1	4.0026 He[2] 2
2 2	6.941 Li[3] 1	9.0122 Be[4] 2											10.81 B[5] 3	12.011 C[6] 4	14.007 N[7] 5	16.000 O[8] 6	19.000 F[9] 7	21.179 Ne[10] 8
3 2, 8	22.99 Na[11] 1	24.305 Mg[12] 2											26.982 Al[13] 3	28.086 Si[14] 4	30.974 P[15] 5	32.06 S[16] 6	35.453 Cl[17] 7	39.948 Ar[18] 8
4 2, 8	39.098 K[19] 8, 1	40.08 Ca[20] 8, 2	44.956 Sc[21] 9, 2	47.90 Ti[22] 10, 2	50.941 V[23] 11, 2	51.996 Cr[24] 13, 1	54.938 Mn[25] 13, 2	55.847 Fe[26] 14, 2	58.933 Co[27] 15, 2	58.70 Ni[28] 16, 2	63.546 Cu[29] 18, 1	65.38 Zn[30] 18, 2	69.72 Ga[31] 18, 3	72.59 Ge[32] 18, 4	74.922 As[33] 18, 5	78.96 Se[34] 18, 6	79.904 Br[35] 18, 7	83.80 Kr[36] 18, 8
5 2,8,18	85.468 Rb[37] 8, 1	87.62 Sr[38] 8, 2	88.906 Y[39] 9, 2	91.22 Zr[40] 10, 2	92.906 Nb[41] 12, 1	95.94 Mo[42] 13, 1	(97) Tc[43] 14, 1	101.07 Ru[44] 15, 1	102.906 Rh[45] 16, 1	106.4 Pd[46] 18	107.868 Ag[47] 18, 1	112.41 Cd[48] 18, 2	114.82 In[49] 18, 3	118.69 Sn[50] 18, 4	121.75 Sb[51] 18, 5	127.60 Te[52] 18, 6	126.905 I[53] 18, 7	131.30 Xe[54] 18, 8
6 2,8,18	132.905 Cs[55] 18,8,1	137.33 Ba[56] 18,8,2	* [57-71]	178.49 Hf[72] 32,10,2	180.948 Ta[73] 32,11,2	183.85 W[74] 32,12,2	186.207 Re[75] 32,13,2	190.2 Os[76] 32,14,2	192.22 Ir[77] 32,15,2	Pt[78] 32,17,1	196.967 Au[79] 32,18,1	200.59 Hg[80] 32,18,2	204.37 Tl[81] 32,18,3	207.2 Pb[82] 32,18,4	208.980 Bi[83] 32,18,5	(209) Po[84] 32,18,6	(210) At[85] 32,18,7	(222) Rn[86] 32,18,8
7 2,8,18,32	(223) Fr[87] 18,8,1	226.025 Ra[88] 18,8,2	† [89-103]	Rf[104] 32,10,2	Ha[105] 32,11,2	[106] 32,12,2	[107]	[108]										

* LANTHANIDE SERIES	138.906 La[57] 18,9,2	140.12 Ce[58] 20,8,2	140.908 Pr[59] 21,8,2	144.24 Nd[60] 22,8,2	(145) Pm[61] 23,8,2	150.4 Sm[62] 24,8,2	151.96 Eu[63] 25,8,2	157.25 Gd[64] 25,9,2	158.925 Tb[65] 27,8,2	162.50 Dy[66] 28,8,2	164.930 Ho[67] 29,8,2	167.26 Er[68] 30,8,2	168.934 Tm[69] 31,8,2	173.04 Yb[70] 32,8,2	174.97 Lu[71] 32,9,2
† ACTINIDE SERIES	(227) Ac[89] 18,9,2	232.038 Th[90] 18,10,2	231.036 Pa[91] 20,9,2	238.029 U[92] 21,9,2	237.048 Np[93] 23,8,2	(244) Pu[94] 24,8,2	(243) Am[95] 25,8,2	(247) Cm[96] 25,9,2	(247) Bk[97] 26,9,2	(251) Cf[98] 28,8,2	(254) Es[99] 29,8,2	(257) Fm[100] 30,8,2	(258) Md[101] 31,8,2	(255) No[102] 32,8,2	(260) Lr[103] 32,9,2

MATH

PRACTICE PROBLEMS

1. A stirred tank, initially containing 20 ft^3 of water, is fed brine at a rate of 3 ft^3/min and a concentration of 2 lbm/ft^3. The tank discharges at 2 ft^3/min. What is the salt concentration when the tank contains exactly 30 ft^3 of brine? How long will it take for the brine in the tank to reach 99% of the input concentration if the tank overflows at 50 ft^3? Note: The answer requires the solution of a first order linear differential equation.

2. Sulfuric acid (specific heat = 0.36) flowing at 5 tons/hour is cooled in a two-stage, counter-current cooler consisting of two tanks with a capacity of five tons each. At steady state, the hot acid (174 °C) is fed to the first tank where it is stirred and exposed to the cooling coils. Continuous discharge from the first tank (88 °C) goes to the second tank where it is stirred and exposed to the cooling coils. The acid leaves the second tank at 45 °C. Cooling water enters the coils of the second tank at 20 °C and 7740 lbm/hour. The cooling water leaves the coil of the second tank at 40 °C on its way to the coil of the first tank, where it leaves at 80 °C. The heat transfer coefficient in the hot tank is 200 CHU/ft^2-hr-°C, and the colder tank is 130 CHU/ft^2-hr-°C.

(a) What is the temperature in each tank after one hour if the cooling water is suddenly shut off? (b) Upon restoration of the water supply at a rate of five tons/hour, what is the discharge acid temperature after one hour? One CHU = heat required to heat 1 lbm water 1 °C.

Note: Part (a) requires the solution of two simultaneous first-order differential equations. Part (b) requires the solution of four simultaneous equations, two of which are linear first-order differential equations. By elimination of variables, part (b) results in a second-order linear differential equation with constant coefficients.

3. It is known that the heat capacity of a particular substance follows the empirical relationship: $c_p = a + bT + \frac{c}{T^2}$, where c_p is in BTU/lbm-°F, and T is in °R. Find the values of a, b, and c if tabular values for heat capacity of the substance are:

T, °F	c_p, BTU/lbm-°F
70	1.3701
100	1.3779
180	1.3996
320	1.4391

What is the heat capacity at 250 °F?

4. A reaction, $2A \rightarrow B$, is known to be second-order, (i.e., $kt = \frac{1}{c_A} - \frac{1}{c_{A_o}}$). Using the following data and linear regression analysis, find the initial concentration of A and the rate constant k.

Note: the data must be transformed to a linear form.

t, min	c_A, moles/liter
2	2.2502
7	1.5888
9	1.4217
13	1.1745
41	0.5298
62	0.3753

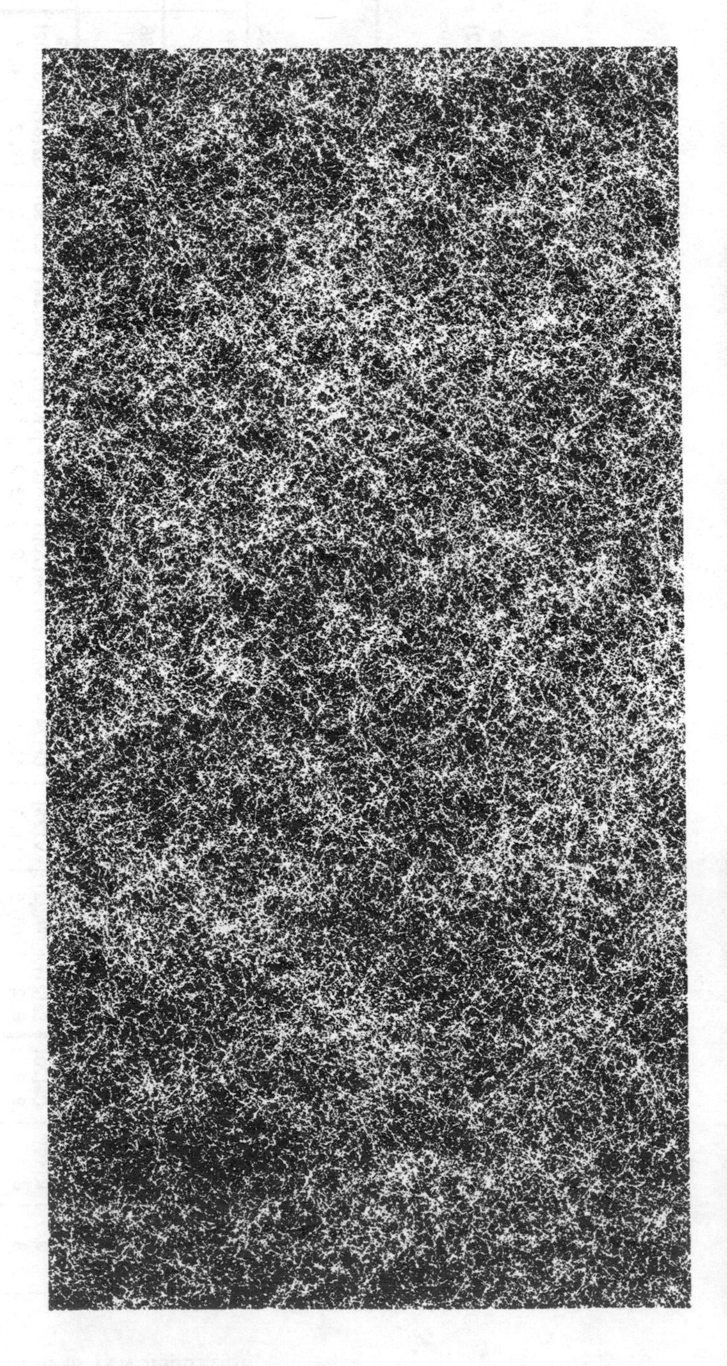

2 STOICHIOMETRY, HEAT AND MATERIAL BALANCES

STOICH

Nomenclature

A_x	excess air	%
c	heat capacity, or concentration	BTU/lbm-°F, lbm/gal (typ)
E	energy	BTU
g	acceleration of gravity (32.2)	ft/sec^2
g_c	gravitational constant (32.2)	lbm-ft/lbf-sec^2
h	enthalpy	BTU/lbm
HHV	higher heating value	BTU/lbm
J	Joule's constant (778)	ft-lbf/BTU
K_p	equilibrium constant	–
LHV	lower heating value	BTU/lbm
m	mass	lbm
M	molecular weight	lbm/lbmole
p	pressure	lbf/ft^2 (typ)
P	total pressure	atm
Q	heat, or volumetric flow rate	BTU, ft^3/min
TFT	theoretical flame temperature	°F
u	internal energy	BTU/lbm
v	velocity	ft/sec
V	volume	ft^3
w	weight	lbf
W	work	ft-lbf
z	elevation	ft

Symbols

Δ	change (e.g., $z_2 - z_1$)	various
Λ	latent heat	BTU/lbm
υ	specific volume	ft^3/lbm

Subscripts

f	final
i	input, initial, or component i
o	output
p	constant pressure
P	products
R	reactants
R_x	reaction
s	steam
t	tank, or total
v	vaporization
w	water

1 PHYSICAL AND CHEMICAL LAWS OF STOICHIOMETRY

Stoichiometry deals with combining proportions of elements and compounds in chemical reactions. The basic principle of stoichiometry is given by the *Law of Definite Multiple Proportions*, which states: (a) a pure chemical substance always contains its elements in the same proportions by weight; and, (b) when two or more elements combine to form a compound, the weights of the elements are simple multiples of the atomic weights. The law's meaning has broadened, through common use, to include the conservation of mass and energy, thereby encompassing operations such as crystallation, where no chemical reactions occur.

For ordinary non-nuclear processes, material and energy balances can be written as

$$\text{input} - \text{output} = \text{accumulation} \quad 2.1$$

This balance applies to (a) the total material and/or energy involved, (b) any element or compound whose quantity has not changed by chemical reaction, and (c) any chemical element. It can also be written for any piece of equipment, several pieces, or an entire process.

A *process* can be classified as either unsteady or steady state. A *steady state* process is one in which conditions at any point in the system do not change with time. All other processes are *unsteady*. The material balance of a steady state process can be written as

$$\text{input} = \text{output} \qquad 2.2$$

A chemical reaction equation is used to (a) relate what reacts to what is produced, (b) show the compositions of the materials involved, and (c) show the quantitative relationships between the materials involved.

A chemical equation does *not* indicate:

- whether a reaction will proceed or not. (The equation simply indicates the reacting proportions if it does proceed.)
- the rate of reaction
- the degree of completion of the reaction
- the best conditions for the reaction

The following definitions make it possible to interpret and record the information conveyed by a chemical equation.

- *limiting reactant* – a reactant present in stoichiometrically limited amounts, whose disappearance would terminate the reaction
- *excess reactant* – a reactant that is present in excess of an amount stoichiometrically equivalent to the quantity of the limiting reactant
- *theoretically required amount* – amount of reactant stoichiometrically equivalent to that of the limiting reactant
- *percent excess* – excess reactant expressed as a percent of the theoretically required reactant
- *degree of completion* – percent of the limiting reactant that reacts
- *percent conversion* – percent of the reactant that reacts

Example 2.1

Nitrogen and hydrogen react under certain conditions to form ammonia according to the equation

$$N_2 + 3H_2 \longleftrightarrow 2NH_3$$

280 pounds of nitrogen and 64.0 pounds of hydrogen are brought together and are allowed to react at 515 °C and 300 atmospheres pressure. Afterwards, it is found that there are 38 lbmoles of gases present at equilibrium.

(a) How many lbmoles of nitrogen, hydrogen, and ammonia are present at equilibrium? (b) Which is the limiting reactant, and which is the excess reactant? (c) How much excess hydrogen is there? (d) What is the amount of theoretically required hydrogen, and what is the percent excess hydrogen? (e) What is the degree of completion of the reaction? (f) What is the percentage conversion of hydrogen to ammonia? (g) What is the equilibrium constant at the reaction conditions?

(a) Let x represent the number of lbmoles of nitrogen used in the reaction. Then, the number of lbmoles of hydrogen consumed is $3x$, and the number of lbmoles of ammonia formed is $2x$.

After the reaction, the nitrogen remaining is

$$N_2 = 10 - x \text{ (lbmoles)}$$

Similarly, the hydrogen remaining is

$$H_2 = 32 - 3x \text{ (lbmoles)}$$

Therefore, at equilibrium,

$$(32 - 3x) + (10 - x) + 2x = 38$$
$$x = 2$$

At equilibrium,

$$N_2 = 8 \text{ lbmoles}$$
$$H_2 = 26 \text{ lbmoles}$$
$$NH_3 = 4 \text{ lbmoles}$$

(b) Since it takes only 30 lbmoles of hydrogen to react with 10 lbmoles of nitrogen, hydrogen is in excess, and nitrogen is the limiting reactant.

(c) Since 30 lbmoles of hydrogen are equivalent to 10 lbmoles of nitrogen, hydrogen is in excess by 2 moles.

(d) The theoretically required hydrogen is 30 lbmoles. The percent excess hydrogen is

$$\frac{2}{30} \times 100\% = 6.67\%$$

(e) The degree of completion is

$$\frac{10 - 8}{10} \times 100\% = 20\%$$

(f) The percentage conversion of hydrogen is

$$\frac{32 - 26}{32} \times 100\% = 18.75\%$$

(g) The equilibrium constant based on partial pressures p_i (atm) and total pressure P (atm) is

$$K_p = \frac{p_{NH_3}^2}{p_{H_2}^3 p_{N_2}}$$

$$= \frac{P^2 y_{NH_3}^2}{P^3 y_{H_2}^3 P y_{N_2}}$$

$$= \frac{y_{NH_3}^2}{y_{H_2}^3 y_{N_2}} P^{-2}$$

$$K_p = \frac{\left(\frac{4}{38}\right)^2}{\left(\frac{26}{38}\right)^3 \left(\frac{8}{38}\right)} (300)^{-2}$$

$$= \frac{(4)^2}{(26)^3 (8)} \left(\frac{38}{300}\right)^2$$

2 MATERIAL BALANCES

If a chemical reaction occurs, the Law of Definite Multiple Proportions must be applied in addition to the input-output relationship. Steady state material balance problems involving masses and compositions can be classified into three types:

Type 1:

- Initial mass and composition are known (input stream: 1).
- A material with known mass and composition is added or removed (add or remove stream: 2).
- The final mass and composition are unknown (output stream: 3).

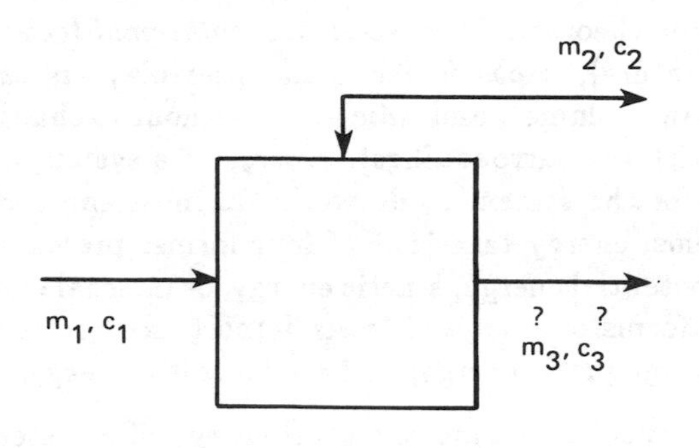

Figure 2.1 Type 1 Process

Example 2.2

A mixer receives two streams, each containing substances A and B. What is the composition of the third stream?

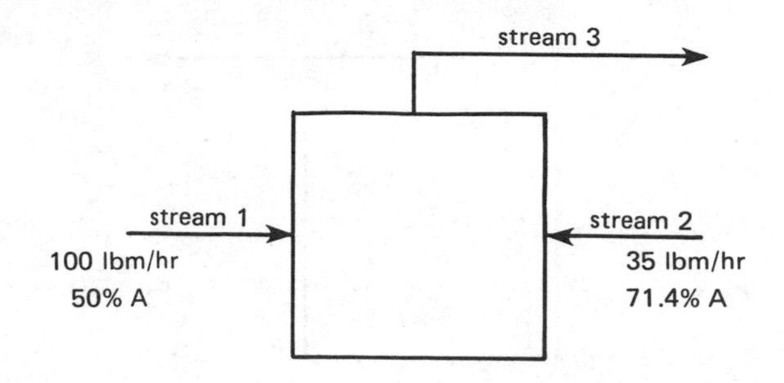

This is a type 1 process. The solution is purely an accounting procedure.

component	stream 1	stream 2	stream 3
A	50	25	75
B	50	10	60
	100	35	135

$$\text{stream 3:} \quad A = \frac{75}{135} = 55.6\%$$

$$B = 44.4\%$$

Type 2:

- The initial composition is known.
- The final composition is known.
- The initial and final masses are unknown.

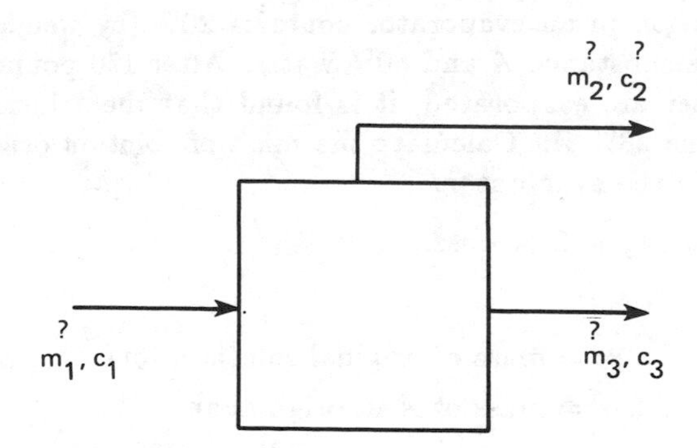

Figure 2.2 Type 2 Process

In type 2 problems, a *tie substance* is required to solve the problem. The tie substance passes through the process unchanged. It may be an element (e.g., nitrogen), a compound (e.g., methane), a group of substances (e.g., coal ash), or the sum of a number of substances in the mixture. The tie substance ties together the initial and final materials and gives a quantitative relationship between the two.

Type 3:

- The final composition is known.
- The initial and final masses are unknown.

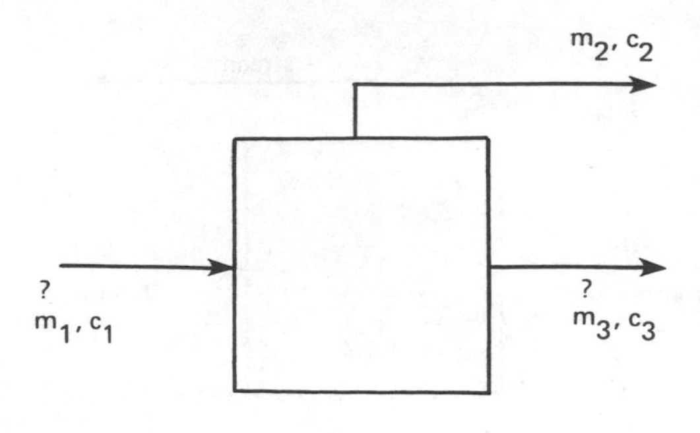

Figure 2.3 Type 3 Process

The solution to this type of system is simplified by the existence of a tie substance. The solution then begins by calculating the amount of substance added or removed per 100 units of initial material. From the amount of material actually added or removed, the initial and final masses are found. If no tie element exists, a material balance can be written using the initial mass, m_i, as an unknown.

Example 2.3

A solution in an evaporator contains 20% (by weight) soluble substance *A* and 80% water. After 120 pounds of water are evaporated, it is found that the solution contains 28% *A*. Calculate the mass of solution originally in the evaporator.

This is a type 2 process.

$$m = \text{mass of original solution, lbm}$$

$$0.2\,m = \text{mass of } A \text{ in original and final solution, (tie)}$$

$$m - 120 = \text{mass of final solution, lbm}$$

$$0.28 = 0.2\frac{m}{m-120} = \text{concentration of } A \text{ in final solution}$$

$$m = 420 \text{ lbm}$$

Example 2.4

It is impossible to measure the amount of sulfuric acid solution in a tank, but the acid is found to contain 8.7% (by mass) H_2SO_4. Ten pounds of 50% H_2SO_4 are added to the tank, and the final mixed composition in the tank is found to be 9.87% H_2SO_4. What is the final mass of the solution in the tank?

This is a type 3 process.

$$m = \text{final mass in tank}$$

$$m - 10 = \text{initial mass in tank}$$

$$(m-10)(0.087) + (10)(0.50) = m(0.0987)$$

$$m = 353 \text{ lbm}$$

3 ENERGY BALANCES

The three laws of thermodynamics are:

First Law (Law of Conservation of Energy): In any ordinary chemical or physical process, energy is neither created nor destroyed.

Second Law (Limiting Law): Energy cannot flow spontaneously from lower to higher temperatures. It is impossible for a machine to derive energy from any portion of matter by cooling it below the temperature of the coldest of surrounding objects.

Third Law: It is impossible to reduce a system to absolute zero temperature.

A *system* is any portion of the universe which is set aside to study. The rest of the universe is called the *surroundings*. A system is described by specifying its state or properties and its boundaries (open, closed, or isolated).

The state of a system is reproducible; the destiny of a gas of specified composition will always be the same at the same temperature and pressure. A system can be changed to another state by varying the properties. The *path*, or series of values the property values assume in passing from one state to another, defines a *process*.

Common theoretical processes are *isothermal* (constant temperature), *isobaric* (constant pressure), *isochoric* (constant volume), and *adiabatic* (without exchange of heat with the surroundings). *Energy* of a system is the ability of the system to do work. In most engineering problems, energy takes one of four forms: pressure energy, potential energy, kinetic energy, or internal energy. Other forms of energy of lesser importance are surface energy, magnetic energy, and electrostatic energy.

At a particular instant, the total energy of a system is

$$E_1 = \frac{m}{J}\left(p_1 v_1 + \frac{z_1 g}{g_c} + \frac{\mathrm{v}_1^2}{2g_c}\right) + m u_1 \qquad 2.3$$

If the system has undergone a change in state without mass change, the new energy state can be written

$$E_2 = \frac{m}{J}\left(p_2 v_2 + \frac{z_2 g}{g_c} + \frac{\mathrm{v}_2^2}{2g_c}\right) + m u_2 \qquad 2.4$$

The change in energy is

$$\Delta E = E_2 - E_1 \tag{2.5}$$

The first law states that energy is neither created nor destroyed, so the only way for the energy of the system to change is for the surroundings to change energy by an equal but opposite amount. When constant mass of the system is assumed, work and heat are the only ways in which the system and surroundings can interchange energy.

Work and *heat* are defined as energy in transit, and they cannot be stored. Two common forms of work in engineering are mechanical work and electrical work. Heat is energy in transit as a result of a temperature difference. If the system does an amount of work, W, on the surroundings and receives an amount of heat, Q, from the surroundings, then the total energy change of the system is defined by the *non-flow* or *static form* of the first law.

$$\Delta E = Q - \frac{W}{J} \tag{2.6}$$

The terms W and Q are algebraically positive. When a chemical reaction gives off heat, Q is negative. In an expansion process,

$$\Delta E = m\,\Delta u = Q - \frac{m}{J}\int p\,dv \tag{2.7}$$

At constant volume, the first law becomes

$$\Delta E = Q \tag{2.8}$$

At constant pressure, the first law becomes

$$\Delta E = Q - \frac{m}{J}\,p\,\Delta v \tag{2.9}$$

In many cases, the operation is continuous. An analysis of continuous problems leads to the *flow equation* form of the first law. In a flow situation, material entering the equipment brings in a certain amount of energy. Material leaving the equipment removes a certain amount of energy, and energy is also interchanged between the surroundings and the system by work and heat.

$$mu_1 + \frac{m}{J}\left(p_1 v_1 + \frac{z_1 g}{g_c} + \frac{\mathrm{v}_1^2}{2g_c}\right) + Q =$$

$$mu_2 + \frac{m}{J}\left(p_2 v_2 + \frac{z_2 g}{g_c} + \frac{\mathrm{v}_2^2}{2g_c}\right) + \frac{W}{J} \tag{2.10}$$

Enthalpy is defined as the sum of the internal energy and pressure energy terms.

$$H = U + \frac{pv}{J} \tag{2.11}$$

Thus, the flow equation becomes

$$Q - \frac{W}{J} = \frac{m}{J}\left(\frac{\Delta z g}{g_c} + \frac{\Delta \mathrm{v}^2}{2g_c}\right) + m\,\Delta h \tag{2.12}$$

The flow equation can be modified to fit a particular energy balance. Consider a boiler, where no work is done to the surroundings ($W = 0$), and where there are no changes in potential and kinetic energy ($\Delta z = 0$ and $\Delta \mathrm{v} = 0$). The energy balance becomes

$$Q = m\Delta h = mc_p\Delta T + m\Lambda \tag{2.13}$$

Consider a well-insulated (adiabatic) steam turbine where $Q = 0$, $\Delta z = 0$, and $\Delta \mathrm{v} = 0$. The energy balance becomes

$$m\,\Delta h = \frac{-W}{J} \tag{2.14}$$

Heat transferred due to a difference in temperature is called *sensible heat*.

$$Q = m\,c_p\,\Delta T \tag{2.15}$$

At constant temperature, heat can also be transferred by a phase change. The heat transferred is called *latent heat*.

$$Q = m\Lambda \tag{2.16}$$

Latent heats are known as *heat of fusion* (for melting a solid), *heat of vaporization* (for vaporizing a liquid), *heat of transition* (for changing one solid phase to another, such as a monoclinic to rhombic sulfur), and *heat of sublimation* (for vaporizing a solid without going through a liquid phase).

Example 2.5

Find the amount of 212 °F steam which must be mixed with 400 pounds of 70 °F water in order to make the final liquid water temperature 212 °F. Heat losses amount to 2000 BTU. The latent heat of vaporization for water at 212 °F is 970.3 BTU/lbm.

A simple heat balance solves the problem.

$$m_s\Lambda = \text{losses} + m_w c_p(T_2 - T_1)$$

$$m_s(970.3) = 2000 + (400)(1)(212 - 70)$$

$$m_s = 60.6 \text{ pounds of steam}$$

4 CHOOSING THE BASIS OF THE CALCULATION

In solving any stoichiometry problem, the first step is always to choose one of the components as a *basis*. One independent material balance can then be written for each component. From this, along with other data relevant to the problem, the corresponding quantities of the other components can be calculated. The final answer to the problem will be obtained by the use of an obvious scale factor.

Generally, a convenient basis substance will be:

- one of the streams entering or leaving the process
- one of the active components entering or leaving the process
- one of the inert components entering or leaving the process

Some useful hints on choosing a basis are given here.

- Since gas analyses are generally reported on a volumetric basis, problems involving mixtures of only gases are suited to taking 100 moles of the gas stream as a basis.
- For solid streams, a basis of 100 pounds or one mole of a component stream is convenient. For solid mixtures containing such miscellaneous things as ash, bark, and rubber, a weight basis is the only reasonable choice, since the analysis is invariably in percentage by weight.
- Liquid compositions are usually reported in weight or mole percent. These should be handled like solid streams. Liquid compositions are sometimes reported in volume percent, in which case, choose one cubic foot of the liquid stream as the basis, multiply the volume of each component by the density, and convert the resulting weight to moles by dividing by the respective molecular weight.
- If the problem calls for an answer involving time rate, it is often convenient to choose the basis as the amount of material flowing in a stream in that unit of time.

When solving problems involving solid-liquid mixtures, such as slurries or wet solids, conversion from a wet basis to a dry basis makes calculations more convenient. Wet-to-dry conversion is accomplished utilizing equations 2.17 and 2.18.

$$\text{wet percent} = \frac{\text{dry\%}}{1 + \frac{\text{dry\%}}{100}} \qquad 2.17$$

$$\text{dry percent} = \frac{\text{wet\%}}{1 - \frac{\text{wet\%}}{100}} \qquad 2.18$$

Example 2.6

A 100 pound batch of clay contains 20% water by weight. It is dried to a water content of 5%. How much water is removed?

$$m_f = \text{total final mass of the 5\% mixture}$$
$$100 - m_f = \text{mass of water removed}$$
$$100(0.80) = 80 = 0.95\,m_f = \text{mass of clay}$$
$$m_f = 84.21 \text{ pounds}$$
$$100 - 84.21 = 15.79 \text{ pounds of water}$$

(Note: The answer is not 15 pounds because the 20% and the 5% do not refer to the same basis.)

5 UNSTEADY STATE BALANCES

In steady state problems, the material entering the system is equated to the material leaving the system. In unsteady state problems, time is a variable, and some properties of the system are functions of time. It is no longer true that input equals output, since there is an allowance for the materials which accumulate or deplete in the system.

Input and output terms in balance equations are computed as rate terms (i.e., pounds per minute), and the accumulation term is computed as the differential of the quantity accumulating in the system (with respect to time). The unsteady state balance for a flowing system with instantaneous volume V_t and concentration c_t, and input flow Q_i at concentration c_i, the output flow Q_o at c_o, is

$$Q_i c_i - Q_o c_o = \frac{d}{dt}(V_t c_t) \qquad 2.19$$

The time-dependent relationship of the variables in equation 2.19 must be known before the problem can be solved. Three steps must be taken to make this equation usable.

step 1: The time dependency nature of the variables has to be determined.

step 2: The interdependence of the variables must be determined and inserted into the balance equation so that there is only one dependent variable and only one independent variable (time).

step 3: The boundary (initial, final, or other) conditions for the equation must be determined so that the constant of integration can be computed.

Most unsteady state material and energy balance problems in chemical engineering can be modeled by simple first- or second-order differential equations.

Consider the situation where the input and output flow rates, Q_i and Q_o are equal and constant. Then the input concentration and system volume are also constant. The balance equation becomes

$$Q(c_i - c_o) = V_t \frac{d}{dt} c_t \qquad 2.20$$

Equation 2.20 is a differential equation having two dependent variables (c_t and c_o) and one independent variable (time). Ideal mixing, where the output concentration and the system concentration are always the same, is the last simplifying assumption to eliminate one of the dependent variables. If c_t and c_o are equal, the balance equation becomes

$$\left(\frac{Q}{V_t}\right) dt = \frac{dc_o}{c_i - c_o} \qquad 2.21$$

This is a first-order, linear differential equation. If the initial concentration in the system at $t = 0$ is $c_{t=0}$, then the differential equation can be integrated. The final solution is

$$c_o = c_i + (c_{t=0} - c_i)e^{\frac{-Qt}{V_t}} \qquad 2.22$$

Example 2.7

A tank initially containing 20 gallons of pure water receives 3 gpm of a brine containing 2 lbm/gal of salt. Liquid flows out of the tank at 2 gpm. What is the brine concentration in the tank after 9 minutes? Assume the tank is well-agitated and does not overflow.

$$\begin{aligned} \text{salt input} &= Q_i c_i = (3)(2) = 6 \text{ lbm/min} \\ \text{salt output} &= Q_o c_o = (2)(c_o) \text{ lbm/min} \\ \text{perfect mixture} &\quad c_o = c_t \\ \text{volume} &= V_t = V_{t=0} + (3-2)t \text{ gal} \end{aligned}$$

The material balance is

$$6 - 2c_o = \frac{d}{dt}[(V_{t=0} + t)c_o]$$

Differentiating,

$$6 - 2c_o = (V_{t=0} + t)\frac{dc_o}{dt} + c_o$$

Rearranging and integrating,

$$ln\,(V_{t=0} + t) = ln\,(2 - c_o)^{-\frac{1}{3}} + C$$

The boundary conditions are

$$\begin{aligned} t &= 0 \\ c_o &= 0 \\ C &= ln\left(\frac{V_{t=0}}{2^{-\frac{1}{3}}}\right) \\ c_o &= 2 - 2\left(1 + \frac{t}{V_{t=0}}\right)^{-3} \end{aligned}$$

When $t = 9$ minutes and $V_{t=0} = 20$ gal,

$$c_o = 2 - 2\left(1 + \frac{9}{20}\right)^{-3} = 1.344 \text{ lbm/gal}$$

6 COMBUSTION STOICHIOMETRY

The basic equations for the combustion of fuels containing carbon, hydrogen, and sulfur are listed in equations 2.23 through 2.26 in their theoretical order of occurrence.

$$2H_2 + O_2 \rightarrow 2H_2O \qquad 2.23$$

$$2C + O_2 \rightarrow 2CO \qquad 2.24$$

$$2CO + O_2 \rightarrow 2CO_2 \qquad 2.25$$

$$S + O_2 \rightarrow SO_2 \qquad 2.26$$

The amount of oxygen (or air) just sufficient to completely burn the carbon, hydrogen, and sulfur in a fuel is the *theoretical oxygen* (or air). The general expression for combustion of a sulfur-free hydrocarbon in oxygen is given by equation 2.27. m and n are general stoichiometric constants.

$$C_mH_n + \left(\frac{4m+n}{4}\right) O_2 \rightarrow mCO_2 + \frac{n}{2} H_2O \qquad 2.27$$

The approximate volume of theoretical oxygen needed to burn any fuel can be calculated from the analysis of the fuel as follows, where C, H_2, O_2 and S are the decimal weights of these elements in one pound of fuel.

$$\frac{\text{volume oxygen}}{\text{pound fuel}} = 359\left(\frac{C}{12} + \frac{H_2}{4} + \frac{S}{32} - \frac{O_2}{32}\right) \qquad 2.28$$

The coefficient 359 is the volume (in cubic feet) of one lbmole of O_2 at 32 °F and one atmosphere. The volume of theoretical air is obtained by using the coefficient 1710 instead of 359.

STOICH

7 FUEL HEATING VALUES

Refer to table 2.1 for fuel heating values. The difference between the *gross* or *higher heating value* (HHV), and the *net* or *lower heating value* (LHV), is due to the condensation of water formed during combustion. The available heat is

$$Q = m(\text{LHV} - \text{enthalpy of combustion products}) \quad 2.29$$

8 AIR PROPERTIES

For most calculations, the following properties of air are adequate:

$$O_2 = 21\% \text{ by mole or volume}$$
$$N_2 = 79\% \text{ by mole or volume}$$
$$\frac{N_2}{O_2} = 3.76 \text{ (mole ratio)}$$
$$\text{molecular weight of air} = 29$$
$$\frac{\text{air}}{N_2} = 1.265 \text{ (mole ratio)}$$

9 EXCESS AIR

The amount of excess air for combustion can be calculated from a dry analysis of the products of combustion. The percent excess air, A_x, is calculated using equation 2.30, which assumes there is a negligible amount of nitrogen in the fuel.

$$A_x = \frac{\%O_2}{0.266\,(\%N_2) - \%O_2} \times 100\% \quad 2.30$$

This equation should only be used if there are no combustible gases (e.g., CO or H_2) in the products of combustion. If these gases are present, the percent excess air is calculated from the *Orsat formula*:

$$A_x = \frac{\%O_2 - 0.5\,(\%CO + \%H_2)}{0.266(\%N_2) - \%O_2 + 0.5\,(\%CO + \%H_2)} \times 100\% \quad 2.31$$

Example 2.8

Coke containing 93% carbon, 7% ash, and no hydrocarbons or hydrogen is burned to produce a flue gas containing 78.1% N_2, 12.5% CO_2, 2.1% CO, and 7.3% O_2. What is the percent excess of air?

Using the Orsat formula,

$$A_x = \frac{7.3 - 0.5\,(2.1 - 0)}{0.266\,(78.1) - (7.3) + 0.5\,(2.1 - 0)} \times 100\% = 43\%$$

10 COMBUSTION CALCULATIONS

There are three types of combustion problems. In the first type, the composition of the fuel is given; the oxygen (or air) required to obtain a given flue gas composition is unknown.

In the second type of combustion problem, you must find the composition of the flue gas, given the amount of air used and the fuel composition. If the composition of the flue gas is given, the analysis is usually made by an *Orsat analyzer.* This analyzer only determines the dry volumetric percentages of CO_2, CO, and O_2. The volume of these gases will not add up to 100%; the difference is usually interpreted as nitrogen. The analyzer cannot analyze for water vapor in the flue gas; this can only be done with knowledge of the hydrogen content of the fuel. (If hydrogen is in the fuel, the additional oxygen usage must be accounted for.) Liquid and gaseous fuels usually contain hydrogen.

In the third type of combustion problem, the flue gas composition is given; the fuel composition is unknown. If a solid (e.g., coal) is in the fuel, the amount of ash will also be given. Either the total fuel mass or the carbon mass will be the tie element in this type of problem.

Use the following steps when solving combustion problems:

step 1: Break the fuel into its basic elements on a mole basis (C, O_2, N_2, H_2, S).

step 2: Equate the quantities of carbon in the flue gas and fuel.

step 3: Determine the air requirements from a nitrogen balance.

step 4: Determine the excess air from an oxygen balance.

step 5: Determine the moisture content using a hydrogen balance.

Example 2.9

From the following flue gas analysis, determine the amount of air used and the carbon burned (per 100 lbmoles of dry flue gas).

component	Orsat analysis mole %
CO_2	12.0%
O_2	7.0%
N_2	81.0%
	100.0%

Use N_2 as the tie element, and take 100 lbmoles of dry flue gas as a basis:

$$N_2 \text{ in air} = N_2 \text{ in flue gas}$$

$$\begin{aligned} \text{moles of air} &= \frac{81.0}{0.79} \\ &= 102.5 \text{ lbmoles air per 100 lbmoles flue gas} \\ &= 2972.5 \text{ pounds of air per 100 lbmoles flue gas} \end{aligned}$$

$$\begin{aligned} O_2 \text{ in air} &= O_2 \text{ in } CO_2,\ O_2, \text{ and } H_2O \\ O_2 \text{ in } H_2O &= (102.5)(0.21) - 12 - 7 \\ &= 2.525 \text{ lbmoles } O_2 \text{ in } H_2O \end{aligned}$$

Thus, the reaction is

$$H_2 + \frac{1}{2}O_2 \rightarrow H_2O$$

Assuming that all hydrogen is consumed, the wet gas analysis becomes

component	flue gas lbmoles	flue gas mole %
CO_2	12.0	11.4%
O_2	7.0	6.7%
N_2	81.0	77.1%
H_2O	5.05	4.8%
	105.05 moles	100.0%

$$\begin{aligned} \text{C in flue gas} &= \text{C in fuel} \\ (12 \text{ lbmoles})(12 \text{ lbm/lbmoles}) &= 144 \text{ pounds of C in fuel} \end{aligned}$$

11 THEORETICAL FLAME TEMPERATURE

A fuel's *theoretical flame temperature* (TFT) is the temperature which could be reached if combustion were complete, and no heat losses occurred (an adiabatic process).

The method used to determine TFT is to equate the sum of the heats of reaction (H^0_{Rx}) and enthalpies of the reactants (H_R) to the enthalpy of the products (H_p). (Note that ΔH^0_{Rx} is negative for exothermic reactions.)

$$\sum H_R + \Delta H^\circ_{Rx} = \sum H_P \qquad 2.32$$

The heats of reaction are taken at some standard temperature (usually 25 °C). The products should include nitrogen from the air, excess oxygen, and other diluents. A trial-and-error solution is typical.

step 1: Compute the heats of reaction.

step 2: Assume a temperature.

step 3: Compute the enthalpies of the reactants and products at the assumed temperature.

step 4: Plot the difference between the enthalpy of the products against the sum of the heat of reaction and enthalpy of the reactants. Plot this difference on the y axis with the assumed temperature on the x axis.

step 5: Assume another temperature, and repeat steps 3 and 4.

step 6: When the y axis points change sign, the TFT has been straddled. Extrapolate or interpolate the TFT from the plot and check the answer.

A plot of the mean specific heats of combustion gases is given in figure 2.4.

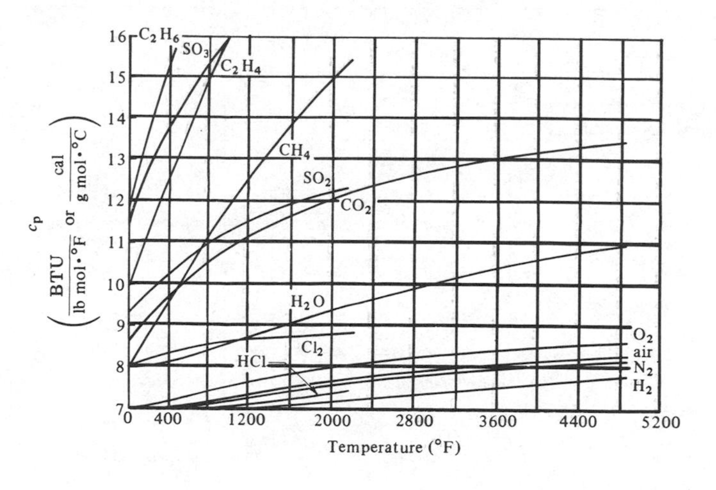

Figure 2.4 Mean Specific Heats of Gases

As an approximation, the TFT can be found from the lower heating value (LHV) of the fuel from equation 2.33.

$$\text{TFT} = T_R + \frac{m_f(\text{LHV})}{\sum m_i c_{pi}} \qquad 2.33$$

T_R is the temperature of the reactants. The LHV can be computed, if necessary, from equation 2.34.

$$\text{LHV} = \text{HHV} - m_w \Lambda \qquad 2.34$$

The values of HHV and LHV are given in table 2.1.

It should be noted that the dissociation of CO_2 and H_2O should be taken into account at temperatures above 3000 °F.

Example 2.10

What is the theoretical flame temperature of methane burned with a theoretical amount of air? The gross heating value of methane is 23,880 BTU/lbm.

	CH_4	$+2O_2$	$+N_2$	$\rightarrow CO_2$	$+2H_2O$	$+N_2$
V	1	2	7.52	1	2	7.52
M	16	32	28	44	18	28
m	1	4	13.18	2.75	2.25	13.18

Assume $T_R = 77$ °F. From equation 2.34,

$$\begin{aligned} \text{LHV} &= \text{HHV} - m_w \Lambda \\ &= 23{,}880 - 2.25\,(1050) \\ &= 21{,}520 \text{ BTU/lbm} \end{aligned}$$

	c_p at 4000 °F	c_p at 3700 °F
CO_2	0.301	0.299
H_2O	0.585	0.578
N_2	0.288	0.286

The first approximation is TFT = 4000 °F.

$$\text{TFT} = \frac{(1)(21{,}520)}{(2.75)(0.301) + (2.25)(0.585) + (13.18)(0.288)} + 77 = 3700 \text{ °F}$$

The second approximation is 3700 °F.

$$\text{TFT} = \frac{(1)(21{,}520)}{(2.75)(0.299) + (2.25)(0.578) + (13.18)(0.286)} + 77 = 3729 \text{ °F}$$

The theoretical flame temperature is

$$\text{TFT} = \frac{1}{2}(3700 + 3729) = 3715 \text{ °F}$$

Table 2.1
Combustion Data for Gases

						Heat of Combustion[c]				Cu Ft per Cu Ft of Combustible						Lb per Lb of Combustible						Experimental Error in Heat of Combustion Percent
						Btu per Cu Ft		Btu per Lb		Required for Combustion			Flue Products			Required for Combustion			Flue Products			
No. Substance	Formula	Molecular Weight[a]	Lb per Cu Ft[b]	Cu Ft per Lb[b]	Sp Gr Air = 1.000[b]	Gross	Net[d]	Gross	Net[d]	O_2	N_2	Air	CO_2	H_2O	N_2	O_2	N_2	Air	CO_2	H_2O	N_2	+ or −
1 Carbon	C	12.01	–	–	–	–	–	14,093[g]	14,093[g]	–	–	–	–	–	–	2.664	8.863	11.527	3.664	–	8.863	0.012
2 Hydrogen	H_2	2.016	0.005327	187.723	0.06959	325.0	275.0	61,100	51,623	0.5	1.882	2.382	–	1.0	1.882	7.937	26.407	34.344	–	8.937	26.407	0.015
3 Oxygen	O_2	32.000	0.08461	11.819	1.1053	–	–	–	–	–	–	–	–	–	–	–	–	–	–	–	–	–
4 Nitrogen (atm)	N_2	28.016	0.07439[e]	13.443[e]	0.9718[e]	–	–	–	–	–	–	–	–	–	–	–	–	–	–	–	–	–
5 Carbon monoxide	CO	28.01	0.07404	13.506	0.9672	321.8	321.8	4,347	4,347	0.5	1.882	2.382	1.0	–	1.882	0.571	1.900	2.471	1.571	–	1.900	0.045
6 Carbon dioxide	CO_2	44.01	0.1170	8.548	1.5282	–	–	–	–	–	–	–	–	–	–	–	–	–	–	–	–	–
Paraffin series C_nH_{2n+2}																						
7 Methane	CH_4	16.041	0.04243	23.565	0.5543	1013.2	913.1	23,879	21,520	2.0	7.528	9.528	1.0	2.0	7.528	3.990	13.275	17.265	2.744	2.246	13.275	0.033
8 Ethane	C_2H_6	30.067	0.08029[e]	12.455[e]	1.04882[e]	1792	1641	22,320	20,432	3.5	13.175	16.675	2.0	3.0	13.175	3.725	12.394	16.119	2.927	1.798	12.394	0.030
9 Propane	C_3H_8	44.092	0.1196[e]	8.365[e]	1.5617[e]	2590	2385	21,661	19,944	5.0	18.821	23.821	3.0	4.0	18.821	3.629	12.074	15.703	2.994	1.634	12.074	0.023
10 n-Butane	C_4H_{10}	58.118	0.1582[e]	6.321[e]	2.06654[e]	3370	3113	21,308	19,680	6.5	24.467	30.967	4.0	5.0	24.467	3.579	11.908	15.487	3.029	1.550	11.908	0.022
11 Isobutane	C_4H_{10}	58.118	0.1582[e]	6.321[e]	2.06654[e]	3363	3105	21,257	19,629	6.5	24.467	30.967	4.0	5.0	24.467	3.579	11.908	15.487	3.029	1.550	11.908	0.019
12 n-Pentane	C_5H_{12}	72.144	0.1904[e]	5.252[e]	2.4872[e]	4016	3709	21,091	19,517	8.0	30.114	38.114	5.0	6.0	30.114	3.548	11.805	15.353	3.050	1.498	11.805	0.025
13 Isopentane	C_5H_{12}	72.144	0.1904[e]	5.252[e]	2.4872[e]	4008	3716	21,052	19,478	8.0	30.114	38.114	5.0	6.0	30.114	3.548	11.805	15.353	3.050	1.498	11.805	0.071
14 Neopentane	C_5H_{12}	72.144	0.1904[e]	5.252[e]	2.4872[e]	3993	3693	20,970	19,396	8.0	30.114	38.114	5.0	6.0	30.114	3.548	11.805	15.353	3.050	1.498	11.805	0.11
15 n-Hexane	C_6H_{14}	86.169	0.2274[e]	4.398[e]	2.9704[e]	4762	4412	20,940	19,403	9.5	35.760	45.260	6.0	7.0	35.760	3.528	11.738	15.266	3.064	1.464	11.738	0.05
Olefin series C_nH_{2n}																						
16 Ethylene	C_2H_4	28.051	0.07456	13.412	0.9740	1613.8	1513.2	21,644	20,295	3.0	11.293	14.293	2.0	2.0	11.293	3.422	11.385	14.807	3.138	1.285	11.385	0.021
17 Propylene	C_3H_6	42.077	0.1110[e]	9.007[e]	1.4504[e]	2336	2186	21,041	19,691	4.5	16.939	21.439	3.0	3.0	16.939	3.422	11.385	14.807	3.138	1.285	11.385	0.031
18 n-Butene (Butylene)	C_4H_8	56.102	0.1480[e]	6.756[e]	1.9336[e]	3084	2885	20,840	19,496	6.0	22.585	28.585	4.0	4.0	22.585	3.422	11.385	14.807	3.138	1.285	11.385	0.031
19 Isobutene	C_4H_8	56.102	0.1480[e]	6.756[e]	1.9336[e]	3068	2869	20,730	19,382	6.0	22.585	28.585	4.0	4.0	22.585	3.422	11.385	14.807	3.138	1.285	11.385	0.031
20 n-Pentene	C_5H_{10}	70.128	0.1852[e]	5.400[e]	2.4190[e]	3836	3586	20,712	19,363	7.5	28.232	35.732	5.0	5.0	28.232	3.422	11.385	14.807	3.138	1.285	11.385	0.037
Aromatic series C_nH_{2n-6}																						
21 Benzene	C_6H_6	78.107	0.2060[e]	4.852[e]	2.6920[e]	3751	3601	18,210	17,480	7.5	28.232	35.732	6.0	3.0	28.232	3.073	10.224	13.297	3.381	0.692	10.224	0.12
22 Toluene	C_7H_8	92.132	0.2431[e]	4.113[e]	3.1760[e]	4484	4284	18,440	17,620	9.0	33.878	42.878	7.0	4.0	33.878	3.126	10.401	13.527	3.344	0.782	10.401	0.21
23 Xylene	C_8H_{10}	106.158	0.2803[e]	3.567[e]	3.6618[e]	5230	4980	18,650	17,760	10.5	39.524	50.024	8.0	5.0	39.524	3.165	10.530	13.695	3.317	0.849	10.530	0.36
Miscellaneous gases																						
24 Acetylene	C_2H_2	26.036	0.06971	14.344	0.9107	1499	1448	21,500	20,776	2.5	9.411	11.911	2.0	1.0	9.411	3.073	10.224	13.297	3.381	0.692	10.224	0.16
25 Naphthalene	$C_{10}H_8$	128.162	0.3384[e]	2.955[e]	4.4208[e]	5854[f]	5654[f]	17,298[f]	16,708[f]	12.0	45.170	57.170	10.0	4.0	45.170	2.996	9.968	12.964	3.434	0.562	9.968	–[f]
26 Methyl alcohol	CH_3OH	32.041	0.0846[e]	11.820[e]	1.1052[e]	867.9	768.0	10,259	9,078	1.5	5.646	7.146	1.0	2.0	5.646	1.498	4.984	6.482	1.374	1.125	4.984	0.027
27 Ethyl alcohol	C_2H_5OH	46.067	0.1216[e]	8.221[e]	1.5890[e]	1600.3	1450.5	13,161	11,929	3.0	11.293	14.293	2.0	3.0	11.293	2.084	6.934	9.018	1.922	1.170	6.934	0.030
28 Ammonia	NH_3	17.031	0.0456[e]	21.914[e]	0.5961[e]	441.1	365.1	9,668	8,001	0.75	2.823	3.573	–	1.5	3.323	1.409	4.688	6.097	–	1.587	5.511	0.088
29 Sulfur	S	32.06	–	–	–	–	–	3,983	3,983	–	–	–	–	–	–	0.998	3.287	4.285	SO_2 1.998	–	3.287	0.071
30 Hydrogen sulfide	H_2S	34.076	0.09109[e]	10.979[e]	1.1898[e]	647	596	7,100	6,545	1.5	5.646	7.146	SO_2 1.0	1.0	5.646	1.409	4.688	6.097	SO_2 1.880	0.529	4.688	0.30
31 Sulfur dioxide	SO_2	64.06	0.1733	5.770	2.264	–	–	–	–	–	–	–	–	–	–	–	–	–	–	–	–	–
32 Water vapor	H_2O	18.016	0.04758[e]	21.017[e]	0.6215[e]	–	–	–	–	–	–	–	–	–	–	–	–	–	–	–	–	–
33 Air	–	28.9	0.07655	13.063	1.0000	–	–	–	–	–	–	–	–	–	–	–	–	–	–	–	–	–

All gas volumes corrected to 60F and 30 in. Hg dry. For gases saturated with water at 60F, 1.73% of the Btu value must be deducted.

[a] Calculated from atomic weights given in "Journal of the American Chemical Society", February 1937.

[b] Densities calculated from values given in grams per liter at 0C and 760 mm in the International Critical Tables allowing for the known deviations from the gas laws. Where the coefficient of expansion was not available, the assumed value was taken as 0.0037 per °C. Compare this with 0.003662 which is the coefficient for a perfect gas. Where no densities were available the volume of the mol was taken as 22.4115 liters.

[c] Converted to mean Btu per lb (1/180 of the heat per lb of water from 32F to 212F) from data by Frederick D. Rossini, National Bureau of Standards, letter of April 10, 1937, except as noted.

[d] Deduction from gross to net heating value determined by deducting 18,919 Btu per pound mol of water in the products of combustion. Osborne, Stimson, and Ginnings, "Mechanical Engineering", p. 163, March 1935, and Osborne, Stimson, and Fiock, National Bureau of Standards Research Paper 209.

[e] Denotes that either the density or the coefficient of expansion has been assumed. Some of the materials cannot exist as gases at 60F and 30 in. Hg pressure, in which case the values are theoretical ones given for ease of calculation of gas problems. Under the actual concentrations in which these materials are present their partial pressure is low enough to keep them as gases.

[f] From Third Edition of "Combustion."

[g] National Bureau of Standards, RP 1141.

Reprinted from "Fuel Flue Gases", 1941 Edition, courtesy of American Gas Association.

12 FLUE GAS DEWPOINT

Flue gas dewpoint is the temperature at which vapor starts to condense in a constant pressure stack. For the cases where SO_2 and SO_3 are absent in the flue gas, the dewpoint can be found from steam tables as the temperature corresponding to the partial pressure of the water vapor.

13 RECYCLE STREAMS

Recycle streams are often part of the process being evaluated. In order to describe processes in general terms, the concept of the *concentration transformation function* (CTF) must be presented. Usually a chemical process alters the concentration of at least one species in the feed stream. The species concentration is changed either by direct addition (mixing) or subtraction (line manifolding or splitting), by chemical reaction, or by some separation process (ion exchange, solvent extraction, etc.). The CTF is a mathematical representation of the output concentration as a function of the input concentration. Figure 2.5 illustrates the concentration transformation function.

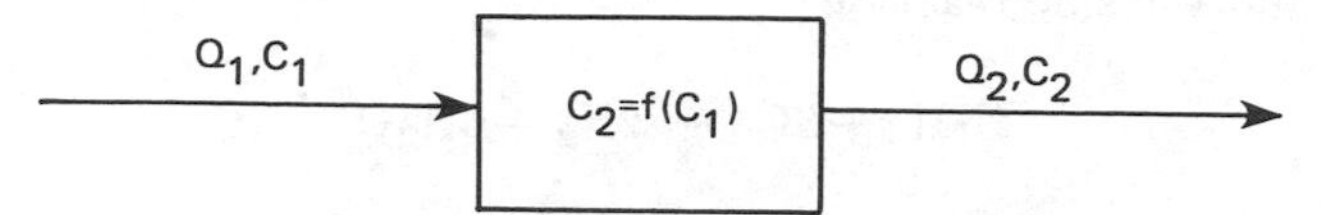

Figure 2.5 A Generalized Process Which Alters One or More Inlet Concentrations

The generalized process in figure 2.5 has no recycle stream, and the CTF relates the inlet to outlet concentration. The CTF, in general, takes the form

$$c_2 = \text{CTF} = f(c_1) \qquad 2.35$$

Some examples of common CTFs are:

$$c_2 = 1.36c_1 \qquad 2.36$$

$$c_2 = c_1 e^{-Q_1 t/V} \qquad 2.37$$

The most important property of the CTF is that it is explicit in c_2.

In certain processes, it is advantageous to divide the outlet stream and return a portion of it to the entrance of the process. The *recycle ratio*, R, is defined as the volume of material returning to the process entrance divided by the volume of material leaving the system. Figure 2.6 illustrates the definition of recycle ratio.

$$R = \frac{Q_3}{Q_2} \qquad 2.38$$

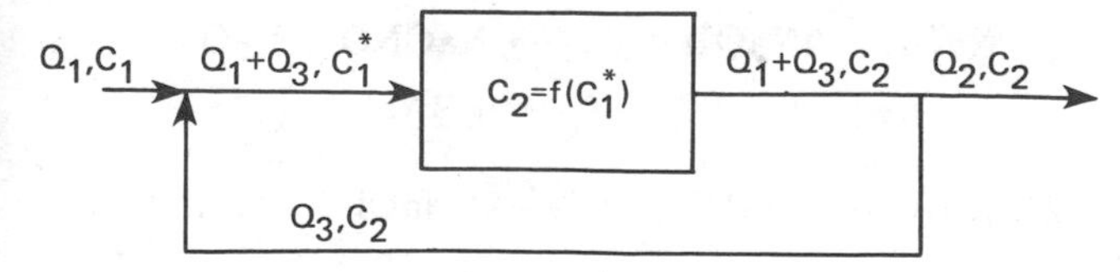

Figure 2.6 A Generalized Process With Recycle

The recycle ratio can vary from zero to infinity. A recycle ratio of four means that the flow entering the process is five times greater than that which would enter it without recycle. For steady state, constant volume systems $Q_1 = Q_2$, and the recycle ratio becomes

$$R = \frac{Q_3}{Q_1} \qquad 2.39$$

For steady state processes, writing a material balance for the mixing point yields

$$c_1 Q_1 + c_2 R Q_1 = (Q_1 + RQ_1)c_1^* \qquad 2.40$$

$$c_1^* = \frac{c_1 + c_2 R}{1 + R} \qquad 2.41$$

Substituting equation 2.41 into the CTF for a process without recycle converts the CTF to one with recycle. The CTF is rearranged to make it explicit in c_2. Care must be taken to transform all variables in the CTF that are functions of the recycle stream. Example 2.11 illustrates how to convert a CTF to one with a recycle stream.

Example 2.11

A plug flow reactor with a reaction involving first order kinetics has a CTF $c_2 = c_1 e^{-kt}$, where k is a constant and t is the residence time in the process. What is the recycle CTF for this process?

Since recycle will decrease residence time in the process due to increased flow rate, it must be accounted for in the recycle CTF. Without recycle, the residence time of a process is defined as

$$t = \frac{V}{v_1}$$

where V is the volume of the process and v_1 is the volumetric flow rate through the process without recycle. With recycle, the residence time becomes

$$t_{recycle} = \frac{V}{v_1(1+R)} = \frac{t_1}{1+R}$$

Substituting equation 2.41 and the equation for the recycle residence time into the CTF yields

$$c_2 = c_1^* e^{-kt/(1+R)} = \frac{c_1 + c_2 R}{1+R} e^{-kt/(1+R)}$$

This equation must be made explicit in c_2, which results in

$$c_2 = \frac{c_1 e^{-kt/(1+R)}}{1 + R - Re^{-kt/(1+R)}}$$

This is the recycle CTF for a plug flow reactor with first-order kinetics.

STOICH

PRACTICE PROBLEMS

1. Coal containing 79% carbon by weight and 6% ash by weight is burned in air. The clinker from the furnace pit contains 90% ash and 10% carbon. How many pounds of clinker are produced per 100 pounds of coal? What is the percent coal burned?

2. Wood containing 40% moisture (wet basis) is dried to 20% moisture (wet basis). How many pounds of water are evaporated per pound of dry wood?

3. When one mole of N_2 gas and 3 moles of H_2 gas are heated to 400 °C, and then allowed to come to equilibrium at 10 atmospheres, 0.148 moles of NH_3 are formed. The reaction is

$$N_2 + 3H_2 \rightarrow 2NH_3$$

The equilibrium constant is

$$K_p = \frac{(p_{NH_3})^2}{(p_{N_2})(p_{H_2})^3}$$

p_i is the partial pressure (atm) of component i. What is the equilibrium constant? What is the partial pressure of H_2 in the mixture?

4. A filter medium is tested by suspending 50 grams of filter material in 100 grams of water. When filtered, the filtrate is clear. The wet cake is weighed and found to be at 53.9 grams. After drying, the cake is found to be at 49.0485 grams. What is the solubility of the filter media in grams per 100 grams of water? Assume that the concentration of the filtrate is the same as the concentration of the liquid in the filter cake.

5. Two streams are mixed to form a single stream. Only the flow in the mixed stream is known. A soluble salt is added to one of the original streams at a steady state, and samples show the stream to have 4.76% salt by weight. Samples of the combined stream show it to be 0.62% salt by weight. What is the flow ratio of the two original streams?

6. Pure oxygen is mixed with air to produce enriched air containing 50% oxygen by volume. What ratio of oxygen to air should be used?

7. 100 grams of a mixture of $Na_2SO_4 \cdot 10H_2O$, and $Na_2CO_3 \cdot 10H_2O$ are heated to drive off the water of hydration. The final mass of the mixed salts is 39.6 grams. What is the molar ratio of Na_2CO_3 to Na_2SO_4 in the original hydrated salts?

8. A gas mixture containing 25% CO_2 and 75% NH_3 by volume is scrubbed with an acid solution to remove the ammonia. The gas mixture leaving the scrubber contains 37.5% NH_3. Assuming that the CO_2 remains unaffected and no acid solution vaporizes, what percent of the original ammonia is removed?

9. One pound of a mixture of NaCl and KCl was treated with H_2SO_4, and 1.2 pounds of potassium and sodium sulfate were recovered. What weights of NaCl and KCl were in the original mixture?

10. Kaolinite clays are produced from the weathering of igneous rocks, of which feldspar is a good example.

$$K_2O \cdot Al_2O_3 \cdot 6SiO_2 + CO_2 + 2H_2O \rightarrow K_2CO_3 + Al_2O_3 \cdot 2SiO_2 \cdot 2H_2O + 4SiO_2$$

How many tons of kaolinite clay are produced from a ton of feldspar?

11. Nitric oxide is made in a reactor by the oxidation of ammonia with the reaction

$$4NH_3 + 5O_2 \rightarrow 4NO + 6H_2O$$

and the side reaction

$$4NH_3 + 3O_2 \rightarrow 2N_2 + 6H_2O$$

The gases leave the reactor at 750 °C, and go to a pressurized scrubber where all the water and 85% of the NO are absorbed. The tail gases from the scrubber are reported as 2.2% O_2, 2.4% NO, and 95.4% N_2. (a) What percent of the ammonia is lost in the side reaction? (b) How much excess air is used?

12. A 73% caustic soda liquor at 150 °F (heat content 340 BTU/lbm) is diluted with 15% NaOH liquor at 70 °F (heat content 30 BTU/lbm) to give a solution of 40% NaOH. If the heat content of 40% NaOH at 120 °F is 107 BTU/lbm, how much heat per 100 pounds of product must be removed to maintain this temperature?

13. A plant with both cyanide-bearing and alkali-bearing effluents is considering the following method of waste treatment: cyanide effluent is to be treated with chlorine in the presence of some alkali effluent to decompose the cyanide. The resulting solution will then be mixed with the balance of alkali-bearing effluent for a final neutralization. The alkali available is NaOH. The reactions proceed according to the equations

$$NaCN + 2NaOH + Cl_2 \rightarrow NaCNO + H_2O + 2NaCl$$

$$2NaCNO + 4NaOH + 3Cl_2 \rightarrow 6NaCl + 2CO_2 + N_2 + 2H_2O$$

Support data is as follows:

cyanide effluent flow	3600 gallons/day
alkali effluent flow	9600 gallons/day
cyanide effluent temperature	90 °F
alkali effluent temperature	120 °F
cyanide content of cyanide effluent	1.64% as NaCN (by weight)

Consider all specific gravities as 1.0. Consider all specific heats as 1.0.

(a) How many pounds per day of Cl_2 and NaOH are required to decompose the cyanide if the chlorine is 20% excess and the caustic soda is 15% excess? What minimum weight percentage of caustic soda must be maintained in the alkali effluent? (b) How much process cooling water (at 75 °F) is required to cool the final mixed stream by direct mixing to 85 °F, ignoring the heats of dilution, neutralization, and reaction? (c) Ignoring the gases evolved from the reactions, what will be the concentration of the NaCl in the final cooled effluent, expressed as parts per million by weight?

14. A carburated water gas has the composition 16.0% C_2H_4, 19.9% CH_4, 32.3% H_2, 26.1% CO_2, 2.9% CO, and 2.8% N_2. After combustion, the flue gas analysis is 11.83% CO_2, 0.4% CO, 4.53% O_2, and 83.44% N_2. (a) How many cubic feet of air are needed per cubic foot of gas fired, if both are measured at 68 °F when the barometric pressure is 28.7 inches of mercury? (b) How many cubic feet of flue gas are produced at 670 °F per cubic feet of gas fired?

15. A solution containing 10% sodium sulfate by weight is cooled to 32 °F (where the solid phase is the decahydrate and the solubility of the sodium sulfate is 0.634 mole/100 moles H_2O). What is the resulting slurry density in pounds per 100 pounds of total mix?

16. A fuel composed of 50% C_7H_{16} and 50% C_8H_{18} is burned with 10% excess air. (a) Determine the volume (in cubic feet) of dry air at 70 °F and 14.9 psia required for 100 pounds of fuel. (b) Determine the volumetric analysis of the combustion products assuming complete oxidation.

17. It is necessary to filter a slurry consisting of the dihydrate $Ca(OCl)_2 \cdot 2H_2O$, and its mother liquor. The following are the pertinent data:

	mother liquor	slurry	cake	actual filtrate
% water	71.28	58.36	47.50	69.60
% $Ca(OCl)_2$	10.20	27.77	42.35	12.92

(a) What is the percent $Ca(OCl)_2$ in the original slurry? (b) It is noted that the actual filtrate is cloudy. What percent dihydrate is in the actual filtrate? (c) What fraction of the total dihydrate is being lost?

18. Gaseous propane at 77 °F is mixed with air at 260 °F and then burned. If 300% theoretical air is used, what is the adiabatic flame temperature?

19. A flat-roofed cylindrical storage tank is to be designed to store 20,000 gallons of a liquid that is twice as dense as water. The roof and bottom of the tank cost $7.00 per square foot, and the vertical surface costs 25tA$, where t is the thickness in inches, and A is the area in ft^2. Note: the vertical wall thickness can be determined by the equation

$$t = \frac{pD}{24,000}$$

where p is in psi, and D is the diameter in inches.

If the tank operates at atmospheric pressure, determine its dimensions for lowest cost. Due to structural considerations, the sides must be at least $\frac{1}{4}''$ thick.

20. A calcium carbonate sludge is burned in a rotary kiln to regenerate lime during a countercurrent process. The flue gas leaving the cold end of the kiln and the sludge entering the same end have known compositions.

flue gas	% by volume	sludge	% by weight
CO_2	20.4	$CaCO_3$	44.7
CO	0.4	H_2O	49.0
N_2	77.1	inerts	6.3

The kiln is fired with methane at 29,000 cubic feet per hour (dry at 60 °F and 14.7 psia). Lime conversion is 90% complete. What is the rate of CaO production?

21. A new organic chemical is prepared by precipitation from an aqueous solution containing excess H_2SO_4. It is proposed to separate the solid from the solution by centrifugal filtration, wash it to remove the excess acid, and then dry it. Pilot tests have shown that the centrifuge can be expected to give a residual mother liquor content of 0.08 lbm/lbm of dry product, and that a single stage of displacement washing, with one pound H_2O per pound of residual mother liquor, will give 80% efficiency in removal of acid.

Assume that after each stage of washing, the residual mother liquor content remains at 0.08 lbm/lbm of dry solid. The design feed is: 2500 lbm/hr of solids; 15,000 lbm/hr of H_2O; and 500 lbm/hr of H_2SO_4.

(a) What product acidity, expressed as percent H_2SO_4 by weight in the dry product, can be expected from a single stage of washing? (b) If it is necessary to reduce the acidity to 0.015% H_2SO_4 (dry weight basis) or less, how many washing stages are required? (c) At the end of the washings in (b), what is the H_2SO_4 content of the combined wash liquors and the acidity of the product (dry basis)?

22. A 10,000 gallon tank originally containing 5000 gallons of a 5% brine wash solution is fed 10 gpm of 22% salt solution. At the same time, valve A is opened to remove 5 gpm. (a) What is the exit concentration after 5 hours? (b) If the overflow can take up to 15 gpm, what is the exit concentration after 12 hours?

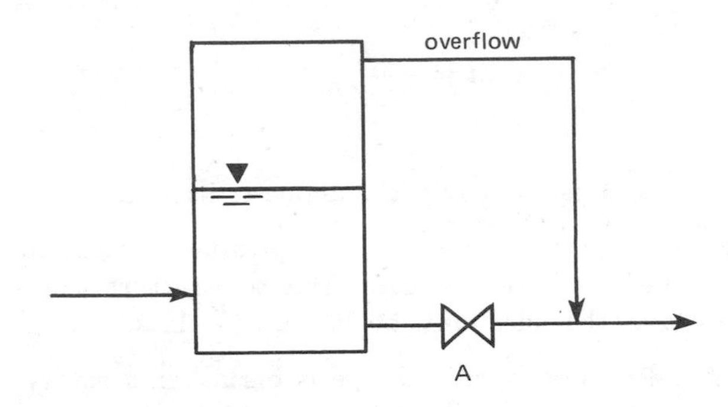

23. A reverse osmosis (RO) process removes a constant fraction of rubidium from a waste stream at all flow rates. If the fraction is q, then the concentration transformation function for the RO is

$$c_2 = c_1(1 - q)$$

where c_1 is the concentration rubidium entering the RO, and c_2 is the concentration leaving. What improvement in overall efficiency in rubidium removal is obtained if the fraction removed is 2/3, and a recycle ratio of 3 is used? What recycle ratio is needed to remove 95% of the rubidium?

3 ENGINEERING ECONOMICS

Nomenclature

A	annual amount or annuity	\$
B	present worth of all benefits	\$
BV_j	book value at the end of the jth year	\$
C	cost, or present worth of all costs	\$
d	declining balance depreciation rate	decimal
D_j	depreciation in year j	\$
$D.R.$	present worth of after-tax depreciation recovery	\$
e	natural logarithm base (2.718)	–
EAA	equivalent annual amount	\$
EUAC	equivalent uniform annual cost	\$
f	federal income tax rate	decimal
F	future amount, or future worth	\$
G	uniform gradient amount	\$
i	effective rate per period (usually per year)	decimal
k	number of compounding periods per year	–
n	number of compounding periods, or life of asset	–
P	present worth, or present value	\$
P_t	present worth after taxes	\$
ROR	rate of return	decimal
ROI	return on investments	\$
r	nominal rate per year (rate per annum)	decimal
s	state income tax rate	decimal
S_n	expected salvage value in year n	\$
t	composite tax rate, or time	decimal
z	a factor equal to $\frac{1+i}{1-d}$	decimal
ϕ	effective rate per period	decimal

1 EQUIVALENCE

Industrial decision-makers using engineering economics are concerned with the timing of a project's cash flows as well as with the total profitability of that project. In this situation, a method is required to compare projects involving receipts and disbursements occurring at different times.

By way of illustration, consider \$100 placed in a bank account which pays 5% effective annual interest at the end of each year. After the first year, the account will have grown to \$105. After the second year, the account will have grown to \$110.25.

Assume that you will have no need for money during the next two years and that any money received would immediately go into your 5% bank account. Then, which of the following options would be more desirable?

option a: \$100 now
option b: \$105 to be delivered in one year
option c: \$110.25 to be delivered in two years

In light of the previous illustration, none of the options is superior under the assumptions given. If the first option is chosen, you will immediately place \$100 into a 5% account, and in two years the account will have grown to \$110.25. In fact, the account will contain \$110.25 at the end of two years regardless of the option chosen. Therefore, these alternatives are said to be *equivalent*.

2 CASH FLOW DIAGRAMS

Although they are not always necessary in simple problems (and they are often unwieldy in very complex problems), *cash flow diagrams* may be drawn to help visualize and simplify problems having diverse receipts and disbursements.

The conventions below are used to standardize cash flow diagrams.

- The horizontal (time) axis is marked off in equal increments, one per period, up to the duration or horizon of the project.

- All disbursements and receipts (cash flows) are assumed to take place at the end of the year in which they occur. This is known as the *year-end convention.* The exception to the year-end convention is any initial cost (purchase cost) which occurs at $t = 0$.
- Two or more transfers in the same year are placed end-to-end, and these may be combined.
- Expenses incurred before $t = 0$ are called *sunk costs.* Sunk costs are not relevant to the problem.
- Receipts are represented by arrows directed upward. Disbursements are represented by arrows directed downward. The arrow length is proportional to the magnitude of the cash flow.

Example 3.1

A mechanical device will cost \$20,000 when purchased. Maintenance will cost \$1000 each year. The device will generate revenues of \$5000 each year for five years after which the salvage value is expected to be \$7000. Draw and simplify the cash flow diagram.

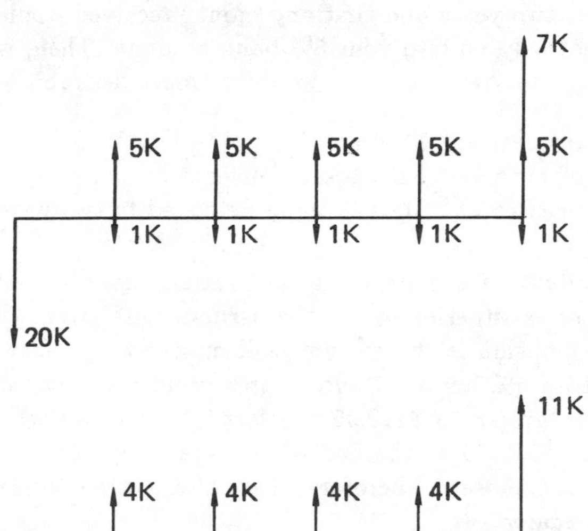

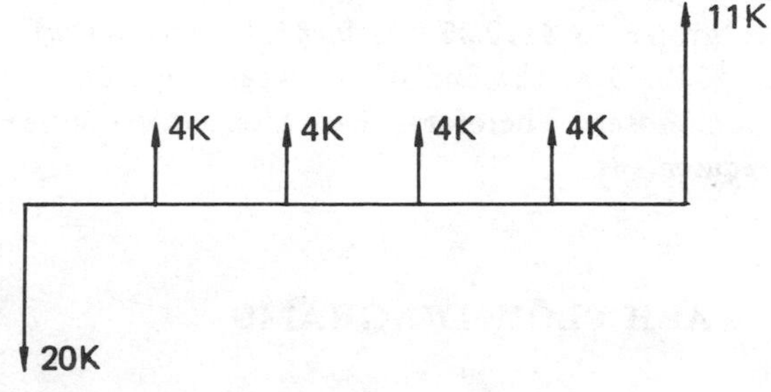

3 TYPICAL PROBLEM FORMAT

With the exception of some investment and rate of return problems, the typical problem involving engineering economics will have the following characteristics:

- An interest rate will be given.
- Two or more alternatives will be competing for funding.
- Each alternative will have its own cash flows.
- It is necessary to select the best alternative.

Example 3.2

Investment A costs \$10,000 today and pays back \$11,500 two years from now. Investment B costs \$8000 today and pays back \$4500 each year for two years. If an interest rate of 5% is used, which alternative is superior?

The solution to this example is not difficult, but it will be postponed until methods of calculating equivalence have been covered.

4 CALCULATING EQUIVALENCE

It was previously illustrated that \$100 now is equivalent at 5% interest to \$105 in one year. The equivalence of any present amount, P, at $t = 0$ to any future amount, F, at $t = n$ is called the *future worth* and can be calculated from equation 3.1.

$$F = P(1+i)^n \qquad 3.1$$

The factor $(1+i)^n$ is known as the *compound amount factor* and has been tabulated at the end of this chapter for various combinations of i and n. Rather than actually writing the formula for the compound amount factor, the convention is to use the standard functional notation $(F/P, i\%, n)$. Thus,

$$F = P(F/P, i\%, n) \qquad 3.2$$

Similarly, the equivalence of any future amount to any present amount is called the *present worth* and can be calculated from

$$P = F(1+i)^{-n} = F(P/F, i\%, n) \qquad 3.3$$

The factor $(1+i)^{-n}$ is known as the *present worth factor*, with functional notation $(P/F, i\%, n)$. Tabulated values are also given for this factor at the end of this chapter.

Example 3.3

How much should you put into a 10% savings account in order to have \$10,000 in five years?

This problem could also be stated: What is the equivalent present worth of $10,000 five years from now if money is worth 10%?

$$P = F(1+i)^{-n} = 10{,}000(1+0.10)^{-5}$$
$$= 6209$$

The factor 0.6209 would usually be obtained from the tables.

A cash flow which repeats regularly each year is known as an *annual amount.* When annual costs are incurred due to the functioning of a piece of equipment, they are often known as *operating and maintenance* (O&M) *costs.* The annual costs associated with operating a business in general are known as *general, selling, and administrative* (GS&A) *expenses.* Although the equivalent value for each of the n annual amounts could be calculated and then summed, it is much easier to use one of the *uniform series factors,* as illustrated in example 3.4.

Example 3.4

Maintenance costs for a machine are $250 each year. What is the present worth of these maintenance costs over a 12-year period if the interest rate is 8% ?

Notice that

$$(P/A, 8\%, 12) = (P/F, 8\%, 1) + (P/F, 8\%, 2) + \cdots + (P/F, 8\%, 12)$$

Then,

$$P = A(P/A, i\%, n) = -250(7.5361)$$
$$= -1884$$

A common complication involves a uniformly increasing cash flow. Such an increasing cash flow should be handled with the *uniform gradient factor,* $(P/G, i\%, n)$. The uniform gradient factor finds the present worth of a uniformly increasing cash flow which starts in year 2 (not year 1) as shown in example 3.5.

Example 3.5

Maintenance on an old machine is $100 this year but is expected to increase by $25 each year thereafter. What is the present worth of five years of maintenance? Use an interest rate of 10%.

In this problem, the cash flow must be broken down into parts. Notice that the five-year gradient factor is used even though there are only four non-zero gradient cash flows.

$$P = A(P/A, 10\%, 5) + G(P/G, 10\%, 5)$$
$$= -100(3.7908) - 25(6.8618)$$
$$= -551$$

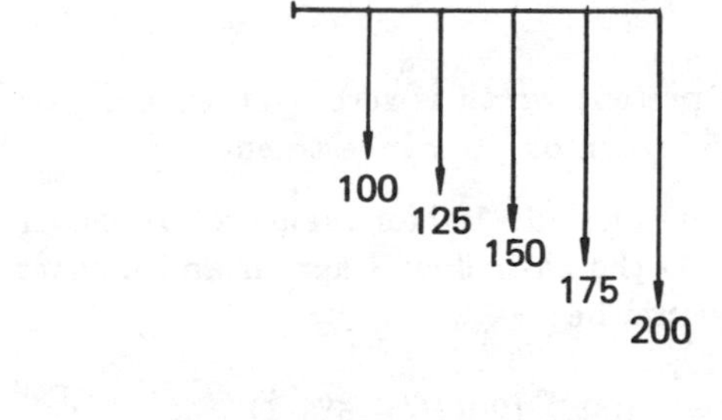

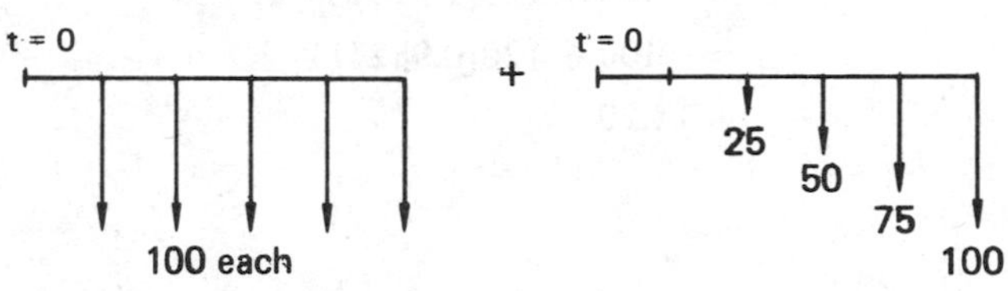

Various combinations of the compounding and discounting factors are possible. For instance, the annual cash flow that would be equivalent to a uniform gradient may be found from

$$A = G(P/G, i\%, n)(A/P, i\%, n) \qquad 3.4$$

Formulas for all of the compounding and discounting factors are contained in table 3.1. Normally, it will not be necessary to calculate factors from the formulas. The tables at the end of this chapter are adequate for solving most problems.

5 THE MEANING OF 'PRESENT WORTH' AND 'i'

It is clear that $100 invested in a 5% bank account will allow you to remove $105 one year from now. If this investment is made, you will clearly receive a *return on investment* (ROI) of $5. The cash flow diagram and the present worth of the two transactions are

$$P = -100 + 105(P/F, 5\%, 1)$$
$$= -100 + 105(0.9524)$$
$$= 0$$

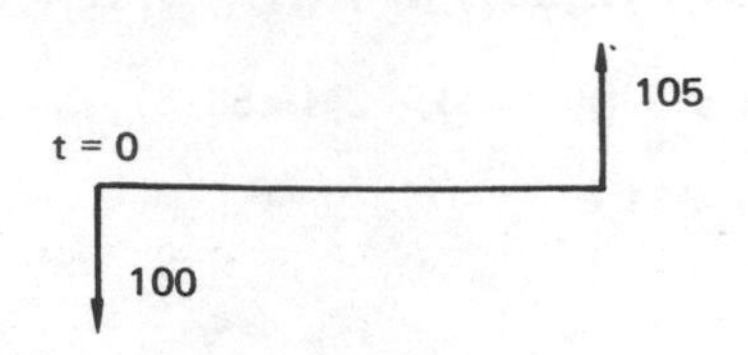

Figure 3.1 Cash Flow Diagram

Notice that the present worth is zero even though you did receive a 5% return on your investment.

However, if you are offered \$120 for the use of \$100 over a one-year period, the cash flow diagram and present worth (at 5%) would be

$$\begin{aligned} P &= -100 + 120(P/F, 5\%, 1) \\ &= -100 + 120(0.9524) \\ &= 14.29 \end{aligned}$$

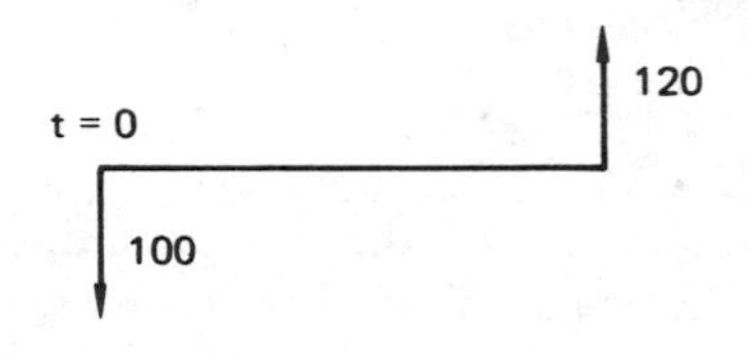

Figure 3.2 Cash Flow Diagram

Therefore, it appears that the present worth of an alternative is equal to the equivalent value at $t = 0$ of the increase in return above that which you would be able to earn in an investment offering $i\%$ per period. In the above case, \$14.29 is the present worth of (\$20−\$5), the difference in the two ROI's.

Alternatively, the actual earned interest rate, called *rate of return*, ROR, can be defined as the rate which makes the present worth of the alternative zero.

The *present worth* is also the amount that you would have to be given to dissuade you from making an investment, since placing the initial investment amount along with the present worth into a bank account earning $i\%$ will yield the same eventual ROI. Relating this to the previous paragraphs, you could be dissuaded against investing \$100 in an alternative which would return \$120 in one year by a $t = 0$ payment of \$14.29. Clearly, (\$100 + \$14.29) invested at $t = 0$ will also yield \$120 in one year at 5%.

The selection of the interest rate is difficult in engineering economics problems. Usually it is taken as the average rate of return that an individual or business organization has realized in past investments. Fortunately, an interest rate is usually given. A company may not know what effective interest rate to use in an economic analysis. In such a case, the company can establish a minimum acceptable return on its investment. This *minimum attractive rate of return* (MARR) should be used as the effective interest rate i in economic analyses.

Table 3.1
Discount Factors for Discrete Compounding

factor name	converts	symbol	formula
single payment compound amount	P to F	$(F/P, i\%, n)$	$(1+i)^n$
present worth	F to P	$(P/F, i\%, n)$	$(1+i)^{-n}$
uniform series Sinking Fund	F to A	$(A/F, i\%, n)$	$\frac{i}{(1+i)^n - 1}$
capital recovery	P to A	$(A/P, i\%, n)$	$\frac{i(1+i)^n}{(1+i)^n - 1}$
compound amount	A to F	$(F/A, i\%, n)$	$\frac{(1+i)^n - 1}{i}$
equal series present worth	A to P	$(P/A, i\%, n)$	$\frac{(1+i)^n - 1}{i(1+1)^n}$
uniform gradient	G to P	$(P/G, i\%, n)$	$\frac{(1+i)^n - 1}{i^2(1+i)^n} - \frac{n}{i(1+i)^n}$

It should be obvious that alternatives with negative present worths are undesirable, and that alternatives with positive present worths are desirable because they increase the average earning power of invested capital.

6 CHOICE BETWEEN ALTERNATIVES

A variety of methods exist for selecting a superior alternative from among a group of proposals. Each method has its own merits and applications.

A. PRESENT WORTH METHOD

The *present worth method* has already been implied. When two or more alternatives are capable of performing the same functions, the superior alternative will have the largest present worth. This method is suitable for ranking the desirability of alternatives. The present worth method is restricted to evaluating alternatives that are mutually exclusive and which have the same lives.

Returning to example 3.2, the present worth of each alternative should be found in order to determine which alternative is superior.

Example 3.2, continued

$$\begin{aligned} P(A) &= -10{,}000 + 11{,}500(P/F, 5\%, 2) \\ &= 431 \\ P(B) &= -8000 + 4500(P/A, 5\%, 2) \\ &= 367 \end{aligned}$$

Alternative A is superior and should be chosen.

B. CAPITALIZED COST METHOD

The present worth of a project with an infinite life is known as the *capitalized cost* or *life cycle cost*. Capitalized cost is the amount of money at $t = 0$ needed to perpetually support the project on the earned interest only. Capitalized cost is a positive number when expenses exceed income.

$$\text{capitalized cost} = \text{initial cost} + \frac{\text{annual costs}}{i} \qquad 3.5$$

Capitalized cost is the present worth of an infinitely-lived project. Normally, it would be difficult to work with an infinite stream of cash flows since most economics tables don't list factors for periods in excess of 100 years. However, the (A/P) discounting factor approaches the interest rate as n becomes large. Since the (P/A) and (A/P) factors are reciprocals of each other, we would expect to divide an infinite series of equal cash flows by the interest rate in order to calculate the present worth of the infinite series. This is the basis of equation 3.5.

Equation 3.5 can be used when the annual costs are equal in every year. The "annual cost" in that equation is assumed to be the same each year. If the operating and maintenance costs occur irregularly instead of annually, or if the costs vary from year to year, it will be necessary to somehow determine a cash flow of *equal annual amounts* (EAA) which is equivalent to the stream of original costs.

The equal annual amount may be calculated in the usual manner by first finding the present worth of all the actual costs, and then multiplying the present worth by the interest rate (the (A/P) factor for an infinite series). However, it is not necessary to convert the present worth to an equal annual amount, since equation 3.6 will convert the equal annual amount back to the present worth.

$$\text{capitalized cost} = \text{initial cost} + \frac{\text{EAA}}{i} \qquad 3.6$$

In comparing two alternatives, each of which is infinitely lived, the superior alternative will have the lowest capitalized cost.

C. ANNUAL COST METHOD

Alternatives which accomplish the same purpose but which have unequal lives must be compared by the *annual cost method*. The annual cost method assumes that each alternative will be replaced by an identical twin at the end of its useful life (infinite renewal). This method, which may also be used to rank alternatives according to their desirability, is also called the *annual return method* and *capital recovery method*.

Restrictions are that the alternatives must be mutually exclusive and infinitely renewed up to the duration of the longest-lived alternative. The calculated annual cost is known as the *equivalent uniform annual cost* (EUAC). Cost is a positive number when expenses exceed income.

Example 3.6

Which of the following alternatives is superior over a 30-year period if the interest rate is 7%?

ECON

	A	B
type	brick	wood
life	30 years	10 years
cost	\$1800	\$450
maintenance	\$5/year	\$20/year

$$\text{EUAC}(A) = 1800(A/P, 7\%, 30) + 5 = 150$$
$$\text{EUAC}(B) = 450(A/P, 7\%, 10) + 20 = 84$$

Alternative B is superior since its annual cost of operation is the lowest. It is assumed that three wood facilities, each with a life of 10 years and a cost of \$450, will be built to span the 30-year period.

ECON

D. BENEFIT-COST RATIO METHOD

The *benefit-cost ratio method* is often used in municipal project evaluations where benefits and costs accrue to different segments of the community. With this method, the present worth of all benefits (regardless of the beneficiary) is divided by the present worth of all costs. The project is considered acceptable if the ratio exceeds *one*.

When the benefit-cost ratio method is used, disbursements by the initiators or sponsors are *costs*. Disbursements by the users of the project are known as *disbenefits*. It is often difficult to determine whether a cash flow is a cost or a disbenefit (whether to place it in the numerator or denominator of the benefit-cost ratio calculation).

Regardless of where the cash flow is placed, an acceptable project will always have a benefit-cost ratio greater than one, although the actual numerical result will depend on the placement. For this reason, the benefit-cost ratio method should not be used to rank competing projects.

The benefit-cost ratio method may be used to rank alternative proposals only if an *incremental analysis* is used. First, determine that the ratio is greater than one for each alternative. Then, calculate the ratio of benefits to costs

$$\frac{B_2 - B_1}{C_2 - C_1}$$

for each possible pair of alternatives. If the ratio exceeds one, alternative 2 is superior to alternative 1. Otherwise, alternative 1 is superior.

E. RATE OF RETURN METHOD

Perhaps no method of analysis is less understood than the *rate of return method*, (ROR). As was stated previously, the ROR is the interest rate that would yield identical profits if all money were invested at that rate. The present worth of any such investment is zero.

The ROR is defined as the interest rate that will discount all cash flows to a total present worth equal to the initial required investment. This definition is used to determine the ROR of an alternative. The advantage of the ROR method is that no knowledge of an interest rate is required.

To find the ROR of an alternative, proceed as follows:

step 1: Set up the problem as if to calculate the present worth.

step 2: Arbitrarily select a reasonable value for i. Calculate the present worth.

step 3: Choose another value of i (not too close to the original value) and again solve for the present worth.

step 4: Interpolate or extrapolate the value of i which gives a zero present worth.

step 5: For increased accuracy, repeat steps (2) and (3) with two more values that straddle the value found in step (4).

A common, although incorrect, method of calculating the ROR involves dividing the annual receipts or returns by the initial investment. This technique ignores such items as salvage, depreciation, taxes, and the time value of money. This technique also fails when the annual returns vary.

Once a rate of return is known for an investment alternative, it is typically compared to the *minimum attractive rate of return* (MARR) specified by a company. However, ROR should not be used to rank alternatives. When two alternatives have ROR's exceeding the MARR, it is not sufficient to select the alternative with the higher ROR.

An *incremental analysis*, also known as a *rate of return on added investment study*, should be performed if ROR is to be used to select between investments. In an incremental analysis, the cash flows for the investment with the lower initial cost are subtracted from the cash flows for the higher-priced alternative on a year-by-year basis. This produces, in effect, a third alternative representing the cost and benefits of the added investment. The added expense of the higher-priced investment is not warranted unless the ROR of this third alternative exceeds the MARR as well.

Example 3.7

What is the return on invested capital if \$1000 is invested now with \$500 being returned in year 4 and \$1000 being returned in year 8?

First, set up the problem as a present worth calculation.

$$P = -1000 + 500(P/F, i\%, 4) + 1000(P/F, i\%, 8)$$

Arbitrarily select $i = 5\%$. The present worth is then found to be \$88.15. Next take a higher value of i to reduce the present worth. If $i = 10\%$, the present worth is -\$192. The ROR is found from simple interpolation to be approximately 6.6%.

7 TREATMENT OF SALVAGE VALUE IN REPLACEMENT STUDIES

An investigation into the retirement of an existing process or piece of equipment is known as a *replacement study*. Replacement studies are similar in most respects to other alternative comparison problems: an interest rate is given, two alternatives exist, and one of the previously mentioned methods of comparing alternatives is used to choose the superior alternative.

In replacement studies, the existing process or piece of equipment is known as the *defender*. The new process or piece of equipment being considered for purchase is known as the *challenger*.

Because most defenders still have some market value when they are retired, the problem of what to do with the salvage arises. It seems logical to use the salvage value of the defender to reduce the initial purchase cost of the challenger. This is consistent with what would actually happen if the defender were to be retired.

By convention, however, the salvage value is subtracted from the defender's present value. This does not seem logical, but it is done to keep all costs and benefits related to the defender with the defender. In this case, the salvage value is treated as an opportunity cost which would be incurred if the defender is not retired.

If the defender and the challenger have the same lives and a present worth study is used to choose the superior alternative, the placement of the salvage value will have no effect on the net difference between present worths for the challenger and defender. Although the values of the two present worths will be different depending on the placement, the difference in present worths will be the same.

If the defender and the challenger have different lives, an annual cost comparison must be made. Since the salvage value would be spread over a different number of years depending on its placement, it is important to abide by the conventions listed in this section.

There are a number of ways to handle salvage value. The best way is to think of the EUAC of the defender as the cost of keeping the defender from now until next year. In addition to the usual operating and maintenance costs, that cost would include an opportunity interest cost incurred by not selling the defender and also a drop in the salvage value if the defender is kept for one additional year. Specifically,

$$\begin{aligned}\text{EUAC(defender)} &= \text{maintenance costs} \\ &\quad + i\,(\text{current salvage value}) \\ &\quad + (\text{current salvage--next} \\ &\qquad \text{year's salvage})\end{aligned} \qquad 3.7$$

It is important in retirement studies not to double count the salvage value. That is, it would be incorrect to add the salvage value to the defender and at the same time subtract it from the challenger.

8 BASIC INCOME TAX CONSIDERATIONS

Assume that an organization pays $f\%$ of its profits to the federal government as income taxes. If the organization also pays a state income tax of $s\%$, and if state taxes paid are recognized by the federal government as expenses, then the composite tax rate is

$$t = s + f - sf \qquad 3.8$$

The basic principles used to incorporate taxation into economic analyses are listed below.

- Initial purchase cost is unaffected by income taxes.
- Salvage value is unaffected by income taxes.
- Deductible expenses, such as operating costs, maintenance costs, and interest payments, are reduced by $t\%$ (i.e., multiplied by the quantity $(1-t)$).
- Revenues are reduced by $t\%$ (i.e., multiplied by the quantity $(1-t)$).
- Depreciation is multiplied by t and added to the appropriate year's cash flow, increasing that year's present worth.

Income taxes and depreciation have no bearing on municipal or governmental projects since municipalities, states, and the U.S. government pay no taxes.

ECON

Example 3.8

A corporation which pays 53% of its revenue in income taxes invests \$10,000 in a project which will result in \$3000 annual revenue for eight years. If the annual expenses are \$700, salvage after eight years is \$500, and 9% interest is used, what is the after-tax present worth? Disregard depreciation.

$$\begin{aligned} P_t &= -10{,}000 + 3000(P/A, 9\%, 8)(1 - 0.53) \\ &\quad - 700(P/A, 9\%, 8)(1 - 0.53) \\ &\quad + 500(P/F, 9\%, 8) \\ &= -3766 \end{aligned}$$

ECON

9 DEPRECIATION

Although depreciation calculations may be considered independently, it is important to recognize that depreciation has no effect on engineering economic calculations unless income taxes are also considered.

Generally, tax regulations do not allow the cost of equipment[1] to be treated as a deductible expense in the year of purchase. Rather, portions of the cost may be allocated to each year of the item's economic life (which may be different from the actual useful life). Each year, the book value (which is initially equal to the purchase price) is reduced by the depreciation in that year. Theoretically, the book value of an item will equal the market value at any time within the economic life of that item.

Since tax regulations allow the depreciation in any year to be handled as if it were an actual operating expense, and since operating expenses are deductible from the income base prior to taxation, the after-tax profits will be increased. If D is the depreciation, the net result to the after-tax cash flow will be the addition of tD.

The present worth of all depreciation over the economic life of the item is called the *depreciation recovery*. Although originally established to do so, depreciation recovery can never fully replace an item at the end of its life.

Depreciation is often confused with amortization and depletion. While depreciation spreads the cost of a fixed asset over a number of years, *amortization* spreads the cost of an intangible asset (e.g., a patent) over some basis such as time or expected units of production.

Depletion is another artificial deductible operating expense designed to compensate mining organizations for decreasing mineral reserves. Since original and remaining quantities of minerals are seldom known accurately, the *depletion allowance* is calculated as a fixed percentage of the organization's gross income. These percentages are usually in the 10% to 20% range and apply to such mineral deposits as oil, natural gas, coal, uranium, and most metal ores.

There are four common methods of calculating depreciation. The book value of an asset depreciated with the *straight line* (SL) *method* (also known as the *fixed percentage method*) decreases linearly from the initial purchase at $t = 0$ to the estimated salvage at $t = n$. The depreciated amount is the same each year. The quantity $(C - S_n)$ in equation 3.9 is known as the *depreciation base*.

$$D_j = \frac{C - S_n}{n} \qquad 3.9$$

Double declining balance[2] (DDB) depreciation is independent of salvage value. Furthermore, the book value never stops decreasing, although the depreciation decreases in magnitude. Usually, any remaining book value is written off in the last year of the asset's estimated life. Unlike any of the other depreciation methods, DDB depends on accumulated depreciation.

$$D_j = \frac{2\left(C - \sum_{i=1}^{j-1} D_i\right)}{n} \qquad 3.10$$

In *sum-of-the-years'-digits* (SOYD) depreciation, the digits from 1 to n inclusive, are summed. The total, T, can also be calculated from

$$T = \frac{1}{2} n(n+1) \qquad 3.11$$

The depreciation can be found from

$$D_j = \frac{(C - S_n)(n - j + 1)}{T} \qquad 3.12$$

[1] The IRS tax regulations allow depreciation on almost all forms of *property* except land. The following types of property are distinguished: *real* (e.g., buildings used for business, etc.), *residential* (e.g., buildings used as rental property), and *personal* (e.g., equipment used for business). Personal property does *not* include items for personal use, despite its name. *Tangible* personal property is distinguished from *intangible property* (e.g., goodwill, copyrights, patents, trademarks, franchises, and agreements not to compete).

[2] Double declining balance depreciation is a particular form of *declining balance depreciation*, as defined by the IRS tax regulations. Declining balance depreciation also includes 125% declining balance and 150% declining balance depreciations which can be calculated by substituting 1.25 and 1.50, respectively for the 2 in equation 3.10.

The *sinking fund method* is seldom used in industry because the initial depreciation is low. The formula for sinking fund depreciation (which increases each year) is

$$D_j = (C - S_n)(A/F, i\%, n)(F/P, i\%, j\text{-}1) \qquad 3.13$$

The above discussion gives the impression that any form of depreciation may be chosen regardless of the nature and circumstances of the purchase. In reality, the IRS tax regulations place restrictions on the higher-rate ("accelerated") methods such as DDB and SOYD. Furthermore, the *Economic Recovery Act of 1981* substantially changed the laws relating to personal and corporate income taxes.

Property placed into service in 1981 or after must use the *accelerated cost recovery system* (ACRS and MACRS). Other methods (straight-line, declining balance, etc.) cannot be used except in special cases.

Property placed into service in 1980 or before must continue to be depreciated according to the method originally chosen (e.g., straight-line, declining balance, or sum-of-the-years'-digits). ACRS and MACRS cannot be used.

Under ACRS and MACRS, the cost recovery amount in the jth year of an asset's cost recovery period is calculated by multiplying the initial cost by a factor.

$$D_j = (\text{initial cost})(\text{factor}) \qquad 3.14$$

The initial cost used is not reduced by the asset's salvage value for either the regular or alternate ACRS and MACRS calculations. The factor used depends on the asset's cost recovery period. Such factors are subject to continuing legislation changes. Current tax publications should be consulted before using the ACRS and MACRS method.

Three other depreciation methods should be mentioned, not because they are currently accepted or in widespread use, but because they are occasionally called for by name.

The *sinking-fund plus interest on first cost depreciation method*, like the following two methods, is an attempt to include the *opportunity interest cost* on the purchase price with the depreciation. That is, the purchasing company not only incurs an annual loss due to the drop in book value, but it also loses the interest on the purchase price. The formula for this method is

$$D_j = (C - S_n)(A/F, i\%, n) + (C)(i) \qquad 3.15$$

The *straight-line plus interest on first cost method* is similar. Its formula is

$$D_j = \left(\frac{1}{n}\right)(C - S_n) + (C)(i) \qquad 3.16$$

The *straight-line plus average interest method* assumes that the opportunity interest cost should be based on the book value only, not on the full purchase price. Since the book value changes each year, an average value is used. The depreciation formula is

$$D_j = \left(\frac{1}{n}\right)(C - S_n) + \frac{1}{2}(i)(C - S_n)\left(\frac{n+1}{n}\right) + iS_n \qquad 3.17$$

These three depreciation methods are not to be used in the usual manner (e.g., in conjunction with the income tax rate). These methods are attempts to calculate a more accurate annual cost of an alternative. Sometimes they work, and sometimes they give misleading answers. Their use cannot be recommended. They are included in this chapter only for the sake of completeness.

ECON

Example 3.9

An asset is purchased for \$9000. Its estimated economic life is 10 years, after which it will be sold for \$1000. Find the depreciation in the first three years using SL, DDB, and SOYD.

SL:

$$D = \frac{9000 - 1000}{10} = 800 \text{ each year}$$

DDB:

$$D_1 = \frac{2(9000)}{10} = 1800 \text{ in year 1}$$

$$D_2 = \frac{2(9000 - 1800)}{10} = 1440 \text{ in year 2}$$

$$D_3 = \frac{2(9000 - 3240)}{10} = 1152 \text{ in year 3}$$

SOYD:

$$T = \frac{1}{2}(10)(11) = 55$$

$$D_1 = \left(\frac{10}{55}\right)(9000 - 1000) = 1455 \text{ in year 1}$$

$$D_2 = \left(\frac{9}{55}\right)(8000) = 1309 \text{ in year 2}$$

$$D_3 = \left(\frac{8}{55}\right)(8000) = 1164 \text{ in year 3}$$

Example 3.10

For the asset described in example 3.9, calculate the book value during the first three years if SOYD depreciation is used.

The book value at the beginning of year 1 is $9000. Then,

$$BV_1 = 9000 - 1455 = 7545$$
$$BV_2 = 7545 - 1309 = 6236$$
$$BV_3 = 6236 - 1164 = 5072$$

Example 3.11

For the asset described in example 3.9, calculate the after-tax depreciation recovery with SL and SOYD depreciation methods. Use 6% interest with 48% income taxes.

SL:

$$D.R. = 0.48(800)(P/A, 6\%, 10) = 2826$$

SOYD: The depreciation series can be thought of as a constant 1,454 term with a negative 145 gradient.

$$D.R. = 0.48(1454)(P/A, 6\%, 10) - 0.48(145)(P/G, 6\%, 10) = 3076$$

Finding book values, depreciation, and depreciation recovery is particularly difficult with DDB depreciation, since all previous years' quantities seem to be required. It appears that the depreciation in the sixth year cannot be calculated unless the values of depreciation for the first five years are calculated first. Questions asking for depreciation or book value in the middle or at the end of an asset's economic life may be solved from the following equations

$$d = \frac{2}{n} \qquad 3.18$$

$$z = \frac{1+i}{1-d} \qquad 3.19$$

$$(P/EG) = \frac{z^n - 1}{z^n(z-1)} \qquad 3.20$$

Then, the present worth of the depreciation recovery is

$$D.R. = t\left[\frac{(d)(C)}{(1-d)}(P/EG)\right] \qquad 3.21$$

$$D_j = (d)(C)(1-d)^{j-1} \qquad 3.22$$

$$BV_j = C(1-d)^j \qquad 3.23$$

Example 3.12

What is the after-tax present worth of the asset described in example 3.8, if SL, SOYD, and DDB depreciation methods are used?

The after-tax present worth, neglecting depreciation, was previously found to be −3766.

Using SL, the depreciation recovery is

$$D.R. = (0.53)\left(\frac{10{,}000 - 500}{8}\right)(P/A, 9\%, 8) = 3483$$

Using SOYD, the depreciation recovery is calculated as follows:

$$T = \frac{1}{2}(8)(9) = 36$$

$$\text{Depreciation base} = (10{,}000 - 500) = 9500$$

$$D_1 = \frac{8}{36}(9500) = 2111$$

$$G = \text{gradient} = \frac{1}{36}(9500) = 264$$

$$D.R. = (0.53)\left[2111(P/A, 9\%, 8) - 264(P/G, 9\%, 8)\right] = 3829$$

Using DDB, the depreciation recovery is calculated as follows:

$$d = \frac{2}{8} = 0.25$$

$$z = \frac{1.09}{0.75} = 1.4533$$

$$(P/EG) = \frac{(1.4533)^8 - 1}{(1.4533)^8(0.4533)} = 2.095$$

$$D.R. = 0.53\left[\frac{(0.25)(10{,}000)}{0.75}\right](2.095) = 3701$$

The after-tax present worths including depreciation recovery are:

SL: $P_t = -3766 + 3483 = -283$
SOYD: $P_t = -3766 + 3829 = 63$
DDB: $P_t = -3766 + 3701 = -65$

10 ADVANCED INCOME TAX CONSIDERATIONS

There are a number of specialized techniques that are needed infrequently. These techniques are related more to the accounting profession than to the engineering profession. Nevertheless, it is occasionally necessary to use these techniques.

A. INVESTMENT TAX CREDIT

An *investment tax credit* (also known as a *tax credit* or an *investment credit*) is a one-time credit against income taxes. The investment tax credit is calculated as a fraction of the initial purchase price of certain types of equipment purchased for industrial, commercial, and manufacturing use.

$$\text{credit} = (\text{initial cost})(\text{fraction}) \qquad 3.24$$

The fraction is subject to continuing legislation changing its value and applicability. The fraction, which typically is taken as 10% for initial estimates, actually depends on the asset life, year of acquisition, and number of years the asset is held before being disposed of. The current tax laws should be studied before using the investment tax credit.[3]

Since the investment tax credit reduces the buyer's tax liability, the credit should only be used in after-tax analyses.

B. GAIN ON THE SALE OF A DEPRECIATED ASSET

If an asset is sold for more than its current book value, the difference between selling price and book value is taxable income. The gain is taxed at capital gains rates. Excluded from this preferential treatment is non-residential real property depreciated under regular ACRS provisions. However, non-residential real property depreciated under the straight-line alternate method qualifies for the capital gains rate.

C. CAPITAL GAINS AND LOSSES

A *gain* is defined as the difference between selling and purchase prices of a capital asset. The gain is called a *regular gain* if the item sold has been kept less than one year. The gain is called a *capital gain* if the item sold has been kept for longer than one year. Capital gains are taxed at the taxpayer's usual rate, but 60% of the gain is excluded from taxation.

Regular (as defined above) *losses* are fully deductible in the year of their occurrence. The IRS tax regulations should be consulted to determine the treatment of *capital losses*.

[3] The 10% investment tax credit for capital equipment purchases was repealed for assets obtained after 1985.

11 RATE AND PERIOD CHANGES

All of the foregoing calculations were based on compounding once a year at an *effective interest rate*, i. However, some problems specify compounding more frequently than annually. In such cases, a *nominal interest rate*, r, will be given. The nominal rate does not include the effect of compounding and is not the same as the effective rate, i. A nominal rate may be used to calculate the effective rate by using equation 3.25 or 3.26.

$$i = \left(1 + \frac{r}{k}\right)^k - 1 \qquad 3.25$$

$$= (1 + \phi)^k - 1 \qquad 3.26$$

A problem may also specify an effective rate per period, ϕ, (e.g., per month). However, that will be a simple problem since compounding for n periods at an effective rate per period is not affected by the definition or length of the period.

The following rules may be used to determine which interest rate is given in a problem:

- Unless specifically qualified in the problem, the interest rate given is an annual rate.
- If the compounding is annually, the rate given is the effective rate. If compounding is other than annually, the rate given is the nominal rate.
- If the type of compounding is not specified, assume annual compounding.

In the case of continuous compounding, the appropriate discount factors may be calculated from the formulas in table 3.2.

Table 3.2
Discount Factors for
Continuous Compounding

(F/P)	e^{rn}
(P/F)	e^{-rn}
(A/F)	$\frac{e^r - 1}{e^{rn} - 1}$
(F/A)	$\frac{e^{rn} - 1}{e^r - 1}$
(A/P)	$\frac{e^r - 1}{1 - e^{-rn}}$
(P/A)	$\frac{1 - e^{-rn}}{e^r - 1}$

ECON

Example 3.13

A savings and loan offers $5\frac{1}{4}\%$ compounded daily. What is the annual effective rate?

method 1:

$$r = 0.0525, \ k = 365$$

$$i = \left(1 + \frac{0.0525}{365}\right)^{365} - 1 = 0.0539$$

method 2: Assume daily compounding is the same as continuous compounding.

$$i = (F/P) - 1$$
$$= e^{0.0525} - 1 = 0.0539$$

12 PROBABILISTIC PROBLEMS

Thus far, all of the cash flows included in the examples have been known exactly. If the cash flows are not known exactly but are given by some implicit or explicit probability distribution, the problem is *probabilistic.*

Probabilistic problems typically possess the following characteristics:

- There is a chance of extreme loss that must be minimized.
- There are multiple alternatives that must be chosen from. Each alternative gives a different degree of protection against the loss or failure.
- The outcome is independent of the alternative chosen. Thus, as illustrated in example 3.15, the size of the dam that is chosen for construction will not alter the rainfall in successive years. However, it will alter the effects on the down-stream watershed areas.

Probabilistic problems are typically solved using annual costs and expected values. An *expected value* is similar to an 'average value' since it is calculated as the mean of the given probability distribution. If cost 1 has a probability of occurrence of p_1, cost 2 has a probability of occurrence of p_2, and so on, the expected value is

$$E(\text{cost}) = p_1(\text{cost 1}) + p_2(\text{cost 2}) + \cdots \qquad 3.27$$

Example 3.14

Flood damage in any year is given according to the table below. What is the present worth of flood damage for a 10-year period? Use 6%.

damage	probability
0	0.75
$10,000	0.20
$20,000	0.04
$30,000	0.01

The expected value of flood damage is

$$E(\text{damage}) = (0)(0.75) + (10{,}000)(0.20) + (20{,}000)(0.04) + (30{,}000)(0.01)$$
$$= 3100$$
$$\text{present worth} = 3100(P/A,6\%,10)$$
$$= 22{,}816$$

Probabilities in probabilistic problems may be given to you in the problem (as in the example above) or you may have to obtain them from some named probability distribution. In either case, the probabilities are known explicitly and such problems are known as *explicit probability problems.*

Example 3.15

A dam is being considered on a river which periodically overflows and causes $600,000 damage. The damage is essentially the same each time the river causes flooding. The project horizon is 40 years. A 10% interest rate is being used.

Three different designs are available, each with different costs and storage capacities.

design alternative	cost	maximum capacity
A	500,000	1 unit
B	625,000	1.5 units
C	900,000	2.0 units

The U.S. Weather Service has provided a statistical analysis of annual rainfall in the area draining into the river.

units annual rainfall	probability
0	0.10
0.1–0.5	0.60
0.6–1.0	0.15
1.1–1.5	0.10
1.6–2.0	0.04
2.1 or more	0.01

Which design alternative would you choose assuming the dam is essentially empty at the start of each rainfall season?

The sum of the construction cost and the expected damage needs to be minimized. If alternative A is chosen, it will have a capacity of 1 unit. Its capacity will be exceeded (causing \$600,000 damage) when the annual rainfall exceeds 1 unit. Therefore, the annual cost of A is

$$\begin{aligned} \text{EUAC}(A) &= 500{,}000(A/P, 10\%, 40) \\ &\quad + 600{,}000(0.10 + 0.04 + 0.01) \\ &= 141{,}150 \end{aligned}$$

Similarly,

$$\begin{aligned} \text{EUAC}(B) &= 625{,}000(A/P, 10\%, 40) \\ &\quad + 600{,}000(0.04 + 0.01) \\ &= 93{,}940 \end{aligned}$$

$$\begin{aligned} \text{EUAC}(C) &= 900{,}000(A/P, 10\%, 40) \\ &\quad + 600{,}000(0.01) \\ &= 98{,}070 \end{aligned}$$

Alternative B should be chosen.

In other problems, a probability distribution will not be given even though some parameter (such as the life of an alternative) is not known with certainty. Such problems are known as ***implicit probability problems*** since they require a reasonable assumption about the probability distribution.

Implicit probability problems typically involve items whose ***expected time to failure*** are known. The key to such problems is in recognizing that an expected time to failure is not the same as a fixed life.

Reasonable assumptions can be made about the form of probability distributions in implicit probability problems.

One such reasonable assumption is that of a ***rectangular distribution***. A rectangular distribution is one which is assumed to give an equal probability of failure in each year. Such an assumption is illustrated in example 3.16.

Example 3.16

A bridge is needed for 20 years. Failure of the bridge at any time will require a 50% reinvestment. Assume that each alternative has an annual probability of failure that is inversely proportional to its expected time to failure. Evaluate the two design alternatives below using 6% interest.

design alternative	initial cost	expected time to failure	annual costs	salvage at $t = 20$
A	15,000	9 years	1200	0
B	22,000	43 years	1000	0

For alternative A, the probability of failure in any year is $\left(\frac{1}{9}\right)$. Similarly, the annual failure probability for alternative B is $\left(\frac{1}{43}\right)$.

$$\begin{aligned} \text{EUAC}(A) &= 15{,}000(A/P, 6\%, 20) \\ &\quad + 15{,}000(0.5)\left(\frac{1}{9}\right) + 1200 \\ &= 3341 \end{aligned}$$

$$\begin{aligned} \text{EUAC}(B) &= 22{,}000(A/P, 6\%, 20) \\ &\quad + 22{,}000(0.5)\left(\frac{1}{43}\right) + 1000 \\ &= 3174 \end{aligned}$$

Alternative B should be chosen.

13 ESTIMATING ECONOMIC LIFE

As assets grow older, their operating and maintenance costs typically increase each year. Eventually, the cost to keep an asset in operation becomes prohibitive, and the asset is retired or replaced. However, it is not always obvious when an asset should be retired or replaced.

As the asset's maintenance is increasing each year, the amortized cost of its initial purchase is decreasing. It is the sum of these two costs that should be evaluated to determine the point at which the asset should be retired or replaced. Since an asset's initial purchase price is likely to be high, the amortized cost will be the controlling factor in those years when the maintenance costs are low. Therefore, the EUAC of the asset will decrease in the initial part of its life.

However, as the asset grows older, the change in its amortized cost decreases while maintenance increases. Eventually the sum of the two costs reaches a minimum and then starts to increase. The age of the asset at the minimum cost point is known as the ***economic life*** of the asset. The economic life is, generally, less than the mission and technological lifetimes of the asset.

The determination of an asset's economic life is illustrated by example 3.17.

Example 3.17

A bus in a municipal transit system has the characteristics listed below. When should the city replace its buses if money can be borrowed at 8%?

Initial cost: \$120,000

year	maintenance cost	salvage value
1	35,000	60,000
2	38,000	55,000
3	43,000	45,000
4	50,000	25,000
5	65,000	15,000

ECON

If the bus is kept for one year and then sold, the annual cost will be

$$\begin{aligned}\text{EUAC}(1) &= 120{,}000(A/P, 8\%, 1) + 35{,}000(A/F, 8\%, 1) \\ &\quad - 60{,}000(A/F, 8\%, 1) \\ &= 104{,}600\end{aligned}$$

If the bus is kept for two years and then sold, the annual cost will be

$$\begin{aligned}\text{EUAC}(2) &= [120{,}000+35{,}000(P/F, 8\%, 1)](A/P, 8\%, 2) \\ &\quad + (38{,}000 - 55{,}000)(A/F, 8\%, 2) \\ &= 77{,}300\end{aligned}$$

If the bus is kept for three years and then sold, the annual cost will be

$$\begin{aligned}\text{EUAC}(3) &= [120{,}000 + 35{,}000(P/F, 8\%, 1) \\ &\quad + 38{,}000(P/F, 8\%, 2)](A/P, 8\%, 3) \\ &\quad + (43{,}000 - 45{,}000)(A/F, 8\%, 3) \\ &= 71{,}200\end{aligned}$$

This process is continued until EUAC begins to increase. In this example, EUAC(4) is 71,700. Therefore, the bus should be retired after three years.

14 BASIC COST ACCOUNTING

Cost accounting is the system which determines the cost of manufactured products. Cost accounting is called *job cost accounting* if costs are accumulated by part number or contract. It is called *process cost accounting* if costs are accumulated by departments or manufacturing processes.

Three types of costs (direct material, direct labor, and all indirect costs) make up the total manufacturing cost of a product.

Direct material costs are the costs of all materials that go into the product, priced at the original purchase cost.

Indirect material and labor costs are generally limited to costs incurred in the factory, excluding costs incurred in the office area. Examples of indirect materials are cleaning fluids, assembly lubricants, and temporary routing tags. Examples of indirect labor are stock-picking, inspection, expediting, and supervision labor.

Here are some important points concerning basic cost accounting:

- The sum of direct material and direct labor costs is known as the *prime cost.*
- Indirect costs may be called *indirect manufacturing expenses* (IME).
- Indirect costs may also include the overhead sector of the company (e.g., secretaries, engineers, and corporate administration). In this case, the indirect cost is usually called *burden* or *overhead.* Burden may also include the EUAC of non-regular costs which must be spread evenly over several years.
- The cost of a product is usually known in advance from previous manufacturing runs or by estimation. Any deviation from this known cost is called a *variance.* Variance may be broken down into *labor variance* and *material variance.*
- Indirect cost per item is not easily measured. The method of allocating indirect costs to a product is as follows:

  step 1: Estimate the total expected indirect (and overhead) costs for the upcoming year.

  step 2: Decide on some convenient vehicle for allocating the overhead to production. Usually, this vehicle is either the number of units expected to be produced or the number of direct hours expected to be worked in the upcoming year.

  step 3: Estimate the quantity or size of the overhead vehicle.

  step 4: Divide expected overhead costs by the expected overhead vehicle to obtain the unit overhead.

  step 5: Regardless of the true size of the overhead vehicle during the upcoming year, one unit of overhead cost is allocated per product.

- Although estimates of production for the next year are always somewhat inaccurate, the cost of the product is assumed to be independent of forecasting errors. Any difference between true cost and calculated cost goes into a variance account.
- *Burden (overhead) variance* will be caused by errors in forecasting both the actual overhead for the upcoming year and the vehicle size. In the former case, the variance is called *burden budget variance;* in the latter, it is called *burden capacity variance.*

Example 3.18

A small company expects to produce 8000 items in the upcoming year. The current material cost is \$4.54 each. 16 minutes of direct labor are required per unit. Workers are paid \$7.50 per hour. 2133 direct labor hours are forecast for the product. Miscellaneous overhead costs are estimated at \$45,000.

Find the expected direct material cost, the direct labor cost, the prime cost, the burden as a function of production and direct labor, and the total cost.

ECON

The direct material cost was given as \$4.54.

The direct labor cost is $\left(\frac{16}{60}\right)$ (\$7.50) = \$2.00.

The prime cost is \$4.54 + \$2.00 = \$6.54.

If the burden vehicle is production, the burden rate is $\$\frac{45{,}000}{8000} = \$5.63$ per item, making the total cost \$4.54 + \$2.00 + \$5.63 = \$12.17.

If the burden vehicle is direct labor hours, the burden rate is $\left(\frac{45{,}000}{2133}\right) = \$21.10$ per hour, making the total cost $\$4.54 + \$2.00 + \frac{16}{60}(\$21.10) = \$12.17$.

Example 3.19

The actual performance of the company in example 3.18 is given by the following figures:

actual production: 7560
actual overhead costs: \$47,000

What are the burden budget variance and the burden capacity variance?

The burden capacity variance is

$$\$45{,}000 - 7560(\$5.63) = \$2437$$

The burden budget variance is

$$\$47{,}000 - \$45{,}000 = \$2000$$

The overall burden variance is

$$\$47{,}000 - 7560(\$5.63) = \$4437$$

15 BREAK-EVEN ANALYSIS

Break-even analysis is a method of determining when costs exactly equal revenue. If the manufactured quantity is less than the break-even quantity, a loss is incurred. If the manufactured quantity is greater than the break-even quantity, a profit is incurred.

Consider the following special variables:

- f a fixed cost which does not vary with production
- a an incremental cost which is the cost to produce one additional item. It may also be called the *marginal cost* or *differential cost.*
- Q the quantity sold
- p the incremental revenue
- R the total revenue
- C the total cost

Assuming no change in the inventory, the *break-even point* can be found from $C = R$, where

$$C = f + aQ \quad 3.28$$

$$R = pQ \quad 3.29$$

An alternate form of the break-even problem is to find the number of units per period for which two alternatives have the same total costs. Fixed costs are to be spread over a period longer than one year. One of the alternatives will have a lower cost if production is less than the break-even point. The other will have a lower cost for production greater than the break-even point.

The *cost per unit* problem is a variation of the breakeven problem. In the typical cost per unit problem, data will be available to determine the direct labor and material costs per unit, but some method is needed to additionally allocate part of the annual overhead (burden) and initial facility purchase/construction costs.

Annual overhead is allocated to the unit cost simply by dividing the overhead by the number of units produced each year. The initial purchase/construction cost is multiplied by the appropriate (A/P) factor before similarly dividing by the production rate. The total unit cost is the sum of the direct labor, direct material, prorata share of overhead, and prorata share of the equivalent annual facility investment costs.

Example 3.20

Two plans are available for a company to obtain automobiles for its salesmen. How many miles must the cars be driven each year for the two plans to have the same costs? Use an interest rate of 10%.

Plan A Lease the cars and pay \$0.15 per mile

Plan B Purchase the cars for \$5000. Each car has an economic life of three years, after which it can be sold for \$1,200. Gas and oil cost \$0.04 per mile. Insurance is \$500 per year. (Assume the year-end convention for the insurance.)

Let x be the number of miles driven per year. Then, the EUAC for both alternatives is:

$$\begin{aligned} \text{EUAC}(A) &= 0.15x \\ \text{EUAC}(B) &= 0.04x + 500 + 5000(A/P, 10\%, 3) \\ &\quad - 1200(A/F, 10\%, 3) \\ &= 0.04x + 2148 \end{aligned}$$

Setting EUAC(A) and EUAC(B) equal and solving for x yields 19,527 miles per year as the break-even point.

ECON

16 HANDLING INFLATION

It is important to perform economic studies in terms of *constant value dollars.* One method of converting all cash flows to constant value dollars is to divide the flows by some annual *economic indicator* or price index. Such indicators would normally be given to you as part of a problem.

If indicators are not available, this method can still be used by assuming that inflation is relatively constant at a decimal rate e per year. Then, all cash flows can be converted to $t = 0$ dollars by dividing by $(1+e)^n$ where n is the year of the cash flow.

Example 3.21

What is the uninflated present worth of $2000 in two years if the average inflation rate is 6% and i is 10% ?

$$P = \frac{\$2000}{(1.10)^2(1.06)^2} = \$1471.07$$

An alternative is to replace i with a value corrected for inflation. This corrected value, i', is

$$i' = i + e + ie \qquad 3.30$$

This method has the advantage of simplifying the calculations. However, pre-calculated factors may not be available for the non-integer values of i'. Therefore, table 3.1 will have to be used to calculate the factors.

Example 3.22

Repeat example 3.21 using i'.

$$i' = 0.10 + 0.06 + (0.10)(0.06)$$

$$= 0.166$$

$$P = \frac{\$2000}{(1.166)^2} = \$1471.07$$

17 LEARNING CURVES

The more products that are made, the more efficient the operation becomes due to experience gained. Therefore, direct labor costs decrease. Usually, a *learning curve* is specified by the decrease in cost each time the quantity produced doubles. If there is a 20% decrease per doubling, the curve is said to be an 80% learning curve.

Consider the following special variables:

- T_1 time or cost for the first item
- T_n time or cost for the nth item
- n total number of items produced
- b learning curve constant

Then, the time to produce the nth item is given by

$$T_n = T_1(n)^{-b} \qquad 3.31$$

Table 3.3
Learning Curve Constants

learning curve	b
80%	0.322
85%	0.234
90%	0.152
95%	0.074

The total time to produce units from quantity n_1 to n_2 inclusive is

$$\int_{n_1}^{n_2} T_n dn \approx \frac{T_1}{(1-b)} \left[\left(n_2 + \frac{1}{2}\right)^{1-b} - \left(n_1 - \frac{1}{2}\right)^{1-b}\right] \qquad 3.32$$

The average time per unit over the production from n_1 to n_2 is the above total time from equation 3.32 divided by the quantity produced, $(n_2 - n_1 + 1)$.

It is important to remember that learning curve reductions apply only to direct labor costs. They are not applied to indirect labor or direct material costs.

Example 3.23

A 70% learning curve is used with an item whose first production time was 1.47 hours. How long will it take to produce the 11th item? How long will it take to produce the 11th through 27th items?

First, find b.

$$\frac{T_2}{T_1} = 0.7 = (2)^{-b}$$

$$b = 0.515$$

Then,

$$T_{11} = 1.47(11)^{-0.515} = 0.428 \text{ hours}$$

The time to produce the 11th item through 27th item is approximately

$$T = \frac{1.47}{1-0.515}\left[(27.5)^{1-0.515} - (10.5)^{1-0.515}\right]$$

$$= 5.643 \text{ hours}$$

18 ECONOMIC ORDER QUANTITY

The *economic order quantity* (EOQ) is the order quantity which minimizes the inventory costs per unit time. Although there are many different EOQ models, the simplest is based on the following assumptions:

- Reordering is instantaneous. The time between order placement and receipt is zero.
- Shortages are not allowed.
- Demand for the inventory item is deterministic (i.e., is not a random variable).
- Demand is constant with respect to time.
- An order is placed when the on-hand quantity is zero.

The following special variables are used:

a the constant depletion rate $\left(\frac{\text{items}}{\text{unit time}}\right)$
h the inventory storage cost $\left(\frac{\$}{\text{item-unit time}}\right)$
H the total inventory storage cost between orders ($)
K the fixed cost of placing an order ($)
Q_0 the order quantity

If the original quantity on hand is Q_0, the stock will be depleted at

$$t^* = \frac{Q_0}{a}$$

The total inventory storage cost between t_0 and t^* is

$$H = \frac{1}{2}h\frac{Q_o^2}{a} \quad 3.33$$

The total inventory and ordering cost per unit time is

$$C_t = \frac{aK}{Q_0} + \frac{1}{2}hQ_0 \quad 3.34$$

C_t can be minimized with respect to Q_0. The EOQ and time between orders are:

$$Q_0^* = \sqrt{2\frac{aK}{h}} \quad 3.35$$

$$t^* = \frac{Q_0^*}{a} \quad 3.36$$

19 CONSUMER LOANS

Many consumer loans cannot be handled by the equivalence formulas presented up to this point. Many different arrangements can be made between lender and borrower. Four of the most common consumer loan arrangements are presented below. Refer to a real estate or investment analysis book for more complex loans.

A. SIMPLE INTEREST

Interest due does not compound with a *simple interest* loan. The interest due is merely proportional to the length of time the principal is outstanding. Because of this, simple interest loans are seldom made for long periods (e.g., longer than one year).

Example 3.24

A $12,000 simple interest loan is taken out at 16% per year. The loan matures in one year with no intermediate payments. How much will be due at the end of the year?

$$\text{Amount due} = (1 + 0.16)(\$12{,}000) = \$13{,}920$$

For loans less than one year, it is commonly assumed that a year consists of 12 months of 30 days each.

Example 3.25

$4000 is borrowed for 75 days at 16% per annum simple interest. How much will be due at the end of 75 days?

$$\text{Amount due} = \$4000 + (0.16)\left(\frac{75}{360}\right)(4000) = \$4133$$

B. LOANS WITH CONSTANT AMOUNT PAID TOWARDS PRINCIPAL

With this loan type, the payment is not the same each period. The amount paid towards the principal is constant, but the interest varies from period to period. The following special symbols are used.

BAL_j principal balance after the jth payment
LV total value loaned (cost minus down payment)
j payment or period number
N total number of payments to pay off the loan
PI_j jth interest payment
PP_j jth principal payment
PT_j jth total payment
ϕ effective rate per period (r/k)

The equations which govern this type of loan are

$$BAL_j = LV - (j)(PP) \quad 3.37$$

$$PI_j = \phi(BAL_{j-1}) \quad 3.38$$

$$PT_j = PP + PI_j \quad 3.39$$

C. DIRECT REDUCTION LOANS

This is the typical 'interest paid on unpaid balance' loan. The amount of the periodic payment is constant, but the amounts paid towards the principal and interest both vary.

The same symbols are used with this type of loan as are listed above.

$$N = -\frac{ln\left[\frac{-\phi(LV)}{PT} + 1\right]}{ln(1+\phi)} \quad 3.40$$

$$BAL_{j-1} = PT\left[\frac{1-(1+\phi)^{j-1-N}}{\phi}\right] \quad 3.41$$

$$PI_j = \phi(BAL_{j-1}) \quad 3.42$$

$$PP_j = PT - PI_j \quad 3.43$$

$$BAL_j = BAL_{j-1} - PP_j \quad 3.44$$

Example 3.26

A $45,000 loan is financed at 9.25% per annum. The monthly payment is $385. What are the amounts paid toward interest and principal in the 14th period? What is the remaining principal balance after the 14th payment has been made?

The effective rate per month is

$$\phi = \frac{r}{k} = \frac{0.0925}{12}$$

$$= 0.007708$$

$$N = \frac{-ln\left[\frac{-(0.007708)(45{,}000)}{385} + 1\right]}{ln(1+0.007708)} = 301$$

$$BAL_{13} = 385\left[\frac{1-(1+0.007708)^{14-1-301}}{0.007708}\right]$$

$$= \$44{,}476.39$$

$$PI_{14} = (0.007708)(\$44{,}476.39) = \$342.82$$

$$PP_{14} = \$385 - \$342.82 = \$42.18$$

$$BAL_{14} = \$44{,}476.39 - \$42.18 = \$44{,}434.20$$

Equation 3.40 calculates the number of payments necessary to pay off a loan. This equation can be solved with effort for the total periodic payment (PT) or the initial value of the loan (LV). It is easier, however, to use the $(A/P, \phi, n)$ factor to find the payment and loan value.

$$PT = (LV)(A/P, \phi\%, n)$$

If the loan is repaid in yearly installments, then i is the effective annual rate. If the loan is paid off monthly, then i should be replaced by the effective rate per month (ϕ from equation 3.26). For monthly payments, n is the number of months in the payback period.

D. DIRECT REDUCTION LOAN WITH BALLOON PAYMENT

This type of loan has a constant periodic payment, but the duration of the loan is insufficient to completely pay back the principal. Therefore, all remaining unpaid principal must be paid back in a lump sum when the loan matures. This large payment is known as a *balloon payment.*

Equations 3.40 through 3.44 can also be used with this type of loan. The remaining balance after the last payment is the balloon payment. This balloon payment must be repaid along with the last regular payment calculated.

20 SENSITIVITY ANALYSIS

Data analysis and forecasts in economic studies represent judgment on costs which will occur in the future. There are always uncertainties about these costs. However, these uncertainties are insufficient reason not to make the best possible estimates of the costs. Nevertheless, a decision between alternatives often can be made more confidently if it is known whether or not the conclusion is sensitive to moderate changes in data forecasts. Sensitivity analysis provides this extra dimension to an economic analysis.

The sensitivity of a decision is determined by inserting a range of estimates for critical cash flows. If radical changes can be made to a cash flow without changing the decision, the decision is said to be insensitive to uncertainties regarding that cash flow. However, if a small change in the estimate of a cash flow will alter the decision, that decision is said to be very sensitive to changes in the estimate.

An established semantic tradition distinguishes between risk analysis and uncertainty analysis. Risk analysis addresses variables which have a known or estimated probability distribution. In this regard, statistics and probability theory can be used to determine the probability of a cash flow varying between given limits. On the other hand, uncertainty analysis is concerned with situations in which there is not enough information to determine the probability or frequency distribution for the variables involved.

As a first step, sensitivity analysis should be applied one at a time to the dominant cost factors. Dominant cost factors are those which have the most significant impact on the present value of the alternative. If warranted, additional investigation can be used to determine the sensitivity to several cash flows varying simultaneously. Significant judgment is needed, however, to successfully determine the proper combinations of cash flows to vary.

It is common to plot the dependency of the present value on the cash flow being varied on a two-dimensional graph. Simple linear interpolation is used (within reason) to determine the critical value of the cash flow being varied.

STANDARD CASH FLOW FACTORS

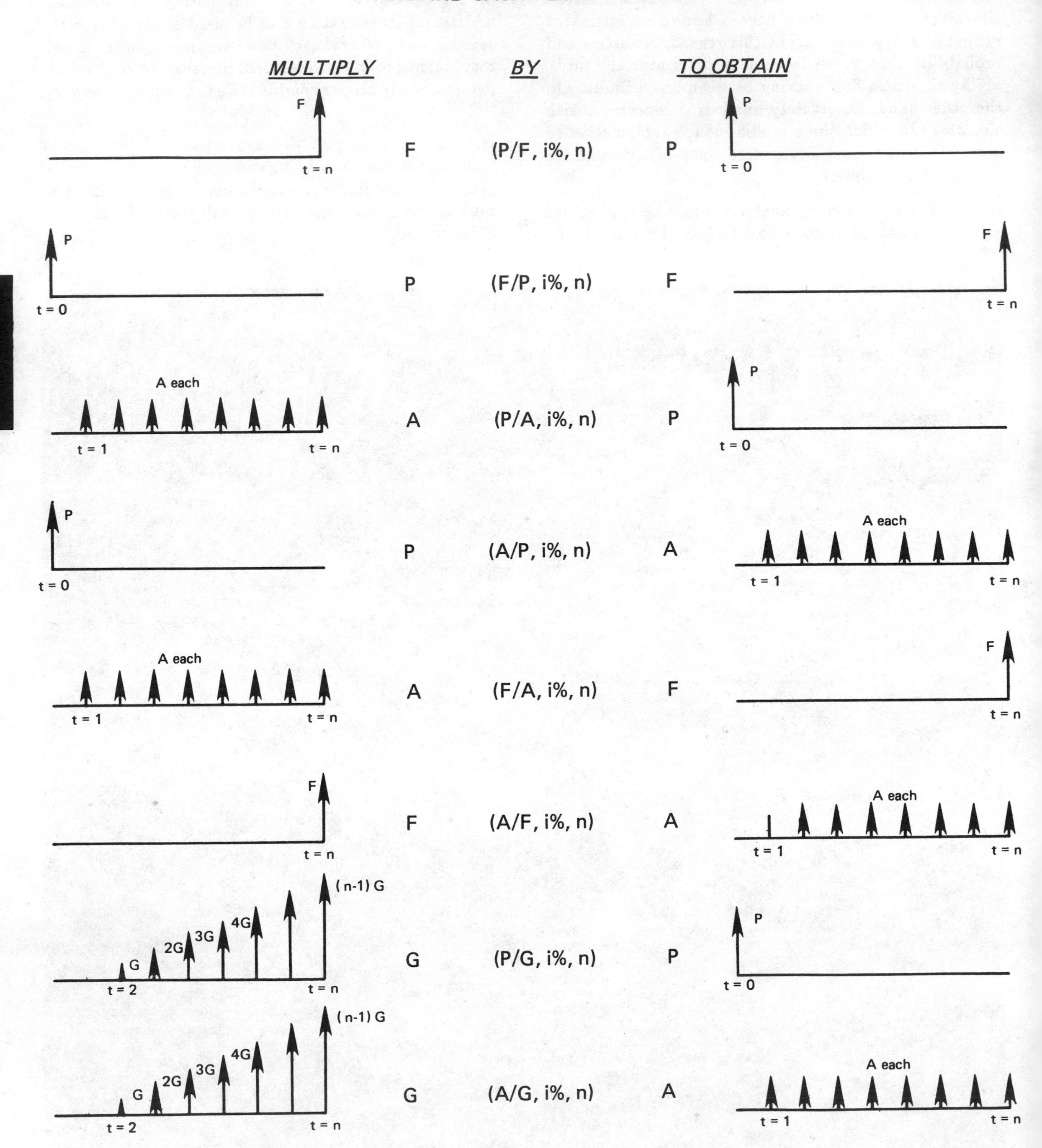

I = 0.50 %

N	(P/F)	(P/A)	(P/G)	(F/P)	(F/A)	(A/P)	(A/F)	(A/G)	N
1	.9950	0.9950	−0.0000	1.0050	1.0000	1.0050	1.0000	−0.0000	1
2	.9901	1.9851	0.9901	1.0100	2.0050	0.5038	0.4988	0.4988	2
3	.9851	2.9702	2.9604	1.0151	3.0150	0.3367	0.3317	0.9967	3
4	.9802	3.9505	5.9011	1.0202	4.0301	0.2531	0.2481	1.4938	4
5	.9754	4.9259	9.8026	1.0253	5.0503	0.2030	0.1980	1.9900	5
6	.9705	5.8964	14.6552	1.0304	6.0755	0.1696	0.1646	2.4855	6
7	.9657	6.8621	20.4493	1.0355	7.1059	0.1457	0.1407	2.9801	7
8	.9609	7.8230	27.1755	1.0407	8.1414	0.1278	0.1228	3.4738	8
9	.9561	8.7791	34.8244	1.0459	9.1821	0.1139	0.1089	3.9668	9
10	.9513	9.7304	43.3865	1.0511	10.2280	0.1028	0.0978	4.4589	10
11	.9466	10.6770	52.8526	1.0564	11.2792	0.0937	0.0887	4.9501	11
12	.9419	11.6189	63.2136	1.0617	12.3356	0.0861	0.0811	5.4406	12
13	.9372	12.5562	74.4602	1.0670	13.3972	0.0796	0.0746	5.9302	13
14	.9326	13.4887	86.5835	1.0723	14.4642	0.0741	0.0691	6.4190	14
15	.9279	14.4166	99.5743	1.0777	15.5365	0.0694	0.0644	6.9069	15
16	.9233	15.3399	113.4238	1.0831	16.6142	0.0652	0.0602	7.3940	16
17	.9187	16.2586	128.1231	1.0885	17.6973	0.0615	0.0565	7.8803	17
18	.9141	17.1728	143.6634	1.0939	18.7858	0.0582	0.0532	8.3658	18
19	.9096	18.0824	160.0360	1.0994	19.8797	0.0553	0.0503	8.8504	19
20	.9051	18.9874	177.2322	1.1049	20.9791	0.0527	0.0477	9.3342	20
21	.9006	19.8880	195.2434	1.1104	22.0840	0.0503	0.0453	9.8172	21
22	.8961	20.7841	214.0611	1.1160	23.1944	0.0481	0.0431	10.2993	22
23	.8916	21.6757	233.6768	1.1216	24.3104	0.0461	0.0411	10.7806	23
24	.8872	22.5629	254.0820	1.1272	25.4320	0.0443	0.0393	11.2611	24
25	.8828	23.4456	275.2686	1.1328	26.5591	0.0427	0.0377	11.7407	25
26	.8784	24.3240	297.2281	1.1385	27.6919	0.0411	0.0361	12.2195	26
27	.8740	25.1980	319.9523	1.1442	28.8304	0.0397	0.0347	12.6975	27
28	.8697	26.0677	343.4332	1.1499	29.9745	0.0384	0.0334	13.1747	28
29	.8653	26.9330	367.6625	1.1556	31.1244	0.0371	0.0321	13.6510	29
30	.8610	27.7941	392.6324	1.1614	32.2800	0.0360	0.0310	14.1265	30
31	.8567	28.6508	418.3348	1.1672	33.4414	0.0349	0.0299	14.6012	31
32	.8525	29.5033	444.7618	1.1730	34.6086	0.0339	0.0289	15.0750	32
33	.8482	30.3515	471.9055	1.1789	35.7817	0.0329	0.0279	15.5480	33
34	.8440	31.1955	499.7583	1.1848	36.9606	0.0321	0.0271	16.0202	34
35	.8398	32.0354	528.3123	1.1907	38.1454	0.0312	0.0262	16.4915	35
36	.8356	32.8710	557.5598	1.1967	39.3361	0.0304	0.0254	16.9621	36
37	.8315	33.7025	587.4934	1.2027	40.5328	0.0297	0.0247	17.4317	37
38	.8274	34.5299	618.1054	1.2087	41.7354	0.0290	0.0240	17.9006	38
39	.8232	35.3531	649.3883	1.2147	42.9441	0.0283	0.0233	18.3686	39
40	.8191	36.1722	681.3347	1.2208	44.1588	0.0276	0.0226	18.8359	40
41	.8151	36.9873	713.9372	1.2269	45.3796	0.0270	0.0220	19.3022	41
42	.8110	37.7983	747.1886	1.2330	46.6065	0.0265	0.0215	19.7678	42
43	.8070	38.6053	781.0815	1.2392	47.8396	0.0259	0.0209	20.2325	43
44	.8030	39.4082	815.6087	1.2454	49.0788	0.0254	0.0204	20.6964	44
45	.7990	40.2072	850.7631	1.2516	50.3242	0.0249	0.0199	21.1595	45
46	.7950	41.0022	886.5376	1.2579	51.5758	0.0244	0.0194	21.6217	46
47	.7910	41.7932	922.9252	1.2642	52.8337	0.0239	0.0189	22.0831	47
48	.7871	42.5803	959.9188	1.2705	54.0978	0.0235	0.0185	22.5437	48
49	.7832	43.3635	997.5116	1.2768	55.3683	0.0231	0.0181	23.0035	49
50	.7793	44.1428	1035.6966	1.2832	56.6452	0.0227	0.0177	23.4624	50
51	.7754	44.9182	1074.4670	1.2896	57.9284	0.0223	0.0173	23.9205	51
52	.7716	45.6897	1113.8162	1.2961	59.2180	0.0219	0.0169	24.3778	52
53	.7677	46.4575	1153.7372	1.3026	60.5141	0.0215	0.0165	24.8343	53
54	.7639	47.2214	1194.2236	1.3091	61.8167	0.0212	0.0162	25.2899	54
55	.7601	47.9814	1235.2686	1.3156	63.1258	0.0208	0.0158	25.7447	55
60	.7414	51.7256	1448.6458	1.3489	69.7700	0.0193	0.0143	28.0064	60
65	.7231	55.3775	1675.0272	1.3829	76.5821	0.0181	0.0131	30.2475	65
70	.7053	58.9394	1913.6427	1.4178	83.5661	0.0170	0.0120	32.4680	70
75	.6879	62.4136	2163.7525	1.4536	90.7265	0.0160	0.0110	34.6679	75
80	.6710	65.8023	2424.6455	1.4903	98.0677	0.0152	0.0102	36.8474	80
85	.6545	69.1075	2695.6389	1.5280	105.5943	0.0145	0.0095	39.0065	85
90	.6383	72.3313	2976.0769	1.5666	113.3109	0.0138	0.0088	41.1451	90
95	.6226	75.4757	3265.3298	1.6061	121.2224	0.0132	0.0082	43.2633	95
100	.6073	78.5426	3562.7934	1.6467	129.3337	0.0127	0.0077	45.3613	100

I = 0.75 %

N	(P/F)	(P/A)	(P/G)	(F/P)	(F/A)	(A/P)	(A/F)	(A/G)	N
1	.9926	0.9926	-0.0000	1.0075	1.0000	1.0075	1.0000	-0.0000	1
2	.9852	1.9777	0.9852	1.0151	2.0075	0.5056	0.4981	0.4981	2
3	.9778	2.9556	2.9408	1.0227	3.0226	0.3383	0.3308	0.9950	3
4	.9706	3.9261	5.8525	1.0303	4.0452	0.2547	0.2472	1.4907	4
5	.9633	4.8894	9.7058	1.0381	5.0756	0.2045	0.1970	1.9851	5
6	.9562	5.8456	14.4866	1.0459	6.1136	0.1711	0.1636	2.4782	6
7	.9490	6.7946	20.1808	1.0537	7.1595	0.1472	0.1397	2.9701	7
8	.9420	7.7366	26.7747	1.0616	8.2132	0.1293	0.1218	3.4608	8
9	.9350	8.6716	34.2544	1.0696	9.2748	0.1153	0.1078	3.9502	9
10	.9280	9.5996	42.6064	1.0776	10.3443	0.1042	0.0967	4.4384	10
11	.9211	10.5207	51.8174	1.0857	11.4219	0.0951	0.0876	4.9253	11
12	.9142	11.4349	61.8740	1.0938	12.5076	0.0875	0.0800	5.4110	12
13	.9074	12.3423	72.7632	1.1020	13.6014	0.0810	0.0735	5.8954	13
14	.9007	13.2430	84.4720	1.1103	14.7034	0.0755	0.0680	6.3786	14
15	.8940	14.1370	96.9876	1.1186	15.8137	0.0707	0.0632	6.8606	15
16	.8873	15.0243	110.2973	1.1270	16.9323	0.0666	0.0591	7.3413	16
17	.8807	15.9050	124.3887	1.1354	18.0593	0.0629	0.0554	7.8207	17
18	.8742	16.7792	139.2494	1.1440	19.1947	0.0596	0.0521	8.2989	18
19	.8676	17.6468	154.8671	1.1525	20.3387	0.0567	0.0492	8.7759	19
20	.8612	18.5080	171.2297	1.1612	21.4912	0.0540	0.0465	9.2516	20
21	.8548	19.3628	188.3253	1.1699	22.6524	0.0516	0.0441	9.7261	21
22	.8484	20.2112	206.1420	1.1787	23.8223	0.0495	0.0420	10.1994	22
23	.8421	21.0533	224.6682	1.1875	25.0010	0.0475	0.0400	10.6714	23
24	.8358	21.8891	243.8923	1.1964	26.1885	0.0457	0.0382	11.1422	24
25	.8296	22.7188	263.8029	1.2054	27.3849	0.0440	0.0365	11.6117	25
26	.8234	23.5422	284.3888	1.2144	28.5903	0.0425	0.0350	12.0800	26
27	.8173	24.3595	305.6387	1.2235	29.8047	0.0411	0.0336	12.5470	27
28	.8112	25.1707	327.5416	1.2327	31.0282	0.0397	0.0322	13.0128	28
29	.8052	25.9759	350.0867	1.2420	32.2609	0.0385	0.0310	13.4774	29
30	.7992	26.7751	373.2631	1.2513	33.5029	0.0373	0.0298	13.9407	30
31	.7932	27.5683	397.0602	1.2607	34.7542	0.0363	0.0288	14.4028	31
32	.7873	28.3557	421.4675	1.2701	36.0148	0.0353	0.0278	14.8636	32
33	.7815	29.1371	446.4746	1.2796	37.2849	0.0343	0.0268	15.3232	33
34	.7757	29.9128	472.0712	1.2892	38.5646	0.0334	0.0259	15.7816	34
35	.7699	30.6827	498.2471	1.2989	39.8538	0.0326	0.0251	16.2387	35
36	.7641	31.4468	524.9924	1.3086	41.1527	0.0318	0.0243	16.6946	36
37	.7585	32.2053	552.2969	1.3185	42.4614	0.0311	0.0236	17.1493	37
38	.7528	32.9581	580.1511	1.3283	43.7798	0.0303	0.0228	17.6027	38
39	.7472	33.7053	608.5451	1.3383	45.1082	0.0297	0.0222	18.0549	39
40	.7416	34.4469	637.4693	1.3483	46.4465	0.0290	0.0215	18.5058	40
41	.7361	35.1831	666.9144	1.3585	47.7948	0.0284	0.0209	18.9556	41
42	.7306	35.9137	696.8709	1.3686	49.1533	0.0278	0.0203	19.4040	42
43	.7252	36.6389	727.3297	1.3789	50.5219	0.0273	0.0198	19.8513	43
44	.7198	37.3587	758.2815	1.3893	51.9009	0.0268	0.0193	20.2973	44
45	.7145	38.0732	789.7173	1.3997	53.2901	0.0263	0.0188	20.7421	45
46	.7091	38.7823	821.6283	1.4102	54.6898	0.0258	0.0183	21.1856	46
47	.7039	39.4862	854.0056	1.4207	56.1000	0.0253	0.0178	21.6280	47
48	.6986	40.1848	886.8404	1.4314	57.5207	0.0249	0.0174	22.0691	48
49	.6934	40.8782	920.1243	1.4421	58.9521	0.0245	0.0170	22.5089	49
50	.6883	41.5664	953.8486	1.4530	60.3943	0.0241	0.0166	22.9476	50
51	.6831	42.2496	988.0050	1.4639	61.8472	0.0237	0.0162	23.3850	51
52	.6780	42.9276	1022.5852	1.4748	63.3111	0.0233	0.0158	23.8211	52
53	.6730	43.6006	1057.5810	1.4859	64.7859	0.0229	0.0154	24.2561	53
54	.6680	44.2686	1092.9842	1.4970	66.2718	0.0226	0.0151	24.6898	54
55	.6630	44.9316	1128.7869	1.5083	67.7688	0.0223	0.0148	25.1223	55
60	.6387	48.1734	1313.5189	1.5657	75.4241	0.0208	0.0133	27.2665	60
65	.6153	51.2963	1507.0910	1.6253	83.3709	0.0195	0.0120	29.3801	65
70	.5927	54.3046	1708.6065	1.6872	91.6201	0.0184	0.0109	31.4634	70
75	.5710	57.2027	1917.2225	1.7514	100.1833	0.0175	0.0100	33.5163	75
80	.5500	59.9944	2132.1472	1.8180	109.0725	0.0167	0.0092	35.5391	80
85	.5299	62.6838	2352.6375	1.8873	118.3001	0.0160	0.0085	37.5318	85
90	.5104	65.2746	2577.9961	1.9591	127.8790	0.0153	0.0078	39.4946	90
95	.4917	67.7704	2807.5694	2.0337	137.8225	0.0148	0.0073	41.4277	95
100	.4737	70.1746	3040.7453	2.1111	148.1445	0.0143	0.0068	43.3311	100

I = 1.00 %

N	(P/F)	(P/A)	(P/G)	(F/P)	(F/A)	(A/P)	(A/F)	(A/G)	N
1	.9901	0.9901	-0.0000	1.0100	1.0000	1.0100	1.0000	-0.0000	1
2	.9803	1.9704	0.9803	1.0201	2.0100	0.5075	0.4975	0.4975	2
3	.9706	2.9410	2.9215	1.0303	3.0301	0.3400	0.3300	0.9934	3
4	.9610	3.9020	5.8044	1.0406	4.0604	0.2563	0.2463	1.4876	4
5	.9515	4.8534	9.6103	1.0510	5.1010	0.2060	0.1960	1.9801	5
6	.9420	5.7955	14.3205	1.0615	6.1520	0.1725	0.1625	2.4710	6
7	.9327	6.7282	19.9168	1.0721	7.2135	0.1486	0.1386	2.9602	7
8	.9235	7.6517	26.3812	1.0829	8.2857	0.1307	0.1207	3.4478	8
9	.9143	8.5660	33.6959	1.0937	9.3685	0.1167	0.1067	3.9337	9
10	.9053	9.4713	41.8435	1.1046	10.4622	0.1056	0.0956	4.4179	10
11	.8963	10.3676	50.8067	1.1157	11.5668	0.0965	0.0865	4.9005	11
12	.8874	11.2551	60.5687	1.1268	12.6825	0.0888	0.0788	5.3815	12
13	.8787	12.1337	71.1126	1.1381	13.8093	0.0824	0.0724	5.8607	13
14	.8700	13.0037	82.4221	1.1495	14.9474	0.0769	0.0669	6.3384	14
15	.8613	13.8651	94.4810	1.1610	16.0969	0.0721	0.0621	6.8143	15
16	.8528	14.7179	107.2734	1.1726	17.2579	0.0679	0.0579	7.2886	16
17	.8444	15.5623	120.7834	1.1843	18.4304	0.0643	0.0543	7.7613	17
18	.8360	16.3983	134.9957	1.1961	19.6147	0.0610	0.0510	8.2323	18
19	.8277	17.2260	149.8950	1.2081	20.8109	0.0581	0.0481	8.7017	19
20	.8195	18.0456	165.4664	1.2202	22.0190	0.0554	0.0454	9.1694	20
21	.8114	18.8570	181.6950	1.2324	23.2392	0.0530	0.0430	9.6354	21
22	.8034	19.6604	198.5663	1.2447	24.4716	0.0509	0.0409	10.0998	22
23	.7954	20.4558	216.0660	1.2572	25.7163	0.0489	0.0389	10.5626	23
24	.7876	21.2434	234.1800	1.2697	26.9735	0.0471	0.0371	11.0237	24
25	.7798	22.0232	252.8945	1.2824	28.2432	0.0454	0.0354	11.4831	25
26	.7720	22.7952	272.1957	1.2953	29.5256	0.0439	0.0339	11.9409	26
27	.7644	23.5596	292.0702	1.3082	30.8209	0.0424	0.0324	12.3971	27
28	.7568	24.3164	312.5047	1.3213	32.1291	0.0411	0.0311	12.8516	28
29	.7493	25.0658	333.4863	1.3345	33.4504	0.0399	0.0299	13.3044	29
30	.7419	25.8077	355.0021	1.3478	34.7849	0.0387	0.0287	13.7557	30
31	.7346	26.5423	377.0394	1.3613	36.1327	0.0377	0.0277	14.2052	31
32	.7273	27.2696	399.5858	1.3749	37.4941	0.0367	0.0267	14.6532	32
33	.7201	27.9897	422.6291	1.3887	38.8690	0.0357	0.0257	15.0995	33
34	.7130	28.7027	446.1572	1.4026	40.2577	0.0348	0.0248	15.5441	34
35	.7059	29.4086	470.1583	1.4166	41.6603	0.0340	0.0240	15.9871	35
36	.6989	30.1075	494.6207	1.4308	43.0769	0.0332	0.0232	16.4285	36
37	.6920	30.7995	519.5329	1.4451	44.5076	0.0325	0.0225	16.8682	37
38	.6852	31.4847	544.8835	1.4595	45.9527	0.0318	0.0218	17.3063	38
39	.6784	32.1630	570.6616	1.4741	47.4123	0.0311	0.0211	17.7428	39
40	.6717	32.8347	596.8561	1.4889	48.8864	0.0305	0.0205	18.1776	40
41	.6650	33.4997	623.4562	1.5038	50.3752	0.0299	0.0199	18.6108	41
42	.6584	34.1581	650.4514	1.5188	51.8790	0.0293	0.0193	19.0424	42
43	.6519	34.8100	677.8312	1.5340	53.3978	0.0287	0.0187	19.4723	43
44	.6454	35.4555	705.5853	1.5493	54.9318	0.0282	0.0182	19.9006	44
45	.6391	36.0945	733.7037	1.5648	56.4811	0.0277	0.0177	20.3273	45
46	.6327	36.7272	762.1765	1.5805	58.0459	0.0272	0.0172	20.7524	46
47	.6265	37.3537	790.9938	1.5963	59.6263	0.0268	0.0168	21.1758	47
48	.6203	37.9740	820.1460	1.6122	61.2226	0.0263	0.0163	21.5976	48
49	.6141	38.5881	849.6237	1.6283	62.8348	0.0259	0.0159	22.0178	49
50	.6080	39.1961	879.4176	1.6446	64.4632	0.0255	0.0155	22.4363	50
51	.6020	39.7981	909.5186	1.6611	66.1078	0.0251	0.0151	22.8533	51
52	.5961	40.3942	939.9175	1.6777	67.7689	0.0248	0.0148	23.2686	52
53	.5902	40.9844	970.6057	1.6945	69.4466	0.0244	0.0144	23.6823	53
54	.5843	41.5687	1001.5743	1.7114	71.1410	0.0241	0.0141	24.0945	54
55	.5785	42.1472	1032.8148	1.7285	72.8525	0.0237	0.0137	24.5049	55
60	.5504	44.9550	1192.8061	1.8167	81.6697	0.0222	0.0122	26.5333	60
65	.5237	47.6266	1358.3903	1.9094	90.9366	0.0210	0.0110	28.5217	65
70	.4983	50.1685	1528.6474	2.0068	100.6763	0.0199	0.0099	30.4703	70
75	.4741	52.5871	1702.7340	2.1091	110.9128	0.0190	0.0090	32.3793	75
80	.4511	54.8882	1879.8771	2.2167	121.6715	0.0182	0.0082	34.2492	80
85	.4292	57.0777	2059.3701	2.3298	132.9790	0.0175	0.0075	36.0801	85
90	.4084	59.1609	2240.5675	2.4486	144.8633	0.0169	0.0069	37.8724	90
95	.3886	61.1430	2422.8811	2.5735	157.3538	0.0164	0.0064	39.6265	95
100	.3697	63.0289	2605.7758	2.7048	170.4814	0.0159	0.0059	41.3426	100

ECON

I = 1.50 %

N	(P/F)	(P/A)	(P/G)	(F/P)	(F/A)	(A/P)	(A/F)	(A/G)	N
1	.9852	0.9852	-0.0000	1.0150	1.0000	1.0150	1.0000	-0.0000	1
2	.9707	1.9559	0.9707	1.0302	2.0150	0.5113	0.4963	0.4963	2
3	.9563	2.9122	2.8833	1.0457	3.0452	0.3434	0.3284	0.9901	3
4	.9422	3.8544	5.7098	1.0614	4.0909	0.2594	0.2444	1.4814	4
5	.9283	4.7826	9.4229	1.0773	5.1523	0.2091	0.1941	1.9702	5
6	.9145	5.6972	13.9956	1.0934	6.2296	0.1755	0.1605	2.4566	6
7	.9010	6.5982	19.4018	1.1098	7.3230	0.1516	0.1366	2.9405	7
8	.8877	7.4859	25.6157	1.1265	8.4328	0.1336	0.1186	3.4219	8
9	.8746	8.3605	32.6125	1.1434	9.5593	0.1196	0.1046	3.9008	9
10	.8617	9.2222	40.3675	1.1605	10.7027	0.1084	0.0934	4.3772	10
11	.8489	10.0711	48.8568	1.1779	11.8633	0.0993	0.0843	4.8512	11
12	.8364	10.9075	58.0571	1.1956	13.0412	0.0917	0.0767	5.3227	12
13	.8240	11.7315	67.9454	1.2136	14.2368	0.0852	0.0702	5.7917	13
14	.8118	12.5434	78.4994	1.2318	15.4504	0.0797	0.0647	6.2582	14
15	.7999	13.3432	89.6974	1.2502	16.6821	0.0749	0.0599	6.7223	15
16	.7880	14.1313	101.5178	1.2690	17.9324	0.0708	0.0558	7.1839	16
17	.7764	14.9076	113.9400	1.2880	19.2014	0.0671	0.0521	7.6431	17
18	.7649	15.6726	126.9435	1.3073	20.4894	0.0638	0.0488	8.0997	18
19	.7536	16.4262	140.5084	1.3270	21.7967	0.0609	0.0459	8.5539	19
20	.7425	17.1686	154.6154	1.3469	23.1237	0.0582	0.0432	9.0057	20
21	.7315	17.9001	169.2453	1.3671	24.4705	0.0559	0.0409	9.4550	21
22	.7207	18.6208	184.3798	1.3876	25.8376	0.0537	0.0387	9.9018	22
23	.7100	19.3309	200.0006	1.4084	27.2251	0.0517	0.0367	10.3462	23
24	.6995	20.0304	216.0901	1.4295	28.6335	0.0499	0.0349	10.7881	24
25	.6892	20.7196	232.6310	1.4509	30.0630	0.0483	0.0333	11.2276	25
26	.6790	21.3986	249.6065	1.4727	31.5140	0.0467	0.0317	11.6646	26
27	.6690	22.0676	267.0002	1.4948	32.9867	0.0453	0.0303	12.0992	27
28	.6591	22.7267	284.7958	1.5172	34.4815	0.0440	0.0290	12.5313	28
29	.6494	23.3761	302.9779	1.5400	35.9987	0.0428	0.0278	12.9610	29
30	.6398	24.0158	321.5310	1.5631	37.5387	0.0416	0.0266	13.3883	30
31	.6303	24.6461	340.4402	1.5865	39.1018	0.0406	0.0256	13.8131	31
32	.6210	25.2671	359.6910	1.6103	40.6883	0.0396	0.0246	14.2355	32
33	.6118	25.8790	379.2691	1.6345	42.2986	0.0386	0.0236	14.6555	33
34	.6028	26.4817	399.1607	1.6590	43.9331	0.0378	0.0228	15.0731	34
35	.5939	27.0756	419.3521	1.6839	45.5921	0.0369	0.0219	15.4882	35
36	.5851	27.6607	439.8303	1.7091	47.2760	0.0362	0.0212	15.9009	36
37	.5764	28.2371	460.5822	1.7348	48.9851	0.0354	0.0204	16.3112	37
38	.5679	28.8051	481.5954	1.7608	50.7199	0.0347	0.0197	16.7191	38
39	.5595	29.3646	502.8576	1.7872	52.4807	0.0341	0.0191	17.1246	39
40	.5513	29.9158	524.3568	1.8140	54.2679	0.0334	0.0184	17.5277	40
41	.5431	30.4590	546.0814	1.8412	56.0819	0.0328	0.0178	17.9284	41
42	.5351	30.9941	568.0201	1.8688	57.9231	0.0323	0.0173	18.3267	42
43	.5272	31.5212	590.1617	1.8969	59.7920	0.0317	0.0167	18.7227	43
44	.5194	32.0406	612.4955	1.9253	61.6889	0.0312	0.0162	19.1162	44
45	.5117	32.5523	635.0110	1.9542	63.6142	0.0307	0.0157	19.5074	45
46	.5042	33.0565	657.6979	1.9835	65.5684	0.0303	0.0153	19.8962	46
47	.4967	33.5532	680.5462	2.0133	67.5519	0.0298	0.0148	20.2826	47
48	.4894	34.0426	703.5462	2.0435	69.5652	0.0294	0.0144	20.6667	48
49	.4821	34.5247	726.6884	2.0741	71.6087	0.0290	0.0140	21.0484	49
50	.4750	34.9997	749.9636	2.1052	73.6828	0.0286	0.0136	21.4277	50
51	.4680	35.4677	773.3629	2.1368	75.7881	0.0282	0.0132	21.8047	51
52	.4611	35.9287	796.8774	2.1689	77.9249	0.0278	0.0128	22.1794	52
53	.4543	36.3830	820.4986	2.2014	80.0938	0.0275	0.0125	22.5517	53
54	.4475	36.8305	844.2184	2.2344	82.2952	0.0272	0.0122	22.9217	54
55	.4409	37.2715	868.0285	2.2679	84.5296	0.0268	0.0118	23.2894	55
60	.4093	39.3803	988.1674	2.4432	96.2147	0.0254	0.0104	25.0930	60
65	.3799	41.3378	1109.4752	2.6320	108.8028	0.0242	0.0092	26.8393	65
70	.3527	43.1549	1231.1658	2.8355	122.3638	0.0232	0.0082	28.5290	70
75	.3274	44.8416	1352.5600	3.0546	136.9728	0.0223	0.0073	30.1631	75
80	.3039	46.4073	1473.0741	3.2907	152.7109	0.0215	0.0065	31.7423	80
85	.2821	47.8607	1592.2095	3.5450	169.6652	0.0209	0.0059	33.2676	85
90	.2619	49.2099	1709.5439	3.8189	187.9299	0.0203	0.0053	34.7399	90
95	.2431	50.4622	1824.7224	4.1141	207.6061	0.0198	0.0048	36.1602	95
100	.2256	51.6247	1937.4506	4.4320	228.8030	0.0194	0.0044	37.5295	100

I = 2.00 %

N	(P/F)	(P/A)	(P/G)	(F/P)	(F/A)	(A/P)	(A/F)	(A/G)	N
1	.9804	0.9804	-0.0000	1.0200	1.0000	1.0200	1.0000	-0.0000	1
2	.9612	1.9416	0.9612	1.0404	2.0200	0.5150	0.4950	0.4950	2
3	.9423	2.8839	2.8458	1.0612	3.0604	0.3468	0.3268	0.9868	3
4	.9238	3.8077	5.6173	1.0824	4.1216	0.2626	0.2426	1.4752	4
5	.9057	4.7135	9.2403	1.1041	5.2040	0.2122	0.1922	1.9604	5
6	.8880	5.6014	13.6801	1.1262	6.3081	0.1785	0.1585	2.4423	6
7	.8706	6.4720	18.9035	1.1487	7.4343	0.1545	0.1345	2.9208	7
8	.8535	7.3255	24.8779	1.1717	8.5830	0.1365	0.1165	3.3961	8
9	.8368	8.1622	31.5720	1.1951	9.7546	0.1225	0.1025	3.8681	9
10	.8203	8.9826	38.9551	1.2190	10.9497	0.1113	0.0913	4.3367	10
11	.8043	9.7868	46.9977	1.2434	12.1687	0.1022	0.0822	4.8021	11
12	.7885	10.5753	55.6712	1.2682	13.4121	0.0946	0.0746	5.2642	12
13	.7730	11.3484	64.9475	1.2936	14.6803	0.0881	0.0681	5.7231	13
14	.7579	12.1062	74.7999	1.3195	15.9739	0.0826	0.0626	6.1786	14
15	.7430	12.8493	85.2021	1.3459	17.2934	0.0778	0.0578	6.6309	15
16	.7284	13.5777	96.1288	1.3728	18.6393	0.0737	0.0537	7.0799	16
17	.7142	14.2919	107.5554	1.4002	20.0121	0.0700	0.0500	7.5256	17
18	.7002	14.9920	119.4581	1.4282	21.4123	0.0667	0.0467	7.9681	18
19	.6864	15.6785	131.8139	1.4568	22.8406	0.0638	0.0438	8.4073	19
20	.6730	16.3514	144.6003	1.4859	24.2974	0.0612	0.0412	8.8433	20
21	.6598	17.0112	157.7959	1.5157	25.7833	0.0588	0.0388	9.2760	21
22	.6468	17.6580	171.3795	1.5460	27.2990	0.0566	0.0366	9.7055	22
23	.6342	18.2922	185.3309	1.5769	28.8450	0.0547	0.0347	10.1317	23
24	.6217	18.9139	199.6305	1.6084	30.4219	0.0529	0.0329	10.5547	24
25	.6095	19.5235	214.2592	1.6406	32.0303	0.0512	0.0312	10.9745	25
26	.5976	20.1210	229.1987	1.6734	33.6709	0.0497	0.0297	11.3910	26
27	.5859	20.7069	244.4311	1.7069	35.3443	0.0483	0.0283	11.8043	27
28	.5744	21.2813	259.9392	1.7410	37.0512	0.0470	0.0270	12.2145	28
29	.5631	21.8444	275.7064	1.7758	38.7922	0.0458	0.0258	12.6214	29
30	.5521	22.3965	291.7164	1.8114	40.5681	0.0446	0.0246	13.0251	30
31	.5412	22.9377	307.9538	1.8476	42.3794	0.0436	0.0236	13.4257	31
32	.5306	23.4683	324.4035	1.8845	44.2270	0.0426	0.0226	13.8230	32
33	.5202	23.9886	341.0508	1.9222	46.1116	0.0417	0.0217	14.2172	33
34	.5100	24.4986	357.8817	1.9607	48.0338	0.0408	0.0208	14.6083	34
35	.5000	24.9986	374.8826	1.9999	49.9945	0.0400	0.0200	14.9961	35
36	.4902	25.4888	392.0405	2.0399	51.9944	0.0392	0.0192	15.3809	36
37	.4806	25.9695	409.3424	2.0807	54.0343	0.0385	0.0185	15.7625	37
38	.4712	26.4406	426.7764	2.1223	56.1149	0.0378	0.0178	16.1409	38
39	.4619	26.9026	444.3304	2.1647	58.2372	0.0372	0.0172	16.5163	39
40	.4529	27.3555	461.9931	2.2080	60.4020	0.0366	0.0166	16.8885	40
41	.4440	27.7995	479.7535	2.2522	62.6100	0.0360	0.0160	17.2576	41
42	.4353	28.2348	497.6010	2.2972	64.8622	0.0354	0.0154	17.6237	42
43	.4268	28.6616	515.5253	2.3432	67.1595	0.0349	0.0149	17.9866	43
44	.4184	29.0800	533.5165	2.3901	69.5027	0.0344	0.0144	18.3465	44
45	.4102	29.4902	551.5652	2.4379	71.8927	0.0339	0.0139	18.7034	45
46	.4022	29.8923	569.6621	2.4866	74.3306	0.0335	0.0135	19.0571	46
47	.3943	30.2866	587.7985	2.5363	76.8172	0.0330	0.0130	19.4079	47
48	.3865	30.6731	605.9657	2.5871	79.3535	0.0326	0.0126	19.7556	48
49	.3790	31.0521	624.1557	2.6388	81.9406	0.0322	0.0122	20.1003	49
50	.3715	31.4236	642.3606	2.6916	84.5794	0.0318	0.0118	20.4420	50
51	.3642	31.7878	660.5727	2.7454	87.2710	0.0315	0.0115	20.7807	51
52	.3571	32.1449	678.7849	2.8003	90.0164	0.0311	0.0111	21.1164	52
53	.3501	32.4950	696.9900	2.8563	92.8167	0.0308	0.0108	21.4491	53
54	.3432	32.8383	715.1815	2.9135	95.6731	0.0305	0.0105	21.7789	54
55	.3365	33.1748	733.3527	2.9717	98.5865	0.0301	0.0101	22.1057	55
60	.3048	34.7609	823.6975	3.2810	114.0515	0.0288	0.0088	23.6961	60
65	.2761	36.1975	912.7085	3.6225	131.1262	0.0276	0.0076	25.2147	65
70	.2500	37.4986	999.8343	3.9996	149.9779	0.0267	0.0067	26.6632	70
75	.2265	38.6771	1084.6393	4.4158	170.7918	0.0259	0.0059	28.0434	75
80	.2051	39.7445	1166.7868	4.8754	193.7720	0.0252	0.0052	29.3572	80
85	.1858	40.7113	1246.0241	5.3829	219.1439	0.0246	0.0046	30.6064	85
90	.1683	41.5869	1322.1701	5.9431	247.1567	0.0240	0.0040	31.7929	90
95	.1524	42.3800	1395.1033	6.5617	278.0850	0.0236	0.0036	32.9189	95
100	.1380	43.0984	1464.7527	7.2446	312.2323	0.0232	0.0032	33.9863	100

I = 3.00 %

N	(P/F)	(P/A)	(P/G)	(F/P)	(F/A)	(A/P)	(A/F)	(A/G)	N
1	.9709	0.9709	-0.0000	1.0300	1.0000	1.0300	1.0000	-0.0000	1
2	.9426	1.9135	0.9426	1.0609	2.0300	0.5226	0.4926	0.4926	2
3	.9151	2.8286	2.7729	1.0927	3.0909	0.3535	0.3235	0.9803	3
4	.8885	3.7171	5.4383	1.1255	4.1836	0.2690	0.2390	1.4631	4
5	.8626	4.5797	8.8888	1.1593	5.3091	0.2184	0.1884	1.9409	5
6	.8375	5.4172	13.0762	1.1941	6.4684	0.1846	0.1546	2.4138	6
7	.8131	6.2303	17.9547	1.2299	7.6625	0.1605	0.1305	2.8819	7
8	.7894	7.0197	23.4806	1.2668	8.8923	0.1425	0.1125	3.3450	8
9	.7664	7.7861	29.6119	1.3048	10.1591	0.1284	0.0984	3.8032	9
10	.7441	8.5302	36.3088	1.3439	11.4639	0.1172	0.0872	4.2565	10
11	.7224	9.2526	43.5330	1.3842	12.8078	0.1081	0.0781	4.7049	11
12	.7014	9.9540	51.2482	1.4258	14.1920	0.1005	0.0705	5.1485	12
13	.6810	10.6350	59.4196	1.4685	15.6178	0.0940	0.0640	5.5872	13
14	.6611	11.2961	68.0141	1.5126	17.0863	0.0885	0.0585	6.0210	14
15	.6419	11.9379	77.0002	1.5580	18.5989	0.0838	0.0538	6.4500	15
16	.6232	12.5611	86.3477	1.6047	20.1569	0.0796	0.0496	6.8742	16
17	.6050	13.1661	96.0280	1.6528	21.7616	0.0760	0.0460	7.2936	17
18	.5874	13.7535	106.0137	1.7024	23.4144	0.0727	0.0427	7.7081	18
19	.5703	14.3238	116.2788	1.7535	25.1169	0.0698	0.0398	8.1179	19
20	.5537	14.8775	126.7987	1.8061	26.8704	0.0672	0.0372	8.5229	20
21	.5375	15.4150	137.5496	1.8603	28.6765	0.0649	0.0349	8.9231	21
22	.5219	15.9369	148.5094	1.9161	30.5368	0.0627	0.0327	9.3186	22
23	.5067	16.4436	159.6566	1.9736	32.4529	0.0608	0.0308	9.7093	23
24	.4919	16.9355	170.9711	2.0328	34.4265	0.0590	0.0290	10.0954	24
25	.4776	17.4131	182.4336	2.0938	36.4593	0.0574	0.0274	10.4768	25
26	.4637	17.8768	194.0260	2.1566	38.5530	0.0559	0.0259	10.8535	26
27	.4502	18.3270	205.7309	2.2213	40.7096	0.0546	0.0246	11.2255	27
28	.4371	18.7641	217.5320	2.2879	42.9309	0.0533	0.0233	11.5930	28
29	.4243	19.1885	229.4137	2.3566	45.2189	0.0521	0.0221	11.9558	29
30	.4120	19.6004	241.3613	2.4273	47.5754	0.0510	0.0210	12.3141	30
31	.4000	20.0004	253.3609	2.5001	50.0027	0.0500	0.0200	12.6678	31
32	.3883	20.3888	265.3993	2.5751	52.5028	0.0490	0.0190	13.0169	32
33	.3770	20.7658	277.4642	2.6523	55.0778	0.0482	0.0182	13.3616	33
34	.3660	21.1318	289.5437	2.7319	57.7302	0.0473	0.0173	13.7018	34
35	.3554	21.4872	301.6267	2.8139	60.4621	0.0465	0.0165	14.0375	35
36	.3450	21.8323	313.7028	2.8983	63.2759	0.0458	0.0158	14.3688	36
37	.3350	22.1672	325.7622	2.9852	66.1742	0.0451	0.0151	14.6957	37
38	.3252	22.4925	337.7956	3.0748	69.1594	0.0445	0.0145	15.0182	38
39	.3158	22.8082	349.7942	3.1670	72.2342	0.0438	0.0138	15.3363	39
40	.3066	23.1148	361.7499	3.2620	75.4013	0.0433	0.0133	15.6502	40
41	.2976	23.4124	373.6551	3.3599	78.6633	0.0427	0.0127	15.9597	41
42	.2890	23.7014	385.5024	3.4607	82.0232	0.0422	0.0122	16.2650	42
43	.2805	23.9819	397.2852	3.5645	85.4839	0.0417	0.0117	16.5660	43
44	.2724	24.2543	408.9972	3.6715	89.0484	0.0412	0.0112	16.8629	44
45	.2644	24.5187	420.6325	3.7816	92.7199	0.0408	0.0108	17.1556	45
46	.2567	24.7754	432.1856	3.8950	96.5015	0.0404	0.0104	17.4441	46
47	.2493	25.0247	443.6515	4.0119	100.3965	0.0400	0.0100	17.7285	47
48	.2420	25.2667	455.0255	4.1323	104.4084	0.0396	0.0096	18.0089	48
49	.2350	25.5017	466.3031	4.2562	108.5406	0.0392	0.0092	18.2852	49
50	.2281	25.7298	477.4803	4.3839	112.7969	0.0389	0.0089	18.5575	50
51	.2215	25.9512	488.5535	4.5154	117.1808	0.0385	0.0085	18.8258	51
52	.2150	26.1662	499.5191	4.6509	121.6962	0.0382	0.0082	19.0902	52
53	.2088	26.3750	510.3742	4.7904	126.3471	0.0379	0.0079	19.3507	53
54	.2027	26.5777	521.1157	4.9341	131.1375	0.0376	0.0076	19.6073	54
55	.1968	26.7744	531.7411	5.0821	136.0716	0.0373	0.0073	19.8600	55
60	.1697	27.6756	583.0526	5.8916	163.0534	0.0361	0.0061	21.0674	60
65	.1464	28.4529	631.2010	6.8300	194.3328	0.0351	0.0051	22.1841	65
70	.1263	29.1234	676.0869	7.9178	230.5941	0.0343	0.0043	23.2145	70
75	.1089	29.7018	717.6978	9.1789	272.6309	0.0337	0.0037	24.1634	75
80	.0940	30.2008	756.0865	10.6409	321.3630	0.0331	0.0031	25.0353	80
85	.0811	30.6312	791.3529	12.3357	377.8570	0.0326	0.0026	25.8349	85
90	.0699	31.0024	823.6302	14.3005	443.3489	0.0323	0.0023	26.5667	90
95	.0603	31.3227	853.0742	16.5782	519.2720	0.0319	0.0019	27.2351	95
100	.0520	31.5989	879.8540	19.2186	607.2877	0.0316	0.0016	27.8444	100

I = 4.00 %

N	(P/F)	(P/A)	(P/G)	(F/P)	(F/A)	(A/P)	(A/F)	(A/G)	N
1	.9615	0.9615	-0.0000	1.0400	1.0000	1.0400	1.0000	-0.0000	1
2	.9246	1.8861	0.9246	1.0816	2.0400	0.5302	0.4902	0.4902	2
3	.8890	2.7751	2.7025	1.1249	3.1216	0.3603	0.3203	0.9739	3
4	.8548	3.6299	5.2670	1.1699	4.2465	0.2755	0.2355	1.4510	4
5	.8219	4.4518	8.5547	1.2167	5.4163	0.2246	0.1846	1.9216	5
6	.7903	5.2421	12.5062	1.2653	6.6330	0.1908	0.1508	2.3857	6
7	.7599	6.0021	17.0657	1.3159	7.8983	0.1666	0.1266	2.8433	7
8	.7307	6.7327	22.1806	1.3686	9.2142	0.1485	0.1085	3.2944	8
9	.7026	7.4353	27.8013	1.4233	10.5828	0.1345	0.0945	3.7391	9
10	.6756	8.1109	33.8814	1.4802	12.0061	0.1233	0.0833	4.1773	10
11	.6496	8.7605	40.3772	1.5395	13.4864	0.1141	0.0741	4.6090	11
12	.6246	9.3851	47.2477	1.6010	15.0258	0.1066	0.0666	5.0343	12
13	.6006	9.9856	54.4546	1.6651	16.6268	0.1001	0.0601	5.4533	13
14	.5775	10.5631	61.9618	1.7317	18.2919	0.0947	0.0547	5.8659	14
15	.5553	11.1184	69.7355	1.8009	20.0236	0.0899	0.0499	6.2721	15
16	.5339	11.6523	77.7441	1.8730	21.8245	0.0858	0.0458	6.6720	16
17	.5134	12.1657	85.9581	1.9479	23.6975	0.0822	0.0422	7.0656	17
18	.4936	12.6593	94.3498	2.0258	25.6454	0.0790	0.0390	7.4530	18
19	.4746	13.1339	102.8933	2.1068	27.6712	0.0761	0.0361	7.8342	19
20	.4564	13.5903	111.5647	2.1911	29.7781	0.0736	0.0336	8.2091	20
21	.4388	14.0292	120.3414	2.2788	31.9692	0.0713	0.0313	8.5779	21
22	.4220	14.4511	129.2024	2.3699	34.2480	0.0692	0.0292	8.9407	22
23	.4057	14.8568	138.1284	2.4647	36.6179	0.0673	0.0273	9.2973	23
24	.3901	15.2470	147.1012	2.5633	39.0826	0.0656	0.0256	9.6479	24
25	.3751	15.6221	156.1040	2.6658	41.6459	0.0640	0.0240	9.9925	25
26	.3607	15.9828	165.1212	2.7725	44.3117	0.0626	0.0226	10.3312	26
27	.3468	16.3296	174.1385	2.8834	47.0842	0.0612	0.0212	10.6640	27
28	.3335	16.6631	183.1424	2.9987	49.9676	0.0600	0.0200	10.9909	28
29	.3207	16.9837	192.1206	3.1187	52.9663	0.0589	0.0189	11.3120	29
30	.3083	17.2920	201.0618	3.2434	56.0849	0.0578	0.0178	11.6274	30
31	.2965	17.5885	209.9556	3.3731	59.3283	0.0569	0.0169	11.9371	31
32	.2851	17.8736	218.7924	3.5081	62.7015	0.0559	0.0159	12.2411	32
33	.2741	18.1476	227.5634	3.6484	66.2095	0.0551	0.0151	12.5396	33
34	.2636	18.4112	236.2607	3.7943	69.8579	0.0543	0.0143	12.8324	34
35	.2534	18.6646	244.8768	3.9461	73.6522	0.0536	0.0136	13.1198	35
36	.2437	18.9083	253.4052	4.1039	77.5983	0.0529	0.0129	13.4018	36
37	.2343	19.1426	261.8399	4.2681	81.7022	0.0522	0.0122	13.6784	37
38	.2253	19.3679	270.1754	4.4388	85.9703	0.0516	0.0116	13.9497	38
39	.2166	19.5845	278.4070	4.6164	90.4091	0.0511	0.0111	14.2157	39
40	.2083	19.7928	286.5303	4.8010	95.0255	0.0505	0.0105	14.4765	40
41	.2003	19.9931	294.5414	4.9931	99.8265	0.0500	0.0100	14.7322	41
42	.1926	20.1856	302.4370	5.1928	104.8196	0.0495	0.0095	14.9828	42
43	.1852	20.3708	310.2141	5.4005	110.0124	0.0491	0.0091	15.2284	43
44	.1780	20.5488	317.8700	5.6165	115.4129	0.0487	0.0087	15.4690	44
45	.1712	20.7200	325.4028	5.8412	121.0294	0.0483	0.0083	15.7047	45
46	.1646	20.8847	332.8104	6.0748	126.8706	0.0479	0.0079	15.9356	46
47	.1583	21.0429	340.0914	6.3178	132.9454	0.0475	0.0075	16.1618	47
48	.1522	21.1951	347.2446	6.5705	139.2632	0.0472	0.0072	16.3832	48
49	.1463	21.3415	354.2689	6.8333	145.8337	0.0469	0.0069	16.6000	49
50	.1407	21.4822	361.1638	7.1067	152.6671	0.0466	0.0066	16.8122	50
51	.1353	21.6175	367.9289	7.3910	159.7738	0.0463	0.0063	17.0200	51
52	.1301	21.7476	374.5638	7.6866	167.1647	0.0460	0.0060	17.2232	52
53	.1251	21.8727	381.0686	7.9941	174.8513	0.0457	0.0057	17.4221	53
54	.1203	21.9930	387.4436	8.3138	182.8454	0.0455	0.0055	17.6167	54
55	.1157	22.1086	393.6890	8.6464	191.1592	0.0452	0.0052	17.8070	55
60	.0951	22.6235	422.9966	10.5196	237.9907	0.0442	0.0042	18.6972	60
65	.0781	23.0467	449.2014	12.7987	294.9684	0.0434	0.0034	19.4909	65
70	.0642	23.3945	472.4789	15.5716	364.2905	0.0427	0.0027	20.1961	70
75	.0528	23.6804	493.0408	18.9453	448.6314	0.0422	0.0022	20.8206	75
80	.0434	23.9154	511.1161	23.0498	551.2450	0.0418	0.0018	21.3718	80
85	.0357	24.1085	526.9384	28.0436	676.0901	0.0415	0.0015	21.8569	85
90	.0293	24.2673	540.7369	34.1193	827.9833	0.0412	0.0012	22.2826	90
95	.0241	24.3978	552.7307	41.5114	1012.7846	0.0410	0.0010	22.6550	95
100	.0198	24.5050	563.1249	50.5049	1237.6237	0.0408	0.0008	22.9800	100

ECON

I = 5.00 %

N	(P/F)	(P/A)	(P/G)	(F/P)	(F/A)	(A/P)	(A/F)	(A/G)	N
1	.9524	0.9524	-0.0000	1.0500	1.0000	1.0500	1.0000	-0.0000	1
2	.9070	1.8594	0.9070	1.1025	2.0500	0.5378	0.4878	0.4878	2
3	.8638	2.7232	2.6347	1.1576	3.1525	0.3672	0.3172	0.9675	3
4	.8227	3.5460	5.1028	1.2155	4.3101	0.2820	0.2320	1.4391	4
5	.7835	4.3295	8.2369	1.2763	5.5256	0.2310	0.1810	1.9025	5
6	.7462	5.0757	11.9680	1.3401	6.8019	0.1970	0.1470	2.3579	6
7	.7107	5.7864	16.2321	1.4071	8.1420	0.1728	0.1228	2.8052	7
8	.6768	6.4632	20.9700	1.4775	9.5491	0.1547	0.1047	3.2445	8
9	.6446	7.1078	26.1268	1.5513	11.0266	0.1407	0.0907	3.6758	9
10	.6139	7.7217	31.6520	1.6289	12.5779	0.1295	0.0795	4.0991	10
11	.5847	8.3064	37.4988	1.7103	14.2068	0.1204	0.0704	4.5144	11
12	.5568	8.8633	43.6241	1.7959	15.9171	0.1128	0.0628	4.9219	12
13	.5303	9.3936	49.9879	1.8856	17.7130	0.1065	0.0565	5.3215	13
14	.5051	9.8986	56.5538	1.9799	19.5986	0.1010	0.0510	5.7133	14
15	.4810	10.3797	63.2880	2.0789	21.5786	0.0963	0.0463	6.0973	15
16	.4581	10.8378	70.1597	2.1829	23.6575	0.0923	0.0423	6.4736	16
17	.4363	11.2741	77.1405	2.2920	25.8404	0.0887	0.0387	6.8423	17
18	.4155	11.6896	84.2043	2.4066	28.1324	0.0855	0.0355	7.2034	18
19	.3957	12.0853	91.3275	2.5270	30.5390	0.0827	0.0327	7.5569	19
20	.3769	12.4622	98.4884	2.6533	33.0660	0.0802	0.0302	7.9030	20
21	.3589	12.8212	105.6673	2.7860	35.7193	0.0780	0.0280	8.2416	21
22	.3418	13.1630	112.8461	2.9253	38.5052	0.0760	0.0260	8.5730	22
23	.3256	13.4886	120.0087	3.0715	41.4305	0.0741	0.0241	8.8971	23
24	.3101	13.7986	127.1402	3.2251	44.5020	0.0725	0.0225	9.2140	24
25	.2953	14.0939	134.2275	3.3864	47.7271	0.0710	0.0210	9.5238	25
26	.2812	14.3752	141.2585	3.5557	51.1135	0.0696	0.0196	9.8266	26
27	.2678	14.6430	148.2226	3.7335	54.6691	0.0683	0.0183	10.1224	27
28	.2551	14.8981	155.1101	3.9201	58.4026	0.0671	0.0171	10.4114	28
29	.2429	15.1411	161.9126	4.1161	62.3227	0.0660	0.0160	10.6936	29
30	.2314	15.3725	168.6226	4.3219	66.4388	0.0651	0.0151	10.9691	30
31	.2204	15.5928	175.2333	4.5380	70.7608	0.0641	0.0141	11.2381	31
32	.2099	15.8027	181.7392	4.7649	75.2988	0.0633	0.0133	11.5005	32
33	.1999	16.0025	188.1351	5.0032	80.0638	0.0625	0.0125	11.7566	33
34	.1904	16.1929	194.4168	5.2533	85.0670	0.0618	0.0118	12.0063	34
35	.1813	16.3742	200.5807	5.5160	90.3203	0.0611	0.0111	12.2498	35
36	.1727	16.5469	206.6237	5.7918	95.8363	0.0604	0.0104	12.4872	36
37	.1644	16.7113	212.5434	6.0814	101.6281	0.0598	0.0098	12.7186	37
38	.1566	16.8679	218.3378	6.3855	107.7095	0.0593	0.0093	12.9440	38
39	.1491	17.0170	224.0054	6.7048	114.0950	0.0588	0.0088	13.1636	39
40	.1420	17.1591	229.5452	7.0400	120.7998	0.0583	0.0083	13.3775	40
41	.1353	17.2944	234.9564	7.3920	127.8398	0.0578	0.0078	13.5857	41
42	.1288	17.4232	240.2389	7.7616	135.2318	0.0574	0.0074	13.7884	42
43	.1227	17.5459	245.3925	8.1497	142.9933	0.0570	0.0070	13.9857	43
44	.1169	17.6628	250.4175	8.5572	151.1430	0.0566	0.0066	14.1777	44
45	.1113	17.7741	255.3145	8.9850	159.7002	0.0563	0.0063	14.3644	45
46	.1060	17.8801	260.0844	9.4343	168.6852	0.0559	0.0059	14.5461	46
47	.1009	17.9810	264.7281	9.9060	178.1194	0.0556	0.0056	14.7226	47
48	.0961	18.0772	269.2467	10.4013	188.0254	0.0553	0.0053	14.8943	48
49	.0916	18.1687	273.6418	10.9213	198.4267	0.0550	0.0050	15.0611	49
50	.0872	18.2559	277.9148	11.4674	209.3480	0.0548	0.0048	15.2233	50
51	.0831	18.3390	282.0673	12.0408	220.8154	0.0545	0.0045	15.3808	51
52	.0791	18.4181	286.1013	12.6428	232.8562	0.0543	0.0043	15.5337	52
53	.0753	18.4934	290.0184	13.2749	245.4990	0.0541	0.0041	15.6823	53
54	.0717	18.5651	293.8208	13.9387	258.7739	0.0539	0.0039	15.8265	54
55	.0683	18.6335	297.5104	14.6356	272.7126	0.0537	0.0037	15.9664	55
60	.0535	18.9293	314.3432	18.6792	353.5837	0.0528	0.0028	16.6062	60
65	.0419	19.1611	328.6910	23.8399	456.7980	0.0522	0.0022	17.1541	65
70	.0329	19.3427	340.8409	30.4264	588.5285	0.0517	0.0017	17.6212	70
75	.0258	19.4850	351.0721	38.8327	756.6537	0.0513	0.0013	18.0176	75
80	.0202	19.5965	359.6460	49.5614	971.2288	0.0510	0.0010	18.3526	80
85	.0158	19.6838	366.8007	63.2544	1245.0871	0.0508	0.0008	18.6346	85
90	.0124	19.7523	372.7488	80.7304	1594.6073	0.0506	0.0006	18.8712	90
95	.0097	19.8059	377.6774	103.0347	2040.6935	0.0505	0.0005	19.0689	95
100	.0076	19.8479	381.7492	131.5013	2610.0252	0.0504	0.0004	19.2337	100

I = 6.00 %

N	(P/F)	(P/A)	(P/G)	(F/P)	(F/A)	(A/P)	(A/F)	(A/G)	N
1	.9434	0.9434	-0.0000	1.0600	1.0000	1.0600	1.0000	-0.0000	1
2	.8900	1.8334	0.8900	1.1236	2.0600	0.5454	0.4854	0.4854	2
3	.8396	2.6730	2.5692	1.1910	3.1836	0.3741	0.3141	0.9612	3
4	.7921	3.4651	4.9455	1.2625	4.3746	0.2886	0.2286	1.4272	4
5	.7473	4.2124	7.9345	1.3382	5.6371	0.2374	0.1774	1.8836	5
6	.7050	4.9173	11.4594	1.4185	6.9753	0.2034	0.1434	2.3304	6
7	.6651	5.5824	15.4497	1.5036	8.3938	0.1791	0.1191	2.7676	7
8	.6274	6.2098	19.8416	1.5938	9.8975	0.1610	0.1010	3.1952	8
9	.5919	6.8017	24.5768	1.6895	11.4913	0.1470	0.0870	3.6133	9
10	.5584	7.3601	29.6023	1.7908	13.1808	0.1359	0.0759	4.0220	10
11	.5268	7.8869	34.8702	1.8983	14.9716	0.1268	0.0668	4.4213	11
12	.4970	8.3838	40.3369	2.0122	16.8699	0.1193	0.0593	4.8113	12
13	.4688	8.8527	45.9629	2.1329	18.8821	0.1130	0.0530	5.1920	13
14	.4423	9.2950	51.7128	2.2609	21.0151	0.1076	0.0476	5.5635	14
15	.4173	9.7122	57.5546	2.3966	23.2760	0.1030	0.0430	5.9260	15
16	.3936	10.1059	63.4592	2.5404	25.6725	0.0990	0.0390	6.2794	16
17	.3714	10.4773	69.4011	2.6928	28.2129	0.0954	0.0354	6.6240	17
18	.3503	10.8276	75.3569	2.8543	30.9057	0.0924	0.0324	6.9597	18
19	.3305	11.1581	81.3062	3.0256	33.7600	0.0896	0.0296	7.2867	19
20	.3118	11.4699	87.2304	3.2071	36.7856	0.0872	0.0272	7.6051	20
21	.2942	11.7641	93.1136	3.3996	39.9927	0.0850	0.0250	7.9151	21
22	.2775	12.0416	98.9412	3.6035	43.3923	0.0830	0.0230	8.2166	22
23	.2618	12.3034	104.7007	3.8197	46.9958	0.0813	0.0213	8.5099	23
24	.2470	12.5504	110.3812	4.0489	50.8156	0.0797	0.0197	8.7951	24
25	.2330	12.7834	115.9732	4.2919	54.8645	0.0782	0.0182	9.0722	25
26	.2198	13.0032	121.4684	4.5494	59.1564	0.0769	0.0169	9.3414	26
27	.2074	13.2105	126.8600	4.8223	63.7058	0.0757	0.0157	9.6029	27
28	.1956	13.4062	132.1420	5.1117	68.5281	0.0746	0.0146	9.8568	28
29	.1846	13.5907	137.3096	5.4184	73.6398	0.0736	0.0136	10.1032	29
30	.1741	13.7648	142.3588	5.7435	79.0582	0.0726	0.0126	10.3422	30
31	.1643	13.9291	147.2864	6.0881	84.8017	0.0718	0.0118	10.5740	31
32	.1550	14.0840	152.0901	6.4534	90.8898	0.0710	0.0110	10.7988	32
33	.1462	14.2302	156.7681	6.8406	97.3432	0.0703	0.0103	11.0166	33
34	.1379	14.3681	161.3192	7.2510	104.1838	0.0696	0.0096	11.2276	34
35	.1301	14.4982	165.7427	7.6861	111.4348	0.0690	0.0090	11.4319	35
36	.1227	14.6210	170.0387	8.1473	119.1209	0.0684	0.0084	11.6298	36
37	.1158	14.7368	174.2072	8.6361	127.2681	0.0679	0.0079	11.8213	37
38	.1092	14.8460	178.2490	9.1543	135.9042	0.0674	0.0074	12.0065	38
39	.1031	14.9491	182.1652	9.7035	145.0585	0.0669	0.0069	12.1857	39
40	.0972	15.0463	185.9568	10.2857	154.7620	0.0665	0.0065	12.3590	40
41	.0917	15.1380	189.6256	10.9029	165.0477	0.0661	0.0061	12.5264	41
42	.0865	15.2245	193.1732	11.5570	175.9505	0.0657	0.0057	12.6883	42
43	.0816	15.3062	196.6017	12.2505	187.5076	0.0653	0.0053	12.8446	43
44	.0770	15.3832	199.9130	12.9855	199.7580	0.0650	0.0050	12.9956	44
45	.0727	15.4558	203.1096	13.7646	212.7435	0.0647	0.0047	13.1413	45
46	.0685	15.5244	206.1938	14.5905	226.5081	0.0644	0.0044	13.2819	46
47	.0647	15.5890	209.1681	15.4659	241.0986	0.0641	0.0041	13.4177	47
48	.0610	15.6500	212.0351	16.3939	256.5645	0.0639	0.0039	13.5485	48
49	.0575	15.7076	214.7972	17.3775	272.9584	0.0637	0.0037	13.6748	49
50	.0543	15.7619	217.4574	18.4202	290.3359	0.0634	0.0034	13.7964	50
51	.0512	15.8131	220.0181	19.5254	308.7561	0.0632	0.0032	13.9137	51
52	.0483	15.8614	222.4823	20.6969	328.2814	0.0630	0.0030	14.0267	52
53	.0456	15.9070	224.8525	21.9387	348.9783	0.0629	0.0029	14.1355	53
54	.0430	15.9500	227.1316	23.2550	370.9170	0.0627	0.0027	14.2402	54
55	.0406	15.9905	229.3222	24.6503	394.1720	0.0625	0.0025	14.3411	55
60	.0303	16.1614	239.0428	32.9877	533.1282	0.0619	0.0019	14.7909	60
65	.0227	16.2891	246.9450	44.1450	719.0829	0.0614	0.0014	15.1601	65
70	.0169	16.3845	253.3271	59.0759	967.9322	0.0610	0.0010	15.4613	70
75	.0126	16.4558	258.4527	79.0569	1300.9487	0.0608	0.0008	15.7058	75
80	.0095	16.5091	262.5493	105.7960	1746.5999	0.0606	0.0006	15.9033	80
85	.0071	16.5489	265.8096	141.5789	2342.9817	0.0604	0.0004	16.0620	85
90	.0053	16.5787	268.3946	189.4645	3141.0752	0.0603	0.0003	16.1891	90
95	.0039	16.6009	270.4375	253.5463	4209.1042	0.0602	0.0002	16.2905	95
100	.0029	16.6175	272.0471	339.3021	5638.3681	0.0602	0.0002	16.3711	100

I = 7.00 %

N	(P/F)	(P/A)	(P/G)	(F/P)	(F/A)	(A/P)	(A/F)	(A/G)	N
1	.9346	0.9346	-0.0000	1.0700	1.0000	1.0700	1.0000	-0.0000	1
2	.8734	1.8080	0.8734	1.1449	2.0700	0.5531	0.4831	0.4831	2
3	.8163	2.6243	2.5060	1.2250	3.2149	0.3811	0.3111	0.9549	3
4	.7629	3.3872	4.7947	1.3108	4.4399	0.2952	0.2252	1.4155	4
5	.7130	4.1002	7.6467	1.4026	5.7507	0.2439	0.1739	1.8650	5
6	.6663	4.7665	10.9784	1.5007	7.1533	0.2098	0.1398	2.3032	6
7	.6227	5.3893	14.7149	1.6058	8.6540	0.1856	0.1156	2.7304	7
8	.5820	5.9713	18.7889	1.7182	10.2598	0.1675	0.0975	3.1465	8
9	.5439	6.5152	23.1404	1.8385	11.9780	0.1535	0.0835	3.5517	9
10	.5083	7.0236	27.7156	1.9672	13.8164	0.1424	0.0724	3.9461	10
11	.4751	7.4987	32.4665	2.1049	15.7836	0.1334	0.0634	4.3296	11
12	.4440	7.9427	37.3506	2.2522	17.8885	0.1259	0.0559	4.7025	12
13	.4150	8.3577	42.3302	2.4098	20.1406	0.1197	0.0497	5.0648	13
14	.3878	8.7455	47.3718	2.5785	22.5505	0.1143	0.0443	5.4167	14
15	.3624	9.1079	52.4461	2.7590	25.1290	0.1098	0.0398	5.7583	15
16	.3387	9.4466	57.5271	2.9522	27.8881	0.1059	0.0359	6.0897	16
17	.3166	9.7632	62.5923	3.1588	30.8402	0.1024	0.0324	6.4110	17
18	.2959	10.0591	67.6219	3.3799	33.9990	0.0994	0.0294	6.7225	18
19	.2765	10.3356	72.5991	3.6165	37.3790	0.0968	0.0268	7.0242	19
20	.2584	10.5940	77.5091	3.8697	40.9955	0.0944	0.0244	7.3163	20
21	.2415	10.8355	82.3393	4.1406	44.8652	0.0923	0.0223	7.5990	21
22	.2257	11.0612	87.0793	4.4304	49.0057	0.0904	0.0204	7.8725	22
23	.2109	11.2722	91.7201	4.7405	53.4361	0.0887	0.0187	8.1369	23
24	.1971	11.4693	96.2545	5.0724	58.1767	0.0872	0.0172	8.3923	24
25	.1842	11.6536	100.6765	5.4274	63.2490	0.0858	0.0158	8.6391	25
26	.1722	11.8258	104.9814	5.8074	68.6765	0.0846	0.0146	8.8773	26
27	.1609	11.9867	109.1656	6.2139	74.4838	0.0834	0.0134	9.1072	27
28	.1504	12.1371	113.2264	6.6488	80.6977	0.0824	0.0124	9.3289	28
29	.1406	12.2777	117.1622	7.1143	87.3465	0.0814	0.0114	9.5427	29
30	.1314	12.4090	120.9718	7.6123	94.4608	0.0806	0.0106	9.7487	30
31	.1228	12.5318	124.6550	8.1451	102.0730	0.0798	0.0098	9.9471	31
32	.1147	12.6466	128.2120	8.7153	110.2182	0.0791	0.0091	10.1381	32
33	.1072	12.7538	131.6435	9.3253	118.9334	0.0784	0.0084	10.3219	33
34	.1002	12.8540	134.9507	9.9781	128.2588	0.0778	0.0078	10.4987	34
35	.0937	12.9477	138.1353	10.6766	138.2369	0.0772	0.0072	10.6687	35
36	.0875	13.0352	141.1990	11.4239	148.9135	0.0767	0.0067	10.8321	36
37	.0818	13.1170	144.1441	12.2236	160.3374	0.0762	0.0062	10.9891	37
38	.0765	13.1935	146.9730	13.0793	172.5610	0.0758	0.0058	11.1398	38
39	.0715	13.2649	149.6883	13.9948	185.6403	0.0754	0.0054	11.2845	39
40	.0668	13.3317	152.2928	14.9745	199.6351	0.0750	0.0050	11.4233	40
41	.0624	13.3941	154.7892	16.0227	214.6096	0.0747	0.0047	11.5565	41
42	.0583	13.4524	157.1807	17.1443	230.6322	0.0743	0.0043	11.6842	42
43	.0545	13.5070	159.4702	18.3444	247.7765	0.0740	0.0040	11.8065	43
44	.0509	13.5579	161.6609	19.6285	266.1209	0.0738	0.0038	11.9237	44
45	.0476	13.6055	163.7559	21.0025	285.7493	0.0735	0.0035	12.0360	45
46	.0445	13.6500	165.7584	22.4726	306.7518	0.0733	0.0033	12.1435	46
47	.0416	13.6916	167.6714	24.0457	329.2244	0.0730	0.0030	12.2463	47
48	.0389	13.7305	169.4981	25.7289	353.2701	0.0728	0.0028	12.3447	48
49	.0363	13.7668	171.2417	27.5299	378.9990	0.0726	0.0026	12.4387	49
50	.0339	13.8007	172.9051	29.4570	406.5289	0.0725	0.0025	12.5287	50
51	.0317	13.8325	174.4915	31.5190	435.9860	0.0723	0.0023	12.6146	51
52	.0297	13.8621	176.0037	33.7253	467.5050	0.0721	0.0021	12.6967	52
53	.0277	13.8898	177.4447	36.0861	501.2303	0.0720	0.0020	12.7751	53
54	.0259	13.9157	178.8173	38.6122	537.3164	0.0719	0.0019	12.8500	54
55	.0242	13.9399	180.1243	41.3150	575.9286	0.0717	0.0017	12.9215	55
60	.0173	14.0392	185.7677	57.9464	813.5204	0.0712	0.0012	13.2321	60
65	.0123	14.1099	190.1452	81.2729	1146.7552	0.0709	0.0009	13.4760	65
70	.0088	14.1604	193.5185	113.9894	1614.1342	0.0706	0.0006	13.6662	70
75	.0063	14.1964	196.1035	159.8760	2269.6574	0.0704	0.0004	13.8136	75
80	.0045	14.2220	198.0748	224.2344	3189.0627	0.0703	0.0003	13.9273	80
85	.0032	14.2403	199.5717	314.5003	4478.5761	0.0702	0.0002	14.0146	85
90	.0023	14.2533	200.7042	441.1030	6287.1854	0.0702	0.0002	14.0812	90
95	.0016	14.2626	201.5581	618.6697	8823.8535	0.0701	0.0001	14.1319	95
100	.0012	14.2693	202.2001	867.7163	12381.6618	0.0701	0.0001	14.1703	100

I = 8.00 %

N	(P/F)	(P/A)	(P/G)	(F/P)	(F/A)	(A/P)	(A/F)	(A/G)	N
1	.9259	0.9259	-0.0000	1.0800	1.0000	1.0800	1.0000	-0.0000	1
2	.8573	1.7833	0.8573	1.1664	2.0800	0.5608	0.4808	0.4808	2
3	.7938	2.5771	2.4450	1.2597	3.2464	0.3880	0.3080	0.9487	3
4	.7350	3.3121	4.6501	1.3605	4.5061	0.3019	0.2219	1.4040	4
5	.6806	3.9927	7.3724	1.4693	5.8666	0.2505	0.1705	1.8465	5
6	.6302	4.6229	10.5233	1.5869	7.3359	0.2163	0.1363	2.2763	6
7	.5835	5.2064	14.0242	1.7138	8.9228	0.1921	0.1121	2.6937	7
8	.5403	5.7466	17.8061	1.8509	10.6366	0.1740	0.0940	3.0985	8
9	.5002	6.2469	21.8081	1.9990	12.4876	0.1601	0.0801	3.4910	9
10	.4632	6.7101	25.9768	2.1589	14.4866	0.1490	0.0690	3.8713	10
11	.4289	7.1390	30.2657	2.3316	16.6455	0.1401	0.0601	4.2395	11
12	.3971	7.5361	34.6339	2.5182	18.9771	0.1327	0.0527	4.5957	12
13	.3677	7.9038	39.0463	2.7196	21.4953	0.1265	0.0465	4.9402	13
14	.3405	8.2442	43.4723	2.9372	24.2149	0.1213	0.0413	5.2731	14
15	.3152	8.5595	47.8857	3.1722	27.1521	0.1168	0.0368	5.5945	15
16	.2919	8.8514	52.2640	3.4259	30.3243	0.1130	0.0330	5.9046	16
17	.2703	9.1216	56.5883	3.7000	33.7502	0.1096	0.0296	6.2037	17
18	.2502	9.3719	60.8426	3.9960	37.4502	0.1067	0.0267	6.4920	18
19	.2317	9.6036	65.0134	4.3157	41.4463	0.1041	0.0241	6.7697	19
20	.2145	9.8181	69.0898	4.6610	45.7620	0.1019	0.0219	7.0369	20
21	.1987	10.0168	73.0629	5.0338	50.4229	0.0998	0.0198	7.2940	21
22	.1839	10.2007	76.9257	5.4365	55.4568	0.0980	0.0180	7.5412	22
23	.1703	10.3711	80.6726	5.8715	60.8933	0.0964	0.0164	7.7786	23
24	.1577	10.5288	84.2997	6.3412	66.7648	0.0950	0.0150	8.0066	24
25	.1460	10.6748	87.8041	6.8485	73.1059	0.0937	0.0137	8.2254	25
26	.1352	10.8100	91.1842	7.3964	79.9544	0.0925	0.0125	8.4352	26
27	.1252	10.9352	94.4390	7.9881	87.3508	0.0914	0.0114	8.6363	27
28	.1159	11.0511	97.5687	8.6271	95.3388	0.0905	0.0105	8.8289	28
29	.1073	11.1584	100.5738	9.3173	103.9659	0.0896	0.0096	9.0133	29
30	.0994	11.2578	103.4558	10.0627	113.2832	0.0888	0.0088	9.1897	30
31	.0920	11.3498	106.2163	10.8677	123.3459	0.0881	0.0081	9.3584	31
32	.0852	11.4350	108.8575	11.7371	134.2135	0.0875	0.0075	9.5197	32
33	.0789	11.5139	111.3819	12.6760	145.9506	0.0869	0.0069	9.6737	33
34	.0730	11.5869	113.7924	13.6901	158.6267	0.0863	0.0063	9.8208	34
35	.0676	11.6546	116.0920	14.7853	172.3168	0.0858	0.0058	9.9611	35
36	.0626	11.7172	118.2839	15.9682	187.1021	0.0853	0.0053	10.0949	36
37	.0580	11.7752	120.3713	17.2456	203.0703	0.0849	0.0049	10.2225	37
38	.0537	11.8289	122.3579	18.6253	220.3159	0.0845	0.0045	10.3440	38
39	.0497	11.8786	124.2470	20.1153	238.9412	0.0842	0.0042	10.4597	39
40	.0460	11.9246	126.0422	21.7245	259.0565	0.0839	0.0039	10.5699	40
41	.0426	11.9672	127.7470	23.4625	280.7810	0.0836	0.0036	10.6747	41
42	.0395	12.0067	129.3651	25.3395	304.2435	0.0833	0.0033	10.7744	42
43	.0365	12.0432	130.8998	27.3666	329.5830	0.0830	0.0030	10.8692	43
44	.0338	12.0771	132.3547	29.5560	356.9496	0.0828	0.0028	10.9592	44
45	.0313	12.1084	133.7331	31.9204	386.5056	0.0826	0.0026	11.0447	45
46	.0290	12.1374	135.0384	34.4741	418.4261	0.0824	0.0024	11.1258	46
47	.0269	12.1643	136.2739	37.2320	452.9002	0.0822	0.0022	11.2028	47
48	.0249	12.1891	137.4428	40.2106	490.1322	0.0820	0.0020	11.2758	48
49	.0230	12.2122	138.5480	43.4274	530.3427	0.0819	0.0019	11.3451	49
50	.0213	12.2335	139.5928	46.9016	573.7702	0.0817	0.0017	11.4107	50
51	.0197	12.2532	140.5799	50.6537	620.6718	0.0816	0.0016	11.4729	51
52	.0183	12.2715	141.5121	54.7060	671.3255	0.0815	0.0015	11.5318	52
53	.0169	12.2884	142.3923	59.0825	726.0316	0.0814	0.0014	11.5875	53
54	.0157	12.3041	143.2229	63.8091	785.1141	0.0813	0.0013	11.6403	54
55	.0145	12.3186	144.0065	68.9139	848.9232	0.0812	0.0012	11.6902	55
60	.0099	12.3766	147.3000	101.2571	1253.2133	0.0808	0.0008	11.9015	60
65	.0067	12.4160	149.7387	148.7798	1847.2481	0.0805	0.0005	12.0602	65
70	.0046	12.4428	151.5326	218.6064	2720.0801	0.0804	0.0004	12.1783	70
75	.0031	12.4611	152.8448	321.2045	4002.5566	0.0802	0.0002	12.2658	75
80	.0021	12.4735	153.8001	471.9548	5886.9354	0.0802	0.0002	12.3301	80
85	.0014	12.4820	154.4925	693.4565	8655.7061	0.0801	0.0001	12.3772	85
90	.0010	12.4877	154.9925	1018.9151	12723.9386	0.0801	0.0001	12.4116	90
95	.0007	12.4917	155.3524	1497.1205	18701.5069	0.0801	0.0001	12.4365	95
100	.0005	12.4943	155.6107	2199.7613	27484.5157	0.0800	0.0000	12.4545	100

ECON

I = 9.00 %

N	(P/F)	(P/A)	(P/G)	(F/P)	(F/A)	(A/P)	(A/F)	(A/G)	N
1	.9174	0.9174	-0.0000	1.0900	1.0000	1.0900	1.0000	-0.0000	1
2	.8417	1.7591	0.8417	1.1881	2.0900	0.5685	0.4785	0.4785	2
3	.7722	2.5313	2.3860	1.2950	3.2781	0.3951	0.3051	0.9426	3
4	.7084	3.2397	4.5113	1.4116	4.5731	0.3087	0.2187	1.3925	4
5	.6499	3.8897	7.1110	1.5386	5.9847	0.2571	0.1671	1.8282	5
6	.5963	4.4859	10.0924	1.6771	7.5233	0.2229	0.1329	2.2498	6
7	.5470	5.0330	13.3746	1.8280	9.2004	0.1987	0.1087	2.6574	7
8	.5019	5.5348	16.8877	1.9926	11.0285	0.1807	0.0907	3.0512	8
9	.4604	5.9952	20.5711	2.1719	13.0210	0.1668	0.0768	3.4312	9
10	.4224	6.4177	24.3728	2.3674	15.1929	0.1558	0.0658	3.7978	10
11	.3875	6.8052	28.2481	2.5804	17.5603	0.1469	0.0569	4.1510	11
12	.3555	7.1607	32.1590	2.8127	20.1407	0.1397	0.0497	4.4910	12
13	.3262	7.4869	36.0731	3.0658	22.9534	0.1336	0.0436	4.8182	13
14	.2992	7.7862	39.9633	3.3417	26.0192	0.1284	0.0384	5.1326	14
15	.2745	8.0607	43.8069	3.6425	29.3609	0.1241	0.0341	5.4346	15
16	.2519	8.3126	47.5849	3.9703	33.0034	0.1203	0.0303	5.7245	16
17	.2311	8.5436	51.2821	4.3276	36.9737	0.1170	0.0270	6.0024	17
18	.2120	8.7556	54.8860	4.7171	41.3013	0.1142	0.0242	6.2687	18
19	.1945	8.9501	58.3868	5.1417	46.0185	0.1117	0.0217	6.5236	19
20	.1784	9.1285	61.7770	5.6044	51.1601	0.1095	0.0195	6.7674	20
21	.1637	9.2922	65.0509	6.1088	56.7645	0.1076	0.0176	7.0006	21
22	.1502	9.4424	68.2048	6.6586	62.8733	0.1059	0.0159	7.2232	22
23	.1378	9.5802	71.2359	7.2579	69.5319	0.1044	0.0144	7.4357	23
24	.1264	9.7066	74.1433	7.9111	76.7898	0.1030	0.0130	7.6384	24
25	.1160	9.8226	76.9265	8.6231	84.7009	0.1018	0.0118	7.8316	25
26	.1064	9.9290	79.5863	9.3992	93.3240	0.1007	0.0107	8.0156	26
27	.0976	10.0266	82.1241	10.2451	102.7231	0.0997	0.0097	8.1906	27
28	.0895	10.1161	84.5419	11.1671	112.9682	0.0989	0.0089	8.3571	28
29	.0822	10.1983	86.8422	12.1722	124.1354	0.0981	0.0081	8.5154	29
30	.0754	10.2737	89.0280	13.2677	136.3075	0.0973	0.0073	8.6657	30
31	.0691	10.3428	91.1024	14.4618	149.5752	0.0967	0.0067	8.8083	31
32	.0634	10.4062	93.0690	15.7633	164.0370	0.0961	0.0061	8.9436	32
33	.0582	10.4644	94.9314	17.1820	179.8003	0.0956	0.0056	9.0718	33
34	.0534	10.5178	96.6935	18.7284	196.9823	0.0951	0.0051	9.1933	34
35	.0490	10.5668	98.3590	20.4140	215.7108	0.0946	0.0046	9.3083	35
36	.0449	10.6118	99.9319	22.2512	236.1247	0.0942	0.0042	9.4171	36
37	.0412	10.6530	101.4162	24.2538	258.3759	0.0939	0.0039	9.5200	37
38	.0378	10.6908	102.8158	26.4367	282.6298	0.0935	0.0035	9.6172	38
39	.0347	10.7255	104.1345	28.8160	309.0665	0.0932	0.0032	9.7090	39
40	.0318	10.7574	105.3762	31.4094	337.8824	0.0930	0.0030	9.7957	40
41	.0292	10.7866	106.5445	34.2363	369.2919	0.0927	0.0027	9.8775	41
42	.0268	10.8134	107.6432	37.3175	403.5281	0.0925	0.0025	9.9546	42
43	.0246	10.8380	108.6758	40.6761	440.8457	0.0923	0.0023	10.0273	43
44	.0226	10.8605	109.6456	44.3370	481.5218	0.0921	0.0021	10.0958	44
45	.0207	10.8812	110.5561	48.3273	525.8587	0.0919	0.0019	10.1603	45
46	.0190	10.9002	111.4103	52.6767	574.1860	0.0917	0.0017	10.2210	46
47	.0174	10.9176	112.2115	57.4176	626.8628	0.0916	0.0016	10.2780	47
48	.0160	10.9336	112.9625	62.5852	684.2804	0.0915	0.0015	10.3317	48
49	.0147	10.9482	113.6661	68.2179	746.8656	0.0913	0.0013	10.3821	49
50	.0134	10.9617	114.3251	74.3575	815.0836	0.0912	0.0012	10.4295	50
51	.0123	10.9740	114.9420	81.0497	889.4411	0.0911	0.0011	10.4740	51
52	.0113	10.9853	115.5193	88.3442	970.4908	0.0910	0.0010	10.5158	52
53	.0104	10.9957	116.0593	96.2951	1058.8349	0.0909	0.0009	10.5549	53
54	.0095	11.0053	116.5642	104.9617	1155.1301	0.0909	0.0009	10.5917	54
55	.0087	11.0140	117.0362	114.4083	1260.0918	0.0908	0.0008	10.6261	55
60	.0057	11.0480	118.9683	176.0313	1944.7921	0.0905	0.0005	10.7683	60
65	.0037	11.0701	120.3344	270.8460	2998.2885	0.0903	0.0003	10.8702	65
70	.0024	11.0844	121.2942	416.7301	4619.2232	0.0902	0.0002	10.9427	70
75	.0016	11.0938	121.9646	641.1909	7113.2321	0.0901	0.0001	10.9940	75
80	.0010	11.0998	122.4306	986.5517	10950.5741	0.0901	0.0001	11.0299	80
85	.0007	11.1038	122.7533	1517.9320	16854.8003	0.0901	0.0001	11.0551	85
90	.0004	11.1064	122.9758	2335.5266	25939.1842	0.0900	0.0000	11.0726	90
95	.0003	11.1080	123.1287	3593.4971	39916.6350	0.0900	0.0000	11.0847	95
100	.0002	11.1091	123.2335	5529.0408	61422.6755	0.0900	0.0000	11.0930	100

I = 10.00 %

N	(P/F)	(P/A)	(P/G)	(F/P)	(F/A)	(A/P)	(A/F)	(A/G)	N
1	.9091	0.9091	−0.0000	1.1000	1.0000	1.1000	1.0000	−0.0000	1
2	.8264	1.7355	0.8264	1.2100	2.1000	0.5762	0.4762	0.4762	2
3	.7513	2.4869	2.3291	1.3310	3.3100	0.4021	0.3021	0.9366	3
4	.6830	3.1699	4.3781	1.4641	4.6410	0.3155	0.2155	1.3812	4
5	.6209	3.7908	6.8618	1.6105	6.1051	0.2638	0.1638	1.8101	5
6	.5645	4.3553	9.6842	1.7716	7.7156	0.2296	0.1296	2.2236	6
7	.5132	4.8684	12.7631	1.9487	9.4872	0.2054	0.1054	2.6216	7
8	.4665	5.3349	16.0287	2.1436	11.4359	0.1874	0.0874	3.0045	8
9	.4241	5.7590	19.4215	2.3579	13.5795	0.1736	0.0736	3.3724	9
10	.3855	6.1446	22.8913	2.5937	15.9374	0.1627	0.0627	3.7255	10
11	.3505	6.4951	26.3963	2.8531	18.5312	0.1540	0.0540	4.0641	11
12	.3186	6.8137	29.9012	3.1384	21.3843	0.1468	0.0468	4.3884	12
13	.2897	7.1034	33.3772	3.4523	24.5227	0.1408	0.0408	4.6988	13
14	.2633	7.3667	36.8005	3.7975	27.9750	0.1357	0.0357	4.9955	14
15	.2394	7.6061	40.1520	4.1772	31.7725	0.1315	0.0315	5.2789	15
16	.2176	7.8237	43.4164	4.5950	35.9497	0.1278	0.0278	5.5493	16
17	.1978	8.0216	46.5819	5.0545	40.5447	0.1247	0.0247	5.8071	17
18	.1799	8.2014	49.6395	5.5599	45.5992	0.1219	0.0219	6.0526	18
19	.1635	8.3649	52.5827	6.1159	51.1591	0.1195	0.0195	6.2861	19
20	.1486	8.5136	55.4069	6.7275	57.2750	0.1175	0.0175	6.5081	20
21	.1351	8.6487	58.1095	7.4002	64.0025	0.1156	0.0156	6.7189	21
22	.1228	8.7715	60.6893	8.1403	71.4027	0.1140	0.0140	6.9189	22
23	.1117	8.8832	63.1462	8.9543	79.5430	0.1126	0.0126	7.1085	23
24	.1015	8.9847	65.4813	9.8497	88.4973	0.1113	0.0113	7.2881	24
25	.0923	9.0770	67.6964	10.8347	98.3471	0.1102	0.0102	7.4580	25
26	.0839	9.1609	69.7940	11.9182	109.1818	0.1092	0.0092	7.6186	26
27	.0763	9.2372	71.7773	13.1100	121.0999	0.1083	0.0083	7.7704	27
28	.0693	9.3066	73.6495	14.4210	134.2099	0.1075	0.0075	7.9137	28
29	.0630	9.3696	75.4146	15.8631	148.6309	0.1067	0.0067	8.0489	29
30	.0573	9.4269	77.0766	17.4494	164.4940	0.1061	0.0061	8.1762	30
31	.0521	9.4790	78.6395	19.1943	181.9434	0.1055	0.0055	8.2962	31
32	.0474	9.5264	80.1078	21.1138	201.1378	0.1050	0.0050	8.4091	32
33	.0431	9.5694	81.4856	23.2252	222.2515	0.1045	0.0045	8.5152	33
34	.0391	9.6086	82.7773	25.5477	245.4767	0.1041	0.0041	8.6149	34
35	.0356	9.6442	83.9872	28.1024	271.0244	0.1037	0.0037	8.7086	35
36	.0323	9.6765	85.1194	30.9127	299.1268	0.1033	0.0033	8.7965	36
37	.0294	9.7059	86.1781	34.0039	330.0395	0.1030	0.0030	8.8789	37
38	.0267	9.7327	87.1673	37.4043	364.0434	0.1027	0.0027	8.9562	38
39	.0243	9.7570	88.0908	41.1448	401.4478	0.1025	0.0025	9.0285	39
40	.0221	9.7791	88.9525	45.2593	442.5926	0.1023	0.0023	9.0962	40
41	.0201	9.7991	89.7560	49.7852	487.8518	0.1020	0.0020	9.1596	41
42	.0183	9.8174	90.5047	54.7637	537.6370	0.1019	0.0019	9.2188	42
43	.0166	9.8340	91.2019	60.2401	592.4007	0.1017	0.0017	9.2741	43
44	.0151	9.8491	91.8508	66.2641	652.6408	0.1015	0.0015	9.3258	44
45	.0137	9.8628	92.4544	72.8905	718.9048	0.1014	0.0014	9.3740	45
46	.0125	9.8753	93.0157	80.1795	791.7953	0.1013	0.0013	9.4190	46
47	.0113	9.8866	93.5372	88.1975	871.9749	0.1011	0.0011	9.4610	47
48	.0103	9.8969	94.0217	97.0172	960.1723	0.1010	0.0010	9.5001	48
49	.0094	9.9063	94.4715	106.7190	1057.1896	0.1009	0.0009	9.5365	49
50	.0085	9.9148	94.8889	117.3909	1163.9085	0.1009	0.0009	9.5704	50
51	.0077	9.9226	95.2761	129.1299	1281.2994	0.1008	0.0008	9.6020	51
52	.0070	9.9296	95.6351	142.0429	1410.4293	0.1007	0.0007	9.6313	52
53	.0064	9.9360	95.9679	156.2472	1552.4723	0.1006	0.0006	9.6586	53
54	.0058	9.9418	96.2763	171.8719	1708.7195	0.1006	0.0006	9.6840	54
55	.0053	9.9471	96.5619	189.0591	1880.5914	0.1005	0.0005	9.7075	55
60	.0033	9.9672	97.7010	304.4816	3034.8164	0.1003	0.0003	9.8023	60
65	.0020	9.9796	98.4705	490.3707	4893.7073	0.1002	0.0002	9.8672	65
70	.0013	9.9873	98.9870	789.7470	7887.4696	0.1001	0.0001	9.9113	70
75	.0008	9.9921	99.3317	1271.8954	12708.9537	0.1001	0.0001	9.9410	75
80	.0005	9.9951	99.5606	2048.4002	20474.0021	0.1000	0.0000	9.9609	80
85	.0003	9.9970	99.7120	3298.9690	32979.6903	0.1000	0.0000	9.9742	85
90	.0002	9.9981	99.8118	5313.0226	53120.2261	0.1000	0.0000	9.9831	90
95	.0001	9.9988	99.8773	8556.6760	85556.7605	0.1000	0.0000	9.9889	95
100	.0001	9.9993	99.9202	13780.6123	137796.1234	0.1000	0.0000	9.9927	100

I = 12.00 %

N	(P/F)	(P/A)	(P/G)	(F/P)	(F/A)	(A/P)	(A/F)	(A/G)	N
1	.8929	0.8929	-0.0000	1.1200	1.0000	1.1200	1.0000	-0.0000	1
2	.7972	1.6901	0.7972	1.2544	2.1200	0.5917	0.4717	0.4717	2
3	.7118	2.4018	2.2208	1.4049	3.3744	0.4163	0.2963	0.9246	3
4	.6355	3.0373	4.1273	1.5735	4.7793	0.3292	0.2092	1.3589	4
5	.5674	3.6048	6.3970	1.7623	6.3528	0.2774	0.1574	1.7746	5
6	.5066	4.1114	8.9302	1.9738	8.1152	0.2432	0.1232	2.1720	6
7	.4523	4.5638	11.6443	2.2107	10.0890	0.2191	0.0991	2.5515	7
8	.4039	4.9676	14.4714	2.4760	12.2997	0.2013	0.0813	2.9131	8
9	.3606	5.3282	17.3563	2.7731	14.7757	0.1877	0.0677	3.2574	9
10	.3220	5.6502	20.2541	3.1058	17.5487	0.1770	0.0570	3.5847	10
11	.2875	5.9377	23.1288	3.4785	20.6546	0.1684	0.0484	3.8953	11
12	.2567	6.1944	25.9523	3.8960	24.1331	0.1614	0.0414	4.1897	12
13	.2292	6.4235	28.7024	4.3635	28.0291	0.1557	0.0357	4.4683	13
14	.2046	6.6282	31.3624	4.8871	32.3926	0.1509	0.0309	4.7317	14
15	.1827	6.8109	33.9202	5.4736	37.2797	0.1468	0.0268	4.9803	15
16	.1631	6.9740	36.3670	6.1304	42.7533	0.1434	0.0234	5.2147	16
17	.1456	7.1196	38.6973	6.8660	48.8837	0.1405	0.0205	5.4353	17
18	.1300	7.2497	40.9080	7.6900	55.7497	0.1379	0.0179	5.6427	18
19	.1161	7.3658	42.9979	8.6128	63.4397	0.1358	0.0158	5.8375	19
20	.1037	7.4694	44.9676	9.6463	72.0524	0.1339	0.0139	6.0202	20
21	.0926	7.5620	46.8188	10.8038	81.6987	0.1322	0.0122	6.1913	21
22	.0826	7.6446	48.5543	12.1003	92.5026	0.1308	0.0108	6.3514	22
23	.0738	7.7184	50.1776	13.5523	104.6029	0.1296	0.0096	6.5010	23
24	.0659	7.7843	51.6929	15.1786	118.1552	0.1285	0.0085	6.6406	24
25	.0588	7.8431	53.1046	17.0001	133.3339	0.1275	0.0075	6.7708	25
26	.0525	7.8957	54.4177	19.0401	150.3339	0.1267	0.0067	6.8921	26
27	.0469	7.9426	55.6369	21.3249	169.3740	0.1259	0.0059	7.0049	27
28	.0419	7.9844	56.7674	23.8839	190.6989	0.1252	0.0052	7.1098	28
29	.0374	8.0218	57.8141	26.7499	214.5828	0.1247	0.0047	7.2071	29
30	.0334	8.0552	58.7821	29.9599	241.3327	0.1241	0.0041	7.2974	30
31	.0298	8.0850	59.6761	33.5551	271.2926	0.1237	0.0037	7.3811	31
32	.0266	8.1116	60.5010	37.5817	304.8477	0.1233	0.0033	7.4586	32
33	.0238	8.1354	61.2612	42.0915	342.4294	0.1229	0.0029	7.5302	33
34	.0212	8.1566	61.9612	47.1425	384.5210	0.1226	0.0026	7.5965	34
35	.0189	8.1755	62.6052	52.7996	431.6635	0.1223	0.0023	7.6577	35
36	.0169	8.1924	63.1970	59.1356	484.4631	0.1221	0.0021	7.7141	36
37	.0151	8.2075	63.7406	66.2318	543.5987	0.1218	0.0018	7.7661	37
38	.0135	8.2210	64.2394	74.1797	609.8305	0.1216	0.0016	7.8141	38
39	.0120	8.2330	64.6967	83.0812	684.0102	0.1215	0.0015	7.8582	39
40	.0107	8.2438	65.1159	93.0510	767.0914	0.1213	0.0013	7.8988	40
41	.0096	8.2534	65.4997	104.2171	860.1424	0.1212	0.0012	7.9361	41
42	.0086	8.2619	65.8509	116.7231	964.3595	0.1210	0.0010	7.9704	42
43	.0076	8.2696	66.1722	130.7299	1081.0826	0.1209	0.0009	8.0019	43
44	.0068	8.2764	66.4659	146.4175	1211.8125	0.1208	0.0008	8.0308	44
45	.0061	8.2825	66.7342	163.9876	1358.2300	0.1207	0.0007	8.0572	45
46	.0054	8.2880	66.9792	183.6661	1522.2176	0.1207	0.0007	8.0815	46
47	.0049	8.2928	67.2028	205.7061	1705.8838	0.1206	0.0006	8.1037	47
48	.0043	8.2972	67.4068	230.3908	1911.5898	0.1205	0.0005	8.1241	48
49	.0039	8.3010	67.5929	258.0377	2141.9806	0.1205	0.0005	8.1427	49
50	.0035	8.3045	67.7624	289.0022	2400.0182	0.1204	0.0004	8.1597	50
51	.0031	8.3076	67.9169	323.6825	2689.0204	0.1204	0.0004	8.1753	51
52	.0028	8.3103	68.0576	362.5243	3012.7029	0.1203	0.0003	8.1895	52
53	.0025	8.3128	68.1856	406.0273	3375.2272	0.1203	0.0003	8.2025	53
54	.0022	8.3150	68.3022	454.7505	3781.2545	0.1203	0.0003	8.2143	54
55	.0020	8.3170	68.4082	509.3206	4236.0050	0.1202	0.0002	8.2251	55
60	.0011	8.3240	68.8100	897.5969	7471.6411	0.1201	0.0001	8.2664	60
65	.0006	8.3281	69.0581	1581.8725	13173.9374	0.1201	0.0001	8.2922	65
70	.0004	8.3303	69.2103	2787.7998	23223.3319	0.1200	0.0000	8.3082	70
75	.0002	8.3316	69.3031	4913.0558	40933.7987	0.1200	0.0000	8.3181	75
80	.0001	8.3324	69.3594	8658.4831	72145.6925	0.1200	0.0000	8.3241	80
85	.0001	8.3328	69.3935	15259.2057	127151.7140	0.1200	0.0000	8.3278	85
90	.0000	8.3330	69.4140	26891.9342	224091.1185	0.1200	0.0000	8.3300	90
95	.0000	8.3332	69.4263	47392.7766	394931.4719	0.1200	0.0000	8.3313	95
100	.0000	8.3332	69.4336	83522.2657	696010.5477	0.1200	0.0000	8.3321	100

I = 15.00 %

N	(P/F)	(P/A)	(P/G)	(F/P)	(F/A)	(A/P)	(A/F)	(A/G)	N
1	.8696	0.8696	-0.0000	1.1500	1.0000	1.1500	1.0000	-0.0000	1
2	.7561	1.6257	0.7561	1.3225	2.1500	0.6151	0.4651	0.4651	2
3	.6575	2.2832	2.0712	1.5209	3.4725	0.4380	0.2880	0.9071	3
4	.5718	2.8550	3.7864	1.7490	4.9934	0.3503	0.2003	1.3263	4
5	.4972	3.3522	5.7751	2.0114	6.7424	0.2983	0.1483	1.7228	5
6	.4323	3.7845	7.9368	2.3131	8.7537	0.2642	0.1142	2.0972	6
7	.3759	4.1604	10.1924	2.6600	11.0668	0.2404	0.0904	2.4498	7
8	.3269	4.4873	12.4807	3.0590	13.7268	0.2229	0.0729	2.7813	8
9	.2843	4.7716	14.7548	3.5179	16.7858	0.2096	0.0596	3.0922	9
10	.2472	5.0188	16.9795	4.0456	20.3037	0.1993	0.0493	3.3832	10
11	.2149	5.2337	19.1289	4.6524	24.3493	0.1911	0.0411	3.6549	11
12	.1869	5.4206	21.1849	5.3503	29.0017	0.1845	0.0345	3.9082	12
13	.1625	5.5831	23.1352	6.1528	34.3519	0.1791	0.0291	4.1438	13
14	.1413	5.7245	24.9725	7.0757	40.5047	0.1747	0.0247	4.3624	14
15	.1229	5.8474	26.6930	8.1371	47.5804	0.1710	0.0210	4.5650	15
16	.1069	5.9542	28.2960	9.3576	55.7175	0.1679	0.0179	4.7522	16
17	.0929	6.0472	29.7828	10.7613	65.0751	0.1654	0.0154	4.9251	17
18	.0808	6.1280	31.1565	12.3755	75.8364	0.1632	0.0132	5.0843	18
19	.0703	6.1982	32.4213	14.2318	88.2118	0.1613	0.0113	5.2307	19
20	.0611	6.2593	33.5822	16.3665	102.4436	0.1598	0.0098	5.3651	20
21	.0531	6.3125	34.6448	18.8215	118.8101	0.1584	0.0084	5.4883	21
22	.0462	6.3587	35.6150	21.6447	137.6316	0.1573	0.0073	5.6010	22
23	.0402	6.3988	36.4988	24.8915	159.2764	0.1563	0.0063	5.7040	23
24	.0349	6.4338	37.3023	28.6252	184.1678	0.1554	0.0054	5.7979	24
25	.0304	6.4641	38.0314	32.9190	212.7930	0.1547	0.0047	5.8834	25
26	.0264	6.4906	38.6918	37.8568	245.7120	0.1541	0.0041	5.9612	26
27	.0230	6.5135	39.2890	43.5353	283.5688	0.1535	0.0035	6.0319	27
28	.0200	6.5335	39.8283	50.0656	327.1041	0.1531	0.0031	6.0960	28
29	.0174	6.5509	40.3146	57.5755	377.1697	0.1527	0.0027	6.1541	29
30	.0151	6.5660	40.7526	66.2118	434.7451	0.1523	0.0023	6.2066	30
31	.0131	6.5791	41.1466	76.1435	500.9569	0.1520	0.0020	6.2541	31
32	.0114	6.5905	41.5006	87.5651	577.1005	0.1517	0.0017	6.2970	32
33	.0099	6.6005	41.8184	100.6998	664.6655	0.1515	0.0015	6.3357	33
34	.0086	6.6091	42.1033	115.8048	765.3654	0.1513	0.0013	6.3705	34
35	.0075	6.6166	42.3586	133.1755	881.1702	0.1511	0.0011	6.4019	35
36	.0065	6.6231	42.5872	153.1519	1014.3457	0.1510	0.0010	6.4301	36
37	.0057	6.6288	42.7916	176.1246	1167.4975	0.1509	0.0009	6.4554	37
38	.0049	6.6338	42.9743	202.5433	1343.6222	0.1507	0.0007	6.4781	38
39	.0043	6.6380	43.1374	232.9248	1546.1655	0.1506	0.0006	6.4985	39
40	.0037	6.6418	43.2830	267.8635	1779.0903	0.1506	0.0006	6.5168	40
41	.0032	6.6450	43.4128	308.0431	2046.9539	0.1505	0.0005	6.5331	41
42	.0028	6.6478	43.5286	354.2495	2354.9969	0.1504	0.0004	6.5478	42
43	.0025	6.6503	43.6317	407.3870	2709.2465	0.1504	0.0004	6.5609	43
44	.0021	6.6524	43.7235	468.4950	3116.6334	0.1503	0.0003	6.5725	44
45	.0019	6.6543	43.8051	538.7693	3585.1285	0.1503	0.0003	6.5830	45
46	.0016	6.6559	43.8778	619.5847	4123.8977	0.1502	0.0002	6.5923	46
47	.0014	6.6573	43.9423	712.5224	4743.4824	0.1502	0.0002	6.6006	47
48	.0012	6.6585	43.9997	819.4007	5456.0047	0.1502	0.0002	6.6080	48
49	.0011	6.6596	44.0506	942.3108	6275.4055	0.1502	0.0002	6.6146	49
50	.0009	6.6605	44.0958	1083.6574	7217.7163	0.1501	0.0001	6.6205	50
51	.0008	6.6613	44.1360	1246.2061	8301.3737	0.1501	0.0001	6.6257	51
52	.0007	6.6620	44.1715	1433.1370	9547.5798	0.1501	0.0001	6.6304	52
53	.0006	6.6626	44.2031	1648.1075	10980.7167	0.1501	0.0001	6.6345	53
54	.0005	6.6631	44.2311	1895.3236	12628.8243	0.1501	0.0001	6.6382	54
55	.0005	6.6636	44.2558	2179.6222	14524.1479	0.1501	0.0001	6.6414	55
60	.0002	6.6651	44.3431	4383.9987	29219.9916	0.1500	0.0000	6.6530	60
65	.0001	6.6659	44.3903	8817.7874	58778.5826	0.1500	0.0000	6.6593	65
70	.0001	6.6663	44.4156	17735.7200	118231.4669	0.1500	0.0000	6.6627	70
75	.0000	6.6665	44.4292	35672.8680	237812.4532	0.1500	0.0000	6.6646	75
80	.0000	6.6666	44.4364	71750.8794	478332.5293	0.1500	0.0000	6.6656	80
85	.0000	6.6666	44.4402	144316.6470	962104.3133	0.1500	0.0000	6.6661	85
90	.0000	6.6666	44.4422	290272.3252	1935142.1680	0.1500	0.0000	6.6664	90
95	.0000	6.6667	44.4433	583841.3276	3892268.8509	0.1500	0.0000	6.6665	95
100	.0000	6.6667	44.4438	1174313.4507	7828749.6713	0.1500	0.0000	6.6666	100

ECON

I = 20.00 %

N	(P/F)	(P/A)	(P/G)	(F/P)	(F/A)	(A/P)	(A/F)	(A/G)	N
1	.8333	0.8333	-0.0000	1.2000	1.0000	1.2000	1.0000	-0.0000	1
2	.6944	1.5278	0.6944	1.4400	2.2000	0.6545	0.4545	0.4545	2
3	.5787	2.1065	1.8519	1.7280	3.6400	0.4747	0.2747	0.8791	3
4	.4823	2.5887	3.2986	2.0736	5.3680	0.3863	0.1863	1.2742	4
5	.4019	2.9906	4.9061	2.4883	7.4416	0.3344	0.1344	1.6405	5
6	.3349	3.3255	6.5806	2.9860	9.9299	0.3007	0.1007	1.9788	6
7	.2791	3.6046	8.2551	3.5832	12.9159	0.2774	0.0774	2.2902	7
8	.2326	3.8372	9.8831	4.2998	16.4991	0.2606	0.0606	2.5756	8
9	.1938	4.0310	11.4335	5.1598	20.7989	0.2481	0.0481	2.8364	9
10	.1615	4.1925	12.8871	6.1917	25.9587	0.2385	0.0385	3.0739	10
11	.1346	4.3271	14.2330	7.4301	32.1504	0.2311	0.0311	3.2893	11
12	.1122	4.4392	15.4667	8.9161	39.5805	0.2253	0.0253	3.4841	12
13	.0935	4.5327	16.5883	10.6993	48.4966	0.2206	0.0206	3.6597	13
14	.0779	4.6106	17.6008	12.8392	59.1959	0.2169	0.0169	3.8175	14
15	.0649	4.6755	18.5095	15.4070	72.0351	0.2139	0.0139	3.9588	15
16	.0541	4.7296	19.3208	18.4884	87.4421	0.2114	0.0114	4.0851	16
17	.0451	4.7746	20.0419	22.1861	105.9306	0.2094	0.0094	4.1976	17
18	.0376	4.8122	20.6805	26.6233	128.1167	0.2078	0.0078	4.2975	18
19	.0313	4.8435	21.2439	31.9480	154.7400	0.2065	0.0065	4.3861	19
20	.0261	4.8696	21.7395	38.3376	186.6880	0.2054	0.0054	4.4643	20
21	.0217	4.8913	22.1742	46.0051	225.0256	0.2044	0.0044	4.5334	21
22	.0181	4.9094	22.5546	55.2061	271.0307	0.2037	0.0037	4.5941	22
23	.0151	4.9245	22.8867	66.2474	326.2369	0.2031	0.0031	4.6475	23
24	.0126	4.9371	23.1760	79.4968	392.4842	0.2025	0.0025	4.6943	24
25	.0105	4.9476	23.4276	95.3962	471.9811	0.2021	0.0021	4.7352	25
26	.0087	4.9563	23.6460	114.4755	567.3773	0.2018	0.0018	4.7709	26
27	.0073	4.9636	23.8353	137.3706	681.8528	0.2015	0.0015	4.8020	27
28	.0061	4.9697	23.9991	164.8447	819.2233	0.2012	0.0012	4.8291	28
29	.0051	4.9747	24.1406	197.8136	984.0680	0.2010	0.0010	4.8527	29
30	.0042	4.9789	24.2628	237.3763	1181.8816	0.2008	0.0008	4.8731	30
31	.0035	4.9824	24.3681	284.8516	1419.2579	0.2007	0.0007	4.8908	31
32	.0029	4.9854	24.4588	341.8219	1704.1095	0.2006	0.0006	4.9061	32
33	.0024	4.9878	24.5368	410.1863	2045.9314	0.2005	0.0005	4.9194	33
34	.0020	4.9898	24.6038	492.2235	2456.1176	0.2004	0.0004	4.9308	34
35	.0017	4.9915	24.6614	590.6682	2948.3411	0.2003	0.0003	4.9406	35
36	.0014	4.9929	24.7108	708.8019	3539.0094	0.2003	0.0003	4.9491	36
37	.0012	4.9941	24.7531	850.5622	4247.8112	0.2002	0.0002	4.9564	37
38	.0010	4.9951	24.7894	1020.6747	5098.3735	0.2002	0.0002	4.9627	38
39	.0008	4.9959	24.8204	1224.8096	6119.0482	0.2002	0.0002	4.9681	39
40	.0007	4.9966	24.8469	1469.7716	7343.8578	0.2001	0.0001	4.9728	40
41	.0006	4.9972	24.8696	1763.7259	8813.6294	0.2001	0.0001	4.9767	41
42	.0005	4.9976	24.8890	2116.4711	10577.3553	0.2001	0.0001	4.9801	42
43	.0004	4.9980	24.9055	2539.7653	12693.8263	0.2001	0.0001	4.9831	43
44	.0003	4.9984	24.9196	3047.7183	15233.5916	0.2001	0.0001	4.9856	44
45	.0003	4.9986	24.9316	3657.2620	18281.3099	0.2001	0.0001	4.9877	45
46	.0002	4.9989	24.9419	4388.7144	21938.5719	0.2000	0.0000	4.9895	46
47	.0002	4.9991	24.9506	5266.4573	26327.2863	0.2000	0.0000	4.9911	47
48	.0002	4.9992	24.9581	6319.7487	31593.7436	0.2000	0.0000	4.9924	48
49	.0001	4.9993	24.9644	7583.6985	37913.4923	0.2000	0.0000	4.9935	49
50	.0001	4.9995	24.9698	9100.4382	45497.1908	0.2000	0.0000	4.9945	50
51	.0001	4.9995	24.9744	10920.5258	54597.6289	0.2000	0.0000	4.9953	51
52	.0001	4.9996	24.9783	13104.6309	65518.1547	0.2000	0.0000	4.9960	52
53	.0001	4.9997	24.9816	15725.5571	78622.7856	0.2000	0.0000	4.9966	53
54	.0001	4.9997	24.9844	18870.6685	94348.3427	0.2000	0.0000	4.9971	54
55	.0000	4.9998	24.9868	22644.8023	113219.0113	0.2000	0.0000	4.9976	55
60	.0000	4.9999	24.9942	56347.5144	281732.5718	0.2000	0.0000	4.9989	60
65	.0000	5.0000	24.9975	140210.6469	701048.2346	0.2000	0.0000	4.9995	65
70	.0000	5.0000	24.9989	348888.9569	1744439.7847	0.2000	0.0000	4.9998	70
75	.0000	5.0000	24.9995	868147.3693	4340731.8466	0.2000	0.0000	4.9999	75

I = 25.00 %

N	(P/F)	(P/A)	(P/G)	(F/P)	(F/A)	(A/P)	(A/F)	(A/G)	N
1	.8000	0.8000	0.0	1.2500	1.0000	1.2500	1.0000	0.0	1
2	.6400	1.4400	0.6400	1.5625	2.2500	0.6944	0.4444	0.4444	2
3	.5120	1.9520	1.6640	1.9531	3.8125	0.5123	0.2623	0.8525	3
4	.4096	2.3616	2.8928	2.4414	5.7656	0.4234	0.1734	1.2249	4
5	.3277	2.6893	4.2035	3.0518	8.2070	0.3718	0.1218	1.5631	5
6	.2621	2.9514	5.5142	3.8147	11.2588	0.3388	0.0888	1.8683	6
7	.2097	3.1611	6.7725	4.7684	15.0735	0.3163	0.0663	2.1424	7
8	.1678	3.3289	7.9469	5.9605	19.8419	0.3004	0.0504	2.3872	8
9	.1342	3.4631	9.0207	7.4506	25.8023	0.2888	0.0388	2.6048	9
10	.1074	3.5705	9.9870	9.3132	33.2529	0.2801	0.0301	2.7971	10
11	.0859	3.6564	10.8460	11.6415	42.5661	0.2735	0.0235	2.9663	11
12	.0687	3.7251	11.6020	14.5519	54.2077	0.2684	0.0184	3.1145	12
13	.0550	3.7801	12.2617	18.1899	68.7596	0.2645	0.0145	3.2437	13
14	.0440	3.8241	12.8334	22.7374	86.9495	0.2615	0.0115	3.3559	14
15	.0352	3.8593	13.3260	28.4217	109.6868	0.2591	0.0091	3.4530	15
16	.0281	3.8874	13.7482	35.5271	138.1085	0.2572	0.0072	3.5366	16
17	.0225	3.9099	14.1085	44.4089	173.6357	0.2558	0.0058	3.6084	17
18	.0180	3.9279	14.4147	55.5112	218.0446	0.2546	0.0046	3.6698	18
19	.0144	3.9424	14.6741	69.3889	273.5558	0.2537	0.0037	3.7222	19
20	.0115	3.9539	14.8932	86.7362	342.9447	0.2529	0.0029	3.7667	20
21	.0092	3.9631	15.0777	108.4202	429.6809	0.2523	0.0023	3.8045	21
22	.0074	3.9705	15.2326	135.5253	538.1011	0.2519	0.0019	3.8365	22
23	.0059	3.9764	15.3625	169.4066	673.6264	0.2515	0.0015	3.8634	23
24	.0047	3.9811	15.4711	211.7582	843.0329	0.2512	0.0012	3.8861	24
25	.0038	3.9849	15.5618	264.6978	1054.7912	0.2509	0.0009	3.9052	25
26	.0030	3.9879	15.6373	330.8722	1319.4890	0.2508	0.0008	3.9212	26
27	.0024	3.9903	15.7002	413.5903	1650.3612	0.2506	0.0006	3.9346	27
28	.0019	3.9923	15.7524	516.9879	2063.9515	0.2505	0.0005	3.9457	28
29	.0015	3.9938	15.7957	646.2349	2580.9394	0.2504	0.0004	3.9551	29
30	.0012	3.9950	15.8316	807.7936	3227.1743	0.2503	0.0003	3.9628	30
31	.0010	3.9960	15.8614	1009.7420	4034.9678	0.2502	0.0002	3.9693	31
32	.0008	3.9968	15.8859	1262.1774	5044.7098	0.2502	0.0002	3.9746	32
33	.0006	3.9975	15.9062	1577.7218	6306.8872	0.2502	0.0002	3.9791	33
34	.0005	3.9980	15.9229	1972.1523	7884.6091	0.2501	0.0001	3.9828	34
35	.0004	3.9984	15.9367	2465.1903	9856.7613	0.2501	0.0001	3.9858	35
36	.0003	3.9987	15.9481	3081.4879	12321.9516	0.2501	0.0001	3.9883	36
37	.0003	3.9990	15.9574	3851.8599	15403.4396	0.2501	0.0001	3.9904	37
38	.0002	3.9992	15.9651	4814.8249	19255.2994	0.2501	0.0001	3.9921	38
39	.0002	3.9993	15.9714	6018.5311	24070.1243	0.2500	0.0000	3.9935	39
40	.0001	3.9995	15.9766	7523.1638	30088.6554	0.2500	0.0000	3.9947	40
41	.0001	3.9996	15.9809	9403.9548	37611.8192	0.2500	0.0000	3.9956	41
42	.0001	3.9997	15.9843	11754.9435	47015.7740	0.2500	0.0000	3.9964	42
43	.0001	3.9997	15.9872	14693.6794	58770.7175	0.2500	0.0000	3.9971	43
44	.0001	3.9998	15.9895	18367.0992	73464.3969	0.2500	0.0000	3.9976	44
45	.0000	3.9998	15.9915	22958.8740	91831.4962	0.2500	0.0000	3.9980	45
46	.0000	3.9999	15.9930	28698.5925	114790.3702	0.2500	0.0000	3.9984	46
47	.0000	3.9999	15.9943	35873.2407	143488.9627	0.2500	0.0000	3.9987	47
48	.0000	3.9999	15.9954	44841.5509	179362.2034	0.2500	0.0000	3.9989	48
49	.0000	3.9999	15.9962	56051.9386	224203.7543	0.2500	0.0000	3.9991	49
50	.0000	3.9999	15.9969	70064.9232	280255.6929	0.2500	0.0000	3.9993	50
51	.0000	4.0000	15.9975	87581.1540	350320.6161	0.2500	0.0000	3.9994	51
52	.0000	4.0000	15.9980	109476.4425	437901.7701	0.2500	0.0000	3.9995	52
53	.0000	4.0000	15.9983	136845.5532	547378.2126	0.2500	0.0000	3.9996	53
54	.0000	4.0000	15.9986	171056.9414	684223.7658	0.2500	0.0000	3.9997	54
55	.0000	4.0000	15.9989	213821.1768	855280.7072	0.2500	0.0000	3.9997	55
60	.0000	4.0000	15.9996	652530.4468	2610117.7872	0.2500	0.0000	3.9999	60

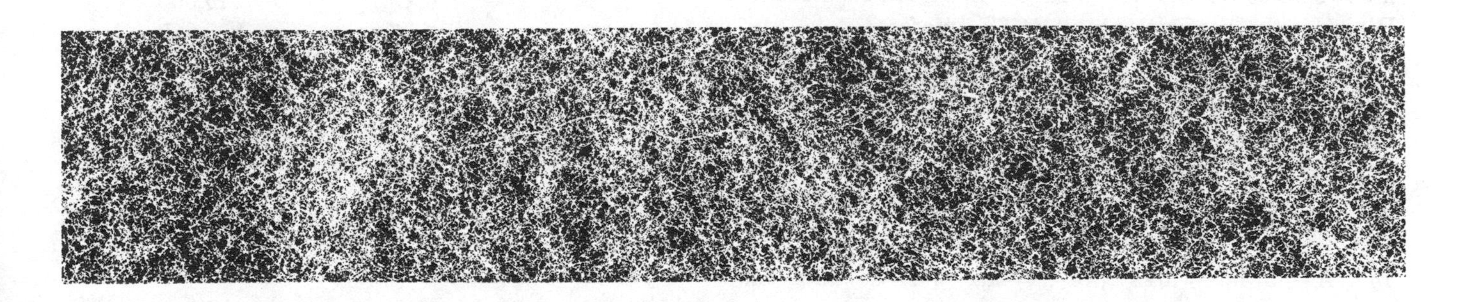

PRACTICE PROBLEMS

1. How much will be accumulated at 6% if $1000 is invested for 10 years?

2. What is the present worth of $2000 at 6% which becomes available in four years?

3. How much will it take to accumulate $2000 in 20 years at 6%?

4. What year-end annual amount over seven years at 6% is equivalent to $500 invested now?

5. $50 is invested at the end of each year for 10 years. What will be the accumulated amount at the end of 10 years at 6%?

6. How much should be deposited at 6% at the start of each year for 10 years in order to empty the fund by drawing out $200 at the end of each year for 10 years?

7. You need $2000 on the date of your last deposit. How much should be deposited at the start of each year for five years at 6%?

8. How much will be accumulated at 6% in 10 years if you deposit three payments of $100 every other year for four years, with the first payment occuring at $t = 0$?

9. $500 is compounded monthly at 6% annual rate. How much will you have in five years?

10. What is the rate of return of an $80 investment that pays back $120 in seven years?

11. A new machine will cost $17,000 and will have a value of $14,000 in five years. Special tooling will cost $5000 and it will have a resale value of $2500 after five years. Maintenance will be $200 per year. What will the average cost of ownership be during the next five years if interest is 6%?

12. An old highway bridge can be strengthened at a cost of $9000, or it can be replaced for $40,000. The present salvage value of the old bridge is $13,000. It is estimated that the reinforced bridge will last for 20 years with an annual cost of $500 and will have a salvage value of $10,000 at the end of 20 years. The estimated salvage of the new bridge after 25 years is $15,000. The maintenance for the new bridge will be $100 annually. Which is the best alternative at 8% interest?

13. A firm expects to receive $32,000 each year for 15 years from the sale of a product. It will require an initial investment of $150,000. Expenses will run $7530 per year. Salvage is zero and straight-line depreciation is used. The tax rate is 48%. What is the after-tax rate of return?

14. A public works project has initial costs of $1,000,000, benefits of $1,500,000, and disbenefits of $300,000. (a) What is the benefit/cost ratio? (b) What is the excess of benefits over costs?

15. A speculator in land pays $14,000 for property that he expects to hold for 10 years. $1000 is spent in renovation and a monthly rent of $75 is collected from the tenants. Taxes are $150 per year and maintenance costs are $250. What must be the sale price in 10 years to realize a 10% rate of return? Use the year-end convention.

16. What is the effective interest rate for a payment plan of 30 equal payments of $89.30 per month when a lump sum of $2000 would have been an outright purchase?

17. An apartment complex is purchased for $500,000. What is the depreciation in each of the first three years if the salvage value is $100,000 in 25 years? Use (a) straight line, (b) sum-of-the-year's digits, and (c) double declining balance.

18. Equipment is purchased for $12,000 which is expected to be sold after 10 years for $2000. The estimated maintenance is $1000 for the first year, but is expected to increase $200 each year thereafter. Using 10%, find the present worth and the annual cost.

19. One of five grades of pipe with average lives (in years) and costs (in dollars) of (9, 1500), (14, 1600), (30, 1750), (52, 1900), and (86, 2100) is to be chosen for a 20-year project. A failure of the pipe at any time during the project will result in a cost equal to 35% of the original cost. Annual costs are 4% of the initial cost, and the pipes are not recoverable. At 6%, which pipe is superior? Note: The lives are average values, not absolute replacement times.

20. A grain combine with a 20-year life can remove seven pounds of rock from its harvest per hour. Any rocks left in its ouptut will cause $25,000 damage in subsequent processes. Several investments are available to increase its removal capacity. At 10%, what should be done?

rock rate (lb/hr)	probability of exceeding rock rate	required investment to meet rock rate
7	0.15	no cost
8	0.10	$15,000
9	0.07	$20,000
10	0.03	$30,000

21. A structure which costs $10,000 has the operating costs and salvage values given.

year 1: maintenance $2000, salvage $8000
year 2: maintenance $3000, salvage $7000

year 3: maintenance $4000, salvage $6000
year 4: maintenance $5000, salvage $5000
year 5: maintenance $6000, salvage $4000

(a) What is the economic life of the structure? (b) Assuming that the structure has been owned and operated for four years, what is the cost of owning the structure for exactly one more year? Use 20% as the interest rate.

22. A man purchases a car for $5000 for personal use, intending to drive 15,000 miles per year. It costs him $200 per year for insurance and $150 per year for maintenance. He gets 15 mpg and gasoline costs $.60 per gallon. The resale value after five years is $1000. Because of unexpected business driving (5000 miles per year extra), his insurance is increased to $300 per year and maintenance to $200. Salvage is reduced to $500. Use 10% interest to answer the following questions. (a) The man's company offers $.10 per mile reimbursement. Is that adequate? (b) How many miles must be driven per year at $.10 per mile to justify the company buying a car for the man's use? The cost would be $5000, but insurance, maintenance, and salvage would be $250, $200, and $800, respectively.

ECON

4 THERMODYNAMICS

Nomenclature

A_{ij}	van Laar and Margules coefficients	–
B	second virial coefficient	in^2/lbf, or ft^2/lbf
c_p	specific heat at constant pressure	BTU/lbm-°R
C_p	specific heat at constant pressure	BTU/lbmole-°R
C_v	heat capacity at constant volume	BTU/lbmole-°R
C	third virial coefficient	in^4/lbf^2, or ft^4/lbf^2
COP	coefficient of performance	–
D	fourth virial coefficient	in^6/lbf^3, or ft^6/lbf^3
EER	energy efficiency ratio	–
g_c	gravitational constant (32.2)	lbm-ft/sec^2-lbf
gef	Gibbs energy function	–
ΔG	Gibbs function	cal/gmole
h	enthalpy	BTU/lbm
H	enthalpy	BTU/lbmole
ΔH_{rx}	heat of reaction	cal/gmole
I	integration constant	–
J	Joule's constant (778.2)	ft-lbf/BTU
k	ratio of specific heat	–
K	equilibrium constant	–
m	mass	lbm
M	molecular weight	g/gmole, or lbm/lbmole
M	generalized thermodynamic function	–
n	number of moles, or polytropic exponent	–
p	pressure	psf, or psi
P	power	hp
q	specific energy	BTU/lbm
Q	heat	BTU/lbmole, or cal/gmole
R	universal gas constant (1545.33)	ft-lbf/°R-lbmole
s	entropy	BTU/lbm-°R
S	entropy	BTU/lbmole-°R, or cal/gmole-°K
t	temperature	°C, or °F
T	absolute temperature	°R, or °K
u	internal energy	BTU/lbm
U	internal energy	BTU/lbmole
v	velocity	ft/sec
V	volume, or specific molar volume	ft^3, or ft^3/lbmole
w	weight	lbf
W	work	BTU/lbmole, or BTU
x	liquid phase mole fraction, or quality	–
y	gas phase mole fraction (mole fraction in vapor)	–
z	elevation	ft
Z	compressibility factor	–

Symbols

λ	latent heat (same as h_{fg})	BTU/lbm
α	activity	–
v	specific volume	ft^3/lbm
β	second virial coefficient	ft^3
γ	third virial coefficient	ft^6
δ	fourth virial coefficient	$ft^9/lbmole^3$
η	efficiency	–
τ	temperature ratio (T/T_o)	–

Subscripts

c	critical, or cold
f	formation, or fusion, or liquid
fg	vaporization
g	gas, or vapor
h	enthalpy, or hot

THERMO

i	component i
in	inlet
m	mixture
out	outlet
p	constant pressure, or equilibrium
r	reduced
s	entropy
sat	saturation
t	total
th	thermal
v	constant volume, or vapor
y	mole fraction
α	activity
γ	activity coefficient
ϕ	fugacity

1 DEFINITIONS

Batch: see *Non-Flow.*

Change of state: any operation in which the properties of the system change.

Critical pressure: the pressure above which a substance may not be distinguished between vapor and liquid; the point at which the specific volumes for the vapor and liquid become equal.

Enthalpy: a usable form of energy, defined as $H = U + \frac{pV}{J}$; the amount of energy (BTU) that may be extracted in the form of useful work. Measurement of enthalpy is made with respect to a reference temperature. For water, enthalpy is zero at 32 °F.

Entropy: a measure of unusable energy; the measure of a material's disorder. Entropy is unchanged in reversible, adiabatic processes. Irreversible processes include:

- stirring of a viscous liquid
- gas expanding into a vacuum
- throttling
- diffusion of two dissimilar inert ideal gases
- freezing of a supercooled liquid
- condensation of a supersaturated vapor
- solution of a solute in a solvent
- osmosis

Entropy is measured with respect to an arbitrary reference temperature, 32 °F for liquid water. Formulas for entropy and its changes are shown in equations 4.1 through 4.3. (Equation 4.3 should be used with ideal gases only.)

$$dS = \frac{dQ}{T} \tag{4.1}$$

$$dS = C_p\left(\frac{dT}{T}\right) - R\left(\frac{dp}{p}\right) \tag{4.2}$$

$$S_2 - S_1 = C_p \, ln\left(\frac{T_2}{T_1}\right) - R \, ln\left(\frac{p_2}{p_1}\right) \tag{4.3}$$

Equilibrium: a state of a system whose intensive properties are uniform and have no tendency to change with time.

Extensive properties: properties which depend on the quantity of the material in the system.

Gas: a substance, above its critical temperature, whose behavior usually follows the perfect gas law. (Also, see *Ideal gas.*)

Heat: an energy transfer caused by a temperature difference. The amount of sensible heat required to raise the temperature of a substance is related to the specific heat as shown in equation 4.4.

$$dQ = C_p dT \tag{4.4}$$

Ideal gas: a gas for which molecular interactions do not exist, and molecular volumes are infinitely small.

Intensive properties: properties which are independent of the quantity of the material in the system.

Internal energy: the energy possessed by virtue of the configuration of its molecules. A function of temperature (only) for an ideal gas. Internal energy is comprised mainly of molecular, kinetic, and rotational energy.

Latent heat: the heat added to a substance which changes its phase (as opposed to changing its temperature). Approximate values for the latent heat of water at one atmosphere are

process	phase change	BTU/lbm	cal/gm	kcal/gmole
fusion	ice to water	143.4	79.71	1.434
vaporization	water to vapor	970.3	539.55	9.703
sublimation	ice to vapor	1293.5	719.26	12.935

Non-flow: a condition in which flow is negligible; also known as batch.

Phase: an indication of the separation between molecules of a substance: solid, liquid, or gas.[1]

[1] High temperature plasma is considered to be a fourth phase.

Process: the events which a substance undergoes when any property changes. The various types of thermodynamic processes are:

- adiabatic - no heat transfer to or from the surroundings
  - irreversible - energy loss but no heat loss
  - isentropic - reversible adiabatic
  - throttling - irreversible with $\Delta H = 0$
- isenthalpic - constant enthalpy
- isobaric - constant pressure
- isochoric - constant volume
- isometric - constant volume
- isothermal - constant temperature
- polytropic - constant pV^n

A *cyclic process* consists of several of these pure processes in a definite sequence.

Property: any quantity of a system or a substance which has a distinct value for each distinct state of the system; also known as *point function.* Common properties are pressure, density, specific volume, temperature, internal energy, enthalpy, and entropy.

Reversible process: a process where the state of a substance may be recovered without loss or absorption of energy from the surroundings, and by retracing the original path.

Sensible heat: heat which is used to change the temperature of a substance.

Specific heat: the amount of heat required to change the temperature of a unit mass by one degree.

Steady flow: a system in which there is no variation in mass flow rate.

STP: Standard Temperature and Pressure. Standard temperature is 32 °F. Standard pressure is 1 atm, 760 mm Hg, or 14.696 psia. Standard gas measurements are usually at 60 °F or 70 °F.

System: a body of matter with finite boundaries. There are three types of systems:

- Open: exchange of mass and energy
- Closed: exchange of energy only, not mass
- Isolated: no exchange of mass or energy

Temperature scales: Most thermodynamic calculations are carried out in absolute scale temperature (Rankine or Kelvin). Calculations involving only changes in temperature may be carried out in absolute or relative scales. The conversions for the relative scales are given in equations 4.5 through 4.8.

$$°R = °F + 460 = \left(\frac{9}{5}\right) °K \qquad 4.5$$

$$°F = \left(\frac{9}{5}\right) °C + 32 \qquad 4.6$$

$$°K = °C + 273 = \left(\frac{5}{9}\right) °R \qquad 4.7$$

$$°C = \left(\frac{5}{9}\right) (°F - 32) \qquad 4.8$$

Vapor: a substance below the critical temperature, (particularly near the saturation line), which does not follow the perfect gas law.

2 IDEAL GASES AND THE PERFECT GAS LAW

A. IDEAL GASES

An *ideal* or *perfect gas* is a substance that would have neither energy nor volume at absolute zero temperature. This means that the gas must have neither molecular volume nor interactions between molecules. Most gases that are supercritical (i.e., are above their critical points) behave sufficiently well to be labeled as ideal. Common gases such as air, nitrogen, oxygen, and carbon dioxide behave ideally at low pressures (below two atmospheres) and at temperatures below 500 °F. The *equation of state* (the equation that relates pressure, temperature, and volume) of an ideal gas is

$$pV = nRT \qquad 4.9$$

Various values for R are

$$\begin{aligned} R &= 0.0821 \text{ liter-atm/gmole-°K} \\ &= 10.73 \text{ psi-ft}^3\text{/lbmole-°R} \\ &= 1.987 \text{ BTU/lbmole-°R} \\ &= 1.987 \text{ cal/gmole-°K} \\ &= 1545.33 \text{ ft-lbf/lbmole-°R} \\ &= 0.73023 \text{ atm-ft}^3\text{/lbmole-°R} \end{aligned}$$

Process calculations for ideal gases are

$p_1V_1 = p_2V_2$	Boyle's Law[2]	4.10
$\frac{T_1}{V_1} = \frac{T_2}{V_2}$	Charles' Law/Isobaric	4.11
$\frac{T_1}{p_1} = \frac{T_2}{p_2}$	Charles' Law/Isometric	4.12
$\frac{p_1V_1}{T_1} = \frac{p_2V_2}{T_2}$	General Law	4.13

[2] Care should be taken when using Boyle's Law. If temperature is not mentioned, the process may be polytropic.

THERMO

For any process, relationships which hold for ideal gases are

$$\Delta U = C_v(T_2 - T_1) \quad 4.14$$
$$\Delta H = C_p(T_2 - T_1) \quad 4.15$$
$$C_p - C_v = R \quad 4.16$$

3 OTHER GAS RELATIONSHIPS AND NON-IDEAL GASES

There are certain named laws which relate to various states of gases.

Amagat's Law: The total volume of a mixture of gases which do not react with each other is equal to the sum of the partial volumes of the gases. Thus, $V = \Sigma V_i$ and $V_i = y_i V$.

Avogadro's Hypothesis: Equal volumes at the same pressure and temperature contain the same number of gas molecules. One gmole of any gas at STP has 6.02 EE23 molecules. One gmole occupies 22.4 liters at STP. One lbmole occupies 359 ft^3 at STP.

Gay-Lussac's Law: At a given temperature and pressure, reacting gases combine with each other in simple whole number proportions of volume.

Gibbs-Dalton Law: The total pressure of a mixture of gases which do not react with each other is equal to the sum of the partial pressures of each of the constituent gases, all evaluated at the same temperature, pressure, and volume. Thus, $p_t = \Sigma p_i$ and $p_i = y_i p_t$.

Graham's Law: At the same temperature and pressure, the rates of diffusion of two gases are inversely proportional to the square roots of their densities.

The equations of state for non-ideal (real) gases take into consideration the effects of molecular volume, intermolecular forces, hydrogen bonding, and polarity. The simplest equations usually have two additional, gas-specific parameters. Real gas, two-parameter equations of state include

$$\left(p + \frac{a}{V^2}\right)(V - b) = nRT$$

van der Waals (1) 4.17

$$\frac{pV}{nRT} = \frac{V}{V-b} - \frac{a}{RTV} = Z$$

van der Waals (2) 4.18

$$p(V-b) = RTe^{\frac{-a}{VRT}} \quad \text{Dieterici} \quad 4.19$$

$$p = \frac{RT}{V-b} - \frac{a}{TV^2} \quad \text{Berthelot} \quad 4.20$$

$$\frac{pV}{RT} = 1 + \frac{\beta}{V} + \frac{\gamma}{V^2} + \ldots \quad \text{Virial}$$

4.21

$$\frac{pV}{RT} = \frac{V}{V-b} - \frac{a}{RT^{\frac{3}{2}}(V+b)} \quad \text{Redlich-Kwong} \quad 4.22$$

A. NON-IDEAL GASES

All equations of state can be related back to the virial form. Since $\frac{pV}{RT} = 1$ for any gas as p approaches zero, the empirical relationship of the vapor state can be conveniently expressed by simply considering that $\frac{pV}{RT}$ is a function of temperature and pressure (or temperature and density), with the power series expansion being

$$\frac{pV}{RT} = 1 + \frac{\beta}{V} + \frac{\gamma}{V^2} + \frac{\delta}{V^3} + \ldots \quad 4.23$$

or

$$\frac{pV}{RT} = 1 + Bp + Cp^2 + Dp^3 + \ldots \quad 4.24$$

The terms β, B, γ, C, δ, and D are called the *virial coefficients* and are generally considered functions of temperature and composition only. β and B are the *second* virial coefficients; γ and C are the *third* virial coefficients, etc. The 1 is considered the *first* virial coefficient. It can be shown that the virial coefficients of the two power series are related:

$$B = \frac{\beta}{RT} \quad 4.25$$

$$C = \frac{\gamma - \beta^2}{(RT)^2} \quad 4.26$$

$$D = \frac{\delta + 2\gamma^3 - 3\beta\gamma}{(RT)^3} \quad 4.27$$

The virial form of the Redlich-Kwong equation of state is

$$\frac{pV}{RT} = 1 + \frac{b - \frac{a}{RT^{\frac{3}{2}}}}{V} + \frac{b^2 + \frac{ab}{RT^{\frac{3}{2}}}}{V^2} + \frac{b^3 - \frac{ab^2}{RT^{\frac{3}{2}}}}{V^3} + \ldots \quad 4.28$$

The *compressibility factor* is a simple method of predicting non-ideal behavior.

$$\frac{pV}{RT} = Z \quad 4.29$$

The compressibility factor is a fractional measure of the deviation from ideal behavior. The virial coefficients of many gases can be determined from their critical properties. Generalized charts for compressibilities of almost all known gases have been devised from functions of dimensionless critical variables known as *reduced properties*. Redućed pressure, p_r, and reduced temperature, T_r, are defined

The values of the *reduced pressure* and *reduced temperature* used to obtain the *compressibility factor*, Z, from the generalized compressibility charts can be calculated if the critical properties are known.

$$p_r = \frac{p}{p_c} \qquad 4.30$$

$$T_r = \frac{T}{T_c} \qquad 4.31$$

The critical properties of several common gases are given in table 4.1. Generalized compressibility charts are given in Appendixes A, B, and C.

gas	critical temperature (°R)	critical pressure (psia)
air	235.8	547.0
ammonia	730.1	1639.0
argon	272.2	705.0
carbon dioxide	547.8	1071.0
carbon monoxide	242.2	508.2
chlorine	751.0	1116.0
ethane	549.8	717.0
ethylene	509.5	745.0
helium	10.0	33.8
hydrogen	60.5	188.0
mercury	2109.0	2646.0
methane	343.9	673.3
neon	79.0	377.8
nitrogen	227.2	492.5
oxygen	278.1	730.9
propane	666.3	617.0
sulfur dioxide	775.0	1141.0
water vapor	1165.4	3206.0
xenon	521.9	855.3

Table 4.1
Critical Properties of Common Gases

4 PHYSICAL THERMODYNAMICS

A. FIRST LAW OF THERMODYNAMICS

The total energy change of a system is the sum of the potential, kinetic, and internal energy changes. The internal energy, U, of a system represents the kinetic and potential energies of the molecules and atoms of the system. Most of the time, the kinetic and potential energies of thermodynamic systems do not change. The *first law of thermodynamics* for a non-flow process is

$$Q = W + \Delta U = W + U_2 - U_1 \qquad 4.32$$

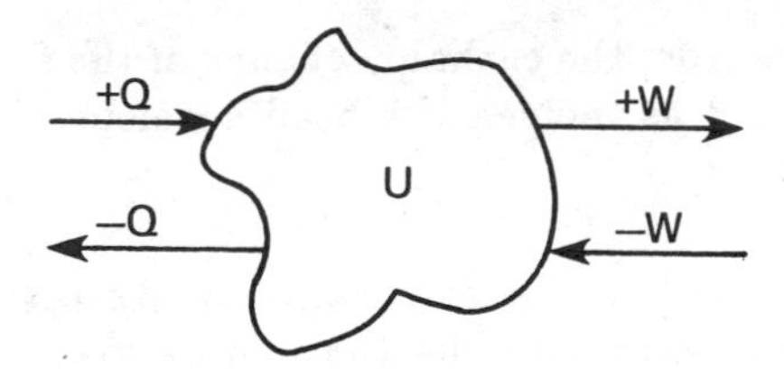

Figure 4.1 Generalized Open Flow System

For flow processses, the first law is

$$Q = W + \Delta U + \left(\frac{\Delta \mathrm{v}^2}{2g_c} + \Delta z \frac{g}{g_c} + \Delta pV\right)\frac{\dot{m}}{J} \qquad 4.33$$

As an example of the application of the first law for steady flow processes, consider a well-insulated turbine. If the process is adiabatic ($Q = 0$), if the velocity component contributes little to the total energy change $\left(\frac{\Delta \mathrm{v}^2}{2g_c} \approx 0\right)$, and if there is no elevation change ($\Delta z = 0$), then the first law can be written

$$U_1 + (p_1V_1)\frac{\dot{m}}{J} = U_2 + (p_2V_2)\frac{\dot{m}}{J} + W \qquad 4.34$$

Since $H = U + \frac{pV}{J}$, performance of the turbine process can be written

$$W = -\Delta H = H_1 - H_2 \qquad 4.35$$

This means that the total work done by the turbine can be approximated by the enthalpy change of the steam.

As another example of the application of the first law for a steady flow process, consider flow through a nozzle. The nozzle does no work ($W = 0$), there is no heat loss ($Q = 0$), there is no elevation change ($\Delta z = 0$), and the upstream velocity is low compared to the velocity in the nozzle throat $\left(\frac{\mathrm{v}_1^2}{2g_c} \approx 0\right)$. The first law for this process can be written

$$U_1 + \frac{p_1V_1}{J} = U_2 + \left(p_2V_2 + \frac{\mathrm{v}_2^2}{2g_c}\right)\frac{\dot{m}}{J} \qquad 4.36$$

Using the definition of enthalpy ($H = U + pV$), the nozzle performance can be predicted by

$$H_2 - H_1 = \frac{-v_2^2}{2g_c}\left(\frac{\dot{m}}{J}\right) \quad 4.37$$

In other words, the enthalpy change of the flowing fluid is calculated as the velocity head developed in the nozzle.

As an example of the application of the first law for a non-flow process, consider the compression process of a reciprocating compressor, or the expansion process in a steam engine.

$$U_1 + Q = W + U_2 \quad 4.38$$

B. SECOND LAW OF THERMODYNAMICS

Energy tends to seek an equilibrium state where the system can do less work. For example, energy is like water which runs downhill to the lowest level. In the form of heat, energy flows from high temperatures to low temperatures. For a reversible constant temperature process this can be written

$$S_2 - S_1 = \frac{Q_2 - Q_1}{T} \quad 4.39$$

Entropy is an intrinsic property of matter defined such that an increase in unavailability of the system corresponds to an increase in its entropy.

The relationship between internal energy and available work may be seen by combining the first law and equation 4.39.

$$U_2 - U_1 = Q - W = T(S_2 - S_1) - W \quad 4.40$$

C. ENTHALPY CHANGES

Enthalpy is a common thermodynamic function used in energy balances. It is a measure of useful work, and it is used to measure the heat absorbed or liberated from a process. The following processes are representative of processes where enthalpy changes are significant.

- *Latent heating or cooling* (no chemical reaction):

$$h_2 - h_1 = \lambda \quad 4.41$$

- *Sensible heating or cooling:*

$$h_2 - h_1 = \int_{T_1}^{T_2} dq \quad 4.42$$

For heating and cooling at constant pressure,

$$h_2 - h_1 = \int_{T_1}^{T_2} c_p dT \quad 4.43$$

If c_p is independent of temperature,

$$h_2 - h_1 = c_p(T_2 - T_1) \quad 4.44$$

If c_p varies with temperature such that

$$c_p = a + bT + cT^2 + dT^3$$

then,

$$h_2 - h_1 = \int_{T_1}^{T_2} c_p dT \quad 4.45$$

$$= \int_{T_1}^{T_2} (a + bT + cT^2 + dT^3)dT \quad 4.46$$

$$= a(T_2 - T_1) + b\left(\frac{T_2^2 - T_1^2}{2}\right) + c\left(\frac{T_2^3 - T_1^3}{3}\right) + d\left(\frac{T_2^4 - T_1^4}{4}\right) \quad 4.47$$

- *Constant temperature bath:*

$$h_2 - h_1 = q \quad 4.48$$

- *Ideal gas* (from p_1, V_1, T_1 to p_2, V_2, T_2):

$$h_2 - h_1 = \int_{T_1}^{T_2} c_p dT = c_p(T_2 - T_1) \quad 4.49$$

- *Adiabatic mixing of hot and cold fluids:*

$$h_2 - h_1 = (h_{2,h} - h_{1,h}) + (h_{2,c} - h_{1,c}) = 0 \quad 4.50$$

$$h_{2,h} - h_{1,h} = \int_{T_1}^{T_2} (c_p)_h dT \quad 4.51$$

$$\Delta h_c = \int_{T_1}^{T_2} (c_p)_c dT \quad 4.52$$

$$\Delta h_c = \Delta h_h \quad 4.53$$

If c_p is constant for both fluids,

$$T_m = \frac{m_h(c_p)_h T_h + m_c(c_p)_c T_c}{m_h(c_p)_h + m_c(c_p)_c} \quad 4.54$$

D. ENTROPY CHANGES

Entropy changes commonly need to be evaluated for the following processes.

- *Latent heating or cooling* (no chemical reaction):

$$s_2 - s_1 = \frac{\lambda}{T} \quad 4.55$$

- *Sensible heating or cooling:*

$$s_2 - s_1 = \int_{T_1}^{T_2} \frac{dq}{T} \quad 4.56$$

If the heating or cooling is at constant pressure,

$$s_2 - s_1 = \int_{T_1}^{T_2} c_p \frac{dT}{T} \quad 4.57$$

If c_p is independent of temperature,

$$s_2 - s_1 = c_p \ ln\left(\frac{T_2}{T_1}\right) \quad 4.58$$

If c_p varies with temperature, such that

$$c_p = a + bT + cT^2 + dT^3$$

then,

$$\begin{aligned} s_2 - s_1 &= \int_{T_1}^{T_2} \left(\frac{a + bT + cT^2 + dT^3}{T}\right) dT \\ &= a \ ln\left(\frac{T_2}{T_1}\right) + b(T_2 - T_1) \\ &\quad + c\left(\frac{T_2^2 - T_1^2}{2}\right) + d\left(\frac{T_2^3 - T_1^3}{3}\right) \end{aligned} \quad 4.59$$

- *Constant temperature bath* (mixer):

$$s_2 - s_1 = \frac{q}{T} \quad 4.60$$

- *Ideal gas* (from p_1, V_1, T_1 to p_2, V_2, T_2) at constant pressure:

$$s_2 - s_1 = \int_{T_1}^{T_2} c_p \frac{dT}{T} \quad 4.61$$

Since $dQ = dU + dW$, and dU is a function of only temperature for ideal gases, then $dU = 0$. Therefore,

$$dW = pdV + Vdp = 0 \quad 4.62$$

$$pdV = Vdp = \frac{RTdp}{p} \quad 4.63$$

- *Ideal gas at constant temperature:*

$$s_2 - s_1 = \int \frac{dQ}{T} = -\int_{p_1}^{p_2} \frac{R}{M}\frac{dp}{p} \quad 4.64$$

- *Ideal gas* (any process):

$$s_2 - s_1 = \int_{T_1}^{T_2} c_p \frac{dT}{T} - \frac{R}{M} \ ln\frac{p_2}{p_1} \quad 4.65$$

- *Adiabatic mixing of hot and cold fluids:*

$$s_{2,h} - s_{1,h} = \int_{T_h}^{T_m} (c_p)_h \frac{dT}{T} \quad 4.66$$

$$s_{2,c} - s_{1,c} = \int_{T_c}^{T_m} (c_p)_c \frac{dT}{T} \quad 4.67$$

The total entropy change is

$$\Delta s = \Delta s_h + \Delta s_c \quad 4.68$$

If c_p is constant for both liquids,

$$s_2 - s_1 = (c_p)_h \ ln\left(\frac{T_m}{T_h}\right) + (c_p)_c \ ln\left(\frac{T_m}{T_c}\right) \quad 4.69$$

THERMO

E. IDEAL GAS PROCESSES

Table 4.2 and figure 4.2 list ideal gas thermodynamic property changes for five basic processes. The sign conventions are:

ΔQ is positive if energy goes into the system.

W is positive if the system does work on the surroundings.

ΔU is positive if internal energy increases within the system.

ΔS is positive if entropy increases within the system.

Example 4.1

How much power can an isentropic turbine produce while receiving 10^6 pounds per hour of an ideal gas entering at 1200 °F and leaving at 600 °F? Assume the gas does not condense and has a constant heat capacity of 0.4 BTU/lbm-°R.

The flow work is calculated from table 4.2:

$$W = mc_p(T_2 - T_1)$$

Since the mass flow rate is given, the power is

$$P = \dot{m}c_pJ\,(T_2 - T_1) \quad \text{(ft-lbf/hr)}$$

$$P = 10^6(0.4)(778)(1200 - 600)$$

$$= 1.87 \times 10^{11} \text{ ft-lbf/hr}$$

Since 1 hp = 550 ft-lbf/sec,

$$P = \frac{1.87 \times 10^{11}}{(60)(60)(550)} = 94{,}444 \text{ hp}$$

F. CYCLIC PROCESSES

A steady state, cyclic, thermodynamic process is one in which a substance returns to its original properties at the same point in the process. Fluids such as steam, ammonia, and freon are commonly used. Cyclic processes usually operate in and around the vapor dome region.

Table 4.2
Thermodynamic Processes Chart

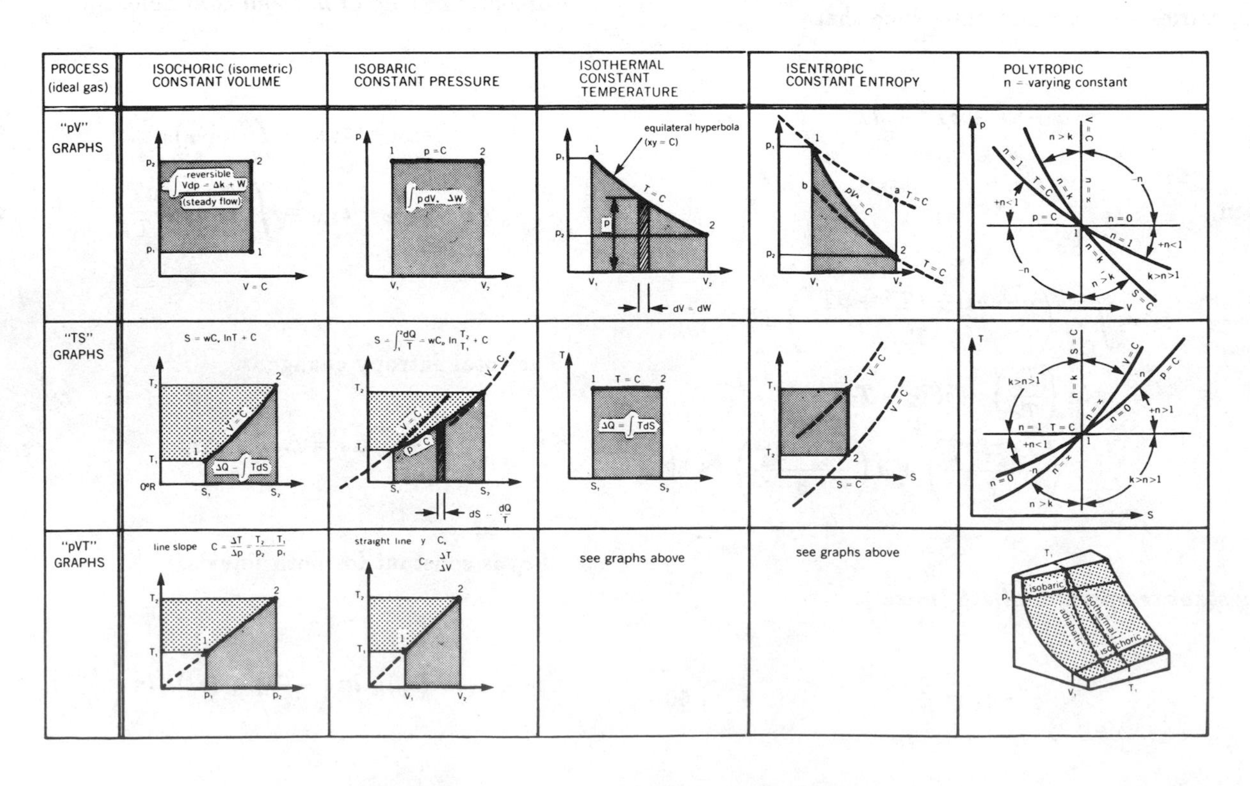

PROCESS (ideal gas) →	ISOCHORIC (isometric) CONSTANT VOLUME	ISOBARIC CONSTANT PRESSURE	ISOTHERMAL CONSTANT TEMPERATURE	ISENTROPIC CONSTANT ENTROPY	POLYTROPIC n = varying constant
LAW	Guy-Lussac V = C	Charles p = C	Boyle T = C	Adiabatic (Q = 0) Reversible (0 = TDS) S = C	$pV^n = C$ any fluid
n POLYTROPIC EXPONENT	$n = \infty$	$n = 0$	$n = 1$	$n = k$	$n = -\infty$ to $+\infty$
GENERAL PATH EQUATION	$p^{1/n}V = p^0V = C = \frac{T}{p}$	$pV^0 = C = \frac{T}{V}$	$pV^1 = C$ (note: $T \neq pV$)	$pV^k = C$	$pV^n = C$ reversible or irreversible
p,V,T RELATIONS $\frac{p_1V_1}{T_1} = \frac{p_2V_2}{T_2} = wR$ [w lb$_f$]	$\frac{p_1}{p_2} = \frac{T_1}{T_2} = C$	$\frac{V_1}{V_2} = \frac{T_1}{T_2} = C$	$\frac{p_1}{p_2} = \frac{V_2}{V_1} = C$	$p_1V_1^k = p_2V_2^k$ $\frac{T_2}{T_1} = \left(\frac{p_2}{p_1}\right)^{(k-1)k} = \left(\frac{V_1}{V_2}\right)^{k-1}$	$p_1V_1^n = p_2V_2^n$ $\frac{T_2}{T_1} = \left(\frac{p_2}{p_1}\right)^{(n-1)/n} = \left(\frac{V_1}{V_2}\right)^{n-1}$
(non-flow reversible) WORK $W = \int_1^2 pdV$ $W = Q - \Delta U$(BTU)	0 (since $dV = 0$)	$p\int dV = p(V_2 - V_1)$ (any fluid) $wR(T_2 - T_1)$ ft − lb (ideal gas)	$C\int \frac{dV}{V} = p_1V_1 \ln \frac{V_2}{V_1}$ ft-lb $wRT \ln \frac{p_1}{p_2}$ ft-lb $\frac{p_1V_1}{J} \ln \frac{V_2}{V_1}$ BTU	ft-lb BTU $\frac{p_2V_2 - p_1V_1}{1-k} = \frac{p_2V_2 - p_1V_1}{J(1-k)}$ $W = -\Delta U = U_1 - U_2$ (any fluid) $W = wC_v(T_1 - T_2)$ BTU (ideal gas)	$\frac{p_2V_2 - p_1V_1}{1-n} = \frac{wR(T_2 - T_1)}{1-n}$ (any fluid) (ideal gas)
(steady-flow reversible) WORK $\Delta W = -\int_1^2 Vdp$ $\Delta p = 0, \Delta K = 0$	$-V\int_1^2 dp = V(p_2 - p_1)$	0 (since $dp = 0$)	$p_1V_1 \ln \frac{V_2}{V_1}$ ft-lb	$\frac{k(p_2V_2 - p_1V_1)}{1-k}$ ft-lb $W = -\Delta H = H_1 - H_2$ (any fluid) $W = wC_p(T_1 - T_2)$ BTU (ideal gas)	$\frac{n(p_2V_2 - p_1V_1)}{1-n} = \frac{nwR(T_2 - T_1)}{1-n}$ $\frac{nRT_1}{1-n}\left[\left(\frac{p_2}{p_1}\right)^{(n-1)/n} - 1\right]$
INTERNAL ENERGY $\Delta U = U_2 - U_1$	$wC_v(T_2 - T_1)$ $\frac{C_v(p_2V_2 - p_1V_1)}{R}$ $\frac{p_2V_2 - p_1V_1}{k-1}$	$wC_v(T_2 - T_1)$ $\frac{C_v(p_2V_2 - p_1V_1)}{R}$ $\frac{p_2V_2 - p_1V_1}{k-1}$	0	$wC_v(T_2 - T_1)$ $\frac{C_v(p_2V_2 - p_1V_1)}{R}$ $\frac{p_2V_2 - p_1V_1}{k-1}$	$wC_v(T_2 - T_1)$ $\frac{C_v(p_2V_2 - p_1V_1)}{R}$ $\frac{p_2V_2 - p_1V_1}{k-1}$
HEAT $Q = \Delta U + \int \frac{pdV}{J}$ $\Delta Q = \int Tds$ $\Delta Q = wC_v(\Delta T) + \frac{wR\Delta T}{1-n}$	$wC_v(T_2 - T_1) = \Delta U$ $\left(h_2 - \frac{p_2V_2}{J}\right) - \left(h_1 - \frac{p_1V_1}{J}\right)$	$wC_p\,\Delta T = wC_v\,\Delta T + \frac{p\Delta V}{J}$ $U_2 - U_1 + \frac{p_2V_2 - p_1V_1}{J}$ $\Delta H = H_2 - H_1$	$\frac{1}{J}\int pdV = \frac{p_1V_1}{J} \ln \frac{V_2}{V_1}$ $\frac{wRT_1}{J} \ln \frac{p_1}{p_2} = -\frac{1}{J}\int Vdp$ $T(S_2 - S_1)$	0 = Tds (since $ds = 0$)	$wC_n(T_2 - T_1) =$ $U_2 - U_2 + \frac{p_2V_2 - p_1V_1}{J(1-n)}$
SPECIFIC HEAT C	$C_v = \left(\frac{\partial u}{\partial T}\right)_v = C_p - R$ $\frac{1}{k-1}(R)$	$C_p = \left(\frac{\partial h}{\partial T}\right)_p = C_v + R$ $\frac{k}{k-1}(R)$	∞	0	$C_n = C_v\left(\frac{k-n}{1-n}\right)$
ENTHALPY $H_2 - H_1$	$wC_p(T_2 - T_1)$	$wC_p(T_2 - T_1)$	0 (since $\Delta T = 0$)	$wC_p(T_2 - T_1)$	$wC_p(T_2 - T_1)$
ENTROPY $S_2 - S_1 = \int_1^2 \frac{dQ}{T}$ (reversible only) $\Delta S = w\left[C_v \ln\left(\frac{T_2}{T_1}\right) + R \ln\left(\frac{V_2}{V_1}\right)\right]$ (general)	$wC_v \int_1^2 \frac{dT}{T} = wC_v \ln \frac{T_2}{T_1}$ $wC_v \ln \frac{p_2}{p_1}$	$wC_p \ln \frac{T_2}{T_1}$ $wC_p \ln \frac{V_2}{V_1}$	reversible $\frac{Q}{T} = \frac{w(h_2 - h_1)}{T}$ $\frac{wR}{J} \ln \frac{V_2}{V_1} = \frac{wR}{J} \ln \frac{p_1}{p_2}$	0 (since $ds = 0$)	$\int \frac{dQ}{T} = w\int \frac{C_n dT}{T} = wC_n \ln\left(\frac{T_2}{T_1}\right)$ $wC_v\left(\frac{k-n}{1-n}\right) \ln\left(\frac{V_1}{V_2}\right)^{n-1}$ $= C_n \ln\left(\frac{p_2}{p_1}\right)^{(n-1)/n}$
FINAL PRESSURE p_2	$p_1\left(\frac{T_2}{T_1}\right)$	p_1	$p_1\left(\frac{V_2}{V_1}\right)$	$p_1\left(\frac{V_1}{V_2}\right)^k$ or $p_1\left(\frac{T_2}{T_1}\right)^{k/(k-1)}$	$p_1\left(\frac{V_1}{V_2}\right)^n$ or $p_1\left(\frac{T_2}{T_1}\right)^{n/(n-1)}$
FINAL VOLUME V_2	V_1	$V_1\left(\frac{T_2}{T_1}\right)$	$V_1\left(\frac{p_1}{p_2}\right)$	$V_1\left(\frac{p_1}{p_2}\right)^{1/k}$ or $V_1\left(\frac{T_1}{T_2}\right)^{1/(k-1)}$	$V_1\left(\frac{p_1}{p_2}\right)^{1/n}$ or $V_1\left(\frac{T_1}{T_2}\right)^{1/(n-1)}$
FINAL TEMPERATURE T_2	$T_1\left(\frac{p_2}{p_1}\right)$	$T_1\left(\frac{V_2}{V_1}\right)$	T_1	$T_1\left(\frac{V_1}{V_2}\right)^{k-1}$ or $T_1\left(\frac{p_2}{p_1}\right)^{(k-1)/k}$	$T_1\left(\frac{V_1}{V_2}\right)^{n-1}$ or $T_1\left(\frac{p_2}{p_1}\right)^{(n-1)/n}$

Nomenclature:

C = Constant, dimensionless

C_p = Molar specific heat at constant pressure,

$$C_p = C_v + \frac{R}{J}, \frac{BTU}{lb_m deg R}$$

C_v = Molar specific heat at constant volume,

$$C_v = C_p - \frac{R}{J}, \frac{BTU}{lb_m deg R}$$

C_n = Polytropic specific heat,

$$C_n = C_v\frac{(k\text{-}n)}{(1\text{-}n)}, \frac{BTU}{lb_m deg R}$$

H = Total enthalpy of "w" lb_m of substance, BTU

h = Specific enthalpy, $\frac{BTU}{lb_m}$

J = Joules constant, 778.172 $\frac{ft\text{-}lb_f}{BTU}$

ΔK = Change of kinetic energy, ft-lb_f

k = Dimensionless ratio, $\frac{C_p}{C_v} = \frac{C_v + R}{C_p - R}$

k = 1.667 monatomic gas

k = 1.400 diatomic gas

k = 1.333 polyatomic gas

k = dimensionless

n = Polytropic exponent

n = $\frac{C_p - C_n}{C_v - C_n}$, dimensionless

p = Unit pressure, $\frac{lb_f}{in^2}$ or $\frac{lb_f}{ft^2}$

ΔP = Change of potential energy, ft-lb_f

Q = Heat, Q = $\int dQ$

$= Q_{in} - Q_{out} = Q_{net}$, BTU

R = Specific gas constant,

R = $\frac{pv}{T}, \frac{ft\text{-}lb_f}{lb_m R}$ or $\frac{BTU}{lb_m deg R}$

R = Deg Rankine, R = F + 460, deg R

S = Total entropy, $\frac{BTU}{deg R}$

T = Absolute temperature (usually deg Rankine), deg R

U = Total internal energy, BTU

u = Specific internal energy, $\frac{BTU}{lb_m}$

V = Total volume, V = $\frac{w}{\rho}$, ft³

v = Specific volume, v = $\frac{V}{w} = \frac{1}{\rho}, \frac{ft^3}{lb_f}$

W = Work, ft-lb

w = Mass, lb, w = $\frac{g}{g_0}$ w_0 = gm, lb_f

ρ = Density, $\rho = \frac{w}{V}, \frac{lb_f}{ft^3}$

Figure 4.2 Ideal Gas Thermodynamic Property Changes

THERMO

Vapor-Liquid Mixtures

When a fluid changes phase at constant temperature and pressure (e.g., boiling), the properties of each phase do not change. If the fluid starts in phase A (e.g., liquid) with properties v_A, H_A, U_A, S_A, etc., and a phase change is slowly (so as to be reversible) carried out to phase B (e.g., vapor), then the vapor will have properties v_B, H_B, U_B, S_B, while the remaining liquid stays at condition A. Intermediate properties of the two-phase substance require a weighted sum of the properties of the individual phases. If x is the (mass or mole) fraction that exists in vapor phase, then the average properties of the two-phase region are

$$v_m = (1-x)v_f + xv_y \quad 4.70$$

$$H_m = (1-x)H_f + xH_g \quad 4.71$$

$$U_m = (1-x)U_f + xU_g \quad 4.72$$

$$S_m = (1-x)S_f + xS_g \quad 4.73$$

In general, if A and B are two phases and if M represents any thermodynamic property,

$$M_m = (1-x)M_A + xM_B \quad 4.74$$

Rearranging equation 4.74,

$$M_m = M_A + x\Delta M_{AB} \quad 4.75$$

ΔM_{AB} is the difference $(M_B - M_A)$between the property at phase A and phase B. In cyclic processes, phase A is usually liquid (or fluid, f), and phase B is vapor (or gas, g). The thermodynamic property in a two-phase region (liquid-vapor) is

$$M_m = M_f + xM_{fg} \quad 4.76$$

For vapor-liquid systems, the mole fraction x, which is vapor, is called *quality*.

Example 4.2

Saturated steam enters an isentropic expansion process at 1000 psia, and exits at 3 psia. What is the final quality of the steam?

From the saturated steam table (Appendix E), the entropy of saturated steam at 1000 psia is

$$s = 1.3897 \text{ BTU/lbm-°R}$$

Since this process is isentropic (i.e., at constant entropy), the exit steam-water mixture must have the same total entropy. At 3 psia,

$$s_f = 0.2008 \text{ BTU/lbm-°R}$$

$$s_{fg} = 1.6855 \text{ BTU/lbm-°R}$$

From equation 4.76

$$s = s_f + xs_{fg}$$

$$1.3897 = 0.2008 + x(1.6855)$$

$$x = 0.7054 \quad (70.54\%)$$

The Ideal Carnot Cycle

The *Carnot cycle* is a theoretical cycle for converting heat into work. It is the most efficient thermodynamic cycle, and serves as a standard for evaluating the efficiency and performance of all actual other cycles. The Carnot cycle consists of four steps.

step 1: (a to b) Saturated fluid enters a heat source (i.e., a boiler) at a constant temperature, T_h, and is changed to saturated vapor. The fluid absorbs heat, Q_h, at a constant temperature. In this step the fluid goes through an isothermal expansion phase change. This step is represented by the process A to B in figure 4.3.

step 2: (b to c) This step is adiabatic (isentropic) expansion of the gas through a machine that extracts work (e.g., a turbine). It is represented by the process B to C. The T-S diagram shows this process as a vertical line, indicating that expansion is taking place at constant entropy.

step 3: (c to d) The expanded vapor-liquid mixture is cooled by removing energy, Q_c, at constant temperature, T_c. In this step, most of the vapor phase is condensed as indicated on the T-S diagram. This step is indicated by the process C to D.

step 4: (d to a) This step is an isentropic compression (as in a pump) of the fluid to the original saturation point. It is represented by a vertical line on the T-S diagram, and is the process represented by D to A. The fluid returns to its original state, saturated liquid, at the end of this step, making this whole process cyclic.

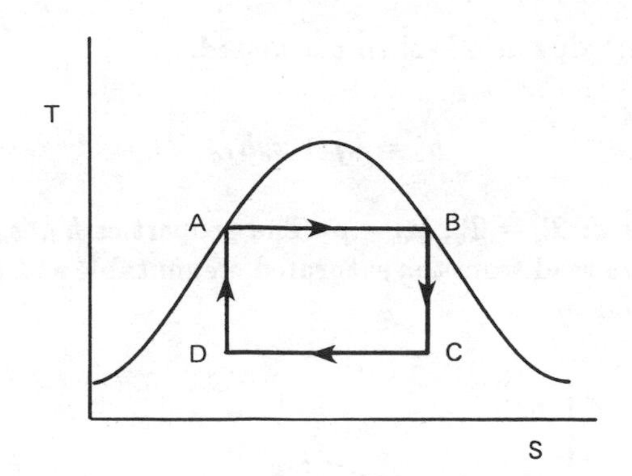

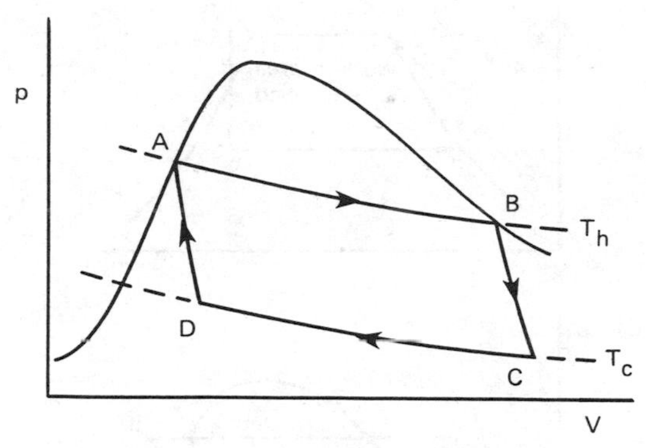

Figure 4.3 The Carnot Cycle

The net work of the cycle is represented by the enclosed area on the P-V diagram. To calculate the efficiency of this cycle, first compute the net work of the cycle from each step

$$W = W_{AB} + W_{BC} + W_{CD} + W_{DA} \tag{4.77}$$

The processes b to c and d to a are adiabatic, so from the first law of thermodynamics, $Q = 0$. For ideal gases,

$$W_{BC} = -\Delta U_{BC} = -\int_{T_h}^{T_c} C_v dT \tag{4.78}$$

$$W_{DA} = \Delta U_{DA} = \int_{T_h}^{T_c} C_v dT \tag{4.79}$$

Therefore, W_{BC} and W_{DA} are equal numerically, but of opposite sign. The net work reduces to

$$W = W_{AB} + W_{CD} \tag{4.80}$$

Since processes A to B and C to D are isothermal, the internal energy change is zero. From the first law of thermodynamics,

$$W_{AB} = Q_h \tag{4.81}$$

$$W_{CD} = Q_c \tag{4.82}$$

The work on an ideal gas in an isothermal process is

$$W_{AB} = RT_h \ln\left(\frac{p_A}{p_B}\right) \tag{4.83}$$

Therefore,

$$\frac{Q_h}{Q_c} = \frac{T_h \ln\left(\frac{p_A}{p_B}\right)}{T_c \ln\left(\frac{p_C}{p_D}\right)} \tag{4.84}$$

For an adiabatic expansion of an ideal gas, the pressure ratios are equal:

$$\frac{T_h}{T_c} = \left(\frac{p_B}{p_C}\right)^{\frac{k-1}{k}} \tag{4.85}$$

$$= \left(\frac{p_A}{p_D}\right)^{\frac{k-1}{k}} \tag{4.86}$$

Therefore,

$$\frac{p_A}{p_B} = \frac{p_D}{p_C} \tag{4.87}$$

$$\frac{Q_c}{Q_h} = \frac{T_c}{T_h} \tag{4.88}$$

The thermal efficiency of the process, η, is defined as

$$\eta = \frac{Q_{in} - Q_{out}}{Q_{in}} \tag{4.89}$$

In this case,

$$Q_{in} = Q_h$$

$$Q_{out} = Q_c \tag{4.90}$$

Therefore,

$$\eta = \frac{Q_h - Q_h \times \frac{T_c}{T_h}}{Q_h} \tag{4.91}$$

$$\eta = \frac{T_h - T_c}{T_h} \tag{4.92}$$

$$\eta = \frac{W}{Q_h} = \frac{T_h - T_c}{T_h} \tag{4.93}$$

Therefore, the efficiency of the Carnot engine is determined entirely by the heating and cooling temperatures of the cycle.

Example 4.3

An engine receives heat at 700 °F, and rejects it to an environment at 180 °F. What is the maximum thermodynamic efficiency for these operating conditions?

From equation 4.93, converting all temperatures to absolute,

$$\eta = \frac{T_h - T_c}{T_h} = \frac{(700 + 460) - (180 + 460)}{700 + 460}$$

$$= 0.44 \quad (44\%)$$

Rankine Cycle with Superheat

The Rankine cycle is similar to the theoretical Carnot cycle, except that the condenser converts the fluid to a single phase: liquid. The compression process occurs entirely in the liquid region. This cycle is closely approximated in actual power plants. The superheating of the steam increases the efficiency of the cycle by raising the mean effective temperature at which heat is added, which has the practical effect of making the turbine last longer. The flow diagram and property charts are shown in figure 4.4. The Rankine cycle is less efficient than the Carnot cycle because the mean temperature at which heat is added is less than T_h of the Carnot cycle. However, the Rankine cycle is a practical cycle, while the Carnot cycle is not.

The Rankine cycle consists of four steps.

step 1: Fluid is heated, vaporized, and superheated in a boiler at constant pressure from T_a to T_b.

step 2: The superheated vapor is expanded isentropically through the turbine where work is extracted.

step 3: The two-phase mixture is condensed to saturated liquid at constant pressure and temperature T_c.

step 4: A pump isentropically compresses the liquid phase from p_d to boiler pressure p_a.

To investigate a Rankine cycle, determine the properties of the fluid at each point in the cycle:

- At a: The properties at point a can be determined by noting that the compression of the liquid phase from point d to a is isentropic, and the exhaust pressure of the turbine p_c (which equals p_d) is usually given. T_a is the saturation temperature for the liquid having enthalpy h_a. To find h_a, use the following relationships:

$$h_a = h_d + v_d \frac{p_a - p_d}{J} \tag{4.94}$$

$$s_a = s_d \tag{4.95}$$

- At b: T_b is usually known. This temperature may be given as *degrees of superheat*, the number of degrees above the saturation temperature at the boiler pressure. Read the values of s_b and h_b at T_b and p_b from the superheated steam property tables (Appendix F).

- At c: The exhaust pressure of the turbine, p_c, is usually given ($s_c = s_b$). T_c is read from the saturated property tables. To calculate h_c, the quality of the vapor in c must first be determined:

$$x_c = \frac{s_c - s_f}{s_{fg}} \tag{4.96}$$

The enthalpy at c is then calculated:

$$h_c = h_f + x_c h_{fg} \tag{4.97}$$

- At d: $T_d = T_c$, $p_d = p_c$ The properties h_d, s_d, v_d, are read from the saturated steam table as h_f, s_f, and v_f.

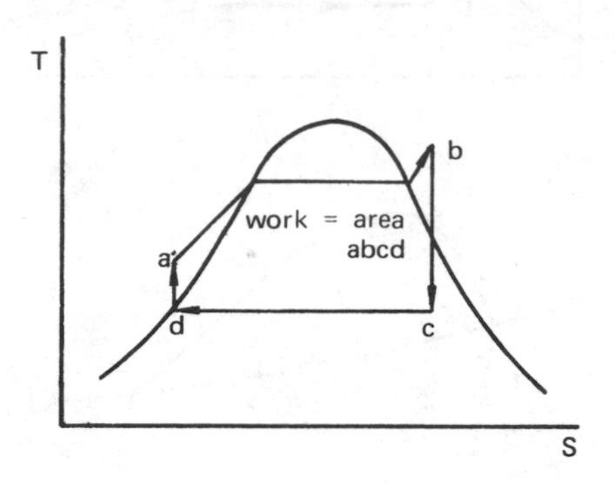

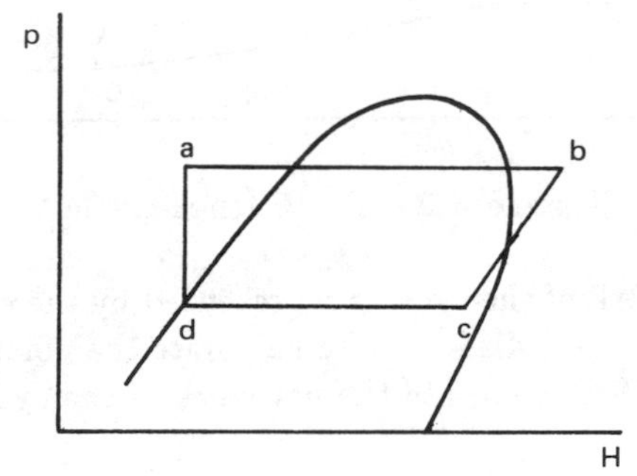

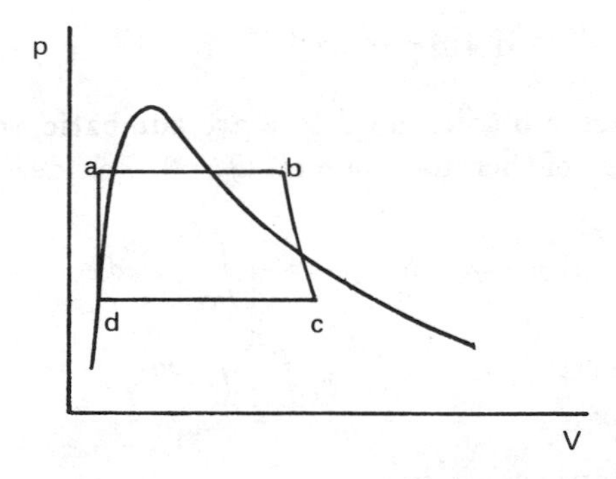

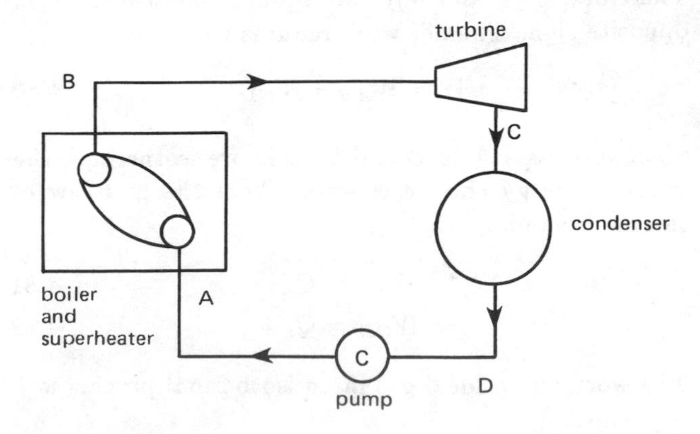

Figure 4.4 Steam Heat Engine Cycles

THERMO

The heat and work terms for each process in the Rankine cycle are:

$$Q_{absorbed} = (h_b - h_a) \times m \quad 4.98$$

$$Q_{rejected} = (h_c - h_d) \times m \quad 4.99$$

$$W_{turbine} = (h_b - h_c) \times m \quad 4.100$$

$$W_{pump} = (h_a - h_d) \times m \quad 4.101$$

The theoretical efficiency is

$$\eta_{th} = \frac{W_{turbine} - W_{pump}}{Q_{absorbed}}$$

$$= \frac{Q_{absorbed} - Q_{rejected}}{Q_{absorbed}}$$

$$= \frac{(h_b - h_a) - (h_c - h_d)}{h_b - h_a} \quad 4.102$$

Example 4.4

A power plant operating on a Rankine cycle has a boiler pressure of 1000 psia, an exhaust turbine pressure of 3 psia, 150 °F superheat, and no subcooling at the condenser. What is the thermal efficiency of the cycle?

From the saturated steam table (Appendix E) at 3 psia,

$$h_d = 109.37 \text{ BTU/lbm}$$

$$v_d = 0.01630 \text{ ft}^3\text{/lbm}$$

$$h_a = 109.37 + 0.01630\ (1000 - 3) \times \frac{144}{778.2}$$

$$= 112.38 \text{ BTU/lbm}$$

From the superheated steam table (Appendix F) at 1000 psia and 150 °F superheat ($T_{sat} + 150$),

$$h_b = 1325 \text{ BTU/lbm}$$

$$s_b = 1.5141 \text{ BTU/lbm-°R}$$

At point *c*, the two-phase properties are calculated from the properties of the saturated phases determined from the saturated steam table (Appendix E) at 3 psia:

$$h_f = 109.37 \text{ BTU/lbm}$$

$$h_{fg} = 1013.2 \text{ BTU/lbm}$$

$$s_f = 0.2008 \text{ BTU/lbm-°R}$$

$$s_{fg} = 1.6855 \text{ BTU/lbm-°R}$$

$$x_c = \frac{1.5141 - 0.2008}{1.6855} = 0.7792$$

$$h_c = 109.37 + 0.7792(1013.2) = 898.86 \text{ BTU/lbm}$$

$$\eta_{th} = \frac{(h_b - h_a) - (h_c - h_d)}{h_b - h_a}$$

$$= \frac{(1325 - 112.38) - (898.86 - 109.37)}{1325 - 112.38}$$

$$= 0.3489 \quad (34.89\%)$$

Refrigeration Cycles

Since it is a reversible cycle, the Carnot cycle can be reversed. All the quantities are the same as the Carnot cycle acting as a power cycle, and the efficiency or *coefficient of performance*, (COP), is again a function of the absolute temperatures of the reservoirs alone.

The primary difference between the Carnot power cycle and the Carnot refrigeration cycle is that, instead of extracting work, work is used to pump heat from the hotter temperature reservoir to the cooler temperature reservoir. Refrigerators work on this principle. The reservoir at T_c is the *cold box* or refrigerator temperature. The reservoir T_h is the environment temperature to which the heat is discarded.

It takes work to produce a refrigeration effect. This requirement is formalized by the *Clausius form* of the second law: you cannot construct a cyclical device which produces no effect other than the transfer of heat from a cooler to a hotter body.

A measure of usefulness in refrigeration cycles is the coefficient of performance

$$\mathrm{COP} = \frac{Q_{in}}{W_{in}} = \frac{Q_{in}}{Q_{out} - Q_{in}} \quad \text{(refrigerators)}$$
$$= \frac{Q_{in} + W_{in}}{Q_{in}} \quad \text{(heat pumps)} \qquad 4.103$$

The *capacity of a refrigerator* is the rate, expressed in tons, at which heat is removed from the cold box. A *ton* is 12,000 BTU/hr or 200 BTU/min (named for the heat flow required to melt a ton of ice in 24 hours). The relationship between COP, capacity in tons, and horsepower is

$$\mathrm{COP} = \frac{4.715\ \text{(tonnage)}}{\text{horsepower}} \qquad 4.104$$

A new term used with appliance products to specify the performance of cooling devices is the *energy efficiency ratio* (EER).

$$\mathrm{EER} = \frac{\text{cooling, BTU/hr}}{\text{watts}} \qquad 4.105$$

Example 4.5

A 100-watt motor powers a small refrigerator which discards heat to the surroundings at 70 °F. The heat absorbed in the refrigerator is 19.9 BTU/min. What is the theoretical minimum temperature that can be maintained in the refrigerator?

The maximum coefficient of performance of a refrigerator operating on a reversed Carnot cycle is

$$\mathrm{COP} = \frac{Q_{in}}{W} = \frac{T_c}{T_h - T_c}$$

Rearranging,

$$T_c = \frac{T_h}{\frac{W}{Q_c} + 1}$$

$$T_h = 70 + 460 = 530\,^\circ\mathrm{R}$$
$$W = 100 \text{ watts} = 100(0.05687) = 5.687 \text{ BTU/min}$$
$$Q_c = 19.9 \text{ BTU/min}$$
$$T_c = \frac{530}{\frac{5.687}{19.9} + 1} = 412.2\,^\circ\mathrm{R} \qquad (-47.8\,^\circ\mathrm{F})$$

5 CHEMICAL THERMODYNAMICS

A. PROBLEM TYPES

There are two common types of chemical thermodynamic problems:

- calculating the heats of reaction at non-standard conditions
- calculating the equilibrium concentrations of the reactants and products, given the equilibrium constants or other thermodynamic properties

B. STANDARD HEATS OF REACTION

Heat effects generally accompany chemical reactions due to the differences in molecular structures of the products and the reactants. Heat can be either evolved or absorbed in a reaction. *Heat of reaction* is the heat *absorbed* by the system after the products have been restored to the same temperature as the reactants.

The energy change of the reaction overall is known as the heat of reaction, ΔH_{rx}, and is the difference between the enthalpy of the products and the enthalpy of the reactants.

$$\Delta H_{rx} = \Delta H_{products} - \Delta H_{reactants} \qquad 4.106$$

Since the heat of reaction is dependent on temperature, it is customary to define a standard heat of reaction based on 25 °C (or occasionally 18 °C) and 1 atmosphere pressure. Standard heat of reaction is designated here as ΔH°_{rx}.

C. HEATS OF REACTION FROM HEATS OF FORMATION AND HEATS OF COMBUSTION

Since enthalpy is a function of the state of the system only, the heat of reaction for any reaction can be evaluated by adding the heats of reaction for the individual reactions whose sum gives the overall desired change. The heats of reaction can be obtained from either the heat of formation or the heat of combustion. The sign convention is different with these two sources, however. Using the heat of formation data,

$$\Delta H_{rx} = \sum_{products} n\Delta H_{formation} - \sum_{reactants} n\Delta H_{formation} \qquad 4.107$$

Using heat of combustion data,

$$H_{rx} = \sum_{reactants} nH_{combustion} - \sum_{products} nH_{combustion} \qquad 4.108$$

The heat of formation of an element is zero.

Example 4.6

Using data from Appendix H, determine the heat of reaction for the following reaction.

$$C(s) + 2S(s) \rightarrow CS_2(g)$$
$$\Delta H_{rx} = 28.11 - 0 = 28.11 \text{ kcal/gmole}$$

Example 4.7

Calculate the heat of reaction for the following reaction. Assume $T = 25\,°C$.

$$C_2H_4(g) + 3O_2(g) \rightarrow 2CO_2(g) + 2H_2O(l)$$

From equation 4.107,

$$\Delta H_{rx} = 2(-94.05) + 2(-68.32) - 12.50$$
$$= -337.24 \text{ kcal/gmole}$$

D. HEAT OF REACTION AT NON-STANDARD CONDITIONS

Heat of reaction must be corrected for the difference in temperature between 25 °C and the actual reaction temperature. (This assumes T is in °C.) This is accomplished by using the average heat capacities of the reactants and products.

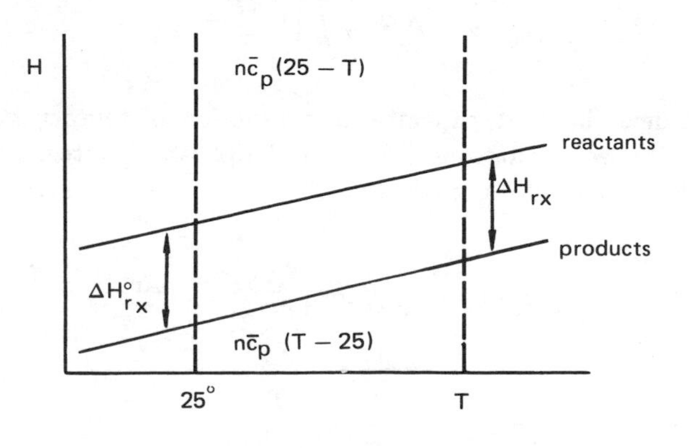

Figure 4.5 Heat of Reaction Relationships at Two Temperatures

$$\Delta H_{rx} = \sum_{reactants} nC_p(25 - T) + \Delta H_{rx,\ 25°} + \sum_{products} nC_p(T - 25) \qquad 4.109$$

or

$$\Delta H_{rx} = \Delta H_{rx,\ 25°} + \left(\sum_{products} nC_p - \sum_{reactants} nC_p \right) (T - 25) \qquad 4.110$$

E. GIBBS FREE ENERGY (Gibbs Function)

Gibbs function is a thermodynamic property that measures the net reversible work for a process. Gibbs function is defined

$$G = H - TS = \left(U + \frac{pV}{J} \right) - TS \qquad 4.111$$

For changes taking place at constant temperature, or for changes where, regardless of path, the initial and final temperatures are the same,

$$\Delta G = \Delta H - T\Delta S \qquad 4.112$$

From the first law,

$$Q = \Delta U + W$$
$$= \Delta U + W_{net\ reversible} + W_{expansion} \qquad 4.113$$

For work done at constant pressure and temperature,

$$W_{expansion} = \frac{p\Delta V}{J} \qquad 4.114$$

Then, for a reversible process at constant pressure and temperature,

$$Q = \Delta U - \Delta G + \frac{p\Delta V}{J} = -\Delta G + \Delta H \qquad 4.115$$
$$= T\Delta S \qquad 4.116$$

For an irreversible process, $T\Delta S > Q$, and the useful work is always less than the decrease in the free energy of the system. The differential relationship between Gibbs function, temperature, and pressure is

$$dG = V\,dp - S\,dT \qquad 4.117$$

THERMO

When a process is restricted to constant temperature,

$$dG = V\,dp \qquad 4.118$$

$$\Delta G = \int V\,dp \qquad 4.119$$

For an ideal gas, since $V = \frac{RT}{p}$, the Gibbs function is

$$\Delta G = RT \int \frac{dp}{p} = RT\, ln\left(\frac{p_2}{p_1}\right) \qquad 4.120$$

The Gibbs function is an extensive property, and G for any process is determined only by the final and initial conditions, not by the intermediate path, (i.e., ΔG is treated in a similar manner to ΔH).

There are two types of free energy change calculations. Both are taken at the temperature of the system and one atmosphere pressure.

- *Actual* ΔG: This corresponds to the actual concentrations of products and reactants in the system. At reaction equilibrium $\Delta G = 0$.
- *Standard* ΔG°: This refers to each reactant in the standard state of unit activity, and with ending products each at unit activity. These standard states do not correspond to the equilibrium states.

THERMO

Example 4.8

Calculate ΔG° in gas phase alkylation of isobutane with C_2H_4 to form neohexane. Use the following information.

	ΔG° cal/gmole
$iC_4H_{10(g)}$	-5000
$C_2H_{4(g)}$	$+16,282$
$C_6H_{14(g)}$	-2370

$$iC_4H_{10} + C_2H_4 \rightarrow C_6H_{14}$$

$$\Delta G^\circ_{25} = \sum_{products} \Delta G^\circ - \sum_{reactants} \Delta G^\circ$$

$$-2370 - (-5000 + 16,282)$$

$$= -13,652 \text{ cal/gmole}$$

The value of ΔG is of theoretical interest only. Most calculations use ΔG° at the temperature of the system. The free energy and equilibrium constant for a reaction can be calculated at any temperature if enough information is given about the change in heat capacity with temperature for the reactants and products, and the value of the heats of formation and absolute entropy at one temperature.

Consider the following generalized reaction with stoichiometric amounts of reactants.

$$rR + sS \rightarrow pP + qQ \qquad 4.121$$

The generalized form of the molal heat capacity is

$$C_p = a + bT + cT^2 + dT^3 + \frac{e}{T^2} \text{ cal/gmole-}^\circ\text{K} \qquad 4.122$$

Let the subscripts R, S, P, and Q stand for the components in equation 4.121. If the heats of formation and absolute entropy are given for some reference temperature, say 298 °K, then the entropy change, heat of reaction, and the coefficient differences are defined by equations 4.123 through 4.129.

$$\Delta S^\circ_{298} = p(S^\circ_{298})_P + q(S^\circ_{298})_Q - r(S^\circ_{298})_R - s(S^\circ_{298})_S \qquad 4.123$$

$$\Delta H^\circ_{298} = p(H^\circ_{298})_P + q(H^\circ_{298})_Q - r(H^\circ_{298})_R - s(H^\circ_{298})_S \qquad 4.124$$

$$\Delta a = pa_P + qa_Q - ra_R - sa_S \qquad 4.125$$

$$\Delta b = pb_P + qb_Q - rb_R - sb_S \qquad 4.126$$

$$\Delta c = pc_P + qc_Q - rc_R - sc_S \qquad 4.127$$

$$\Delta d = pd_P + qd_Q - rd_R - sd_S \qquad 4.128$$

$$\Delta e = pe_P + qe_Q - re_R - se_S \qquad 4.129$$

Free energy is defined as

$$\Delta G = \Delta H - T\Delta S \qquad 4.130$$

The equations for enthalpy and entropy change are

$$\Delta H = \int C_p dT \qquad 4.131$$

$$\Delta S = \int C_p \frac{dT}{T} \qquad 4.132$$

Since the heat capacity as a function of temperature is known, equations 4.131 and 4.132 can be integrated, with the following results:

$$\Delta H^\circ_T = I_h + \Delta aT + \frac{\Delta bT^2}{2} + \frac{\Delta cT^3}{3} + \frac{\Delta dT^4}{4} - \frac{\Delta e}{T} \qquad 4.133$$

$$\Delta S^\circ_T = I_S + \Delta a\, lnT + \Delta bT + \frac{\Delta cT^2}{2} + \frac{\Delta dT^3}{3} - \frac{\Delta e}{2T^2} \qquad 4.134$$

The constants of integration, I_h and I_S, can be evaluated by using the values of ΔS^o_{298} and ΔH^o_{298} calculated in equations 4.123 and 4.124 (since $T = 298$ °K). Combining equations 4.130, 4.133, and 4.134, the free energy is

$$\frac{\Delta G^o_T}{T} = \frac{I_h}{T} + (\Delta a - I_S) - \frac{\Delta bT}{2} - \frac{\Delta cT^2}{6} - \frac{\Delta dT^3}{12} - \frac{\Delta e}{2T^2} - \Delta a\ lnT \qquad 4.135$$

The equilibrium constant can be calculated since the free energy and equilibrium constant are related:

$$\frac{\Delta G^o_T}{T} = -R\ lnK \qquad 4.136$$

If, at standard temperature, T_o, the value of H^o_o and the quantities $\frac{G_o - H_o}{T_o}$ are known, the constant of integration in equation 4.133 can be evaluated and equations 4.133 and 4.34 can be rearranged. τ is a dimensionless temperature, $\tau = \frac{T}{T_o}$.

$$\Delta H_o = \Delta H^o_o + T_o\Delta a(\tau - 1) + \frac{T_o^2\Delta b}{2}(\tau^2 - 1) + \frac{T_o^3\Delta c}{3}(\tau^3 - 1) + \frac{T_o^4\Delta d}{4}(\tau^4 - 1) + \frac{\Delta e(\tau - 1)}{T_o\tau} \qquad 4.137$$

$$\Delta G_o = \Delta H^o_o + (\Delta G^o_o - \Delta H^o_o)\tau + T_o\Delta a(\tau - 1 - \tau ln\tau) - \frac{T_o^2\Delta b}{2}(\tau^2 - 2\tau + 1) - \frac{T_o^3\Delta c}{6}(\tau^3 - 3\tau + 2) - \frac{T_o^4\Delta d}{12}(\tau^4 - 4\tau + 3) - \frac{\Delta e(\tau^2 - 2\tau + 1)}{2T_o\tau} \qquad 4.138$$

Example 4.9

Determine the equilibrium constant for the reaction that converts calcium carbonate to slaked lime at 725 °C.

$$CaCO_3 \rightarrow CaO + CO_2$$

The molal heat capacity of all the reactants and products follow the generalized equation

$$C_p = a + bT + \frac{e}{T^2}\ \text{cal/gmole°K}$$

Use the following data:

	a	b × 10^3	e × 10^{-5}	ΔH^o_{298} (cal/gmole)	ΔG^o_{298} (cal/gmole)
$CaCO_3$	24.98	5.24	−6.20	−288,450	−269,780
CaO	11.67	1.08	−1.56	−151,900	−144,400
CO_2	10.55	2.16	−2.04	-94,052	-94,260

Since S^o_{298} for the components is not given, it must be calculated from equation 4.130.

$$\Delta S^o_{298} = -\frac{\Delta G^o_{298} - \Delta H^o_{298}}{T}$$

$$\Delta G^o_{298} = (-144,400) + (-94,260) - (-269,780) = 31,120\ \text{cal/gmole}$$

$$\Delta H^o_{298} = (-151,900) + (-94,052) - (-288,450) = 42,498\ \text{cal/gmole}$$

$$\Delta S^o_{298} = \frac{-31,120 - 42,498}{298} = 38.181\ \text{cal/gmole-°K}$$

$$\Delta a = 11.67 + 10.55 - 24.98 = -2.76$$

$$\Delta b = (1.08 + 2.16 - 5.24) \times 10^{-3} = -0.002$$

$$\Delta e = (-1.56 - 2.04 + 6.20) \times 10^5 = 260,000$$

Now, evaluate the integration constants:

$$\Delta H^o_{298} = 42,498 = I_h + (-2.76)(298) - \frac{(0.02)(298)^2}{2} - \frac{260,000}{298}$$

$$I_h = 44,281.8$$

$$\Delta S^o_{298} = 38.181 = I_s + (-2.76)ln(298) + (-0.002)(298)^2 - \frac{260,000}{2(298)^2}$$

$$I_s = 55.965$$

THERMO

At $T = 273 + 750 = 1023\ ^\circ K$, the free energy is calculated as

$$\frac{\Delta G_T^o}{T} = \frac{44,281.8}{1023} + (-2.76 - 55.965) - (-0.002)\left(\frac{1023}{2}\right) - \frac{260,000}{2(1023)^2} - (-2.76\ ln\ 1023)$$

$$\frac{\Delta G_T^o}{T} = 4.588$$

$$-\left(\frac{\Delta G_T^o}{T}\right)\left(\frac{1}{R}\right) = lnK$$

$$= -4.588\left(\frac{1}{1.987}\right) = -2.309$$

The free energy can be calculated for any temperature, provided any two of the three thermodynamic properties, G, H, or S, are given at one temperature, and the heat capacity as a function of temperature is known. If the values of ΔG_T^o are given for each component, ΔG_T^o for the reaction is calculated by the summation principle similar to that used in the calculation for the heat of reaction in example 4.9.

THERMO

Sometimes, along with the heats of formation, H_f at 298 °K, the data may be given as the *Gibbs energy function (gef)* defined

$$gef = \frac{-(\Delta G_T^o - \Delta H_{298}^o)}{T} \qquad 4.139$$

To find the free energy at any temperature from this data, the gef must be given at the temperature in question. However, the gef changes slowly with temperature changes, and linear interpolations can be made for intermediate temperatures. The free energy of the reaction is calculated with the data in equations 4.140 and 4.141.

$$\Delta\ gef = \sum_{products} gef_{products} - \sum_{reactants} gef_{reactants} \qquad 4.140$$

$$\Delta H_{298}^o = \sum(\Delta H_f^o)_{products} - \sum(\Delta H_f^o)_{reactants} \qquad 4.141$$

The free energy is

$$\frac{-\Delta G_T^o}{T} = \Delta gef - \frac{\Delta H_{298}^o}{T} \qquad 4.142$$

Example 4.10

Compute the equilibrium constant and conversion at 1200 °K for the reaction given, when one mole of steam per mole of methane are reacted at one atmosphere.

$$CH_4(g) + H_2O(g) \leftrightarrow 3H_2(g) + CO(g)$$

Use the following data:

	$-\left[\frac{\Delta G_T^o - \Delta H_{298}^o}{T}\right]$	$\Delta H_{f,298}^o$ (cal/gmole)
CH_4	51.814	-17,895
H_2O	50.575	-57,797.9
H_2	35.636	0
CO	51.834	-26,420

The *Gibbs free energy function* is given at 1200 °K with units of cal/gmole-°K. Using equations 4.138 and 4.139,

$$\Delta gef = 51.834 + 3(35.636) - 50.575 - 51.814 = 56.353\ \text{cal/gmole-}^\circ\text{K}$$

$$\Delta H_f = -26,420 - (-17,895) - (-57,797.9) = 49,272.9\ \text{cal/gmole}$$

$$-\frac{\Delta G_T^o}{T} = 56.353 - \frac{49,272.9}{1200} = 15.292$$

$$lnK = \frac{15.292}{1.987} = 7.696$$

$$K = 2200$$

Let h represent the number of moles of CH_4 at equilibrium. Assume there are 100 moles of CH_4 at the start.

	moles at equilibrium	mole fraction
CH_4	h	$\frac{h}{500-2h}$
H_2O	$100 + h$	$\frac{100+h}{500-2h}$
CO	$100 - h$	$\frac{100-h}{500-2h}$
H_2	$300 - 3h$	$\frac{300-3h}{500-2h}$
	$500 - 2h$	

$$K = \frac{(100-h)(300-3h)^3}{h(100+h)(500-2h)^2}(1)^{3+1-1-1} = 2200$$

As a first approximation, assume that h is small compared to 100. Then, the equation for the equilibrium constant reduces to

$$K = 2200 = \frac{108}{h}$$

$$h = 0.0491$$

As a check, insert h into the original equation:

$$K = \frac{(100 - 0.0491)(300 - (3)0.0491)^3}{0.0491(100 + 0.0491)(500 - (2)0.0491)^2} = 2195.5$$

(This solution would normally be sufficient.)

Several more iterations produce

$$h = 0.04899 \quad (\text{check: yields } K = 2200.00)$$

Since we chose 100 moles to start, conversions can be calculated directly:

$$\text{conversion of } CH_4 = 100 - 0.04899 = 99.95\%$$

6 CHEMICAL EQUILIBRIUM

Consider the typical reaction

$$aA + bB + \ldots \leftrightarrow pP + qQ \ldots \qquad 4.143$$

The *equilibrium constant*, K (actually a ratio which varies with temperature only), expresses the equilibrium of the reaction with the final concentrations of the products and the reactants. The general form of the equilibrium constant for liquids and gases is

$$K_\alpha = \frac{(\alpha_P)^p(\alpha_Q)^q}{(\alpha_A)^a(\alpha_B)^b} \qquad 4.144$$

Activities are approximately equal to concentrations (for liquids), and can be expressed either as mole/liter or as mole fractions. For gases only, the equivalent form of the equilibrium constant is

$$K_\alpha = K_\gamma K_\phi K_p \qquad 4.145$$

$$K_p = \frac{(p_P)^p(p_Q)^q}{(p_A)^a(p_B)^b} \qquad 4.146$$

The partial pressures in equation 4.146 must be expressed in atmospheres.

For ideal gases, $K_\gamma = K_\phi = 1$, so that $K_\alpha = K_p$. *Dalton's law of partial pressures* states that the sum of the partial pressures is equal to the total pressure, so the ratio K_p can be rewritten. Since the partial pressure is $p_i = y_i p_t$,

$$K_p = \frac{(y_P p)^p(y_Q p)^q}{(y_A p)^a(y_B p)^b} = K_y p^{p+q-a-b} \qquad 4.147$$

$$K_y = \frac{(y_P)^p(y_Q)^q}{(y_A)^a(y_B)^b} \qquad 4.148$$

The relationship between the equilibrium constant and free energy is

$$\Delta G - \Delta G^o = RT\ lnK \qquad 4.149$$

At equilibrium, $\Delta G = 0$. The working equation for free energy is

$$-\Delta G^o = RT\ lnK_\alpha \qquad 4.150$$

$$= 1.987\,T\ lnK_\alpha \qquad 4.151$$

$$= 4.576\,T\ \log_{10}K_\alpha \qquad 4.152$$

Given the equilibrium constant, the free energy can be calculated by equations 4.151 and 4.152. The equilibrium constant can be calculated from

$$K_\alpha = e^{\frac{-\Delta G^o}{RT}} \qquad 4.153$$

At constant pressure and temperature, ΔG^o is used as a rough criterion of equilibrium chemical conversion.

Table 4.3
Gibbs Function vs. Probability of Reaction

ΔG^o	probability of reaction
<0	probable/high
0 – 10,000	doubtful/low
> 10,000	very unfavorable/very low

The values of ΔG^o should be evaluated at system temperature (not 25 °C). *Le Chatelier's Principle* can be used to predict the effects of reaction conditions on equilibrium.

THERMO

The change of equilibrium constant with temperature can be determined by the *van't Hoff equation,*

$$\frac{d(lnK)}{dT} = \frac{\Delta H^o}{RT^2} \qquad 4.154$$

$$\frac{d(R\ ln\ K)}{d\left(\frac{1}{T}\right)} = \frac{d\left(\frac{-\Delta G^o}{T}\right)}{d\left(\frac{1}{T}\right)} = -\Delta H^o \qquad 4.155$$

A plot of $R\ lnK$ or $\frac{-\Delta G^o}{T}$ versus $\frac{1}{T}$ results in a straight line with slope $-\Delta H^o$.

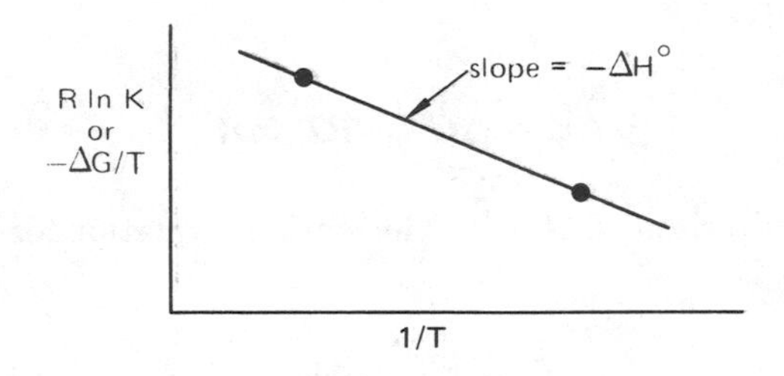

Figure 4.6 Graph of van't Hoff Equation

The van't Hoff equation allows the calculation of the equilibrium constant at any temperature, given the equilibrium constant at two different temperatures. The importance of this relationship is illustrated by examples 4.11 and 4.12.

Example 4.11

The equilibrium constant for a reaction is 10 at 25 °C and 2000 at 927 °C. What is the equilibrium constant at 200 °C?

Let the subscripts 1, 2, and 3 stand for the conditions at 25, 200, and 927 °C, respectively. Then,

$$\frac{lnK_2 - lnK_1}{\frac{1}{T_2} - \frac{1}{T_1}} = \frac{lnK_3 - lnK_1}{\frac{1}{T_3} - \frac{1}{T_1}}$$

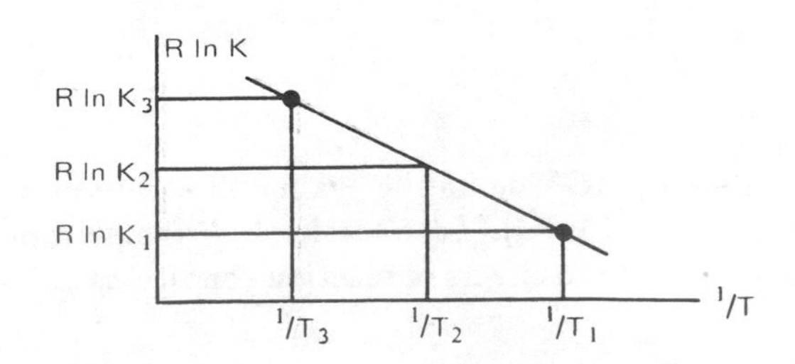

Therefore,

$$\frac{lnK_2 - ln(10)}{\frac{1}{473} - \frac{1}{298}} = \frac{ln(2000) - ln(10)}{\frac{1}{1200} - \frac{1}{298}}$$

$$lnK_2 = 4.91047$$
$$K_2 = 135.7$$

Example 4.12

What is the enthalpy change at 267 °C in example 4.11 at equilibrium?

$$-\Delta H^o = \frac{ln(2000) - ln(10)}{\frac{1}{1200} - \frac{1}{298}}(R)$$
$$-\Delta H^o = (-2100)(1.987) = -4172.7\ \text{cal/gmole}$$
$$\Delta H^o = 4172.7\ \text{cal/gmole}$$

Use of the van't Hoff equation implies the change in enthalpy at equilibrium is constant at any temperature.

A. SOLUTION THERMODYNAMICS

When a gas is in equilibrium with a non-ideal liquid, the resulting calculations become complex. To simplify the equations involved, it is necessary to assume that one of the phases is ideal. At common pressures and temperatures, the vapor phase can usually be considered ideal, and the vapor-liquid equilibrium expression can be simplified to a modified version of the ***Lewis-Randall relationship.*** For an activity coefficient, γ, and partial pressure, p.

$$\gamma_1 x_1 p_1^o = y_1 p_t \qquad 4.156$$

The subscript 1 denotes the component. In non-ideal liquid problems, the vapor pressure of the pure component is usually assumed to behave according to

$$ln\ p^o = A + \frac{B}{T} \qquad 4.157$$

The vapor pressure is a function of temperature only; the activities are functions of composition only. The usual problem dealing with non-ideal liquids is to find the composition of the vapor-liquid equilibrium at one state, given the composition at another state. To do this, you must know the form of the ***activity coefficients.***

Two common types of equations for activity coefficients are the ***van Laar equations*** for binary mixtures and the ***three-suffix Margules equations.***

The van Laar equations are

$$ln\ \gamma_1 = \frac{A_{12}}{\left[1+\left(\frac{A_{12}x_1}{A_{21}x_2}\right)\right]^2} \qquad 4.158$$

$$ln\ \gamma_2 = \frac{A_{21}}{\left[1+\left(\frac{A_{21}x_2}{A_{12}x_1}\right)\right]^2} \qquad 4.159$$

The three-suffix Margules equations are

$$\ln\gamma_1 = x_2^2[A_{12} + 2x_1(A_{21} - A_{12})] \qquad 4.160$$

$$\ln\gamma_2 = x_1^2[A_{21} + 2x_2(A_{12} - A_{21})] \qquad 4.161$$

In order to determine the coefficients, A_{12} and A_{21}, something must be known about the composition at equilibrium. For example, the liquid and vapor compositions are identical at the *azeotrope*. Another case would be if the activity were known at infinite dilution. (Setting $x = 0$ would result in the determination of the coefficients.)

Substituting the van Laar or Margules equation and the relationship for vapor pressure into equation 4.156 requires a very difficult mathematical solution. Trial and error solutions are often required.

If the azeotropic composition is given, the van Laar or Margules constant can be calculated. Since $x_1 = y_1$ at the azeotrope, equation 4.156 becomes

$$\gamma_1 = \frac{p_t}{p_1^o} \qquad 4.162$$

Equations 4.158 through 4.161 can be rearranged to compute the van Laar or Margules constants.

$$\text{van Laar}: \quad A_{12} = ln\gamma_1\left(1+\frac{x_2\ ln\gamma_2}{x_1\ ln\gamma_1}\right)^2 \qquad 4.162$$

$$A_{21} = ln\gamma_2\left(1+\frac{x_1\ ln\gamma_1}{x_2\ ln\gamma_2}\right)^2 \qquad 4.164$$

$$\text{Margules}: \quad A_{12} = \frac{1-2x_1}{x_2^2}\ ln\gamma_1 + \frac{2}{x_1}\ ln\gamma_2 \qquad 4.165$$

$$A_{21} = \frac{1-2x_2}{x_1^2}\ ln\gamma_2 - ln\gamma_1 \qquad 4.166$$

Example 4.13

Compute the van Laar constants for the binary solution of hydrofluoric acid and water. An azeotrope exists at 120 °C with 37% HF by weight.

Use the following vapor pressure data:

	A	B	C
HF	19.2965	4495.9	335.52
H_2O	18.6686	4030.2	235.00

$$\ln p^o = A - \frac{B}{C+t} \qquad p^o,\ \text{mm Hg} \quad t\ ^\circ\text{C}$$

Assume 1 atmosphere. At the azeotrope,

$$x_{HF} = \frac{\frac{0.37}{20}}{\frac{0.37}{20}+\frac{0.63}{18}} = 0.346$$

From vapor pressure at 120 °C,

$$\ln p_{HF}^o = 19.2965 - \frac{4495.9}{335.52+120}$$

$$p_{HF}^o = 12,415.2\ \text{mm Hg}$$

$$\ln p_{H_2O}^o = 18.6686 - \frac{4030.2}{235+120}$$

$$p_{H_2O}^o = 1504.06\ \text{mm Hg}$$

Using equation 4.156,

$$\gamma_{HF} = \frac{760}{12,415.2} = 0.0612$$

$$\gamma_{H_2O} = \frac{760}{1504.06} = 0.5053$$

$$A_{HF} = ln\ 0.0612\left(\frac{1+0.654\times ln\ 0.5053}{0.354\times ln\ 0.0612}\right)^2$$

$$= -5.885$$

$$A_{H_2O} = ln\ 0.5053\left(\frac{1+0.354\times ln\ 0.0612}{0.654\times ln\ 0.5053}\right)^2$$

$$= -7.057$$

7 MATHEMATHICS OF THERMODYNAMICS

The thermodynamic relationships dealt with in introductory sections of thermodynamic process include the quantities Q and W, which are not properties but the products of such processes. For reversible processes, Q and W can be replaced by thermodynamic state variables by mathematical techniques. The state variables such as U, H, and S, for example, can be expressed entirely in terms of T, V, and p or other measurable quantities. Except for expansion, compression, and piston or balloon processes, volume is typically a constant in chemical engineering processes. Therefore, thermodynamic state functions in terms of pressure and temperature are usually desired. The mathematical properties of thermodynamic state functions allow for the development of a great number of relationships interconnecting them.

If a thermodynamic function Y is a function of independent variables X_i, then

$$Y = f(X_1, X_2, X_3 \cdots X_n) \qquad 4.167$$

Thermodynamic functions can be expressed differentially as

$$dY = C_1 dX_1 + C_2 dX_2 + \cdots + C_n dX_n \qquad 4.168$$

where C_1 is a function of X_1 only, C_2 is a function of X_2 only, etc. Equation 4.168 is then defined mathematically as an exact differential. Exact differentials have the following property when expressed in partial differential form:

$$dY = \left(\frac{\partial Y}{\partial X_1}\right)_{X_j} dX_1 + \left(\frac{\partial Y}{\partial X_2}\right)_{X_j} dX_2 + \cdots + \left(\frac{\partial Y}{\partial X_n}\right)_{X_j} dX_n \qquad 4.169$$

where the subscript X_j means that all the independent variables X_i are held constant except the ones in the derivative itself. As a consequence, for any pair of independent variables X_i and X_k in an exact differential, the cross derivatives are:

$$\frac{\partial^2 Y}{\partial X_i \partial X_k} = \frac{\partial^2 Y}{\partial X_k \partial X_i} \qquad 4.170$$

Using equation 4.170 in 4.168 results in the relationship

$$\left(\frac{\partial C_k}{\partial X_i}\right)_{X_j} = \left(\frac{\partial C_i}{\partial X_k}\right)_{X_j} \qquad 4.171$$

This equation is true for any pair of independent variables. To illustrate the importance of equation 4.171, the equation for internal energy of a reversible process is

$$dU = TdS - pdV \qquad 4.172$$

Since U is a function of temperature and pressure, and dU is an exact differential, equation 4.171 implies that

$$\left(\frac{\partial T}{\partial V}\right)_S = -\left(\frac{\partial p}{\partial S}\right)_V \qquad 4.173$$

From other thermodynamic definitions,

$$\left(\frac{\partial T}{\partial p}\right)_S = \left(\frac{\partial V}{\partial S}\right)_p \qquad 4.174$$

$$\left(\frac{\partial p}{\partial T}\right)_V = \left(\frac{\partial S}{\partial V}\right)_T \qquad 4.175$$

$$\left(\frac{\partial V}{\partial T}\right)_p = -\left(\frac{\partial S}{\partial p}\right)_T \qquad 4.176$$

Equations 4.173, 4.174, 4.175, and 4.176 are known as the *Maxwell relations*. For common systems where a thermodynamic function, for example X, can be fully specified by fixing two state functions, for example Y and Z, such that

$$f(X, Y, Z) = 0 \qquad 4.177$$

The total differentials can be written

$$dX = \left(\frac{\partial X}{\partial Y}\right)_Z dY + \left(\frac{\partial X}{\partial Z}\right)_Y dZ \qquad 4.178$$

$$dY = \left(\frac{\partial Y}{\partial X}\right)_Z dX + \left(\frac{\partial Y}{\partial Z}\right)_X dZ \qquad 4.179$$

Eliminating dY from equations 4.178 and 4.179, and equating coefficients of the resulting equation, has the following result:

$$\left(\frac{\partial X}{\partial Y}\right)_Z = \frac{1}{\left(\frac{\partial Y}{\partial X}\right)_Z} \qquad 4.180$$

Equation 4.180 is important when algebraically manipulating differentials. The chain rule for differentiating provides still yet another important relationship:

$$\left(\frac{\partial X}{\partial Z}\right)_Y = -\left(\frac{\partial X}{\partial Y}\right)_Z \left(\frac{\partial Y}{\partial Z}\right)_X \qquad 4.181$$

Using equation 4.180, equation 4.181 can be rearranged as:

$$\left(\frac{\partial X}{\partial Y}\right)_Z = -\left(\frac{\partial X}{\partial Z}\right)_Y \left(\frac{\partial Z}{\partial Y}\right)_X \qquad 4.182$$

The introduction of a fourth state variable, W, results in the last of two important mathematical relationships derived from equations 4.180, 4.181, and 4.182:

$$\left(\frac{\partial X}{\partial Y}\right)_Z = \left(\frac{\partial X}{\partial W}\right)_Z \left(\frac{\partial W}{\partial Y}\right)_Z \qquad 4.183$$

$$\left(\frac{\partial X}{\partial Y}\right)_Z = \left(\frac{\partial X}{\partial Y}\right)_W + \left(\frac{\partial X}{\partial W}\right)_Y \left(\frac{\partial W}{\partial Y}\right)_Z \qquad 4.184$$

Using the relationships in equations 4.180 through 4.184 and the Maxwell relations, the thermodynamic functions can be arranged in countless forms. The following three examples show the use of the techniques of the mathematics of thermodynamics to relate thermodynamic functions in terms of T, p, and V.

Example 4.14

Express internal energy and enthalpy in terms of heat capacity, pressure and temperature.

The definition of heat capacity is:

$$c_v = \left(\frac{\partial U}{\partial T}\right)_V \qquad 4.185$$

$$dU = TdS - pdV \qquad 4.186$$

$$dU = \left(\frac{\partial U}{\partial T}\right)_V dT + \left(\frac{\partial U}{\partial V}\right)_T dV \qquad 4.187$$

Substituting 4.185 in 4.187,

$$dU = c_v dT + \left(\frac{\partial U}{\partial V}\right)_T dV \qquad 4.188$$

Dividing equation 4.186 by $(\partial V)_T$

$$\left(\frac{\partial U}{\partial V}\right)_T = T\left(\frac{\partial S}{\partial V}\right)_T - p \qquad 4.189$$

Therefore,

$$dU = c_v dT + \left[T\left(\frac{\partial S}{\partial V}\right)_T - p\right] dV \qquad 4.190$$

Using equation 4.175, internal energy is calculated as follows:

$$dU = c_v dT + \left[T\left(\frac{\partial p}{\partial T}\right)_V - p\right] dV \qquad 4.191$$

The definition of heat capacity is:

$$c_p \equiv \left(\frac{\partial H}{\partial T}\right)_p \qquad 4.192$$

$$dH = TdS + Vdp \qquad 4.193$$

$$dH = \left(\frac{\partial H}{\partial T}\right)_p dT + \left(\frac{\partial H}{\partial p}\right)_T dp \qquad 4.194$$

Substituting 4.192 in 4.194

$$dH = c_p dT + \left(\frac{\partial H}{\partial p}\right)_T dp \qquad 4.195$$

Divide 4.195 by $(\partial p)_T$

$$\left(\frac{\partial H}{\partial p}\right)_T = -T\left(\frac{\partial S}{\partial p}\right)_T + V \qquad 4.196$$

Therefore,

$$dH = c_p dT + \left[T\left(\frac{\partial S}{\partial p}\right)_T + V\right] dp \qquad 4.197$$

Using equation 4.176, enthalpy is calculated:

$$dH = c_p dT + \left[-T\left(\frac{\partial V}{\partial T}\right)_p + V\right] dp \qquad 4.198$$

Example 4.15

Express enthalpy in terms of variables of the equation of state for an ideal gas (p, V,and T).

For an ideal gas,

$$pV = RT$$

$$\left(\frac{\partial V}{\partial T}\right)_p = \frac{R}{p}$$

$$dH = c_p dt + \left[-T\left(\frac{R}{p}\right) + V\right] dp = c_p dT$$

Example 4.16

Express thermal expansivity, isothermal compressibility and Joule-Thompson coefficient in terms of the variables of the equation of state for an ideal gas and Van der Waals gas.

thermal expansivity:

$$\beta = \frac{1}{V}\left(\frac{\partial V}{\partial T}\right)_p$$

For an ideal gas,

$$\left(\frac{\partial V}{\partial T}\right)_p = \frac{\partial}{\partial T_p}\left(\frac{RT}{p}\right) = \frac{R}{p}$$

$$\beta_{ideal} = \frac{1}{V}\left(\frac{R}{p}\right) = \frac{1}{T}$$

For Van der Waals gas:

$$\frac{pV}{RT} = \frac{V}{V-b} - \frac{a}{RTV}$$

$$\left(\frac{\partial V}{\partial T}\right)_p = \frac{RV^3(V-b)}{RTV^3 - 2a(V-b)^2}$$

$$\beta_{Van\ der\ Waals} = \frac{RV^2(V-b)}{RTV^3 - 2a(V-b)^2}$$

thermal compressibility:

$$K \equiv \frac{-1}{V}\left(\frac{\partial V}{\partial p}\right)_T$$

for ideal gases:

$$\left(\frac{\partial V}{\partial p}\right)_T = \frac{-V}{p}$$

$$K_{ideal} = \frac{-1}{V}\left(\frac{-V}{p}\right) = \frac{1}{p}$$

THERMO

for Van der Waals gases:

$$\left(\frac{\partial V}{\partial p}\right)_T = \frac{V^3(V-b)^2}{2a(V-b)^2 - RTV^3}$$

$$K_{Van\ der\ Waals} = \frac{V^2(V-b)}{2a(V-b)^2 - RTV^3}$$

Joule-Thompson coefficient:

$$\mu \equiv \left(\frac{\partial T}{\partial p}\right)_H$$

but,

$$\left(\frac{\partial T}{\partial p}\right)_H = -\left(\frac{\partial T}{\partial H}\right)_p \left(\frac{\partial H}{\partial p}\right)_T$$

$$\left(\frac{\partial T}{\partial p}\right)_H = \frac{-1}{c_p}\left(\frac{\partial H}{\partial p}\right)_T$$

therefore:

$$\mu = \frac{-1}{c_p}\left[V - T\left(\frac{\partial V}{\partial T}\right)_p\right]$$

for ideal gases:

$$\left(\frac{\partial V}{\partial T}\right)_p = \frac{R}{p}$$

$$\mu_{ideal} = \frac{-1}{c_p}\left[V - T\left(\frac{R}{p}\right)\right] = 0$$

for Van der Waals gases:

$$\mu = \frac{-1}{c_p}\left[V - \frac{RTV^3(V-b)}{RTV^3 - 2a(V-b)^2}\right]$$

Appendix A

Nelson-Obert Generalized Compressibility Chart
Low Pressure Range

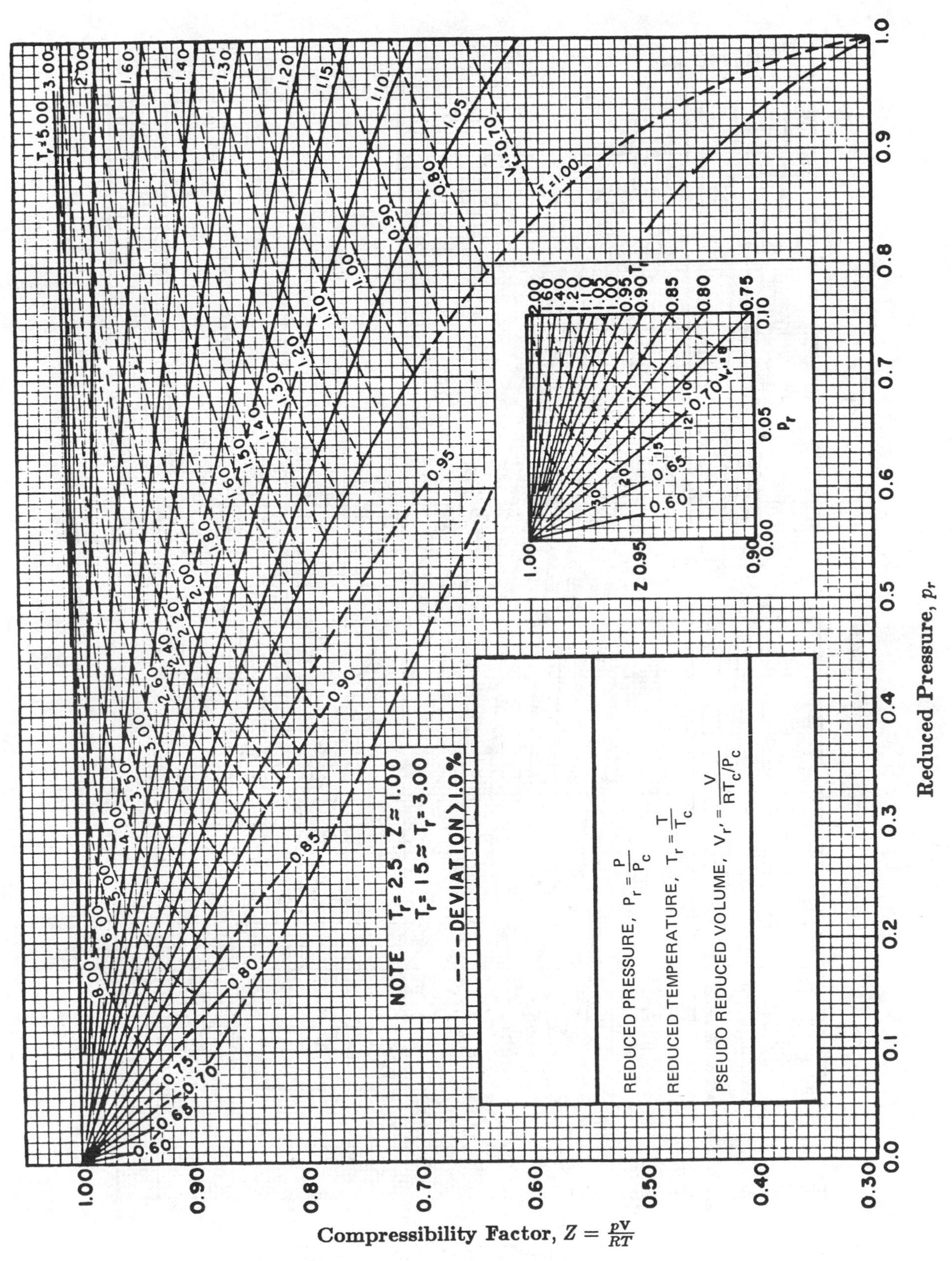

Appendix B
Nelson-Obert Generalized Compressibility Chart
Intermediate Pressure Range

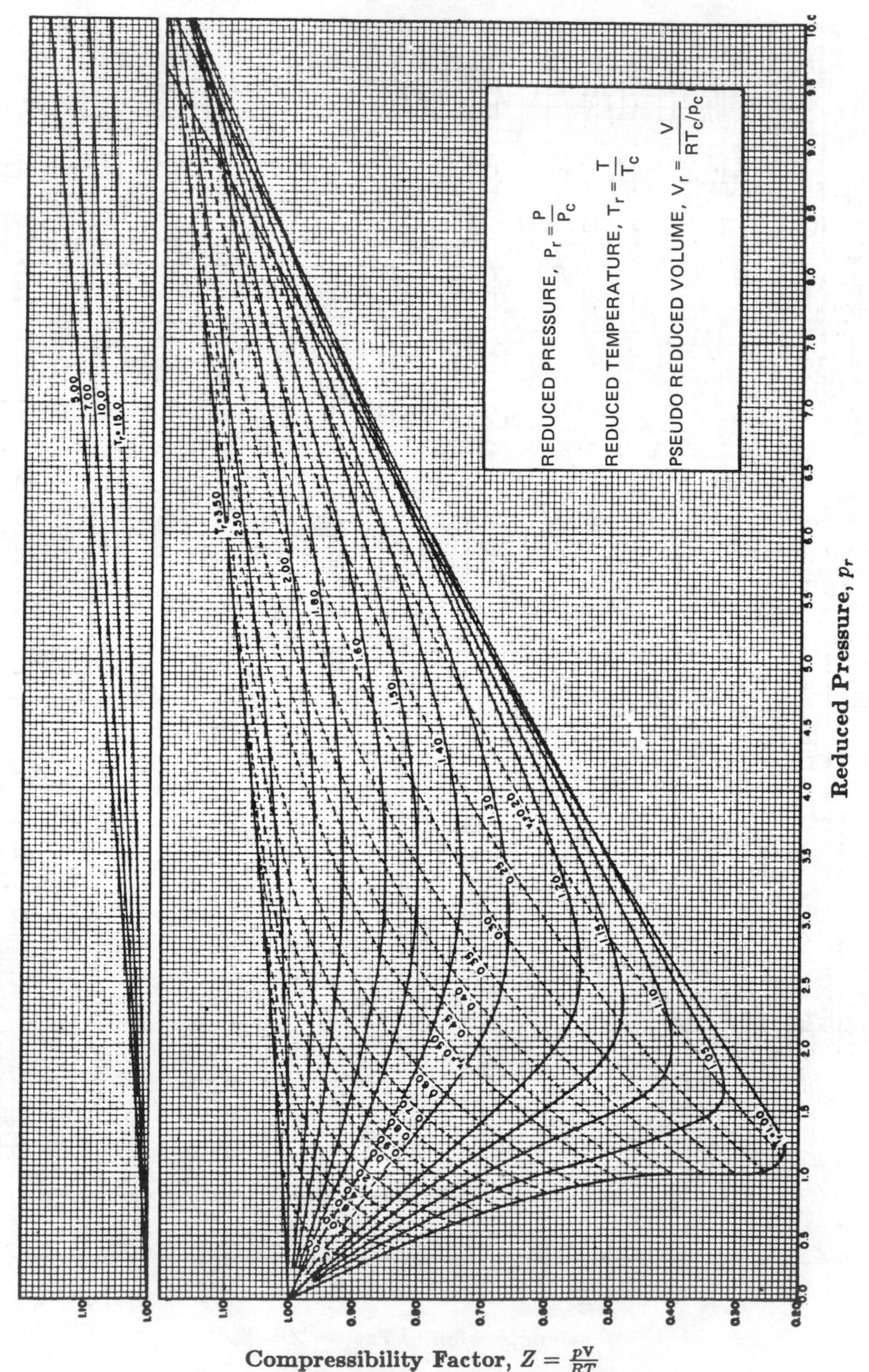

Appendix C

Nelson-Obert Generalized Compressibility Chart
High Pressure Range

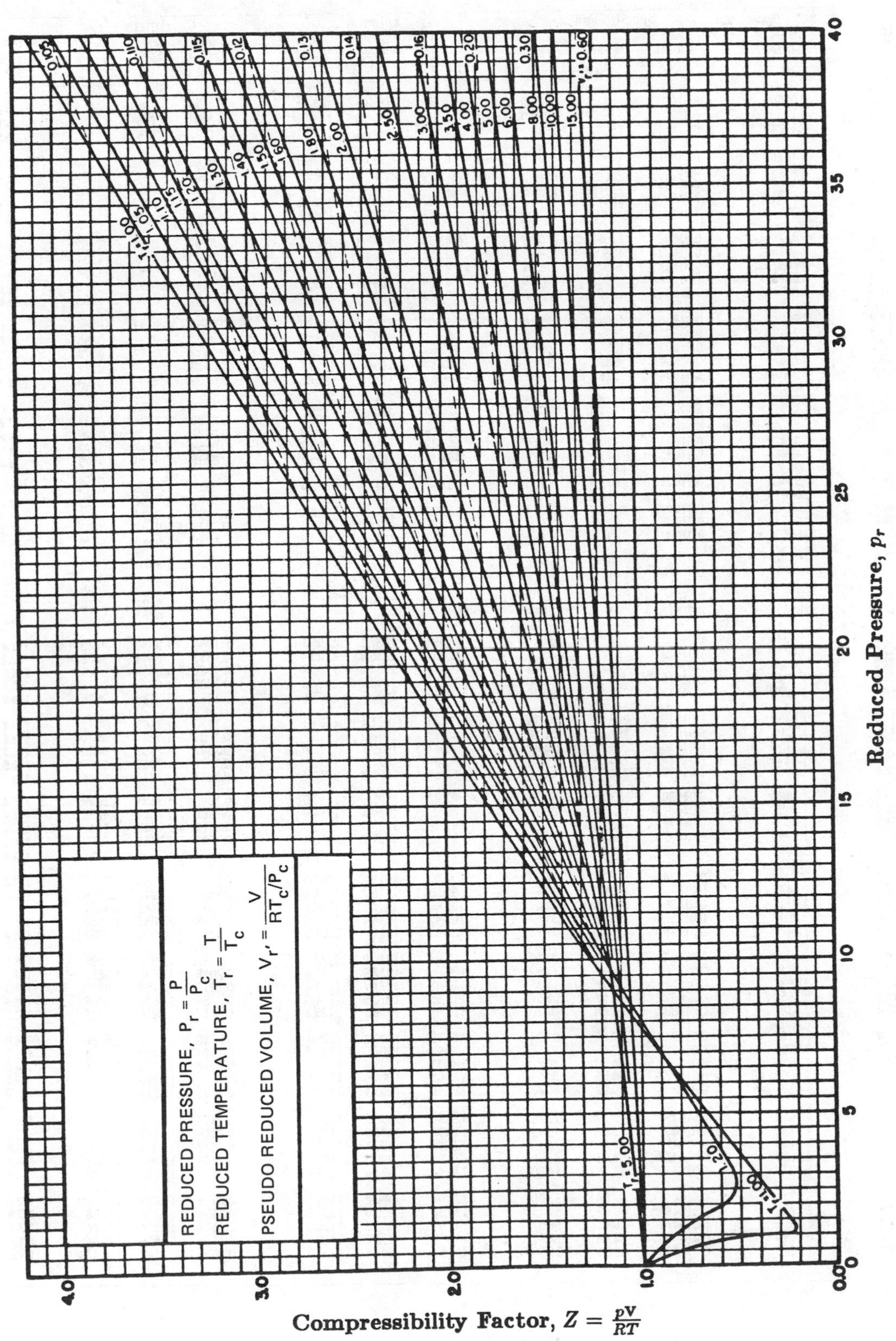

Appendix D
Saturated Steam: Temperature

Temp Fahr t	Abs Press. Lb per Sq In. p	Specific Volume Sat. Liquid v_f	Specific Volume Evap v_{fg}	Specific Volume Sat. Vapor v_g	Enthalpy Sat. Liquid h_f	Enthalpy Evap h_{fg}	Enthalpy Sat. Vapor h_g	Entropy Sat. Liquid s_f	Entropy Evap s_{fg}	Entropy Sat. Vapor s_g	Temp Fahr t
32.0*	0.08859	0.016022	3304.7	3304.7	−0.0179	1075.5	1075.5	0.0000	2.1873	2.1873	32.0*
34.0	0.09600	0.016021	3061.9	3061.9	1.996	1074.4	1076.4	0.0041	2.1762	2.1802	34.0
36.0	0.10395	0.016020	2839.0	2839.0	4.008	1073.2	1077.2	0.0081	2.1651	2.1732	36.0
38.0	0.11249	0.016019	2634.1	2634.2	6.018	1072.1	1078.1	0.0122	2.1541	2.1663	38.0
40.0	0.12163	0.016019	2445.8	2445.8	8.027	1071.0	1079.0	0.0162	2.1432	2.1594	40.0
42.0	0.13143	0.016019	2272.4	2272.4	10.035	1069.8	1079.9	0.0202	2.1325	2.1527	42.0
44.0	0.14192	0.016019	2112.8	2112.8	12.041	1068.7	1080.7	0.0242	2.1217	2.1459	44.0
46.0	0.15314	0.016020	1965.7	1965.7	14.047	1067.6	1081.6	0.0282	2.1111	2.1393	46.0
48.0	0.16514	0.016021	1830.0	1830.0	16.051	1066.4	1082.5	0.0321	2.1006	2.1327	48.0
50.0	0.17796	0.016023	1704.8	1704.8	18.054	1065.3	1083.4	0.0361	2.0901	2.1262	50.0
52.0	0.19165	0.016024	1589.2	1589.2	20.057	1064.2	1084.2	0.0400	2.0798	2.1197	52.0
54.0	0.20625	0.016026	1482.4	1482.4	22.058	1063.1	1085.1	0.0439	2.0695	2.1134	54.0
56.0	0.22183	0.016028	1383.6	1383.6	24.059	1061.9	1086.0	0.0478	2.0593	2.1070	56.0
58.0	0.23843	0.016031	1292.2	1292.2	26.060	1060.8	1086.9	0.0516	2.0491	2.1008	58.0
60.0	0.25611	0.016033	1207.6	1207.6	28.060	1059.7	1087.7	0.0555	2.0391	2.0946	60.0
62.0	0.27494	0.016036	1129.2	1129.2	30.059	1058.5	1088.6	0.0593	2.0291	2.0885	62.0
64.0	0.29497	0.016039	1056.5	1056.5	32.058	1057.4	1089.5	0.0632	2.0192	2.0824	64.0
66.0	0.31626	0.016043	989.0	989.1	34.056	1056.3	1090.4	0.0670	2.0094	2.0764	66.0
68.0	0.33889	0.016046	926.5	926.5	36.054	1055.2	1091.2	0.0708	1.9996	2.0704	68.0
70.0	0.36292	0.016050	868.3	868.4	38.052	1054.0	1092.1	0.0745	1.9900	2.0645	70.0
72.0	0.38844	0.016054	814.3	814.3	40.049	1052.9	1093.0	0.0783	1.9804	2.0587	72.0
74.0	0.41550	0.016058	764.1	764.1	42.046	1051.8	1093.8	0.0821	1.9708	2.0529	74.0
76.0	0.44420	0.016063	717.4	717.4	44.043	1050.7	1094.7	0.0858	1.9614	2.0472	76.0
78.0	0.47461	0.016067	673.8	673.9	46.040	1049.5	1095.6	0.0895	1.9520	2.0415	78.0
80.0	0.50683	0.016072	633.3	633.3	48.037	1048.4	1096.4	0.0932	1.9426	2.0959	80.0
82.0	0.54093	0.016077	595.5	595.5	50.033	1047.3	1097.3	0.0969	1.9334	2.0303	82.0
84.0	0.57702	0.016082	560.3	560.3	52.029	1046.1	1098.2	0.1006	1.9242	2.0248	84.0
86.0	0.61518	0.016087	227.5	527.5	54.026	1045.0	1099.0	0.1043	1.9151	2.0193	86.0
88.0	0.65551	0.016093	496.8	496.8	56.022	1043.9	1099.9	0.1079	1.9060	2.0139	88.0
90.0	0.69813	0.016099	468.1	468.1	58.018	1042.7	1100.8	0.1115	1.8970	2.0086	90.0
92.0	0.74313	0.016105	441.3	441.3	60.014	1041.6	1101.6	0.1152	1.8881	2.0033	92.0
94.0	0.79062	0.016111	416.3	416.3	62.010	1040.5	1102.5	0.1188	1.8792	1.9980	94.0
96.0	0.84072	0.016117	392.8	392.9	64.006	1039.3	1103.3	0.1224	1.8704	1.9928	96.0
98.0	0.89356	0.016123	370.9	370.9	66.003	1038.2	1104.2	0.1260	1.8617	1.9876	98.0
100.0	0.94924	0.016130	350.4	350.4	67.999	1037.1	1105.1	0.1295	1.8530	1.9825	100.0
102.0	1.00789	0.016137	331.1	331.1	69.995	1035.9	1105.9	0.1331	1.8444	1.9775	102.0
104.0	1.06965	0.016144	313.1	313.1	71.992	1034.8	1106.8	0.1366	1.8358	1.9725	104.0
106.0	1.1347	0.016151	296.16	296.18	73.99	1033.6	1107.6	0.1402	1.8273	1.9675	106.0
108.0	1.2030	0.016158	280.28	280.30	75.98	1032.5	1108.5	0.1437	1.8188	1.9626	108.0
110.0	1.2750	0.016165	265.37	265.39	77.98	1031.4	1109.3	0.1472	1.8105	1.9577	110.0
112.0	1.3505	0.016173	251.37	251.38	79.98	1030.2	1110.2	0.1507	1.8021	1.9528	112.0
114.0	1.4299	0.016180	238.21	238.22	81.97	1029.1	1111.0	0.1542	1.7938	1.9480	114.0
116.0	1.5133	0.016188	225.84	225.85	83.97	1027.9	1111.9	0.1577	1.7856	1.9433	116.0
118.0	1.6009	0.016196	214.20	214.21	85.97	1026.8	1112.7	0.1611	1.7774	1.9386	118.0
120.0	1.6927	0.016204	203.25	203.26	87.97	1025.6	1113.6	0.1646	1.7693	1.9339	120.0
122.0	1.7891	0.016213	192.94	192.95	89.96	1024.5	1114.4	0.1680	1.7613	1.9293	122.0
124.0	1.8901	0.016221	183.23	183.24	91.96	1023.3	1115.3	0.1715	1.7533	1.9247	124.0
126.0	1.9959	0.016229	174.08	174.09	93.96	1022.2	1116.1	0.1749	1.7453	1.9202	126.0
128.0	2.1068	0.016238	165.45	165.47	95.96	1021.0	1117.0	0.1783	1.7374	1.9157	128.0
130.0	2.2230	0.016247	157.32	157.33	97.96	1019.8	1117.8	0.1817	1.7295	1.9112	130.0
132.0	2.3445	0.016256	149.64	149.66	99.95	1018.7	1118.6	0.1851	1.7217	1.9068	132.0
134.0	2.4717	0.016265	142.40	142.41	101.95	1017.5	1119.5	0.1884	1.7140	1.9024	134.0
136.0	2.6047	0.016274	135.55	135.57	103.95	1016.4	1120.3	0.1918	1.7063	1.8980	136.0
138.0	2.7438	0.016284	129.09	129.11	105.95	1015.2	1121.1	0.1951	1.6986	1.8937	138.0
140.0	2.8892	0.016293	122.98	123.00	107.95	1014.0	1122.0	0.1985	1.6910	1.8895	140.0
142.0	3.0411	0.016303	117.21	117.22	109.95	1012.9	1122.8	0.2018	1.6534	1.8852	142.0
144.0	3.1997	0.016312	111.74	111.76	111.95	1011.7	1123.6	0.2051	1.6759	1.8810	144.0
146.0	3.3653	0.016322	106.58	106.59	113.95	1010.5	1124.5	0.2084	1.6684	1.8769	146.0
148.0	3.5381	0.016332	101.68	101.70	115.95	1009.3	1125.3	0.2117	1.6610	1.8727	148.0
150.0	3.7184	0.016343	97.05	97.07	117.95	1008.2	1126.1	0.2150	1.6536	1.8686	150.0
152.0	3.9065	0.016353	92.66	92.68	119.95	1007.0	1126.9	0.2183	1.6463	1.8646	152.0
154.0	4.1025	0.016363	88.50	88.52	121.95	1005.8	1127.7	0.2216	1.6390	1.8606	154.0
156.0	4.3068	0.016374	84.56	84.57	123.95	1004.6	1128.6	0.2248	1.6318	1.8566	156.0
158.0	4.5197	0.016384	80.82	80.83	125.96	1003.4	1129.4	0.2281	1.6245	1.8526	158.0
160.0	4.7414	0.016395	77.27	77.29	127.96	1002.2	1130.2	0.2313	1.6174	1.8487	160.0
162.0	4.9722	0.016406	73.90	73.92	129.96	1001.0	1131.0	0.2345	1.6103	1.8448	162.0
164.0	5.2124	0.016417	70.70	70.72	131.96	999.8	1131.8	0.2377	1.6032	1.8409	164.0
166.0	5.4623	0.016428	67.67	67.68	133.97	998.6	1132.6	0.2409	1.5961	1.8371	166.0
168.0	5.7223	0.016440	64.78	64.80	135.97	997.4	1133.4	0.2441	1.5892	1.8333	168.0
170.0	5.9926	0.016451	62.04	62.06	137.97	996.2	1134.2	0.2473	1.5822	1.8295	170.0
172.0	6.2736	0.016463	59.43	59.45	139.98	995.0	1135.0	0.2505	1.5753	1.8258	172.0
174.0	6.5656	0.016474	56.95	56.97	141.98	993.8	1135.8	0.2537	1.5684	1.8221	174.0
176.0	6.8690	0.016486	54.59	54.61	143.99	992.6	1136.6	0.2568	1.5616	1.8184	176.0
178.0	7.1840	0.016498	52.35	52.36	145.99	991.4	1137.4	0.2600	1.5548	1.8147	178.0

*The states shown are metastable

THERMO

Appendix D (continued)

Temp Fahr t	Abs Press. Lb per Sq In. p	Specific Volume Sat. Liquid v_f	Evap v_{fg}	Sat. Vapor v_g	Enthalpy Sat. Liquid h_f	Evap h_{fg}	Sat. Vapor h_g	Entropy Sat. Liquid s_f	Evap s_{fg}	Sat. Vapor s_g	Temp Fahr t
180.0	7.5110	0.016510	50.21	50.22	148.00	990.2	1138.2	0.2631	1.5480	1.8111	180.0
182.0	7.850	0.016522	48.172	48.189	150.01	989.0	1139.0	0.2662	1.5413	1.8075	182.0
184.0	8.203	0.016534	46.232	46.249	152.01	987.8	1139.8	0.2694	1.5346	1.8040	184.0
186.0	8.568	0.016547	44.383	44.400	154.02	986.5	1140.5	0.2725	1.5279	1.8004	186.0
188.0	8.947	0.016559	42.621	42.638	156.03	985.3	1141.3	0.2756	1.5213	1.7969	188.0
190.0	9.340	0.016572	40.941	40.957	158.04	984.1	1142.1	0.2787	1.5148	1.7934	190.0
192.0	9.747	0.016585	39.337	39.354	160.05	982.8	1142.9	0.2818	1.5082	1.7900	192.0
194.0	10.168	0.016598	37.808	37.824	162.05	981.6	1143.7	0.2848	1.5017	1.7865	194.0
196.0	10.605	0.016611	36.348	36.364	164.06	980.4	1144.4	0.2879	1.4952	1.7831	196.0
198.0	11.058	0.016624	34.954	34.970	166.08	979.1	1145.2	0.2910	1.4888	1.7798	198.0
200.0	11.526	0.016637	33.622	33.639	168.09	977.9	1146.0	0.2940	1.4824	1.7764	200.0
204.0	12.512	0.016664	31.135	31.151	172.11	975.4	1147.5	0.3001	1.4697	1.7698	204.0
208.0	13.568	0.016691	28.862	28.878	176.14	972.8	1149.0	0.3061	1.4571	1.7632	208.0
212.0	14.696	0.016719	26.782	26.799	180.17	970.3	1150.5	0.3121	1.4447	1.7568	212.0
216.0	15.901	0.016747	24.878	24.894	184.20	967.8	1152.0	0.3181	1.4323	1.7505	216.0
220.0	17.186	0.016775	23.131	23.148	188.23	965.2	1153.4	0.3241	1.4201	1.7442	220.0
224.0	18.556	0.016805	21.529	21.545	192.27	962.6	1154.9	0.3300	1.4081	1.7380	224.0
228.0	20.015	0.016834	20.056	20.073	196.31	960.0	1156.3	0.3359	1.3961	1.7320	228.0
232.0	21.567	0.016864	18.701	18.718	200.35	957.4	1157.8	0.3417	1.3842	1.7260	232.0
236.0	23.216	0.016895	17.454	17.471	204.40	954.8	1159.2	0.3476	1.3725	1.7201	236.0
240.0	24.968	0.016926	16.304	16.321	208.45	952.1	1160.6	0.3533	1.3609	1.7142	240.0
244.0	26.826	0.016958	15.243	15.260	212.50	949.5	1162.0	0.3591	1.3494	1.7085	244.0
248.0	28.796	0.016990	14.264	14.281	216.56	946.8	1163.4	0.3649	1.3379	1.7028	248.0
252.0	30.883	0.017022	13.358	13.375	220.62	944.1	1164.7	0.3706	1.3266	1.6972	252.0
256.0	33.091	0.017055	12.520	12.538	224.69	941.4	1166.1	0.3763	1.3154	1.6917	256.0
260.0	35.427	0.017089	11.745	11.762	228.76	938.6	1167.4	0.3819	1.3043	1.6862	260.0
264.0	37.894	0.017123	11.025	11.042	232.83	935.9	1168.7	0.3876	1.2933	1.6808	264.0
268.0	40.500	0.017157	10.358	10.375	236.91	933.1	1170.0	0.3932	1.2823	1.6755	268.0
272.0	43.249	0.017193	9.738	9.755	240.99	930.3	1171.3	0.3987	1.2715	1.6702	272.0
276.0	46.147	0.017228	9.162	9.180	245.08	927.5	1172.5	0.4043	1.2607	1.6650	276.0
280.0	49.200	0.017264	8.627	8.644	249.17	924.6	1173.8	0.4098	1.2501	1.6599	280.0
284.0	52.414	0.01730	8.1280	8.1453	253.3	921.7	1175.0	0.4154	1.2395	1.6548	284.0
288.0	55.795	0.01734	7.6634	7.6807	257.4	918.8	1176.2	0.4208	1.2290	1.6498	288.0
292.0	59.350	0.01738	7.2301	7.2475	261.5	915.9	1177.4	0.4263	1.2186	1.6449	292.0
296.0	63.084	0.01741	6.8259	6.8433	265.6	913.0	1178.6	0.4317	1.2082	1.6400	296.0
300.0	67.005	0.01745	6.4483	6.4658	269.7	910.0	1179.7	0.4372	1.1979	1.6351	300.0
304.0	71.119	0.01749	6.0955	6.1130	273.8	907.0	1180.9	0.4426	1.1877	1.6303	304.0
308.0	75.433	0.01753	5.7655	5.7830	278.0	904.0	1182.0	0.4479	1.1776	1.6256	308.0
312.0	79.953	0.01757	5.4566	5.4742	282.1	901.0	1183.1	0.4533	1.1676	1.6209	312.0
316.0	84.688	0.01761	5.1673	5.1849	286.3	897.9	1184.1	0.4586	1.1576	1.6162	316.0
320.0	89.643	0.01766	4.8961	4.9138	290.4	894.8	1185.2	0.4640	1.1477	1.6116	320.0
324.0	94.826	0.01770	4.6418	4.6595	294.6	891.6	1186.2	0.4692	1.1378	1.6071	324.0
328.0	100.245	0.01774	4.4030	4.4208	298.7	888.5	1187.2	0.4745	1.1280	1.6025	328.0
332.0	105.907	0.01779	4.1788	4.1966	302.9	885.3	1188.2	0.4798	1.1183	1.5981	332.0
336.0	111.820	0.01783	3.9681	3.9859	307.1	882.1	1189.1	0.4850	1.1086	1.5936	336.0
340.0	117.992	0.01787	3.7699	3.7878	311.3	878.8	1190.1	0.4902	1.0990	1.5892	340.0
344.0	124.430	0.01792	3.5834	3.6013	315.5	875.5	1191.0	0.4954	1.0894	1.5849	344.0
348.0	131.142	0.01797	3.4078	3.4258	319.7	872.2	1191.1	0.5006	1.0799	1.5806	348.0
352.0	138.138	0.01801	3.2423	3.2603	323.9	868.9	1192.7	0.5058	1.0705	1.5763	352.0
356.0	145.424	0.01806	3.0863	3.1044	328.1	865.5	1193.6	0.5110	1.0611	1.5721	356.0
360.0	153.010	0.01811	2.9392	2.9573	332.3	862.1	1194.4	0.5161	1.0517	1.5678	360.0
364.0	160.903	0.01816	2.8002	2.8184	336.5	858.6	1195.2	0.5212	1.0424	1.5637	364.0
368.0	169.113	0.01821	2.6691	2.6873	340.8	855.1	1195.9	0.5263	1.0332	1.5595	368.0
372.0	177.648	0.01826	2.5451	2.5633	345.0	851.6	1196.7	0.5314	1.0240	1.5554	372.0
376.0	186.517	0.01831	2.4279	2.4462	349.3	848.1	1197.4	0.5365	1.0148	1.5513	376.0
380.0	195.729	0.01836	2.3170	2.3353	353.6	844.5	1198.0	0.5416	1.0057	1.5473	380.0
384.0	205.294	0.01842	2.2120	2.2304	357.9	840.8	1198.7	0.5466	0.9966	1.5432	384.0
388.0	215.220	0.01847	2.1126	2.1311	362.2	837.2	1199.3	0.5516	0.9876	1.5392	388.0
392.0	225.516	0.01853	2.0184	2.0369	366.5	833.4	1199.9	0.5567	0.9786	1.5352	392.0
396.0	236.193	0.01858	1.9291	1.9477	370.8	829.7	1200.4	0.5617	0.9696	1.5313	396.0
400.0	247.259	0.01864	1.8444	1.8630	375.1	825.9	1201.0	0.5667	0.9607	1.5274	400.0
404.0	258.725	0.01870	1.7640	1.7827	379.4	822.0	1201.5	0.5717	0.9518	1.5234	404.0
408.0	270.600	0.01875	1.6877	1.7064	383.8	818.2	1201.9	0.5766	0.9429	1.5195	408.0
412.0	282.894	0.01881	1.6152	1.6340	388.1	814.2	1202.4	0.5816	0.9341	1.5157	412.0
416.0	295.617	0.01887	1.5463	1.5651	392.5	810.2	1202.8	0.5866	0.9253	1.5118	416.0
420.0	308.780	0.01894	1.4808	1.4997	396.9	806.2	1203.1	0.5915	0.9165	1.5080	420.0
424.0	322.391	0.01900	1.4184	1.4374	401.3	802.2	1203.5	0.5964	0.9077	1.5042	424.0
428.0	336.463	0.01906	1.3591	1.3782	405.7	798.0	1203.7	0.6014	0.8990	1.5004	428.0
432.0	351.00	0.01913	1.30266	1.32179	410.1	793.9	1204.0	0.6063	0.8903	1.4966	432.0
436.0	366.03	0.01919	1.24887	1.26806	414.6	789.7	1204.2	0.6112	0.8816	1.4928	436.0
440.0	381.54	0.01926	1.19761	1.21687	419.0	785.4	1204.4	0.6161	0.8729	1.4890	440.0
444.0	397.56	0.01933	1.14874	1.16806	423.5	781.1	1204.6	0.6210	0.8643	1.4853	444.0
448.0	414.09	0.01940	1.10212	1.12152	428.0	776.7	1204.7	0.6259	0.8557	1.4815	448.0
452.0	431.14	0.01947	1.05764	1.07711	432.5	772.3	1204.8	0.6308	0.8471	1.4778	452.0
456.0	448.73	0.01954	1.01518	1.03472	437.0	767.8	1204.8	0.6356	0.8385	1.4741	456.0

THERMO

Appendix D (continued)

Temp Fahr t	Abs Press. Lb per Sq In. p	Specific Volume Sat. Liquid v_f	Specific Volume Evap v_{fg}	Specific Volume Sat. Vapor v_g	Enthalpy Sat. Liquid h_f	Enthalpy Evap h_{fg}	Enthalpy Sat. Vapor h_g	Entropy Sat. Liquid s_f	Entropy Evap s_{fg}	Entropy Sat. Vapor s_g	Temp Fahr t
460.0	466.87	0.01961	0.97463	0.99424	441.5	763.2	1204.8	0.6405	0.8299	1.4704	**460.0**
464.0	485.56	0.01969	0.93588	0.95557	446.1	758.6	1204.7	0.6454	0.8213	1.4667	**464.0**
468.0	504.83	0.01976	0.89885	0.91862	450.7	754.0	1204.6	0.6502	0.8127	1.4629	**468.0**
472.0	524.67	0.01984	0.86345	0.88329	455.2	749.3	1204.5	0.6551	0.8042	1.4592	**472.0**
476.0	545.11	0.01992	0.82958	0.84950	459.9	744.5	1204.3	0.6599	0.7956	1.4555	**476.0**
480.0	566.15	0.02000	0.79716	0.81717	464.5	739.6	1204.1	0.6648	0.7871	1.4518	**480.0**
484.0	587.81	0.02009	0.76613	0.78622	469.1	734.7	1203.8	0.6696	0.7785	1.4481	**484.0**
488.0	610.10	0.02017	0.73641	0.75658	473.8	729.7	1203.5	0.6745	0.7700	1.4444	**488.0**
492.0	633.03	0.02026	0.70794	0.72820	478.5	724.6	1203.1	0.6793	0.7614	1.4407	**492.0**
496.0	656.61	0.02034	0.68065	0.70100	483.2	719.5	1202.7	0.6842	0.7528	1.4370	**496.0**
500.0	680.86	0.02043	0.65448	0.67492	487.9	714.3	1202.2	0.6890	0.7443	1.4333	**500.0**
504.0	705.78	0.02053	0.62938	0.64991	492.7	709.0	1201.7	0.6939	0.7357	1.4296	**504.0**
508.0	731.40	0.02062	0.60530	0.62592	497.5	703.7	1201.1	0.6987	0.7271	1.4258	**508.0**
512.0	757.72	0.02072	0.58218	0.60289	502.3	698.2	1200.5	0.7036	0.7185	1.4221	**512.0**
516.0	784.76	0.02081	0.55997	0.58079	507.1	692.7	1199.8	0.7085	0.7099	1.4183	**516.0**
520.0	812.53	0.02091	0.53864	0.55956	512.0	687.0	1199.0	0.7133	0.7013	1.4146	**520.0**
524.0	841.04	0.02102	0.51814	0.53916	516.9	681.3	1198.2	0.7182	0.6926	1.4108	**524.0**
528.0	870.31	0.02112	0.49843	0.51955	521.8	675.5	1197.3	0.7231	0.6839	1.4070	**528.0**
532.0	900.34	0.02123	0.47947	0.50070	526.8	669.6	1196.4	0.7280	0.6752	1.4032	**532.0**
536.0	931.17	0.02134	0.46123	0.48257	531.7	663.6	1195.4	0.7329	0.6665	1.3993	**536.0**
540.0	962.79	0.02146	0.44367	0.46513	536.8	657.5	1194.3	0.7378	0.6577	1.3954	**540.0**
544.0	995.22	0.02157	0.42677	0.44834	541.8	651.3	1193.1	0.7427	0.6489	1.3915	**544.0**
548.0	1028.49	0.02169	0.41048	0.43217	546.9	645.0	1191.9	0.7476	0.6400	1.3876	**548.0**
552.0	1062.59	0.02182	0.39479	0.41660	552.0	638.5	1190.6	0.7525	0.6311	1.3837	**552.0**
556.0	1097.55	0.02194	0.37966	0.40160	557.2	632.0	1189.2	0.7575	0.6222	1.3797	**556.0**
560.0	1133.38	0.02207	0.36507	0.38714	562.4	625.3	1187.7	0.7625	0.6132	1.3757	**560.0**
564.0	1170.10	0.02221	0.35099	0.37320	567.6	618.5	1186.1	0.7674	0.6041	1.3716	**564.0**
568.0	1207.72	0.02235	0.33741	0.35975	572.9	611.5	1184.5	0.7725	0.5950	1.3675	**568.0**
572.0	1246.26	0.02249	0.32429	0.34678	578.3	604.5	1182.7	0.7775	0.5859	1.3634	**572.0**
576.0	1285.74	0.02264	0.31162	0.33426	583.7	597.2	1180.9	0.7825	0.5766	1.3592	**576.0**
580.0	1326.17	0.02279	0.29937	0.32216	589.1	589.9	1179.0	0.7876	0.5673	1.3550	**580.0**
584.0	1367.7	0.02295	0.28753	0.31048	594.6	582.4	1176.9	0.7927	0.5580	1.3507	**584.0**
588.0	1410.0	0.02311	0.27608	0.29919	600.1	574.7	1174.8	0.7978	0.5485	1.3464	**588.0**
592.0	1453.3	0.02328	0.26499	0.28827	605.7	566.8	1172.6	0.8030	0.5390	1.3420	**592.0**
596.0	1497.8	0.02345	0.25425	0.27770	611.4	558.8	1170.2	0.8082	0.5293	1.3375	**596.0**
600.0	1543.2	0.02364	0.24384	0.26747	617.1	550.6	1167.7	0.8134	0.5196	1.3330	**600.0**
604.0	1589.7	0.02382	0.23374	0.25757	622.9	542.2	1165.1	0.8187	0.5097	1.3284	**604.0**
608.0	1637.3	0.02402	0.22394	0.24796	628.8	533.6	1162.4	0.8240	0.4997	1.3238	**608.0**
612.0	1686.1	0.02422	0.21442	0.23865	634.8	524.7	1159.5	0.8294	0.4896	1.3190	**612.0**
616.6	1735.9	0.02444	0.20516	0.22960	640.8	515.6	1156.4	0.8348	0.4794	1.3141	**616.0**
620.0	1786.9	0.02466	0.19615	0.22081	646.9	506.3	1153.2	0.8403	0.4689	1.3092	**620.0**
624.0	1839.0	0.02489	0.18737	0.21226	653.1	496.6	1149.8	0.8458	0.4583	1.3041	**624.0**
628.0	1892.4	0.02514	0.17880	0.20394	659.5	486.7	1146.1	0.8514	0.4474	1.2988	**628.0**
632.0	1947.0	0.02539	0.17044	0.19583	665.9	476.4	1142.2	0.8571	0.4364	1.2934	**632.0**
636.0	2002.8	0.02566	0.16226	0.18792	672.4	465.7	1138.1	0.8628	0.4251	1.2879	**636.0**
640.0	2059.9	0.02595	0.15427	0.18021	679.1	454.6	1133.7	0.8686	0.4134	1.2821	**640.0**
644.0	2118.3	0.02625	0.14644	0.17269	685.9	443.1	1129.0	0.8746	0.4015	1.2761	**644.0**
648.0	2178.1	0.02657	0.13876	0.16534	692.9	431.1	1124.0	0.8806	0.3893	1.2699	**648.0**
652.0	2239.2	0.02691	0.13124	0.15816	700.0	418.7	1118.7	0.8868	0.3767	1.2634	**652.0**
656.0	2301.7	0.02728	0.12387	0.15115	707.4	405.7	1113.1	0.8931	0.3637	1.2567	**656.0**
660.0	2365.7	0.02768	0.11663	0.14431	714.9	392.1	1107.0	0.8995	0.3502	1.2498	**660.0**
664.0	2431.1	0.02811	0.10947	0.13757	722.9	377.7	1100.6	0.9064	0.3361	1.2425	**664.0**
668.0	2498.1	0.02858	0.10229	0.13087	731.5	362.1	1093.5	0.9137	0.3210	1.2347	**668.0**
672.0	2566.6	0.02911	0.09514	0.12424	740.2	345.7	1085.9	0.9212	0.3054	1.2266	**672.0**
676.0	2636.8	0.02970	0.08799	0.11769	749.2	328.5	1077.6	0.9287	0.2892	1.2179	**676.0**
680.0	2708.6	0.03037	0.08080	0.11117	758.5	310.1	1068.5	0.9365	0.2720	1.2086	**680.0**
684.0	2782.1	0.03114	0.07349	0.10463	768.2	290.2	1058.4	0.9447	0.2537	1.1984	**684.0**
688.0	2857.4	0.03204	0.06595	0.09799	778.8	268.2	1047.0	0.9535	0.2337	1.1872	**688.0**
692.0	2934.5	0.03313	0.05797	0.09110	790.5	243.1	1033.6	0.9634	0.2110	1.1744	**692.0**
696.0	3013.4	0.03455	0.04916	0.08371	804.4	212.8	1017.2	0.9749	0.1841	1.1591	**696.0**
700.0	3094.3	0.03662	0.03857	0.07519	822.4	172.7	995.2	0.9901	0.1490	1.1390	**700.0**
702.0	3135.5	0.03824	0.03173	0.06997	835.0	144.7	979.7	1.0006	0.1246	1.1252	**702.0**
704.0	3177.2	0.04108	0.02192	0.06300	854.2	102.0	956.2	1.0169	0.0876	1.1046	**704.0**
705.0	3198.3	0.04427	0.01304	0.05730	873.0	61.4	934.4	1.0329	0.0527	1.0856	**705.0**
705.47*	3208.2	0.05078	0.00000	0.05078	906.0	0.0	906.0	1.0612	0.0000	1.0612	**705.47***

*Critical temperature

Appendix E
Saturated Steam: Pressure

Abs Press. Lb/Sq In. p	Temp Fahr t	Specific Volume: Sat. Liquid v_f	Specific Volume: Evap v_{fg}	Specific Volume: Sat. Vapor v_g	Enthalpy: Sat. Liquid h_f	Enthalpy: Evap h_{fg}	Enthalpy: Sat. Vapor h_g	Entropy: Sat. Liquid s_f	Entropy: Evap s_{fg}	Entropy: Sat. Vapor s_g	Abs Press. Lb/Sq In. p
0.08865	32.018	0.016022	3302.4	3302.4	0.0003	1075.5	1075.5	0.0000	2.1872	2.1872	**0.08865**
0.25	59.323	0.016032	1235.5	1235.5	27.382	1060.1	1087.4	0.0542	2.0425	2.0967	**0.25**
0.50	79.586	0.016071	641.5	641.5	47.623	1048.6	1096.3	0.0925	1.9446	2.0370	**0.50**
1.0	101.74	0.016136	333.59	333.60	69.73	1036.1	1105.8	0.1326	1.8455	1.9781	**1.0**
5.0	162.24	0.016407	73.515	73.532	130.20	1000.9	1131.1	0.2349	1.6094	1.8443	**5.0**
10.0	193.21	0.016592	38.404	38.420	161.26	982.1	1143.3	0.2836	1.5043	1.7879	**10.0**
14.696	212.00	0.016719	26.782	26.799	180.17	970.3	1150.5	0.3121	1.4447	1.7568	**14.696**
15.0	213.03	0.016726	26.274	26.290	181.21	969.7	1150.9	0.3137	1.4415	1.7552	**15.0**
20.0	227.96	0.016834	20.070	20.087	196.27	960.1	1156.3	0.3358	1.3962	1.7320	**20.0**
30.0	250.34	0.017009	13.7266	13.7436	218.9	945.2	1164.1	0.3682	1.3313	1.6995	**30.0**
40.0	267.25	0.017151	10.4794	10.4965	236.1	933.6	1169.8	0.3921	1.2844	1.6765	**40.0**
50.0	281.02	0.017274	8.4967	8.5140	250.2	923.9	1174.1	0.4112	1.2474	1.6586	**50.0**
60.0	292.71	0.017383	7.1562	7.1736	262.2	915.4	1177.6	0.4273	1.2167	1.6440	**60.0**
70.0	302.93	0.017482	6.1875	6.2050	272.7	907.8	1180.6	0.4411	1.1905	1.6316	**70.0**
80.0	312.04	0.017573	5.4536	5.4711	282.1	900.9	1183.1	0.4534	1.1675	1.6208	**80.0**
90.0	320.28	0.017659	4.8779	4.8953	290.7	894.6	1185.3	0.4643	1.1470	1.6113	**90.0**
100.0	327.82	0.017740	4.4133	4.4310	298.5	888.6	1187.2	0.4743	1.1284	1.6027	**100.0**
110.0	334.79	0.01782	4.0306	4.0484	305.8	883.1	1188.9	0.4834	1.1115	1.5950	**110.0**
120.0	341.27	0.01789	3.7097	3.7275	312.6	877.8	1190.4	0.4919	1.0960	1.5879	**120.0**
130.0	347.33	0.01796	3.4364	3.4544	319.0	872.8	1191.7	0.4998	1.0815	1.5813	**130.0**
140.0	353.04	0.01803	3.2010	3.2190	325.0	868.0	1193.0	0.5071	1.0681	1.5752	**140.0**
150.0	358.43	0.01809	2.9958	3.0139	330.6	863.4	1194.1	0.5141	1.0554	1.5695	**150.0**
160.0	363.55	0.01815	2.8155	2.8336	336.1	859.0	1195.1	0.5206	1.0435	1.5641	**160.0**
170.0	368.42	0.01821	2.6556	2.6738	341.2	854.8	1196.0	0.5269	1.0322	1.5591	**170.0**
180.0	373.08	0.01827	2.5129	2.5312	346.2	850.7	1196.9	0.5328	1.0215	1.5543	**180.0**
190.0	377.53	0.01833	2.3847	2.4030	350.9	846.7	1197.6	0.5384	1.0113	1.5498	**190.0**
200.0	381.80	0.01839	2.2689	2.2873	355.5	842.8	1198.3	0.5438	1.0016	1.5454	**200.0**
210.0	385.91	0.01844	2.16373	2.18217	359.9	839.1	1199.0	0.5490	0.9923	1.5413	**210.0**
220.0	389.88	0.01850	2.06779	2.08629	364.2	835.4	1199.6	0.5540	0.9834	1.5374	**220.0**
230.0	393.70	0.01855	1.97991	1.99846	368.3	831.8	1200.1	0.5588	0.9748	1.5336	**230.0**
240.0	397.39	0.01860	1.89909	1.91769	372.3	828.4	1200.6	0.5634	0.9665	1.5299	**240.0**
250.0	400.97	0.01865	1.82452	1.84317	376.1	825.0	1201.1	0.5679	0.9585	1.5264	**250.0**
260.0	404.44	0.01870	1.75548	1.77418	379.9	821.6	1201.5	0.5722	0.9508	1.5230	**260.0**
270.0	407.80	0.01875	1.69137	1.71013	383.6	818.3	1201.9	0.5764	0.9433	1.5197	**270.0**
280.0	411.07	0.01880	1.63169	1.65049	387.1	815.1	1202.3	0.5805	0.9361	1.5166	**280.0**
290.0	414.25	0.01885	1.57597	1.59482	390.6	812.0	1202.6	0.5844	0.9291	1.5135	**290.0**
300.0	417.35	0.01889	1.52384	1.54274	394.0	808.9	1202.9	0.5882	0.9223	1.5105	**300.0**
350.0	431.73	0.01912	1.30642	1.32554	409.8	794.2	1204.0	0.6059	0.8909	1.4968	**350.0**
400.0	444.60	0.01934	1.14162	1.16095	424.2	780.4	1204.6	0.6217	0.8630	1.4847	**400.0**
450.0	456.28	0.01954	1.01224	1.03179	437.3	767.5	1204.8	0.6360	0.8378	1.4738	**450.0**
500.0	467.01	0.01975	0.90787	0.92762	449.5	755.1	1204.7	0.6490	0.8148	1.4639	**500.0**
550.0	476.94	0.01994	0.82183	0.84177	460.9	743.3	1204.3	0.6611	0.7936	1.4547	**550.0**
600.0	486.20	0.02013	0.74962	0.76975	471.7	732.0	1203.7	0.6723	0.7738	1.4461	**600.0**
650.0	494.89	0.02032	0.68811	0.70843	481.9	720.9	1202.8	0.6828	0.7552	1.4381	**650.0**
700.0	503.08	0.02050	0.63505	0.65556	491.6	710.2	1201.8	0.6928	0.7377	1.4304	**700.0**
750.0	510.84	0.02069	0.58880	0.60949	500.9	699.8	1200.7	0.7022	0.7210	1.4232	**750.0**
800.0	518.21	0.02087	0.54809	0.56896	509.8	689.6	1199.4	0.7111	0.7051	1.4163	**800.0**
850.0	525.24	0.02105	0.51197	0.53302	518.4	679.5	1198.0	0.7197	0.6899	1.4096	**850.0**
900.0	531.95	0.02123	0.47968	0.50091	526.7	669.7	1196.4	0.7279	0.6753	1.4032	**900.0**
950.0	538.39	0.02141	0.45064	0.47205	534.7	660.0	1194.7	0.7358	0.6612	1.3970	**950.0**
1000.0	544.58	0.02159	0.42436	0.44596	542.6	650.4	1192.9	0.7434	0.6476	1.3910	**1000.0**
1050.0	550.53	0.02177	0.40047	0.42224	550.1	640.9	1191.0	0.7507	0.6344	1.3851	**1050.0**
1100.0	556.28	0.02195	0.37863	0.40058	557.5	631.5	1189.1	0.7578	0.6216	1.3794	**1100.0**
1150.0	561.82	0.02214	0.35859	0.38073	564.8	622.2	1187.0	0.7647	0.6091	1.3738	**1150.0**
1200.0	567.19	0.02232	0.34013	0.36245	571.9	613.0	1184.8	0.7714	0.5969	1.3683	**1200.0**
1250.0	572.38	0.02250	0.32306	0.34556	578.8	603.8	1182.6	0.7780	0.5850	1.3630	**1250.0**
1300.0	577.42	0.02269	0.30722	0.32991	585.6	594.6	1180.2	0.7843	0.5733	1.3577	**1300.0**
1350.0	582.32	0.02288	0.29250	0.31537	592.3	585.4	1177.8	0.7906	0.5620	1.3525	**1350.0**
1400.0	587.07	0.02307	0.27871	0.30178	598.8	576.5	1175.3	0.7966	0.5507	1.3474	**1400.0**
1450.0	591.70	0.02327	0.26584	0.28911	605.3	567.4	1172.8	0.8026	0.5397	1.3423	**1450.0**
1500.0	596.20	0.02346	0.25372	0.27719	611.7	558.4	1170.1	0.8085	0.5288	1.3373	**1500.0**
1550.0	600.59	0.02366	0.24235	0.26601	618.0	549.4	1167.4	0.8142	0.5182	1.3324	**1550.0**
1600.0	604.87	0.02387	0.23159	0.25545	624.2	540.3	1164.5	0.8199	0.5076	1.3274	**1600.0**
1650.0	609.05	0.02407	0.22143	0.24551	630.4	531.3	1161.6	0.8254	0.4971	1.3225	**1650.0**
1700.0	613.13	0.02428	0.21178	0.23607	636.5	522.2	1158.6	0.8309	0.4867	1.3176	**1700.0**
1750.0	617.12	0.02450	0.20263	0.22713	642.5	513.1	1155.6	0.8363	0.4765	1.3128	**1750.0**
1800.0	621.02	0.02472	0.19390	0.21861	648.5	503.8	1152.3	0.8417	0.4662	1.3079	**1800.0**
1850.0	624.83	0.02495	0.18558	0.21052	654.5	494.6	1149.0	0.8470	0.4561	1.3030	**1850.0**
1900.0	628.56	0.02517	0.17761	0.20278	660.4	485.2	1145.6	0.8522	0.4459	1.2981	**1900.0**
1950.0	632.22	0.02541	0.16999	0.19540	666.3	475.8	1142.0	0.8574	0.4358	1.2931	**1950.0**
2000.0	635.80	0.02565	0.16266	0.18831	672.1	466.2	1138.3	0.8625	0.4256	1.2881	**2000.0**
2100.0	642.76	0.02615	0.14885	0.17501	683.8	446.7	1130.5	0.8727	0.4053	1.2780	**2100.0**
2200.0	649.45	0.02669	0.13603	0.16272	695.5	426.7	1122.2	0.8828	0.3848	1.2676	**2200.0**
2300.0	655.89	0.02727	0.12406	0.15133	707.2	406.0	1113.2	0.8929	0.3640	1.2569	**2300.0**
2400.0	662.11	0.02790	0.11287	0.14076	719.0	384.8	1103.7	0.9031	0.3430	1.2460	**2400.0**
2500.0	668.11	0.02859	0.10209	0.13068	731.7	361.6	1093.3	0.9139	0.3206	1.2345	**2500.0**
2600.0	673.91	0.02938	0.09172	0.12110	744.5	337.6	1082.0	0.9247	0.2977	1.2225	**2600.0**
2700.0	679.53	0.03029	0.08165	0.11194	757.3	312.3	1069.7	0.9356	0.2741	1.2097	**2700.0**
2800.0	684.96	0.03134	0.07171	0.10305	770.7	285.1	1055.8	0.9468	0.2491	1.1958	**2800.0**
2900.0	690.22	0.03262	0.06158	0.09420	785.1	254.7	1039.8	0.9588	0.2215	1.1803	**2900.0**
3000.0	695.33	0.03428	0.05073	0.08500	801.8	218.4	1020.3	0.9728	0.1891	1.1619	**3000.0**
3100.0	700.28	0.03681	0.03771	0.07452	824.0	169.3	993.3	0.9914	0.1460	1.1373	**3100.0**
3200.0	705.08	0.04472	0.01191	0.05663	875.5	56.1	931.6	1.0351	0.0482	1.0832	**3200.0**
3208.2*	705.47	0.05078	0.00000	0.05078	906.0	0.0	906.0	1.0612	0.0000	1.0612	**3208.2***

*Critical pressure

Appendix F
Superheated Steam

Abs Press. Lb/Sq In. (Sat. Temp)		Sat. Water	Sat. Steam	Temperature – Degrees Fahrenheit 200	250	300	350	400	450	500	600	700	800	900	1000	1100	1200
1 (101.74)	Sh			98.26	148.26	198.26	248.26	298.26	348.26	398.26	498.26	598.26	698.26	798.26	898.26	998.26	1098.26
	v	0.01614	333.6	392.5	422.4	452.3	482.1	511.9	541.7	571.5	631.1	690.7	750.3	809.8	869.4	929.0	988.6
	h	69.73	1105.8	1150.2	1172.9	1195.7	1218.7	1241.8	1265.1	1288.6	1336.1	1384.5	1433.7	1483.8	1534.9	1586.8	1639.7
	s	0.1326	1.9781	2.0509	2.0841	2.1152	2.1445	2.1722	2.1985	2.2237	2.2708	2.3144	2.3551	2.3934	2.4296	2.4640	2.4969
5 (162.24)	Sh			37.76	87.76	137.76	187.76	237.76	287.76	337.76	437.76	537.76	637.76	737.76	837.76	937.76	1037.76
	v	0.01641	73.53	78.14	84.21	90.24	96.25	102.24	108.23	114.21	126.15	138.08	150.01	161.94	173.86	185.78	197.70
	h	130.20	1131.1	1148.6	1171.7	1194.8	1218.0	1241.3	1264.7	1288.2	1335.9	1384.3	1433.6	1483.7	1534.7	1586.7	1639.6
	s	0.2349	1.8443	1.8716	1.9054	1.9369	1.9664	1.9943	2.0208	2.0460	2.0932	2.1369	2.1776	2.2159	2.2521	2.2866	2.3194
10 (193.21)	Sh			6.79	56.79	106.79	156.79	206.79	256.79	306.79	406.79	506.79	606.79	706.79	806.79	906.79	1006.79
	v	0.01659	38.42	38.84	41.93	44.98	48.02	51.03	54.04	57.04	63.03	69.00	74.98	80.94	86.91	92.87	98.84
	h	161.26	1143.3	1146.6	1170.2	1193.7	1217.1	1240.6	1264.1	1287.8	1335.5	1384.0	1433.4	1483.5	1534.6	1586.6	1639.5
	s	0.2836	1.7879	1.7928	1.8273	1.8593	1.8892	1.9173	1.9439	1.9692	2.0166	2.0603	2.1011	2.1394	2.1757	2.2101	2.2430
14.696 (212.00)	Sh				38.00	88.00	138.00	188.00	238.00	288.00	388.00	488.00	588.00	688.00	788.00	888.00	988.00
	v	.0167	26.799		28.42	30.52	32.60	34.67	36.72	38.77	42.86	46.93	51.00	55.06	59.13	63.19	67.25
	h	180.17	1150.5		1168.8	1192.6	1216.3	1239.9	1263.6	1287.4	1335.2	1383.8	1433.2	1483.4	1534.5	1586.5	1639.4
	s	.3121	1.7568		1.7833	1.8158	1.8459	1.8743	1.9010	1.9265	1.9739	2.0177	2.0585	2.0969	2.1332	2.1676	2.2005
15 (213.03)	Sh				36.97	86.97	136.97	186.97	236.97	286.97	386.97	486.97	586.97	686.97	786.97	886.97	986.97
	v	0.01673	26.290		27.837	29.899	31.939	33.963	35.977	37.985	41.986	45.978	49.964	53.946	57.926	61.905	65.882
	h	181.21	1150.9		1168.7	1192.5	1216.2	1239.9	1263.6	1287.3	1335.2	1383.8	1433.2	1483.4	1534.5	1586.5	1639.4
	s	0.3137	1.7552		1.7809	1.8134	1.8437	1.8720	1.8988	1.9242	1.9717	2.0155	2.0563	2.0946	2.1309	2.1653	2.1982
20 (227.96)	Sh				22.04	72.04	122.04	172.04	222.04	272.04	372.04	472.04	572.04	672.04	772.04	872.04	972.04
	v	0.01683	20.087		20.788	22.356	23.900	25.428	26.946	28.457	31.466	34.465	37.458	40.447	43.435	46.420	49.405
	h	196.27	1156.3		1167.1	1191.4	1215.4	1239.2	1263.0	1286.9	1334.9	1383.5	1432.9	1483.2	1534.3	1586.3	1639.3
	s	0.3358	1.7320		1.7475	1.7805	1.8111	1.8397	1.8666	1.8921	1.9397	1.9836	2.0244	2.0628	2.0991	2.1336	2.1665
25 (240.07)	Sh				9.93	59.93	109.93	159.93	209.93	259.93	359.93	459.93	559.93	659.93	759.93	859.93	959.93
	v	0.01693	16.301		16.558	17.829	19.076	20.307	21.527	22.740	25.153	27.557	29.954	32.348	34.740	37.130	39.518
	h	208.52	1160.6		1165.6	1190.2	1214.5	1238.5	1262.5	1286.4	1334.6	1383.3	1432.7	1483.0	1534.2	1586.2	1639.2
	s	0.3535	1.7141		1.7212	1.7547	1.7856	1.8145	1.8415	1.8672	1.9149	1.9588	1.9997	2.0381	2.0744	2.1089	2.1418
30 (250.34)	Sh					49.66	99.66	149.66	199.66	249.66	349.66	449.66	549.66	649.66	749.66	849.66	949.66
	v	0.01701	13.744			14.810	15.859	16.892	17.914	18.929	20.945	22.951	24.952	26.949	28.943	30.936	32.927
	h	218.93	1164.1			1189.0	1213.6	1237.8	1261.9	1286.0	1334.2	1383.0	1432.5	1482.8	1534.0	1586.1	1639.0
	s	0.3682	1.6995			1.7334	1.7647	1.7937	1.8210	1.8467	1.8946	1.9386	1.9795	2.0179	2.0543	2.0888	2.1217
35 (259.29)	Sh					40.71	90.71	140.71	190.71	240.71	340.71	440.71	540.71	640.71	740.71	840.71	940.71
	v	0.01708	11.896			12.654	13.562	14.453	15.334	16.207	17.939	19.662	21.379	23.092	24.803	26.512	28.220
	h	228.03	1167.1			1187.8	1212.7	1237.1	1261.3	1285.5	1333.9	1382.8	1432.3	1482.7	1533.9	1586.0	1638.9
	s	0.3809	1.6872			1.7152	1.7468	1.7761	1.8035	1.8294	1.8774	1.9214	1.9624	2.0009	2.0372	2.0717	2.1046
40 (267.25)	Sh					32.75	82.75	132.75	182.75	232.75	332.75	432.75	532.75	632.75	732.75	832.75	932.75
	v	0.01715	10.497			11.036	11.838	12.624	13.398	14.165	15.685	17.195	18.699	20.199	21.697	23.194	24.689
	h	236.14	1169.8			1186.6	1211.7	1236.4	1260.8	1285.0	1333.6	1382.5	1432.1	1482.5	1533.7	1585.8	1638.8
	s	0.3921	1.6765			1.6992	1.7312	1.7608	1.7883	1.8143	1.8624	1.9065	1.9476	1.9860	2.0224	2.0569	2.0899
45 (274.44)	Sh					25.56	75.56	125.56	175.56	225.56	325.56	425.56	525.56	625.56	725.56	825.56	925.56
	v	0.01721	9.399			9.777	10.497	11.201	11.892	12.577	13.932	15.276	16.614	17.950	19.282	20.613	21.943
	h	243.49	1172.1			1185.4	1210.4	1235.7	1260.2	1284.6	1333.3	1382.3	1431.9	1482.3	1533.6	1585.7	1638.7
	s	0.4021	1.6671			1.6849	1.7173	1.7471	1.7748	1.8010	1.8492	1.8934	1.9345	1.9730	2.0093	2.0439	2.0768
50 (281.02)	Sh					18.98	68.98	118.98	168.98	218.98	318.98	418.98	518.98	618.98	718.98	818.98	918.98
	v	0.01727	8.514			8.769	9.424	10.062	10.688	11.306	12.529	13.741	14.947	16.150	17.350	18.549	19.746
	h	250.21	1174.1			1184.1	1209.9	1234.9	1259.6	1284.1	1332.9	1382.0	1431.7	1482.2	1533.4	1585.6	1638.6
	s	0.4112	1.6586			1.6720	1.7048	1.7349	1.7628	1.7890	1.8374	1.8816	1.9227	1.9613	1.9977	2.0322	2.0652
55 (287.07)	Sh					12.93	62.93	112.93	162.93	212.93	312.93	412.93	512.93	612.93	712.93	812.93	912.93
	v	0.01733	7.787			7.945	8.546	9.130	9.702	10.267	11.381	12.485	13.583	14.677	15.769	16.859	17.948
	h	256.43	1176.0			1182.9	1208.9	1234.2	1259.1	1283.6	1332.6	1381.8	1431.5	1482.0	1533.3	1585.5	1638.5
	s	0.4196	1.6510			1.6601	1.6933	1.7237	1.7518	1.7781	1.8266	1.8710	1.9121	1.9507	1.987	2.022	2.055
60 (292.71)	Sh					7.29	57.29	107.29	157.29	207.29	307.29	407.29	507.29	607.29	707.29	807.29	907.29
	v	0.01738	7.174			7.257	7.815	8.354	8.881	9.400	10.425	11.438	12.446	13.450	14.452	15.452	16.450
	h	262.21	1177.6			1181.6	1208.0	1233.5	1258.5	1283.2	1332.3	1381.5	1431.3	1481.8	1533.2	1585.3	1638.4
	s	0.4273	1.6440			1.6492	1.6829	1.7134	1.7417	1.7681	1.8168	1.8612	1.9024	1.9410	1.9774	2.0120	2.0450
65 (297.98)	Sh					2.02	52.02	102.02	152.02	202.02	302.02	402.02	502.02	602.02	702.02	802.02	902.02
	v	0.01743	6.653			6.675	7.195	7.697	8.186	8.667	9.615	10.552	11.484	12.412	13.337	14.261	15.183
	h	267.63	1179.1			1180.3	1207.0	1232.7	1257.9	1282.7	1331.9	1381.3	1431.1	1481.6	1533.0	1585.2	1638.3
	s	0.4344	1.6375			1.6390	1.6731	1.7040	1.7324	1.7590	1.8077	1.8522	1.8935	1.9321	1.9685	2.0031	2.0361
70 (302.93)	Sh						47.07	97.07	147.07	197.07	297.07	397.07	497.07	597.07	697.07	797.07	897.07
	v	0.01748	6.205				6.664	7.133	7.590	8.039	8.922	9.793	10.659	11.522	12.382	13.240	14.097
	h	272.74	1180.6				1206.0	1232.0	1257.3	1282.2	1331.6	1381.0	1430.9	1481.5	1532.9	1585.1	1638.2
	s	0.4411	1.6316				1.6640	1.6951	1.7237	1.7504	1.7993	1.8439	1.8852	1.9238	1.9603	1.9949	2.0279
75 (307.61)	Sh						42.39	92.39	142.39	192.39	292.39	392.39	492.39	592.39	692.39	792.39	892.39
	v	0.01753	5.814				6.204	6.645	7.074	7.494	8.320	9.135	9.945	10.750	11.553	12.355	13.155
	h	277.56	1181.9				1205.0	1231.2	1256.7	1281.7	1331.3	1380.7	1430.7	1481.3	1532.7	1585.0	1638.1
	s	0.4474	1.6260				1.6554	1.6868	1.7156	1.7424	1.7915	1.8361	1.8774	1.9161	1.9526	1.9872	2.0202

Sh = superheat, F
v = specific volume, cu ft per lb
h = enthalpy, Btu per lb
s = entropy, Btu per R per lb

THERMO

Appendix F (continued)

Abs Press. Lb/Sq In. (Sat. Temp)		Sat. Water	Sat. Steam	Temperature – Degrees Fahrenheit 350	400	450	500	550	600	700	800	900	1000	1100	1200	1300	1400
	Sh			37.96	87.96	137.96	187.96	237.96	287.96	387.96	487.96	587.96	687.96	787.96	887.96	987.96	1087.96
80	v	0.01757	5.471	5.801	6.218	6.622	7.018	7.408	7.794	8.560	9.319	10.075	10.829	11.581	12.331	13.081	13.829
(312.04)	h	282.15	1183.1	1204.0	1230.5	1256.1	1281.3	1306.2	1330.9	1380.5	1430.5	1481.1	1532.6	1584.9	1638.0	1692.0	1746.8
	s	0.4534	1.6208	1.6473	1.6790	1.7080	1.7349	1.7602	1.7842	1.8289	1.8702	1.9089	1.9454	1.9800	2.0131	2.0446	2.0750
	Sh			33.74	83.74	133.74	183.74	233.74	283.74	383.74	483.74	583.74	683.74	783.74	883.74	983.74	1083.74
85	v	0.01762	5.167	5.445	5.840	6.223	6.597	6.966	7.330	8.052	8.768	9.480	10.190	10.898	11.604	12.310	13.014
(316.26)	h	286.52	1184.2	1203.0	1229.7	1255.5	1280.8	1305.8	1330.6	1380.2	1430.3	1481.0	1532.4	1584.7	1637.9	1691.9	1746.8
	s	0.4590	1.6159	1.6396	1.6716	1.7008	1.7279	1.7532	1.7772	1.8220	1.8634	1.9021	1.9386	1.9733	2.0063	2.0379	2.0682
	Sh			29.72	79.72	129.72	179.72	229.72	279.72	379.72	479.72	579.72	679.72	779.72	879.72	979.72	1079.72
90	v	0.01766	4.895	5.128	5.505	5.869	6.223	6.572	6.917	7.600	8.277	8.950	9.621	10.290	10.958	11.625	12.290
(320.28)	h	290.69	1185.3	1202.0	1228.9	1254.9	1280.3	1305.4	1330.2	1380.0	1430.1	1480.8	1532.3	1584.6	1637.8	1691.8	1746.7
	s	0.4643	1.6113	1.6323	1.6646	1.6940	1.7212	1.7467	1.7707	1.8156	1.8570	1.8957	1.9323	1.9669	2.0000	2.0316	2.0619
	Sh			25.87	75.87	125.87	175.87	225.87	275.87	375.87	475.87	575.87	675.87	775.87	875.87	975.87	1075.87
95	v	0.01770	4.651	4.845	5.205	5.551	5.889	6.221	6.548	7.196	7.838	8.477	9.113	9.747	10.380	11.012	11.643
(324.13)	h	294.70	1186.2	1200.9	1228.1	1254.3	1279.8	1305.0	1329.9	1379.7	1429.9	1480.6	1532.1	1584.5	1637.7	1691.7	1746.6
	s	0.4694	1.6069	1.6253	1.6580	1.6876	1.7149	1.7404	1.7645	1.8094	1.8509	1.8897	1.9262	1.9609	1.9940	2.0256	2.0559
	Sh			22.18	72.18	122.18	172.18	222.18	272.18	372.18	472.18	572.18	672.18	772.18	872.18	972.18	1072.18
100	v	0.01774	4.431	4.590	4.935	5.266	5.588	5.904	6.216	6.833	7.443	8.050	8.655	9.258	9.860	10.460	11.060
(327.82)	h	298.54	1187.2	1199.9	1227.4	1253.7	1279.3	1304.6	1329.6	1379.5	1429.7	1480.4	1532.0	1584.4	1637.6	1691.6	1746.5
	s	0.4743	1.6027	1.6187	1.6516	1.6814	1.7088	1.7344	1.7586	1.8036	1.8451	1.8839	1.9205	1.9552	1.9883	2.0199	2.0502
	Sh			18.63	68.63	118.63	168.63	218.63	268.63	368.63	468.63	568.63	668.63	768.63	868.63	968.63	1068.63
105	v	0.01778	4.231	4.359	4.690	5.007	5.315	5.617	5.915	6.504	7.086	7.665	8.241	8.816	9.389	9.961	10.532
(331.37)	h	302.24	1188.0	1198.8	1226.6	1253.1	1278.8	1304.2	1329.2	1379.2	1429.4	1480.3	1531.8	1584.2	1637.5	1691.5	1746.4
	s	0.4790	1.5988	1.6122	1.6455	1.6755	1.7031	1.7288	1.7530	1.7981	1.8396	1.8785	1.9151	1.9498	1.9828	2.0145	2.0448
	Sh			15.21	65.21	115.21	165.21	215.21	265.21	365.21	465.21	565.21	665.21	765.21	865.21	965.21	1065.21
110	v	0.01782	4.048	4.149	4.468	4.772	5.068	5.357	5.642	6.205	6.761	7.314	7.865	8.413	8.961	9.507	10.053
(334.79)	h	305.80	1188.9	1197.7	1225.8	1252.5	1278.3	1303.8	1328.9	1379.0	1429.2	1480.1	1531.7	1584.1	1637.4	1691.4	1746.4
	s	0.4834	1.5950	1.6061	1.6396	1.6698	1.6975	1.7233	1.7476	1.7928	1.8344	1.8732	1.9099	1.9446	1.9777	2.0093	2.0397
	Sh			11.92	61.92	111.92	161.92	211.92	261.92	361.92	461.92	561.92	661.92	761.92	861.92	961.92	1061.92
115	v	0.01785	3.881	3.957	4.265	4.558	4.841	5.119	5.392	5.932	6.465	6.994	7.521	8.046	8.570	9.093	9.615
(338.08)	h	309.25	1189.6	1196.7	1225.0	1251.8	1277.9	1303.3	1328.6	1378.7	1429.0	1479.9	1531.6	1584.0	1637.2	1691.4	1746.3
	s	0.4877	1.5913	1.6001	1.6340	1.6644	1.6922	1.7181	1.7425	1.7877	1.8294	1.8682	1.9049	1.9396	1.9727	2.0044	2.0347
	Sh			8.73	58.73	108.73	158.73	208.73	258.73	358.73	458.73	558.73	658.73	758.73	858.73	958.73	1058.73
120	v	0.01789	3.7275	3.7815	4.0786	4.3610	4.6341	4.9009	5.1637	5.6813	6.1928	6.7006	7.2060	7.7096	8.2119	8.7130	9.2134
(341.27)	h	312.58	1190.4	1195.6	1224.1	1251.2	1277.4	1302.9	1328.2	1378.4	1428.8	1479.8	1531.4	1583.9	1637.1	1691.3	1746.2
	s	0.4919	1.5879	1.5943	1.6286	1.6592	1.6872	1.7132	1.7376	1.7829	1.8246	1.8635	1.9001	1.9349	1.9680	1.9996	2.0300
	Sh			2.67	52.67	102.67	152.67	202.67	252.67	352.67	452.67	552.67	652.67	752.67	852.67	952.67	1052.67
130	v	0.01796	3.4544	3.4699	3.7489	4.0129	4.2672	4.5151	4.7589	5.2384	5.7118	6.1814	6.6486	7.1140	7.5781	8.0411	8.5033
(347.33)	h	318.95	1191.7	1193.4	1222.5	1249.9	1276.4	1302.1	1327.5	1377.9	1428.4	1479.4	1531.1	1583.6	1636.9	1691.1	1746.1
	s	0.4998	1.5813	1.5833	1.6182	1.6493	1.6775	1.7037	1.7283	1.7737	1.8155	1.8545	1.8911	1.9259	1.9591	1.9907	2.0211
	Sh				46.96	96.96	146.96	196.96	246.96	346.96	446.96	546.96	646.96	746.96	846.96	946.96	1046.96
140	v	0.01803	3.2190		3.4661	3.7143	3.9526	4.1844	4.4119	4.8588	5.2995	5.7364	6.1709	6.6036	7.0349	7.4652	7.8946
(353.04)	h	324.96	1193.0		1220.8	1248.7	1275.3	1301.3	1326.8	1377.4	1428.0	1479.1	1530.8	1583.4	1636.7	1690.9	1745.9
	s	0.5071	1.5752		1.6085	1.6400	1.6686	1.6949	1.7196	1.7652	1.8071	1.8461	1.8828	1.9176	1.9508	1.9825	2.0129
	Sh				41.57	91.57	141.57	191.57	241.57	341.57	441.57	541.57	641.57	741.57	841.57	941.57	1041.57
150	v	0.01809	3.0139		3.2208	3.4555	3.6799	3.8978	4.1112	4.5298	4.9421	5.3507	5.7568	6.1612	6.5642	6.9661	7.3671
(358.43)	h	330.65	1194.1		1219.1	1247.4	1274.3	1300.5	1326.1	1376.9	1427.6	1478.7	1530.5	1583.1	1636.5	1690.7	1745.7
	s	0.5141	1.5695		1.5993	1.6313	1.6602	1.6867	1.7115	1.7573	1.7992	1.8383	1.8751	1.9099	1.9431	1.9748	2.0052
	Sh				36.45	86.45	136.45	186.45	236.45	336.45	436.45	536.45	636.45	736.45	836.45	936.45	1036.45
160	v	0.01815	2.8336		3.0060	3.2288	3.4413	3.6469	3.8480	4.2420	4.6295	5.0132	5.3945	5.7741	6.1522	6.5293	6.9055
(363.55)	h	336.07	1195.1		1217.4	1246.0	1273.3	1299.6	1325.4	1376.4	1427.2	1478.4	1530.3	1582.9	1636.3	1690.5	1745.6
	s	0.5206	1.5641		1.5906	1.6231	1.6522	1.6790	1.7039	1.7499	1.7919	1.8310	1.8678	1.9027	1.9359	1.9676	1.9980
	Sh				31.58	81.58	131.58	181.58	231.58	331.58	431.58	531.58	631.58	731.58	831.58	931.58	1031.58
170	v	0.01821	2.6738		2.8162	3.0288	3.2306	3.4255	3.6158	3.9879	4.3536	4.7155	5.0749	5.4325	5.7888	6.1440	6.4983
(368.42)	h	341.24	1196.0		1215.6	1244.7	1272.2	1298.8	1324.7	1375.8	1426.8	1478.0	1530.0	1582.6	1636.1	1690.4	1745.4
	s	0.5269	1.5591		1.5823	1.6152	1.6447	1.6717	1.6968	1.7428	1.7850	1.8241	1.8610	1.8959	1.9291	1.9608	1.9913
	Sh				26.92	76.92	126.92	176.92	226.92	326.92	426.92	526.92	626.92	726.92	826.92	926.92	1026.92
180	v	0.01827	2.5312		2.6474	2.8508	3.0433	3.2286	3.4093	3.7621	4.1084	4.4508	4.7907	5.1289	5.4657	5.8014	6.1363
(373.08)	h	346.19	1196.9		1213.8	1243.4	1271.2	1297.9	1324.0	1375.3	1426.3	1477.7	1529.7	1582.4	1635.9	1690.2	1745.3
	s	0.5328	1.5543		1.5743	1.6078	1.6376	1.6647	1.6900	1.7362	1.7784	1.8176	1.8545	1.8894	1.9227	1.9545	1.9849
	Sh				22.47	72.47	122.47	172.47	222.47	322.47	422.47	522.47	622.47	722.47	822.47	922.47	1022.47
190	v	0.01833	2.4030		2.4961	2.6915	2.8756	3.0525	3.2246	3.5601	3.8889	4.2140	4.5365	4.8572	5.1766	5.4949	5.8124
(377.53)	h	350.94	1197.6		1212.0	1242.0	1270.1	1297.1	1323.3	1374.8	1425.9	1477.4	1529.4	1582.1	1635.7	1690.0	1745.1
	s	0.5384	1.5498		1.5667	1.6006	1.6307	1.6581	1.6835	1.7299	1.7722	1.8115	1.8484	1.8834	1.9166	1.9484	1.9789
	Sh				18.20	68.20	118.20	168.20	218.20	318.20	418.20	518.20	618.20	718.20	818.20	918.20	1018.20
200	v	0.01839	2.2873		2.3598	2.5480	2.7247	2.8939	3.0583	3.3783	3.6915	4.0008	4.3077	4.6128	4.9165	5.2191	5.5209
(381.80)	h	355.51	1198.3		1210.1	1240.6	1269.0	1296.2	1322.6	1374.3	1425.5	1477.0	1529.1	1581.9	1635.4	1689.8	1745.0
	s	0.5438	1.5454		1.5593	1.5938	1.6242	1.6518	1.6773	1.7239	1.7663	1.8057	1.8426	1.8776	1.9109	1.9427	1.9732

Sh = superheat, F
v = specific volume, cu ft per lb
h = enthalpy, Btu per lb
s = entropy, Btu per F per lb

THERMO

Appendix F (continued)

Abs Press. Lb/Sq In. (Sat. Temp)		Sat. Water	Sat. Steam	400	450	500	550	600	700	800	900	1000	1100	1200	1300	1400	1500
	Sh			14.09	64.09	114.09	164.09	214.09	314.09	414.09	514.09	614.09	714.09	814.09	914.09	1014.09	1114.09
210	v	0.01844	2.1822	2.2364	2.4181	2.5880	2.7504	2.9078	3.2137	3.5128	3.8080	4.1007	4.3915	4.6811	4.9695	5.2571	5.5440
(385.91)	h	359.91	1199.0	1208.2	1239.2	1268.0	1295.3	1321.9	1373.7	1425.1	1476.7	1528.8	1581.6	1635.2	1689.6	1744.8	1800.8
	s	0.5490	1.5413	1.5522	1.5872	1.6180	1.6458	1.6715	1.7182	1.7607	1.8001	1.8371	1.8721	1.9054	1.9372	1.9677	1.9970
	Sh			10.12	60.12	110.12	160.12	210.12	310.12	410.12	510.12	610.12	710.12	810.12	910.12	1010.12	1110.12
220	v	0.01850	2.0863	2.1240	2.2999	2.4638	2.6199	2.7710	3.0642	3.3504	3.6327	3.9125	4.1905	4.4671	4.7426	5.0173	5.2913
(389.88)	h	364.17	1199.6	1206.3	1237.8	1266.9	1294.5	1321.2	1373.2	1424.7	1476.3	1528.5	1581.4	1635.0	1689.4	1744.7	1800.6
	s	0.5540	1.5374	1.5453	1.5808	1.6120	1.6400	1.6658	1.7128	1.7553	1.7948	1.8318	1.8668	1.9002	1.9320	1.9625	1.9919
	Sh			6.30	56.30	106.30	156.30	206.30	306.30	406.30	506.30	606.30	706.30	806.30	906.30	1006.30	1106.30
230	v	0.01855	1.9985	2.0212	2.1919	2.3503	2.5008	2.6461	2.9276	3.2020	3.4726	3.7406	4.0068	4.2717	4.5355	4.7984	5.0606
(393.70)	h	368.28	1200.1	1204.4	1236.3	1265.7	1293.6	1320.4	1372.7	1424.2	1476.0	1528.2	1581.1	1634.8	1689.3	1744.5	1800.5
	s	0.5588	1.5336	1.5385	1.5747	1.6062	1.6344	1.6604	1.7075	1.7502	1.7897	1.8268	1.8618	1.8952	1.9270	1.9576	1.9869
	Sh			2.61	52.61	102.61	152.61	202.61	302.61	402.61	502.61	602.61	702.61	802.61	902.61	1002.61	1102.61
240	v	0.01860	1.9177	1.9268	2.0928	2.2462	2.3915	2.5316	2.8024	3.0661	3.3259	3.5831	3.8385	4.0926	4.3456	4.5977	4.8492
(397.39)	h	372.27	1200.6	1202.4	1234.9	1264.6	1292.7	1319.7	1372.1	1423.8	1475.6	1527.9	1580.9	1634.6	1689.1	1744.3	1800.4
	s	0.5634	1.5299	1.5320	1.5687	1.6006	1.6291	1.6552	1.7025	1.7452	1.7848	1.8219	1.8570	1.8904	1.9223	1.9528	1.9822
	Sh				49.03	99.03	149.03	199.03	299.03	399.03	499.03	599.03	699.03	799.03	899.03	999.03	1099.03
250	v	0.01865	1.8432		2.0016	2.1504	2.2909	2.4262	2.6872	2.9410	3.1909	3.4382	3.6837	3.9278	4.1709	4.4131	4.6546
(400.97)	h	376.14	1201.1		1233.4	1263.5	1291.8	1319.0	1371.6	1423.4	1475.3	1527.6	1580.6	1634.4	1688.9	1744.2	1800.2
	s	0.5679	1.5264		1.5629	1.5951	1.6239	1.6502	1.6976	1.7405	1.7801	1.8173	1.8524	1.8858	1.9177	1.9482	1.9776
	Sh				45.56	95.56	145.56	195.56	295.56	395.56	495.56	595.56	695.56	795.56	895.56	995.56	1095.56
260	v	0.01870	1.7742		1.9173	2.0619	2.1981	2.3289	2.5808	2.8256	3.0663	3.3044	3.5408	3.7758	4.0097	4.2427	4.4750
(404.44)	h	379.90	1201.5		1231.9	1262.4	1290.9	1318.2	1371.1	1423.0	1474.9	1527.3	1580.4	1634.2	1688.7	1744.0	1800.1
	s	0.5722	1.5230		1.5573	1.5899	1.6189	1.6453	1.6930	1.7359	1.7756	1.8128	1.8480	1.8814	1.9133	1.9439	1.9732
	Sh				42.20	92.20	142.20	192.20	292.20	392.20	492.20	592.20	692.20	792.20	892.20	992.20	1092.20
270	v	0.01875	1.7101		1.8391	1.9799	2.1121	2.2388	2.4824	2.7186	2.9509	3.1806	3.4084	3.6349	3.8603	4.0849	4.3087
(407.80)	h	383.56	1201.9		1230.4	1261.2	1290.0	1317.5	1370.5	1422.6	1474.6	1527.1	1580.1	1634.0	1688.5	1743.9	1800.0
	s	0.5764	1.5197		1.5518	1.5848	1.6140	1.6406	1.6885	1.7315	1.7713	1.8085	1.8437	1.8771	1.9090	1.9396	1.9690
	Sh				38.93	88.93	138.93	188.93	288.93	388.93	488.93	588.93	688.93	788.93	888.93	988.93	1088.93
280	v	0.01880	1.6505		1.7665	1.9037	2.0322	2.1551	2.3909	2.6194	2.8437	3.0655	3.2855	3.5042	3.7217	3.9384	4.1543
(411.07)	h	387.12	1202.3		1228.8	1260.0	1289.1	1316.8	1370.0	1422.1	1474.2	1526.8	1579.9	1633.8	1688.4	1743.7	1799.8
	s	0.5805	1.5166		1.5464	1.5798	1.6093	1.6361	1.6841	1.7273	1.7671	1.8043	1.8395	1.8730	1.9050	1.9356	1.9649
	Sh				35.75	85.75	135.75	185.75	285.75	385.75	485.75	585.75	685.75	785.75	885.75	985.75	1085.75
290	v	0.01885	1.5948		1.6988	1.8327	1.9578	2.0772	2.3058	2.5269	2.7440	2.9585	3.1711	3.3824	3.5926	3.8019	4.0106
(414.25)	h	390.60	1202.6		1227.3	1258.9	1288.1	1316.0	1369.5	1421.7	1473.9	1526.5	1579.6	1633.5	1688.2	1743.6	1799.7
	s	0.5844	1.5135		1.5412	1.5750	1.6048	1.6317	1.6799	1.7232	1.7630	1.8003	1.8356	1.8690	1.9010	1.9316	1.9610
	Sh				32.65	82.65	132.65	182.65	282.65	382.65	482.65	582.65	682.65	782.65	882.65	982.65	1082.65
300	v	0.01889	1.5427		1.6356	1.7665	1.8883	2.0044	2.2263	2.4407	2.6509	2.8585	3.0643	3.2688	3.4721	3.6746	3.8764
(417.35)	h	393.99	1202.9		1225.7	1257.7	1287.2	1315.2	1368.9	1421.3	1473.6	1526.2	1579.4	1633.3	1688.0	1743.4	1799.6
	s	0.5882	1.5105		1.5361	1.5703	1.6003	1.6274	1.6758	1.7192	1.7591	1.7964	1.8317	1.8652	1.8972	1.9278	1.9572
	Sh				29.64	79.64	129.64	179.64	279.64	379.64	479.64	579.64	679.64	779.64	879.64	979.64	1079.64
310	v	0.01894	1.4939		1.5763	1.7044	1.8233	1.9363	2.1520	2.3600	2.5638	2.7650	2.9644	3.1625	3.3594	3.5555	3.7509
(420.36)	h	397.30	1203.2		1224.1	1256.5	1286.3	1314.5	1368.4	1420.9	1473.2	1525.9	1579.2	1633.1	1687.8	1743.3	1799.4
	s	0.5920	1.5076		1.5311	1.5657	1.5960	1.6233	1.6719	1.7153	1.7553	1.7927	1.8280	1.8615	1.8935	1.9241	1.9536
	Sh				26.69	76.69	126.69	176.69	276.69	376.69	476.69	576.69	676.69	776.69	876.69	976.69	1076.69
320	v	0.01899	1.4480		1.5207	1.6462	1.7623	1.8725	2.0823	2.2843	2.4821	2.6774	2.8708	3.0628	3.2538	3.4438	3.6332
(423.31)	h	400.53	1203.4		1222.5	1255.2	1285.3	1313.7	1367.8	1420.5	1472.9	1525.6	1578.9	1632.9	1687.6	1743.1	1799.3
	s	0.5956	1.5048		1.5261	1.5612	1.5918	1.6192	1.6680	1.7116	1.7516	1.7890	1.8243	1.8579	1.8899	1.9206	1.9500
	Sh				23.82	73.82	123.82	173.82	273.82	373.82	473.82	573.82	673.82	773.82	873.82	973.82	1073.82
330	v	0.01903	1.4048		1.4684	1.5915	1.7050	1.8125	2.0168	2.2132	2.4054	2.5950	2.7828	2.9692	3.1545	3.3389	3.5227
(426.18)	h	403.70	1203.6		1220.9	1254.0	1284.4	1313.0	1367.3	1420.0	1472.5	1525.3	1578.7	1632.7	1687.5	1742.9	1799.2
	s	0.5991	1.5021		1.5213	1.5568	1.5876	1.6153	1.6643	1.7079	1.7480	1.7855	1.8208	1.8544	1.8864	1.9171	1.9466
	Sh				21.01	71.01	121.01	171.01	271.01	371.01	471.01	571.01	671.01	771.01	871.01	971.01	1071.01
340	v	0.01908	1.3640		1.4191	1.5399	1.6511	1.7561	1.9552	2.1463	2.3333	2.5175	2.7000	2.8811	3.0611	3.2402	3.4186
(428.99)	h	406.80	1203.8		1219.2	1252.8	1283.4	1312.2	1366.7	1419.6	1472.2	1525.0	1578.4	1632.5	1687.3	1742.8	1799.0
	s	0.6026	1.4994		1.5165	1.5525	1.5836	1.6114	1.6606	1.7044	1.7445	1.7820	1.8174	1.8510	1.8831	1.9138	1.9432
	Sh				18.27	68.27	118.27	168.27	268.27	368.27	468.27	568.27	668.27	768.27	868.27	968.27	1068.27
350	v	0.01912	1.3255		1.3725	1.4913	1.6002	1.7028	1.8970	2.0832	2.2652	2.4445	2.6219	2.7980	2.9730	3.1471	3.3205
(431.73)	h	409.83	1204.0		1217.5	1251.5	1282.4	1311.4	1366.2	1419.2	1471.8	1524.7	1578.2	1632.3	1687.1	1742.6	1798.9
	s	0.6059	1.4968		1.5119	1.5483	1.5797	1.6077	1.6571	1.7009	1.7411	1.7787	1.8141	1.8477	1.8798	1.9105	1.9400
	Sh				15.59	65.59	115.59	165.59	265.59	365.59	465.59	565.59	665.59	765.59	865.59	965.59	1065.59
360	v	0.01917	1.2891		1.3285	1.4454	1.5521	1.6525	1.8421	2.0237	2.2009	2.3755	2.5482	2.7196	2.8898	3.0592	3.2279
(434.41)	h	412.81	1204.1		1215.8	1250.3	1281.5	1310.6	1365.6	1418.7	1471.5	1524.4	1577.9	1632.1	1686.9	1742.5	1798.8
	s	0.6092	1.4943		1.5073	1.5441	1.5758	1.6040	1.6536	1.6976	1.7379	1.7754	1.8109	1.8445	1.8766	1.9073	1.9368
	Sh				10.39	60.39	110.39	160.39	260.39	360.39	460.39	560.39	660.39	760.39	860.39	960.39	1060.39
380	v	0.01925	1.2218		1.2472	1.3606	1.4635	1.5598	1.7410	1.9139	2.0825	2.2484	2.4124	2.5750	2.7366	2.8973	3.0572
(439.61)	h	418.59	1204.4		1212.4	1247.7	1279.5	1309.0	1364.5	1417.9	1470.8	1523.8	1577.4	1631.6	1686.5	1742.2	1798.5
	s	0.6156	1.4894		1.4982	1.5360	1.5683	1.5969	1.6470	1.6911	1.7315	1.7692	1.8047	1.8384	1.8705	1.9012	1.9307

Abs Press. Lb/Sq In. (Sat. Temp)		Sat. Water	Sat. Steam	450	500	550	600	650	700	800	900	1000	1100	1200	1300	1400	1500
	Sh			5.40	55.40	105.40	155.40	205.40	255.40	355.40	455.40	555.40	655.40	755.40	855.40	955.40	1055.40
400	v	0.01934	1.1610	1.1738	1.2841	1.3836	1.4763	1.5646	1.6499	1.8151	1.9759	2.1339	2.2901	2.4450	2.5987	2.7515	2.9037
(444.60)	h	424.17	1204.6	1208.8	1245.1	1277.5	1307.4	1335.9	1363.4	1417.0	1470.1	1523.3	1576.9	1631.2	1686.2	1741.9	1798.2
	s	0.6217	1.4847	1.4894	1.5282	1.5611	1.5901	1.6163	1.6406	1.6850	1.7255	1.7632	1.7988	1.8325	1.8647	1.8955	1.9250

Sh = superheat, F
v = specific volume, cu ft per lb
h = enthalpy, Btu per lb
s = entropy, Btu per F per lb

THERMO

Appendix G
Air Table

T °R	T °F	h BTU/lbm	p_r	u BTU/lbm	v_r	ϕ BTU/lbm-°R
100	−359.67	23.82	.003826	16.96	9684	.19698
120	−339.67	28.60	.007229	20.38	6149	.24061
140	−319.67	33.39	.012382	23.79	4189	.27750
160	−299.67	38.18	.019734	27.21	3004	.30945
180	−279.67	42.96	.029769	30.62	2240	.33764
200	−259.67	47.75	.04300	34.04	1723.0	.36285
220	−239.67	52.53	.05998	37.45	1358.9	.38566
240	−219.67	57.32	.08126	40.87	1094.1	.40648
260	−199.67	62.11	.10746	44.28	896.4	.42564
280	−179.67	66.89	.13919	47.70	745.3	.44338
300	−159.67	71.68	.17709	51.11	627.6	.45989
320	−139.67	76.47	.22184	54.53	534.4	.47533
340	−119.67	81.25	.27414	57.94	459.5	.48984
360	−99.67	86.04	.3347	61.36	398.5	.50353
380	−79.67	90.83	.4042	64.78	348.3	.51647
400	−59.67	95.62	.4835	68.19	306.5	.52875
420	−39.67	100.40	.5733	71.61	271.4	.54043
440	−19.67	105.19	.6745	75.03	241.7	.55157
460	.33	109.98	.7878	78.45	216.31	.56222
480	20.33	114.78	.9142	81.87	194.52	.57241
500	40.33	119.57	1.0544	85.29	175.68	.58220
520	60.33	124.36	1.2094	88.71	159.29	.59160
540	80.33	129.16	1.3801	92.14	144.96	.60065
560	100.33	133.96	1.5675	95.57	132.35	.60938
580	120.33	138.76	1.7725	99.00	121.23	.61781
600	140.33	143.57	1.996	102.44	111.35	.62595
620	160.33	148.38	2.240	105.88	102.56	.63384
640	180.33	153.20	2.504	109.32	94.70	.64148
660	200.33	158.01	2.790	112.77	87.65	.64890
680	220.33	162.84	3.099	116.22	81.30	.65610
700	240.33	167.67	3.432	119.68	75.57	.66310
720	260.33	172.51	3.790	123.14	70.38	.66991
740	280.33	177.35	4.175	126.62	65.66	.67655
760	300.33	182.20	4.588	130.10	61.36	.68301
780	320.33	187.06	5.031	133.58	57.44	.68932
800	340.33	191.92	5.504	137.08	53.85	.69548
820	360.33	196.79	6.008	140.58	50.56	.70150
840	380.33	201.68	6.547	144.09	47.53	.70738
860	400.33	206.57	7.120	147.61	44.75	.71314
880	420.33	211.47	7.730	151.14	42.18	.71877
900	440.33	216.38	8.378	154.68	39.80	.72429
920	460.33	221.30	9.066	158.23	37.60	.72970
940	480.33	226.23	9.795	161.79	35.55	.73500
960	500.33	231.18	10.567	165.36	33.66	.74020
980	520.33	236.13	11.384	168.95	31.89	.74531

THERMO

Appendix G (continued)

T °R	T °F	h BTU/lbm	p_r	u BTU/lbm	v_r	ϕ BTU/lbm-°R
1000	540.33	241.10	12.248	172.54	30.25	.75032
1020	560.33	246.07	13.161	176.15	28.71	.75525
1040	580.33	251.06	14.125	179.76	27.28	.76010
1060	600.33	256.06	15.141	183.39	25.94	.76486
1080	620.33	261.08	16.212	187.03	24.68	.76954
1100	640.33	266.10	17.339	190.69	23.50	.77415
1120	660.33	271.14	18.526	194.35	22.40	.77869
1140	680.33	276.19	19.774	198.03	21.36	.78316
1160	700.33	281.25	21.09	201.72	20.381	.78756
1180	720.33	286.33	22.46	205.43	19.462	.79190
1200	740.33	291.41	23.91	209.15	18.595	.79618
1220	760.33	296.51	25.42	212.88	17.777	.80039
1240	780.33	301.63	27.01	216.62	17.006	.80455
1260	800.33	306.75	28.68	220.37	16.276	.80865
1280	820.33	311.89	30.42	224.14	15.587	.81270
1300	840.33	317.04	32.25	227.92	14.935	.81669
1320	860.33	322.21	34.16	231.72	14.317	.82063
1340	880.33	327.39	36.15	235.52	13.732	.82453
1360	900.33	332.58	38.24	239.34	13.177	.82837
1380	920.33	337.78	40.42	243.17	12.650	.83217
1400	940.33	342.99	42.69	247.02	12.150	.83592
1420	960.33	348.22	45.06	250.87	11.675	.83963
1440	980.33	353.46	47.54	254.74	11.223	.84329
1460	1000.33	358.71	50.11	258.62	10.793	.84691
1480	1020.33	363.98	52.80	262.51	10.384	.85050
1500	1040.33	369.25	55.60	266.42	9.995	.85404
1520	1060.33	374.54	58.52	270.34	9.623	.85754
1540	1080.33	379.84	61.55	274.27	9.270	.86100
1560	1100.33	385.15	64.70	278.21	8.932	.86443
1580	1120.33	390.48	67.98	282.16	8.610	.86782
1600	1140.33	395.81	71.39	286.12	8.303	.87118
1620	1160.33	401.16	74.94	290.10	8.009	.87450
1640	1180.33	406.52	78.62	294.09	7.728	.87778
1660	1200.33	411.89	82.44	298.08	7.460	.88104
1680	1220.33	417.27	86.41	302.09	7.203	.88426
1700	1240.33	422.66	90.52	306.11	6.958	.88745
1720	1260.33	428.06	94.79	310.14	6.723	.89061
1740	1280.33	433.47	99.21	314.18	6.497	.89374
1760	1300.33	438.89	103.80	318.23	6.282	.89684
1780	1320.33	444.32	108.55	322.29	6.075	.89990
1800	1340.33	449.77	113.48	326.37	5.877	.90294
1820	1360.33	455.22	118.57	330.45	5.686	.90596
1840	1380.33	460.68	123.85	334.54	5.504	.90894
1860	1400.33	466.16	129.31	338.64	5.329	.91190
1880	1420.33	471.64	134.96	342.75	5.161	.91483

Appendix G (continued)

T °R	T °F	h BTU/lbm	p_r	u BTU/lbm	v_r	ϕ BTU/lbm-°R
1900	1440.33	477.13	140.81	346.87	4.999	.91774
1920	1460.33	482.63	146.85	351.00	4.844	.92062
1940	1480.33	488.14	153.09	355.14	4.695	.92347
1960	1500.33	493.66	159.54	359.29	4.551	.92630
1980	1520.33	499.19	166.21	363.45	4.413	.92911
2000	1540.33	504.72	173.10	367.61	4.281	.93189
2020	1560.33	510.27	180.20	371.79	4.153	.93465
2040	1580.33	515.82	187.54	375.97	4.030	.93739
2060	1600.33	521.39	195.11	380.16	3.911	.94010
2080	1620.33	526.96	202.93	384.36	3.797	.94279
2100	1640.33	532.54	211.0	388.57	3.687	.94546
2120	1660.33	538.13	219.3	392.79	3.582	.94811
2140	1680.33	543.72	227.9	397.01	3.479	.95074
2160	1700.33	549.32	236.7	401.24	3.381	.95334
2180	1720.33	554.94	245.8	405.48	3.286	.95593
2200	1740.33	560.55	255.2	409.73	3.194	.95850
2220	1760.33	566.18	264.8	413.99	3.106	.96104
2240	1780.33	571.82	274.8	418.25	3.020	.96357
2260	1800.33	577.46	285.0	422.52	2.938	.96607
2280	1820.33	583.10	295.5	426.80	2.858	.96856
2300	1840.33	588.76	306.4	431.08	2.781	.97103
2320	1860.33	594.42	317.5	435.37	2.707	.97348
2340	1880.33	600.09	329.0	439.67	2.635	.97592
2360	1900.33	605.77	340.8	443.98	2.566	.97833
2380	1920.33	611.45	352.9	448.29	2.498	.98073
2400	1940.33	617.14	365.4	452.61	2.433	.98311
2420	1960.33	622.84	378.2	456.93	2.371	.98547
2440	1980.33	628.54	391.4	461.26	2.310	.98782
2460	2000.33	634.25	404.9	465.60	2.251	.99015
2480	2020.33	639.96	418.8	469.94	2.194	.99247
2500	2040.33	645.68	433.1	474.29	2.1386	.99476
2520		651.41	447.7	478.65	2.0852	.99704
2540		657.14	462.8	483.01	2.0334	.99931
2560		662.88	478.2	487.38	1.9832	1.00156
2580		668.63	494.1	491.75	1.9346	1.00380
2600	2140.33	674.38	510.3	496.13	1.8875	1.00602
2620		680.14	527.0	500.52	1.8418	1.00822
2640		685.90	544.1	504.91	1.7974	1.01041
2660		691.67	561.7	509.31	1.7545	1.01259
2680		697.44	579.7	513.71	1.7128	1.01475
2700	2240.33	703.22	598.1	518.12	1.6723	1.01690
2720		709.00	617.0	522.53	1.6331	1.01903
2740		714.79	636.4	526.95	1.5950	1.02116
2760		720.58	656.3	531.37	1.5580	1.02326
2780		726.38	676.6	535.80	1.5221	1.02536

THERMO

Appendix G (continued)

T °R	T °F	h BTU/lbm	p_r	u BTU/lbm	v_r	φ BTU/lbm-°R
2800	2340.33	732.19	697.5	540.23	1.4872	1.02744
2820		738.00	718.9	544.67	1.4533	1.02950
2840		743.81	740.7	549.11	1.4204	1.03156
2860		749.63	763.1	553.56	1.3885	1.03360
2880		755.45	786.0	558.01	1.3574	1.03563
2900	2440.33	761.28	809.5	562.47	1.3272	1.03765
2920		767.12	833.5	566.93	1.2978	1.03965
2940		772.95	858.1	571.40	1.2693	1.04164
2960		778.80	883.3	575.87	1.2416	1.04362
2980		784.64	909.0	580.35	1.2146	1.04559
3000	2540.33	790.49	935.3	584.83	1.1883	1.04755
3020		796.35	962.2	589.31	1.1628	1.04949
3040		802.21	989.8	593.80	1.1379	1.05143
3060		808.07	1017.9	598.29	1.1137	1.05335
3080		813.94	1046.7	602.79	1.0902	1.05526
3100	2640.33	819.82	1076.1	607.29	1.0672	1.05716
3120		825.69	1106.2	611.80	1.0449	1.05905
3140		831.57	1136.9	616.31	1.0232	1.06093
3160		837.46	1168.3	620.82	1.0020	1.06280
3180		843.35	1200.4	625.34	.9814	1.06466
3200	2740.33	849.24	1233.2	629.86	.9613	1.06651
3220		855.14	1266.7	634.39	.9417	1.06834
3240		861.04	1300.9	638.92	.9227	1.07017
3260		866.94	1335.9	643.45	.9041	1.07199
3280		872.85	1371.5	647.99	.8860	1.07379
3300	2840.33	878.77	1408.0	652.53	.8683	1.07559
3320		884.68	1445.2	657.07	.8511	1.07738
3340		890.60	1483.1	661.62	.8343	1.07916
3360		896.52	1521.9	666.18	.8179	1.08092
3380		902.45	1561.4	670.73	.8020	1.08268
3400	2940.33	908.38	1601.8	675.29	.7864	1.08443
3420		914.32	1643.0	679.85	.7712	1.08617
3440		920.25	1685.0	684.42	.7564	1.08790
3460		926.19	1727.8	688.99	.7419	1.08963
3480		932.14	1771.6	693.56	.7278	1.09134
3500	3040.33	938.09	1816.1	698.14	.7140	1.09304
3520		944.04	1861.6	702.72	.7005	1.09474
3540		949.99	1908.0	707.30	.6874	1.09642
3560		955.95	1955.3	711.89	.6745	1.09810
3580		961.91	2003.5	716.48	.6620	1.09977
3600	3140.33	967.88	2052.6	721.07	.6498	1.10143
3620		973.84	2102.7	725.67	.6378	1.10309
3640		979.81	2153.8	730.27	.6261	1.10473
3660		985.79	2205.8	734.87	.6147	1.10637
3680		991.76	2258.8	739.48	.6036	1.10800

Appendix G (continued)

T °R	T °F	h BTU/lbm	p_r	u BTU/lbm	v_r	ϕ BTU/lbm-°R
3700	3240.33	997.74	2312.9	744.09	.5927	1.10962
3720		1003.73	2367.9	748.70	.5820	1.11123
3740		1009.71	2424.0	753.31	.5716	1.11284
3760		1015.70	2481.1	757.93	.5614	1.11443
3780		1021.69	2539.3	762.55	.5515	1.11602
3800	3340.33	1027.69	2598.6	767.18	.5418	1.11760
3820		1033.69	2659.0	771.80	.5322	1.11918
3840		1039.69	2720.4	776.43	.5229	1.12074
3860		1045.69	2783.0	781.06	.5138	1.12230
3880		1051.70	2846.7	785.70	.5049	1.12386
3900	3440.33	1057.71	2911.6	790.34	.4962	1.12540
3920		1063.72	2977.7	794.98	.4877	1.12694
3940		1069.73	3044.9	799.62	.4794	1.12847
3960		1075.75	3113.3	804.27	.4712	1.12999
3980		1081.77	3182.9	808.92	.4632	1.13151
4000	3540.33	1087.79	3254	813.57	.4554	1.13302
4020		1093.82	3326	818.22	.4478	1.13452
4040		1099.85	3399	822.88	.4403	1.13602
4060		1105.88	3474	827.54	.4330	1.13751
4080		1111.91	3550	832.20	.4258	1.13899
4100	3640.33	1117.95	3627	836.87	.4188	1.14046
4120		1123.99	3706	841.54	.4119	1.14193
4140		1130.03	3786	846.21	.4052	1.14340
4160		1136.07	3867	850.88	.3986	1.14485
4180		1142.12	3950	855.55	.3921	1.14630
4200	3740.33	1148.17	4034	860.23	.3858	1.14775
4220		1154.22	4119	864.91	.3796	1.14918
4240		1160.27	4206	869.59	.3735	1.15061
4260		1166.33	4294	874.28	.3675	1.15204
4280		1172.39	4384	878.96	.3617	1.15346
4300	3840.33	1178.45	4475	883.65	.3560	1.15487
4320		1184.51	4568	888.35	.3504	1.15628
4340		1190.57	4662	893.04	.3449	1.15768
4360		1196.64	4758	897.74	.3395	1.15907
4380		1202.71	4856	902.44	.3342	1.16046
4400	3940.33	1208.78	4955	907.14	.3290	1.16185
4420		1214.86	5055	911.84	.3239	1.16322
4440		1220.94	5157	916.55	.3189	1.16459
4460		1227.02	5261	921.25	.3141	1.16596
4480		1233.10	5367	925.97	.3093	1.16732
4500	4040.33	1239.18	5474	930.68	.3046	1.16868
4520		1245.27	5582	935.39	.3000	1.17003
4540		1251.35	5693	940.11	.2954	1.17137
4560		1257.44	5805	944.83	.2910	1.17271
4580		1263.54	5919	949.55	.2867	1.17404

THERMO

Appendix G (continued)

T °R	*T* °F	*h* BTU/lbm	p_r	*u* BTU/lbm	v_r	ϕ BTU/lbm-°R
4600	4140.33	1269.63	6035	954.27	.2824	1.17537
4620		1275.73	6153	959.00	.2782	1.17669
4640		1281.83	6272	963.73	.2741	1.17801
4660		1287.93	6393	968.45	.2700	1.17932
4680		1294.03	6516	973.19	.2661	1.18063
4700	4240.33	1300.13	6641	977.92	.2622	1.18193
4720		1306.24	6768	982.66	.2584	1.18323
4740		1312.35	6897	987.39	.2546	1.18452
4760		1318.46	7027	992.13	.2509	1.18580
4780		1324.57	7160	996.88	.2473	1.18709
4800	4340.33	1330.69	7294	1001.62	.2438	1.18836
4820		1336.81	7431	1006.36	.2403	1.18963
4840		1342.92	7569	1011.11	.2369	1.19090
4860		1349.05	7710	1015.86	.2335	1.19216
4880		1355.17	7853	1020.61	.2302	1.19342
4900	4440.33	1361.29	7998	1025.37	.2270	1.19467
4920		1367.42	8144	1030.12	.2238	1.19592
4940		1373.55	8293	1034.88	.2207	1.19716
4960		1379.68	8445	1039.64	.2176	1.19840
4980		1385.81	8598	1044.40	.2146	1.19963
5000	4540.33	1391.94	8754	1049.16	.21161	1.20086
5020		1398.08	8911	1053.93	.20870	1.20209
5040		1404.22	9071	1058.69	.20583	1.20331
5060		1410.36	9234	1063.46	.20302	1.20452
5080		1416.50	9398	1068.23	.20025	1.20574
5100	4640.33	1422.64	9565	1073.00	.19753	1.20694
5120		1428.79	9734	1077.78	.19486	1.20815
5140		1434.93	9906	1082.55	.19223	1.20934
5160		1441.08	10080	1087.33	.18965	1.21054
5180		1447.23	10257	1092.11	.18711	1.21173
5200	4740.33	1453.38	10435	1096.89	.18461	1.21291
5220		1459.54	10617	1101.67	.18215	1.21409
5240		1465.69	10801	1106.46	.17974	1.21527
5260		1471.85	10987	1111.24	.17736	1.21644
5280		1478.01	11176	1116.03	.17503	1.21761
5300	4840.33	1484.17	11367	1120.82	.17273	1.21878
5320		1490.33	11561	1125.61	.17048	1.21994
5340		1496.49	11758	1130.40	.16825	1.22109
5360		1502.66	11957	1135.20	.16607	1.22225
5380		1508.82	12159	1139.99	.16392	1.22339
5400	4940.33	1514.99	12364	1144.79	.16181	1.22454
5420		1521.16	12571	1149.59	.15973	1.22568
5440		1527.33	12782	1154.39	.15768	1.22682
5460		1533.51	12994	1159.19	.15567	1.22795
5480		1539.68	13210	1163.99	.15368	1.22908

Appendix G (continued)

T °R	T °F	h BTU/lbm	p_r	u BTU/lbm	v_r	ϕ BTU/lbm-°R
5500	5040.33	1545.86	13429	1168.80	.15173	1.23020
5520		1552.04	13650	1173.61	.14982	1.23132
5540		1558.22	13875	1178.41	.14793	1.23244
5560		1564.40	14102	1183.22	.14607	1.23355
5580		1570.58	14332	1188.04	.14424	1.23466
5600	5140.33	1576.76	14565	1192.85	.14244	1.23577
5620		1582.95	14801	1197.66	.14067	1.23687
5640		1589.14	15040	1202.48	.13892	1.23797
5660		1595.32	15283	1207.30	.13721	1.23907
5680		1601.51	15528	1212.11	.13552	1.24016
5700	5240.33	1607.71	15776	1216.94	.13385	1.24125
5720		1613.90	16028	1221.76	.13221	1.24233
5740		1620.09	16283	1226.58	.13060	1.24341
5760		1626.29	16541	1231.40	.12901	1.24449
5780		1632.49	16802	1236.23	.12745	1.24556
5800	5340.33	1638.68	17066	1241.06	.12591	1.24663
5820		1644.88	17334	1245.89	.12439	1.24770
5840		1651.09	17605	1250.72	.12290	1.24877
5860		1657.29	17879	1255.55	.12142	1.24983
5880		1663.49	18157	1260.38	.11997	1.25088
5900	5440.33	1669.70	18438	1265.22	.11855	1.25194
5920		1675.91	18723	1270.05	.11714	1.25299
5940		1682.11	19011	1274.89	.11575	1.25403
5960		1688.32	19303	1279.73	.11439	1.25508
5980		1694.54	19598	1284.57	.11304	1.25612

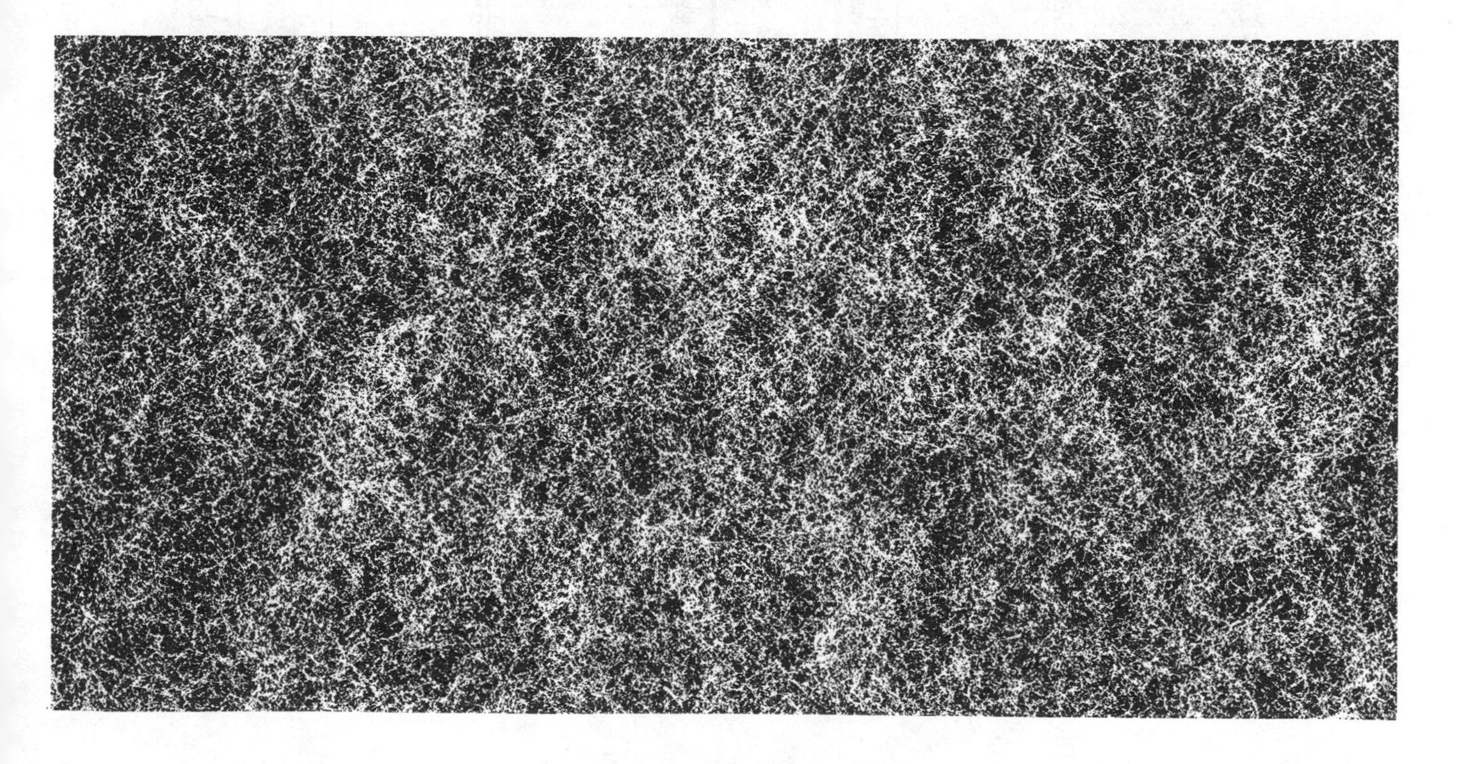

THERMO

Appendix H

Heats and Free Energies of Formation

The values given in the following table for the heats and free energies of formation of inorganic compounds are derived from (*a*) Bichowsky and Rossini, "Thermochemistry of the Chemical Substances," Reinhold, New York, 1936; (*b*) Latimer, "Oxidation States of the Elements and Their Potentials in Aqueous Solution," Prentice-Hall, New York, 1938; (*c*) the tables of the American Petroleum Institute Research Project 44 at the National Bureau of Standards; and (*d*) the tables of Selected Values of Chemical Thermodynamic Properties of the National Bureau of Standards. The reader is referred to the preceding books and tables for additional details as to methods of calculation, standard states, etc.

The organic compounds in the following table are all given under the element carbon. The values for the non-hydrocarbons are largely from E. I. du Pont de Nemours & Co., Ammonia Department, Chemical Division, Experimental Station; and the values for the hydrocarbons are from the tables of the American Petroleum Institute Research Project 44 at the National Bureau of Standards.*

Compound	State†	Heat of formation‡§ ΔH (formation) at 25°C., kcal./mole	Free energy of formation‖¶ ΔF (formation) at 25°C., kcal./mole
Aluminum:			
Al	c	0.00	0.00
$AlBr_3$	c	−123.4	
	aq	−209.5	−189.2
Al_4C_3	c	−30.8	−29.0
$AlCl_3$	c	−163.8	
	aq, 600	−243.9	−209.5
AlF_3	c	−329	
	aq	−360.8	−312.6
AlI_3	c	−72.8	
	aq	−163.4	−152.5
AlN	c	−57.7	−50.4
$Al(NH_4)(SO_4)_2$	c	−561.19	−486.17
$Al(NH_4)(SO_4)_2.12H_2O$	c	−1419.36	−1179.26
$Al(NO_3)_3.6H_2O$	c	−680.89	−526.32
$Al(NO_3)_3.9H_2O$	c	−897.59	
Al_2O_3	c, corundum	−399.09	−376.87
$Al(OH)_3$	c	−304.8	−272.9
$Al_2O_3.SiO_2$	c, sillimanite	−648.7	
$Al_2O_3.SiO_2$	c, disthene	−642.4	
$Al_2O_3.SiO_2$	c, andalusite	−642.0	
$3Al_2O_3.2SiO_2$	c, mullite	−1874	
Al_2S_3	c	−121.6	
$Al_2(SO_4)_3$	c	−820.99	−739.53
	aq	−893.9	−759.3
$Al_2(SO_4)_3.6H_2O$	c	−1268.15	−1103.39
$Al_2(SO_4)_3.18H_2O$	c	−2120	
Antimony:			
Sb	c	0.00	0.00
$SbBr_3$	c	−59.9	
$SbCl_3$	c	−91.3	−77.8
$SbCl_5$	l	−104.8	
SbF_3	c	−216.6	
SbI_3	c	−22.8	
Sb_2O_3	c, I, orthorhombic	−165.4	−146.0
	c, II, octahedral	−166.6	
Sb_2O_4	c	−213.0	−186.6
Sb_2O_5	c	−230.0	−196.1
Sb_2S_3	c, black	−38.2	−36.9
Arsenic:			
As	c	0.00	0.00
$AsBr_3$	c	−45.9	
$AsCl_3$	l	−80.2	−70.5
AsF_3	l	−223.76	−212.27
AsH_3	g	43.6	37.7
AsI_3	c	−13.6	
As_2O_3	c	−154.1	−134.8
As_2O_5	c	−217.9	−183.9
As_2S_3	c	−20	−20
	amorphous	−34.76	
Barium:			
Ba	c	0.00	0.00
$BaBr_2$	c	−180.38	
	aq, 400	−185.67	−183.0
$BaCl_2$	c	−205.25	
	aq, 300	−207.92	−196.5
$Ba(ClO_3)_2$	c	−176.6	
	aq, 1600	−170.0	−134.4
$Ba(ClO_4)_2$	c	−210.2	
	aq, 800	....	−155.3
$Ba(CN)_2$	c	−48	
$Ba(CNO)_2$	c	−212.1	
	aq	....	−180.7
$BaCN_2$	c	−63.6	
$BaCO_3$	c, witherite	−284.2	−271.4
$BaCrO_4$	c	−342.2	
BaF_2	c	−287.9	
	aq, 1600	−284.6	−265.3
BaH_2	c	−40.8	−31.5
$Ba(HCO_2)_2$	aq	−459	−414.4
BaI_2	c	−144.6	
	aq, 400	−155.17	−158.52
$Ba(IO_3)_2$	c	−264.5	
	aq	−237.50	−198.35
$BaMoO_4$	c	−370	
Ba_3N_2	c	−90.7	
$Ba(NO_2)_2$	c	−184.5	
	aq	−179.05	−150.75
Barium (*Cont.*):			
$Ba(NO_3)_2$	c	−236.99	−189.94
	aq, 600	−227.74	
BaO	c	−133.0	
$Ba(OH)_2$	c	−225.9	
	aq, 400	−237.76	−209.02
$BaO.SiO_2$	c	−363	
$Ba_3(PO_4)_2$	c	−992	
$BaPtCl_6$	c	−284.9	
BaS	c	−111.2	
$BaSO_3$	c	−282.5	
$BaSO_4$	c	−340.2	−313.4
$BaWO_4$	c	−402	
Beryllium:			
Be	c	0.00	0.00
$BeBr_2$	c	−79.4	
	aq	−142	−127.9
$BeCl_2$	c	−112.6	
	aq	−163.9	−141.4
BeI_2	c	−39.4	
	aq	−112	−103.4
Be_3N_2	c	−134.5	−122.4
BeO	c	−145.3	−138.3
$Be(OH)_2$	c	−215.6	
BeS	c	−56.1	
$BeSO_4$	c	−281	
	aq	....	−254.8
Bismuth:			
Bi	c	0.00	0.00
$BiCl_3$	c	−90.5	−76.4
	aq	−101.6	
BiI_3	c	−24	
	aq	−27	
BiO	c	−49.5	−43.2
Bi_2O_3	c	−137.1	−117.9
$Bi(OH)_3$	c	−171.1	
Bi_2S_3	c	−43.9	−39.1
$Bi_2(SO_4)_3$	c	−607.1	
Boron:			
B	c	0.00	0.00
BBr_3	l	−52.7	
	g	−44.6	−50.9
BCl_3	g	−94.5	−90.8
BF_3	g	−265.2	−261.0
B_2H_6	g	7.5	19.9
BN	c	−32.1	−27.2
B_2O_3	c	−302.0	−282.9
	gls	−297.6	−280.3
$B(OH)_3$	c	−260.0	−229.4
B_2S_3	c	−56.6	
Bromine:			
Br_2	l	0.00	0.00
	g	7.47	0.931
$BrCl$	g	3.06	−0.63
Cadmium:			
Cd	c	0.00	0.00
$CdBr_2$	c	−75.8	−70.7
	aq, 400	−76.6	−67.6
$CdCl_2$	c	−92.149	−81.889
	aq, 400	−96.44	−81.2
$Cd(CN)_2$	c	36.2	
$CdCO_3$	c	−178.2	−163.2
CdI_2	c	−48.40	
	aq, 400	−47.46	−43.22
Cd_3N_2	c	39.8	
$Cd(NO_3)_2$	aq, 400	−115.67	−71.05
CdO	c	−62.35	−55.28
$Cd(OH)_2$	c	−135.0	−113.7
CdS	c	−34.5	−33.6
$CdSO_4$	c	−222.23	
	aq, 400	−232.635	−194.65
Calcium:			
Ca	c	0.00	0.00
$CaBr_2$	c	−162.20	
	aq, 400	−187.19	−181.86
CaC_2	c	−14.8	−16.0
$CaCl_2$	c	−190.6	−179.8
	aq	−209.15	−195.36

*For footnotes see end of table.

Appendix H (continued)

Compound	State†	Heat of formation‡§ ΔH (formation) at 25°C., kcal./mole	Free energy of formation‖¶ ΔF (formation) at 25°C., kcal./mole
Calcium (*Cont.*):			
$CaCN_2$	c	−85	
$Ca(CN)_2$	c	−43.3	
	aq	....	−54.0
$CaCO_3$	c, calcite	−289.5	−270.8
	c, aragonite	−289.54	−270.57
$CaCO_3.MgCO_3$	c	−558.8	
CaC_2O_4	c	−332.2	
$Ca(C_2H_3O_2)_2$	c	−356.3	
	aq	−364.1	−311.3
CaF_2	c	−290.2	
	aq	−286.5	−264.1
CaH_2	c	−46	−35.7
CaI_2	c	−128.49	
	aq, 400	−156.63	−157.37
Ca_3N_2	c	−103.2	−88.2
$Ca(NO_3)_2$	c	−224.05	−177.38
	aq, 400	−228.29	
$Ca(NO_3)_2.2H_2O$	c	−367.95	−293.57
$Ca(NO_3)_2.3H_2O$	c	−439.05	−351.58
$Ca(NO_3)_2.4H_2O$	c	−509.43	−409.32
CaO	c	−151.7	−144.3
$Ca(OH)_2$	c	−235.58	−213.9
	aq, 800	−239.2	−207.9
$CaO.SiO_2$	c, II, wollastonite	−377.9	−357.5
	c, I, pseudowollastonite	−376.6	−356.6
CaS	c	−114.3	−113.1
$CaSO_4$	c, insoluble form	−338.73	−311.9
	c, soluble form α	−336.58	−309.8
	c, soluble form β	−335.52	−308.8
$CaSO_4.\frac{1}{2}H_2O$	c	−376.13	
$CaSO_4.2H_2O$	c	−479.33	−425.47
$CaWO_4$	c	−387	
Carbon:			
C	c, graphite	0.00	0.00
	c, diamond	0.453	0.685
CO	g	−26.416	−32.808
CO_2	g	−94.052	−94.260
CH_4 methane	g	−17.889	−12.140
C_2H_6 ethane	g	−20.236	−7.860
C_3H_8 propane	g	−24.820	−5.614
C_4H_{10} n-butane	g	−29.812	−3.754
C_4H_{10} isobutane	g	−31.452	−4.296
C_5H_{12} n-pentane	g	−35.00	−1.96
	l	−41.36	−2.21
C_5H_{12} 2-methylbutane	g	−36.92	−3.50
	l	−42.85	−3.59
C_5H_{12} 2,2-dimethylpropane	g	−39.67	−3.64
C_6H_{14} n-hexane	g	−39.96	0.05
	l	−47.52	−0.91
C_6H_{14} 2-methylpentane	g	−41.66	−0.96
	l	−48.82	−1.73
C_6H_{14} 3-methylpentane	g	−41.02	−0.29
	l	−48.28	−1.12
C_6H_{14} 2,2-dimethylbutane	g	−44.35	−2.35
	l	−51.00	−2.88
C_6H_{14} 2,3-dimethylbutane	g	−42.49	−0.73
	l	−49.48	−1.44
C_7H_{16} n-heptane	g	−44.89	2.09
	l	−53.63	0.42
C_7H_{16} 2-methylhexane	g	−46.60	0.98
	l	−54.93	−0.47
C_7H_{16} 3-methylhexane	g	−45.96	1.10
	l	−54.35	−0.39
C_7H_{16} 3-ethylpentane	g	−45.34	2.59
	l	−53.77	1.06
C_7H_{16} 2,2-dimethylpentane	g	−49.29	0.09
	l	−57.05	−1.08
C_7H_{16} 2,3-dimethylpentane	g	−47.62	0.16
	l	−55.81	−1.27
C_7H_{16} 2,4-dimethylpentane	g	−48.30	0.72
	l	−56.17	−0.49
C_7H_{16} 3,3-dimethylpentane	g	−48.17	0.63
	l	−56.07	−0.69
C_7H_{16} 2,2,3-trimethylbutane	g	−48.96	0.76
	l	−56.63	−0.43
C_8H_{18} n-octane	g	−49.82	4.14
	l	−59.74	1.77
C_8H_{18} 2-methylheptane	g	−51.50	3.06
	l	−60.98	0.92
C_8H_{18} 3-methylheptane	g	−50.82	3.29
	l	−60.34	1.12
C_8H_{18} 4-methylheptane	g	−50.69	4.00
	l	−60.17	1.86
C_8H_{18} 3-ethylhexane	g	−50.40	3.95
	l	−59.88	1.80
Carbon (*Cont.*):			
C_8H_{18} 2,2-dimethylhexane	g	−53.71	2.56
	l	−62.63	−0.72
C_8H_{18} 2,3-dimethylhexane	g	−51.13	4.23
	l	−60.40	2.17
C_8H_{18} 2,4-dimethylhexane	g	−52.44	2.80
	l	−61.47	0.89
C_8H_{18} 2,5-dimethylhexane	g	−53.21	2.50
	l	−62.26	0.59
C_8H_{18} 3,3-dimethylhexane	g	−52.61	3.17
	l	−61.58	1.23
C_8H_{18} 3,4-dimethylhexane	g	−50.91	4.97
	l	−60.23	2.86
C_8H_{18} 2-methyl-3-ethylpentane	g	−50.48	5.08
	l	−59.69	3.03
C_8H_{18} 3-methyl-3-ethylpentane	g	−51.38	4.76
	l	−60.46	2.69
C_8H_{18} 2,2,3-trimethylpentane	g	−52.61	4.09
	l	−61.44	2.22
C_8H_{18} 2,2,4-trimethylpentane	g	−53.57	3.13
	l	−61.97	1.51
C_8H_{18} 2,3,3,-trimethylpentane	g	−51.73	4.52
	l	−60.63	2.54
C_8H_{18} 2,3,4-trimethylpentane	g	−51.97	4.32
	l	−60.98	2.34
C_8H_{18} 2,2,3,3,-tetramethylbutane	g	−53.99	4.88
	c	−64.23	2.74
C_2H_4 ethylene	g	12.496	16.282
C_3H_6 propylene	g	4.879	14.964
C_4H_8 1-butene	g	0.280	17.217
C_4H_8 cis-2-butene	g	−1.362	16.007
C_4H_8 trans-2-butene	g	−2.405	15.323
C_4H_8 2-methyl-2-propene	g	−3.343	14.574
C_5H_{10} 1-pentene	g	−5.000	18.787
C_5H_{10} cis-2-pentene	g	−6.710	17.173
C_5H_{10} trans-2-pentene	g	−7.590	16.575
C_5H_{10} 2-methyl-1-butene	g	−8.680	15.509
C_5H_{10} 3-methyl-1-butene	g	−6.920	17.874
C_5H_{10} 2-methyl-2-butene	g	−10.170	14.267
C_2H_2 acetylene	g	54.194	50.000
C_3H_4 methylacetylene	g	44.319	46.313
C_4H_6 1-butyne	g	39.70	48.52
C_4H_6 2-butyne	g	35.374	44.725
C_5H_8 1-pentyne	g	34.50	50.17
C_5H_8 2-pentyne	g	30.80	46.41
C_5H_8 3-methyl-1-butyne	g	32.60	49.12
C_6H_6 benzene	g	19.820	30.989
	l	11.718	29.756
C_7H_8 toluene	g	11.950	29.228
	l	2.867	27.282
C_8H_{10} ethylbenzene	g	7.120	31.208
	l	−2.977	28.614
C_8H_{10} o-xylene	g	4.540	29.177
	l	−5.841	26.370
C_8H_{10} m-xylene	g	4.120	28.405
	l	−6.075	25.730
C_8H_{10} p-xylene	g	4.290	28.952
	l	−5.838	26.310
C_9H_{12} n-propylbenzene	g	1.870	32.810
	l	−9.178	29.600
C_9H_{12} isopropylbenzene	g	0.940	32.738
	l	−9.848	29.708
C_9H_{12} 1-methyl-2-ethylbenzene	g	0.290	31.323
	l	−11.110	27.973
C_9H_{12} 1-methyl-3-ethylbenzene	g	−0.460	30.217
	l	−11.670	26.977
C_9H_{12} 1-methyl-4-ethylbenzene	g	−0.780	30.281
	l	−11.920	27.041
C_9H_{12} 1,2,3-trimethylbenzene	g	−2.290	29.319
	l	−14.013	25.679
C_9H_{12} 1,2,4-trimethylbenzene	g	−3.330	27.912
	l	−14.785	24.462
C_9H_{12} 1,3,5-trimethylbenzene	g	−3.840	28.172
	l	−15.184	24.832
C_5H_{10} cyclopentane	g	−18.46	9.23
	l	−25.31	8.70
C_6H_{12} methylcyclopentane	g	−25.50	8.55
	l	−33.08	7.53
C_7H_{14} ethylcyclopentane	g	−30.38	10.59
	l	−39.09	8.84

Appendix H (continued)

Compound	State†	Heat of formation‡§ ΔH (formation) at 25°C., kcal./mole	Free energy of formation‖¶ ΔF (formation) at 25°C.; kcal./mole
Carbon (*Cont.*):			
C_6H_{12} cyclohexane	g	−29.43	7.59
	l	−37.34	6.39
C_7H_{14} methylcyclohexane	g	−37.00	6.52
	l	−45.46	4.86
C_8H_{16} ethylcyclohexane	g	−41.06	9.38
	l	−50.73	6.96
CH_4O methanol	g	−48.08	−38.62
	l	−57.04	−39.80
C_2H_6O ethanol	g	−52.23	−40.23
	l	−66.35	−41.76
C_3H_8O *n*-propanol	g	−61.17	−38.83
	l	−71.87	−39.84
C_3H_8O isopropanol	g	−62.41	−38.20
	l	−74.32	−38.83
$C_4H_{10}O$ *n*-butanol	g	−67.81	−38.88
	l	−79.61	−40.37
$C_4H_{10}O$ isobutanol	g	−69.05	−38.25
	l	−81.06	−39.36
$C_2H_6O_2$ ethylene glycol	g	−92.53	−71.26
	l	−107.91	−76.44
$C_3H_8O_3$ glycerol	g		
	l	−159.16	−113.65
C_6H_6O phenol	g	−21.71	−6.26
	l	−37.80	−11.02
C_7H_8O cresol	g	....	−13.17
C_2H_4O ethylene oxide	g	−16.1	−6.94
C_2H_6O dimethyl ether	g	−43.06	−26.06
	l	−51.3	
$C_4H_{10}O$ diethyl ether	l	−65.2	−27.75
CH_2O formaldehyde	g	−28.29	−26.88
C_2H_4O acetaldehyde	g	−39.72	−31.46
C_3H_4O acrolein	g	−20.50	−15.57
	l	−27.97	−16.17
C_3H_6O propionaldehyde	g	−49.15	−33.96
C_4H_8O *n*-butyraldehyde	g	−52.40	−73.24
C_7H_6O bensaldehyde	g	−9.57	5.85
	l	−21.23	2.24
C_8H_8O *p*-toluic aldehyde	g	−17.78	4.09
	l	−29.79	0.97
C_2H_2O ketene	g	−14.78	−14.30
	l	−18.78	−13.32
C_3H_6O acetone	g	−51.79	−36.45
	l	−59.32	−37.16
$C_5H_{10}O$ diethylketone	l	−73.8	
CH_2O_2 formic acid	g	−86.67	−80.24
	l	−97.8	−82.7
½$(CH_2O_2)_2$ bimolecular formic acid	g	−93.85	−81.90
$C_2H_4O_2$ acetic acid	g	−104.72	−91.24
	l	−116.2	−93.56
$C_3H_6O_2$ propionic acid	g	−108.75	−88.27
	l	−121.7	−91.65
$C_2H_4O_3$ hydroxyacctic acid	l	−155.33	−125.57
$C_6H_{10}O_4$ adipic acid	g	−216.19	−163.96
	l	−235.51	−177.17
$C_2H_4O_2$ methyl formate	g	−84.69	−71.37
	l	−95.26	−71.53
$C_4H_6O_2$ methyl acrylate	g	−70.10	−56.78
	l	−82.76	−58.13
$C_4H_8O_2$ ethyl acetate	g	−102.02	−74.93
	l	−110.72	−76.11
$C_5H_{10}O_2$ ethyl propionate	g	−112.36	−77.37
	l	−122.16	−79.16
$C_4H_6O_3$ acetic anhydride	g	−148.82	−119.29
	l	−155.16	−121.75
$C_6H_{10}O_3$ propionic anhydride	g	−147.32	−109.78
	l	−161.53	−113.66
CS_2 carbon disulfide	g	28.11	16.13
COS carbonyl sulfide	g	−33.83	−40.85
C_2N_2 cyanogen	g	73.82	71.02
HCN hydrogen cyanide	g	31.1	27.94
	l	25.2	29.0
	aq, 100	25.2	26.8
C_2H_3N acetonitrile	g	19.81	
CH_5N methylamine	g	−6.7	6.6
C_2H_7N ethylamine	g	−12.24	10.01
C_3H_9N propylamine	g	−16.45	14.38
$C_4H_{11}N$ butylamine	g	−15.60	19.55
$C_6H_{13}N$ hexamethyleneimine	g	−14.37	31.52
	l	−24.90	28.84
CH_2N_2 cyanamide	l	11.18	24.30
	c	9.15	24.18
$C_6H_8N_2$ adiponitrile	g	33.34	61.43
	l	19.19	54.63
$C_6H_{16}N_2$ hexamethylenediamine	g	−30.57	28.91
CH_5N_3 guanidine	l	−27.48	7.34
	c	−30.68	6.33
Carbon (*Cont.*):			
$C_3H_6N_6$ melamine	l	−19.33	40.80
CH_3NO formamide	g	−44.64	−36.60
C_2H_7NO ethanolamine	l	−62.52	27.50
CH_4N_2O urea	l	−77.55	−46.45
	c	−79.634	−47.118
Cerium:			
Ce	c	0.00	0.00
CeN	c	−78.2	−70.8
Cesium:			
Cs	c	0.00	0.00
CsBr	c	−97.64	
	aq, 500	−91.39	−94.86
CsCl	c	−106.31	
	aq, 400	−102.01	−101.61
Cs_2CO_3	c	−271.88	
CsF	c	−131.67	
	aq, 400	−140.48	−135.98
CsH	c	−12	−7.30
$CsHCO_3$	c	−230.6	
	aq, 2000	−226.6	−210.56
CsI	c	−83.91	
	aq, 400	−75.74	−82.61
$CsNH_2$	c	−28.2	
$CsNO_3$	c	−121.14	
	aq, 400	−111.54	−96.53
Cs_2O	c	−82.1	
CsOH	c	−100.2	
	aq, 200	−117.0	−107.87
Cs_2S	c	−87	
Cs_2SO_4	c	−344.86	
	aq	−340.12	−316.66
Chlorine:			
Cl_2	g	0.00	0.00
ClF	g	−25.7	
ClO	g	33	
ClO_2	g	24.7	29.5
ClO_3	g	37	
Cl_2O	g	18.20	22.40
Cl_2O_7	g	63	
Chromium:			
Cr	c	0.00	0.00
$CrBr_3$	aq	....	−122.7
Cr_3C_2	c	−21.008	−21.20
Cr_4C	c	−16.378	−16.74
$CrCl_2$	c	−103.1	−93.8
	aq	....	−102.1
CrF_2	c	−152	
CrF_3	c	−231	
CrI_2	c	−63.7	
	aq	....	−64.1
CrO_3	c	−139.3	
Cr_2O_3	c	−268.8	−249.3
$Cr_2(SO_4)_3$	aq	....	−626.3
Cobalt:			
Co	c	0.00	0.00
$CoBr_2$	c	−55.0	
	aq	−73.61	−61.96
Co_3C	c	9.49	7.08
$CoCl_2$	c	−76.9	−66.6
	aq, 400	−95.58	−75.46
$CoCO_3$	c	−172.39	−155.36
CoF_2	aq	−172.98	−144.2
CoI_2	c	−24.2	
	aq	−43.15	−37.4
$Co(NO_3)_2$	c	−102.8	
	aq	−114.9	−65.3
CoO	c	−57.5	
Co_3O_4	c	−196.5	
$Co(OH)_2$	c	−131.5	−108.9
$Co(OH)_3$	c	−177.0	−142.0
CoS	c	−22.3	−19.8
Co_2S_3	c	−40.0	
$CoSO_4$	c	−216.6	
	aq, 400	....	−188.9
Columbium:			
Cb	c	0.00	0.00
Cb_2O_5	c	−462.96	
Copper:			
Cu	c	0.00	0.00
CuBr	c	−26.7	−23.8
$CuBr_2$	c	−34.0	
	aq	−42.4	−33.25
CuCl	c	−31.4	−24.13
$CuCl_2$	c	−48.83	
	aq, 400	−64.7	
$CuClO_4$	aq	−28.3	1.34
$Cu(ClO_3)_2$	aq, 400	....	15.4
$Cu(ClO_4)_2$	aq	....	−5.5

THERMO

Appendix H (continued)

Compound	State†	Heat of formation‡§ ΔH (formation) at 25°C., kcal./mole	Free energy of formation‖¶ ΔF (formation) at 25°C.; kcal./mole
Copper (*Cont.*):			
CuI	c	−17.8	−16.66
CuI_2	c	−4.8	
	aq	−11.9	−8.76
Cu_3N	c	17.78	
$Cu(NO_3)_2$	c	−73.1	
	aq, 200	−83.6	−36.6
CuO	c	−38.5	−31.9
Cu_2O	c	−43.00	−38.13
$Cu(OH)_2$	c	−108.9	−85.5
CuS	c	−11.6	−11.69
Cu_2S	c	−18.97	−20.56
$CuSO_4$	c	−184.7	−158.3
	aq, 800	−200.78	−160.19
Cu_2SO_4	c	−179.6	
	aq	....	−152.0
Erbium:			
Er	c	0.00	0.00
$Er(OH)_3$	c	−326.8	
Fluorine:			
F_2	g	0.00	0.00
F_2O	g	5.5	9.7
Gallium:			
Ga	c	0.00	0.00
$GaBr_3$	c	−92.4	
$GaCl_3$	c	−125.4	
GaN	c	−26.2	
Ga_2O	c	−84.3	
Ga_2O_3	c	−259.9	
Germanium:			
Ge	c	0.00	0.00
Ge_3N_4	c	−15.7	
GeO_2	c	−128.6	
Gold:			
Au	c	0.00	0.00
$AuBr$	c	−3.4	
$AuBr_3$	c	−14.5	
	aq	−11.0	24.47
$AuCl$	c	−8.3	
$AuCl_3$	c	−28.3	
	aq	−32.96	4.21
AuI	c	0.2	−0.76
Au_2O_3	c	11.0	18.71
$Au(OH)_3$	c	−100.6	
Hafnium:			
Hf	c	0.00	0.00
HfO_2	c	−271.1	−258.2
Hydrogen:			
H_3AsO_3	aq	−175.6	−153.04
H_3AsO_4	c	−214.9	
	aq	−214.8	−183.93
HBr	g	−8.66	−12.72
	aq, 400	−28.80	−24.58
$HBrO$	aq	−25.4	−19.90
$HBrO_3$	aq	−11.51	5.00
HCl	g	−22.063	−22.778
	aq, 400	−39.85	−31.330
HCN	g	31.1	27.94
	aq, 100	24.2	26.55
$HClO$	aq, 400	−28.18	−19.11
$HClO_3$	aq	−23.4	−0.25
$HClO_4$	aq, 660	−31.4	−10.70
$HC_2H_3O_2$	l	−116.2	−93.56
	aq, 400	−116.74	−96.8
$H_2C_2O_4$	c	−196.7	
	aq, 300	−194.6	−165.64
$HCOOH$	l	−97.8	−82.7
	aq, 200	−98.0	−85.1
H_2CO_3	aq	−167.19	−149.0
HF	g	−64.2	−64.7
	aq, 200	−75.75	
HI	g	6.27	0.365
	aq, 400	−13.47	−12.35
HIO	aq	−38	−23.33
HIO_3	c	−56.77	
	aq	−54.8	−32.25
HN_3	g	70.3	78.50
HNO_3	g	−31.99	−17.57
	l	−41.35	−19.05
	aq, 400	−49.210	
$HNO_3.H_2O$	l	−112.91	−78.36
$HNO_3.3H_2O$	l	−252.15	−193.70
H_2O	g	−57.7979	−54.6351
	l	−68.3174	−56.6899
H_2O_2	l	−45.16	−28.23
	aq, 200	−45.80	−31.47
H_3PO_2	c	−145.5	
	aq	−145.6	−120.0
H_3PO_3	c	−232.2	
	aq	−232.2	−204.0
Hydrogen (*Cont.*):			
H_3PO_4	c	−306.2	
	aq, 400	−309.32	−270.0
H_2S	g	−4.77	−7.85
	aq, 2000	−9.38	
H_2S_2	l	−3.6	
H_2SO_3	aq, 200	−146.88	−128.54
H_2SO_4	l	−193.69	
	aq, 400	−212.03	
H_2Se	g	20.5	17.0
	aq	18.1	18.4
H_2SeO_3	c	−126.5	
	aq	−122.4	−101.36
H_2SeO_4	c	−130.23	
	aq, 400	−143.4	
H_2SiO_3	c	−267.8	−247.9
H_4SiO_4	c	−340.6	
H_2Te	g	36.9	33.1
H_2TeO_3	c	−145.0	−115.7
	aq	−145.0	
H_2TeO_4	aq	−165.6	
Indium:			
In	c	0.00	0.00
$InBr_3$	c	−97.2	
	aq	−112.9	−97.2
$InCl_3$	c	−128.5	
	aq	−145.6	−117.5
InI_3	c	−56.5	
	aq	−67.2	−60.5
InN	c	−4.8	
In_2O_3	c	−222.47	
Iodine:			
I_2	c	0.00	0.00
	g	14.88	4.63
IBr	g	10.05	1.24
ICl	g	4.20	−1.32
ICl_3	c	−21.8	−6.05
I_2O_5	c	−42.5	
Iridium:			
Ir	c	0.00	0.00
$IrCl$	c	−20.5	−16.9
$IrCl_2$	c	−40.6	−32.0
$IrCl_3$	c	−60.5	−46.5
IrF_6	l	−130	
IrO_2	c	−40.14	
Iron:			
Fe	c, α	0.00	0.00
$FeBr_2$	c	−57.15	
	aq, 540	−78.7	−69.47
$FeBr_3$	aq	−95.5	−76.26
Fe_3C	c	5.69	4.24
$Fe(CO)_5$	l	−187.6	
$FeCO_3$	c, siderite	−172.4	−154.8
$FeCl_2$	c	−81.9	−72.6
	aq	−100.0	−83.0
$FeCl_3$	c	−96.4	
	aq, 2000	−128.5	−96.5
FeF_2	aq, 1200	−177.2	−151.7
FeI_2	c	−24.2	
	aq	−47.7	−45
FeI_3	aq	−49.7	−39.5
Fe_4N	c	−2.55	0.862
$Fe(NO_3)_2$	aq	−118.9	−72.8
$Fe(NO_3)_3$	aq, 800	−156.5	−81.3
FeO	c	−64.62	−59.38
Fe_2O_3	c	−198.5	−179.1
Fe_3O_4	c	−266.9	−242.3
$Fe(OH)_2$	c	−135.9	−115.7
$Fe(OH)_3$	c	−197.3	−166.3
$FeO.SiO_2$	c	−273.5	
Fe_2P	c	−13	
$FeSi$	c	−19.0	
FeS	c	−22.64	−23.23
FeS_2	c, pyrites	−38.62	−35.93
	c, marcasite	−33.0	
$FeSO_4$	c	−221.3	−195.5
	aq, 400	−236.2	−196.4
$Fe_2(SO_4)_3$	aq, 400	−653.3	−533.4
$FeTiO_3$	c, ilmenite	−295.51	−277.06
Lanthanum:			
La	c	0.00	0.00
$LaCl_3$	c	−253.1	
	aq	−284.7	
La_2H_3	c	−160	
LaN	c	−72.0	−64.6
La_2O_3	c	−539	
LaS_2	c	−148.3	
La_2S_3	c	−351.4	
$La_2(SO_4)_3$	aq	−972	

THERMO

Appendix H (continued)

Compound	State†	Heat of formation‡§ ΔH (formation) at 25°C., kcal./mole	Free energy of formation‖¶ ΔF (formation) at 25°C., kcal./mole
Lead:			
Pb	c	0.00	0.00
$PbBr_2$	c	−66.24	−62.06
	aq	−56.4	−54.97
$PbCO_3$	c, cerussite	−167.6	−150.0
$Pb(C_2H_3O_2)_2$	c	−232.6	
	aq, 400	−234.2	−184.40
PbC_2O_4	c	−205.3	
$PbCl_2$	c	−85.68	−75.04
	aq	−82.5	−68.47
PbF_2	c	−159.5	−148.1
PbI_2	c	−41.77	−41.47
$Pb(NO_3)_2$	c	−106.88	
	aq, 400	−99.46	−58.3
PbO	c, red	−51.72	−45.53
	c, yellow	−50.86	−43.88
PbO_2	c	−65.0	−52.0
Pb_3O_4	c	−172.4	−142.2
$Pb(OH)_2$	c	−123.0	−102.2
PbS	c	−22.38	−21.98
$PbSO_4$	c	−218.5	−192.9
Lithium:			
Li	c	0.00	0.00
LiBr	c	−83.75	
	aq, 400	−95.40	−95.28
$LiBrO_3$	aq	−77.9	−65.70
Li_2C_2	c	−13.0	
LiCN	aq	−31.4	−31.35
LiCNO	aq	−101.2	−94.12
$LiC_2H_3O_2$	aq	−183.9	−160.00
Li_2CO_3	c	−289.7	−269.8
	aq, 1900	−293.1	−267.58
LiCl	c	−97.63	
	aq, 278	−106.45	−102.03
$LiClO_3$	aq	−87.5	−70.95
$LiClO_4$	aq	−106.3	−81.4
LiF	c	−145.57	
	aq, 400	−144.85	−136.40
LiH	c	−22.9	
$LiHCO_3$	aq, 2000	−231.1	−210.98
LiI	c	−65.07	
	aq, 400	−80.09	−83.03
$LiIO_3$	aq	−121.3	−102.95
Li_3N	c	−47.45	−37.33
$LiNO_3$	c	−115.350	
	aq, 400	−115.88	−96.95
Li_2O	c	−142.3	
Li_2O_2	c	−151.9	−138.0
	aq	−159	
LiOH	c	−116.58	−106.44
	aq, 400	−121.47	−108.29
$LiOH.H_2O$	c	−188.92	
$Li_2O.SiO_2$	gls	−374	
Li_2Se	c	−84.9	
	aq	−95.5	−105.64
Li_2SO_4	c	−340.23	−314.66
	aq, 400	−347.02	
$Li_2SO_4.H_2O$	c	−411.57	−375.07
Magnesium:			
Mg	c	0.00	0.00
$Mg(AsO_4)_2$	c	−731.3	
	aq	−749	−630.14
$MgBr_2$	c	−123.9	
	aq, 400	−167.33	−156.94
$Mg(CN)_2$	aq	−39.7	−29.08
$MgCN_2$	c	−61	
$Mg(C_2H_3O_2)_2$	aq	−344.6	−286.38
$MgCO_3$	c	−261.7	−241.7
$MgCl_2$	c	−153.220	−143.77
	aq, 400	−189.76	
$MgCl_2.H_2O$	c	−230.970	−205.93
$MgCl_2.2H_2O$	c	−305.810	−267.20
$MgCl_2.4H_2O$	c	−453.820	−387.98
$MgCl_2.6H_2O$	c	−597.240	−505.45
MgF_2	c	−263.8	
MgI_2	c	−86.8	
	aq, 400	−136.79	−132.45
$MgMoO_4$	c	−329.9	
Mg_3N_2	c	−115.2	−100.8
$Mg(NO_3)_2$	c	−188.770	−140.66
	aq, 400	−209.927	−160.28
$Mg(NO_3)_2.2H_2C$	c	−336.625	
$Mg(NO_3)_2.6H_2O$	c	−624.48	−496.03
MgO	c	−143.84	−136.17
$MgO.SiO_2$	c	−347.5	−326.7
$Mg(OH)_2$	c, ppt.	−221.90	−200.17
	c, brucite	−223.9	−193.3
MgS	c	−84.2	
	aq	−108	
Magnesium (*Cont.*):			
$MgSO_4$	c	−304.94	−277.7
	aq, 400	−325.4	−283.88
MgTe	c	−25	
$MgWO_4$	c	−345.2	
Manganese:			
Mn	c, α	0.00	0.00
$MnBr_2$	c	−91	
	aq	−106	−97.8
Mn_3C	c	1.1	1.26
$Mn(C_2H_3O_2)_2$	c	−270.3	
	aq	−282.7	−227.2
$MnCO_3$	c	−211	−192.5
MnC_2O_4	c	−240.9	
$MnCl_2$	c	−112.0	−102.2
	aq, 400	−128.9	
MnF_2	aq, 1200	−206.1	−180.0
MnI_2	c	−49.8	
	aq	−76.2	−73.3
Mn_5N_2	c	−57.77	−46.49
$Mn(NO_3)_2$	c	−134.9	
	aq, 400	−148.0	−101.1
$Mn(NO_3)_2.6H_2O$	c	−557.07	−441.2
MnO	c	−92.04	−86.77
MnO_2	c	−124.58	−111.49
Mn_2O_3	c	−229.5	−209.9
Mn_3O_4	c	−331.65	−306.22
$MnO.SiO_2$	c	−301.3	−282.1
$Mn(OH)_2$	c	−163.4	−143.1
$Mn(OH)_3$	c	−221	−190
$Mn_3(PO_4)_2$	c	−736	
MnSe	c	−26.3	−27.5
MnS	c, green	−47.0	−48.0
$MnSO_4$	c	−254.18	−228.41
	aq, 400	−265.2	
$Mn_2(SO_4)_3$	c	−635	
	aq	−657	
Mercury:			
Hg	l	0.00	0.00
HgBr	g	23	18
$HgBr_2$	c	−40.68	−38.8
	aq	−38.4	−9.74
$Hg(C_2H_3O_2)_2$	c	−196.3	
	aq	−192.5	−139.2
$HgCl_2$	c	−53.4	−42.2
	aq	−50.3	−23.25
HgCl	g	19	14
Hg_2Cl_2	c	−63.13	
$Hg(CN)_2$	c	62.8	
	aq, 1110	66.25	
HgC_2O_4	c	−159.3	
HgH	g	57.1	52.25
HgI_2	c, red	−25.3	−24.0
HgI	g	33	23
Hg_2I_2	c	−28.88	−26.53
$Hg(NO_3)_2$	aq	−56.8	−13.09
$Hg_2(NO_3)_2$	aq	−58.5	−15.65
HgO	c, red	−21.6	−13.94
	c, yellow ppt.	−20.8	
Hg_2O	c	−21.6	−12.80
HgS	c, black	−10.7	−8.80
$HgSO_4$	c	−166.6	
Hg_2SO_4	c	−177.34	−149.12
Molybdenum:			
Mo	c	0.00	0.00
Mo_2C	c	4.36	2.91
Mo_2N	c	−8.3	
MoO_2	c	−130	−118.0
MoO_3	c	−180.39	−162.01
MoS_2	c	−56.27	−54.19
MoS_3	c	−61.48	−57.38
Nickel:			
Ni	c	0.00	0.00
$NiBr_2$	c	−53.4	
	aq	−72.6	−60.7
Ni_3C	c	9.2	8.88
$Ni(C_2H_3O_2)_2$	aq	−249.6	−190.1
$Ni(CN)_2$	aq	230.9	66.3
$NiCl_2$	c	−75.0	
	aq, 400	−94.34	−74.19
NiF_2	c	−157.5	
	aq	−171.6	−142.9
NiI_2	c	−22.4	
	aq	−42.0	−36.2
$Ni(NO_3)_2$	c	−101.5	
	aq, 200	−113.5	−64.0
NiO	c	−58.4	−51.7
$Ni(OH)_2$	c	−129.8	−105.6
$Ni(OH)_3$	c	−163.2	

THERMO

Appendix H (continued)

Compound	State†	Heat of formation‡§ ΔH (formation) at 25°C., kcal./mole	Free energy of formation‖¶ ΔF (formation) at 25°C., kcal./mole
Nickel (*Cont.*):			
NiS	c	−20.4	
$NiSO_4$	c	−216	
	aq, 200	−231.3	−187.6
Nitrogen:			
N_2	g	0.00	0.00
NF_3	g	−27	
NH_3	g	−10.96	−3.903
	aq, 200	−19.27	
NH_4Br	c	−64.57	
	aq	−60.27	−43.54
$NH_4C_2H_3O_2$	c	−148.1	
	aq, 400	−148.58	−108.26
NH_4CN	c	−0.7	
	aq	3.6	20.4
NH_4CNS	c	−17.8	
	aq	−12.3	4.4
$(NH_4)_2CO_3$	aq	−223.4	−164.1
$(NH_4)_2C_2O_4$	c	−266.3	
	aq	−260.6	−196.2
NH_4Cl	c	−75.23	−48.59
	aq, 400	−71.20	
NH_4ClO_4	c	−69.4	
	aq	−63.2	−21.1
$(NH_4)_2CrO_4$	c	−276.9	
	aq	−271.3	−209.3
NH_4F	c	−111.6	
	aq	−110.2	−84.7
NH_4I	c	−48.43	
	aq	−44.97	−31.3
NH_4NO_3	c	−87.40	
	aq, 500	−80.89	
NH_4OH	aq	−87.59	
$(NH_4)_2S$	aq, 400	−55.21	−14.50
$(NH_4)_2SO_4$	c	−281.74	−215.06
	aq, 400	−279.33	−214.02
N_2H_4	l	12.06	
$N_2H_4.H_2O$	l	−57.96	
$N_2H_4.H_2SO_4$	c	−232.2	
N_2O	g	19.55	24.82
NO	g	21.600	20.719
NO_2	g	7.96	12.26
N_2O_4	g	2.23	23.41
N_2O_5	c	−10.0	
$NOBr$	l	11.6	19.26
$NOCl$	g	12.8	16.1
Osmium:			
Os	c	0.00	0.00
OsO_4	c	−93.6	−70.9
	g	−80.1	−68.1
Oxygen:			
O_2	g	0.00	0.00
O_3	g	33.88	38.86
Palladium:			
Pd	c	0.00	0.00
PdO	c	−20.40	
Phosphorus:			
P	c, white ("yellow")	0.00	0.00
	c, red ("violet")	−4.22	−1.80
P	g	150.35	141.88
P_2	g	33.82	24.60
P_4	g	13.2	5.89
PBr_3	l	−45	
PBr_5	c	−60.6	
PCl_3	g	−70.0	−65.2
	l	−76.8	−63.3
PCl_5	g	−91.0	−73.2
PH_3	g	2.21	−1.45
PI_3	c	−10.9	
P_2O_5	c	−360.0	
$POCl_3$	g	−138.4	−127.2
Platinum:			
Pt	c	0.00	0.00
$PtBr_4$	c	−40.6	
	aq	−50.7	
$PtCl_2$	c	−34	
$PtCl_4$	c	−62.6	
	aq	−82.3	
PtI_4	c	−18	
$Pt(OH)_2$	c	−87.5	−67.9
PtS	c	−20.18	−18.55
PtS_2	c	−26.64	−24.28
Potassium:			
K	c	0.00	0.00
K_3AsO_3	aq	−323.0	
K_3AsO_4	aq	−390.3	−355.7
KH_2AsO_4	c	−271.2	−236.7
KBr	c	−94.06	−90.8
	aq, 400	−89.19	−92.0
Potassium (*Cont.*):			
$KBrO_3$	c	−81.58	−60.30
	aq, 1667	−71.68	
$KC_2H_3O_2$	c	−173.80	
	aq, 400	−177.38	−156.73
KCl	c	−104.348	−97.76
	aq, 400	−100.164	−98.76
$KClO_3$	c	−93.5	−69.30
	aq, 400	−81.34	
$KClO_4$	c	−103.8	−72.86
	aq, 400	−101.14	
KCN	c	−28.1	
	aq, 400	−25.3	−28.08
$KCNO$	c	−99.6	
	aq	−94.5	−90.85
$KCNS$	c	−47.0	
	aq, 400	−41.07	−44.08
K_2CO_3	c	−274.01	
	aq, 400	−280.90	−264.04
$K_2C_2O_4$	c	−319.9	
	aq, 400	−315.5	−293.1
K_2CrO_4	c	−333.4	
	aq, 400	−328.2	−306.3
$K_2Cr_2O_7$	c	−488.5	
	aq, 400	−472.1	−440.9
KF	c	−134.50	
	aq, 180	−138.36	−133.13
$K_3Fe(CN)_6$	c	−48.4	
	aq	−34.5	
$K_4Fe(CN)_6$	c	−131.8	
	aq	−119.9	
KH	c	−10	−5.3
$KHCO_3$	c	−229.8	
	aq, 2000	−224.85	−207.71
KI	c	−78.88	−77.37
	aq, 500	−73.95	−79.76
KIO_3	c	−121.69	−101.87
	aq, 400	−115.18	−99.68
KIO_4	aq	−98.1	
$KMnO_4$	c	−192.9	−169.1
	aq, 400	−182.5	−168.0
K_2MoO_4	aq, 880	−364.2	−342.9
KNH_2	c	−28.25	
KNO_2	aq	−86.0	−75.9
KNO_3	c	−118.08	−94.29
	aq, 400	−109.79	−93.68
K_2O	c	−86.2	
$K_2O.Al_2O_3.S_iO_2$)	c, leucite	−1379.6	
	gls	−1368.2	
$K_2O.Al_2O_3.S_iO_2$	c, adularia	−1784.5	
	c, microcline	−1784.5	
	gls	−1747	
KOH	c	−102.02	
	aq, 400	−114.96	−105.0
K_3PO_3	aq	−397.5	
K_3PO_4	aq	−478.7	−443.3
KH_2PO_4	c	−362.7	−326.1
K_2PtCl_4	c	−254.7	
	aq	−242.6	−226.5
K_2PtCl_6	c	−299.5	−263.6
	aq, 9400	−286.1	
K_2Se	c	−74.4	
	aq	−83.4	−99.10
K_2SeO_4	aq	−267.1	−240.0
K_2S	c	−121.5	
	aq, 400	−110.75	−111.44
K_2SO_3	c	−267.7	
	aq	−269.7	−251.3
K_2SO_4	c	−342.65	−314.62
	aq, 400	−336.48	−310.96
$K_2SO_4.Al_2(SO_4)_3$	c	−1178.38	−1068.48
$K_2SO_4.Al_2(SO_4)_3.24H_2O$	c	−2895.44	−2455.68
$K_2S_2O_6$	c	−418.62	
Rhenium:			
Re	c	0.00	0.00
ReF_6	g	−274	
Rhodium:			
Rh	c	0.00	0.00
RhO	c	−21.7	
Rh_2O	c	−22.7	
Rh_2O_3	c	−68.3	
Rubidium:			
Rb	c	0.00	0.00
$RbBr$	c	−95.82	
	g	−45.0	−52.50
	aq, 500	−90.54	−93.38
$RbCN$	aq	−25.9	
Rb_2CO_3	c	−273.22	
	aq, 220	−282.61	−263.78

Appendix H (continued)

Compound	State†	Heat of formation‡§ ΔH (formation) at 25°C., kcal./mole	Free energy of formation‖¶ ΔF (formation) at 25°C., kcal./mole
Rubidium (*Cont.*):			
$RbCl$	c	−105.06	−98.48
	g	−53.6	−57.9
	aq, ∞	−101.06	−100.13
RbF	c	−133.23	
	aq, 400	−139.31	−134.5
$RbHCO_3$	c	−230.01	
	aq, 2000	−225.59	−209.07
RbI	c	−81.04	
	g	−31.2	−40.5
	aq, 400	−74.57	−81.13
$RbNH_2$	c	−27.74	
$RbNO_3$	c	−119.22	
	aq, 400	−110.52	−95.05
Rb_2O	c	−82.9	
Rb_2O_2	c	−107	
$RbOH$	c	−101.3	
	aq, 200	−115.8	−106.39
Ruthenium:			
Ru	c	0.00	0.00
RuS_2	c	−46.99	−44.11
Selenium:			
Se	c, I, hexagonal	0.00	0.00
	c, II, red, monoclinic	0.2	
Se_2Cl_2	l	−22.06	−13.73
SeF_6	g	−246	−222
SeO_2	c	−56.33	
Silicon:			
Si	c	0.00	0.00
$SiBr_4$	l	−93.0	
SiC	c	−28	−27.4
$SiCl_4$	l	−150.0	−133.9
	g	−142.5	−133.0
SiF_4	g	−370	−360
SiH_4	g	−14.8	−9.4
SiI_4	c	−29.8	
Si_3N_4	c	−179.25	−154.74
SiO_2	c, cristobalite, 1600° form	−202.62	
	c, cristobalite, 1100° form	−202.46	
	c, quarts	−203.35	−190.4
	c, tridymite	−203.23	
Silver:			
Ag	c	0.00	0.00
$AgBr$	c	−23.90	−23.02
Ag_2C_2	c	84.5	
$AgC_2H_3O_2$	c	−95.9	
	aq	−91.7	−70.86
$AgCN$	c	33.8	38.70
Ag_2CO_3	c	−119.5	−103.0
$Ag_2C_2O_4$	c	−158.7	
$AgCl$	c	−30.11	−25.98
AgF	c	−48.7	
	aq, 400	−53.1	−47.26
AgI	c	−15.14	−16.17
$AgIO_3$	c	−42.02	−24.08
$AgNO_2$	c	−11.6	3.76
	aq	−2.9	9.99
$AgNO_3$	c	−29.4	−7.66
	aq, 6500	−24.02	−7.81
Ag_2O	c	−6.95	−2.23
Ag_2S	c	−5.5	−7.6
Ag_2SO_4	c	−170.1	−146.8
	aq	−165.8	−139.22
Sodium:			
Na	c	0.00	0.00
Na_3AsO_3	aq, 500	−314.61	
Na_3AsO_4	c	−366	
	aq, 500	−381.97	−341.17
$NaBr$	c	−86.72	
	aq, 400	−86.33	−87.17
$NaBrO$	aq	−78.9	
$NaBrO_3$	aq, 400	−68.89	−57.59
$NaC_2H_3O_2$	c	−170.45	
	aq, 400	−175.450	−152.31
$NaCN$	c	−22.47	
	aq, 200	−22.29	−23.24
$NaCNO$	c	−96.3	
	aq	−91.7	−86.00
$NaCNS$	c	−39.94	
	aq, 400	−38.23	−39.24
Na_2CO_3	c	−269.46	−249.55
	aq, 1000	−275.13	−251.36
$NaCO_2NH_2$	c	−142.17	
$Na_2C_2O_4$	c	−313.8	
	aq, 600	−309.92	−283.42
$NaCl$	c	−98.321	−91.894
	aq, 400	−97.324	−93.92
Sodium (*Cont.*):			
$NaClO_3$	c	−83.59	
	aq, 400	−78.42	−62.84
$NaClO_4$	c	−101.12	
	aq, 476	−97.66	−73.29
Na_2CrO_4	c	−319.8	
	aq, 800	−323.0	−296.58
$Na_2Cr_2O_7$	aq, 1200	−465.9	−431.18
NaF	c	−135.94	−129.0
	aq, 400	−135.711	−128.29
NaH	c	−14	−9.30
$NaHCO_3$	c	−226.0	−202.66
	aq	−222.1	−202.87
NaI	c	−69.28	
	aq, ∞	−71.10	−74.92
$NaIO_3$	aq, 400	−112.300	−94.84
Na_2MoO_4	c	−364	
	aq	−358.7	−333.18
$NaNO_2$	c	−86.6	
	aq	−83.1	−71.04
$NaNO_3$	c	−111.71	−87.62
	aq, 400	−106.880	−88.84
Na_2O	c	−99.45	−90.06
Na_2O_2	c	−119.2	−105.0
$Na_2O.SiO_2$	c	−383.91	−361.49
$Na_2O.Al_2O_3.3SiO_2$	c, natrolite	−1180	
$Na_2O.Al_2O_3.4SiO_2$	c	−1366	
$NaOH$	c	−101.96	−90.60
	aq, 400	−112.193	−100.18
Na_3PO_3	aq, 1000	−389.1	
Na_3PO_4	c	−457	
	aq, 400	−471.9	−428.74
Na_2PtCl_4	aq	−237.2	−216.78
Na_2PtCl_6	c	−272.1	
	aq	−280.9	
Na_2Se	c	−59.1	
	aq, 440	−78.1	−89.42
Na_2SeO_4	c	−254	
	aq, 800	−261.5	−230.30
Na_2S	c	−89.8	
	aq, 400	−105.17	−101.76
Na_2SO_3	c	−261.2	−240.14
	aq, 800	−264.1	−241.58
Na_2SO_4	c	−330.50	−302.38
	aq, 1100	−330.82	−301.28
$Na_2SO_4.10H_2O$	c	−1033.85	−870.52
Na_2WO_4	c	−391	
	aq	−381.5	−345.18
Strontium:			
Sr	c	0.00	0.00
$SrBr_2$	c	−171.0	
	aq, 400	−187.24	−182.36
$Sr(C_2H_3O_2)_2$	c	−358.0	
	aq	−364.4	−311.80
$Sr(CN)_2$	aq	−59.5	−54.50
$SrCO_3$	c	−290.9	−271.9
$SrCl_2$	c	−197.84	
	aq, 400	−209.20	−195.86
SrF_2	c	−289.0	
$Sr(HCO_3)_2$	aq	−459.1	−413.76
SrI_2	c	−136.1	
	aq, 400	−156.70	−157.87
Sr_3N_2	c	−91.4	−76.5
$Sr(NO_3)_2$	c	−233.2	
	aq, 400	−228.73	−185.70
SrO	c	−140.8	−133.7
$SrO.SiO_2$	gls	−364	
SrO_2	c	−153.3	−139.0
Sr_2O	c	−153.6	
$Sr(OH)_2$	c	−228.7	
	aq, 800	−239.4	−208.27
$Sr_3(PO_4)_2$	c	−980	
	aq	−985	−881.54
SrS	c	−113.1	
	aq	−120.4	−109.78
$SrSO_4$	c	−345.3	
	aq, 400	−345.0	−309.30
$SrWO_4$	c	−393	
Sulfur:			
S	c, rhombic	0.00	0.00
	c, monoclinic	−0.071	−0.023
	l, λ	0.257	0.072
	l, λμ equilibrium	....	0.071
	g	53.25	43.57
S_2	g	31.02	19.36
S_6	g	27.78	13.97
S_8	g	27.090	12.770
S_2Br_2	l	−4	
SCl_4	l	−13.7	

Appendix H (continued)

Compound	State†	Heat of formation‡§ ΔH (formation) at 25°C., kcal./mole	Free energy of formation‖¶ ΔF (formation) at 25°C., kcal./mole
Sulfur (*Cont.*):			
S_2Cl_2	l	−14.2	−5.90
S_2Cl_4	l	−24.1	
SF_6	g	−262	−237
SO	g	19.02	12.75
SO_2	g	−70.94	−71.68
SO_3	g	−94.39	−88.59
	l	−103.03	−88.28
	c, α	−105.09	−88.22
	c, β	−105.92	−88.34
	c, γ	−109.34	−88.98
SO_2Cl_2	g	−82.04	−74.06
	l	−89.80	−75.06
Tantalum:			
Ta	c	0.00	0.00
TaN	c	−51.2	−45.11
Ta_2O_5	c	−486.0	−453.7
Tellurium:			
Te	c	0.00	0.00
$TeBr_4$	c	−49.3	
$TeCl_4$	c	−77.4	−57.4
TeF_6	g	−315	−292
TeO_2	c	−77.56	−64.66
Thallium:			
Tl	c	0.00	0.00
TlBr	c	−41.5	−39.43
	aq	−28.0	−32.34
TlCl	c	−49.37	−44.46
	aq	−38.4	−39.09
$TlCl_3$	c	−82.4	
	aq	−91.0	−44.25
TlF	aq	−77.6	−73.46
TlI	c	−31.1	−31.3
	aq	−12.7	−20.09
$TlNO_3$	c	−58.2	−36.32
	aq	−48.4	−34.01
Tl_2O	c	−43.18	
Tl_2O_3	c	−120	
TlOH	c	−57.44	−45.54
	aq	−53.9	−45.35
Tl_2S	c	−22	
Tl_2SO_4	c	−222.8	−197.79
	aq, 800	−214.1	−191.62
Thorium:			
Th	c	0.00	0.00
$ThBr_4$	c	−281.5	
	aq	−352.0	−295.31
ThC_2	c	−45.1	
$ThCl_4$	c	−335	
	aq	−392	−322.32
ThI_4	aq	−292.0	−246.33
Th_3N_4	c	−309.0	−282.3
ThO_2	c	−291.6	−280.1
$Th(OH)_4$	c, "soluble"	−336.1	
$Th(SO_4)_2$	c	−632	
	aq	−668.1	−549.2
Tin:			
Sn	c, II, tetragonal	0.00	0.00
	c, III, "gray," cubic	0.6	1.1
$SnBr_2$	c	−61.4	
	aq	−60.0	−55.43
$SnBr_4$	c	−94.8	
	aq	−110.6	−97.66
$SnCl_2$	c	−83.6	
	aq	−81.7	−68.94
$SnCl_4$	l	−127.3	−110.4
	aq	−157.6	−124.67
SnI_2	c	−38.9	
	aq	−33.3	−30.95
Tin (*Cont.*):			
SnO	c	−67.7	−60.75
SnO_2	c	−138.1	−123.6
$Sn(OH)_2$	c	−136.2	−115.95
$Sn(OH)_4$	c	−268.9	−226.00
SnS	c	−18.61	
Titanium:			
Ti	c	0.00	0.00
TiC	c	−110	−109.2
$TiCl_4$	l	−181.4	−165.5
TiN	c	−80.0	−73.17
TiO_2	c, III, rutil	−225.0	−211.9
	amorphous	−214.1	−201.4
Tungsten:			
W	c	0.00	0.00
WO_2	c	−130.5	−118.3
WO_3	c	−195.7	−177.3
WS_2	c	−84	
Uranium:			
U	c	0.00	0.00
UC_2	c	−29	
UCl_3	c	−213	
UCl_4	c	−251	
U_3N_4	c	−274	−249.6
UO_2	c	−256.6	−242.2
$UO_2(NO_3)_2.6H_2O$	c	−756.8	−617.8
UO_3	c	−291.6	
U_3O_8	c	−845.1	
Vanadium:			
V	c	0.00	0.00
VCl_2	c	−147	
VCl_3	l	−187	
VCl_4	l	−165	
VN	c	−41.43	−35.08
V_2O_2	c	−195	
V_2O_3	c	−296	−277
V_2O_4	c	−342	−316
V_2O_5	c	−373	−342
Zinc:			
Zn	c	0.00	0.00
ZnSb	c	−3.6	−3.88
$ZnBr_2$	c	−77.0	−72.9
	aq, 400	−93.6	
$Zn(C_2H_3O_2)_2$	c	−259.4	
	aq, 400	−269.4	−214.4
$Zn(CN)_2$	c	17.06	
$ZnCO_3$	c	−192.9	−173.5
$ZnCl_2$	c	−99.9	−88.8
	aq, 400	−115.44	
ZnF_2	aq	−192.9	−166.6
ZnI_2	c	−50.50	−49.93
	aq	−61.6	
$Zn(NO_3)_2$	aq, 400	−134.9	−87.7
ZnO	c, hexagonal	−83.36	−76.19
$ZnO.SiO_2$	c	−282.6	
$Zn(OH)_2$	c, rhombic	−153.66	
ZnS	c, wurtzite	−45.3	−44.2
$ZnSO_4$	c	−233.4	
	aq, 400	−252.12	−211.28
Zirconium:			
Zr	c	0.00	0.00
ZrC	c	−29.8	−34.6
$ZrCl_4$	c	−268.9	
ZrN	c	−82.5	−75.9
ZrO_2	c, monoclinic	−258.5	−244.6
$Zr(OH)_4$	c	−411.0	
$ZrO(OH)_2$	c	−337	−307.6

† The physical state is indicated as follows: *c*, crystal (solid); *l*, liquid; *g*, gas; *gls*, glass or solid supercooled liquid; *aq*, in aqueous solution. A number following the symbol *aq* applies only to the values of the heats of formation (not to those of free energies of formation); and indicates the number of moles of water per mole of solute; when no number is given, the solution is understood to be dilute. For the free energy of formation of a substance in aqueous solution, the concentration is always that of the hypothetical solution of unit molality.

‡ The increment in heat content, ΔH, in the reaction of forming the given substance from its elements in their standard states. When ΔH is negative, heat is evolved in the process, and, when positive, heat is absorbed.

§ The heat of solution in water of a given solid, liquid, or gaseous compound is given by the difference in the value for the heat of formation of the given compound in the solid, liquid, or gaseous state and its heat of formation in aqueous solution. The following two examples serve as an illustration of the procedure: (1) For NaCl(*c*) and NaCl(*aq*, 400H_2O), the values of ΔH(formation) are, respectively, −98.321 and −97.324 kg.-cal. per mole. Subtraction of the first value from the second gives $\Delta H = 0.998$ kg.-cal. per mole for the reaction of dissolving crystalline sodium chloride in 400 moles of water. When this process occurs at a constant pressure of 1 atm., 0.998 kg.-cal. of energy are absorbed. (2) For HCl(*g*) and HCl(*aq*, 400H_2O), the values for ΔH(formation) are, respectively, −22.06 and −39.85 kg.-cal. per mole. Subtraction of the first from the second gives $\Delta H = -17.79$ kg.-cal per mole for the reaction of dissolving gaseous hydrogen chloride in 400 moles of water. At a constant pressure of 1 atm. 17.79 kg.-cal. of energy are evolved in this process.

‖ The increment in the free energy, ΔF, in the reaction of forming the given substance in its standard state from its elements in their standard states. The standard states are: for a gas, fugacity (approximately equal to the pressure) of 1 atm.; for a pure liquid or solid, the substance at a pressure of 1 atm.; for a substance in aqueous solution, the hypothetical solution of unit molality, which has all the properties of the infinitely dilute solution except the property of concentration.

¶ The free energy of solution of a given substance from its normal standard state as a solid, liquid, or gas to the hypothetical one molal state in aqueous solution may be calculated in a manner similar to that described in footnote § for calculating the heat of solution.

NOTE: °F = 9/5 °C + 32; to convert kilocalories per gram-mole to British thermal units per pound-mole, multiply by 1.799×10^{-3}.

THERMO

Appendix I
Heats of Combustion

Heats of combustion of additional compounds may be calculated from the heats of formation given in **Appendix H**
The following values are taken from the tables of the American Petroleum Institute Research Project 44 of the National Bureau of Standards on the Collection, Analysis, Calculation, and Compilation of Data on the Properties of Hydrocarbons.

Compound	Formula	State	Heat of combustion, $-\Delta Hc°$, at 25°C. and constant pressure, to form: H_2O (liq.) and CO_2 (gas)			H_2O (gas) and CO_2 (gas)		
			Kcal./mole	Cal./g.	B.t.u./lb.	Kcal./mole	Cal./g.	B.t.u./lb.
Hydrogen	H_2	gas	68.3174	33,887.6	60,957.7	57.7979	28,669.6	51,571.4
Carbon	C	solid, graph.	94.0518	7,831.1	14,086.8			
Carbon monoxide	CO	gas	67.6361	2,414.7	4,343.6			
Paraffins								
Methane	CH_4	gas	212.798	13,265.1	23,861	191.759	11,953.6	21,502
Ethane	C_2H_6	gas	372.820	12,399.2	22,304	341.261	11,349.6	20,416
Propane	C_3H_8	gas	530.605	12,033.5	21,646	488.527	11,079.2	19,929
Propane	C_3H_8	liq.*	526.782	11,946.8	21,490	484.704	10,992.5	19,774
n-Butane	C_4H_{10}	gas	687.982	11,837.3	21,293	635.384	10,932.3	19,665
n-Butane	C_4H_{10}	liq.*	682.844	11,748.9	21,134	630.246	10,843.9	19,506
2-Methylpropane (Isobutane)	C_4H_{10}	gas	686.342	11,809.1	21,242	633.744	10,904.1	19,614
2-Methylpropane (Isobutane)	C_4H_{10}	liq.*	681.625	11,727.9	21,096	629.027	10,822.9	19,468
n-Pentane	C_5H_{12}	gas	845.16	11,714.6	21,072	782.04	10,839.7	19,499
n-Pentane	C_5H_{12}	liq.	838.80	11,626.4	20,914	775.68	10,751.5	19,340
2-Methylbutane (Isopentane)	C_5H_{12}	gas	843.24	11,688.0	21,025	780.12	10,813.1	19,451
2-Methylbutane (Isopentane)	C_5H_{12}	liq.	837.31	11,605.8	20,877	774.19	10,730.9	19,303
2,2-Dimethylpropane (Neopentane)	C_5H_{12}	gas	840.49	11,649.8	20,956	777.37	10,775.0	19,382
2,2-Dimethylpropane (Neopentane)	C_5H_{12}	liq.	835.18	11,576.2	20,824	772.06	10,701.4	19,250
n Hexane	C_6H_{14}	gas	1,002.57	11,634.5	20,928	928.93	10,780.0	19,391
n-Hexane	C_6H_{14}	liq.	995.01	11,546.8	20,771	921.37	10,692.2	19,233
2-Methylpentane	C_6H_{14}	gas	1,000.87	11,614.8	20,893	927.23	10,760.2	19,356
2-Methylpentane	C_6H_{14}	liq.	993.71	11,531.7	20,743	920.07	10,677.1	19,206
3-Methylpentane	C_6H_{14}	gas	1,001.51	11,622.2	20,906	927.87	10,767.6	19,369
3-Methylpentane	C_6H_{14}	liq.	994.25	11,538.0	20,755	920.61	10,683.4	19,218
2,2-Dimethylbutane	C_6H_{14}	gas	998.17	11,583.5	20,837	924.53	10,728.9	19,299
2,2-Dimethylbutane	C_6H_{14}	liq.	991.52	11,506.3	20,698	917.88	10,651.7	19,161
2,3-Dimethylbutane	C_6H_{14}	gas	1,000.04	11,605.2	20,876	926.40	10,750.6	19,338
2,3-Dimethylbutane	C_6H_{14}	liq.	993.05	11,524.0	20,730	919.41	10,669.5	19,192
n-Heptane	C_7H_{16}	gas	1,160.01	11,577.2	20,825	1,075.85	10,737.2	19,314
n-Heptane	C_7H_{16}	liq.	1,151.27	11,489.9	20,668	1,067.11	10,650.0	19,157
2-Methylhexane	C_7H_{16}	gas	1,158.30	11,560.1	20,795	1,074.14	10,720.2	19,284
2-Methylhexane	C_7H_{16}	liq.	1,149.97	11,477.0	20,645	1,065.81	10,637.0	19,134
3-Methylhexane	C_7H_{16}	gas	1,158.94	11,566.5	20,806	1,074.78	10,726.6	19,295
3-Methylhexane	C_7H_{16}	liq.	1,150.55	11,482.8	20,655	1,066.39	10,642.8	19,145
3-Ethylpentane	C_7H_{16}	gas	1,159.56	11,572.7	20,817	1,075.40	10,732.7	19,306
3-Ethylpentane	C_7H_{16}	liq.	1,151.13	11,488.6	20,666	1,066.97	10,648.6	19,155
2,2-Dimethylpentane	C_7H_{16}	gas	1,155.61	11,533.3	20,746	1,071.45	10,693.3	19,235
2,2-Dimethylpentane	C_7H_{16}	liq.	1,147.85	11,455.8	20,607	1,063.69	10,615.9	19,096
2,3-Dimethylpentane	C_7H_{16}	gas	1,157.28	11,549.9	20,776	1,073.12	10,710.0	19,265
2,3-Dimethylpentane	C_7H_{16}	liq.	1,149.09	11,468.2	20,629	1,064.93	10,628.3	19,118
2,4-Dimethylpentane	C_7H_{16}	gas	1,156.60	11,543.1	20,764	1,072.44	10,703.2	19,253
2,4-Dimethylpentane	C_7H_{16}	liq.	1,148.73	11,464.6	20,623	1,064.57	10,624.7	19,112
3,3-Dimethylpentane	C_7H_{16}	gas	1,156.73	11,544.4	20,766	1,072.57	10,704.5	19,255
3,3-Dimethylpentane	C_7H_{16}	liq.	1,148.83	11,465.6	20,625	1,064.67	10,625.7	19,114
2,2,3-Trimethylbutane	C_7H_{16}	gas	1,155.94	11,536.6	20,752	1,071.78	10,696.6	19,241
2,2,3-Trimethylbutane	C_7H_{16}	liq.	1,148.27	11,460.0	20,614	1,064.11	10,620.1	19,104
n-Octane	C_8H_{18}	gas	1,317.45	11,533.9	20,747	1,222.77	10,705.0	19,256
n-Octane	C_8H_{18}	liq.	1,307.53	11,447.1	20,591	1,212.85	10,618.2	19,100
2-Methylheptane	C_8H_{18}	gas	1,315.76	11,519.1	20,721	1,221.08	10,690.2	19,230
2-Methylheptane	C_8H_{18}	liq.	1,306.28	11,436.1	20,572	1,211.60	10,607.2	19,080
3-Methylheptane	C_8H_{18}	gas	1,316.44	11,525.1	20,732	1,221.76	10,696.2	19,240
3-Methylheptane	C_8H_{18}	liq.	1,306.92	11,441.7	20,582	1,212.24	10,612.8	19,091
4-Methylheptane	C_8H_{18}	gas	1,316.57	11,526.2	20,734	1,221.89	10,697.3	19,243
4-Methylheptane	C_8H_{16}	liq.	1,307.09	11,443.2	20,584	1,212.41	10,614.3	19,093
3-Ethylhexane	C_8H_{18}	gas	1,316.87	11,528.8	20,738	1,222.19	10,699.9	19,247
3-Ethylhexane	C_8H_{18}	liq.	1,307.39	11,445.8	20,589	1,212.71	10,616.9	19,098
2,2-Dimethylhexane	C_8H_{18}	gas	1,313.56	11,499.9	20,686	1,218.88	10,671.0	19,195
2,2-Dimethylhexane	C_8H_{18}	liq.	1,304.64	11,421.8	20,546	1,209.96	10,592.9	19,055
2,3-Dimethylhexane	C_8H_{18}	gas	1,316.13	11,522.4	20,727	1,221.45	10,693.5	19,236
2,3-Dimethylhexane	C_8H_{18}	liq.	1,306.86	11,441.2	20,581	1,212.18	10,612.3	19,090
2,4-Dimethylhexane	C_8H_{18}	gas	1,314.83	11,511.0	20,706	1,220.15	10,682.1	19,215
2,4-Dimethylhexane	C_8H_{18}	liq.	1,305.80	11,431.9	20,564	1,211.12	10,603.0	19,073
2,5-Dimethylhexane	C_8H_{18}	gas	1,314.05	11,504.2	20,694	1,219.37	10,675.3	19,203
2,5-Dimethylhexane	C_8H_{18}	liq.	1,305.00	11,424.9	20,551	1,210.32	10,596.0	19,060
3,3-Dimethylhexane	C_8H_{18}	gas	1,314.65	11,509.4	20,703	1,219.97	10,680.5	19,212
3,3-Dimethylhexane	C_8H_{18}	liq.	1,305.68	11,430.9	20,562	1,211.00	10,602.0	19,071
3,4-Dimethylhexane	C_8H_{18}	gas	1,316.36	11,524.4	20,730	1,221.68	10,695.5	19,239
3,4-Dimethylhexane	C_8H_{18}	liq.	1,307.04	11,442.8	20,583	1,212.36	10,613.9	19,092
2-Methyl-3-ethylpentane	C_8H_{18}	gas	1,316.79	11,528.1	20,737	1,222.11	10,699.2	19,246
2-Methyl-3-ethylpentane	C_8H_{18}	liq.	1,307.58	11,447.5	20,592	1,212.90	10,618.6	19,101
3-Methyl-3-ethylpentane	C_8H_{18}	gas	1,315.88	11,520.2	20,723	1,221.20	10,691.3	19,232
3-Methyl-3-ethylpentane	C_8H_{18}	liq.	1,306.80	11,440.7	20,580	1,212.12	10,611.8	19,089
2,2,3-Trimethylpentane	C_8H_{18}	gas	1,314.66	11,509.5	20,703	1,219.98	10,680.6	19,212
2,2,3-Trimethylpentane	C_8H_{18}	liq.	1,305.83	11,432.2	20,564	1,211.15	10,603.3	19,073
2,2,4-Trimethylpentane	C_8H_{18}	gas	1,313.69	11,501.0	20,688	1,219.01	10,672.1	19,197
2,2,4-Trimethylpentane	C_8H_{18}	liq.	1,305.29	11,427.5	20,556	1,210.61	10,598.6	19,065
2,3,3-Trimethylpentane	C_8H_{18}	gas	1,315.54	11,517.2	20,717	1,220.86	10,688.3	19,226
2,3,3-Trimethylpentane	C_8H_{18}	liq.	1,306.64	11,439.3	20,577	1,211.96	10,610.4	19,086
2,3,4-Trimethylpentane	C_8H_{18}	gas	1,315.29	11,515.0	20,713	1,220.61	10,686.1	19,222
2,3,4-Trimethylpentane	C_8H_{18}	liq.	1,306.28	11,436.1	20,572	1,211.60	10,607.2	19,080
2,2,3,3-Tetramethylbutane	C_8H_{18}	gas	1,313.27	11,497.3	20,682	1,218.59	10,668.4	19,191

Appendix I (continued)

Compound	Formula	State	Heat of combustion, $-\Delta Hc°$, at 25°C. and constant pressure, to form: H_2O (liq.) and CO_2 (gas)			H_2O (gas) and CO_2 (gas)		
			Kcal./mole	Cal./g.	B.t.u./lb.	Kcal./mole	Cal./g.	B.t.u./lb.
2,2,3,3-Tetramethylbutane	C_8H_{18}	solid	1,303.03	11,407.7	20,520	1,208.35	10,578.8	19,029
n-Nonane	C_9H_{20}	gas	1,474.90	11,500.2	20,687	1,369.70	10,680.0	19,211
n-Nonane	C_9H_{20}	liq.	1,463.80	11,413.6	20,531	1,358.60	10,593.4	19,056
n-Decane	$C_{10}H_{22}$	gas	1,632.34	11,473.0	20,638	1,516.63	10,659.7	19,175
n-Decane	$C_{10}H_{22}$	liq.	1,620.06	11,386.7	20,483	1,504.35	10,573.4	19,020
n-Undecane	$C_{11}H_{24}$	gas	1,789.78	11,450.8	20,598	1,663.55	10,643.2	19,145
n-Undecane	$C_{11}H_{24}$	liq.	1,776.32	11,364.7	20,443	1,650.09	10,557.0	18,990
n-Dodecane	$C_{12}H_{26}$	gas	1,947.23	11,432.2	20,564	1,810.48	10,629.4	19,120
n-Dodecane	$C_{12}H_{26}$	liq.	1,932.59	11,346.3	20,410	1,795.84	10,543.4	18,966
n-Tridecane	$C_{13}H_{28}$	gas	2,104.67	11,416.5	20,536	1,957.40	10,617.6	19,099
n-Tridecane	$C_{13}H_{28}$	liq.	2,088.85	11,330.6	20,382	1,941.58	10,531.8	18,945
n-Tetradecane	$C_{14}H_{30}$	gas	2,262.11	11,402.9	20,512	2,104.32	10,607.5	19,081
n-Tetradecane	$C_{14}H_{30}$	liq.	2,245.11	11,317.2	20,358	2,087.32	10,521.8	18,927
n-Pentadecane	$C_{15}H_{32}$	gas	2,419.55	11,391.2	20,491	2,251.24	10,598.7	19,065
n-Pentadecane	$C_{15}H_{32}$	liq.	2,401.37	11,305.6	20,337	2,233.06	10,513.2	18,911
n-Hexadecane	$C_{16}H_{34}$	gas	2,577.00	11,380.9	20,472	2,398.17	10,591.1	19,052
n-Hexadecane	$C_{16}H_{34}$	liq.	2,557.64	11,295.4	20,318	2,378.81	10,505.6	18,898
n-Heptadecane	$C_{17}H_{36}$	gas	2,734.44	11,371.8	20,456	2,545.09	10,584.3	19,039
n-Heptadecane	$C_{17}H_{36}$	liq.	2,713.90	11,286.4	20,302	2,524.55	10,498.9	18,886
n-Octadecane	$C_{18}H_{38}$	gas	2,891.88	11,363.7	20,441	2,692.01	10,578.3	19,028
n-Octadecane	$C_{18}H_{38}$	liq.	2,870.16	11,278.4	20,288	2,670.29	10,493.0	18,875
n-Nonadecane	$C_{19}H_{40}$	gas	3,049.33	11,356.5	20,428	2,838.94	10,572.9	19,019
n-Nonadecane	$C_{19}H_{40}$	liq.	3,026.43	11,271.2	20,275	2,816.04	10,487.7	18,865
n-Eicosane	$C_{20}H_{42}$	gas	3,206.77	11,350.0	20,416	2,985.86	10,568.1	19,010
n-Eicosane	$C_{20}H_{42}$	liq.	3,182.69	11,264.7	20,263	2,961.78	10,482.8	18,857
Alkyl benzenes								
Benzene	C_6H_6	gas	789.08	10,102.4	18,172	757.52	9,698.4	17,446
Benzene	C_6H_6	liq.	780.98	9,998.7	17,986	749.42	9,594.7	17,259
Methylbenzene (toluene)	C_7H_8	gas	943.58	10,241.4	18,422	901.50	9,784.7	17,601
Methylbenzene (toluene)	C_7H_8	liq.	934.50	10,142.8	18,245	892.42	9,686.1	17,424
Ethylbenzene	C_8H_{10}	gas	1,101.13	10,372.4	18,658	1,048.53	9,876.9	17,767
Ethylbenzene	C_8H_{10}	liq.	1,091.03	10,277.2	18,487	1,038.43	9,781.7	17,596
1,2-Dimethylbenzene (o-xylene)	C_8H_{10}	gas	1,098.54	10,348.0	18,614	1,045.94	9,852.5	17,723
1,2-Dimethylbenzene (o-xylene)	C_8H_{10}	liq.	1,088.16	10,250.2	18,438	1,035.56	9,754.7	17,547
1,3-Dimethylbenzene (m-xylene)	C_8H_{10}	gas	1,098.12	10,344.0	18,607	1,045.52	9,848.5	17,716
1,3-Dimethylbenzene (m-xylene)	C_8H_{10}	liq.	1,087.92	10,247.9	18,434	1,035.32	9,752.4	17,543
1,4-Dimethylbenzene (p-xylene)	C_8H_{10}	gas	1,098.29	10,345.6	18,610	1,045.69	9,850.1	17,719
1,4-Dimethylbenzene (p-xylene)	C_8H_{10}	liq.	1,088.16	10,250.2	18,438	1,035.56	9,754.7	17,547
n-Propylbenzene	C_9H_{12}	gas	1,258.24	10,469.1	18,832	1,195.12	9,943.9	17,887
n-Propylbenzene	C_9H_{12}	liq.	1,247.19	10,377.2	18,667	1,184.07	9,852.0	17,722
Isopropylbenzene (cumene)	C_9H_{12}	gas	1,257.31	10,461.4	18,818	1,194.19	9,936.2	17,873
Isopropylbenzene (cumene)	C_9H_{12}	liq.	1,246.52	10,371.6	18,657	1,183.40	9,846.4	17,712
1-Methyl-2-ethylbenzene	C_9H_{12}	gas	1,256.66	10,456.0	18,808	1,193.54	9,930.8	17,864
1-Methyl-2-ethylbenzene	C_9H_{12}	liq.	1,245.26	10,361.1	18,638	1,182.14	9,835.9	17,693
1-Methyl-3-ethylbenzene	C_9H_{12}	gas	1,255.92	10,449.8	18,797	1,192.80	9,924.6	17,853
1-Methyl-3-ethylbenzene	C_9H_{12}	liq.	1,244.71	10,356.5	18,630	1,181.59	9,831.3	17,685
1-Methyl-4-ethylbenzene	C_9H_{12}	gas	1,255.59	10,447.1	18,792	1,192.47	9,921.9	17,848
1-Methyl-4-ethylbenzene	C_9H_{12}	liq.	1,244.45	10,354.4	18,626	1,181.33	9,829.2	17,681
1,2,3-Trimethylbenzene (hemimellitene)	C_9H_{12}	gas	1,254.08	10,434.5	18,770	1,190.96	9,909.3	17,825
1,2,3-Trimethylbenzene (hemimellitene)	C_9H_{12}	liq.	1,242.36	10,337.0	18,594	1,179.24	9,811.8	17,650
1,2,4-Trimethylbenzene (pseudocumene)	C_9H_{12}	gas	1,253.04	10,425.8	18,754	1,189.92	9,900.7	17,809
1,2,4-Trimethylbenzene (pseudocumene)	C_9H_{12}	liq.	1,241.58	10,330.5	18,583	1,178.46	9,805.3	17,638
1,3,5-Trimethylbenzene (mesitylene)	C_9H_{12}	gas	1,252.53	10,421.6	18,747	1,189.41	9,896.4	17,802
1,3,5-Trimethylbenzene (mesitylene)	C_9H_{12}	liq.	1,241.19	10,327.2	18,577	1,178.07	9,802.1	17,632
n-Butylbenzene	$C_{10}H_{14}$	gas	1,415.44	10,546.3	18,971	1,341.80	9,997.6	17,984
n-Butylbenzene	$C_{10}H_{14}$	liq.	1,403.46	10,457.0	18,810	1,329.82	9,908.4	17,823
Alkyl cyclopentanes								
Cyclopentane	C_5H_{10}	gas	793.39	11,313.1	20,350	740.79	10,563.1	19,001
Cyclopentane	C_5H_{10}	liq.	786.54	11,215.5	20,175	733.94	10,465.4	18,825
Methylcyclopentane	C_6H_{12}	gas	948.72	11,273.4	20,279	885.60	10,523.3	18,930
Methylcyclopentane	C_6H_{12}	liq.	941.14	11,183.3	20,117	878.02	10,433.2	18,768
Ethylcyclopentane	C_7H_{14}	gas	1,106.21	11,266.9	20,267	1,032.57	10,516.9	18,918
Ethylcyclopentane	C_7H_{14}	liq.	1,097.50	11,178.2	20,108	1,023.86	10,428.2	18,758
n-Propylcyclopentane	C_8H_{16}	gas	1,263.56	11,260.9	20,256	1,179.40	10,510.8	18,907
n-Propylcyclopentane	C_8H_{16}	liq.	1,253.74	11,173.4	20,099	1,169.58	10,423.3	18,750
n-Butylcyclopentane	C_9H_{18}	gas	1,421.10	11,257.7	20,250	1,326.42	10,507.6	18,901
n-Butylcyclopentane	C_9H_{18}	liq.	1,410.10	11,170.5	20,094	1,315.42	10,420.5	18,745
Alkyl cyclohexanes								
Cyclohexane	C_6H_{12}	gas	944.79	11,226.7	20,195	881.67	10,476.7	18,846
Cyclohexane	C_6H_{12}	liq.	936.88	11,132.7	20,026	873.76	10,382.7	18,676
Methylcyclohexane	C_7H_{14}	gas	1,099.59	11,199.5	20,146	1,025.95	10,449.5	18,797
Methylcyclohexane	C_7H_{14}	liq.	1,091.13	11,113.3	19,991	1,017.49	10,363.3	18,642
Ethylcyclohexane	C_8H_{16}	gas	1,257.90	11,210.4	20,166	1,173.74	10,460.4	18,816
Ethylcyclohexane	C_8H_{16}	liq.	1,248.23	11,124.3	20,011	1,164.07	10,374.3	18,661
n-Propylcyclohexane	C_9H_{18}	gas	1,415.12	11,210.3	20,165	1,320.44	10,460.3	18,816
n-Propylcyclohexane	C_9H_{18}	liq.	1,404.34	11,124.9	20,012	1,309.66	10,374.9	18,663
n-Butylcyclohexane	$C_{10}H_{20}$	gas	1,572.74	11,213.0	20,170	1,467.54	10,463.0	18,821
n-Butylcyclohexane	$C_{10}H_{20}$	liq.	1,560.78	11,127.8	20,017	1,455.58	10,377.8	18,668
Monoolefins								
Ethene (ethylene)	C_2H_4	gas	337.234	12,021.7	21,625	316.195	11,271.7	20,276
Propene (propylene)	C_3H_6	gas	491.987	11,692.3	21,032	460.428	10,942.3	19,683
1-Butene	C_4H_8	gas	649.757	11,581.3	20,833	607.679	10,831.3	19,484
cis-2-Butene	C_4H_8	gas	648.115	11,552.0	20,780	606.037	10,802.0	19,431
trans-2-Butene	C_4H_8	gas	647.072	11,533.4	20,747	604.994	10,783.4	19,397
2-Methylpropene (isobutene)	C_4H_8	gas	646.134	11,516.7	20,716	604.056	10,766.7	19,367
1-Pentene	C_5H_{10}	gas	806.85	11,505.1	20,696	754.25	10,755.1	19,346

Appendix I (continued)

Compound	Formula	State	Heat of combustion, $-\Delta Hc°$, at 25°C. and constant pressure, to form					
			H_2O (liq.) and CO_2 (gas)			H_2O (gas) and CO_2 (gas)		
			Kcal./mole	Cal./g.	B.t.u./lb.	Kcal./mole	Cal./g.	B.t.u./lb.
cis-2-Pentene	C_5H_{10}	gas	805.34	11,483.5	20,657	752.74	10,733.5	19,308
trans-2-Pentene	C_5H_{10}	gas	804.26	11,468.1	20,629	751.66	10,718.1	19,280
2-Methyl-1-butene	C_5H_{10}	gas	803.17	11,452.6	20,601	750.57	10,702.6	19,252
3-Methyl-1-butene	C_5H_{10}	gas	804.93	11,477.7	20,646	752.33	10,727.7	19,297
2-Methyl-2-butene	C_5H_{10}	gas	801.68	11,431.3	20,563	749.08	10,681.3	19,214
Acetylenes								
Ethyne (acetylene)	C_2H_2	gas	310.615	11,930.2	21,460	300.096	11,526.2	20,734
Propyne (methylacetylene)	C_3H_4	gas	463.109	11,559.8	20,794	442.070	11,034.6	19,849
1-Butyne (ethylacetylene)	C_4H_6	gas	620.86	11,478.7	20,648	589.302	10,895.2	19,599
2-Butyne (dimethylacetylene)	C_4H_6	gas	616.533	11,398.7	20,504	584.974	10,815.2	19,455
1-Pentyne	C_5H_8	gas	778.03	11,422.5	20,547	735.95	10,804.7	19,436
2-Pentyne	C_5H_8	gas	774.33	11,368.2	20,449	732.25	10,750.4	19,338
3-Methyl-1-butyne	C_5H_8	gas	776.13	11,394.6	20,497	734.05	10,776.8	19,386

NOTE: $°F = \frac{9}{5}\,°C + 32$.
°Saturation pressure.

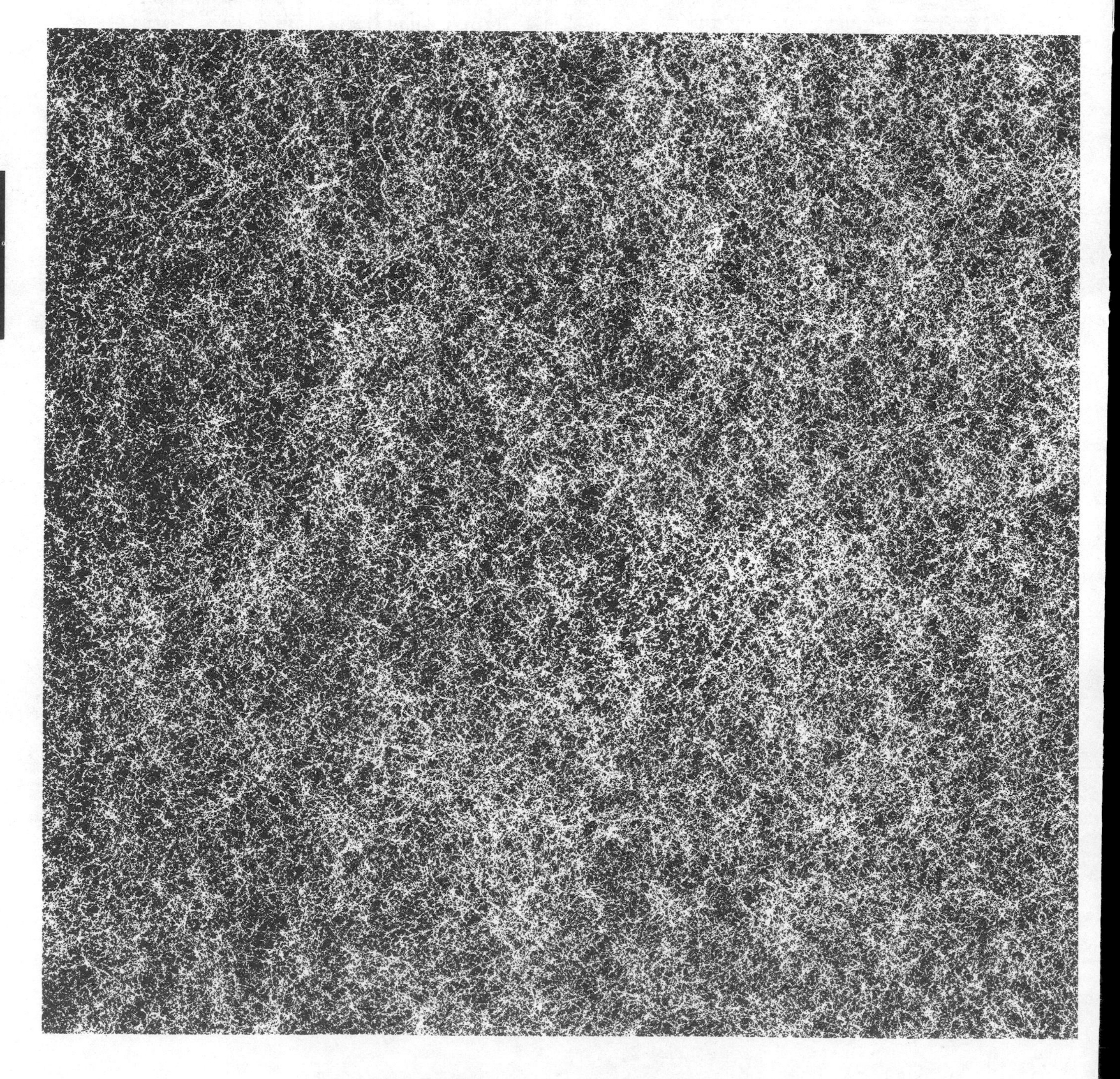

Appendix J
Specific Heats of Miscellaneous Materials

Material	Specific Heat, cal./g. °C.
Alumina	0.2 (100°C.); 0.274 (1500°C.)
Alundum	0.186 (100°C.)
Asbestos	0.25
Asphalt	0.22
Bakelite	0.3 to 0.4
Brickwork	About 0.2
Carbon	0.168 (26° to 76°C.)
	0.314 (40° to 892°C.)
	0.387 (56° to 1450°C.)
(gas retort)	0.204
(See under Graphite)	
Cellulose	0.32
Cement, Portland Clinker	0.186
Charcoal (wood)	0.242
Chrome brick	0.17
Clay	0.224
Coal	0.26 to 0.37
tar oils	0.34 (15° to 90°C.)
Coal tars	0.35 (40°C.); 0.45 (200°C.)
Coke	0.265 (21° to 400°C.)
	0.359 (21° to 800°C.)
	0.403 (21° to 1300°C.)
Concrete	0.156 (70° to 312°F.); 0.219 (72° to 1472°F.)
Cryolite	0.253 (16° to 55°C.)
Diamond	0.147
Fireclay brick	0.198 (100°C.); 0.298 (1500°C.)
Fluorspar	0.21 (30°C.)
Gasoline	0.53
Glass (crown)	0.16 to 0.20
(flint)	0.117
(pyrex)	0.20
(silicate)	0.188 to 0.204 (0 to 100°C.)
	0.24 to 0.26 (0 to 700°C.)
wool	0.157
Granite	0.20 (20° to 100°C.)
Graphite	0.165 (26° to 76°C.); 0.390 (56° to 1450°C.)
Gypsum	0.259 (16° to 46°C.)
Kerosene	0.47
Limestone	0.217
Litharge	0.055
Magnesia	0.234 (100°C.); 0.188 (1500°C.)
Magnesite brick	0.222 (100°C.); 0.195 (1500°C.)
Marble	0.21 (18°C.)
Pyrites (copper)	0.131 (19° to 50°C.)
(iron)	0.136 (15° to 98°C.)
Quartz	0.17 (0°C.); 0.28 (350°C.,
Sand	0.191
Silica	0.316
Steel	0.12
Stone	About 0.2
Turpentine	0.42 (18°C.)
Wood (oak)	0.570
Most woods vary between	0.45 and 0.65

THERMO

Appendix K
Mollier Chart

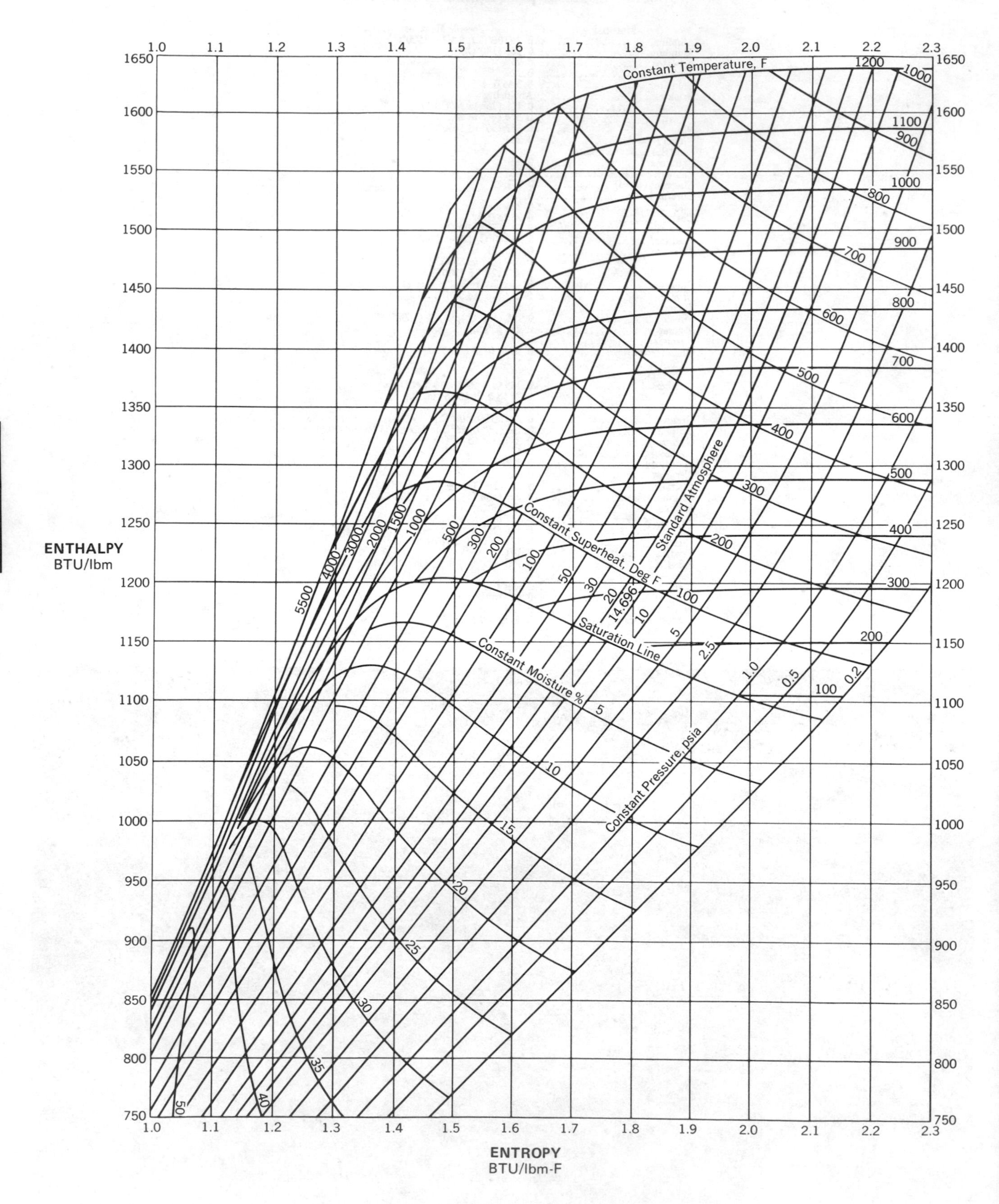

PRACTICE PROBLEMS

1. An insulated, 10 cubic foot compartmented tank contains an ideal (but unknown) corrosive gas. The starting conditions are as follows:

	compartment 1	compartment 2
volume	9 ft^3	1 ft^3
pressure	8 atm	6 atm
temperature	200 °C	300 °C

The thin wall between compartments suddenly dissolves. What are the final pressure, temperature, and internal energy change in the tank?

2. A bicycle tire is initially pumped up with air to 100 psig. After a long ride, the tire pressure is 105 psig. What was the internal energy change of the air in the tire? Assume air is an ideal gas, C_p – 7 BTU/lbmole-°F, and the tire volume is unchanged at 0.01 cubic foot.

3. Which is larger: c_p or c_v? Why?

4. The heat of transition for a change from rhombic to monoclinic sulfur at 368.5 °K is 96 cal/gmole. What are the changes in entropy and internal energy for this transition?

5. Supercooled water at –2 °C is converted to ice at –2 °C. The molal heat capacities of ice and water are 9 and 18 cal/mole °C. The heat of fusion of water is 1436 cal/mole. What are the enthalpy and entropy changes for this process?

6. Iodine vapor dissociates according to

$$I_2 \rightarrow 2I$$

$$K = \frac{(I)^2}{(I_2)}$$

The equilibrium constants for this reaction are 3992 and 0.0002 at 1000 °C and 600 °C, respectively. What is the heat of reaction and percent dissociation at 900 °C?

7. Rewrite van der Waal's equation of state in the virial form.

8. At what approximate reduced pressure is the compressibility factor greater than one?

9. Which four thermodynamic variables of an ideal gas are functions of temperature only?

10. In a binary solution, the activity of component one is 2.00 at infinite dilution; the activity of component two is 0.5 at infinite dilution. What are the constants for the van Laar and Margules equations?

11. Calculate the entropy change when one pound of ammonia is cooled from 500 °F to –150 °F at a constant pressure of 1 atmosphere. Express the results in BTU/lbm-°R. Calculate the absolute entropy of solid NH_3 at its melting point at 1 atmosphere pressure in BTU/lbm-°R.

NH_3 data:

$a = 6.5846$
$b = 6.1251 \times 10^{-3}$
$c = 2.3663 \times 10^{-6}$
$d = -1.5981 \times 10^{-9}$
boiling point = –33.4 °C
melting point = –77.7 °C
$C_p = a + bT + cT^2 + dT^3$ cal/gmole °K with T in °K

S = 46.03 cal/gmole °K (absolute entropy of gas at 25 °C)

λ_v = 5581 cal/gmole
λ_f = 1352 cal/gmole
c_p (solid) = 0.502 cal/g-°K
c_p (liquid) = 1.06 cal/g-°K

12. Dry sodium acetate, $NaC_2H_3O_2$, is stored and charged into a dry material feeder in an area maintained at 68 °F. The dry feed is added to a jacketed mixer agitated by a 5 horsepower, high shear impeller. The mixer also receives a stream of 55 °F water at a rate calculated to produce a 9 mole percent solution when drawn off at a rate of 20 gpm. The solution is then pumped with a $7\frac{1}{2}$ horsepower pump into a reactor where it must enter at 135 °F. The specific gravity of the solution is 1.15; the specific heat of the solid of the solution is 0.94 cal/g-°C; the specific heat of the solid is 0.339 cal/g-°C; the heat of the solution is –3943 cal/g formula weight at 25 °C. Calculate the heat, in BTU/hr, that must be added to or taken from the mixer if this temperature requirement is to be met.

13. Sufficient impact in nitroglycerine causes local hot spots from which detonation proceeds. The temperature of the hot spot must exceed a critical temperature in order for the detonation to occur. If liquid nitroglycerine is degassed, and a drop is placed on a steel table, a certain weight dropped from 12 cm or more will produce a detonation. If a bubble of air is introduced into the drop, the same weight will produce a detonation at 2 cm. The effect in the latter case is that the compression of the air bubble produces a temperature rise above the critical temperature. Assuming that the critical temperature for nitroglycerine is 300 °C, what pressure must be produced in the bubble to cause detonation?

14. Water flowing at 1.2 gpm at 50 °F is heated to 150 °F with saturated steam at 14.7 psia. Calculate the steam flow rate required to heat the water in (a) a mixing type heater, and (b) a parallel type heat exchanger.

15. An equimolecular mixture of hydrogen and carbon monoxide is mixed with the stoichiometric quantity of air in a closed, insulated vessel. The total pressure of the mixture is 5 atmospheres and the temperature is 25 °C. The mixture is ignited by a spark plug. The oxidation of the reactants is assumed complete. Assuming all the gases are ideal, estimate (a) the maximum temperature, and (b) the pressure.

16. A compressor is used to supply 50 psig air to an air lift that is raising 20 gpm of a liquid with a specific gravity of 1.5 to a height of 50′. The pump efficiency is 30%, and it may be assumed that the compressor works isentropically. Assume $k = 1.4$. Calculate the horsepower required for the compressor.

17. Iron oxide, FeO, is reduced to Fe by passing a mixture at 150 atmospheres and 100 °C of 20 mole percent CO and 80 mole percent N_2 over the oxide. $K_a = 40$ at the stated conditions.

$$FeO(s) + CO(g) \rightarrow Fe(s) + CO_2(g)$$

(a) What is the weight of Fe produced per 1000 ft^3 of entering gas measured at 1 atmosphere and 25 °C? (b) Calculate the weight assuming pure CO is used instead of the gas mixture. (Assume equilibrium is reached in both cases.)

18. A Rankine cycle power plant generates dry saturated steam at a gauge pressure of 400 psig which is isentropically expanded through engines to an exhaust gauge pressure of 10 psi for use in subsequent process heating. Assume that the condensate is returned to the boilers at its saturation temperature under exhaust pressure by an isentropic pump. (a) Calculate the mass of dry saturated low pressure steam available for process heating per 100 pounds of high pressure steam generated. (b) Calculate the thermodynamic efficiency of the cycle. (Since the process steam would have to be generated anyway, deduct its heat from the power generation cycle.) (c) Determine the effect on your answers of superheating the steam 100 °F.

19. Acetonitrile, C_2H_3N, is stored at a temperature of 550 °F and a pressure of 4500 psia in a high-pressure vessel, A, with a volume of 0.2 cubic feet. For safety, the 0.2 cubic foot vessel is surrounded by a second pressure vessel. The volume of the second vessel, B, is 2 cubic feet (exclusive of the volume occupied by vessel A which it surrounds). This volume is filled with inert nitrogen at 10 atmosphere pressure. The entire assembly is maintained at 550 °F. If the inner tank ruptures, what is the final pressure?

	T_c (°K)	p_c (atm)
N_2	126.2	33.5
C_2H_3N	548.0	47.7

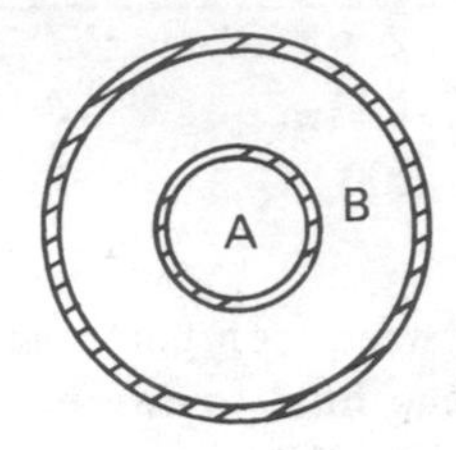

20. Steam is delivered by a boiler at 550 psia and 700 °F. After expansion to 110 psia, the steam is reheated to 700 °F. Expansion is to 1.5 inches of mercury (abs). For an ideal reheat cycle (Rankine with reheating), calculate on the basis of one pound of steam: (a) the heat added, (b) the heat rejected, (c) the net work, (d) thermodynamic efficiency, and (e) the quality of the exhaust steam.

21. In a binary system of acetone (component 1) and chloroform (component 2), the van Laar constants are $A_{12} = -0.44$ and $A_{21} = -0.34$. Assuming an ideal gas, determine the azeotrope composition and temperature at a constant total pressure of 1 atmosphere. p is measured in mm Hg, and T is measured in °C. The vapor pressure for each component is

$$\log_{10} p_{acetone} = 7.02447 - \frac{1161.0}{224 + t}$$

$$\log_{10} p_{chloroform} = 6.90328 - \frac{1163.0}{227 + t}$$

5 FLUID STATICS AND DYNAMICS

Nomenclature

a	acceleration	ft/sec^2
A	area	ft^2
B	rate of flow	bbl/hr
c	velocity of sound, or specific heat	ft/sec, or BTU/lbm-°F
C	compressibility, flow coefficient, or Hazen-Williams constant	in^2/lbf, –, –
d	internal diameter	in
D	internal diameter	ft
E	bulk modulus, or modulus of elasticity	psi, psi
f	Moody friction factor	–
g	acceleration of gravity (32.2)	ft/sec^2
g_c	gravitational conversion factor (32.2)	lbm-ft/lbf-sec^2
G	mass flow rate per unit area	lbm/ft^2-sec
h	head, or height of a fluid column	ft-lbf/lbm or ft
h_f	head loss due to friction	ft-lbf/lbm
H	total head (loss)	ft
k	ratio of specific heats	–
K	resistance coefficient	–
l	length	in
L	length	ft
m	mass density	slug/ft^3
M	mass flow (GA)	lbm/sec
n	polytropic exponent (pV^n = constant)	–
n	pump rpm	min^{-1}
N_{Re}	Reynolds number	–
p	pressure	lbf/ft^2
p'	pressure	lbf/in^2
q	flow rate	ft^3/sec
q'	flow rate at standard conditions	ft^3/hr
Q	flow rate	gal/min
r	radius	ft
R	specific gas constant, or resultant force	ft/°R, or lbf
SG	specific gravity	–
T	absolute temperature	°R
v	velocity	ft/sec
V	volume, or specific volume	ft^3, ft^3/lbm
w	flow rate	lbm/sec
W	flow rate	lbm/hr
z	elevation	ft

Symbols

α	ratio of actual to ideal energy	–
β	angle, or ratio of diameters	degrees, –
ρ	density	lbm/ft^3
η	efficiency	–
ε	specific roughness	ft
μ	relative viscosity	centipoise, or
μ_e	viscosity	lbm/ft-sec lbf-sec/ft^2
ν	kinematic viscosity	centistokes
ν'	kinematic viscosity	ft^2/sec
τ	surface tension	lbf/ft
ω	rotational velocity	radian/sec

Subscripts

0	orifice or nozzle
1	upstream
2	downstream
c	center
d	discharge
e	equivalent
f	friction
m	manometer fluid
p	constant pressure
s	suction
t	total
v	constant volume, or valve
vp	vapor pressure
y	distance from center of conduit

1 FLUID CHARACTERISTICS

There are two types of fluids: ideal and real. *Ideal fluids* are those which have zero viscosity and shearing forces, are incompressible, and exhibit uniform velocity distributions when flowing.

Real fluids exhibit finite viscosity and non-uniform velocity distributions. They can be divided into Newtonian and Non-Newtonian fluids. *Newtonian fluids*, typified by gases and thin liquids, exhibit viscosities which are independent of rate of change of shear stress. *Non-Newtonian fluids*, typified by gels, emulsions, suspensions, and gases near critical points, exhibit viscosities dependent on rate of change of shear stress. In a Newtonian fluid relative viscosity does not change with flow. Most fluid problems assume Newtonian fluid characteristics.

Fluid characteristics include *mass density* which is the mass of a fluid in a unit volume, and specific weight, which is the force exerted by the fluid in a unit volume. American engineering practice usually applies the term *density* to specific weight. However, a look at the units will usually serve to indicate which is required. Mass density in consistent English units will be in slugs per cubic foot. Specific weight will be in pounds per cubic foot.

$$\text{weight density} = \frac{\rho g}{g_c} \tag{5.1}$$

$$m = \text{mass density} = \frac{\rho}{g_c} \tag{5.2}$$

Specific volume is the reciprocal of weight density, and is the volume per unit mass, usually in units of cubic feet per pound.

Specific gravity (SG) is the ratio of the density of a substance to the density of pure water (or air at STP if the fluid is a gas). Specific gravity can be calculated if the density of a fluid is known. The specific gravity for liquids and solids is

$$SG_{Liquids} = \frac{\rho}{62.45} \tag{5.3}$$

Although a similar calculation can be made for gases, usually only the conditions (temperature, pressure, volume) at which the gas exists will be known. The actual density must first be calculated from the ideal gas law. The gas density at STP should be calculated. If ρ is in lbm/ft^3, the specific gravity for gases is

$$SG_{Gases} = \frac{\rho}{0.0749} = \frac{p}{0.0749 \times RT} \tag{5.4}$$

Viscosity is a measure of the resistance to motion of the fluid. Low-viscosity fluids will flow faster and more freely than viscous fluids. Viscosity depends greatly on temperature. Liquids experience decreasing viscosities as temperature increases, since cohesive forces between molecules dominate. Gases experience increasing viscosities as temperature increases, since cohesive forces are negligible when compared to molecular momentum and agitation, both of which increase when temperature increases. Viscosity of liquids and gases is essentially independent of pressure.

Table 5.1
Typical Specific Gravities

liquids		solids		gases	
alcohol	0.79	aluminum	2.56	air	1.00
gasoline	0.67	brass	8.40	acetylene	0.910
mercury	13.56	brick	1.80	ammonia	0.592
motor oil	0.82	cast iron	7.20	carbon dioxide	1.52
water, pure	1.00	cement	3.1	hydrogen	0.069
water, sea	1.03	concrete	2.2	mercury vapor	6.94
		copper	8.82	nitrogen	0.971
		glass	2.6	oxygen	1.106
		gold	19.32	water vapor	0.623
		lead	11.37		
		magnesium	1.74		
		silver	10.53		
		steel	7.80		
		wood (walnut)	0.65		
		zinc	7.15		

Absolute viscosity is formally known as the *coefficient of viscosity* since it is the proportionality constant between shear stress and the velocity distribution.

Two types of viscosities are used — *relative viscosity*, also known as *dynamic viscosity*, and *absolute viscosity* (absolute viscosity is the same as the coefficient of viscosity). *Kinematic viscosity* is relative viscosity divided by mass density.

$$\nu = \frac{\mu g_c}{\rho} \quad 5.5$$

Pressure exerted by free molecules at a liquid surface, or by the fluid on the surface molecules, is known as the *vapor pressure*. Boiling occurs when the vapor pressure is increased to the local ambient pressure. A liquid's boiling point depends on the external pressure. Liquids with low vapor pressure and high stability factors are used in accurate barometers.

Table 5.2
Viscosity Conversions

to obtain	multiply	by	and divide by
ft^2/s	$lbf\text{-}sec/ft^2$	32.2	density
ft^2/s	stokes	1.076 EE−3	1
$lbf\text{-}sec/ft^2$	ft^2/sec	density	32.2
$lbf\text{-}sec/ft^2$	poise	1	478.8
m^2/s	centistokes	1 EE−6	1
m^2/s	stokes	1 EE−4	1
m^2/s	ft^2/sec	9.29 EE−2	1
pascal-sec	centipoise	1 EE−3	1
pascal-sec	lbm/ft-sec	1.488	1
pascal-sec	$lbf\text{-}sec/ft^2$	47.88	1
pascal-sec	poise	.1	1
pascal-sec	slug/ft-sec	47.88	1
poise	$lbf\text{-}sec/ft^2$	478.8	1
stokes	ft^2/sec	929	1
stokes	poise	1	specific gravity

Table 5.3
Miscellaneous Viscosities

material	temperature °F	relative viscosity $lbf\text{-}sec/ft^2 = \mu'_e$
pure water	32	2.746 EE−5
	70	2.050
	100	1.424
	200	0.637
mercury	32	3.55 EE−5
	68	3.28
dry air at 14.7 psia*	32	3.55 EE−7
	68	3.80

* may be calculated from *Sutherland's formula*:
dry air viscosity = 3.02 EE−7 $(T/392)^{0.76}$

FLUIDS

Table 5.4
Typical Surface Tensions

fluid	surface tension
ethyl alcohol	0.001527 lbf/ft
turpentine	0.001857
water	0.004985
mercury	0.03562
n-octane	0.00144
acetone	0.00192
benzene	0.00192
carbon tetrachloride	0.00180

The value of pressure is dependent on the reference point chosen. Two such reference points exist: zero absolute pressure and standard atmospheric pressure.

If *standard atmospheric pressure* (14.7 psia) is chosen as the reference, pressures measured are known as *gage pressures*. Positive gage pressures are pressures above atmospheric. *Vacuum* (negative gage pressure) is the pressure below atmospheric. Maximum vacuum, according to this convention, is −14.7 psig.

If *zero pressure* is chosen as the reference, pressure is said to be *absolute pressure*. There is no distinction between positive and negative pressure.

Equation 5.6 can be used to convert between gage and absolute pressure.

$$\text{absolute pressure} = \text{atmospheric pressure} + \text{gage pressure} \qquad 5.6$$

The term *gage* is somewhat misleading, since a mechanical device (Bourdon pressure gauge) may be used to indicate either gage or absolute pressure. However, the barometer is the most common device used to measure atmospheric pressure.

The *barometer* is constructed by filling a long hollow-bore tube, which is open at one end, with mercury and inverting (placing the open end below the surface level of a mercury-filled container). If the vapor pressure is neglected, the mercury column will be supported only by the atmospheric pressure transmitted through the container fluid at the lower, open end. If an equation is set up balancing the weight of the fluid against the atmospheric force, atmospheric pressure in psia is

$$\text{atm} = 0.491 \times h_{inches} \qquad 5.7$$

Any fluid may be used to measure atmospheric pressure, although vapor pressure should not be significant. For any other fluid, atmospheric pressure is

$$\text{atm} = 0.0361 \times SG \times h_{inches} \qquad 5.8$$

The values 0.491 and 0.0361 are the densities in pounds per cubic inch of mercury and water, respectively. The height, h, must be in inches.

If a fluid which has considerable vapor pressure is used, that vapor pressure should be added to the value of atmospheric pressure obtained from equation 5.7 or equation 5.8. In the case of water, vapor pressure can be found from the saturated pressure column in the steam tables.

Manometers are frequently used to specify pressures or pressure differentials. Figure 5.1 is a simple U-tube manometer whose two ends are connected to two sources of pressure. One end may be open to the atmosphere, which will define that end's pressure.

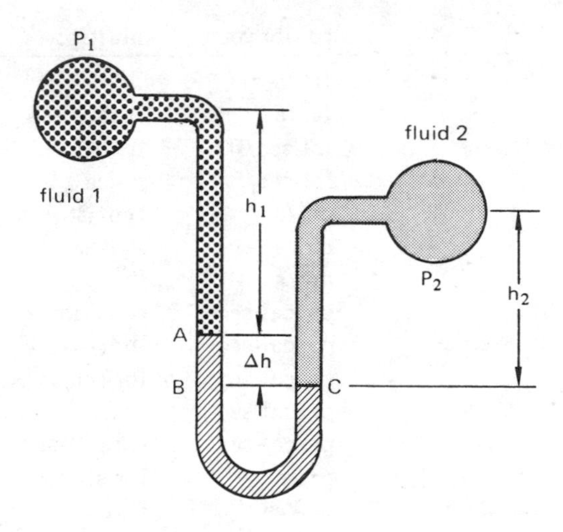

Figure 5.1 A Basic Manometer

Since the pressure at point B is the same as that at point C, the pressure differential must be causing the fluid column height.

$$\Delta p = p_2 - p_1 = \rho \Delta h \qquad 5.9$$

Equation 5.9 assumes that only low-density gases fill the tubes above the measuring fluid. If a fluid, such as water, is present above the measuring fluid (e.g., mercury), or if the gas column h_1 or h_2 is very long, a correction must be made.

$$\Delta p = \rho \Delta h + \rho_1 h_1 - \rho_2 h_2 \qquad 5.10$$

Hydrostatic pressure is the pressure which a fluid exerts on an object or container walls. Hydrostatic pressure acts through the *center of pressure* (CP) normal to the exposed surface, regardless of the object's orientation, and is a function of depth and density only. It varies linearly with depth.

Buoyant force is predicted by *Archimedes' Principle:* the upward buoyant force of an immersed object is equal to the weight of the displaced fluid. A buoyant force due to displaced air is also relevant in the case of partially submerged objects. For lighter-than-air crafts, the buoyant force results entirely from displaced air.

$$\text{buoyant force} = \text{weight of fluid displaced} \qquad 5.11$$

The buoyant force acts upward through the center of gravity of the displaced volume, a point known as the *center of buoyancy*. The weight acts downward through the center of gravity of the object. For totally submerged objects (e.g., balloons and submarines), the center of buoyancy must be above the center of gravity for stability.

2 BASIC DEFINITIONS FOR FLUID DYNAMICS

In general, for any liquid where Δh is in feet,

$$\Delta p' = \frac{\rho \Delta h}{144} = \frac{\Delta h(\text{SG})}{2.31} \qquad 5.12$$

For water, $SG = 1$, so

$$\Delta p' = \frac{\Delta h}{2.31} \qquad 5.13$$

$$\text{where } 2.31 = \frac{144 \text{ lb/ft}^2\text{-psi}}{62.4 \text{ lb/ft}^3} = \text{ft/psi for water}$$

$$\text{and } SG = \frac{\rho}{62.4}$$

A manometer with two immiscible fluids can measure a pressure differential. If ρ_m is the density of the manometer fluid,

$$\Delta p' = \frac{\rho_m \Delta h_{inches}}{144} \qquad 5.14$$

For an inclined U-tube manometer, where R_0 is the zero reading,

$$\Delta p' = \frac{\rho_m (R - R_0) \sin\theta}{144} \qquad 5.15$$

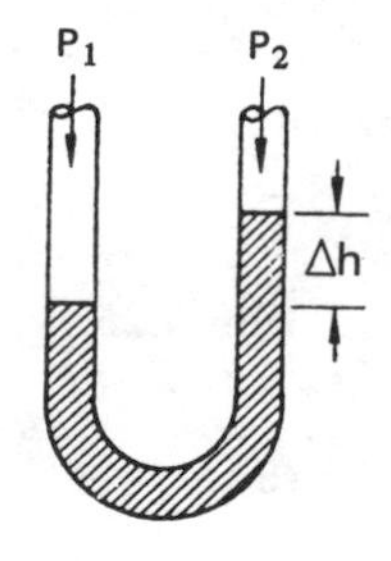

Figure 5.2 U-Tube Manometer

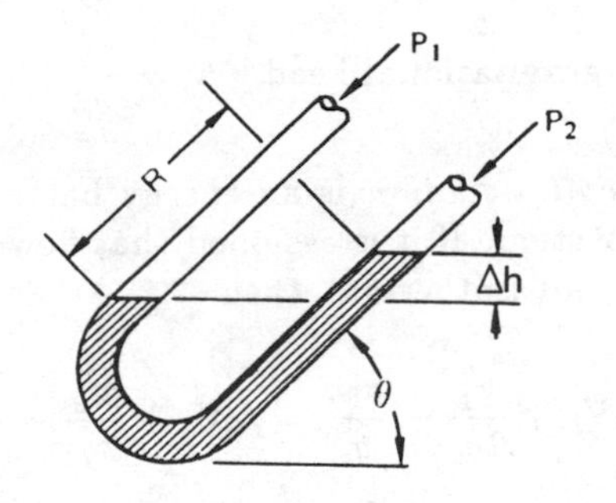

Figure 5.3 Inclined U-Tube Manometer

3 FLUID DYNAMICS

The continuity of flow equation always applies to fluids problems, no matter how complex.

$$M_1 = M_2 \qquad 5.16$$

$$\rho_1 v_1 A_1 = \rho_2 v_2 A_2 \qquad 5.17$$

For incompressible fluids, such as water,

$$\rho_1 = \rho_2 \qquad 5.18$$

$$A_1 v_1 = A_2 v_2 \qquad 5.19$$

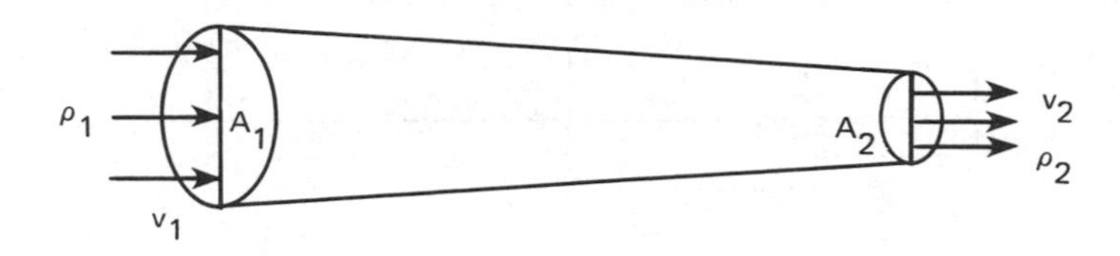

Figure 5.4 Continuity of Flow

There are three ways energy (ft-lbf/lbm) can be stored in a fluid.

$$\text{pressure head} \quad h_p = \frac{p}{\rho} = pV \qquad 5.20$$

$$\text{velocity head} \quad h_v = \frac{v^2}{2g_c} \qquad 5.21$$

$$\text{gravitational head} \quad h_g = \frac{zg}{g_c} \qquad 5.22$$

The *Bernoulli equation* is an energy balance made for a flowing system. If it is assumed that flow is constant, adiabatic, and isothermal, then

$$p_1 V_1 + \frac{v_1^2}{2g_c} + \frac{z_1 g}{g_c} = p_2 V_2 + \frac{v_2^2}{2g_c} + \frac{z_2 g}{g_c} \qquad 5.23$$

For one pound of an incompressible fluid, the specific volume is $V_1 = V_2 = 1/\rho$, and equation 5.23 becomes

$$\frac{p_1}{\rho} + \frac{v_1^2}{2g_c} + \frac{z_1 g}{g_c} = \frac{p_2}{\rho} + \frac{v_2^2}{2g_c} + \frac{z_2 g}{g_c} \qquad 5.24$$

If friction and pump energy are considered, the conservation of energy equation for an incompressible fluid is

$$\frac{p_1}{\rho} + \frac{v_1^2}{2g_c} + \frac{z_1 g}{g_c} + h_{pump} = \frac{p_2}{\rho} + \frac{v_2^2}{2g_c} + \frac{z_2 g}{g_c} + h_f \qquad 5.25$$

Example 5.1

A pump draws a solution with a specific gravity of 1.84 from a storage tank of large cross-section through a 3″ schedule-40 pipe. The velocity in the suction line is 3 feet per second. The pump discharges through a 2″ schedule-40 pipe to an overhead tank. The end of the discharge line is 50′ above the level of the solution in the feed tank. Friction losses in the entire system are 10′ of solution. What pressure, in psi, must the pump develop?

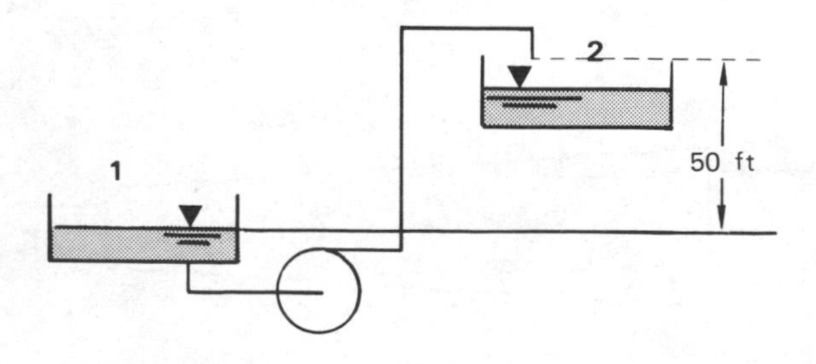

From the pipe schedule appendix, $D_3 = 3.068''$ and $D_2 = 2.067''$.

$$z_1 = 0$$

$$z_2 = 50 \text{ ft}$$

$$v_1 = 0$$

$$v_2 = \frac{(3)(3.068)^2}{(2.067)^2} = 6.61 \text{ ft/sec in 2'' pipe}$$

$$h_f = 10 \text{ ft}$$

$$p_1 = p_2 = \text{atmospheric pressure}$$

$$\Delta p = 0$$

$$\rho_1 = \rho_2 = (1.84)(62.4) = 115 \text{ lbm/ft}^3$$

Substituting in equation 5.25

$$\frac{p}{\rho} + 0 + 0 + h_{pump} = \frac{p}{\rho} + \frac{(6.61)^2}{(2)(32.2)} + 50 + 10$$

$$h_{pump} = 60.68 \text{ ft of solution}$$

$$p_{pump} = \rho h = \frac{(60.68)(115)}{144}$$

$$= 48.38 \text{ psi}$$

Example 5.2

A horizontal ram in a large hydraulic press exerts a net force of 10^6 pounds while moving at a velocity of 4.0 feet per minute. The system used for pumping oil to the ram's cylinder is shown. Neglecting all friction losses, determine the input horsepower to the pump.

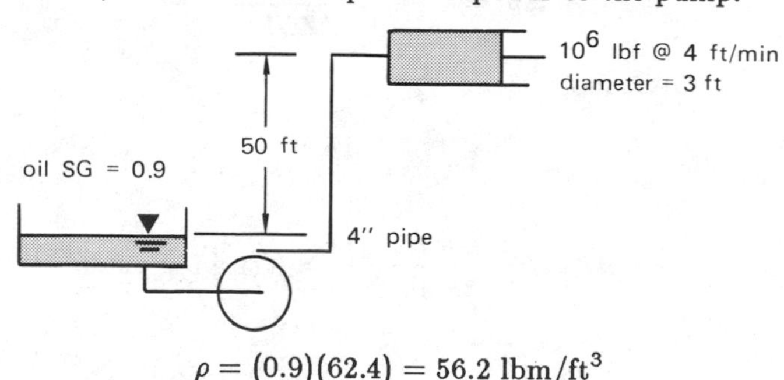

$$\rho = (0.9)(62.4) = 56.2 \text{ lbm/ft}^3$$

Since friction losses are to be neglected, $h_f = 0$, and pump efficiency is 100 percent.

Take locations at (1) oil surface and (2) piston center.

$$z_2 = 50 \text{ ft}$$

$$v_2 = \frac{4}{60} = 0.067 \text{ ft/sec}$$

$$p_2 = \frac{\text{force}}{\text{area}} = \frac{10^6}{\left(\frac{\pi}{4}\right) \times 3^2} = 141{,}540 \text{ psfg}$$

$$z_1 = 0$$

$$v_1 = 0 \text{ (large surface)}$$

$$p_1 = 0 \text{ psfg}$$

Substituting into equation 5.25,

$$h_{pump} = (50 - 0) + \frac{0.067^2}{64.4} + \frac{141{,}540 - 0}{56.2}$$

$$= 50 + 2520 = 2570 \text{ ft-lbf/lbm oil}$$

$$\text{mass of oil flowing} = \text{cfm} \times \rho$$
$$= \left(\frac{\pi}{4}\right)(3)^2(4)(56.2)\ (\text{ft})^2$$
$$= 1588\ \text{lbm/min}$$
$$(\text{ft}^3/\text{min})(\text{lb/ft}^3) = \text{lb/min}$$
$$\text{THP of pump} = \frac{\text{ft-lb/min}}{33{,}000}$$
$$= \frac{(2570)(1588)}{33{,}000}$$
$$= 124\ \text{hp}$$

The ***hydraulic radius*** is defined by equation 5.26. The exposed surface should not be included when calculating the wetted perimeter.

$$r_h = \frac{\text{area flowing}}{\text{wetted perimeter}} \qquad 5.26$$

In full and half-full circular pipes, the hydraulic radius is

$$r_h = \frac{r}{2} = \frac{D}{4} \qquad 5.27$$

In circular pipes filled to $\frac{3}{4}$ of their diameter, the hydraulic radius is

$$r_h = 0.302\,D \qquad 5.28$$

The most efficient cross section will flow the most liquid and will have the smallest wetted perimeter. The circular pipe (closed channel flow) and semi-circular channel (open channel flow) are the most efficient cross sections.

The most efficient trapezoidal open channel is shaped as a half-hexagon with each inclined side making a 30° angle with the horizontal.

With free, unpressurized flow, maximum discharge from a circular cross section occurs at less than full depth, at 0.938 diameters full. A pipe carrying unpressurized fluid will never flow full. Back pressure is usually present, however, and will tend to fill the pipe.

The hydraulic radius can be used to calculate the ***equivalent diameter***.

$$D_e = 4\,r_h \qquad 5.29$$

Osbourne Reynolds showed that whether or not fluid flowing in a pipe is laminar or turbulent, its nature can be determined by a dimensionless number, known as the ***Reynolds number***, which can be thought of as the ratio of the dynamic (inertial) forces of mass flow to the shear stress due to viscosity. The Reynolds number is a dimensionless function of velocity, representing the ratio of inertial to viscous forces.

Table 5.5
Equivalent Diameters for Various Cross Sections

full stream cross-section	D_e
circle	D
annulus	$D_o - D_i$
square	L
rectangle	$\frac{2L_1L_2}{L_1 + L_2}$
partially filled	
rectangle, h deep and L wide	$\frac{4hL}{L + 2h}$
semicircle (half of a circle)	D
wide, shallow stream, h deep	$4h$
triangle, h deep, L broad, s side	$\frac{hL}{s}$
trapezoid, h deep, a top, b bottom, s side	$\frac{2h(a + b)}{b + 2s}$

$$N_{Re} = \frac{D_e \text{v} \rho}{\mu g_c} = \frac{D_e \text{v} m}{\mu}$$
$$= \frac{DG}{\mu g_c} \qquad 5.30$$

It should be noted that equation 5.30 must be dimensionally consistent so that N_{Re} is a dimensionless number.

Laminar and turbulent flow are categorized by the Reynolds number.

$$\text{laminar flow: } N_{Re} < 2100$$
$$\text{transitional region: } 2100 < N_{Re} < 3000$$
$$\text{turbulent flow: } N_{Re} > 3000$$

Usually, flow in the transition range is assumed to be turbulent. A pipe designed for turbulent flow will have excess capacity for laminar flow.

4 FLUID DISTRIBUTION OF VELOCITY

A. GENERAL CASE

In general, the kinetic energy of a flowing fluid can be expressed as

$$KE = \alpha \frac{m\bar{\text{v}}^2}{2} \qquad 5.31$$

α is the ratio of observed kinetic energy to the energy of a uniformly distributed flow. The actual kinetic energy can be found by integration.

FLUIDS

$$KE = \frac{1}{2} \int_{y=0}^{y=d} v^2 \, dM \qquad 5.32$$

M must be expressed in terms of dy and the velocity. If density is assumed constant, then

$$\alpha = \frac{1}{A} \int_0^d \left(\frac{V}{v} \right)^3 dA \qquad 5.33$$

B. LAMINAR FLOW

Due to viscous effects, laminar velocity distributions are parabolic. The actual profile depends on the Reynolds number, but theoretically $\alpha = 2$. Also, the maximum velocity is twice the average velocity found from knowledge of the quantity flowing.

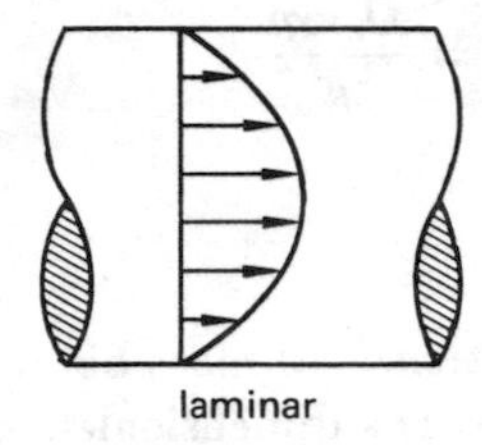

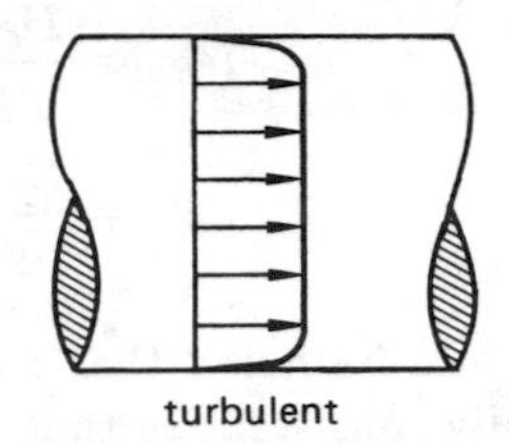

Figure 5.5 Velocity Distributions

C. TURBULENT FLOW

The velocity distribution of fluid in turbulent flow is ellipsoid and approaches a rectangular distribution at high Reynolds numbers. α typically varies between 1.02 and 1.15. Although many empirical formulas exist for the velocity profile, only *Nikuradse's equation* for smooth tubes is given here.

$$v_y = v_c \left(\frac{y}{r_0} \right)^n \qquad 5.34$$

$$v_c = \text{center velocity} \qquad 5.35$$

$$n = \frac{1}{7} \quad N_{Re} \leq 100{,}000$$

$$n = \frac{1}{8} \quad 100{,}000 \leq N_{Re} \leq 400{,}000$$

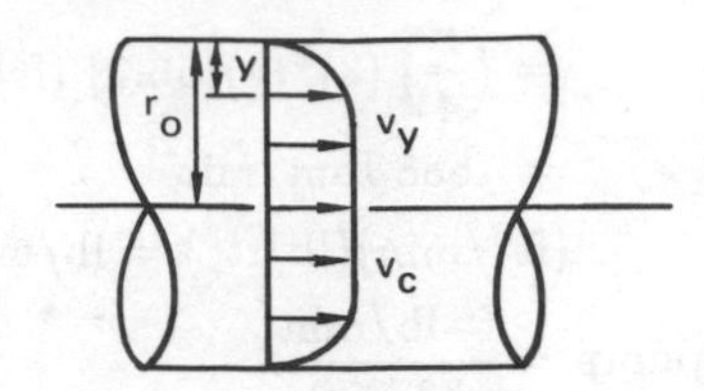

Figure 5.6 Turbulent Flow Velocity Distributions

D. UNIFORM FLOW

Most practical problems assume a uniform distribution ($\alpha = 1$).

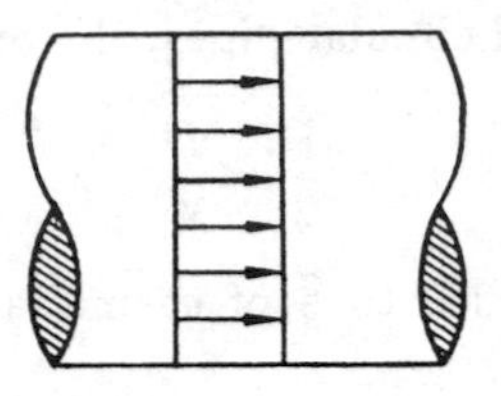

Figure 5.7 Uniform Velocity Distribution

5 FRICTION LOSSES

Fluid flow is always accompanied by friction and an accompanying loss of energy. There is a pressure drop in the direction of flow. If the pressures in a horizontal pipe at two points are measured, higher static pressure would be measured on the upstream point. *Darcy's equation* states that the pressure drop is proportional to the square of the velocity, the length of the pipe, and inversely proportional to the diameter of the pipe. The Darcy equation is

$$h_f = \frac{f L v^2}{2 D g_c} \qquad 5.36$$

The Darcy equation is valid for all regimes of flow, including laminar flow, where extreme velocities cause the downstream pressure to fall below the vapor pressure of the liquid. The equation is for use with straight pipe of constant diameter carrying fluids of reasonably constant properties. The pipes can be horizontal, vertical, or sloping. For inclined pipes, pipes of changing diameter, or vertical pipe, the change in pressure due to elevation changes or fluid properties must be made according to the Bernoulli equation. The head loss due to friction can also be expressed as a pressure drop in psi.

$$\Delta p'_{psi} = \frac{\rho f L v^2}{144 \times 2 D g_c} \qquad 5.37$$

If the flow is laminar ($N_{Re} \leq 2000$), the friction factor is

$$f = \frac{64}{N_{Re}} \qquad 5.38$$

If this quantity is substituted into equation 5.36, the pressure drop is predicted by the *Poiseuille equation* for laminar flow.

$$\Delta p' = 0.000668 \frac{\mu L v}{d^2} \qquad 5.39$$

If the flow is turbulent, the friction factor is a function of the Reynolds number, the *relative roughness* ($\frac{\epsilon}{D}$) (a dimensionless quantity which is the ratio of the absolute roughness of the pipe walls, ϵ, and the pipe diameter, D). This dependence can be seen in figure 5.8, the *Moody Friction Factor Chart.*[1]

[1] Figure 5.8 gives the Moody friction factor. The Fanning friction factor is also used by chemical engineers. The Moody friction factor is four times the Fanning friction factor.

In the case of very smooth pipes, such as glass or drawn brass tubing, the friction factor decreases more rapidly with increasing Reynolds numbers than for pipe with rough walls. Since the roughness of the internal surface of pipe is usually independent of the diameter, the roughness has a greater effect on smaller diameter pipes. As a result, smaller pipes will have much higher friction factors than larger pipes of like material.

The turbulent region of the Moody chart is based on the *Colebrook and White equation.*

$$\frac{1}{\sqrt{f}} = -2 \log \left(\frac{\frac{\epsilon}{D}}{3.7} + \frac{2.51}{N_{Re}\sqrt{f}} \right) \qquad 5.40$$

Equation 5.40 is valid for smooth pipes ($\frac{\epsilon}{D} = 0$) and rough pipes. This equation is difficult to work with since it is implicit in f. For this reason the friction factor chart is used instead of the equation.

For the trial-and-error calculations or calculations of an iterative nature, an explicit equation relating friction factor to Reynolds number and relative roughness would be useful. The Colebrook and White equation relates the friction factor explicitly as a function of the Reynolds number. Equation 5.41 is accurate to $\pm 1\%$ for smooth pipes with turbulent flow.

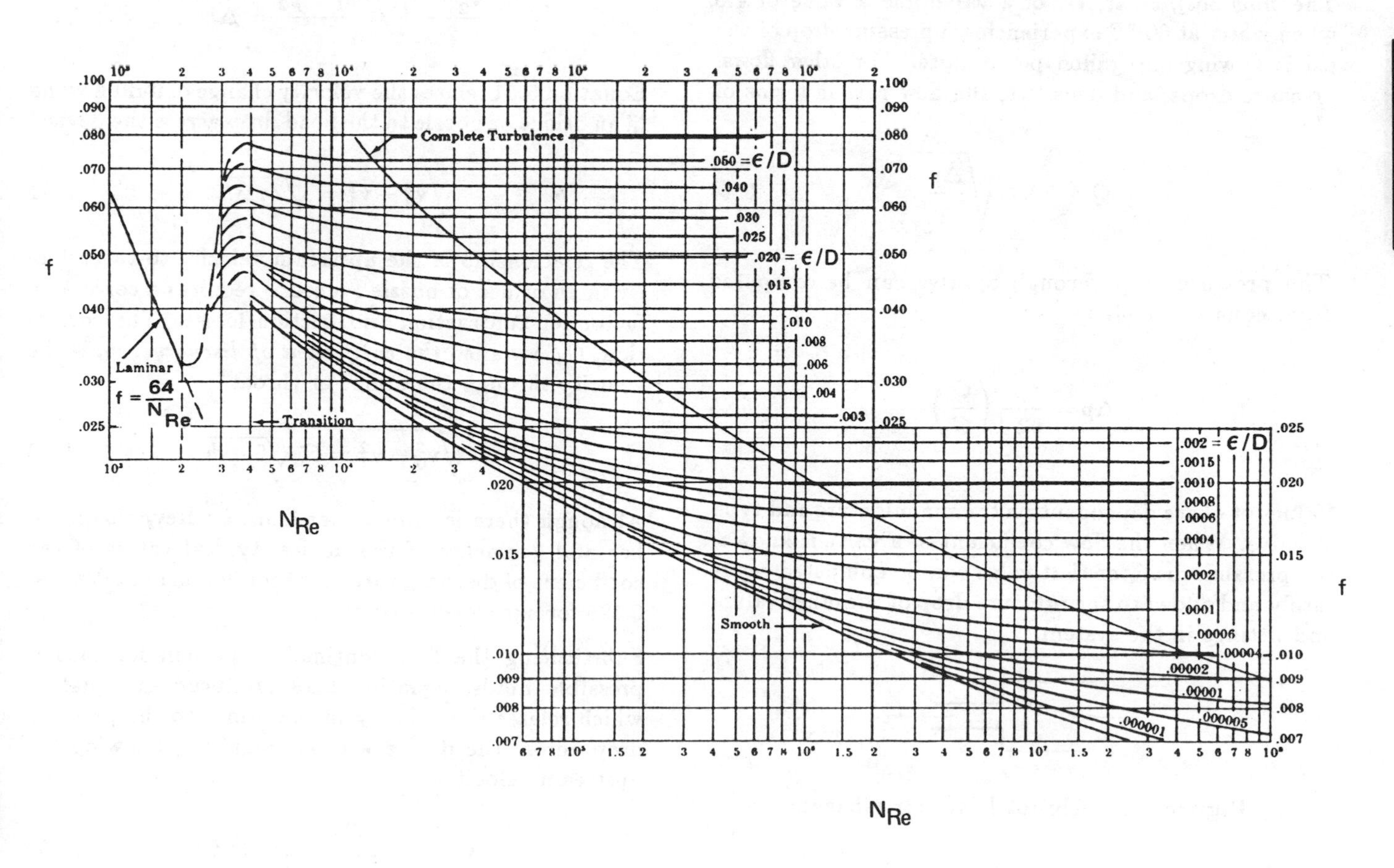

Figure 5.8 Moody Friction Factor Chart

FLUIDS

$$f = \left[1.8 \log \left(\frac{N_{Re}}{7}\right)\right]^{-2} \qquad 5.41$$

The *Sacham equation* is also explicit in f and accurate to $\pm 1\%$.

$$f = \left\{-2 \log \left[\frac{\frac{\epsilon}{D}}{3.7} - \frac{5.02}{N_{Re}} \log \left(\frac{\frac{\epsilon}{D}}{3.7} + \frac{14.5}{N_{Re}}\right)\right]\right\}^{-2} \qquad 5.42$$

Valves and fittings also produce pressure drops in the direction of flow. The head loss through valves and fittings is predicted by equation 5.43.

$$h_f = K\frac{\mathrm{v}^2}{2g_c} \qquad 5.43$$

When h_f, for different sizes of a given valve design, is plotted as a function of the velocity head, $\frac{\mathrm{v}^2}{2g}$, the curve follows the slope of the $f\left(\frac{L}{D}\right)$ curve for straight pipes. Therefore, each valve type can be thought of as having an equivalent length $\left(\frac{L}{D}\right)$ in pipe diameters which will cause the same pressure drop. A list of equivalent lengths of pipe for valves and fittings is presented in table 5.11.

The *flow coefficient*, C, of a valve has a value of 1.0 when water at 60 °F experiencing a pressure drop of one psi is flowing one gallon per minute. For other flows, pressure drops, and densities, the flow rate in terms of C is

$$Q = 7.9\,C\sqrt{\frac{\Delta p}{\rho}} \qquad 5.44$$

The pressure drop through a valve can be computed from equation 5.44:

$$\Delta p = \frac{\rho}{62.4}\left(\frac{Q}{C}\right)^2 \qquad 5.45$$

Values of C are dependent on the manufacture and type of valve. When the flow coefficient of a valve is known, the pressure drop for that valve can be computed separately and added to the pressure drop of the other valves and fittings in the system.

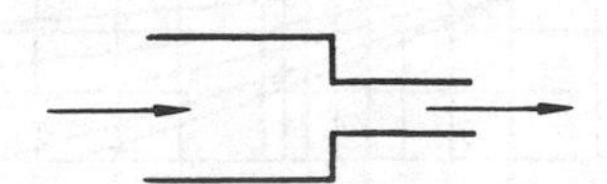

Figure 5.9 Abrupt Diameter Changes

The head loss due to flow through sudden enlargements and contractions follows equation 5.43.

For abrupt diameter changes, as in figure 5.9:

$$K_{entry} = 0.5 \qquad 5.46$$

$$K_{exit} = 1.0 \qquad 5.47$$

A. THE EQUIVALENT DIAMETER

For non-circular cross sections, the equivalent diameter, D_e, can be used in calculating the Reynolds number and friction loss. The equivalent diameter is

$$D_e = \frac{4 \times \text{cross sectional area}}{\text{wetted perimeter}} \qquad 5.48$$

Therefore, for a square channel with side s,

$$D_e = \frac{4s^2}{4s} = s \qquad 5.49$$

B. ORIFICES AND NOZZLES

With the conditions $\Delta z = h_{pump} = h_f = 0$, the Bernoulli equation becomes

$$\frac{\mathrm{v}_2^2 - \mathrm{v}_1^2}{2g_c} = \frac{p_1 - p_2}{\rho} = \Delta h \qquad 5.50$$

Equation 5.51 relates the velocity change of fluid flowing in an orifice or nozzle to the head drop across the device.

$$\sqrt{\mathrm{v}_2^2 - \mathrm{v}_1^2} = \sqrt{2g\,\Delta h} \qquad 5.51$$

The relationship of the upstream velocity to the velocity in an orifice or nozzle (venturi) requires a correction factor for contraction and friction losses. The correction, C_d, is called the *coefficient of discharge*. v_0 is the velocity through the orifice or throat.

$$\sqrt{\mathrm{v}_0^2 - \mathrm{v}_1^2} = C_d\sqrt{2g\Delta h} \qquad 5.52$$

Although there is some dependence on Reynolds number and geometry of the device, typical values of the coefficient of discharge are 0.985 for venturi nozzles and 0.595 for square-edge orifices.

Substituting the flow continuity equation for incompressible fluids, equation 5.19 produces an equation which relates the velocity in the orifice to the pressure drop across the device without requiring knowing the upstream velocity.

$$\mathrm{v}_0 = \frac{C_d}{\sqrt{1 - \left(\frac{d_0}{d_1}\right)^4}}\sqrt{2g\,\Delta h} \qquad 5.53$$

Several terms in equation 5.53 can be combined into the flow coefficient, C_f. If $\frac{d_o}{d_1} \leq 0.2$, then $C_f \cong C_d$.

$$C_f = \frac{C_d}{\sqrt{1-\left(\frac{d_o}{d_1}\right)^4}} \quad 5.54$$

The flow through nozzles and orifices is found by multiplying the velocity by the area and density.

$$q = 0.0438\, d_o^2\, C_f \sqrt{\Delta h}$$

$$= 0.525\, d_o^2\, C_f \sqrt{\frac{\Delta p}{\rho}} \text{ in cfs} \quad 5.55$$

lbf/ft² see p. 5-1

$$Q = 19.65\, d_o^2\, C_f \sqrt{\Delta h}$$

$$= 236\, d_o^2\, C_f \sqrt{\frac{\Delta p}{\rho}} \text{ in gpm} \quad 5.56$$

ditto

$$w = 0.0438\, d_o^2\, C_f \sqrt{\Delta h \rho^2}$$

$$= 0.525\, d_o^2\, C_f \sqrt{\Delta p\, \rho} \quad 5.57$$

$$W = 157.6\, d_o^2\, C_f \sqrt{\Delta h \rho^2}$$

$$= 1891\, d_o^2\, C_f \sqrt{\Delta p\, \rho} \quad 5.58$$

If the problem is to calculate the diameter of the orifice, a trial-and-error technique may be required. Use the following procedure to compute the orifice diameter when the flow, pipe diameter, density, and pressure drop across the orifice are known.

step 1: Assume a ratio $\frac{d_o}{d_1}$.

step 2: Calculate $d_o = d_1\left(\frac{d_o}{d_1}\right)$.

step 3: Calculate C_f using equation 5.54.

step 4: Calculate the flow using equations 5.55 – 5.58.

step 5: Compare the calculated flow with the given flow. If the computed flow is higher (lower) than the given flow, reduce (increase) the previous value of the ratio by the factor

$$\sqrt{\frac{\text{given flow}}{\text{calculated flow}}}$$

step 6: Repeat step 2 as required.

Example 5.3

Find the diameter of a square-edged orifice used to measure a 150 gpm water flow at 60 °F in a 4″, schedule-40 pipe with a pressure differential of 3 psi.

Iteration 1: Assume $\frac{d_o}{d_1} = 0.5$.

$$d_o = (4.026)(0.5) = 2.013 \text{ inches}$$

From equation 5.54,

$$C_f = \frac{0.595}{\sqrt{1-(0.5)^4}} = 0.6145$$

From equation 5.56,

$$\begin{aligned} Q &= (236)(2.013)^2(0.6145)\sqrt{\frac{(3)(144)}{62.34}} \\ &= (621.3)(2.013)^2(0.6145) \\ &= 1547 \text{ gpm} \end{aligned}$$

$1547 > 150$
(too high)

Iteration 2:

$$\frac{d_o}{d_1} = 0.5\sqrt{\frac{150}{1547}} = 0.156$$

$$d_o = (4.026)(0.156) = 0.6268 \text{ inches}$$

$$C_f = \frac{0.595}{\sqrt{1-0.156^4}} = 0.5952$$

$$Q = (621.3)(0.6268)^2\,0.5952 = 145.3 \text{ gpm}$$

$145.3 < 150$
(too low)

Iteration 3:

$$\frac{d_o}{d_1} = 0.156\sqrt{\frac{150}{145.3}} = 0.1585$$

$$d_o = (4.026)(0.1585) = 0.6381 \text{ inches}$$

$$C_f = \frac{0.595}{\sqrt{1-0.1585^4}} = 0.5952$$

$$Q = (621.3)(0.6381)^2\,0.5952 = 150.6 \text{ gpm}$$

(close – try one more time)

Iteration 4:

$$\frac{d_o}{d_1} = 0.1585\sqrt{\frac{150}{150.6}} = 0.1582$$

$$d_o = (4.026)(0.1582) = 0.6368 \text{ inches}$$

$$C_f = 0.5952$$

$$\begin{aligned} Q &= (621.3)(0.6368)^2\,(0.5952) \\ &= 149.95 \text{ gpm} \end{aligned}$$

(close enough)

6 COMPRESSIBLE FLOW THROUGH ORIFICES AND NOZZLES

The flow of compressible fluids through orifices and nozzles can be calculated from the formulas for incompressible flows, except that the *expansion factor*, Y, must also be included. For example, the equations for flow through nozzles and orifices for compressible fluids become

$$q = 0.0438\, d_0^2\, C_f Y \sqrt{\Delta h} = 0.525\, d_0^2\, C_f Y \sqrt{\frac{\Delta p}{\rho}} \qquad 5.59$$

$$Q = 19.65\, d_0^2\, C_f Y \sqrt{\Delta h} = 236\, d_0^2\, C_f Y \sqrt{\frac{\Delta p}{\rho}} \qquad 5.60$$

$$w = 0.0438\, d_0^2\, C_f Y \sqrt{\Delta h \rho^2} = 0.525\, d_0^2\, C_f Y \sqrt{\Delta p\, \rho} \qquad 5.61$$

$$W = 157.6\, d_0^2\, C_f Y \sqrt{\Delta h \rho^2} = 1891\, d_0^2\, C_f Y \sqrt{\Delta p\, \rho} \qquad 5.62$$

The expansion factor is a function of the specific heat ratio, k, the ratio of the orifice to pipe diameters, $\frac{d_0}{d_1}$, and the ratio of downstream to upstream pressures. Plots of the expansion factor as a function of these variables are found in Appendix I. The equations for calculating the expansion factor can be simplified utilizing the quantities $\beta = d_0/d_1$, $\phi = p_2/p_1$, and $k = c_p/c_v$.

For square-edged orifices,

$$Y = \left(1 - \frac{1}{k}\right)(0.41 + 0.35\,\beta^4)(1 - \phi) \qquad 5.63$$

For nozzles,

$$Y = \sqrt{\frac{\left(\frac{k-1}{k}\right)\phi^{\frac{2}{k}}\left(1 - \phi^{\frac{k-1}{k}}\right)(1 - \beta^4)}{(1 - \phi)(1 - \beta^4 \phi^{\frac{2}{k}})}} \qquad 5.64$$

7 STATIC PITOT TUBES

Static pitot tubes measure the velocity head. In a pitot tube, the fluid is brought to rest ($\mathrm{v}_2 = 0$) in a stagnation process. As with equation 5.52, the velocity measured by the pitot tube is

$$\mathrm{v} = C_\mathrm{v} \sqrt{2g\,\Delta h} \qquad 5.65$$

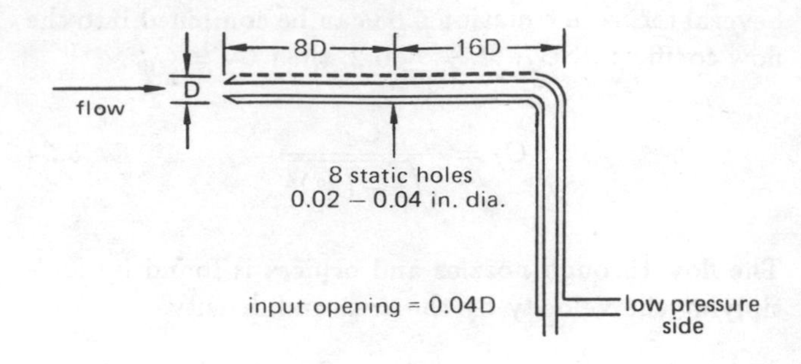

Figure 5.10 Pitot Tube Dimension

For pitot tubes, C_v is the *coefficient of velocity*, which typically has values of 1.00 ± 0.015.

Example 5.4

What is the differential reading on a static pitot tube inserted at the centerline of an 8″ (inside diameter) pipe carrying 1200 gpm of water? The average velocity is 80% of that at the center of the pipe.

$$V_{max} = \left(\frac{1200 \text{ gpm}}{0.8}\right)\left(\frac{1}{60}\ \text{min/sec}\right)\left(\frac{1}{7.48}\ \text{ft}^3/\text{gal}\right)$$
$$= 3.342\ \text{ft}^3/\text{sec}$$
$$\mathrm{v}_{max} = \frac{V}{A} = \frac{3.342}{\left(\frac{\pi}{4}\right)\left(\frac{8}{12}\right)^2}$$
$$= 9.52\ \text{ft/sec}$$
$$C_\mathrm{v} = 1.00$$
$$\Delta h = \frac{\mathrm{v}^2}{2g_c} = \frac{9.52^2}{(2)(32.2)}$$
$$= 1.41\ \text{ft of water}$$

Example 5.5

It is desired to meter the flow of fluid passing through a 6″, schedule-40 pipeline. A sharp-edged orifice gives an 8″ mercury differential at 150 gpm flow. The liquid is at 60 °F, and has a specific gravity of 1.2 and viscosity of 1.1 centistokes. Calculate the size of the orifice.

$$\rho_1 = (1.2)(62.4) = 74.9\ \text{lb/ft}^3$$
$$\mu = (1.1)(1.2) = 1.32\ \text{centipoise}$$
$$D_1 = \frac{6.065}{12} = 0.505\ \text{ft}$$
$$A = 0.2006\ \text{ft}^2$$
$$q = \frac{150}{(60)(7.48)} = 0.334\ \text{ft}^3/\text{sec}$$
$$\mathrm{v}_1 = \frac{q}{A_1} = \frac{0.334}{0.2006} = 1.67\ \text{ft/sec}$$

FLUIDS

$$N_{Re} \text{ in pipeline} = \frac{Dv\rho}{\mu_e} = \frac{(0.505)(1.67)(74.9)}{(1.32)(6.72)(10^{-4})} = 71,210$$

Since N_{Re} in the orifice will be even greater, use 0.61 for the orifice coefficient. There are two ways to solve this.

method 1: The simple approximate method assumes $\frac{D_0}{D_1} < 0.2$.

For a manometer with two fluids:

$$\Delta p = l(\rho_1 - \rho_2) = \Delta h \rho_1$$

$$\text{or } \Delta h = l\frac{(\rho_1 - \rho_2)}{\rho_1} = (8)\frac{(13.6 - 1.2)}{(1.2)(12)} = 6.89 \text{ ft}$$

$$v_0 = C_0\sqrt{2g\Delta h} = 0.61\sqrt{(2)(32.2)(6.89)} = 12.85 \text{ ft/sec orifice only}$$

$$150 \text{ gpm} = 150 \text{ gal/min} \times 1 \text{ min/60 sec} \times 1 \text{ ft}^3/7.48 \text{ gal} = 0.334 \text{ ft}^3/\text{sec}$$

$$\text{area} = \frac{0.334 \text{ ft}^3/\text{sec}}{12.85 \text{ ft/sec}} = 0.0260 \text{ ft}^2$$

$$\text{diameter} = \sqrt{\frac{(4)(0.0260)}{\pi}} = 0.182 \text{ ft}$$

$$(0.182)(12) = 2.18 \text{ inches}$$

method 2: Multiplying equation 5.53 by A_2,

$$w = q_1\rho_1 = C_d A_2\sqrt{\frac{2g(p_1 - p_2)\rho_1}{1 - \beta^4}} = K A_2\sqrt{2g(p_1 - p_2)\rho_1}$$

$$q_1\rho_1 = w = C_0 A_0\sqrt{\frac{2g\rho_1\Delta p}{1 - \beta^4}}$$

$$\text{where } \beta = \frac{D_0}{D_1}$$

$$\text{Then, } \frac{\beta^4}{1 - \beta^4} = \frac{w^2}{(C_0 A_1)^2 2g\rho_1\Delta p}$$

$$w = (150)(8.33)\left(\frac{1.2}{60}\right) = 25 \text{ lb/sec}$$

$$\Delta p = l(\rho_2 - \rho_1) = \left(\frac{8}{12}\right)(13.6 - 1.2)(62.4) = 516 \text{ lb/ft}^2$$

$$\frac{\beta^4}{1 - \beta^4} = \frac{25^2}{[(0.61)(0.02)]^2(64)(1.2)(62.4)(516)} = 0.01698$$

$$\beta^4 = \frac{0.01698}{1.1698} = 0.017$$

$$\beta = \sqrt[4]{0.017} = 0.36$$

$$D_0 = \beta D_1 = 0.36(6.065) = 2.18 \text{ inches}$$

8 COMPOUND PIPE PROBLEMS

A compound pipe consists of two or more parallel pipes connected at each end. The typical problem encountered with this type of network is to find the total flow through the system, the total head loss through the system, or the diameter or length of one of the parallel legs.

With a compound pipe, the head loss across each parallel leg is the same. Using the Darcy form of the head loss equation (equation 5.36), and letting R be the ratio of the flow rates in each parallel leg,

$$R = \frac{Q_3}{Q_2} \qquad 5.66$$

The ratio of the flow rates is

$$R = \sqrt{\left(\frac{f_2}{f_3}\right)\left(\frac{L_2}{L_3}\right)\left(\frac{d_3}{d_2}\right)^5} \qquad 5.67$$

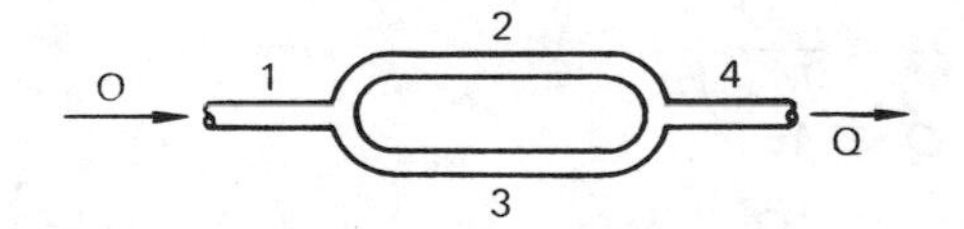

Figure 5.11 Compound Pipe

Equation 5.67 is difficult to use. A trial-and-error technique is required to solve it since the friction factors are functions of the velocities in the pipe. Simplifying assumptions can help solve the problem quickly. There are several pipe loss equations that can be used for the compound pipe problem.

- Darcy

$$h_f = fL\left(\frac{1}{d}\right)\left(\frac{v^2}{2g_c}\right) \qquad 5.68$$

- Lea-Darcy

$$h_f = KL\left(\frac{1}{d^{1.25}}\right)\left(\frac{v^2}{2g_c}\right) \qquad 5.69$$

- Hazen-Williams

$$h_f = CL\left(\frac{1}{d^{1.166}}\right)\left(\frac{v^2}{2g_c}\right)^{0.926} \qquad 5.70$$

FLUIDS

The parallel leg flow ratio can be written with the following simplifying assumptions:

Equation 5.71 is based on the Darcy equation, and assumes $f_2 = f_3$, which is approximately true for similar pipes and turbulent flow.

$$R = \sqrt{\left(\frac{L_2}{L_3}\right)\left(\frac{d_3}{d_2}\right)^5} \quad 5.71$$

Equation 5.72 is based on the Lea-Darcy equation, and assumes $K_2 = K_3$, which is true as long as v_2 and v_3 are not too different.

$$R = \sqrt{\left(\frac{L_2}{L_3}\right)\left(\frac{d_3}{d_2}\right)^{5.25}} \quad 5.72$$

Equation 5.73 is based on the Hazen-Williams equation, and assumes $C_2 = C_3$, which is true for similar pipe.

$$R = \sqrt{\left(\frac{L_2}{L_3}\right)^{1.08}\left(\frac{d_3}{d_2}\right)^{5.26}} \quad 5.73$$

Other useful relationships are

$$Q = Q_2 + Q_3 \quad 5.74$$

$$Q_2 = \frac{Q}{R+1} \quad 5.75$$

$$Q = Av \quad 5.76$$

$$H = \frac{8Q^2}{g\pi^2}\left[\frac{f_1L_1}{d_1^5} + \frac{f_2L_2}{(R+1)^2d_2^5} + \frac{f_4L_4}{d_4^5}\right] \quad 5.77$$

$$H = h_1 + h_2 + h_4 = h_1 + h_3 + h_4 \quad 5.78$$

Type 1 Problem

Given: All lengths, diameters, and H
Find: Q

step 1: Calculate R using equation 5.71, 5.72, or 5.73 depending on which method is used for head loss calculation. If equation 5.71 is used, the final answer should be checked to verify that $f_2 = f_3$.

step 2: Assume a value for v_1.

step 3: Calculate Q using equation 5.76.

step 4: Calculate Q_2 using equation 5.75.

step 5: Calculate v_2 and v_4 or v_3 and v_4 using equation 5.76.

step 6: Calculate f_1, f_2 or f_3 and f_4 by calculating N_{Re} and finding f on the Moody chart (see figure 5.8).

step 7: Calculate H using equation 5.78.

step 8: Use graph paper to plot the value of H versus Q. If H (calculated from step 7) equals the H given, stop. If H is too high, reduce v_1. If the value for H is too low, increase v_1. Go to step 3. If you have bracketed the value for H, draw a line between the two Q points, and use the value for Q found at the intersection for the given value of H. Go to step 4, and calculate H one more time.

Type 2 Problem

Given: All lengths, diameters, and Q
Find: H

step 1: Calculate the velocities v_1 and v_4 from equation 5.76.

step 2: Calculate R using equation 5.71, 5.72, or 5.73, depending on which method is used for the head loss calculation.

step 3: Calculate v_2 or v_3 noting that

$$v_2 = \frac{Q}{A_2(1+R)} \quad 5.79$$

$$v_3 = \frac{QR}{A_3(1+R)} \quad 5.80$$

step 4: Determine the values f_1, f_2 or f_3, and f_4 from the Moody chart (see figure 5.8).

step 5: Calculate H using equation 5.78.

Type 3 Problem

Given: Q, H, all lengths, and diameters of three pipes
Find: diameter of all remaining pipe leg

step 1: Find the head loss h_1 and h_4 using equation 5.68, 5.69, or 5.70, depending on the given head loss calculation method.

step 2: Calculate the head loss in the leg of unknown diameter using equation 5.78.

step 3: Calculate R using either equation 5.71, 5.72, or 5.73.

step 4: Calculate Q_2 and Q_3 using equations 5.74 and 5.75.

FLUIDS

step 5: Calculate the velocity in the leg of unknown diameter using head loss equation 5.68, 5.69, or 5.70. This is a trial-and-error method similar to the method employed in step 8 of problem type 1. Plot head loss versus the assumed velocity in the pipe of known diameter.

When you have bracketed the head loss calculated in step 2, draw a line between the two points and use the value of velocity at the intercept with the head loss calculated in step 2. Note that f must be calculated for each trial.

step 6: Calculate Q in the legs of unknown diameter using equations 5.74 and 5.77.

step 7: Calculate d of the pipe of unknown diameter using another trial-and-error method on the head loss formula. For example, if the leg of unknown diameter is leg 3 and the Darcy pipe loss formula is used, the formula to iterate on is

$$d_3 = \sqrt[5]{\frac{8\, f_3\, L_3\, Q_3^2}{g_c\, \pi^2\, h_3}} \qquad 5.81$$

Plot d versus h until you have bracketed the value of h calculated in step 2. Draw a line between these points and use the value of d found at the intersection of the line and the value of h determined in step 2.

step 8: Calculate the value of h_3 one more time using the value of d_3 in equation 5.68, 5.69, or 5.70 to check the value determined in step 2.

Solving a type 3 problem can be tedious, and should only be attempted using a programmable calculator or a simplifying assumption. You may find, for instance, that the friction factors do not vary, and can be assumed to be constant (true for turbulent flow and similar pipes).

9 SONIC VELOCITY (CRITICAL FLOW)

The speed of sound in a gas is the maximum velocity that a compression wave can attain. Ordinarily, the speed is approximately 1200 feet per second for air. The sonic velocity in an ideal gas is

$$c = \sqrt{\frac{c_p\, gRT}{c_v}} = \sqrt{kg_c RT} \qquad 5.82$$

$$= 41.4\sqrt{\frac{c_p T}{c_v\, SG}} \qquad 5.83$$

$$= 68.1\sqrt{\frac{c_p p}{c_v \rho}} \qquad 5.84$$

All properties (i.e., temperature, pressure, and density) are evaluated at the condition of sonic flow. Sonic flow is achieved if the ratio of exit to upstream pressure is less than the critical pressure ratio (0.5283 for air).

Example 5.6

Steam is flowing in a 3″ diameter pipe. Sonic flow is achieved at the exit where the temperature is 730 °F. What is the maximum velocity that can be expected at the exit?

For steam, the ratio of specific heats is approximately 1.3, and the molecular weight is 18. The specific gas constant can be calculated from the universal gas constant (1545) and the molecular weight. From equation 5.82,

$$c = \sqrt{(1.3)(32.2)\left(\frac{1545}{18}\right)(1190)} = 2065 \text{ ft/sec}$$

10 CENTRIFUGAL PUMPS

Centrifugal pumps are commonly used. They operate by imparting kinetic energy to the fluid flowing into the suction of the pump. The pump captures fluid at the low velocity portion at the impeller center and moves the fluid to the high velocity portion in the casing. The velocity head imparted to the fluid is converted to pressure energy when the fluid is discharged from the pump outlet.

For a centrifugal pump to operate, the pump suction must be flooded. If the section is not flooded, the pump will lose or be unable to maintain its prime and flow will stop.

In addition to the velocity head, h_v, the pressure head, h_p, and the atmospheric head, h_a, previously described, the following head terms apply to centrifugal pumps (refer to figures 5.12 and 5.13).

Static suction head, h_s: the vertical distance *above* the centerline of the suction inlet to the free level of the liquid being pumped.

Static lift head, $-h_s$: the vertical distance *below* the centerline of the suction inlet to the free level of the liquid being pumped.

Dynamic suction head, H_s: the static suction head minus the total suction friction head, $H_s = h_s - h_f$.

FLUIDS

Dynamic lift head, $-H_s$: the static suction lift plus the total suction friction head, $-H_s = -(h_s + h_f)$.

Static discharge head, h_d: the vertical distance above the pump centerline to the free level of the discharge tank or point of free discharge.

Dynamic discharge head, H_d: the static discharge head plus the discharge velocity head plus the discharge friction head, $H_d = h_d + h_f + \frac{v^2}{2g_c}$.

Total static head, h_t: the vertical distance between the free level of the source of supply and the point of free discharge or free level of the discharge tank.

Total dynamic head, H: the total discharge head less the total suction head or plus the total suction lift, $H = H_d - H_s$.

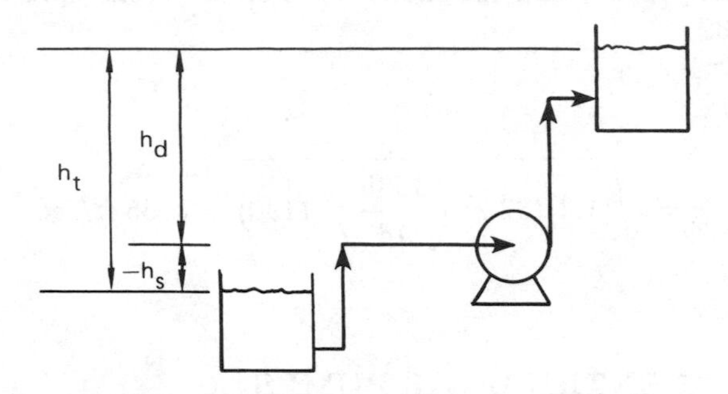

Figure 5.12 Pump Suction Diagram with Negative Lift

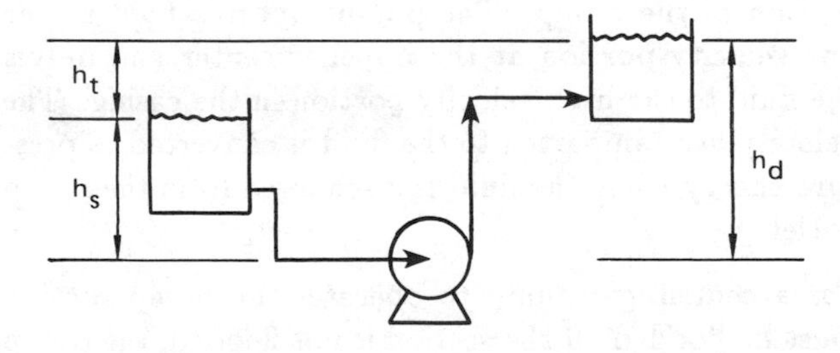

Figure 5.13 Pump Suction Diagram with Positive Lift

A. WORK, POWER, AND EFFICIENCY

The work performed by a pump is the product of the total dynamic head and the weight of the fluid being pumped in a given time.

$$\text{work} = H \times \text{weight (ft-lbf)} \qquad 5.85$$

The pump output power measured as hydraulic (water) horsepower, whp, is

$$\text{whp} = \text{QH}\frac{SG}{3956} \qquad 5.86$$

With other units, the hydraulic horsepower can also be calculated:

$$\text{whp} = \frac{WH}{1{,}980{,}000} = \frac{Q\Delta p'}{1714} = \frac{QH\rho}{247{,}000} \qquad 5.87$$

The power input to a pump having efficiency η_p is

$$\text{bhp} = \frac{\text{whp}}{\eta_p} \qquad 5.88$$

The power input to the motor having efficiency η_m driving the pump is

$$\text{ehp} = \frac{\text{bhp}}{\eta_m} = \frac{\text{whp}}{\eta_p \eta_m} \qquad 5.89$$

To convert the motor input power to kilowatts,

$$\text{kilowatts} = 0.7457 \times \text{ horsepower} \qquad 5.90$$

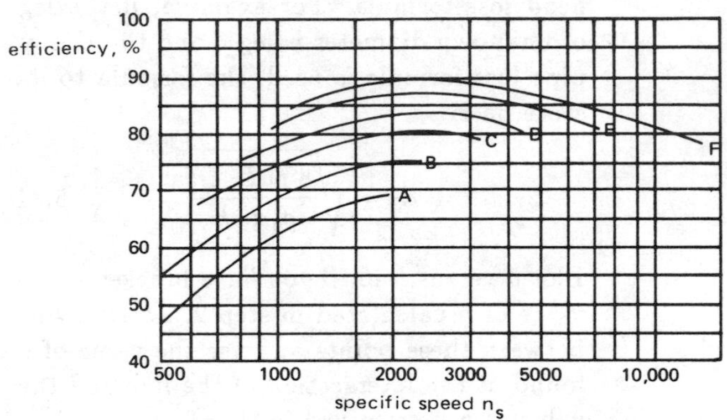

Figure 5.14 Typical Pump Efficiencies

A typical situation would be to find the motor size, given a pump efficiency and motor efficiency. If efficiencies are not known, a motor efficiency of 90 percent should be used. Typical pump efficiencies are plotted in figure 5.14. Standard motor sizes are tabulated in table 5.6, and standard motor speeds for 60 Hertz motors in table 5.7.

Table 5.6
Standard Motor Sizes (bhp)

0.5	7.5	50
0.75	10	60
1	15	75
1.5	20	100
2	25	150
3	30	200
5	40	250

FLUIDS

Table 5.7
Standard Motor Speeds (rpm)
(no load)

3600	1800
1200	900

Example 5.7

What size of motor is required to increase the pressure of 100 gpm of 70 °F water 100 psi? Assume the pump's specific speed is 1800.

From figure 5.14, $\eta_p = 68\%$.

$$\text{Assume } \eta_m = 90\%$$

$$\text{ehp} = \frac{(100)(100)}{(1714)(0.68)(0.9)} = 9.5 \text{ hp}$$

$$\text{ehp} = 10 \text{ hp (next standard size)}$$

B. NET POSITIVE SUCTION HEAD

Net positive suction head available (NPSHA) is the suction head at the pump inlet less the vapor pressure of the liquid. A positive head is normally needed to push the liquid into the pump impeller. NPSHA is the measure (in feet of fluid) of this head.

A high value of NPSHA will ensure that the fluid at the entrance to the pump has enough pressure to overcome the internal friction losses of the pump without the pressure in the pump dropping below the vapor pressure of the liquid.

Each pump has a minimum NPSH requirement (NPSHR). The available head (NPSHA) of the system must always be equal to or greater than the NPSHR. The net positive suction head available, NPSHA, is

$$\text{NPSHA} = h_a + h_s - h_f - h_{vp} \qquad 5.91$$

$$= h_a + \frac{p_i}{\rho} + \frac{v_1^2}{2g_c} - h_{vp} \qquad 5.92$$

In equations 5.91 and 5.92, h_{vp} is the fluid vapor pressure at the pump inlet. h_f is the friction loss (computed from equation 5.36) for the inlet line only.

Table 5.8 serves as a general guide for selecting pumps given the NPSHA.

Table 5.8
General Guide for Selecting Pumps
(Given NPSH)

NPSHA	select pump with NPSH
> 12 ft	at least 4 ft less
10–12 ft	< 8 ft
< 10 ft	< 80% of NPSHA

C. PUMP PERFORMANCE CURVES

The performance of a pump under various conditions can be determined from curves usually supplied by the pump's vendor. There are four basic parameters which define the performance: horsepower, speed, efficiency, and NPSHR. These parameters are plotted against the capacity of the pump, usually in gallons per minute. As the volumetric output increases, the pump cannot provide the total head it can at a lower flow rate.

At low capacities, the overall pump efficiency usually increases with increasing capacity. There are five factors affecting pump efficiency:

- mechanical losses (essentially constant at all capacities)
- impeller losses (essentially constant at all capacities)
- disk friction losses (maximum at low capacity)
- leakage losses (maximum at intermediate capacity)
- casing losses (maximum at high capacity)

The overall result is an efficiency curve that is maximum at intermediate capacity. Typical efficiencies for hydraulic pumps are shown in figure 5.14.

The performance curve shown in figure 5.15 assumes that the impeller diameter and rotational speed are constant.

Homologous pumps are pumps of different sizes having the same basic design. The size of a pump is typically established by specifying the inlet (suction) and outlet (discharge) diameters. A pump designated as a 3 × 2 has a three-inch suction and a two-inch discharge.

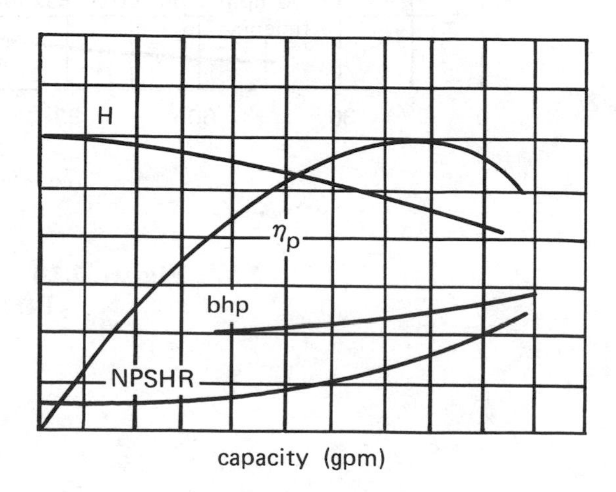

Figure 5.15 Typical Pump Performance Curve

FLUIDS

Figure 5.16 shows typical pump performance curves for homologous pumps.

The curves shown in figure 5.16 cover only a limited range of impeller speeds. Often, it is desired to predict the performance of a pump outside the range of speeds in the performance chart. The affinity laws for homologous pumps assume the impeller diameter is constant and no change in efficiency.

$$\frac{Q_2}{Q_1} = \frac{n_2}{n_1} \quad 5.93$$

$$\frac{H_2}{H_1} = \left(\frac{n_2}{n_1}\right)^2 = \left(\frac{Q_2}{Q_1}\right)^2 \quad 5.94$$

$$\frac{\text{bhp}_2}{\text{bhp}_1} = \left(\frac{n_2}{n_1}\right)^3 = \left(\frac{Q_2}{Q_1}\right)^3 \quad 5.95$$

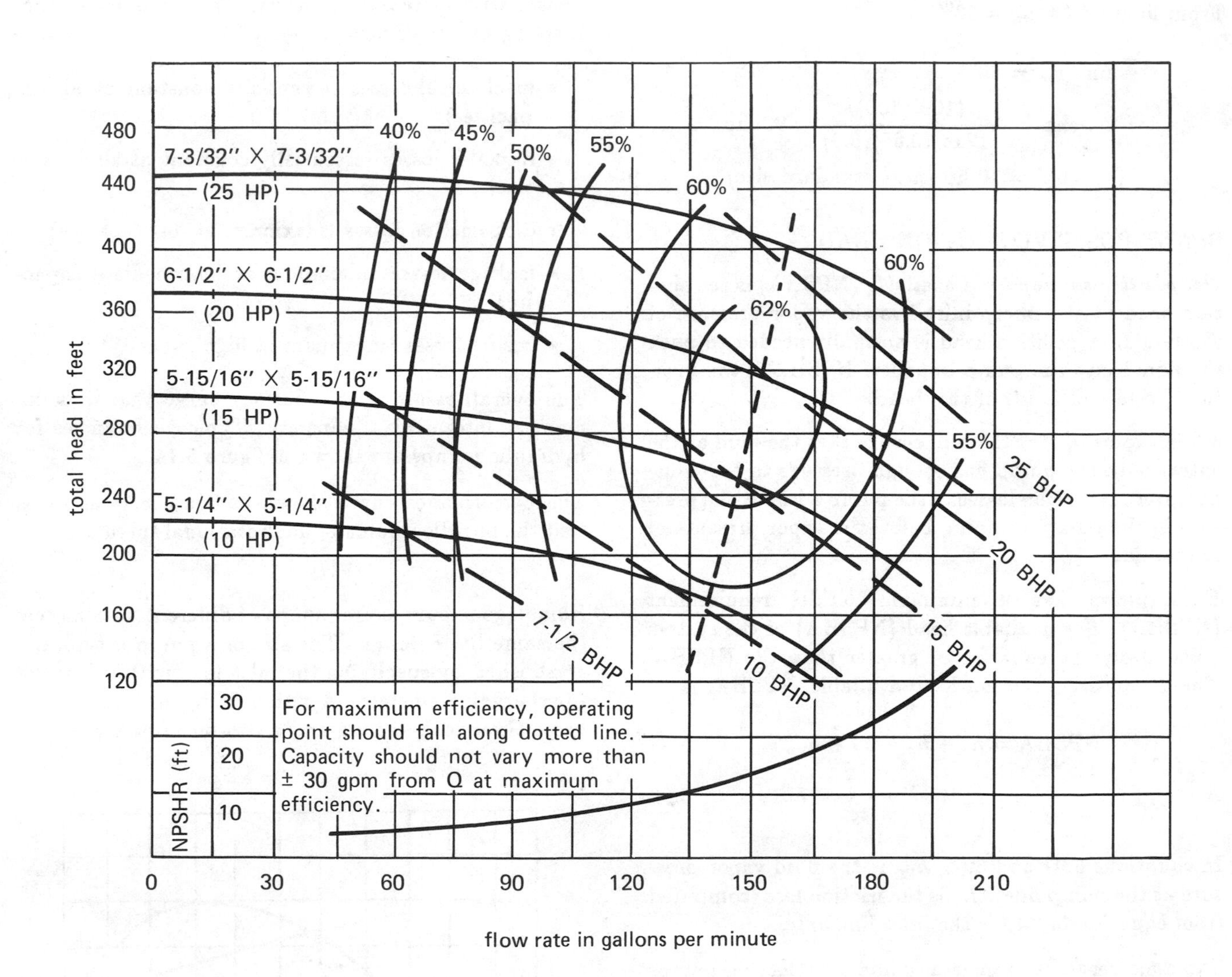

Figure 5.16 Typical Homologous Pump Performance Curves

When the speed is held constant and the impeller size varied, the affinity laws are

$$\frac{Q_2}{Q_1} = \frac{d_2}{d_1} \quad 5.96$$

$$\frac{H_2}{H_1} = \left(\frac{d_2}{d_1}\right)^2 \quad 5.97$$

$$\frac{\text{bhp}_2}{\text{bhp}_1} = \left(\frac{d_2}{d_1}\right)^3 \quad 5.98$$

Example 5.8

A 1770 rpm pump moves at 500 gpm against 200′ total head. How fast must this pump operate to move the same amount of fluid against a total head of 375′?

From equation 5.94,

$$n_2 = n_1\sqrt{\frac{H_2}{H_1}}$$

$$n_2 = 1770\sqrt{\frac{375}{200}}$$

$$= 2424 \text{ rpm}$$

A curve of the system friction losses (from piping, valves, etc.) can be developed. When plotted against the performance curve of the pump, the operating point can be established. The static head plus friction losses of a system can be represented as a system curve. The operating point of the pump is where the system curve intersects the performance curve. This is shown in figure 5.17.

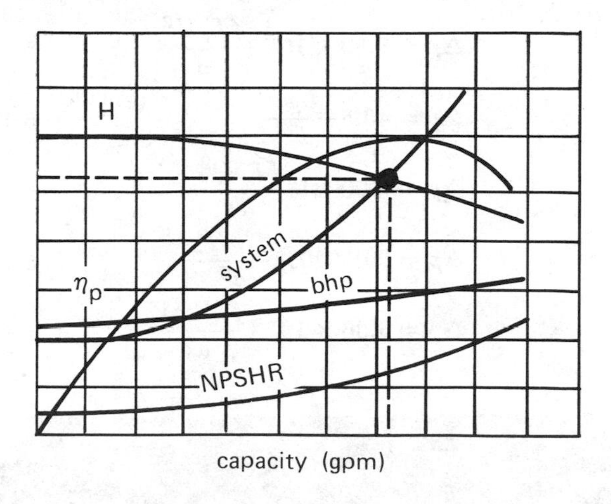

Figure 5.17 Pump Performance and System Curves

A common pump problem involves selecting the best pump for a particular application. Pump performance curves for one or more pumps must be available. The piping system must be detailed and the required flow rate given.

To select a pump, the following steps should be taken:

step 1: Calculate the flow velocity from the pipe dimensions and the given flow.

step 2: Calculate the friction loss in the piping system.

step 3: Use the Bernoulli equation to solve for the required total head, h_{pump}.

step 4: Use the performance curves to eliminate pumps that cannot supply the necessary total head. Select the most efficient pump from those that can supply the required total head.

step 5: Determine the horsepower requirements from the performance curve or by using equations 5.86 and 5.87.

Pumps can be operated in series (the discharge of the first pump feeding the inlet of the second pump), or in parallel (two pumps having a common discharge header). Parallel operation increases the capacity of the system while operating against the same total head. Series operation increases the total head with only a slight increase in capacity. The efficiency of two pumps operated in combination is

$$\eta_{p,parallel} = \frac{(Q_a + Q_b)H}{3956\,(bhp_a + bhp_b)} \quad 5.99$$

$$\eta_{p,series} = \frac{Q(H_a + H_b)}{3956\,(bhp_a + bhp_b)} \quad 5.100$$

D. CAVITATION

If the pressure in the fluid drops below its vapor pressure, the fluid will vaporize. This vaporization and the subsequent bubble implosion at the impeller face is known as *cavitation.* Cavitation usually occurs at pump inlets. Cavitation is destructive to impellers and pump housings, and should be avoided. It usually manifests itself in vibration, noise, impeller pitting, and a drop in pumping efficiency.

Cavitation will not occur as long as the net positive suction head available (NPSHA) is above the NPSHR. A good safety factor is two to three feet of liquid.

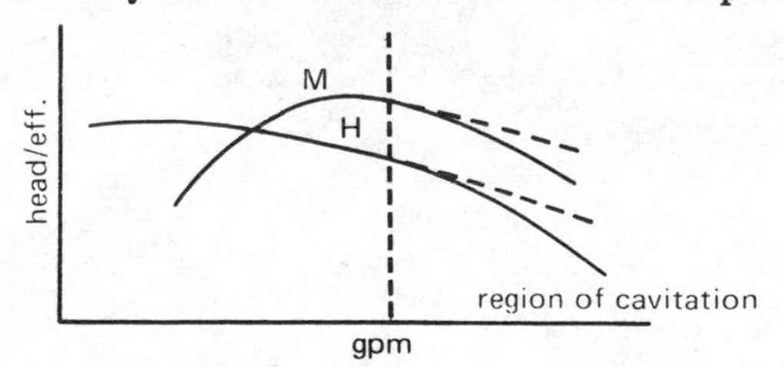

Figure 5.18 Region of Cavitation

FLUIDS

11 MIXED UNIT EQUATIONS

Experienced engineers know that the conversion of variables to proper units is a time-consuming task in fluids problem solving. Mixed unit equations eliminate the need to convert each variable to the proper units in order to obtain the proper units in the result. All conversion factors for proper solution are combined into one factor.

The Reynolds number can be calculated in a variety of ways depending on the units of available data.

$$N_{Re} = \frac{Dv\rho}{\mu_e} \tag{5.101}$$

$$N_{Re} = 123.9\frac{dv\rho}{\mu} \tag{5.102}$$

$$N_{Re} = 22{,}700\frac{q\rho}{d\mu} \tag{5.103}$$

$$N_{Re} = 50.6\frac{Q\rho}{d\mu} \tag{5.104}$$

$$N_{Re} = 35.4\frac{B\rho}{d\mu} \tag{5.105}$$

$$N_{Re} = 6.31\frac{W}{d\mu} \tag{5.106}$$

$$N_{Re} = 0.482\frac{q'S_g}{d\mu} \tag{5.107}$$

$$N_{Re} = \frac{Dv}{\nu'} \tag{5.108}$$

$$N_{Re} = \frac{dv}{12\nu'} \tag{5.109}$$

$$N_{Re} = 7740\,\frac{dv}{\nu} \tag{5.110}$$

$$N_{Re} = 1{,}419{,}000\,\frac{q}{\nu d} \tag{5.111}$$

$$N_{Re} = 3160\,\frac{Q}{\nu d} \tag{5.112}$$

Similarly, useful velocity equivalents are:

$$v = \frac{q}{A} \tag{5.113}$$

$$v = 183.3\frac{q}{d^2} \tag{5.114}$$

$$v = 0.408\frac{Q}{d^2} \tag{5.115}$$

$$v = 0.286\frac{B}{d^2} \tag{5.116}$$

$$v = 183.3\frac{wV}{d^2} \tag{5.117}$$

$$v = 0.0509\frac{WV}{d^2} \tag{5.118}$$

$$v = 0.00389\frac{q'S_g}{\rho d^2} \tag{5.119}$$

Useful equivalents for head loss due to friction are:

$$h_f = 0.1863\frac{fLv^2}{d} \tag{5.120}$$

$$h_f = 6260\frac{fLq^2}{d^5} \tag{5.121}$$

$$h_f = 0.0311\frac{fLQ^2}{d^5} \tag{5.122}$$

$$h_f = 0.01524\frac{fLB^2}{d^5} \tag{5.123}$$

$$h_f = 0.000483\frac{flW^2V^2}{d^5} \tag{5.124}$$

Finally, the following pressure drop equivalents are handy:

$$\Delta p = 0.001294\frac{fL\rho v^2}{d} \tag{5.125}$$

$$\Delta p = 3.59 \times 10^{-7}\frac{fL\rho V^2}{d} \tag{5.126}$$

$$\Delta p = 43.5\frac{fL\rho q^2}{d^5} \tag{5.127}$$

$$\Delta p = 0.000216\frac{fL\rho Q^2}{d^5} \tag{5.128}$$

$$\Delta p = 0.0001058\frac{fl\rho B^2}{d^5} \tag{5.129}$$

$$\Delta p = 3.36 \times 10^{-6}\frac{fLW^2V}{d^5} \tag{5.130}$$

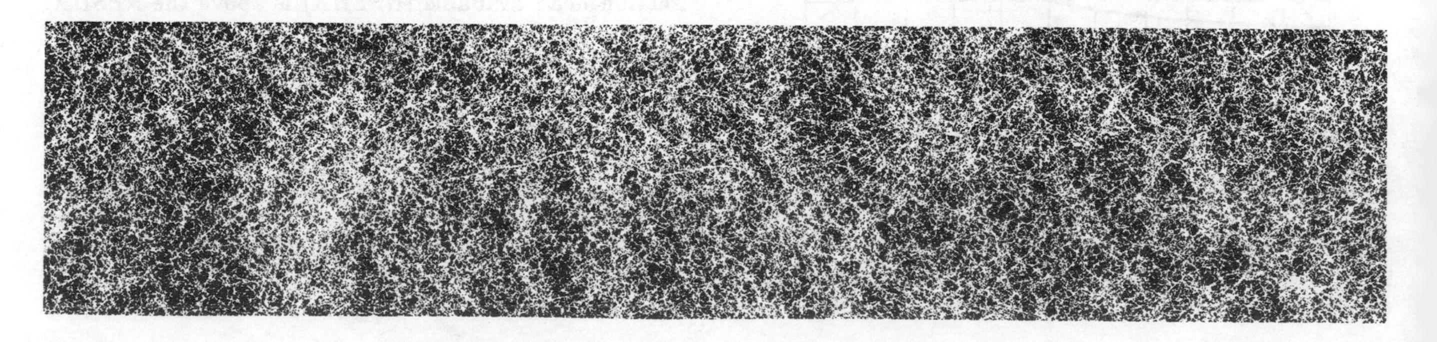

12 FLOW THROUGH PACKED BEDS

Nomenclature

A_p	area of particle	ft^2
D_p	diameter of particle	ft
g_c	gravitational conversion factor (32.2)	$lbm\text{-}ft/lbf\text{-}sec^2$
k_2	constant	–
k_4	constant	–
L	length of bed	ft
N_{Re}	Reynolds number	–
Δp	pressure drop across bed	lbf/ft^2
v_s	superficial velocity	ft/sec
V_p	volume of particle	ft^3

Symbols

ε	porosity of bed	–
ρ	density of fluid	lbm/ft^3
μ	viscosity of fluid	lbm/ft-sec
ψ	sphericity of particle	–

Calculation of the pressure drop of gases or liquids flowing through a particulate-solid bed is required for applications such as filtration, pebble heaters, absorption towers, and solid catalyst reactors. For laminar flow through a packed bed (as defined by a modified Reynold's number based on superficial fluid velocity, particle diameter, fluid viscosity, and fluid density), the pressure drop through a bed of length L, having particulates with diameter D_p and bed porosity ε, is defined by the *Carman-Kozeny equation*:

$$\Delta p_{laminar} = k_2 \frac{(1-\varepsilon)^2 \mu v_s L}{\varepsilon^3 D_p^2 g_c} \qquad 5.131$$

The value of k_2 equals 150 for laminar flow only. In the turbulent region, the pressure drop through a bed is given by the *Burke-Plummer equation*:

$$\Delta p_{turbulent} = k_4 \frac{(1-\varepsilon)\rho v_s^2 L}{\varepsilon^3 D_p g_c} \qquad 5.132$$

By combining both the laminar and turbulent data, the pressure drop and flow conditions at *any* flow region are given by the *Ergun equation*:

$$\Delta p = \frac{(1-\varepsilon)\rho v_s^2 L}{\varepsilon^3 D_p g_c}\left(150\,\frac{1-\varepsilon}{N_{Re}^*} + 1.75\right) \qquad 5.133$$

For irregular particles of known area, A_p, and volume, V_p, the ***characteristic particle diameter*** given in equations 5.131, 5.132, and 5.133 is

$$D_p = \frac{6V_p}{A_p} \qquad 5.134$$

The ***superficial velocity***, v_s, is the velocity that would occur in the container holding the bed if the container was empty. If the flowing fluid is compressible, the superficial velocity is the average bed-free velocity at a density averaged between the inlet and outlet conditions. The modified Reynolds' number is calculated from equation 5.135.

$$N_{Re}^* = \frac{\rho D_p v_s}{\mu} \qquad 5.135$$

Porosity, ε, is a function of particle size and distribution, particle shape, surface roughness, the packing method, and the size of the container relative to the particle diameter. Particle shape is the most important variable in determining porosity. The most common shape factor used is the ***sphericity***, ψ, and is defined as the ratio of the surface area of a sphere of volume equal to that of the particle to the surface area of the particle. Therefore, sphericity is calculated from the equation

$$\psi = \frac{\pi\left(\frac{6V_p}{\pi}\right)^{2/3}}{A_p} \qquad 5.136$$

If the ratio of the particle diameter to the vessel diameter is less than 0.05, the relationship between sphericity and porosity can be calculated using the empirical relationship

$$\varepsilon = 1.08 - 1.12\psi + 0.405\psi^2 \qquad 5.137$$

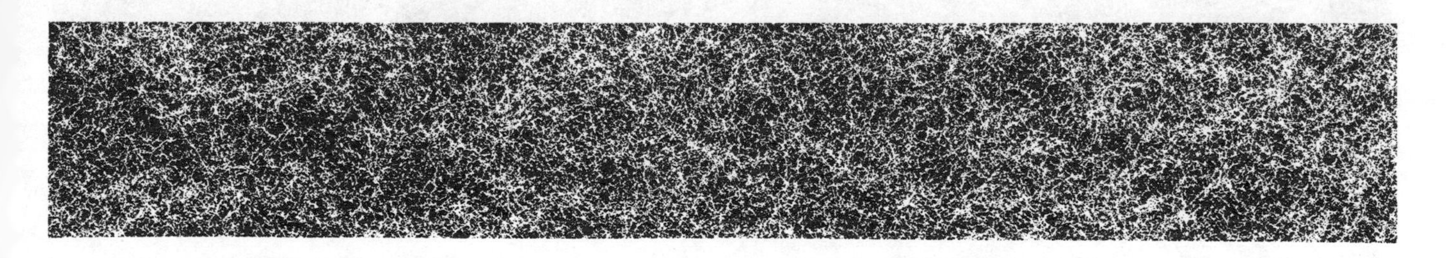

FLUIDS

Table 5.9
Properties of Water at Atmospheric Pressure

temp. °F	density lbm/ft^3	viscosity lbf-sec/ft^2	kinematic viscosity ft^2/sec	surface tension lbf/ft	vapor pressure head ft	bulk modulus lbf/in^2
32	62.42	3.746 EE−5	1.931 EE−5	0.518 EE−5	0.20	293 EE3
40	62.43	3.229 EE−5	1.664 EE−5	0.514 EE−2	0.28	294 EE3
50	62.41	2.735 EE−5	1.410 EE−5	0.509 EE−2	0.41	305 EE3
60	62.37	2.359 EE−5	1.217 EE−5	0.504 EE−2	0.59	311 EE3
70	62.30	2.050 EE−5	1.059 EE−5	0.500 EE−2	0.84	320 EE3
80	62.22	1.799 EE−5	0.930 EE−5	0.492 EE−2	1.17	322 EE3
90	62.11	1.595 EE−5	0.826 EE−5	0.486 EE−2	1.61	323 EE3
100	62.00	1.424 EE−5	0.739 EE−5	0.480 EE−2	2.19	327 EE2
110	61.86	1.284 EE−5	0.667 EE−5	0.473 EE−2	2.95	331 EE3
120	61.71	1.168 EE−5	0.609 EE−5	0.465 EE−2	3.91	333 EE3
130	61.55	1.069 EE−5	0.558 EE−5	0.460 EE−2	5.13	334 EE3
140	61.38	0.981 EE−5	0.514 EE−5	0.454 EE−2	6.67	330 EE3
150	61.20	0.905 EE−5	0.476 EE−5	0.447 EE−2	8.58	328 EE3
160	61.00	0.838 EE−5	0.442 EE−5	0.441 EE−2	10.95	326 EE3
170	60.80	0.780 EE−5	0.413 EE−5	0.433 EE−2	13.83	322 EE3
180	60.58	0.726 EE−5	0.385 EE−5	0.426 EE−2	17.33	313 EE3
190	60.36	0.678 EE−5	0.362 EE−5	0.419 EE−2	21.55	313 EE3
200	60.12	0.637 EE−5	0.341 EE−5	0.412 EE−2	26.59	308 EE3
212	59.83	0.593 EE−5	0.319 EE−5	0.404 EE−2	33.90	300 EE3

Table 5.10
Properties of Air at Atmospheric Pressure

temp. °F	density lbm/ft^3	kinematic viscosity ft^2/sec	dynamic viscosity lbf-sec/ft^2
0	0.0862	12.6 EE−5	3.28 EE−7
20	0.0827	13.6 EE−5	3.50 EE−7
40	0.0794	14.6 EE−5	3.62 EE−7
60	0.0763	15.8 EE−5	3.74 EE−7
68	0.0752	16.0 EE−5	3.75 EE−7
80	0.0735	16.9 EE−5	3.85 EE−7
100	0.0709	18.0 EE−5	3.96 EE−7
120	0.0684	18.9 EE−5	4.07 EE−7
250	0.0559	27.3 EE−5	4.74 EE−7

Table 5.11
Equivalent Length of Straight Pipe for Various Fittings

(turbulent flow only, for any fluid)

fittings			pipe size																				
			1/4	3/8	1/2	3/4	1	1 1/4	1 1/2	2	2 1/2	3	4	5	6	8	10	12	14	16	18	20	24
regular 90°ell	screwed	steel	2.3	3.1	3.6	4.4	5.2	6.6	7.4	8.5	9.3	11.0	13.0										
		c.i.										9.0	11.0										
	flanged	steel			0.92	1.2	1.6	2.1	2.4	3.1	3.6	4.4	5.9	7.3	8.9	12.0	14.0	17.0	18.0	21.0	23.0	25.0	30.0
		c.i.										3.6	4.8		7.2	9.8	12.0	15.0	17.0	19.0	22.0	24.0	28.0
long radius 90°ell	screwed	steel	1.5	2.0	2.2	2.3	2.7	3.2	3.4	3.6	3.6	4.0	4.6										
		c.i.										3.3	3.7										
	flanged	steel			1.1	1.3	1.6	2.0	2.3	2.7	2.9	3.4	4.2	5.0	5.7	7.0	8.0	9.0	9.4	10.0	11.0	12.0	14.0
		c.i.										2.8	3.4		4.7	5.7	6.8	7.8	8.6	9.6	11.0	11.0	13.0
regular 45°ell	screwed	steel	0.34	0.52	0.71	0.92	1.3	1.7	2.1	2.7	3.2	4.0	5.5										
		c.i.										3.3	4.5										
	flanged	steel			0.45	0.59	0.81	1.1	1.3	1.7	2.0	2.6	3.5	4.5	5.6	7.7	9.0	11.0	13.0	15.0	16.0	18.0	22.0
		c.i.										2.1	2.9		4.5	6.3	8.1	9.7	12.0	13.0	15.0	17.0	20.0
tee-line flow	screwed	steel	0.79	1.2	1.7	2.4	3.2	4.6	5.6	7.7	9.3	12.0	17.0										
		c.i.										9.9	14.0										
	flanged	steel			0.69	0.82	1.0	1.3	1.5	1.8	1.9	2.2	2.8	3.3	3.8	4.7	5.2	6.0	6.4	7.2	7.6	8.2	9.6
		c.i.										1.9	2.2		3.1	3.9	4.6	5.2	5.9	6.5	7.2	7.7	8.8
tee-branch flow	screwed	steel	2.4	3.5	4.2	5.3	6.6	8.7	9.9	12.0	13.0	17.0	21.0										
		c.i.										14.0	17.0										
	flanged	steel			2.0	2.6	3.3	4.4	5.2	6.6	7.5	9.4	12.0	15.0	18.0	24.0	30.0	34.0	37.0	43.0	47.0	52.0	62.0
		c.i.										7.7	10.0		15.0	20.0	25.0	30.0	35.0	39.0	44.0	49.0	57.0
180° return bend	screwed	steel	2.3	3.1	3.6	4.4	5.2	6.6	7.4	8.5	9.3	11.0	13.0										
		c.i.										9.0	11.0										
	reg. flanged	steel			0.92	1.2	1.6	2.1	2.4	3.1	3.6	4.4	5.9	7.3	8.9	12.0	14.0	17.0	18.0	21.0	23.0	25.0	30.0
		c.i.										3.6	4.8		7.2	9.8	12.0	15.0	17.0	19.0	22.0	24.0	28.0
	long rad. flanged	steel			1.1	1.3	1.6	2.0	2.3	2.7	2.9	3.4	4.2	5.0	5.7	7.0	8.0	9.0	9.4	10.0	11.0	12.0	14.0
		c.i.										2.8	3.4		4.7	5.7	6.8	7.8	8.6	9.6	11.0	11.0	13.0
globe valve	screwed	steel	21.0	22.0	22.0	24.0	29.0	37.0	42.0	54.0	62.0	79.0	110.0										
		c.i.										65.0	86.0										
	flanged	steel			38.0	40.0	45.0	54.0	59.0	70.0	77.0	94.0	120.0	150.0	190.0	260.0	310.0	390.0					
		c.i.										77.0	99.0		150.0	210.0	270.0	330.0					
gate valve	screwed	steel	0.32	0.45	0.56	0.67	0.84	1.1	1.2	1.5	1.7	1.9	2.5										
		c.i.										1.6	2.0										
	flanged	steel								2.6	2.7	2.8	2.9	3.1	3.2	3.2	3.2	3.2	3.2	3.2	3.2	3.2	3.2
		c.i.										2.3	2.4		2.6	2.7	2.8	2.9	2.9	3.0	3.0	3.0	3.0
angle valve	screwed	steel	12.8	15.0	15.0	15.0	17.0	18.0	18.0	18.0	18.0	18.0	18.0										
		c.i.										15.0	15.0										
	flanged	steel			15.0	15.0	17.0	18.0	18.0	21.0	22.0	28.0	38.0	50.0	63.0	90.0	120.0	140.0	160.0	190.0	210.0	240.0	300.0
		c.i.										23.0	31.0		52.0	74.0	98.0	120.0	150.0	170.0	200.0	230.0	280.0
swing check valve	screwed	steel	7.2	7.3	8.0	8.8	11.0	13.0	15.0	19.0	22.0	27.0	38.0										
		c.i.										22.0	31.0										
	flanged	steel			3.8	5.3	7.2	10.0	12.0	17.0	21.0	27.0	38.0	50.0	63.0	90.0	120.0	140.0					
		c.i.										22.0	31.0		52.0	74.0	98.0	120.0					
coupling or union	screwed	steel	0.14	0.18	0.21	0.24	0.29	0.36	0.39	0.45	0.47	0.53	0.65										
		c.i.										0.44	0.52										
	bell mouth inlet	steel	0.04	0.07	0.10	0.13	0.18	0.26	0.31	0.43	0.52	0.67	0.95	1.3	1.6	2.3	2.9	3.5	4.0	4.7	5.3	6.1	7.6
		c.i.										0.55	0.77		1.3	1.9	2.4	3.0	3.6	4.3	5.0	5.7	7.0
	square mouth inlet	steel	0.44	0.68	0.96	1.3	1.8	2.6	3.1	4.3	5.2	6.7	9.5	13.0	16.0	23.0	29.0	35.0	40.0	47.0	53.0	61.0	76.0
		c.i.										5.5	7.7		13.0	19.0	24.0	30.0	36.0	43.0	50.0	57.0	70.0
	re-entrant pipe	steel	0.88	1.4	1.9	2.6	3.6	5.1	6.2	8.5	10.0	13.0	19.0	25.0	32.0	45.0	58.0	70.0	80.0	95.0	110.0	120.0	150.0
		c.i.										11.0	15.0		26.0	37.0	49.0	61.0	73.0	86.0	100.0	110.0	140.0

Table 5.12
Atmospheric Pressure vs. Altitude

Altitude (above sea level)	Atmospheric Pressure feet of water	psia
0	33.9	14.7
1000	32.8	14.2
2000	31.6	13.7
3000	30.5	13.2
4000	29.4	12.7
5000	28.3	12.3
6000	27.3	11.8
7000	26.2	11.3
8000	25.2	10.9
9000	24.3	10.5
10,000	23.4	10.1

Table 5.13
Vapor Pressure of Water (Absolute)

Temperature °C	°F	Vapor Pressure feet of water	psia
0	32	0.20	0.0886
5	41	0.29	0.126
10	50	0.40	0.173
15	59	0.56	0.245
20	68	0.78	0.339
40	104	2.47	1.07
60	140	6.68	2.89
80	176	15.87	6.87
100	212	33.96	14.7
120	248	66.53	28.8
140	284	121.04	52.4
160	320	206.98	89.6
180	356	334.95	145
200	392	519.75	225
220	428	773.85	335

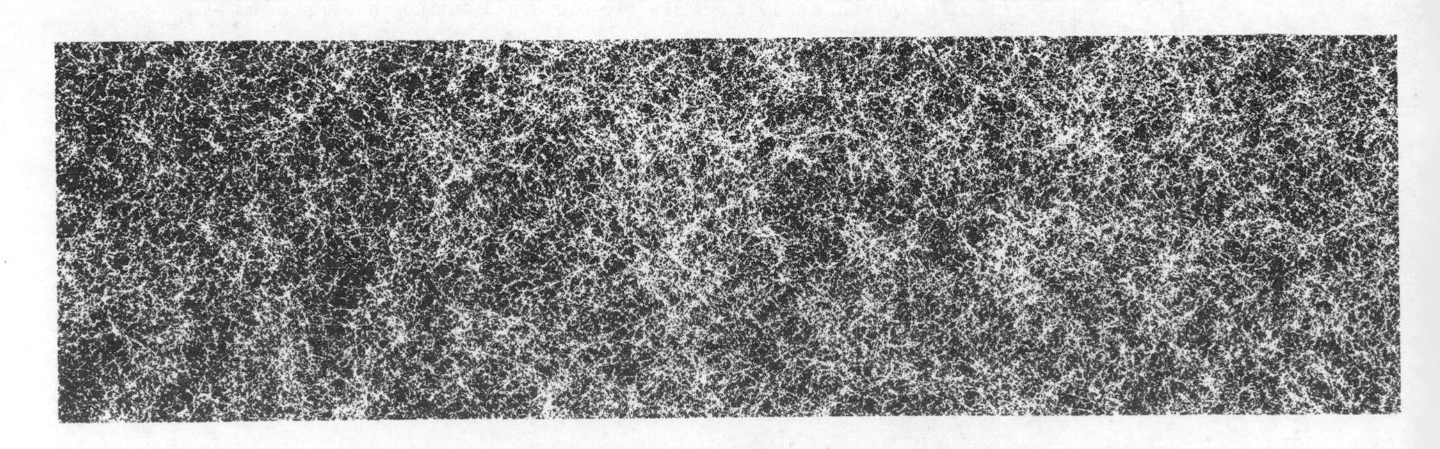

Appendix A
Viscosity of Water

temperature (°F)	absolute viscosity	kinematic viscosity		
	centipoise	centistokes	SSU	ft²/sec
32	1.79	1.79	33.0	0.00001931
50	1.31	1.31	31.6	0.00001410
60	1.12	1.12	31.2	0.00001217
70	0.98	0.98	30.9	0.00001059
80	0.86	0.86	30.6	0.00000930
85	0.81	0.81	30.4	0.00000869
100	0.68	0.69	30.2	0.00000739
120	0.56	0.57	30.0	0.00000609
140	0.47	0.48	29.7	0.00000514
160	0.40	0.41	29.6	0.00000442
180	0.35	0.36	29.5	0.00000385
212	0.28	0.29	29.3	0.00000319

Appendix B
Important Fluid Conversions

multiply	by	to obtain
cubic feet	7.4805	gallons
cfs	448.83	gpm
cfs	0.64632	MGD
gallons	0.1337	cubic feet
gpm	0.002228	cfs
inches of mercury	0.491	psi
inches of mercury	70.7	psf
inches of mercury	13.60	inches of water
inches of water	5.199	psf
inches of water	0.0361	psi
inches of water	0.0735	inches of mercury
psi	144	psf
psi	2.308	feet of water
psi	27.7	inches of water
psi	2.037	inches of mercury
psf	0.006944	psi

FLUIDS

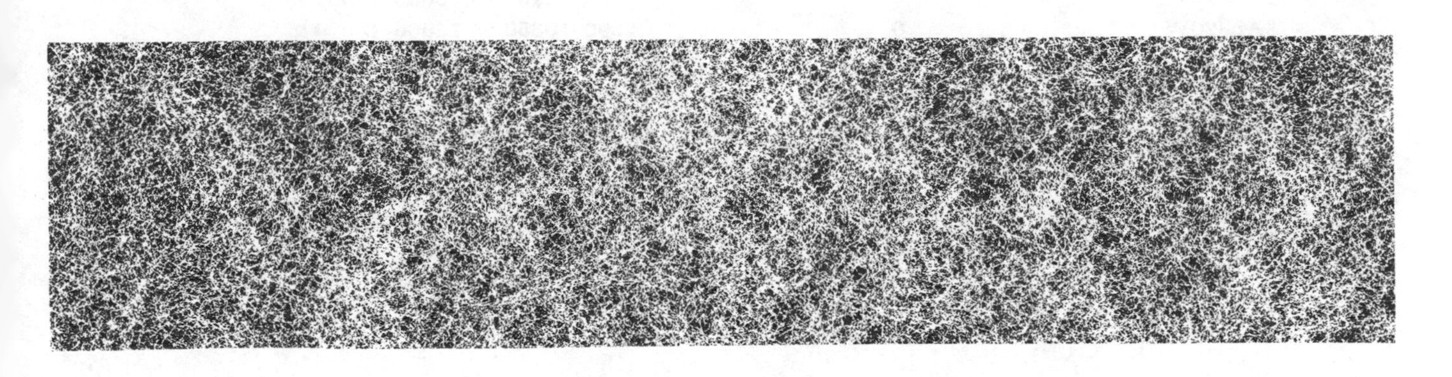

Appendix C
Properties of Liquids

Liquid	Temp, °F	Specific Gravity*	Viscosity centistokes	Viscosity SSU	Viscosity ft^2/sec
Acetone	68	.792	.41		
Alcohol, ethyl	68	.789	1.52	31.7	1.65 EE−5
(C_2H_5OH)	104	.772	1.2	31.5	
Alcohol, methyl	68	.79			
(CH_3OH)	59		.74		
Ammonia	0	.662	.30		
Butane	−50		.52		
	30		.35		
	60	.584			
Castor Oil	68	.96			1110 EE−5
	104	.95	259–325	1200–1500	
	130		98–130	450–600	
Ethylene glycol	60	1.125			
	70		17.8	88.4	
Freon-11	70	1.49	21.1	.21	
Freon-12	70	1.33	21.1	.27	
Fuel oils, #1 to #6	60	.82–.95			
Fuel oil #1	70		2.39–4.28	34–40	
	100		−2.69	32–35	
Fuel oil #2	70		3.0–7.4	36–50	
	100		2.11–4.28	33–40	
Fuel oil #3	70		2.69–5.84	35–45	
	100		2.06–3.97	32.8–39	
Fuel oil #5A	70		7.4–26.4	50–125	
	100		4.91–13.7	42–72	
Fuel oil #5B	70		26.4–	125–	
	100		13.6–67.1	72–310	
Fuel oil #6	122		97.4–660	450–3000	
	160		37.5–172	175–780	
Gasoline (regular)	60	.728			.73 EE−5
	80	.719			.66 EE−5
	100	.710			.60 EE−5
Kerosene	60	.78–.82			
	68		2.17	35	
Jet Fuel	−30		7.9	52	
	60	.82			
Mercury	70	13.6	21.1	.118	
	100	13.6	37.8	.11	
Oils, SAE 5 to 150	60	.88–.94			
SAE-5W	0		1295 max	6000 max	
SAE-10W	0		1295–2590	6000–12000	
SAE-20W	0		2590–10350	12000–48000	
SAE-20	210		5.7–9.6	45–58	
SAE-30	210		9.6–12.9	58–70	
SAE-40	210		12.9–16.8	70–85	
SAE-50	210		16.8–22.7	85–110	
Salt Water (5%)	39	1.037			
	68		1.097	31.1	
Salt Water (25%)	39	1.196			
	60	1.19	2.4	34	

*Measured with respect to 60°F water

FLUIDS

Appendix D

Dimensions of Welded and Seamless Steel Pipe

nominal diameter inches	schedule	outside diameter inches	wall thickness inches	internal diameter inches	internal area sq inches	internal diameter feet	internal area sq feet
$\frac{1}{8}$	40 (S)	0.405	0.068	0.269	0.0568	0.0224	0.00039
	80 (X)		0.095	0.215	0.0363	0.0179	0.00025
$\frac{1}{4}$	40 (S)	0.540	0.088	0.364	0.1041	0.0303	0.00072
	80 (X)		0.119	0.302	0.0716	0.0252	0.00050
$\frac{3}{8}$	40 (S)	0.675	0.091	0.493	0.1909	0.0411	0.00133
	80 (X)		0.126	0.423	0.1405	0.0353	0.00098
$\frac{1}{2}$	40 (S)	0.840	0.109	0.622	0.3039	0.0518	0.00211
	80 (X)		0.147	0.546	0.2341	0.0455	0.00163
	160		0.187	0.466	0.1706	0.0388	0.00118
	(XX)		0.294	0.252	0.499	0.0210	0.00035
$\frac{3}{4}$	40 (S)	1.050	0.113	0.824	0.5333	0.0687	0.00370
	80 (X)		0.154	0.742	0.4324	0.0618	0.00300
	160		0.219	0.612	0.2942	0.0510	0.00204
	(XX)		0.308	0.434	0.1479	0.0362	0.00103
1	40 (S)	1.315	0.133	1.049	0.8643	0.0874	0.00600
	80 (X)		0.179	0.957	0.7193	0.0798	0.00500
	160		0.250	0.815	0.5217	0.0679	0.00362
	(XX)		0.358	0.599	0.2818	0.0499	0.00196
$1\frac{1}{4}$	40 (S)	1.660	0.140	1.380	1.496	0.1150	0.01039
	80 (X)		0.191	1.278	1.283	0.1065	0.00890
	160		0.250	1.160	1.057	0.0967	0.00734
	(XX)		0.382	0.896	0.6305	0.0747	0.00438
$1\frac{1}{2}$	40 (S)	1.900	0.145	1.610	2.036	0.1342	0.01414
	80 (X)		0.200	1.500	1.767	0.1250	0.01227
	160		0.281	1.338	1.406	0.1115	0.00976
	(XX)		0.400	1.100	0.9503	0.0917	0.00660
2	40 (S)	2.375	0.154	2.067	3.356	0.1723	0.02330
	80 (X)		0.218	1.939	2.953	0.1616	0.02051
	160		0.344	1.687	2.235	0.1406	0.01552
	(XX)		0.436	1.503	1.774	0.1253	0.01232
$2\frac{1}{2}$	40 (S)	2.875	0.203	2.469	4.788	0.2058	0.03325
	80 (X)		0.276	2.323	4.238	0.1936	0.02943
	160		0.375	2.125	3.547	0.1771	0.02463
	(XX)		0.552	1.771	2.464	0.1476	0.01711
3	40 (S)	3.500	0.216	3.068	7.393	0.2557	0.05134
	80 (X)		0.300	2.900	6.605	0.2417	0.04587
	160		0.438	2.624	5.408	0.2187	0.03755
	(XX)		0.600	2.300	4.155	0.1917	0.02885
$3\frac{1}{2}$	40 (S)	4.000	0.226	3.548	9.887	0.2957	0.06866
	80 (X)		0.318	3.364	8.888	0.2803	0.06172

nominal diameter		outside diameter	wall thickness	internal diameter	internal area	internal diameter	internal area
inches	schedule	inches	inches	inches	sq inches	feet	sq feet
	40 (S)	4.500	0.237	4.026	12.73	0.3355	0.08841
4	80 (X)		0.337	3.826	11.50	0.3188	0.07984
	120		0.438	3.624	10.32	0.3020	0.07163
	160		0.531	3.438	9.283	0.2865	0.06447
	(XX)		0.674	3.152	7.803	0.2627	0.05419
5	40 (S)	5.563	0.258	5.047	20.01	0.4206	0.1389
	80 (X)		0.375	4.813	18.19	0.4011	0.1263
	120		0.500	4.563	16.35	0.3803	0.1136
	160		0.625	4.313	14.61	0.3594	0.1015
	(XX)		0.750	4.063	12.97	0.3386	0.09004
6	40 (S)	6.625	0.280	6.065	28.89	0.5054	0.2006
	80 (X)		0.432	5.761	26.07	0.4801	0.1810
	120		0.562	5.501	23.77	0.4584	0.1650
	160		0.719	5.187	21.13	0.4323	0.1467
	(XX)		0.864	4.897	18.83	0.4081	0.1308
8	20	8.625	0.250	8.125	51.85	0.6771	0.3601
	30		0.277	8.071	51.16	0.6726	0.3553
	40 (S)		0.322	7.981	50.03	0.6651	0.3474
	60		0.406	7.813	47.94	0.6511	0.3329
	80 (X)		0.500	7.625	45.66	0.6354	0.3171
	100		0.594	7.437	43.44	0.6198	0.3017
	120		0.719	7.187	40.57	0.5989	0.2817
	140		0.812	7.001	38.50	0.5834	0.2673
	(XX)		0.875	6.875	37.12	0.5729	0.2578
	160		0.906	6.813	36.46	0.5678	0.2532
10	20	10.75	0.250	10.250	82.52	0.85417	0.5730
	30		0.307	10.136	80.69	0.84467	0.5604
	40 (S)		0.365	10.020	78.85	0.83500	0.5476
	60 (X)		0.500	9.750	74.66	0.8125	0.5185
	80		0.594	9.562	71.81	0.7968	0.4987
	100		0.719	9.312	68.11	0.7760	0.4730
	120		0.844	9.062	64.50	0.7552	0.4479
	140 (XX)		1.000	8.750	60.13	0.7292	0.4176
	160		1.125	8.500	56.75	0.7083	0.3941
12	20	12.75	0.250	12.250	117.86	1.0208	0.8185
	30		0.330	12.090	114.80	1.0075	0.7972
	(S)		0.375	12.000	113.10	1.0000	0.7854
	40		0.406	11.938	111.93	0.99483	0.7773
	(X)		0.500	11.750	108.43	0.97917	0.7530
	60		0.562	11.626	106.16	0.96883	0.7372
	80		0.688	11.374	101.61	0.94783	0.7056
	100		0.844	11.062	96.11	0.92183	0.6674
	120 (XX)		1.000	10.750	90.76	0.89583	0.6303
	140		1.125	10.500	86.59	0.87500	0.6013
	160		1.312	10.126	80.53	0.84383	0.5592

nominal diameter		outside diameter	wall thickness	internal diameter	internal area	internal diameter	internal area
inches	schedule	inches	inches	inches	sq inches	feet	sq feet
14 OD	10	14.00	0.250	13.500	143.14	1.1250	0.9940
	20		0.312	13.376	140.52	1.1147	0.9758
	30 (S)		0.375	13.250	137.89	1.1042	0.9575
	40		0.438	13.124	135.28	1.0937	0.9394
	(X)		0.500	13.000	132.67	1.0833	0.9213
	60		0.594	12.812	128.92	1.0677	0.8953
	80		0.750	12.500	122.72	1.0417	0.8522
	100		0.938	12.124	115.45	1.0104	0.8017
	120		1.094	11.812	109.58	0.98433	0.7610
	140		1.250	11.500	103.87	0.95833	0.7213
	160		1.406	11.188	98.31	0.93233	0.6827
16 OD	10	16.00	0.250	15.500	188.69	1.2917	1.3104
	20		0.312	15.376	185.69	1.2813	1.2895
	30 (S)		0.375	15.250	182.65	1.2708	1.2684
	40 (X)		0.500	15.000	176.72	1.2500	1.2272
	60		0.656	14.688	169.44	1.2240	1.1767
	80		0.844	14.312	160.88	1.1927	1.1172
	100		1.031	13.938	152.58	1.1615	1.0596
	120		1.219	13.562	144.46	1.1302	1.0032
	140		1.438	13.124	135.28	1.0937	0.9394
	160		1.594	12.812	128.92	1.0677	0.8953
18 OD	10	18.00	0.250	17.500	240.53	1.4583	1.6703
	20		0.312	17.376	237.13	1.4480	1.6467
	(S)		0.375	17.250	233.71	1.4375	1.6230
	30		0.438	17.124	230.00	1.4270	1.5993
	(X)		0.500	17.000	226.98	1.4167	1.5762
	40		0.562	16.876	223.68	1.4063	1.5533
	60		0.750	16.500	213.83	1.3750	1.4849
	80		0.938	16.124	204.19	1.3437	1.4180
	100		1.156	15.688	193.30	1.3073	1.3423
	120		1.375	15.250	182.65	1.2708	1.2684
	140		1.562	14.876	173.81	1.2397	1.2070
	160		1.781	14.438	163.72	1.2032	1.1370
20 OD	10	20.00	0.250	19.500	298.65	1.6250	2.0739
	20 (S)		0.375	19.250	291.04	1.6042	2.0211
	30 (X)		0.500	19.000	283.53	1.5833	1.9689
	40		0.594	18.812	277.95	1.5677	1.9302
	60		0.812	18.376	265.21	1.5313	1.8417
	80		1.031	17.938	252.72	1.4948	1.7550
	100		1.281	17.438	238.83	1.4532	1.6585
	120		1.500	17.000	226.98	1.4167	1.5762
	140		1.750	16.500	213.83	1.3750	1.4849
	160		1.969	16.062	202.62	1.3385	1.4071

nominal diameter inches	schedule	outside diameter inches	wall thickness inches	internal diameter inches	internal area sq inches	internal diameter feet	internal area sq feet
24 OD	10	24.00	0.250	23.500	433.74	1.9583	3.0121
	20 (S)		0.375	23.250	424.56	1.9375	2.9483
	(X)		0.500	23.000	415.48	1.9167	2.8852
	30		0.562	22.876	411.01	1.9063	2.8542
	40		0.688	22.624	402.00	1.8853	2.7917
	60		0.969	22.062	382.28	1.8385	2.6547
	80		1.219	21.562	365.15	1.7802	2.5358
	100		1.531	20.938	344.32	1.7448	2.3911
	120		1.812	20.376	326.92	1.6980	2.2645
	140		2.062	19.876	310.28	1.6563	2.1547
	160		2.344	19.312	292.92	1.6093	2.0342
30 OD	10	30.00	0.312	29.376	677.76	2.4480	4.7067
	(S)		0.375	29.250	671.62	2.4375	4.6640
	20 (X)		0.500	29.000	660.52	2.4167	4.5869
	30		0.625	28.750	649.18	2.3958	4.5082

S = Wall thickness, formerly designated "standard weight"

X = Wall thickness, formerly designated "extra strong"

XX = Wall thickness, formerly designated "double extra strong"

Actual wall thickness may vary slightly.

Extracted from American Standard Wrought Steel and Wrought Iron Pipe (ASA B36, 10—1959), The American Society of Mechanical Engineers.

FLUIDS

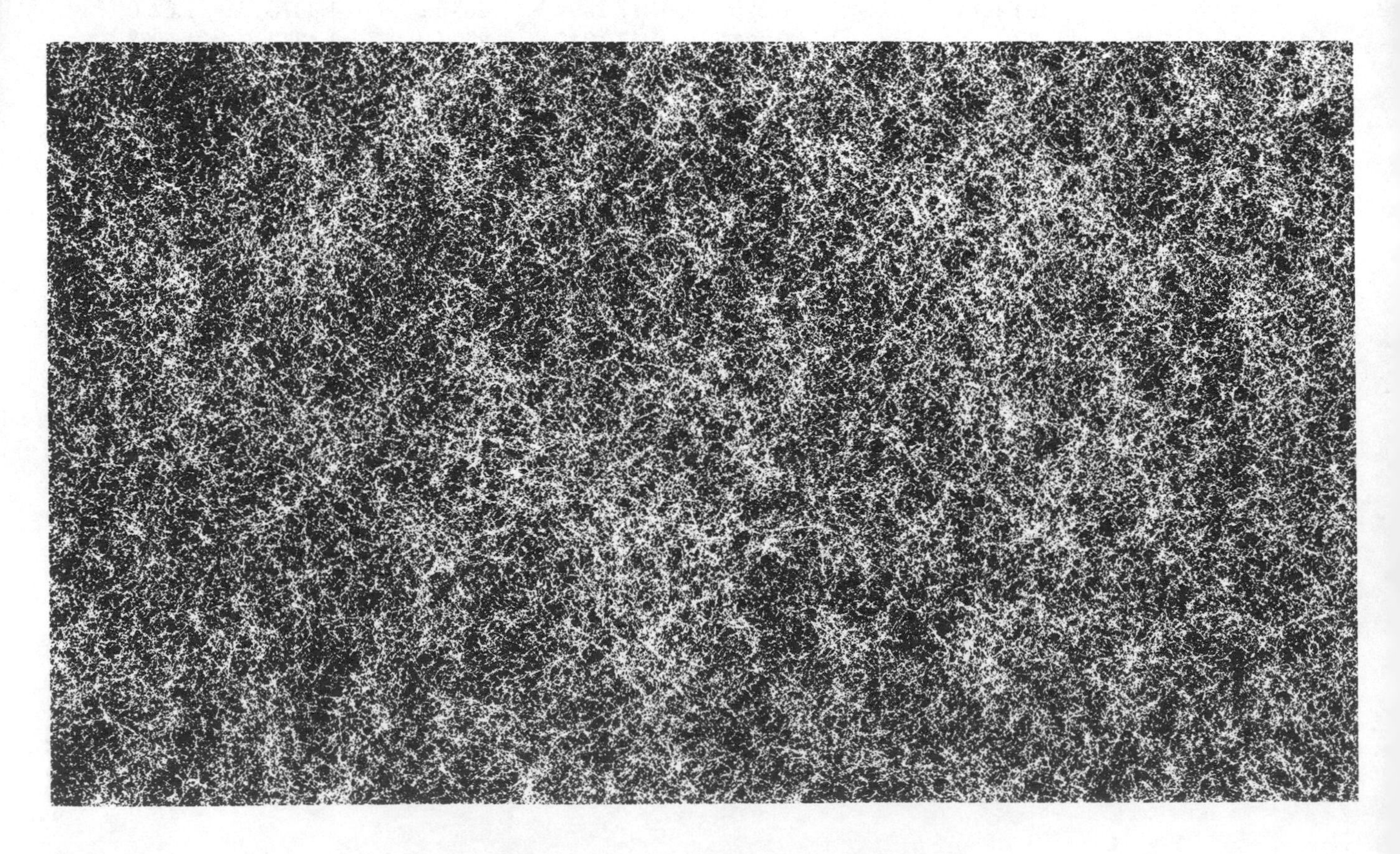

Appendix E

Dimensions of Copper Water Tubing

CLASSIFICATION	NOM. TUBE SIZE (in.)	OUTSIDE DIAM (in.)	WALL THICK-NESS (in.)	INSIDE DIAM (in.)	TRANS-VERSE AREA (sq. in.)	SAFE WORKING PRESSURE (psi)
HARD	1/4	3/8	.025	.325	.083	1000
	3/8	1/2	.025	.450	.159	1000
	1/2	5/8	.028	.569	.254	890
	3/4	7/8	.032	.811	.516	710
	1	1 1/8	.035	1.055	.874	600
	1 1/4	1 3/8	.042	1.291	1.309	590
Type "M" 250 psi Working Pressure	1 1/2	1 5/8	.049	1.527	1.831	580
	2	2 1/8	.058	2.009	3.17	520
	2 1/2	2 5/8	.065	2.495	4.89	470
	3	3 1/8	.072	2.981	6.98	440
	3 1/2	3 5/8	.083	3.459	9.40	430
	4	4 1/8	.095	3.935	12.16	430
	5	5 1/8	.109	4.907	18.91	400
	6	6 1/8	.122	5.881	27.16	375
	8	8 1/8	.170	7.785	47.6	375
HARD	3/8	1/2	.035	.430	.146	1000
	1/2	5/8	.040	.545	.233	1000
	3/4	7/8	.045	.785	.484	1000
	1	1 1/8	.050	1.025	.825	880
	1 1/4	1 3/8	.055	1.265	1.256	780
Type "L" 250 psi Working Pressure	1 1/2	1 5/8	.060	1.505	1.78	720
	2	2 1/8	.070	1.985	3.094	640
	2 1/2	2 5/8	.080	2.465	4.77	580
	3	3 1/8	.090	2.945	6.812	550
	3 1/2	3 5/8	.100	3.425	9.213	530
	4	4 1/8	.110	3.905	11.97	510
	5	5 1/8	.125	4.875	18.67	460
	6	6 1/8	.140	5.845	26.83	430
HARD	1/4	3/8	.032	.311	.076	1000
	3/8	1/2	.049	.402	.127	1000
	1/2	5/8	.049	.527	.218	1000
	3/4	7/8	.065	.745	.436	1000
	1	1 1/8	.065	.995	.778	780
	1 1/4	1 3/8	.065	1.245	1.217	630
Type "K" 400 psi Working Pressure	1 1/2	1 5/8	.072	1.481	1.722	580
	2	2 1/8	.083	1.959	3.014	510
	2 1/2	2 5/8	.095	2.435	4.656	470
	3	3 1/8	.109	2.907	6.637	450
	3 1/2	3 5/8	.120	3.385	8.999	430
	4	4 1/8	.134	3.857	11.68	420
	5	5 1/8	.160	4.805	18.13	400
	6	6 1/8	.192	5.741	25.88	400
SOFT	1/4	3/8	.032	.311	.076	1000
	3/8	1/2	.049	.402	.127	1000
	1/2	5/8	.049	.527	.218	1000
	3/4	7/8	.065	.745	.436	1000
	1	1 1/8	.065	.995	.778	780
	1 1/4	1 3/8	.065	1.245	1.217	630
Type "K" 250 psi Working Pressure	1 1/2	1 5/8	.072	1.481	1.722	580
	2	2 1/8	.083	1.959	3.014	510
	2 1/2	2 5/8	.095	2.435	4.656	470
	3	3 1/8	.109	2.907	6.637	450
	3 1/2	2 5/8	.120	3.385	8.999	430
	4	4 1/8	.134	3.857	11.68	420
	5	5 1/8	.160	4.805	18.13	400
	6	6 1/8	.192	5.741	25.88	400

Appendix F
Dimensions of Brass and Copper Tubing

regular

pipe size in.	nominal dimensions in.			cross sectional area of bore sq. in.	lb per ft	
	O.D.	I.D.	wall		red brass	copper
1/8	.405	.281	.062	.062	.253	.259
1/4	.540	.376	.082	.110	.447	.457
3/8	.675	.495	.090	.192	.627	.641
1/2	.840	.626	.107	.307	.934	.955
3/4	1.050	.822	.114	.531	1.270	1.300
1	1.315	1.063	.126	.887	1.780	1.820
1 1/4	1.660	1.368	.146	1.470	2.630	2.690
1 1/2	1.900	1.600	.150	2.010	3.130	3.200
2	2.375	2.063	.156	3.340	4.120	4.220
2 1/2	2.875	2.501	.187	4.910	5.990	6.120
3	3.500	3.062	.219	7.370	8.560	8.750
3 1/2	4.000	3.500	.250	9.620	11.200	11.400
4	4.500	4.000	.250	12.600	12.700	12.900
5	5.562	5.062	.250	20.100	15.800	16.200
6	6.625	6.125	.250	29.500	19.000	19.400
8	8.625	8.001	.312	50.300	30.900	31.600
10	10.750	10.020	.365	78.800	45.200	46.200
12	12.750	12.000	.375	113.000	55.300	56.500

extra strong

pipe size in.	nominal dimensions in.			cross sectional area of bore sq. in.	lb per ft	
	O.D.	I.D.	wall		red brass	copper
1/8	.405	.205	.100	.033	.363	.371
1/4	.540	.294	.123	.068	.611	.625
3/8	.675	.421	.127	.139	.829	.847
1/2	.840	.542	.149	.231	1.230	1.250
3/4	1.050	.736	.157	.425	1.670	1.710
1	1.315	.951	.182	.710	2.460	2.510
1 1/4	1.660	1.272	.194	1.270	3.390	3.460
1 1/2	1.900	1.494	.203	1.750	4.100	4.190
2	2.375	1.933	.221	2.94	5.670	5.800
2 1/2	2.875	2.315	.280	4.21	8.660	8.850
3	3.500	2.892	.304	6.57	11.600	11.800
3 1/2	4.000	3.358	.321	8.86	14.100	14.400
4	4.500	3.818	.341	11.50	16.900	17.300
5	5.562	4.812	.375	18.20	23.200	23.700
6	6.625	5.751	.437	26.00	32.200	32.900
8	8.625	7.625	.500	45.70	48.400	49.500
10	10.750	9.750	.500	74.70	61.100	62.400

Appendix G
Specific Gravity of Hydrocarbons

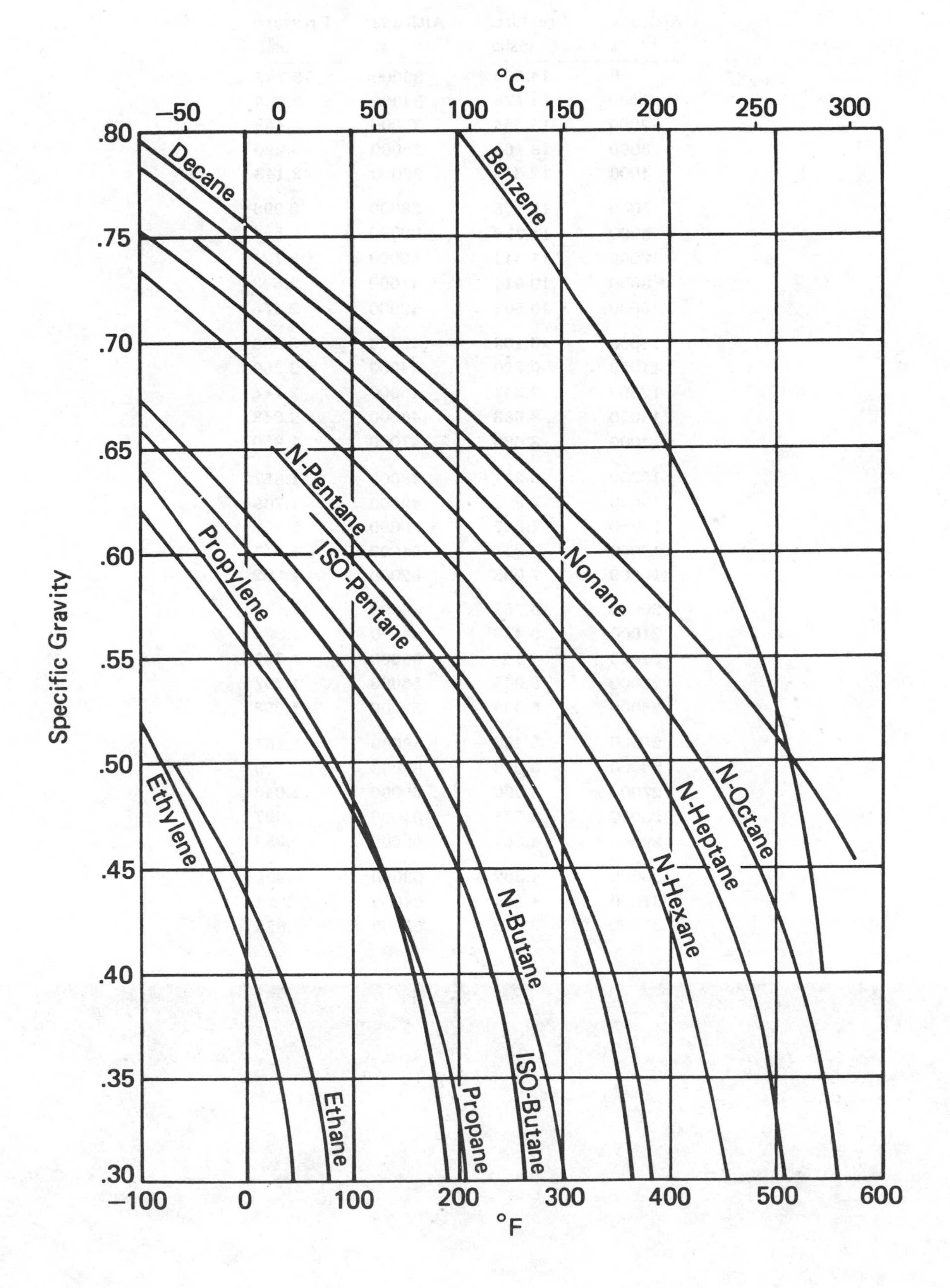

Appendix H
Atmospheric Pressure versus Altitude

Altitude ft	Pressure psia	Altitude ft	Pressure psia
0	14.696	33000	3.797
1000	14.175	34000	3.625
2000	13.664	35000	3.458
3000	13.168	36000	3.296
4000	12.692	37000	3.143
5000	12.225	38000	2.996
6000	11.778	39000	2.854
7000	11.341	40000	2.721
8000	10.914	41000	2.593
9000	10.501	42000	2.475
10000	10.108	43000	2.358
11000	9.720	44000	2.250
12000	9.347	45000	2.141
13000	8.983	46000	2.043
14000	8.630	47000	1.950
15000	8.291	48000	1.857
16000	7.962	49000	1.768
17000	7.642	50000	1.690
18000	7.338	51000	1.611
19000	7.038	52000	1.532
20000	6.753	53000	1.464
21000	6.473	54000	1.395
22000	6.2	55000	1.331
23000	5.943	56000	1.267
24000	5.693	57000	1.208
25000	5.452	58000	1.154
26000	5.216	59000	1.100
27000	4.990	60000	1.046
28000	4.774	61000	.997
29000	4.563	62000	.953
30000	4.362	63000	.909
31000	4.165	64000	.864
32000	3.978	65000	.825

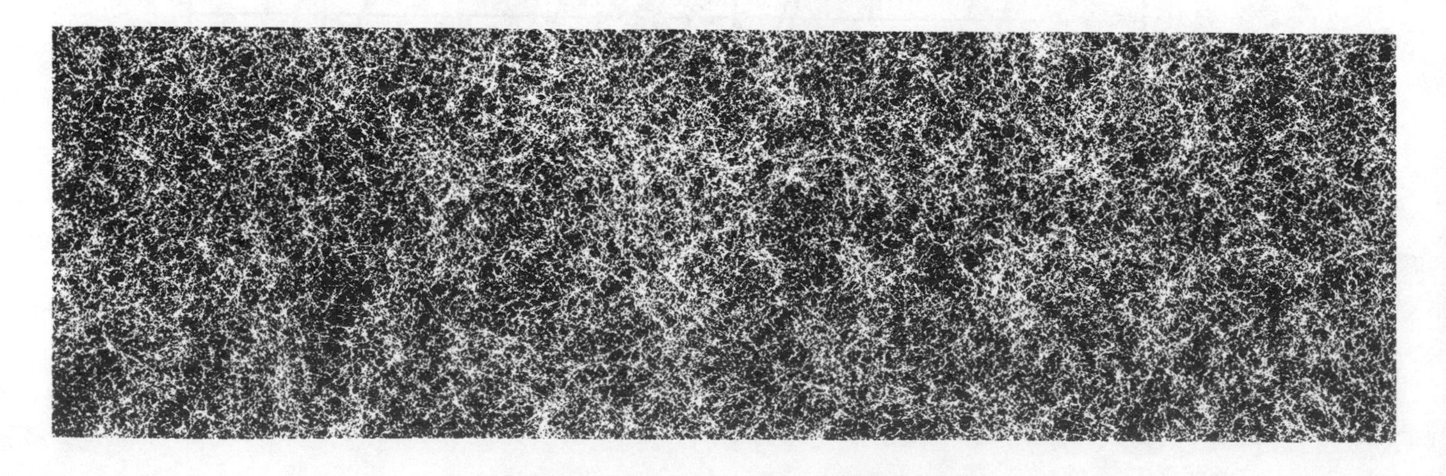

Appendix I
Net Expansion Factor, Y
For Compressible Flow Through Nozzles and Orifices

k = 1.3 approximately
(for CO_2, SO_2, H_2O, H_2S, NH_3, N_2O, Cl_2, CH_4, C_2H_2, and C_2H_4)

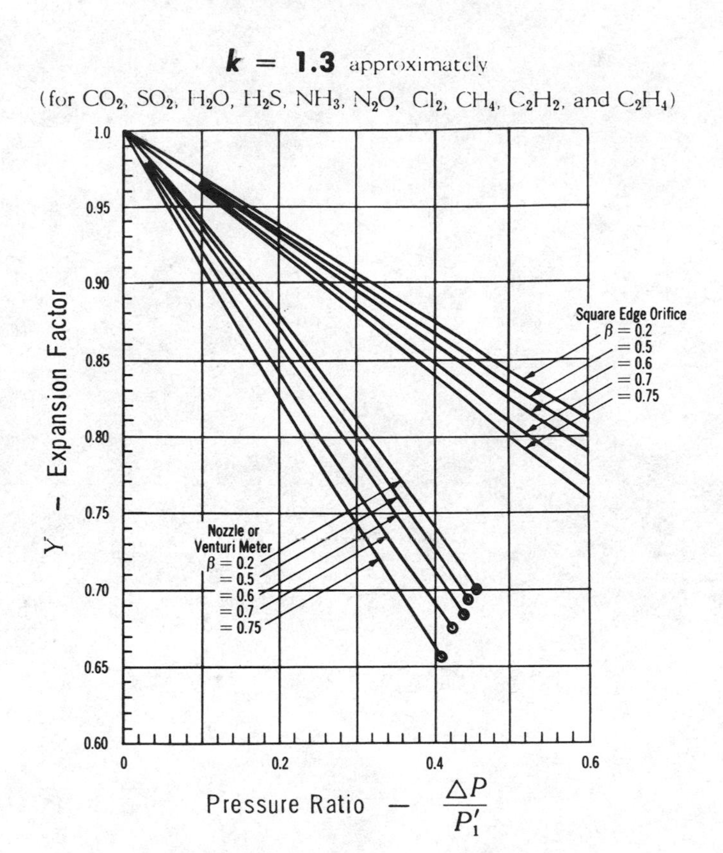

k = 1.4 approximately
(for Air, H_2, O_2, N_2, CO, NO, and HCl)

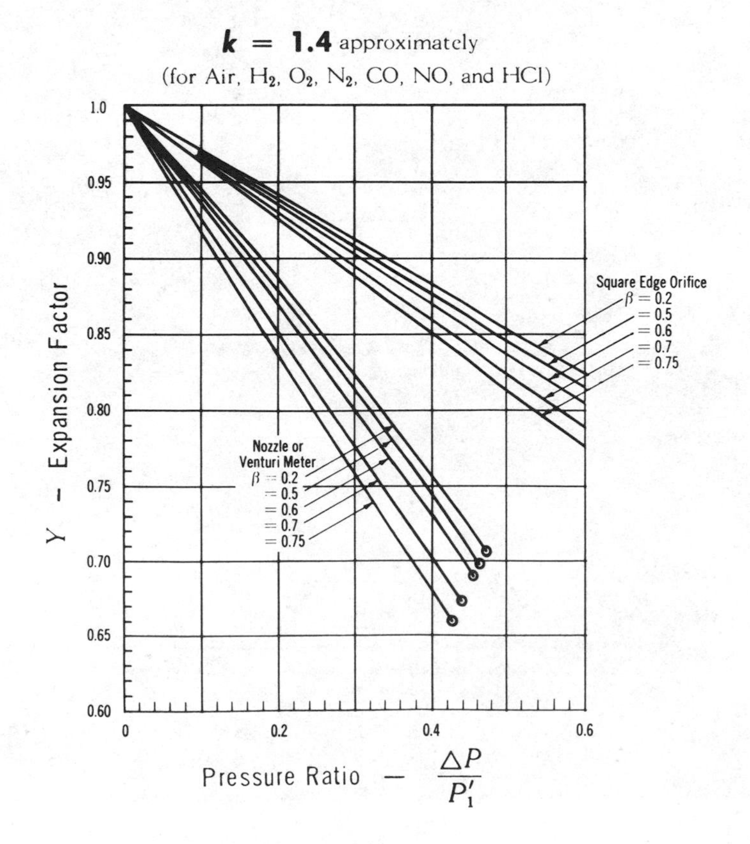

PRACTICE PROBLEMS

1. A water-tight cubical box, one foot on each exterior side, is made from $\frac{1}{4}''$ iron plate. Will the box float? If so, how far above the water surface will the box float?

2. You are in the field-piping area of a refinery, without a calculator or friction factor chart, doing an inspection with your manager. The flow through pipe A is always fully turbulent. While you are watching the gauges on pipe A, which read pressure drop through the length of the pipe, the pressure increases by a factor of two. Your manager wants to know what is happening. You know that no valves closed downstream, and the density and viscosity remain unchanged. In five seconds you figure out the flow rate increase. What do you tell your manager?

3. A pitot-static tube reads 1″ of mercury in a pipe. What is the velocity if the fluid is air at 14.7 psi and 70 °F? What is the velocity if the fluid is water at 110 °F?

4. A 6″ schedule-40 pipe is 3000′ long. The outlet is 100′ above the inlet. Oil ($SG = 0.9$) is pumped through the pipe at 750 gpm, the dynamic viscosity is 0.0015 lbm/ft-sec, and the outlet is maintained at 50 psig. What is the inlet pressure?

5. What is the equivalent length, in diameters, of a coil of $\frac{1}{4}''$ copper tubing with an inside diameter of 0.21″ bent into 12 turns, 6″ in diameter (measured centerline to centerline)?

6. Honey flows through a pipe with $N_{Re} < 10$. The honey is heated to a temperature at which its viscosity decreases to one-half its original value. What effect will the heating have on the friction pressure drop in the pipe?

7. What is the pressure drop, in psi, when 20 gpm of 60 °F water flows through 65′ of horizontal 1″ schedule-40 pipe?

8. What is the pressure drop when 100 standard cubic feet per minute of air at 112 psig and 95 °F flow through 80′ of $1\frac{1}{4}''$ schedule-40 steel pipe?

9. What is the pressure drop through a 4″ schedule-40 pipe with 500 gpm of 70 °F water flowing? The pipe is 400′ long and has three 90° ells, four couplings, one gate valve (half-open), one ball valve (sometimes called a plug cock) closed 20°, and one heat exchanger where friction loss is equivalent to 200 pipe diameters.

10. Water at 68 °F is pumped at a constant rate of 5 cubic feet per minute from a supply tank resting on the floor to the top of an experimental absorber. The point of discharge is 15′ above the floor, and the frictional losses in the 2″ schedule-40 pipe are estimated to be 0.8′. At what height in the supply tank must the water level be kept if the pump can develop only $\frac{1}{8}$ net horsepower?

11. A helical coil made of 3″ O.D. 16-gauge type 316 stainless steel tubing has a total of 15 turns on a 72″ pitch diameter. If water at 80 °F flows through the coil at 150 gpm, calculate the pressure drop through the coil.

12. A 1″ diameter smooth tube 1000′ long is carrying a product with a viscosity of 25 centipoise and a density of 55 pcf at an average velocity of 7.33′ per second. Due to a production change, the product viscosity will be decreased to 12.5 centipoise with no change in density. What will be the percentage change in pump horsepower needed to overcome the friction losses in the smooth tube for the same flow?

13. 450,000 pounds per hour of NaCl solution ($SG = 1.20$) and 50,000 pounds per hour of suspended NaCl crystals ($SG = 2.1$), all at a temperature of 140 °F, are pumped from a vessel operating at 26″ vacuum into an atmospheric slurry tank in which the liquor inlet nozzle is 30′ above the nominal liquor level in the vacuum vessel. The barometer reads 30″. The friction loss in the interconnecting piping (not including the required control valve) has been calculated as 18′ . Without adding any safety factors, calculate the head and capacity for the pump when a control valve having a capacity coefficient of $C = 400$ is used.

14. A gasoline storage tank 15′ in diameter and 30′ tall is mounted on an elevated support so that its base is 15′ above the ground level of an adjacent truck loading dock. A discharge line is connected to the center of the tank bottom and runs horizontally 60′ to the loading dock. When the liquid level in the tank is 5′ above the bottom of the tank, the flow rate in the 2″ schedule-40 steel pipe discharge line has been measured at 75 gpm. Because it is desired to know how long it takes to load a given size truck, information on flow rate versus liquid level is required. Indicate your method of obtaining the desired flow rate level data, indicating assumptions with justifications.

15. Fluid is flowing through a 5-mile length of 4″ schedule-40 steel pipe with an average linear velocity of 10′ per second. Assume that the pipeline is well-insulated to prevent heat loss to the surroundings. The constant liquid properties are: density = 62.4 pounds per cubic foot; viscosity = 4.35 centipoise; and specific heat = 0.2 BTU/lbm-°F. Calculate the temperature rise of the liquid over the 5-mile section.

16. It is necessary to supply 250 standard cubic feet per second of natural gas (CH_4) at 60 psig and 70 °F. The gas must travel through a 12″ inside diameter pipe which is 3 miles long. Calculate the required pressure at the reservoir. Assume the gas is incompressible.

17. A large centrifugal pump has a 10″ diameter inlet and a 5″ diameter outlet. The measured flow rate is 818 gpm. The measured inlet pressure is 5″ of mercury above atmospheric. The discharge pressure measured at 4′ above the pump outlet is 30.7 psia. The pump input horsepower is 10. Find (a) the pump efficiency, and (b) the new flow rate, net head, and brake horsepower if the pump speed is increased from 1750 rpm to 3500 rpm.

18. A reaction vessel designed for 100 psig normally operates at 35 psig and 65 °F. If the temperature rises to 100 °C, a decomposition occurs releasing CO gas. To relieve the pressure, a 6″ schedule-80 steel pipe 23′ long, containing three elbows, is provided as a vent. This pipe contains a lead rupture disc scored to break at 100 psig. What is the maximum possible rate of formation of CO that can be tolerated without exceeding the design pressure of the vessel? The viscosity of the CO is 0.021 centipoise at 100 °C.

19. Water at 60 °F is pumped from a reservoir to the top of a mountain through a 6″ schedule-80 pipe at a velocity of 10 fps. The pipe discharges into the atmosphere at a level 4000′ above the reservoir. The pipeline itself is 5000′ long. If the overall efficiency of the pump and motor driving it is 70%, and the cost of the electrical energy to the motor is $1\frac{1}{2}$ cents per kilowatt hour, what is the hourly energy cost for pumping this water?

20. Boiler feed water is pumped downward through a bed of ion exchange (IX) resin. The resin is spherical in shape, having an average particle diameter of 0.065″. The IX bed is 4′ long. The bed operates at a flow rate of 20 gpm/ft^3 of bed. What is the pressure drop through the bed?

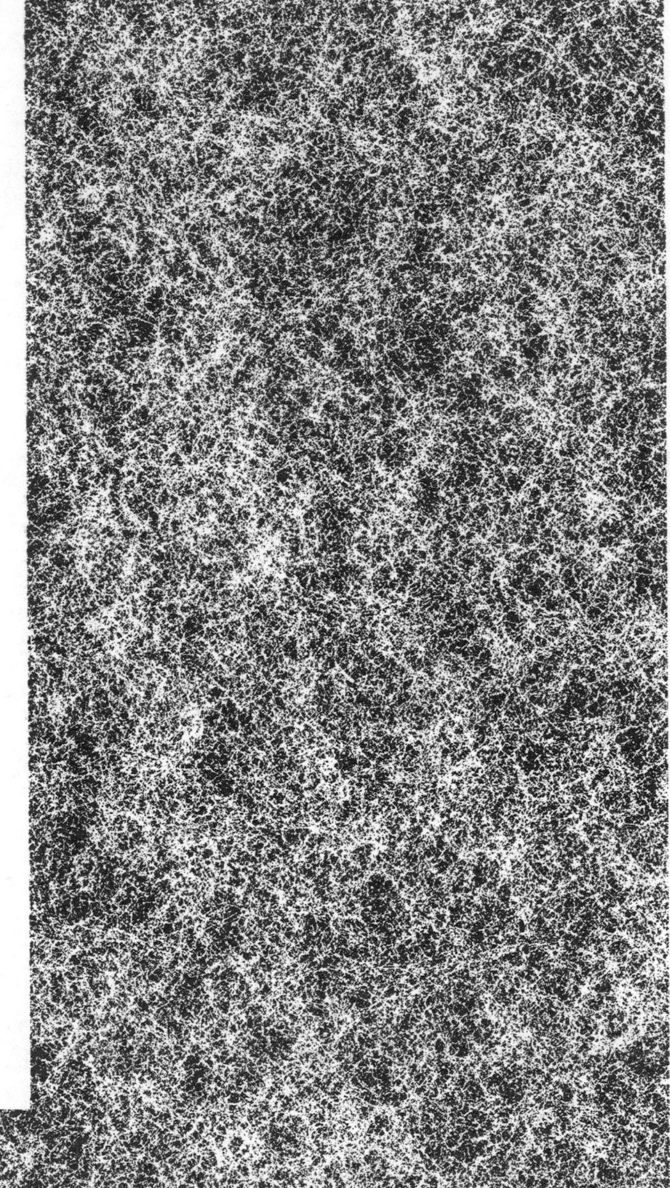

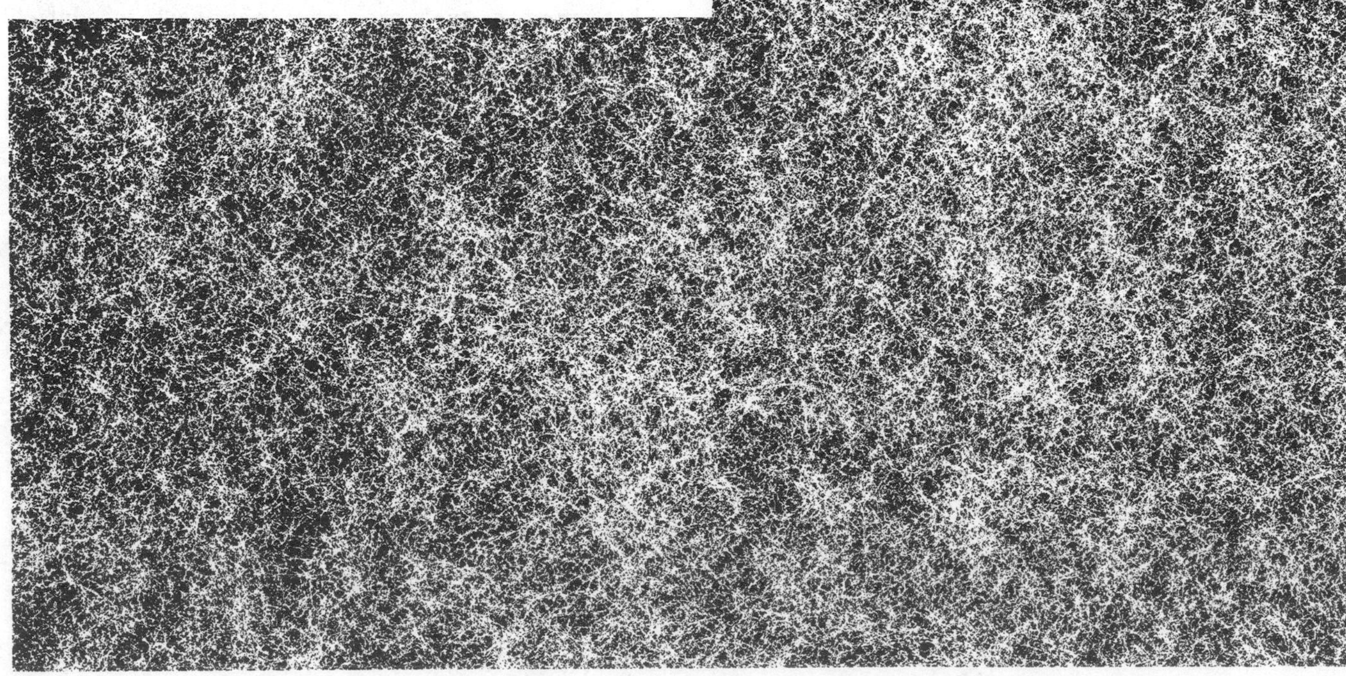

6 HEAT TRANSFER: CONDUCTION AND RADIATION

Nomenclature

a	inside radius	ft
A	area	ft^2
b	outside radius	ft
B	cooling fin half-thickness	ft
c	specific heat	BTU/lbm-°F
D	diameter	ft
F_{12}	direct view factor	–
F_ϵ	emissivity factor	–
F_A	arrangement factor	–
h	film coefficient	BTU/hr-ft^2-°F
k	thermal conductivity	BTU/hr-ft-°F
L	length, or wall thickness	ft
N	length of cylinder	ft
q	rate of heat transfer	BTU/hr
Q	heat transfer	BTU
r	radius	ft
R	resistance	°F-hr/BTU
S	heat source strength	BTU/hr-ft^3
t	time	hr
T	temperature	°R
V	volume	ft^3
w	cooling fin width	ft
x	distance	ft
y	distance	ft
z	distance	ft

Symbols

ε	emissivity	–
η	effectiveness of a fin	–
ρ	density	lbm/ft^3
σ	Stefan-Boltzmann constant (0.1713×10^{-8})	BTU/ft^2-hr-°R^4
ϕ	heat flux	BTU/hr-ft^2
ψ	angle, or cylindrical coordinates	–

Subscripts

A	arrangement
b	body
c	thermocouple
g	gas
i	inner
ins	insulation
m	log mean, or mid-point
o	outer
p	constant pressure
r	radiation
s	surface
T	total
w	wall
∞	surroundings (environment)

1 CONDUCTIVE HEAT TRANSFER

The conductive heat flow, q, through a slab of isotropic material of thickness L, having constant thermal conductivity k, and having a cross sectional area A, can be found from the one-dimensional form of *Fourier's Law*: heat flux, ϕ, is directly proportional to the negative temperature gradient, $\frac{-\partial T}{\partial x}$.

$$q = \frac{dQ}{dt} = -kA\frac{dT}{dx} \qquad 6.1$$

It is important to distinguish between the three types of energy transfer: heat flow rate, q; heat flux, ϕ; and source strength, S. These terms are computed from generalized differential equations.

$$q = \frac{d}{dt}(Q) \qquad 6.2$$

$$\phi = \frac{d}{dt}\left(\frac{Q}{A}\right) \qquad 6.3$$

$$S = \frac{d}{dt}\left(\frac{Q}{V}\right) \qquad 6.4$$

HT XFER

While equation 6.1 is frequently used, it is not the most general equation for conductive energy transfer. For materials experiencing a one-dimensional conductive heat transfer, with uniform constant conductivity, and with a heat source within the material (e.g., a nuclear, electrical, or chemical source), the generalized one-dimensional equation for heat transfer by conduction becomes

$$\frac{\partial^2 T}{\partial x^2} + \frac{S}{k} = \frac{\rho c_p}{k}\frac{\partial T}{\partial t} \qquad 6.5$$

If there are temperature gradients in all three dimensions, the generalized unsteady state equation for an isotropic material having uniform thermal conductivity is

$$\frac{\partial^2 T}{\partial x^2} + \frac{\partial^2 T}{\partial y^2} + \frac{\partial^2 T}{\partial z^2} + \frac{S}{k} = \frac{\rho c_p}{k}\frac{\partial T}{\partial t} \qquad 6.6$$

With no heat source, equation 6.6 reduces to the generalized *Fourier equation*:

$$\frac{\partial^2 T}{\partial x^2} + \frac{\partial^2 T}{\partial y^2} + \frac{\partial^2 T}{\partial z^2} = \frac{\rho c_p}{k}\frac{\partial T}{\partial t} \qquad 6.7$$

At steady state, $\frac{\partial T}{\partial t} = 0$, and equation 6.7 further reduces to the *Laplace equation*:

$$\nabla^2 T = \frac{\partial^2 T}{\partial x^2} + \frac{\partial^2 T}{\partial y^2} + \frac{\partial^2 T}{\partial z^2} = 0 \qquad 6.8$$

At steady state with a source term, the generalized equation reduces to the *Poisson equation*:

$$\frac{\partial^2 T}{\partial x^2} + \frac{\partial^2 T}{\partial y^2} + \frac{\partial^2 T}{\partial z^2} + \frac{S}{k} = 0 \qquad 6.9$$

The solutions to these equations, although initially formidable, can be derived for one-dimensional problems such as slabs, walls, and cylinders since the equations reduce to ordinary differential equations.

A. VALUES OF k

Figures 6.1 and 6.2 are graphs of thermal conductivity versus temperature for several conductors. Note that

A. silver 99.999 percent pure
B. high-purity copper
C. coalesced copper
D. copper, electrolytic tough pitch
E. aluminum, single crystal
F. free-machining tellurium copper
G. aluminum, 1100-F
H. aluminum, 6063-T5
I. copper, phosphorus deoxidized
J. aluminum, 2024-T4
K. free-machining loaded brass

1 watt/(cm)(°K)=57.79 BTU/(hr)(ft)(°R)

Figure 6.1 Thermal Conductivities of Some Good Conductors

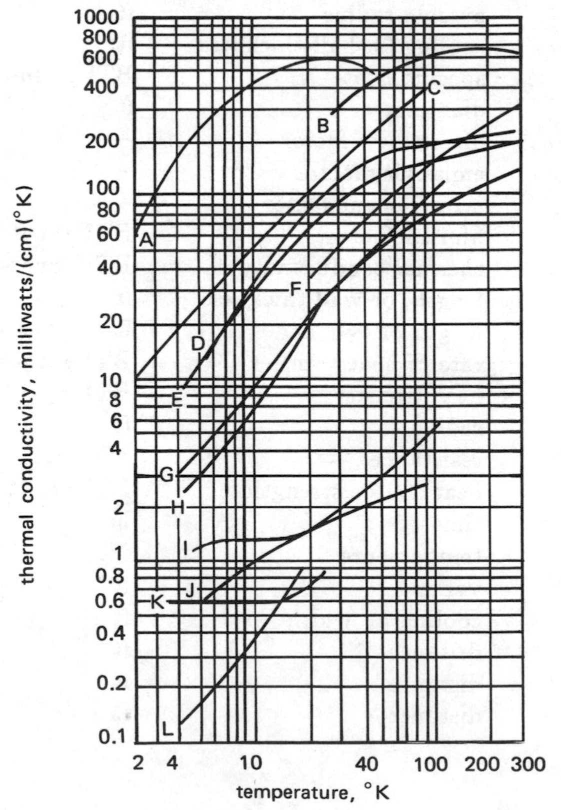

A. 50-50 lead-tin solder
B. steel SAE 1020
C. beryllium copper
D. constantan
E. monel
F. silicon bronze
G. inconel
H. type 347 stainless steel
I. fused quartz
J. pollytetrafluoroethylene (teflon)
K. polymethylmethacrylate (perspex)
L. nylon

1 mW/(cm)(°K)=57.79 x 10^{-8} BTU/(hr)(ft)(°R)

Figure 6.2 Thermal Conductivities of Some Poor Conductors

the ordinate of these figures is in units of watts/cm-°K, and the conversion factor for English units is given below the figure. Table 6.1 lists the thermal conductivities of several metals and alloys.

B. CONDUCTION THROUGH A SLAB WALL

The Fourier equation can be integrated directly for a slab wall. Since the area is independent of length, and k is assumed independent of temperature, the resulting equation for the slab wall illustrated is[1]

$$q = \frac{-kA}{L}\Delta T \qquad 6.10$$

$$= \frac{kA}{L}(T_1 - T_2) \qquad 6.11$$

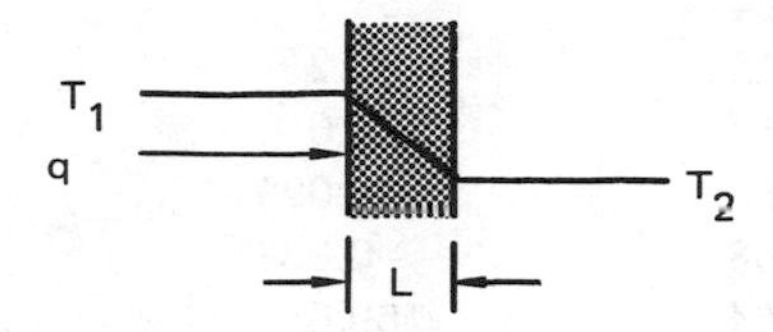

Figure 6.3 Temperature Gradient

The reciprocal of thermal conductance is *thermal resistance.*

$$R = \frac{L}{kA} \qquad 6.12$$

Using an analogy to electrical flow, where current $I = \frac{E}{R}$, heat flow is the ratio of the driving potential (temperature difference) to the thermal resistance. This concept is important when we determine the conduction of heat transfer through composite structures such as layered walls, and insulated pipes, tanks, spheres, and boxes.

C. CONDUCTION THROUGH A COMPOSITE WALL

Thermal resistances can be added for a flow of heat through a composite structure. In the case of the composite slab composed of two materials with different thicknesses and thermal conductivities, each of the materials contributes a thermal resistance. The total thermal resistance is the sum of the individual resistances, and the total driving force is the sum of the individual temperature differences. The equation for a composite wall can be written

$$q = \frac{\Delta T_{total}}{R_{total}} = \frac{\Delta T_1 + \Delta T_2}{R_1 + R_2} = \frac{\Delta T_1 + \Delta T_2}{\frac{L_1}{k_1 A} + \frac{L_2}{k_2 A}} \qquad 6.13$$

D. CONDUCTION THROUGH A HOLLOW CYLINDER (LONG PIPE)

The Fourier equation can be integrated for a hollow cylinder since it is known how the area varies with length. If the direction of flow is radial (perpendicular to the longitudinal axis), the cross section of the path of heat flow is proportional to the distance from the center of the cylinder, so that

$$q = \frac{kA_m}{L}\Delta T = \frac{k\pi N(D_o - D_i)}{\left[\frac{D_o - D_i}{2}\right] ln \frac{D_o}{D_i}}\Delta T$$

$$= \frac{2k\pi N}{ln \frac{D_o}{D_i}}\Delta T \qquad 6.14$$

The *logarithmic mean area* is

$$A_m = \frac{A_o - A_i}{ln \frac{A_o}{A_i}} \qquad 6.15$$

When $\frac{A_o}{A_i} < 2$, the *arithmetic mean area*, $\frac{A_o + A_i}{2}$, can be used instead of the log mean area, with about four percent error. When calculating heat loss from a cylinder of finite length, the loss from the ends must be included.

E. CONDUCTION THROUGH AN INSULATED PIPE

The common insulated pipe problem is an extension of the hollow cylinder solution of Fourier's equation. Using the concepts of driving force and resistance in series, the equation of the insulated pipe is

$$q = \frac{\pi N \Delta T}{\frac{ln\left(\frac{D_o}{D_i}\right)}{2k_{pipe}} + \frac{ln\left(\frac{D_{ins}}{D_o}\right)}{2k_{ins}}} \qquad 6.16$$

[1] Equation 6.10 is the last equation in this chapter which makes a point of including the negative sign in equations for heat transfer. Strictly speaking, the negative sign is required to show that the transfer is against the thermal gradient. The remainder of this chapter assumes the direction of heat transfer can be determined by inspection.

HT XFER

Table 6.1
Thermal Conductivities of Some Materials

substance	T, °F	k, BTU/hr-ft-°F
air	32	0.014
aluminum 6063	68	118
antimony	32	10.6
	212	9.7
asbestos	32	0.087
	398	0.12
bismuth	64	4.7
	212	3.9
brass (yellow)	32-212	69.1
brass (naval)	32-212	68.3
brass (admiralty)	32-212	64.3
bronze	32-212	108.3
cadmium	64	53.7
	212	52.2
copper	32-212	225.0
cork	68	0.025
dacron	68	0.088
gold	64	169.0
	212	170.0
iron (pure)	64	39.0
	212	36.6
iron (wrought)	64	34.9
	212	34.6
iron (cast)	120	27.6
	216	26.8
lead	32-212	20.0
magnesium	32-212	92.0
mercury	32	4.8
mylar	68	0.088
nickel	32-212	45.3
nickel alloy (62% Ni, 12% Cr, 26% Fe)	68	7.8
nylon	68	0.18
platinum	64	40.2
	212	41.9
silver	32-212	242
stainless steel (301, 302, 303, 304, 316)	212	9.4
	932	12.4
stainless steel (321, 347)	212	9.3
	932	12.8
steel (1% C)	64	26.2
	212	25.9
teflon	68	0.14
tin	32-212	1.07
titanium	32-212	9.8
tungsten	32-212	0.2
water	32	0.32
zinc	32-212	1.07
zirconium	32-212	0.33

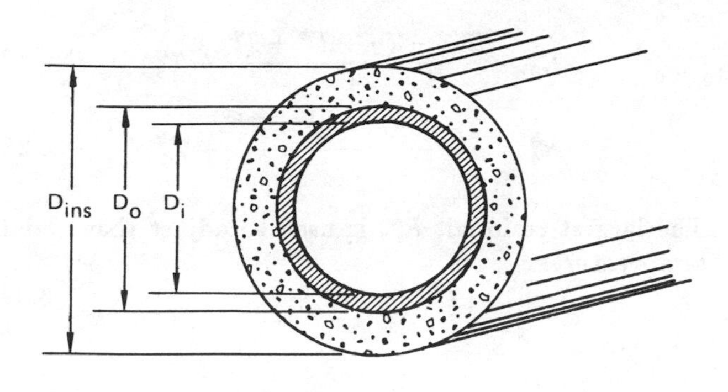

Figure 6.4 An Insulated Pipe

If only the bulk temperatures of the pipe fluid and the air surrounding the insulation are known, more resistances must be known to solve the insulated pipe problem. Specifically, the resistances of heat flow from the bulk fluid to the inside wall of the pipe, $\frac{1}{h_i A_i}$, from the outer wall of the pipe to the inner wall of the insulation, $\frac{1}{h_o A_o}$, and from the outer surface of the insulation to the air, $\frac{1}{h_{ins} A_{ins}}$, must be known. h values are the convection film coefficients for the respective surfaces. Equation 6.16 becomes

$$q = \frac{\pi N \Delta T}{\frac{1}{h_i D_i} + \frac{ln\left(\frac{D_o}{D_i}\right)}{2k_{pipe}} + \frac{1}{h_o D_o} + \frac{ln\left(\frac{D_{ins}}{D_o}\right)}{2k_{ins}} + \frac{1}{h_{ins} D_{ins}}} \qquad 6.17$$

Equation 6.17 is used to calculate heat losses when only the bulk temperatures are known; equation 6.16 is used when the pipe surface temperatures are known.

The addition of insulation to a pipe increases the surface area of the system, causing an increase of heat loss over bare pipe levels until a critical diameter is reached, after which additional thicknesses of insulation will decrease the heat loss. Eventually, the heat loss will again be the same as for the bare pipe. This insulation diameter is known as the *minimum diameter for insulation* or *critical diameter* since any thickness less than this minimum will cause the heat loss to increase. The determination of the critical diameter involves finding q as a function of D in equation 6.17. The partial derivative of q with respect to D_{ins} is

$$\frac{\partial q}{\partial D_{ins}} = \frac{-\pi N \Delta T \left[\frac{1}{2k_{ins} D_{ins}} - \frac{1}{D_{ins}^2 h_{ins}}\right]}{\left[\frac{ln\left(\frac{D_o}{D_i}\right)}{2k_{pipe}} + \frac{ln\left(\frac{D_{ins}}{D_o}\right)}{2k_{ins}} + \frac{1}{h_i D_i} + \frac{1}{h_o D_o} + \frac{1}{h_{ins} D_{ins}}\right]^2} \qquad 6.18$$

The critical insulation diameter is found by setting $\frac{\partial q}{\partial D_{ins}}$ equal to zero and solving for $D_{ins} = D_{critical}$.

$$D_{critical} = \frac{2k_{ins}}{h_{ins}} \qquad 6.19$$

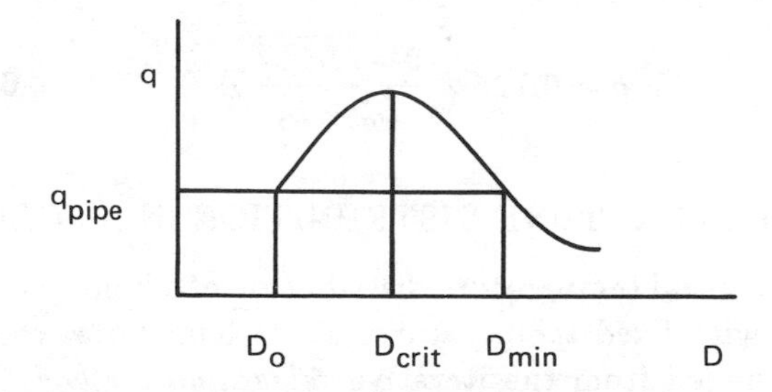

Figure 6.5 Heat Flow Versus Insulation Diameter

F. CONDUCTION THROUGH AN INSULATED SPHERE

The equation for conduction through a hollow sphere can also be derived from Fourier's equation. If the wall temperatures are known,

$$\begin{aligned} q &= \frac{2\pi k\, D_o D_i \Delta T}{D_o - D_i} \\ &= \frac{2k\sqrt{A_o A_i}\Delta T}{D_o - D_i} \\ &= \frac{4\pi k\, r_i r_o \Delta T}{r_o - r_i} \end{aligned} \qquad 6.20$$

For an insulated sphere whose wall temperatures are known, the heat flow equation becomes

$$q = \frac{2\pi \Delta T}{\frac{D_o - D_i}{k_{wall} D_o D_i} + \frac{D_{ins} - D_o}{k_{ins} D_{ins} D_o}} \qquad 6.21$$

If only the bulk temperatures are known, the heat flow is

$$q = \frac{2\pi \Delta T}{\frac{2}{h_i D_i^2} + \frac{D_o - D_i}{k_{wall} D_o D_i} + \frac{2}{h_o D_o^2} + \frac{D_{ins} - D_o}{k_{ins} D_{ins} D_o} + \frac{1}{h_{ins} D_{ins}^2}} \qquad 6.22$$

The *critical thickness* of insulation for a sphere is

$$D_{critical} = \frac{4k_{ins}}{h_{ins}} \qquad 6.23$$

HT XFER

G. CONDUCTION THROUGH A HOLLOW CUBE

For hollow cubes with $\frac{A_o}{A_i} > 2$, equation 6.20 for a sphere can be used to estimate the heat flow from a side of the cube by using the correction factor 0.725, and substituting L for D, and L^2 for A.

$$q = 0.725 \times \frac{2k\sqrt{L_o^2 L_i^2}}{L_o - L_i} \Delta T \qquad 6.24$$

H. TEMPERATURE DISTRIBUTION IN A SOLID

The internal temperature distribution of a homogeneous solid with fixed temperatures at its boundaries can be determined from the iterative *relaxation method.* The solid cross section is drawn with a square grid pattern. The grid intersections, called *nodes*, are the points where the temperatures are to be calculated. To start, the temperature of each node is estimated for the "zeroeth" iteration. The technique involves calculating a residual for each node as the average adjacent nodal temperature minus the node temperature.

After all the *nodal residuals* have been calculated, the node with the largest residual is adjusted by adding the residual temperature to the node's temperature. The residuals of the system are recomputed, and the largest residual is again determined. (Only those nodes immediately affected have to be recomputed.) This procedure continues until the residuals are all nearly equal to zero. For an object with four nodes, with boundary temperatures of T_N, T_W, T_S, and T_E (for north, west, south, and east), the residuals at iteration k are[2]

$$R_1^k = \frac{T_W + T_N + T_2^k + T_3^k}{4} - T_1^k \qquad 6.25$$

$$R_2^k = \frac{T_E + T_N + T_1^k + T_4^k}{4} - T_2^k \qquad 6.26$$

$$R_3^k = \frac{T_W + T_S + T_1^k + T_4^k}{4} - T_3^k \qquad 6.27$$

$$R_4^k = \frac{T_E + T_S + T_2^k + T_3^k}{4} - T_4^k \qquad 6.28$$

The largest residual, R_i^k, is used to adjust that nodal temperature:

$$T_i^{k+1} = T_i^k + R_i^k \qquad 6.29$$

Example 6.1

Assume a solid has the boundary temperatures indicated. Assume all the nodal temperatures for iteration zero to be 200 °F. The iterations of the temperature and residual for each node are tabulated. What is the temperature distribution of the solid?

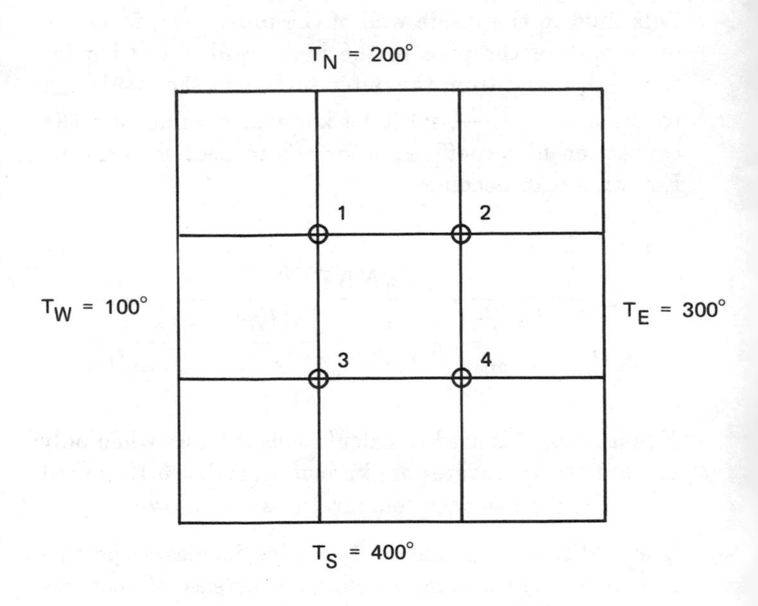

Table for Example 6.1

iteration	node 1		node 2		node 3		node 4	
k	T_1	R_1	T_2	R_2	T_3	R_3	T_4	R_4
0	200	−25	200	−75	200	−75	200	+75
1	200	−25	200	+64	200	+19	275	0
2	200	−8	264	0	200	+19	275	+16
3	200	−4	264	0	219	0	275	+21
4	200	−4	264	−15	219	30	296	0
5	200	+3	264	−15	249	0	296	7
6	200	−1	249	0	249	0	296	3
7	200	−1	249	+1	249	1	299	0

[2] Note that the k indicates the iteration number, and is not a power of the numbers calculated.

Example 6.2

A furnace wall is constructed with a layer of chrome brick (exposed to flame) 6″ thick with a mean thermal conductivity of 0.77 BTU-ft/hr-ft^2-°F, and another layer of fire brick 8″ thick with a mean thermal conductivity of 0.19 BTU-ft/hr-ft^2-°F. The inside furnace wall temperature is 1450 °F, and the outside furnace wall is 85 °F. What is the rate of heat loss through 10 ft^2 of furnace wall? What is the temperature between the two layers of bricks? Disregard inside films.

In this example, the wall temperatures are known, so film data is not needed. Ignoring the resistance between the bricks, the heat loss is

$$q = \frac{\Delta T}{R_1 + R_2} = \frac{\Delta T}{\frac{L_1}{k_1 A_1} + \frac{L_2}{k_2 A_2}}$$

$$q = \frac{1450 - 85}{\frac{6}{(12)(0.77)(10)} + \frac{8}{(12)(0.19)(10)}}$$

$$= \frac{1365}{0.0649 + 0.351}$$

$$q = 3282 \text{ BTU/hr}$$

$$\Delta T_{1-2} = qR_1 = (3282)(0.0649) = 213 \text{ °F}$$

$$T_2 = 1450 - 213 = 1237 \text{ °F}$$

Example 6.3[3]

The temperature of a 6″ steam pipe with an outside diameter of 6.63″, insulated with magnesia, is evaluated with thermocouples, one touching the pipe, and one in the insulation 3″ from the first thermocouple. If the first thermocouple registers 350 °F, and the second registers 150 °F, what is the rate of heat loss per foot of length?

$$k = 0.49 \text{ BTU-in/hr-ft}^2\text{-°F}$$
for magnesia

$$q = \frac{kA_m}{L}\Delta T$$

$$D_i = 6.63''$$

$$D_o = 6.63 + 2(3) = 12.63''$$

$$\frac{D_o}{D_i} = \frac{12.63}{6.63} = 1.9$$

$$A_m = \frac{\pi(1)(12.63 - 6.63)}{(12)\, ln\, \frac{12.63}{6.63}} = 2.44 \text{ ft}^2$$

[3] The arithmetic average method may be used to solve this problem; the log mean method is used here for illustration.

For 12 inches (1 foot) of pipe,

$$q = \left(\frac{0.49}{12}\right)(2.44)\left(\frac{12}{3}\right)(350 - 150)$$

$$= 79.7 \text{ BTU/hr-ft of length}$$

I. TEMPERATURE DISTRIBUTION IN A SLAB WITH HEAT SOURCE

Problems involving slabs or walls with heat sources, (e.g., nuclear fuel plates, curing concrete slabs, etc.) can be solved by simplifying equation 6.9. At steady state, the equation reduces to the one-dimensional *Poisson equation:*

$$\frac{d^2T}{dx^2} + \frac{S}{k} = 0 \qquad 6.30$$

The boundary conditions are: at $x = 0$, $T = T_1$; at $x = L$, $T = T_2$. Using the methods in chapter 1 to solve the differential equation, the first integration produces:

$$\frac{dT}{dx} + \frac{S}{k}x = C_1 \qquad 6.31$$

After the second integration,

$$T + \frac{S}{k}\frac{x^2}{2} = C_1 x + C_2 \qquad 6.32$$

$$C_2 = T_1 \qquad 6.33$$

$$C_1 = \frac{T_2 - T_1}{L} + \frac{SL}{2k} \qquad 6.34$$

$$T = \frac{SL^2}{2k}\left[\frac{x}{L} - \left(\frac{x}{L}\right)^2\right] + (T_2 - T_1)\frac{x}{L} + T_1 \qquad 6.35$$

The heat flux is found from the temperature gradient.

$$\phi = -k\frac{dT}{dx} = -k\left(C_1 - \frac{Sx}{k}\right) \qquad 6.36$$

$$\phi = k\frac{T_1 - T_2}{L} - \frac{SL}{2} + Sx \qquad 6.37$$

Example 6.4

A small concrete wall, idealized as an extensive plate 4′ thick, is to be completely poured in a short period of time. If both wall surfaces are kept at 60 °F and the heat of hydration of the concrete is considered to be a uniformly distributed heat source of 7 BTU/hr-ft^3, what is the maximum temperature in the concrete, assuming steady state conditions and $k = 0.7$ BTU/hr-ft-°F?

HT XFER

The maximum temperature occurs at the center. Using equation 6.35,

$$T_{max} = \frac{7(4)^2}{2(0.7)}\left[\frac{2}{4} - \left(\frac{2}{4}\right)^2\right] + 0 + 60$$

$$T_{max} = 80\ ^\circ\text{F}$$

J. TEMPERATURE DISTRIBUTION IN A CYLINDER WITH HEAT GENERATION

The best way to evaluate a cylinder with internal heat generation is to convert the generalized equation into cylindrical coordinates.

$$\frac{\partial^2 T}{\partial r^2} + \frac{1}{r}\frac{\partial T}{\partial r} + \frac{1}{r^2}\frac{\partial^2 T}{\partial \psi^2} + \frac{\partial^2 T}{\partial z^2} + \frac{S}{k} = \frac{\rho c_p}{k}\frac{\partial T}{\partial t} \qquad 6.38$$

If the cylinder is subjected to a uniformly distributed heat source, and if there are no longitudinal temperature gradients (i.e., $\frac{\partial^2 T}{\partial \psi^2} = 0$ and $\frac{\partial^2 T}{\partial z^2} = 0$), then at steady state the generalized one-dimensional heat transfer equation reduces to

$$\frac{d^2 T}{dr^2} + \frac{1}{r}\frac{dT}{dr} + \frac{S}{k} = 0 \qquad 6.39$$

If the cylinder is hollow and has inside and outside radii a and b, respectively, the boundary conditions are

$$\text{at } r = a, \quad T = T_1$$
$$\text{at } r = b, \quad T = T_2$$

Solving the differential equation 6.39, the temperature distribution of a hollow cylinder with heat source is

$$T = -\frac{S}{4k}r^2 + C_1 \ln r + C_2 \qquad 6.40$$

The constants of integration are determined from the boundary conditions.

$$C_1 = \frac{\left(\frac{S}{4k}\right)(b^2 - a^2) - (T_1 - T_2)}{\ln\left(\frac{b}{a}\right)} \qquad 6.41$$

$$C_2 = T_2 + \frac{S}{4k}b^2 - C_1 \ln b \qquad 6.42$$

$$T = \frac{Sb^2}{4k}\left[\left(1 - \frac{a^2}{b^2}\right)\frac{\ln\left(\frac{r}{b}\right)}{\ln\left(\frac{b}{a}\right)} - \left(1 + \frac{r^2}{b^2}\right)\right] - (T_1 - T_2)\frac{\ln\left(\frac{r}{b}\right)}{\ln\left(\frac{b}{a}\right)} + T_2 \qquad 6.43$$

The heat flux is found from the temperature gradient.

$$\phi = \frac{S}{2}\left[r - \frac{b^2 - a^2}{2r\ln\left(\frac{b}{a}\right)}\right] + \frac{(T_1 - T_2)k}{r\ln\left(\frac{b}{a}\right)} \qquad 6.44$$

If the cylinder is solid, $a = 0$, and the surface temperature at the outer radius is T_o, the constant C_1 is zero. The temperature gradient is

$$T = \frac{Sb^2}{4k}\left(1 - \frac{r^2}{b^2}\right) + T_o \qquad 6.45$$

Example 6.5

What is the temperature at the center of a solid nuclear fuel rod 0.5″ in diameter generating 12,000 BTU/in^3-hr if the surface temperature is held at 540 °F by the reactor coolant? Assume $k = 9$ BTU/hr-ft-°F.

The temperature at the center is calculated from equation 6.45 with $r = 0$ and $b = 0.52/2$.

$$T = \frac{(12{,}000 \times 12^3)\left(\frac{0.5}{12}\right)^2}{(4)(9)}(1 - 0) + 540$$
$$= 790\ ^\circ\text{F}$$

K. HEAT CONDUCTION IN A COOLING FIN

Consider the cooling fin shown in figure 6.6. Assuming that the wall temperature is constant, the temperature in the fin is a function of z alone, and there is no heat loss from the ends or edges of the fin, then the heat flux at the surface of the fin is governed by *Newton's Law of Cooling*:

$$\phi = \frac{q}{A} = h(T_s - T_{air}) \qquad 6.46$$

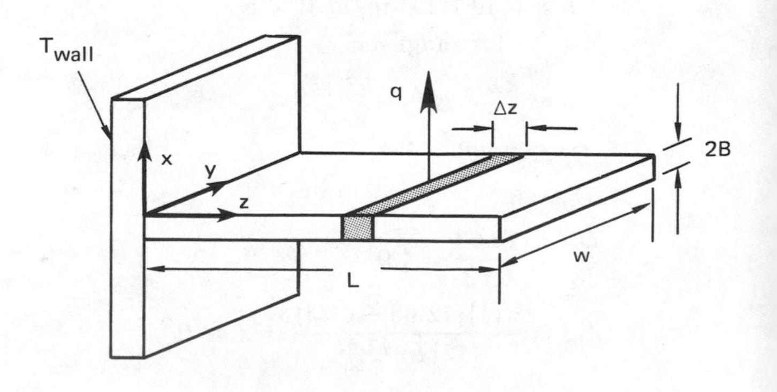

Figure 6.6 Fin Nomenclature

An energy balance for a fin section Δz in length with a constant heat flow area of 2Bw yields the equation

$$\frac{-\partial\phi}{\partial z} = \frac{h}{B}(T - T_{air}) \qquad 6.47$$

Inserting this into Fourier's Law,

$$\frac{-\partial^2 T}{\partial z^2} = \frac{h}{kB}(T - T_{air}) \qquad 6.48$$

The boundary conditions are

$$\text{at } z = 0, \quad T = T_w$$

$$\text{at } z = L, \quad \frac{\partial T}{\partial z} = 0$$

The equation for the temperature profile along the fin is

$$T = \frac{(T_w - T_{air})\cosh\left[\sqrt{\frac{hL^2}{kB}}\left(1 - \frac{z}{L}\right)\right]}{\cosh\left(\sqrt{\frac{hL^2}{kB}}\right)} + T_{air} \qquad 6.49$$

The *effectiveness* of a fin, η, is defined as

$$\eta = \frac{\text{heat actually dissipated by fin}}{\text{heat which would be dissipated if fin were at } T_w} \qquad 6.50$$

$$= \frac{-\tanh\left(\sqrt{\frac{hL^2}{kB}}\right)}{\sqrt{\frac{hL^2}{kB}}} \qquad 6.51$$

Example 6.6

What is the surface temperature of a 1.0″ thick copper fin 3″ from the wall it is attached to, if the wall is at 350 °F, the fin is 1′ wide, and air at 100 °F is flowing over it? Assume that $h = 5$ BTU/hr-ft^2-°R.

From table 6.1, k_{copper} = 225 BTU/hr-ft-°F. From equation 6.49, with $B = 0.5''$,

$$T = \frac{(350 - 100)\cosh\left[\left(1 - \frac{\frac{1}{4}}{1}\right)\sqrt{\frac{5(1)^2}{225\left(\frac{0.5}{12}\right)}}\right]}{\cosh\sqrt{\frac{5(1)^2}{225\left(\frac{0.5}{12}\right)}}} + 100$$

$$= \frac{250\cosh\left(0.75\sqrt{0.5333}\right)}{\cosh\sqrt{0.5333}} + 100$$

$$= \frac{250(1.1538)}{1.2808} + 100$$

$$T = 325.2\ °\text{F}$$

2 RADIATION HEAT TRANSFER

A. BLACKBODY RADIATION

An *ideal emitter* radiates a heat flux of

$$q = \sigma A T^4 = 0.1713\,A\left(\frac{T}{100}\right)^4 \qquad 6.52$$

If the surface is isolated and can receive no radiant energy from any other source, equation 6.52 gives the maximum energy that can be emitted. This upper limit is known as the *blackbody radiation.*

B. ACTUAL RADIATION FROM A SINGLE BODY

In equation 6.53, ε is the *emissivity* of surface. Equation 6.53 reduces the blackbody radiation according to the characteristics of the actual surface.

$$q = \varepsilon\sigma A T^4 \qquad 6.53$$

C. RADIATION INTERCHANGE BETWEEN SURFACES

The net transfer between two surfaces, 1 and 2, is

$$q_1 = 0.1713\,F_{12}A_1\left[\left(\frac{T_1}{100}\right)^4 - \left(\frac{T_2}{100}\right)^4\right] \qquad 6.54$$

F_{12} is the *direct view factor*, which depends on the emissivities and geometric arrangement. F_{12} is sometimes written as $F_\varepsilon F_A$, where F_ε is the combined emissivity factor for both surfaces.

F_A is an *arrangement factor* used to account for the geometric arrangement, which shields some of the emitted radiation from the absorber.

Alternatively, the radiant energy transfer can be calculated by using an equation analogous to the conduction equation.

$$q = h_r A\,(T_1 - T_2) \qquad 6.55$$

h_r is the *coefficient of heat transfer by radiation.*

$$h_r = \sigma F_\varepsilon F_A \frac{T_1^4 - T_2^4}{T_1 - T_2} \qquad 6.56$$

D. APPLICATION OF RADIATION EQUATIONS

Between two large parallel plates:

$$\frac{q}{A} = \frac{\sigma}{\frac{1}{\varepsilon_1} + \frac{1}{\varepsilon_2} - 1}\left(T_1^4 - T_2^4\right) \qquad 6.57$$

For spheres or cylinders inside spherical or cylindrical enclosures:

$$q = \frac{\sigma A_1}{\frac{1}{\varepsilon_1} + \frac{A_1}{A_2}\left(\frac{1}{\varepsilon_2} - 1\right)}(T_1^4 - T_2^4) \qquad 6.58$$

Example 6.7

Two very large parallel walls are at constant temperatures of 800 °F and 1000 °F. (a) Assuming that the walls are blackbodies, how much heat must be removed from the colder wall to maintain a constant temperature? (b) If the two walls have emissivities of 0.6 and 0.8, what is the net exchange of heat?

(a) Converting °F to °R by adding 460, using $F_e = 1$, and assuming $F_A = 1$, equation 6.54 yields

$$\frac{q}{A} = 0.1713\left[\left(\frac{1000+460}{100}\right)^4 - \left(\frac{800+460}{100}\right)^4\right]$$

$$= 3500 \text{ BTU/hr-ft}^2$$

(b) From equation 6.57,

$$\frac{q}{A} = \frac{\sigma}{\frac{1}{\varepsilon_1} + \frac{1}{\varepsilon_2} - 1}(14.6^4 - 12.6^4)$$

$$= \frac{0.173}{\frac{1}{0.6} + \frac{1}{0.8} - 1}(14.6^4 - 12.6^4)$$

$$= 1825 \text{ BTU/hr-ft}^2$$

E. RADIATION ERRORS IN PYROMETRY

Radiation errors in pyrometry occur when a thermocouple (or thermometer) is used to measure the temperature of a gas, and the temperature of the surroundings is different from the temperature of the gas. The thermocouple will indicate a temperature between the temperatures of the gas and the surroundings.

Under conditions of thermal equilibrium, the rate of heat flow by radiation is equal to that by conduction and convection.

$$\frac{q}{A} = 0.1713\,\varepsilon_c\left[\left(\frac{T_w}{100}\right)^4 - \left(\frac{T_c}{100}\right)^4\right] = h(T_c - T_g) \qquad 6.59$$

$$\varepsilon_c \approx 0.90 \text{ to } 0.96 \qquad 6.60$$

3 CONVECTIVE HEAT TRANSFER

Convection is the transfer of heat by moving matter. The fluid used for convection absorbs heat and then physically moves away, carrying the heat with it. It is constantly replaced by fluid at a lower temperature, by the convective flow. The two classes of convection are *natural* and *forced*.

Convective heat transfer is predicted by equation 6.61:

$$q = hA(T_s - T_\infty) \qquad 6.61$$

4 UNSTEADY STATE HEAT TRANSFER

Analytical solutions for unsteady state heat transfer have been developed for solid slabs, cylinders, and spheres. The solutions are plotted as curves involving four ratios. Since the ratios are dimensionless, any consistent set of units may be used.

$$Y = \frac{T_\infty - T}{T_\infty - T_b} = \frac{T_s - T}{T_s - T_o} \qquad 6.62$$

$$X = \frac{kt}{\rho c_p r_m^2} = \frac{4\alpha t}{L^2} \qquad 6.63$$

$$m = \frac{k}{h_T r_m} = \frac{2k}{hL} \qquad 6.64$$

$$n = \frac{r}{r_m} = \frac{2x}{L} \qquad 6.65$$

T_b = initial uniform temperature of the body

T_∞ = temperature of the surroundings

T = temperature in the body at time t measured from the start of heating or cooling operation

h_T = coefficient of total heat transfer between the surroundings and the surface of the body

r = distance, in the direction of heat conduction, from the midpoint or midplane of the body to the point under consideration

r_m = radius of a sphere or cylinder, one-half thickness of a slab heated or cooled from both faces, or the total thickness of a slab heated or cooled from one face and insulated perfectly on the other

x = distance in the direction of heat conduction, from the surface of a semi-infinite body to the point under consideration

The curves in figures 6.7(a) through 6.7(d) have Y as the ordinate on logarithmic scale, and X as the abscissa on an arithmetic scale for various values of m and n.

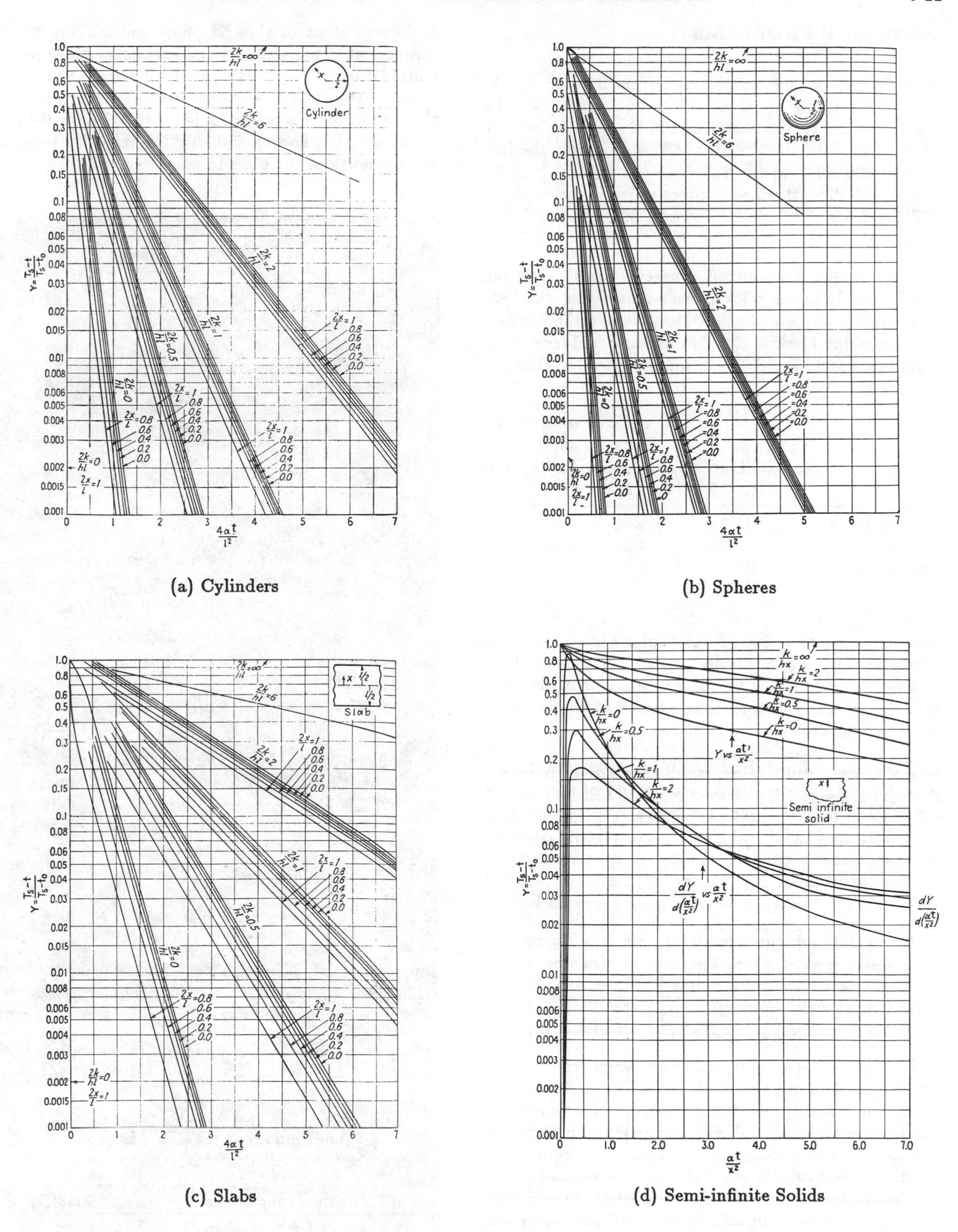

(a) Cylinders

(b) Spheres

(c) Slabs

(d) Semi-infinite Solids

Figure 6.7 Transient Cooling Charts for Solid Objects

PRACTICE PROBLEMS

1. A furnace wall is constructed from 9″ of firebrick, 4.5″ of insulating brick, and 9″ of building brick. The inside surface temperature is 1400 °F, and the outside air temperature is 85 °F. If the heat loss from the furnace is 200 BTU/hr-ft^2, calculate the air heat transfer coefficient. The thermal conductivities of the brick, in BTU-ft/hr-ft^2-°F, are: firebrick, 0.8; insulating brick, 0.15; and, building brick, 0.4.

2. A flat wall is constructed of fireclay brick, insulating brick, and building brick in series such that the heat loss will not exceed 250 BTU/hr-ft^2 when the hot face of the fireclay brick is 2000 °F, and the cold face of the building is 100 °F. (a) What minimum wall thickness is required? (b) What actual heat loss will occur in (a)?

	k, BTU-ft/hr -ft^2-°F	brick thickness, inches	maximum allowable temperature °F
fireclay brick	0.90	4.5	
insulating brick	0.12	3.0	1800
building brick	0.40	4.0	300

3. A hollow metal sphere is heated so that the inside wall temperature is 300 °F. The sphere has an internal diameter of 6″, and has walls 2″ thick. What is the heat loss from the sphere if the outer surface is maintained at 212 °F? The thermal conductivity of the metal is 8 BTU-ft/hr-ft^2-°F.

4. The inside dimensions of an elevated furnace are 4′ × 5′ × 3′. The thermal conductivity of the walls is 0.8 BTU-ft/hr-ft^2-°F. The inner and outer surface temperatures are 500 °F and 100 °F, respectively. All walls are the same material and thickness. All six outer surfaces dissipate heat. For the loss not to exceed 1,440,000 BTU/day (24 hours), how thick must the walls be?

5. A new 2.0″ (outside diameter) steel tube carries saturated steam at 52 psig. The tube is insulated with a 2″ layer of asbestos covered by a $1\frac{1}{2}$″ layer of cork. The measured surface temperature is 85 °F. Calculate the heat loss from the pipe in BTU/hr per foot of pipe.

6. Parallel wooden outer and inner walls of a building are 15′ long, 10′ high, and 4″ apart. The outer surface of the inner wall is at 170 °F, and the inner surface of the outer wall is at 0 °F. (a) Calculate the heat loss in BTU/hr neglecting any leakage of air through the walls. (Note that the calculation of the film coefficient for natural convection is part of this problem.) (b) Suppose the air space is divided in half by a 0.001″ aluminum foil. How much would this affect the heat transfer through the air space?

7. A large sheet of glass 2″ thick and initially 300 °F throughout is plunged into a stream of water having a temperature of 60 °F. How long will it take to cool the center of the glass to a temperature of 100 °F? Assume that $m = 0$. For glass: $k =$ .40 BTU/hr-ft-°F; ρ = 155 lbm/ft^3; and $c_p = 0.20$ BTU/lbm-°F. Disregard any tendency of the glass to crack.

HT XFER

7 HEAT TRANSFER: CONVECTION AND EQUIPMENT

Nomenclature

a	dimensionless parameter in Nusselt equation	–
A	area	ft^2
c	heat capacity	BTU/lbm-°F
d	diameter	in
D	diameter	ft
F_t	temperature correction factor	–
g	gravitational constant (4.17×10^8)	ft/hr^2
G	mass velocity ($V\rho$)	$lbm/hr\text{-}ft^2$
GTTD	greater terminal temperature difference	°F
h	film coefficient of heat transfer, or enthalpy	$BTU/hr\text{-}ft^2\text{-}°F$ BTU/lbm
j_H	Colburn factor for heat transfer	–
k	thermal conductivity	BTU/hr-ft-°F
L	wall thickness, tube length, or characteristic length	ft
L′	wall thickness	ft
LMTD	log mean temperature difference	°F
LTTD	lesser terminal temperature difference	°F
m	mass	lbm
$\dot{m}$	mass rate	lbm/hr
N_{Gr}	Grashof number $\left(\frac{g D^3 \beta \Delta T \rho^2}{\mu^2}\right)$	–
N_{Gz}	Graetz number $\left(\frac{c G D^2}{k L}\right)$	–
N_{Nu}	Nusselt number $\left(\frac{h D}{k}\right)$	–
N_{Pr}	Prandtl number $\left(\frac{c \mu}{k}\right)$	–
N_{Re}	Reynolds number $\left(\frac{D V \rho}{\mu}\right)$	–
N_{St}	Stanton number $\left(\frac{h}{c G}\right)$	–
q	heat transfer	BTU/hr
R	thermal resistance	$ft^2\text{-}hr\text{-}°F/BTU$
t	temperature of cold fluid	°F
T	temperature	°F
U	overall heat transfer coefficient	$BTU/hr\text{-}ft^2\text{-}°F$
v	velocity	ft/sec
V	velocity	ft/hr

Symbols

β	coefficient of thermal expansion	$°F^{-1}$
Λ	latent heat	BTU/lbm
μ	viscosity	lbm/ft-hr
ρ	density	lbm/ft^3

Subscripts

b	bulk
c	cold side, clean, or condensing
d	dirty
e	equivalent
f	film, or fluid (liquid), or fouling
g	gas (vapor)
h	hot side
i	inside
io	inside referred to outside area
m	log mean
o	outside
oi	outside referred to inside area
p	constant pressure
s	shell
w	wall

1 CONVECTION

Convection is the removal of heat from one point within a fluid (gas or liquid) by mixing one portion of the fluid with another. There are two types of convection: natural and forced. *Natural convection* occurs when the movement of the fluid is due to the density gradients resulting from the temperature difference. *Forced convection* occurs when the fluid motion is produced by mechanical means.

A. FILM THEORY

Convective heat transfer is an important factor when a liquid or gas transfers heat. The majority of heat transfer processes involve the transfer of heat from a solid (e.g., a tube wall) to a fluid. Figure 7.1 shows a generalized diagram of the temperature gradient from a hot fluid to a cold fluid through a tube wall.

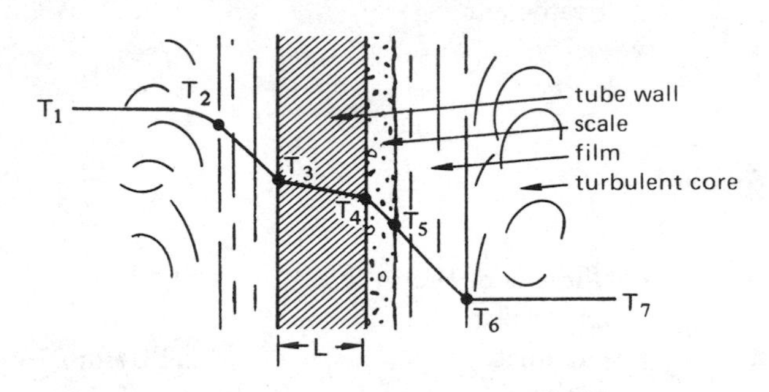

Figure 7.1 Temperature Gradient Through Heat Exchanger Tube Wall

For turbulent flow, there is a laminar zone of fluid adjacent to the surface of the tube wall called the *film*. The resistance to heat flow through the film can vary, depending on its thickness, from one to 95 percent of the total thermal resistance. The turbulent fluid core and the transition zone between the core and the film also introduce resistance to heat flow. The resistance is dependent upon the degree of turbulence and the fluid properties.

B. HEAT TRANSFER COEFFICIENTS

Since it is not convenient to measure temperatures in the film, the convective film heat transfer coefficients are implicitly defined in equation 7.1:

$$q = h_i A_i (T_1 - T_3) = h_o A_o (T_5 - T_7) \quad 7.1$$

The heat flow from the core of one fluid to the core of another fluid with a solid wall between them is

$$q = UA\,\Delta T = UA(T_1 - T_7) \quad 7.2$$

In equation 7.2, the *overall coefficient of heat transfer*, U, includes all the resistances due to turbulence, buffer zones, scale, conduction through the wall, fouling, and films. Analogous to the resistance concept developed in the conductive heat transfer chapter, all of the individual heat transfer coefficients contribute to the overall coefficient.

$$q = UA\,\Delta T = \frac{\Delta T}{\frac{1}{UA}} = \frac{\Delta T}{R_1 + R_2 + \cdots}$$

$$= \frac{\Delta T}{\frac{1}{h_i A_i} + \frac{L'}{kA_m} + \frac{1}{h_o A_o}} \quad 7.3$$

$$\frac{1}{U_i A_i} = \frac{1}{h_i A_i} + \frac{L'}{kA_m} + \frac{1}{h_o A_o} \quad 7.4$$

$$\frac{1}{U_i} = \frac{1}{h_i} + \frac{L' A_i}{kA_m} + \frac{A_i}{h_o A_o} \quad 7.5$$

In equation 7.5, U_i is the overall heat transfer coefficient based on the inside area of the tube. The term A_m is the *log mean area* of the inner and outer surfaces of the tube. Based on the inside area, the overall heat transferred is

$$q = U_i A_i \Delta T \quad 7.6$$

The term h_{oi} is the outside convection coefficient based on the inside area.

$$\frac{1}{h_{oi}} = \frac{A_i}{h_o A_o} \quad 7.7$$

If the heat transfer coefficient is based on the outside area (as is the usual practice), then

$$q = U_o A_o \Delta T \quad 7.8$$

$$\frac{1}{U_o A_o} = \frac{1}{h_i A_i} + \frac{L'}{kA_m} + \frac{1}{h_o A_o} \quad 7.9$$

$$\frac{1}{U_o} = \frac{A_o}{h_i A_i} + \frac{L' A_o}{kA_m} + \frac{1}{h_o} \quad 7.10$$

The term h_{io} can also be used. It is the inside convection coefficient based on the outside area:

$$\frac{1}{h_{io}} = \frac{A_o}{h_i A_i} \quad 7.11$$

The convection coefficients must coincide with the area upon which U is based.

If the tube material is not known, the overall heat transfer coefficient can be approximately determined from the two film coefficients:

$$\frac{1}{U_o} \approx R_{io} + R_o = \frac{1}{h_{io}} + \frac{1}{h_o} = \frac{h_{io} + h_o}{h_{io} h_o} \quad 7.12$$

Equation 7.12 disregards tube wall resistance, tube scaling, and fouling. Where the metal wall is tubular, the area of a tube is proportional to the diameter, so that

$$\frac{1}{U_o} = \frac{1}{h_o} + \frac{L' D_o}{k D_m} + \frac{D_o}{h_i D_i} \quad 7.13$$

With large diameter, thin wall tubes, $\frac{D_o}{D_m}$ and $\frac{D_o}{D_i}$ can be taken as 1.0, so that

$$\frac{1}{U_o} = \frac{1}{h_o} + \frac{L'}{k} + \frac{1}{h_i} \quad 7.14$$

Where h_o is small compared to h_i and $\frac{k}{L'}$, then,

$$U_o = h_o \quad 7.15$$

C. EMPIRICAL FILM COEFFICIENTS FOR NATURAL CONVECTION

Most of the empirical relationships for convection coefficients require simple geometric configurations. The common configurations are horizontal and vertical cylinders (pipes), horizontal and vertical plates (floor or walls), and long thin wires (heating elements). Figure 7.2 illustrates the characteristic dimensions of common geometric shapes. Tables 7.1 and 7.2 show formulas used for these shapes in air and liquids.

The *Nusselt correlation* is used widely for estimating convection heat transfer coefficients. (Equations 7.16 and 7.17 both require use of consistent units.)

$$\frac{hL}{k} = a\left[\left(\frac{g L^3 \beta \Delta T \rho^2}{\mu^2}\right)\left(\frac{c_p \mu}{k}\right)\right]^m \quad 7.16$$

$$N_{Nu} = a\left(N_{Gr} N_{Pr}\right)^m \quad 7.17$$

Table 7.1[1]
Values of Natural Convection Coefficients for Air (14.7 psia)
(ΔT in °F, L in feet)

$N_{Gr} N_{Pr}$	h, BTU/hr-ft^2-°F	surface
$> 10^9$	$0.18\ \Delta T^{0.33}$	all
10^3 to 10^9	$0.27\left(\frac{\Delta T}{L}\right)^{0.25}$	horizontal cylinders
10^3 to 10^9	$0.28\left(\frac{\Delta T}{L}\right)^{0.25}$	vertical plates or cylinders
10^5 to 10^7	$0.27\ \Delta T^{0.25}$	heated horizontal plates facing up
10^5 to 10^7	$0.20\ \Delta T^{0.25}$	cooled horizontal plates facing down

Table 7.2[1]
Values of h for Natural Convection to Liquids at 70°F

liquid	h	surface
water	$43\left(\frac{\Delta T}{L}\right)^{0.25}$	horizontal cylinders
water	$26\left(\frac{\Delta T}{L}\right)^{0.25}$	vertical plates
organic liquids	$20\left(\frac{\Delta T}{L}\right)^{0.25}$	horizontal plates up (heated)
organic liquids	$12\left(\frac{\Delta T}{L}\right)^{0.25}$	horizontal plates down (cooled)

[1] Units in this table are inconsistent. ΔT must be in °F and L must be in feet. (L is the diameter of horizontal pipes, height of vertical pipes, or the longest dimension of horizontal plates.) For horizontal surfaces, the heated surface faces up, the cooled surface faces down.

Table 7.3
Values of a and m in Nusselt Equation for Natural Convection

	vertical surfaces		horizontal cylinders		horizontal plates		horizontal or vertical thin wires	
$N_{Gr}N_{Pr}$	a	m	a	m	a	m	a	m
$> 10^9$	0.13	1/3	0.13	1/3				
10^4 to 10^9	0.59	1/4	0.53	1/4				
$< 10^4$	1.36	1/6	1.09	1/6				
$> 2 \times 10^7$					0.14	1/3		
10^5 to 2×10^7					0.54	1/4		
10^{-7} to 1							1	0.1

D. EMPIRICAL FILM COEFFICIENTS FOR FORCED CONVECTION

Relationships for forced convection coefficients are commonly calculated using one of two correlations: the *Sieder-Tate equation* (also known as the Nusselt correlation) and the *Colburn type equation.* For $N_{Re} > 10{,}000$, $0.7 < N_{Pr} < 700$, $\frac{L}{D} > 60$ (where L is the length of the tube with diameter D), and properties based on bulk temperature, the Sieder-Tate equation is

$$N_{Nu} = 0.023\,(N_{Re})^{0.8}(N_{Pr})^{\frac{1}{3}}\left(\frac{\mu_b}{\mu_w}\right)^{0.14} \qquad 7.18$$

The ratio $\frac{\mu_b}{\mu_w}$ is the ratio of the bulk and wall viscosities. When this ratio is equal to 1.0, the Sieder-Tate equation is known as the *Nusselt form.* Table 7.4 gives some values for two functions of the viscosity ratio, $\frac{\mu_b}{\mu_w}$.

Table 7.4
Functions of $\frac{\mu_b}{\mu_w}$

$\frac{\mu_b}{\mu_w}$	$\left(\frac{\mu_b}{\mu_w}\right)^{0.14}$	$\left(\frac{\mu_b}{\mu_w}\right)^{-0.14}$
10.0	1.380	0.724
8.0	1.338	0.747
6.0	1.285	0.778
5.0	1.253	0.798
4.0	1.214	0.824
2.0	1.102	0.908
1.0	1.000	1.000
0.8	0.969	1.032
0.6	0.931	1.074
0.5	0.908	1.102

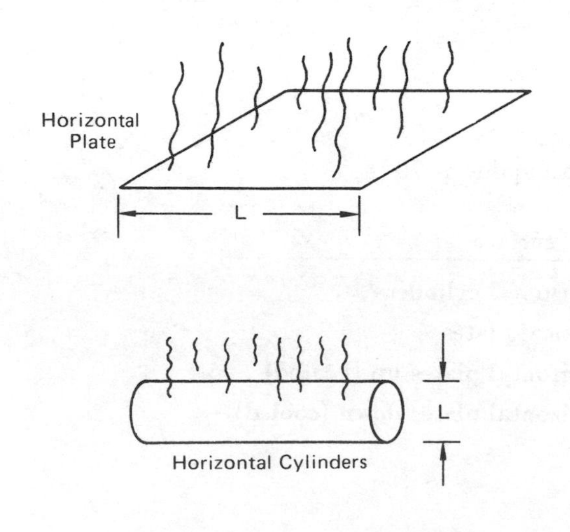

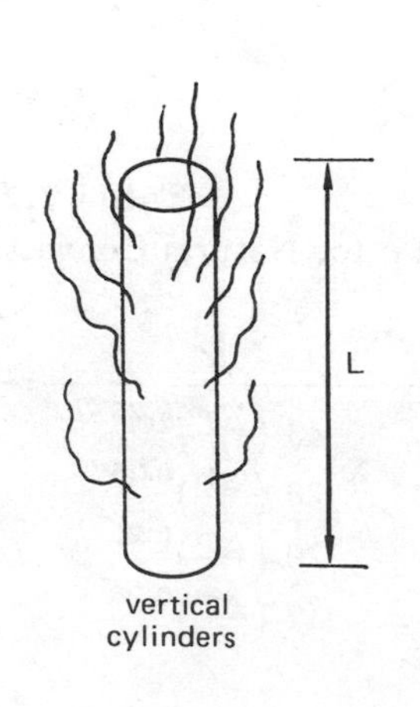

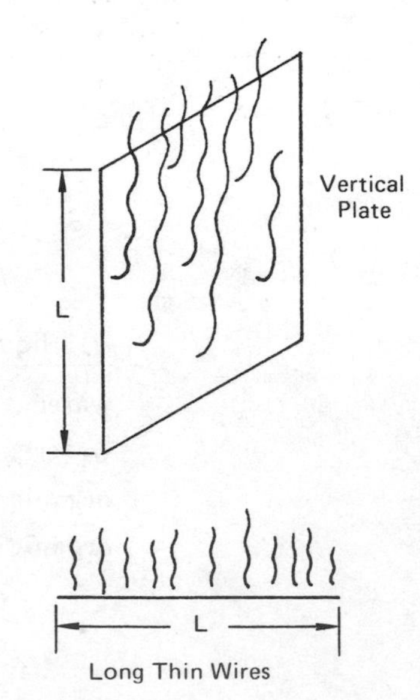

Figure 7.2 Characteristic Dimensions of Common Geometric Shapes

A generalized relationship derived from the Sieder-Tate equation for air between 0 °F and 240 °F is

$$h = \frac{C\,G^{0.8}}{D^{0.2}} \quad 7.19$$

$$C = (351 + 0.1583\,T_{^\circ F}) \times 10^5 \quad 7.20$$

The Colburn type equation for heating or cooling fluids flowing inside the tubes is

$$j_H = \frac{h}{c_p\,G}\left(\frac{c_p\,\mu}{k}\right)^{\frac{2}{3}}\left(\frac{\mu_w}{\mu_b}\right)^{0.14} \approx \frac{0.023}{(N_{Re})^{0.2}} \quad 7.21$$

For fluids flowing outside of but parallel to the tubes, substitute D_e for D in the Reynolds number.

$$D_e = 4\left(\frac{\text{flow area}}{\text{heated perimeter}}\right) \quad 7.22$$

It should be noted that equivalent diameter for heat transfer is not necessarily defined the same as equivalent diameter for fluid flow.

A plot of the Colburn factor, j_H, is shown in figure 7.9. For laminar flow, $N_{Gz} > 100$, small diameters, and small temperature differences, the heat transfer coefficient can be correlated with N_{Gz}:

$$N_{Nu} = \frac{h\,L}{k} = 1.86\,(N_{Gz})^{\frac{1}{3}}\left(\frac{\mu_b}{\mu_w}\right)^{0.14} \quad 7.23$$

E. DIMENSIONLESS NUMBERS

Numerous dimensionless numbers are used in convection heat transfer correlations. Table 7.5 lists these numbers and their associated variables. Because the numbers are dimensionless, consistent units must be used.

Table 7.5
Dimensionless Numbers

name and symbol	formula
Nusselt, N_{Nu}	$\frac{hD}{k}$
Stanton, N_{St}	$\frac{N_{Nu}}{N_{Re}N_{Pr}} = \frac{h}{c_pG}$
Reynolds, N_{Re}	$\frac{Dv\rho}{\mu} = \frac{DG}{\mu}$
Prandtl, N_{Pr}	$\frac{c_p\mu}{k}$
Grashof, N_{Gr}	$\frac{gD^3\beta\Delta T\rho^2}{\mu^2}$
Graetz, N_{Gz}	$N_{Re}N_{Pr}\frac{D}{L} = \frac{c_pGD^2}{kL}$

2 HEAT EXCHANGER DESIGN

There are three main types of heat exchangers. The most commonly used type is the *surface heat exchanger.* Two fluids with different temperatures flow in adjoining spaces separated by a wall, and they exchange heat by convection and conduction.

The other two types of heat exchangers are *regenerators* and *cooling towers.* In these latter two types, the two fluids which exchange heat occupy the same space. The regenerator consists of high-bulk solids (rocks, pebbles, rotating honeycomb, etc.) which alternately store energy taken from the warmer fluid passing around them, and release energy to the colder fluid passing around them during a different period of time. In a regenerator, the heat is transferred under unsteady state conditions, and is a cyclic, repetitive process.

A cooling tower operates by bringing atmospheric air into contact with the hot fluid. A small portion of the hot liquid evaporates. The loss of latent heat shows up as a decrease in liquid temperature. In a cooling tower, heat and mass transfer occur in a steady state process.

There are many different designs for ordinary heat exchangers, but in principle, the only significant difference between them is the directions of flow of the two fluids. In general, there are two designs of surface heat exchangers: *cross flow* and *parallel flow.* Parallel flow can be subdivided into conflow (cocurrent) and counterflow exchangers. Under comparable conditions, more heat is transferred in counterflow exchangers than in parallel exchangers. Figure 7.3 shows the flow arrangement for parallel flow exchangers. There are conditions, however, where cross flow exchangers may transfer more heat than counterflow exchangers.

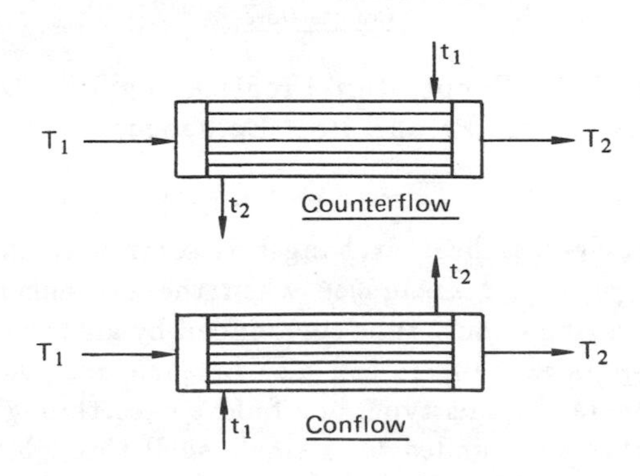

Figure 7.3 Parallel Flow Exchangers

Parallel flow heat exchangers carry two fluids: one hot and one cold. Heat is transferred to the cooled fluid primarily through forced convection. In general, these types of exchangers use tubes to provide the heat transfer area. (Other configurations use fins welded to the tubes.)

In a *single-pass heat exchanger*, each fluid is exposed to the other once. Figure 7.4 shows that, for conflow, the temperature of the warmer fluid decreases and the temperature of the colder fluid increases in the flow direction. The temperature gradient between the fluids becomes smaller, and the heat transfer decreases.

In contrast, the counterflow exchanger temperature profiles shown in figure 7.4 indicate that the temperature difference does not change significantly. For equal flow rates of fluids having identical properties, the temperature gradient is essentially constant over the entire length, the temperature of each fluid varies over the length, and the temperature profiles are parallel lines.

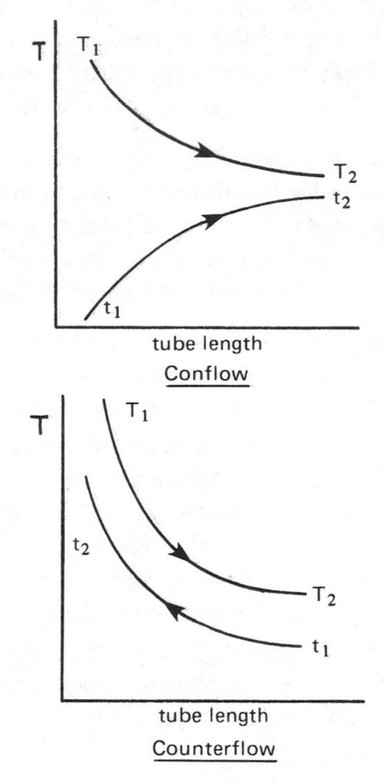

Figure 7.4 Temperature Profile Along Tube Length for Parallel Heat Exchanger

The single-pass heat exchanger is commonly called a *double pipe heat exchanger* when the exchanger consists of a single inner tube surrounded by an outer tube. Single-pass exchangers can also be *shell and tube heat exchangers*. In this type, one fluid passes through several tubes surrounded by a single shell through which the second fluid passes. Even in the simplest form, the shell and tube exchanger operates with mixed cross and parallel flow directions. Using baffles in the shell, the average cross section of the shell side fluid is reduced and its path length is increased.

If the tubes pass through the shell more than once, or if the shell fluid passes through the shell more than once by the use of baffles, then the heat exchanger is called a *multiple-pass heat exchanger*. The multiple-pass heat exchanger is used for increased efficiency. This kind of exchanger experiences conflow, counterflow, and crossflow fluid flow patterns.

In the *crossflow exchanger* (e.g., an automobile radiator), each fluid flows perpendicular to the other.

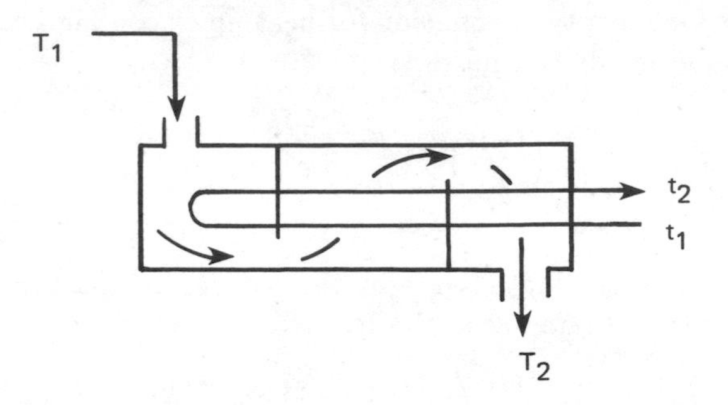

Figure 7.5 Multiple-Pass Shell and Tube Heat Exchanger

A. DEFINITIONS AND CONVENTIONS USED FOR HEAT EXCHANGERS

The following conventions are used so that general equations can be applied to the various types of exchangers. (Refer to figure 7.3.)

- The subscript 1 refers to the inlet, and the subscript 2 refers to the outlet.
- The temperature of the hot fluid is represented by T, and the temperature of the cold fluid is represented by t.
- *Temperature difference* of a heat exchanger refers to the *terminal temperature difference*, the difference in fluid temperature at the terminal ends of an exchanger.

	cold terminal	hot terminal
countercurrent	$T_2 - t_1$	$T_1 - t_2$
concurrent	$T_2 - t_2$	$T_1 - t_1$

- *Range* is the actual temperature rise or fall of one fluid. Thus, $T_1 - T_2$ and $t_2 - t_1$ are the ranges for hot and cold fluids, respectively.
- *Approach* is the smallest terminal temperature difference of the exchanger. A *close approach* is the smallest temperature difference allowed for a certain type of exchanger. A typical close approach for a countercurrent heat exchanger is 5 °F; a typical close approach for a conflow exchanger is 10 °F.

B. TEMPERATURE DIFFERENCES

The temperature gradient between the two fluids in a heat exchanger will, in general, vary along the flow path. The mean temperature difference can be calculated from the terminal temperatures if all the elements of a given fluid stream have the same thermal history, the exchanger operates in steady state, the specific heat is constant, the overall heat transfer coefficient is constant, and the losses are negligible. The basic heat exchanger equation, relating the total heat exchanged as a function of total area, overall heat transfer coefficient, and average log mean temperature difference is

$$q = U_o A_o \Delta T_m \qquad 7.24$$

The *log mean temperature difference* should be used in equation 7.24. Provided GTTD and LTTD are not equal,

$$\Delta T_m = \frac{\text{GTTD} - \text{LTTD}}{ln \dfrac{\text{GTTD}}{\text{LTTD}}} \qquad 7.25$$

Equation 7.25 applies to both counterflow and conflow heat exchangers, where both fluids undergo temperature variations which are not linear with tube length (as shown in figure 7.4). Except where one fluid is isothermal, such as condensing steam, there is a distinct disadvantage to using conflow exchangers. For the same terminal temperatures, heat transfer, and overall heat transfer coefficients, a conflow exchanger requires more area. Conflow exchangers are efficient when used as condensers.

Example 7.1

Calculate the log mean temperature difference for counterflow and conflow heat exchangers, each having a hot fluid entering at 300 °F with a range of 100 °F, and a cold fluid entering at 100 °F with a range of 50 °F.

$$T_1 = 300\ °\text{F}$$
$$T_2 = 200\ °\text{F}$$
$$t_1 = 100\ °\text{F}$$
$$t_2 = 150\ °\text{F}$$

Calculations for Example 7.1

	counterflow			conflow		
	hot fluid	cold fluid		hot fluid	cold fluid	
GTTD	300 −	150	= 150	300 −	100	= 200
LTTD	200 −	100	= 100	200 −	150	= 50
LMTD	$\Delta T_m = \dfrac{150-100}{ln \frac{150}{100}}$		= 123.3 °F	$\Delta T_m = \dfrac{200-50}{ln \frac{200}{50}}$		= 108.2 °F

C. ASSUMPTIONS GENERALLY USED IN HEAT EXCHANGER CALCULATIONS

The following assumptions are usually made when solving heat exchanger problems:

- Heat is transferred between fluids with no losses.
- Heat transfer is steady state.
- The heat transfer coefficient, U, is constant over the length of the exchanger tube.
- Specific heat is constant over the length of the tube.

Based on these assumptions, the heat balance for an exchanger with no phase change is

$$q = (\dot{m}\, c_p)_{hot}\,(T_1 - T_2) = (\dot{m}\, c_p)_{cold}\,(t_2 - t_1) \qquad 7.26$$

With an isothermal phase change in one of the fluids, the heat balance for the exchanger is

$$q = \dot{m}\,(h_g - h_f) = \dot{m}\,\Lambda \qquad 7.27$$

D. FOULING FACTORS

Fouling refers to scaling or deposits of material on a heat transfer surface. Such deposits usually have low thermal conductivity, which produces high resistance to heat transfer. In the expression for the overall heat transfer coefficient, the fouling factors are expressed as *resistances* or *reciprocal film coefficients*. Alternatively, both the inside and outside resistances can be combined into a single factor. The relationships for the overall dirty heat transfer coefficient, U_d, (also called *design* or *service coefficient*) and the *fouling resistances* and *fouling factor* are

$$U_{do} = \frac{1}{\frac{1}{h_o} + R_{do} + \frac{L}{kD_oD_{avg}} + R_{di}\frac{D_o}{D_i} + \frac{1}{h_i}\frac{D_o}{D_i}} \qquad 7.28$$

$$R_{do} = \text{outside fouling resistance} = \frac{1}{h_{do}} \qquad 7.29$$

$$R_{di} = \text{inside fouling resistance} = \frac{1}{h_{di}} \qquad 7.30$$

$$R_d = \text{overall fouling resistance}$$

$$= R_{di} + R_{do} = \frac{1}{h_f} \qquad 7.31$$

$$h_f = \text{fouling coefficient} \qquad 7.32$$

Table 7.6
Heat Transfer Coefficients (Fouling Factors)
for Deposits (BTU/hr-ft^2-°F)

temperature of heating medium	up to 240 °F		240-400 °F	
temperature of water	< 126 °F		≥ 126 °F	
water velocity, ft/sec	< 3	> 3	< 3	> 3
distilled	2000	2000	2000	2000
sea water	2000	2000	1000	1000
treated boiler feed water	1000	2000	1000	1000
treated cooling water makeup	1000	1000	500	500
brackish river water	500	1000	330	500
muddy, silty river water	330	500	250	330
hard water (over 15 grains/gal)	330	330	200	200

The quantity $\frac{1}{h_f}$ is called the *fouling resistance*.

There are no methods for consistently predicting fouling resistances. Many fluids do not foul heat exchanger surfaces at all. In general, fouling rates decrease with increasing flow rate. Fouling is prevalent in vaporizers. Some typical values of fouling resistances in hr-ft^2-°F/BTU, are: clean dry gas, 0; liquids of condensing vapors, 0.001; reboilers or vaporizers, 0.002. Other values of fouling coefficients are given in table 7.6.

The value of U obtained without consideration of fouling factors is called the *clean coefficient*, U_c.

$$\frac{1}{U_d} = \frac{1}{U_c} + R_{di} + R_{do} = \frac{1}{U_c} + R_d \qquad 7.33$$

$$R_d = \frac{U_c - U_d}{U_c U_d} \qquad 7.34$$

E. DESIGN OF DOUBLE PIPE HEAT EXCHANGERS

A double pipe heat exchanger is the purest form of a parallel flow heat exchanger. It consists of two concentric pipes or tubes. The outside pipe acts as the shell; the inside pipe acts as the tube.

The design of double pipe heat exchangers is straightforward. The effect of natural convection is ignored. In laminar flow, the required length of the inner pipe of the double pipe exchanger is computed by determining the area from equation 7.24 and dividing by the tube circumference.

Example 7.2

9820 pounds per hour of cold benzene are heated from 80 °F to 120 °F by passing through the inner pipe of a double pipe heat exchanger. Hot toluene flows countercurrently in the annulus, entering at 160 °F, and leaving at 100 °F. The specific gravities at 68 °F are 0.88 for benzene and 0.87 for toluene. A fouling factor of 0.001 should be provided for each stream. Heat loss to the air from the outer pipe can be neglected. The dimensions of the pipes are:

inside pipe: ID = 1.38 inches, OD = 1.66 inches
outside pipe: ID = 2.067 inches

The physical properties of the fluids are:

	benzene	toluene	
c_p	0.425	0.440	BTU/lbm-°F
μ	0.5	0.41	centipoise
k	0.091	0.085	BTU/hr-ft-°F

Find the length of 0.14″ thick interchanger pipe (0.435 ft^2/ft of outside area for inner pipe) required to produce this transfer of heat.

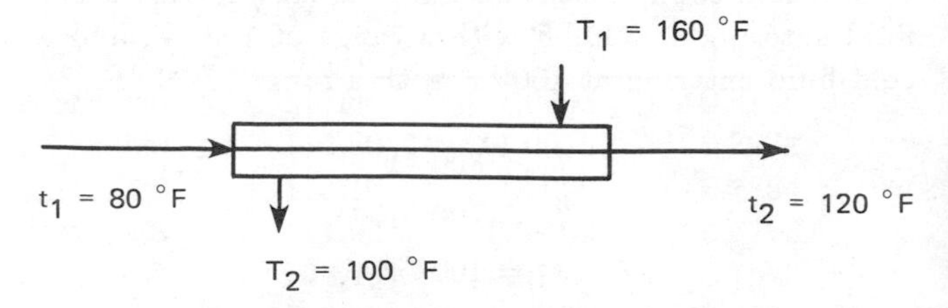

step 1: Heat balance from equation 7.26:

$$\begin{aligned} q &= \dot{m}_c c_{pc} (t_2 - t_1) \\ &= (9820)(0.425)(120 - 80) \\ &= 167{,}000 \text{ BTU/hr} \end{aligned}$$

$$\begin{aligned} \dot{m}_h &= \frac{167{,}000}{(0.44)(160 - 100)} \\ &= 6330 \text{ lbm/hr toluene} \end{aligned}$$

step 2: Log mean temperature difference from equation 7.25:

hot	cold	
160	120	GTTD = 160 − 120 = 40
100	80	LTTD = 100 − 80 = 20

$$\text{LMTD} = \frac{40 - 20}{ln\,\dfrac{40}{20}} = 28.8\ °\text{F}$$

step 3: Inner pipe:

$$D_i = \frac{1.38}{12} = 0.115\ \text{ft}$$
$$A_i = \frac{\pi}{4}(0.115)^2 = 0.0104\ \text{ft}^2$$

step 4: Outer pipe:

The equivalent diameter for heating purposes (equation 7.22) is

$$D_e = \frac{D_o^2 - D_{io}^2}{D_{io}} = \frac{(0.1725)^2 - (0.138)^2}{0.138} = 0.07762\ \text{ft}$$
$$A_o = \frac{\pi}{4}(D_o^2 - D_{io}^2) = 0.00841\ \text{ft}^2$$

step 5: Heat transfer:

benzene, inner pipe, cold fluid

$$G = \frac{\dot{m}}{A_i} = \frac{9820}{0.0104} = 943{,}000\ \text{lbm/hr-ft}^2$$
$$\mu = 0.5\,(2.42) = 1.21\ \text{lbm/ft-hr}$$
$$N_{Re} = \frac{D_i\,G}{\mu} = (0.115)\left(\frac{943{,}000}{1.21}\right) = 89{,}500$$

From figure 7.9,

$$N_{Re}j_H = 236;\ j_H = 0.002636$$
$$\left(\frac{c_p\mu}{k}\right)^{2/3} = \left[(0.425)\left(\frac{1.21}{0.91}\right)\right]^{2/3} = 3.173$$

From equation 7.21,

$$h_i = j_H c_p G\left(\frac{c_p\mu}{k}\right)^{-2/3}\left(\frac{\mu}{\mu_w}\right)^{0.14}$$
$$= (0.002636)(0.425)(943{,}000)\left(\frac{1}{3.173}\right)(\sim 1)$$
$$= 333\ \text{BTU/hr-ft}^2\text{-°F}$$

toluene, outer pipe, hot fluid

$$G = \frac{6330}{0.00826} = 767{,}000\ \text{lbm/hr-ft}^2$$
$$\mu = (0.41)(2.42) = 0.99\ \text{lbm/ft-hr}$$
$$N_{Re} = (0.0762)\left(\frac{767{,}000}{0.99}\right) = 59{,}000$$

From figure 7.9,

$$N_{Re}j_H = 167;\ j_H = 0.00283$$
$$\left(c_p\frac{\mu}{k}\right)^{2/3} = \left[0.44\left(\frac{0.99}{0.085}\right)\right]^{2/3} = 2.972$$
$$h_o = (0.00283)(0.44)(767{,}000)\left(\frac{1}{2.972}\right)(\sim 1)$$
$$= 321\ \text{BTU/hr-ft}^2$$
$$h_o = h_i\ \text{annulus}$$

step 6: The overall (clean) heat transfer coefficient is based on the outside area of the inner pipe:

$$U_c = \frac{h_{io}h_o}{h_{io} + h_o} = \frac{(333)(321)}{333 + 321} = 163\ \text{BTU/hr-ft}^2\text{-°F}$$

The overall (fouled) heat transfer coefficient is

$$U_d = \frac{1}{R_d + \dfrac{1}{U_c}} = \frac{1}{0.001 + 0.001 + \dfrac{1}{163}} = 123\ \text{BTU/hr-ft}^2\text{-°F}$$

step 7: The required area is found from equation 7.24:

$$A = \frac{q}{U \times \Delta T_m} = \frac{167{,}000}{(123)(28.8)} = 47.2\ \text{ft}^2$$

step 8: The pipe length can be calculated from the required area (step 7) and the surface area per foot of pipe. The surface area is

$$A_{surface} = \pi \times \frac{1.66}{12} \times 1 = 0.435\ \text{ft}^2/\text{ft}$$
$$L = \frac{A}{A_{surface}} = \frac{50.5\ \text{ft}^2}{0.435\ \text{ft}^2/\text{ft}} = 108\ \text{ft}$$

CONVECT

F. DESIGN OF MULTIPLE-PASS EXCHANGERS

If the flow path in the exchanger is not completely countercurrent or concurrent, it will be necessary to apply a correction factor, F_t, to the mean temperature difference in equation 7.24. Figure 7.8 plots correction factors for multiple-pass heat exchangers as functions of the terminal temperature differences. Figure 7.6 is a schematic of a one-two exchanger (a one-shell, two-tube pass exchanger).

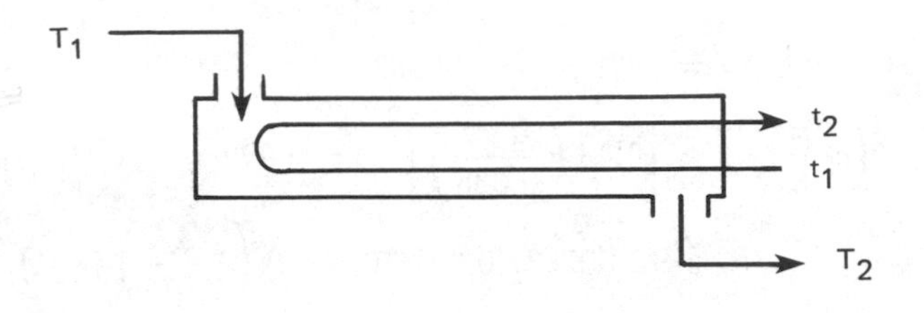

Figure 7.6 A One-Two Exchanger

Certain guidelines should be followed in the design of multiple pass exchangers:

- Corrections to the mean temperature difference are generally required, as shown in figure 7.8, except when either true counterflow or conflow prevails. (If the correction factor is below 0.75, the wrong type of exchanger is being used.)
- If, due to phase changes, there are isothermal sections in the exchanger, assume linear temperature changes along the tube length, and check for *temperature crosses.* If the hot and cold fluid temperatures cross, the exchanger will not operate as designed.
- Calculate the tube side heat transfer coefficient using the *Colburn factor chart* (figure 7.9).
- If water is circulating, determine the tube side heat transfer coefficient from figure 7.10.
- If gases or low viscosity liquids are circulating in the tubes, use figure 7.11 or 7.12 to determine the tube side heat transfer coefficient.
- The shell side heat transfer coefficient is much more difficult to determine due to the many arrangements of tubes, spacing, baffles, etc. A convenient aid for determining shell side heat transfer is figure 7.13. The tube layout (diamond, triangle, or square) and baffle cut must be known.
- The effects of fluid velocity and tubing characteristics can also be determined from figure 7.14 or table 7.7.

G. DESIGN OF CONDENSERS

Condensation occurs when a saturated vapor comes into contact with a surface whose temperature is below the saturation temperature. Normally, a film of condensate is formed on the surface. The thickness of this film can vary with the size and orientation (e.g., horizontal or vertical) of the surface.

Film type condensation (as opposed to *drop-wise condensation*) is ordinarily assumed for design. Physical properties of the fluid, rather than those of the vapor, are used for determining the film coefficient for condensation. A chart of condensing coefficients is given in figure 7.15 for both horizontal and vertical tubes.

The usual approach to condenser design follows the following steps:

step 1: Establish the condensing temperature of the vapor. Use either the vapor pressure to find the saturation temperature or cooling water temperature.

step 2: Use equation 7.27 to calculate the heat load.

step 3: Determine an allowable temperature rise for the cooling water.

step 4: Estimate an overall heat transfer coefficient from tables 7.8 and 7.9. Be sure to use the values for condensing fluids.

step 5: If individual film coefficients are known, establish a film temperature for condensation from equations 7.35 through 7.40. (If the film coefficients are not known, they will have to be assumed and subsequently verified in an iterative process.)

Condensation on outside of tubes:

$$T_w = T_h - \frac{h_{io}}{h_{io} + h_o}(T_h - T_c) \qquad 7.35$$

$$= T_c + \frac{h_o}{h_{io} + h_o}(T_h - T_c) \qquad 7.36$$

Condensation on inside of tubes:

$$T_w = T_c + \frac{h_{io}}{h_{io} + h_o}(T_h - T_c) \qquad 7.37$$

$$= T_h - \frac{h_o}{h_{io} + h_o}(T_h - T_c) \qquad 7.38$$

Estimate the temperature of the film:

$$T_f = \frac{1}{2}(T_h + T_w) \qquad 7.39$$

$$\Delta T_f = T_f - T_w \qquad 7.40$$

CONVECT

Table 7.7
Characteristics of Tubing

O.D. of Tubing	B.W.G. Gauge	Thickness Inches	Internal Area Sq. Inch	Sq. Ft. External Surface Per Foot Length	Sq. Ft. Internal Surface Per Foot Length	Weight Per Ft. Length Steel Lbs.*	I. D. Tubing Inches	Moment of Inertia Inches⁴	Section Modulus Inches³	Radius of Gyration Inches	Constant C**	O. D. / I. D.	Metal Area (Transverse Metal Area) Sq. Inch
1/4	22	.028	.0295	.0655	.0508	.066	.194	.00012	.00098	.0792	46	1.289	.0195
1/4	24	.022	.0333	.0655	.0539	.054	.206	.00011	.00083	.0810	52	1.214	.0159
1/4	26	.018	.0360	.0655	.0560	.045	.214	.00009	.00071	.0824	56	1.168	.0131
3/8	18	.049	.0603	.0982	.0725	.171	.277	.00068	.0036	.1164	94	1.354	.0502
3/8	20	.035	.0731	.0982	.0798	.127	.305	.00055	.0029	.1213	114	1.233	.0374
3/8	22	.028	.0799	.0982	.0835	.104	.319	.00046	.0025	.1227	125	1.176	.0305
3/8	24	.022	.0860	.0982	.0867	.083	.331	.00038	.0020	.1248	134	1.133	.0244
1/2	16	.065	.1075	.1309	.0969	.302	.370	.0022	.0086	.1556	168	1.351	.0888
1/2	18	.049	.1269	.1309	.1052	.236	.402	.0018	.0072	.1606	198	1.244	.0694
1/2	20	.035	.1452	.1309	.1126	.174	.430	.0014	.0056	.1649	227	1.163	.0511
1/2	22	.028	.1548	.1309	.1162	.141	.444	.0012	.0046	.1671	241	1.126	.0415
5/8	12	.109	.1301	.1636	.1066	.602	.407	.0061	.0197	.1864	203	1.536	.177
5/8	13	.095	.1486	.1636	.1139	.537	.435	.0057	.0183	.1903	232	1.437	.158
5/8	14	.083	.1655	.1636	.1202	.479	.459	.0053	.0170	.1938	258	1.362	.141
5/8	15	.072	.1817	.1636	.1259	.425	.481	.0049	.0156	.1971	283	1.299	.125
5/8	16	.065	.1924	.1636	.1296	.388	.495	.0045	.0145	.1993	300	1.263	.114
5/8	17	.058	.2035	.1636	.1333	.350	.509	.0042	.0134	.2016	317	1.228	.103
5/8	18	.049	.2181	.1636	.1380	.303	.527	.0037	.0118	.2043	340	1.186	.089
5/8	19	.042	.2298	.1636	.1416	.262	.541	.0033	.0105	.2068	358	1.155	.077
5/8	20	.035	.2419	.1636	.1453	.221	.555	.0028	.0091	.2089	377	1.126	.065
3/4	10	.134	.1825	.1963	.1262	.884	.482	.0129	.0344	.2229	285	1.556	.260
3/4	11	.120	.2043	.1963	.1335	.809	.510	.0122	.0326	.2267	319	1.471	.238
3/4	12	.109	.2223	.1963	.1393	.748	.532	.0116	.0309	.2299	347	1.410	.220
3/4	13	.095	.2463	.1963	.1466	.666	.560	.0107	.0285	.2340	384	1.339	.196
3/4	14	.083	.2679	.1963	.1529	.592	.584	.0098	.0262	.2376	418	1.284	.174
3/4	15	.072	.2884	.1963	.1587	.520	.606	.0089	.0238	.2410	450	1.238	.153
3/4	16	.065	.3019	.1963	.1623	.476	.620	.0083	.0221	.2433	471	1.210	.140
3/4	17	.058	.3157	.1963	.1660	.428	.634	.0076	.0203	.2455	492	1.183	.126
3/4	18	.049	.3339	.1963	.1707	.367	.652	.0067	.0178	.2484	521	1.150	.108
3/4	20	.035	.3632	.1963	.1780	.269	.680	.0050	.0134	.2532	567	1.103	.079
7/8	10	.134	.2892	.2291	.1589	1.061	.607	.0221	.0505	.2662	451	1.441	.312
7/8	11	.120	.3166	.2291	.1662	.969	.635	.0208	.0475	.2703	494	1.378	.285
7/8	12	.109	.3390	.2291	.1720	.891	.657	.0196	.0449	.2736	529	1.332	.262
7/8	13	.095	.3685	.2291	.1793	.792	.685	.0180	.0411	.2778	575	1.277	.233
7/8	14	.083	.3948	.2291	.1856	.704	.709	.0164	.0374	.2815	616	1.234	.207
7/8	16	.065	.4359	.2291	.1950	.561	.745	.0137	.0312	.2873	680	1.174	.165
7/8	18	.049	.4742	.2291	.2034	.432	.777	.0109	.0249	.2925	740	1.126	.127
7/8	20	.035	.5090	.2291	.2107	.313	.805	.0082	.0187	.2972	794	1.087	.092
1	8	.165	.3526	.2618	.1754	1.462	.670	.0392	.0784	.3009	550	1.493	.430
1	10	.134	.4208	.2618	.1916	1.237	.732	.0350	.0700	.3098	656	1.366	.364
1	11	.120	.4536	.2618	.1990	1.129	.760	.0327	.0654	.3140	708	1.316	.332
1	12	.109	.4803	.2618	.2047	1.037	.782	.0307	.0615	.3174	749	1.279	.305
1	13	.095	.5153	.2618	.2121	.918	.810	.0280	.0559	.3217	804	1.235	.270
1	14	.083	.5463	.2618	.2183	.813	.834	.0253	.0507	.3255	852	1.199	.239
1	15	.072	.5755	.2618	.2241	.714	.856	.0227	.0455	.3291	898	1.167	.210
1	16	.065	.5945	.2618	.2278	.649	.870	.0210	.0419	.3314	927	1.149	.191
1	18	.049	.6390	.2618	.2361	.496	.902	.0166	.0332	.3366	997	1.109	.146
1	20	.035	.6793	.2618	.2435	.360	.930	.0124	.0247	.3414	1060	1.075	.106
1-1/4	7	.180	.6221	.3272	.2330	2.057	.890	.0890	.1425	.3836	970	1.404	.605
1-1/4	8	.165	.6648	.3272	.2409	1.921	.920	.0847	.1355	.3880	1037	1.359	.565
1-1/4	10	.134	.7574	.3272	.2571	1.598	.982	.0741	.1186	.3974	1182	1.273	.470
1-1/4	11	.120	.8012	.3272	.2644	1.448	1.010	.0688	.1100	.4018	1250	1.238	.426
1-1/4	12	.109	.8365	.3272	.2702	1.329	1.032	.0642	.1027	.4052	1305	1.211	.391
1-1/4	13	.095	.8825	.3272	.2775	1.173	1.060	.0579	.0926	.4097	1377	1.179	.345
1-1/4	14	.083	.9229	.3272	.2838	1.033	1.084	.0521	.0833	.4136	1440	1.153	.304
1-1/4	16	.065	.9852	.3272	.2932	.823	1.120	.0426	.0682	.4196	1537	1.116	.242
1-1/4	18	.049	1.042	.3272	.3016	.629	1.152	.0334	.0534	.4250	1626	1.085	.185
1-1/4	20	.035	1.094	.3272	.3089	.456	1.180	.0247	.0395	.4297	1707	1.059	.134
1-1/2	10	.134	1.192	.3927	.3225	1.955	1.232	.1354	.1806	4853	1860	1.218	.575
1-1/2	12	.109	1.291	.3927	.3356	1.618	1.282	.1159	.1546	.4933	2014	1.170	.476
1-1/2	14	.083	1.398	.3927	.3492	1.258	1.334	.0931	.1241	.5018	2181	1.124	.370
1-1/2	16	.065	1.474	.3927	.3587	.996	1.370	.0756	.1008	.5079	2299	1.095	.293
2	11	.120	2.433	.5236	.4608	2.410	1.760	.3144	.3144	.6660	3795	1.136	.709
2	13	.095	2.573	.5236	.4739	1.934	1.810	.2586	.2586	.6744	4014	1.105	.569
2-1/2	9	.148	3.815	.6540	.5770	3.719	2.204	.7592	.6074	.8332	5951	1.134	1.094

*Weights are based on low carbon steel with a density of 0.2833 lb/inch³. For other metals multiply by the following factors:

Metal	Factor	Metal	Factor
Aluminum	0.35	Nickel-Chrome-Iron	1.07
A.I.S.I. 400 Series Stainless Steels	0.99	Admiralty	1.09
A.I.S.I. 300 Series Stainless Steels	1.02	Nickel and Nickel-Copper	1.13
Aluminum Bronze	1.04	Copper and Cupro-Nickels	1.14
Aluminum Brass	1.06		

**Liquid Velocity $= \frac{\text{Lbs. Per Tube Per Hour}}{\text{C x SP. GR. of Liquid}}$ in feet per sec. (Sp. Gr. of Water at 60°F. = 1.0)

step 6: Use figure 7.15 to determine the condensing film coefficients if the loading G'' (or G') is known. If loading is not known, use equations 7.41 or 7.42.

$$h = 0.943 \left(\frac{k_f^3 \rho_f^2 \Lambda g}{\mu_f L \Delta T_f} \right)^{\frac{1}{4}} \qquad 7.41$$

(vertical condensers)

$$h = 0.73 \left(\frac{k_f^3 \rho_f^2 \Lambda g}{\mu_f D \Delta T_f} \right)^{\frac{1}{4}} \qquad 7.42$$

(horizontal condensers)

step 7: For cooling water, use figure 7.10 to calculate the tube side heat transfer coefficient, or use figure 7.13 for shell side coefficients.

step 8: Calculate U from h values. Calculate the area from equation 7.24. (No multiple-pass corrections are needed for isothermal condensation.)

step 9: If necessary, validate the values of h assumed in step 5. Iterate as needed.

H. TEMPERATURE CROSSES

Temperature crosses prevent an exchanger from operating as it is intended. Figure 7.7 illustrates two situations involving phase changes in an exchanger.

The first situation occurs when the vapor is desuperheated and condensed, causing a non-isothermal region and an isothermal area to exist in the exchanger. The second condition is where sensible heat and vaporization of a liquid also exhibit non-isothermal and isothermal sections. It cannot be assumed the solution of these two situations involves the shape of the profile curves. To be certain of performance, the heating, cooling, condensing, or vaporizing curves for all fluids should be established.

If the exchanger design is based on the terminal temperatures, and a cross exists inside, the expected heat exchange will not occur. A temperature profile of an exchanger can be estimated by assuming that the overall heat transfer coefficient is constant along the exchanger tube. Then, the total heat exchanged is directly proportional to the length of the tube. For example, if 10 percent of the total heat transferred in the exchanger is for desuperheating a vapor, then the isothermal condensation section comprises 90 percent of the tube length. Thus, the diagrams shown in figure 7.7 can be drawn.

In vaporization, one fluid vaporizes at constant temperature, T_2, while the second fluid is cooled from t_1 to t_2. When a refrigerant vaporizes while another fluid condenses, the unit actually operates at a constant temperature difference for its entire length. In that case, vaporization and condensation both occur along the tube.

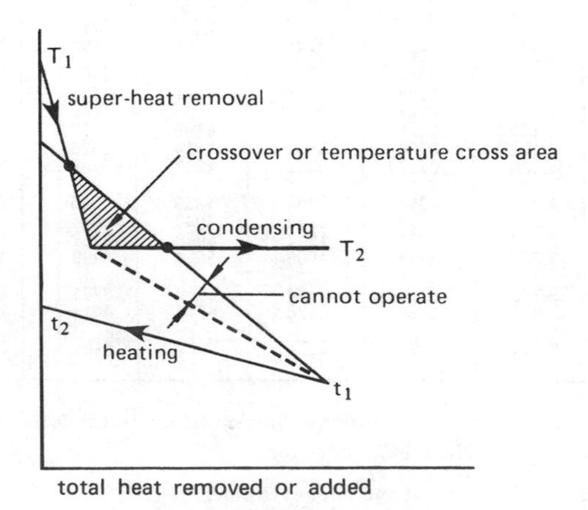

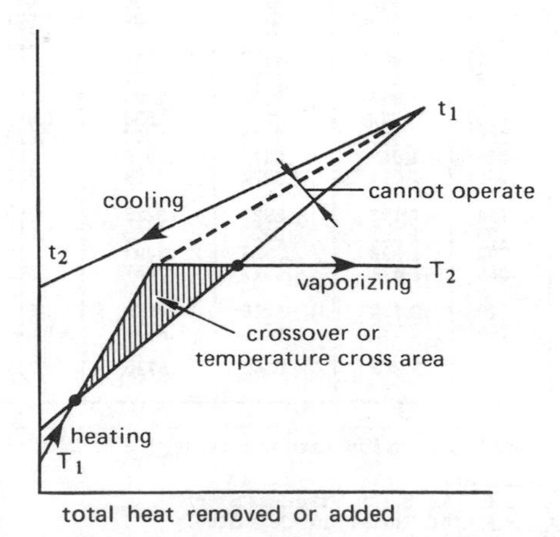

Figure 7.7 Phase Changes in an Exchanger

CONVECT

Table 7.8
Typical Overall Heat Transfer Coefficients in Tubular Heat Exchangers

Shell side	Tube side	Design U	Includes total dirt
Liquid-liquid media			
Aroclor 1248	Jet fuels	100–150	0.0015
Cutback asphalt	Water	10–20	.01
Demineralized water	Water	300–500	.001
Ethanol amine (MEA or DEA) 10–25% solutions	Water or DEA, or MEA solutions	140–200	.003
Fuel oil	Water	15–25	.007
Fuel oil	Oil	10–15	.008
Gasoline	Water	60–100	.003
Heavy oils	Heavy oils	10–40	.004
Heavy oils	Water	15–50	.005
Hydrogen-rich reformer stream	Hydrogen-rich reformer stream	90–120	.002
Kerosene or gas oil	Water	25–50	.005
Kerosene or gas oil	Oil	20–35	.005
Kerosene or jet fuels	Trichlorethylene	40–50	.0015
Jacket water	Water	230–300	.002
Lube oil (low viscosity)	Water	25–50	.002
Lube oil (high viscosity)	Water	40–80	.003
Lube oil	Oil	11–20	.006
Naphtha	Water	50–70	.005
Naphtha	Oil	25–35	.005
Organic solvents	Water	50–150	.003
Organic solvents	Brine	35–90	.003
Organic solvents	Organic solvents	20–60	.002
Tall oil derivatives, vegetable oil, etc.	Water	20–50	.004
Water	Caustic soda solutions (10–30%)	100–250	.003
Water	Water	200–250	.003
Wax distillate	Water	15–25	.005
Wax distillate	Oil	13–23	.005
Condensing vapor-liquid media			
Alcohol vapor	Water	100–200	.002
Asphalt (450°F.)	Dowtherm vapor	40–60	.006
Dowtherm vapor	Tall oil and derivatives	60–80	.004

Shell side	Tube side	Design U	Includes total dirt
Dowtherm vapor	Dowtherm liquid	80–120	.0015
Gas-plant tar	Steam	40–50	.0055
High-boiling hydrocarbons V	Water	20–50	.003
Low-boiling hydrocarbons A	Water	80–200	.003
Hydrocarbon vapors (partial condenser)	Oil	25–40	.004
Organic solvents A	Water	100–200	.003
Organic solvents high NC, A	Water or brine	20–60	.003
Organic solvents low NC, V	Water or brine	50–120	.003
Kerosene	Water	30–65	.004
Kerosene	Oil	20–30	.005
Naphtha	Water	50–75	.005
Naphtha	Oil	20–30	.005
Stabilizer reflux vapors	Water	80–120	.003
Steam	Feed water	400–1000	.0005
Steam	No. 6 fuel oil	15–25	.0055
Steam	No. 2 fuel oil	60–90	.0025
Sulfur dioxide	Water	150–200	.003
Tall-oil derivatives, vegetable oils (vapor)	Water	20–50	.004
Water	Aromatic vapor-stream azeotrope	40–80	.005
Gas-liquid media			
Air, N_2, etc. (compressed)	Water or brine	40–80	.005
Air, N_2, etc., A	Water or brine	10–50	.005
Water or brine	Air, N_2 (compressed)	20–40	.005
Water or brine	Air, N_2, etc., A	5–20	.005
Water	Hydrogen containing natural-gas mixtures	80–125	.003
Vaporizers			
Anhydrous ammonia	Steam condensing	150–300	.0015
Chlorine	Steam condensing	150–300	.0015
Chlorine	Light heat-transfer oil	40–60	.0015
Propane, butane, etc.	Steam condensing	200–300	.0015
Water	Steam condensing	250–400	.0015

NC = noncondensable gas present.
V = vacuum.
A = atmospheric pressure.
Dirt (or fouling factor) units are hr-ft^2-°F/BTU
To convert British thermal units per hour–square foot–degrees Fahrenheit to joules per square meter–second–kelvins, multiply by 5.6783; to convert hours per square foot–degree Fahrenheit–British thermal units to square meters per second–kelvin–joules, multiply by 0.1761.

Table 7.9
Typical Overall Heat Transfer Coefficients in Refinery Service

	Fluid	API gravity	Fouling factor (one stream)	Reboiler, steam-heated	Condenser, water-cooled*	Exchangers, liquid to liquid (tube-side fluid designation appears below)			Reboiler (heating liquid designated below)			Condenser (cooling liquid designated below)			
						C	G	H	C	G†	K	D	F	G	J
A	Propane	...	0.001	160	95	85	85	80	110	95	35				
B	Butane	...	.001	155	90	80	75	75	105	90	35	80	55	40	30
C	400°F. end-point gasoline	50	.001	120	80	70	65	60	65	50	30				
D	Virgin light naphtha	70	.001	140	85	70	55	55	75	60	35	75			
E	Virgin heavy naphtha	45	.001	95	75	65	55	50	55	45	30	70	50	35	30
F	Kerosene	40	.001	85	60	60	55	50	...	45	25	...	50	35	30
G	Light gas oil	30	.002	70	50	60	50	50	...	40	25	70	45	30	30
H	Heavy gas oil	22	.003	60	45	55	50	45	50	40	20	70	40	30	20
J	Reduced crude	17	.005	...	...	55	45	40							
K	Heavy fuel oil (tar)	10	.005	...	...	50	40	35							

Fouling factor, water side 0.0002; heating or cooling streams are shown at top of columns as C, D, F, G, etc.; to convert British thermal units per hour-square foot–degrees Fahrenheit to joules per square meter–second–kelvins, multiply by 5.6783; to convert hours per square foot–degree Fahrenheit–British thermal units to square meters per second–kelvin–joules, multiply by 0.1761.
*Cooler, water-cooled, rates are about 5 percent lower.
†With heavy gas oil (H) as heating medium, rates are about 5 percent lower.

CONVECT

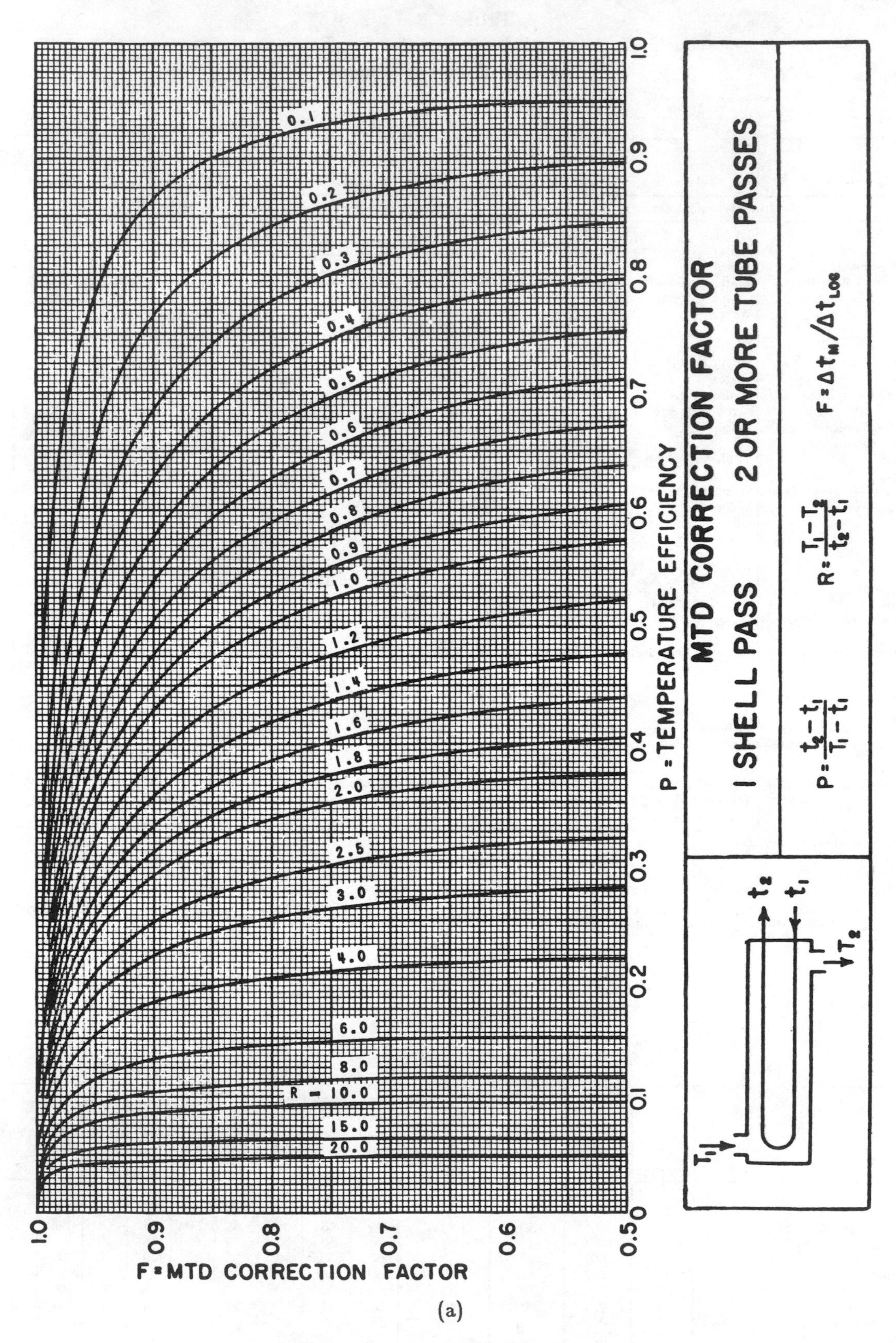

(a)

Figure 7.8 Correction Factors for Multiple-Pass Heat Exchangers

Figure 7.8 continued

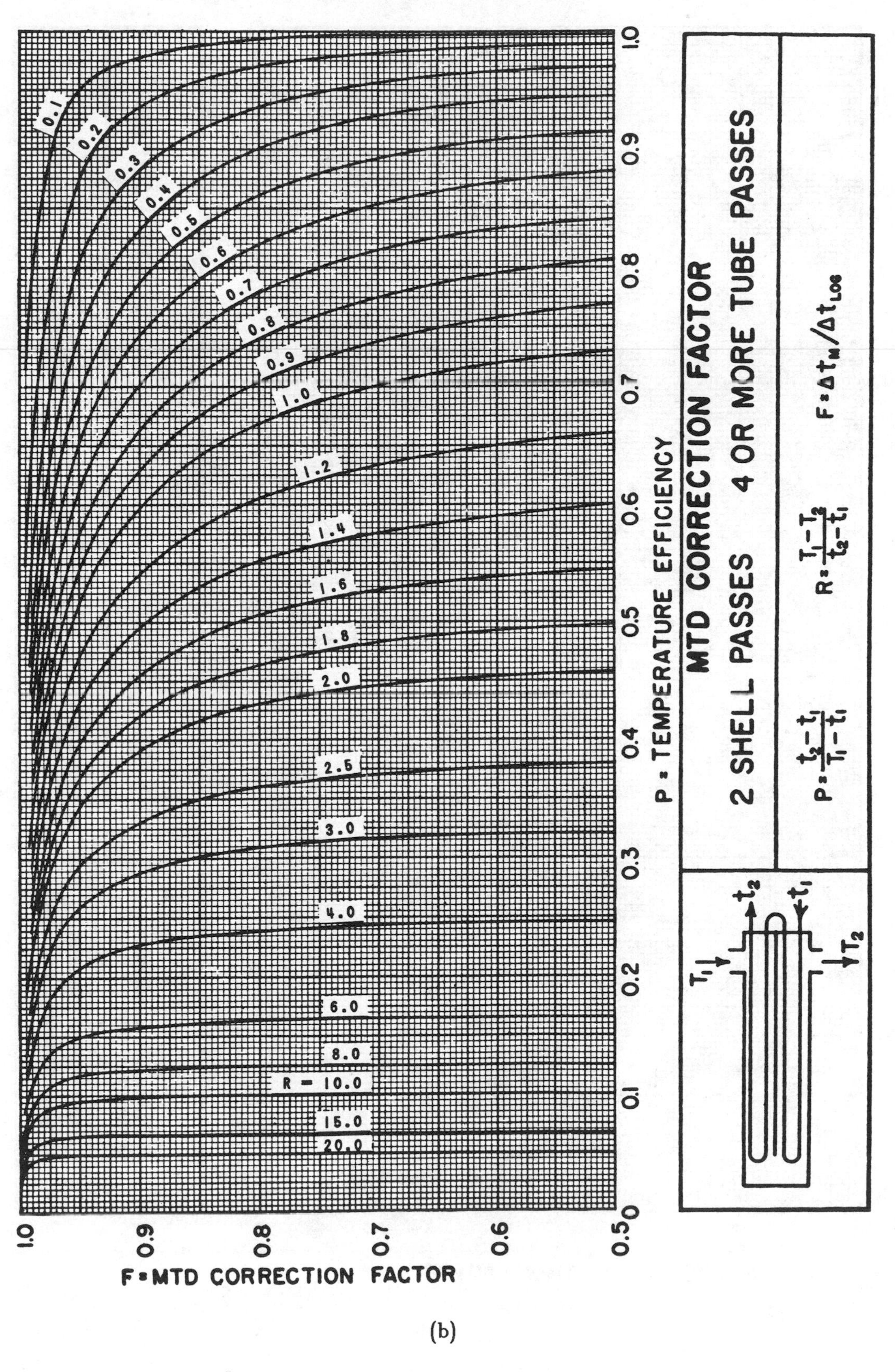

(b)

Figure 7.8 continued

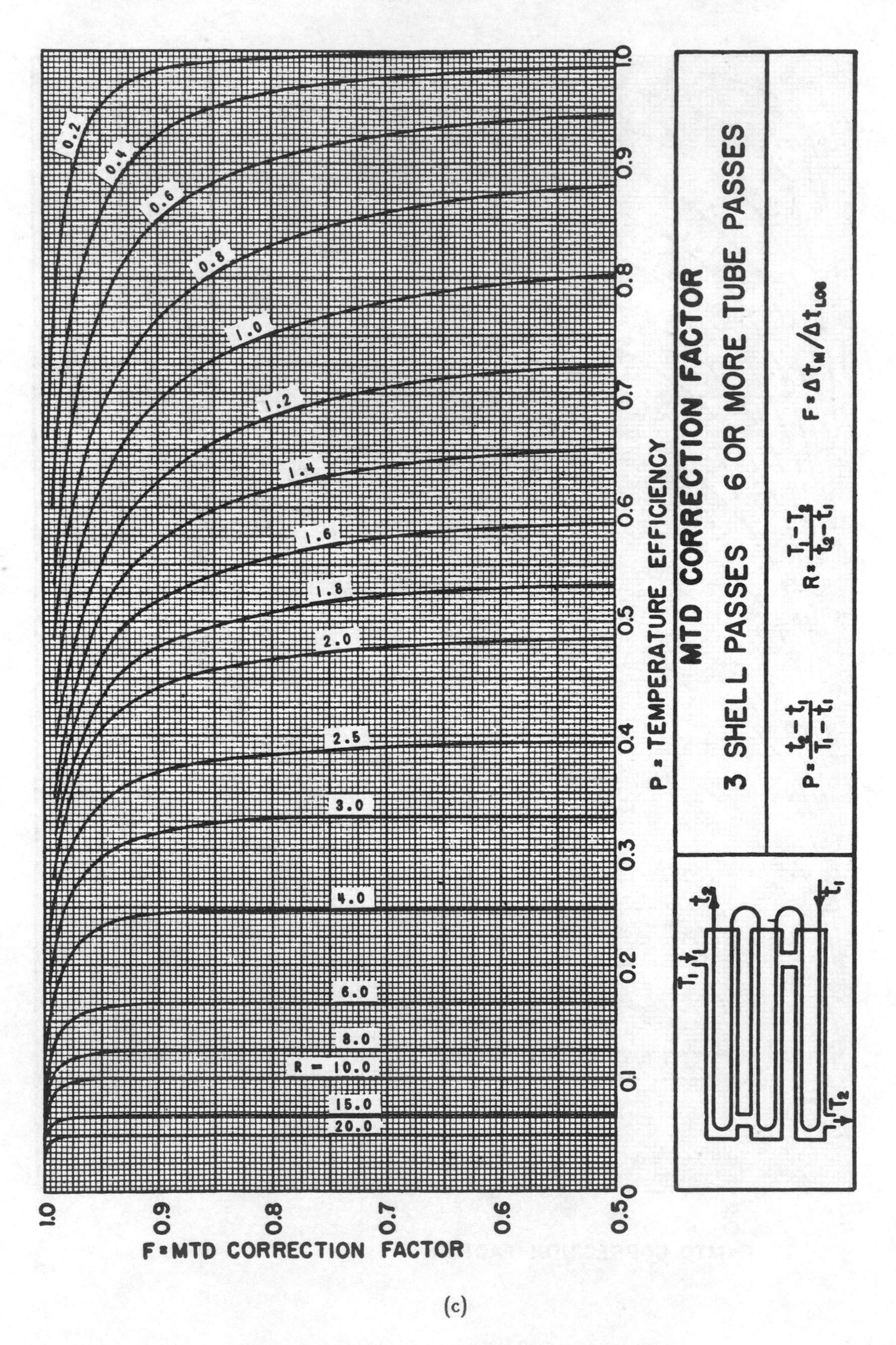

(c)

Figure 7.8 continued

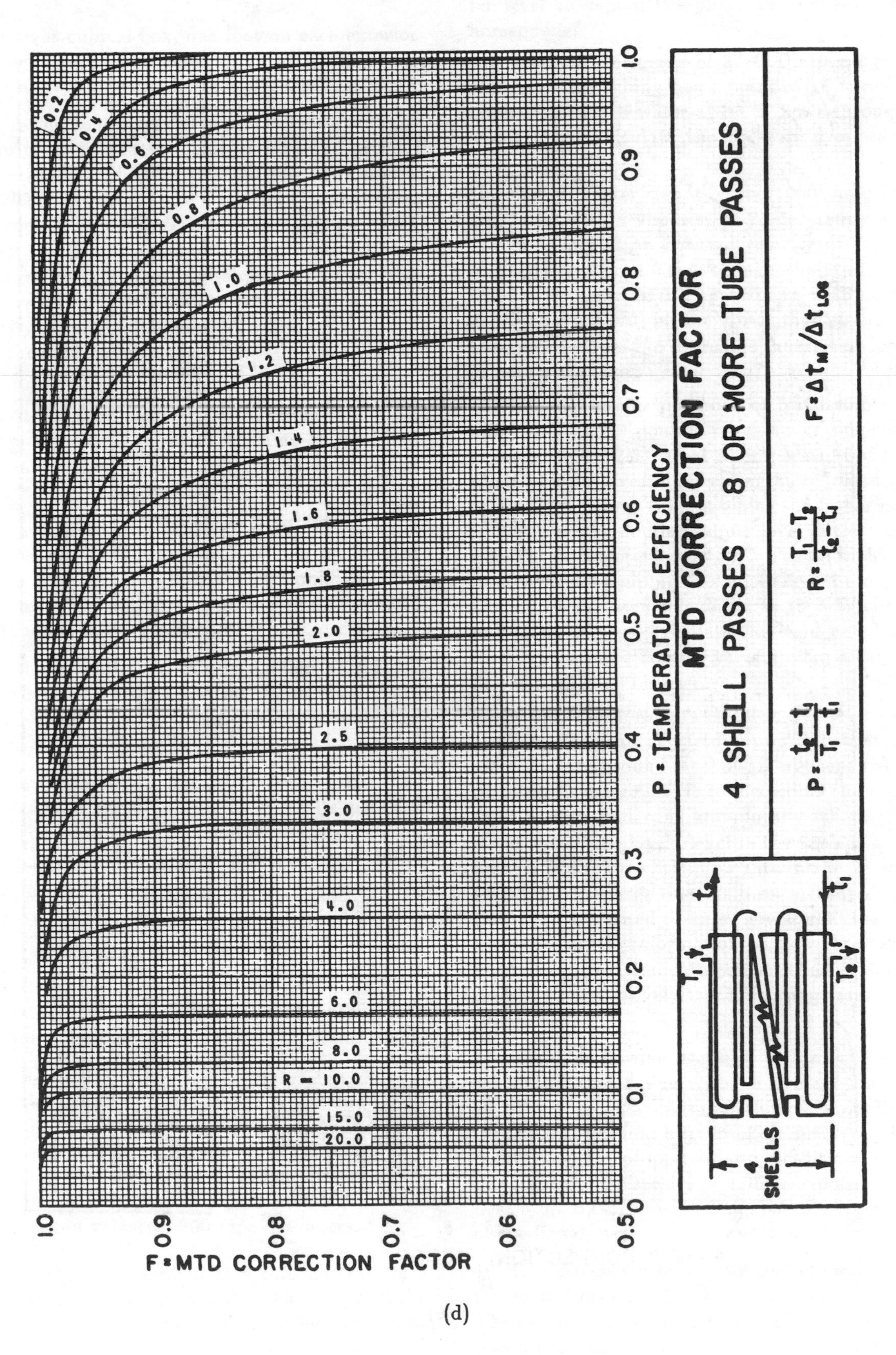

(d)

Figure 7.8 continued

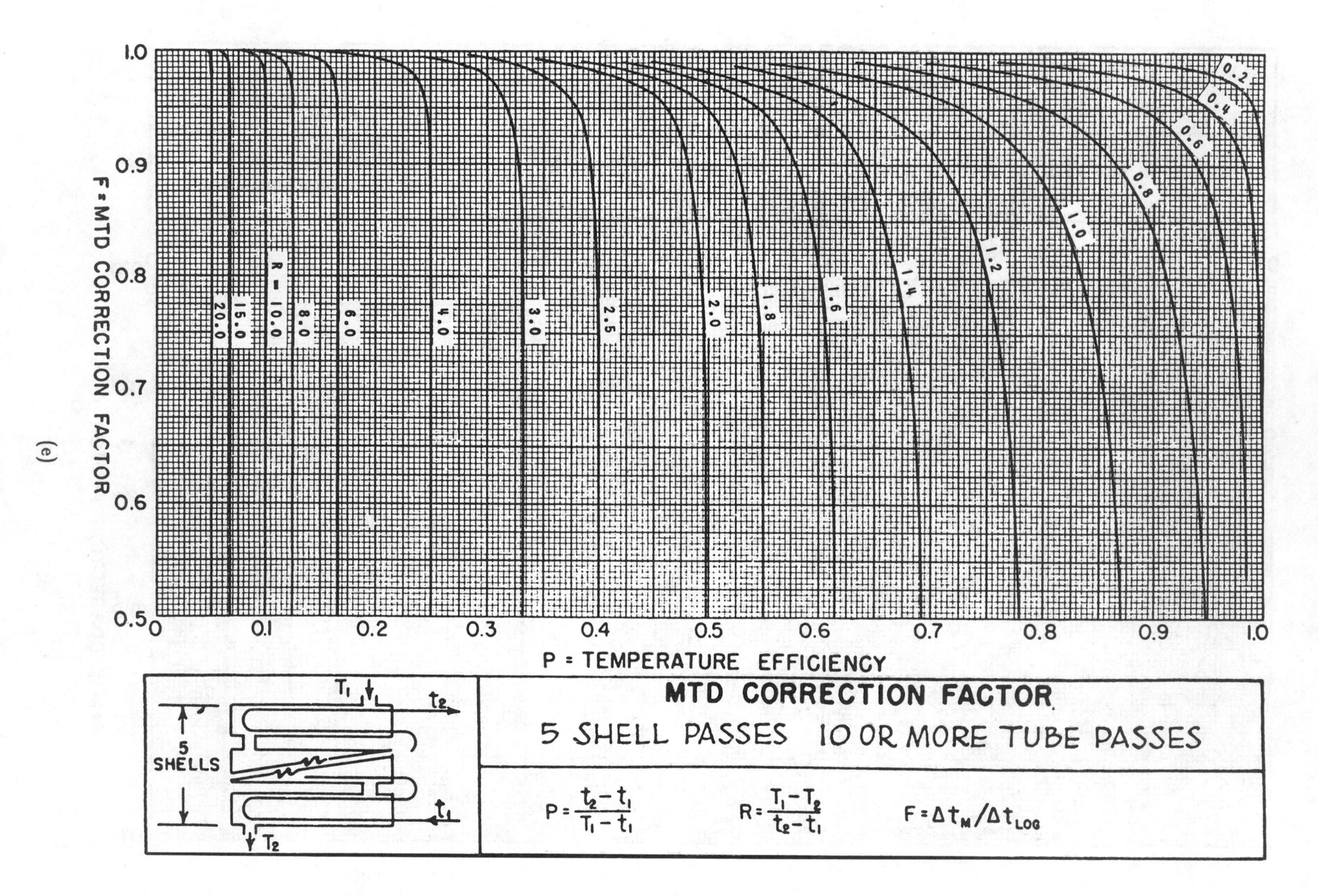

(e)

Figure 7.8 continued

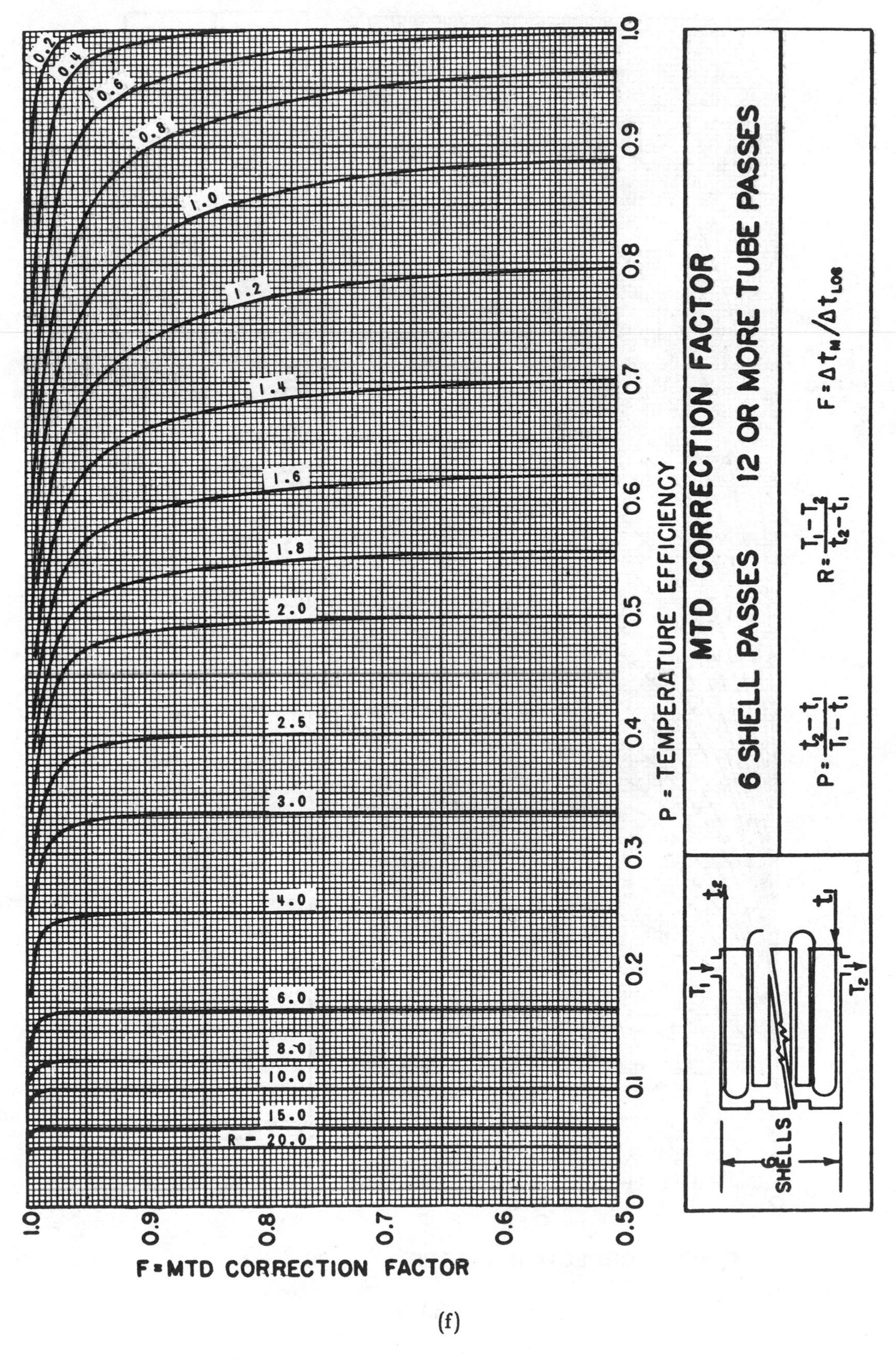

(f)

Figure 7.8 continued

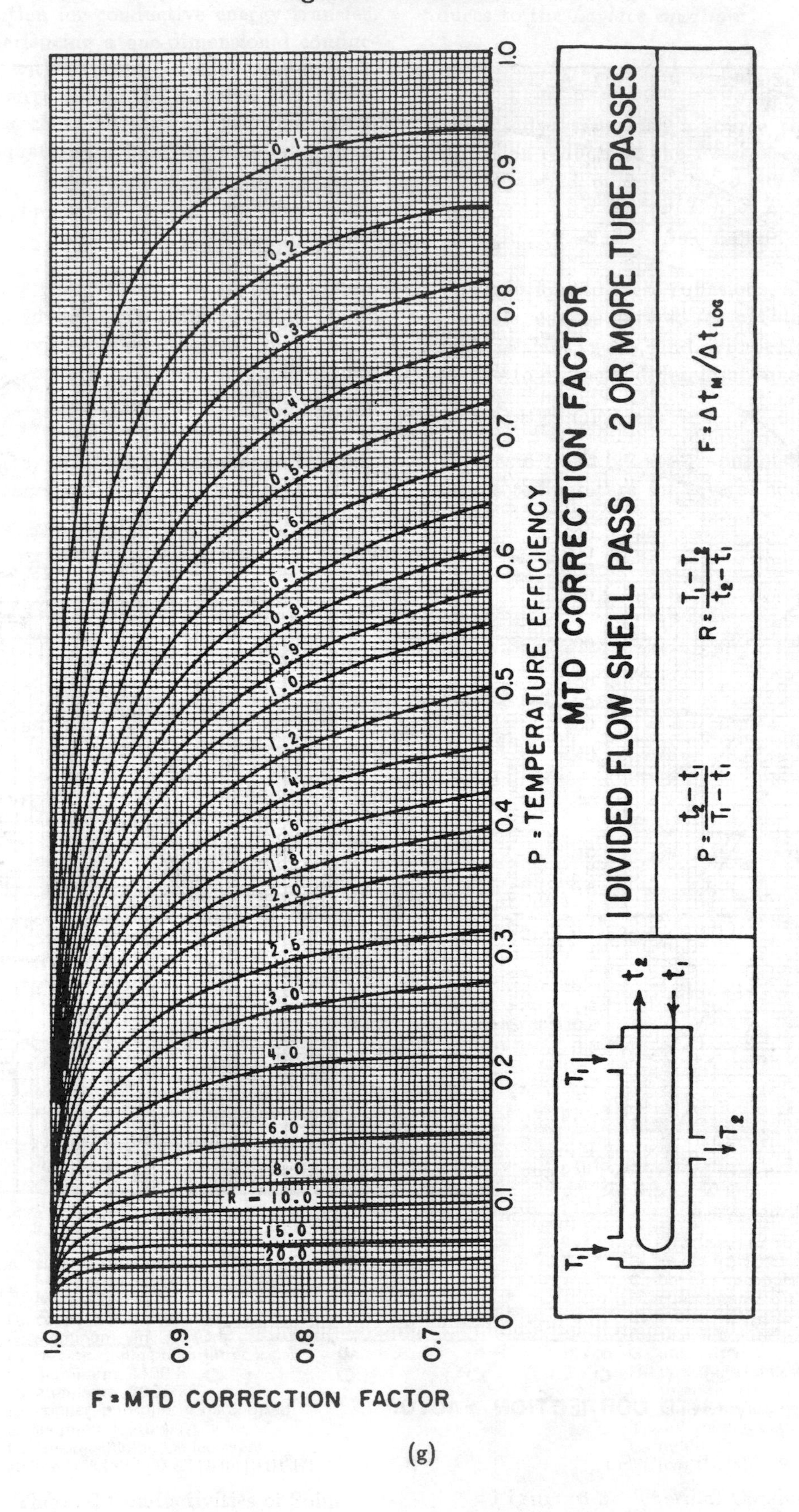

(g)

Figure 7.8 continued

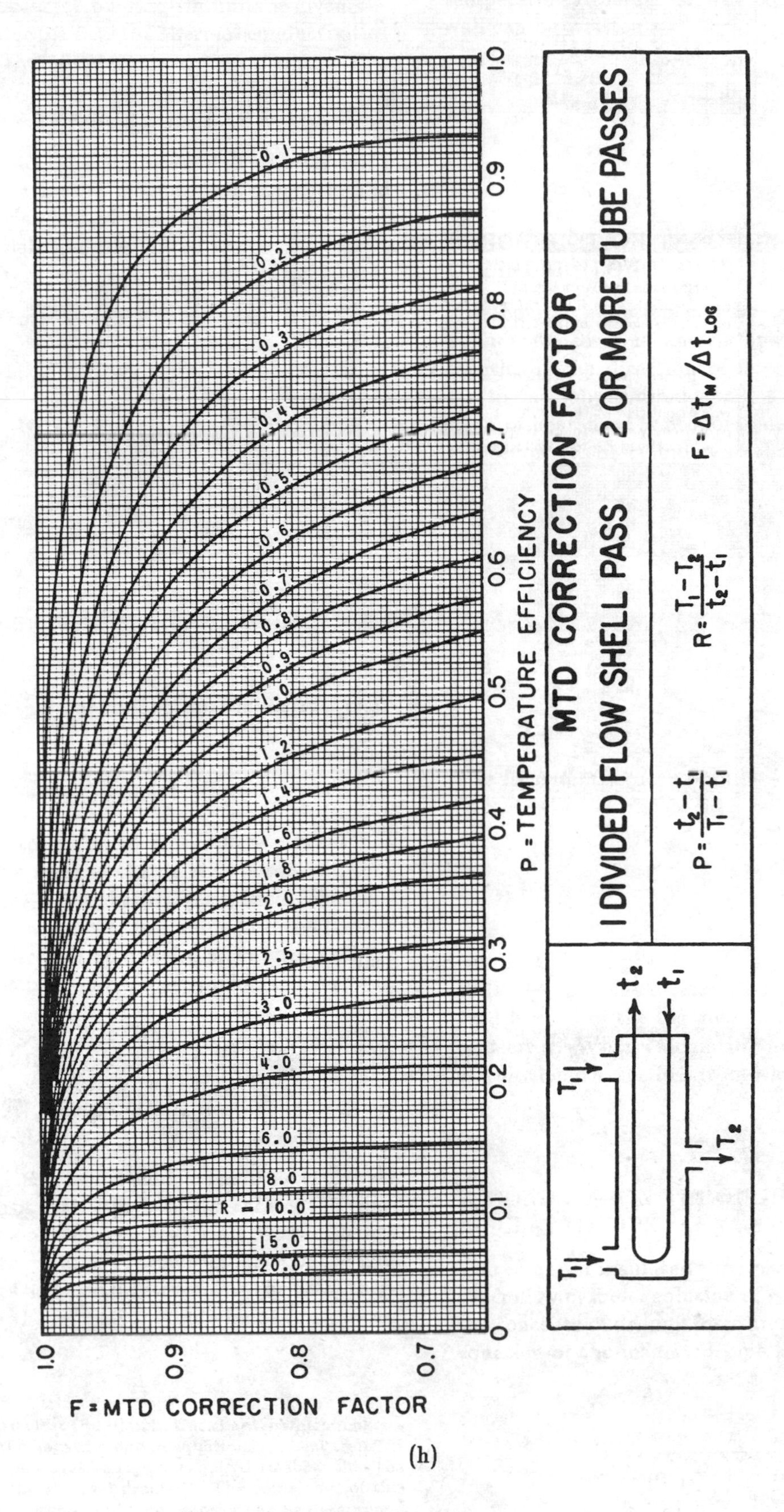

(h)

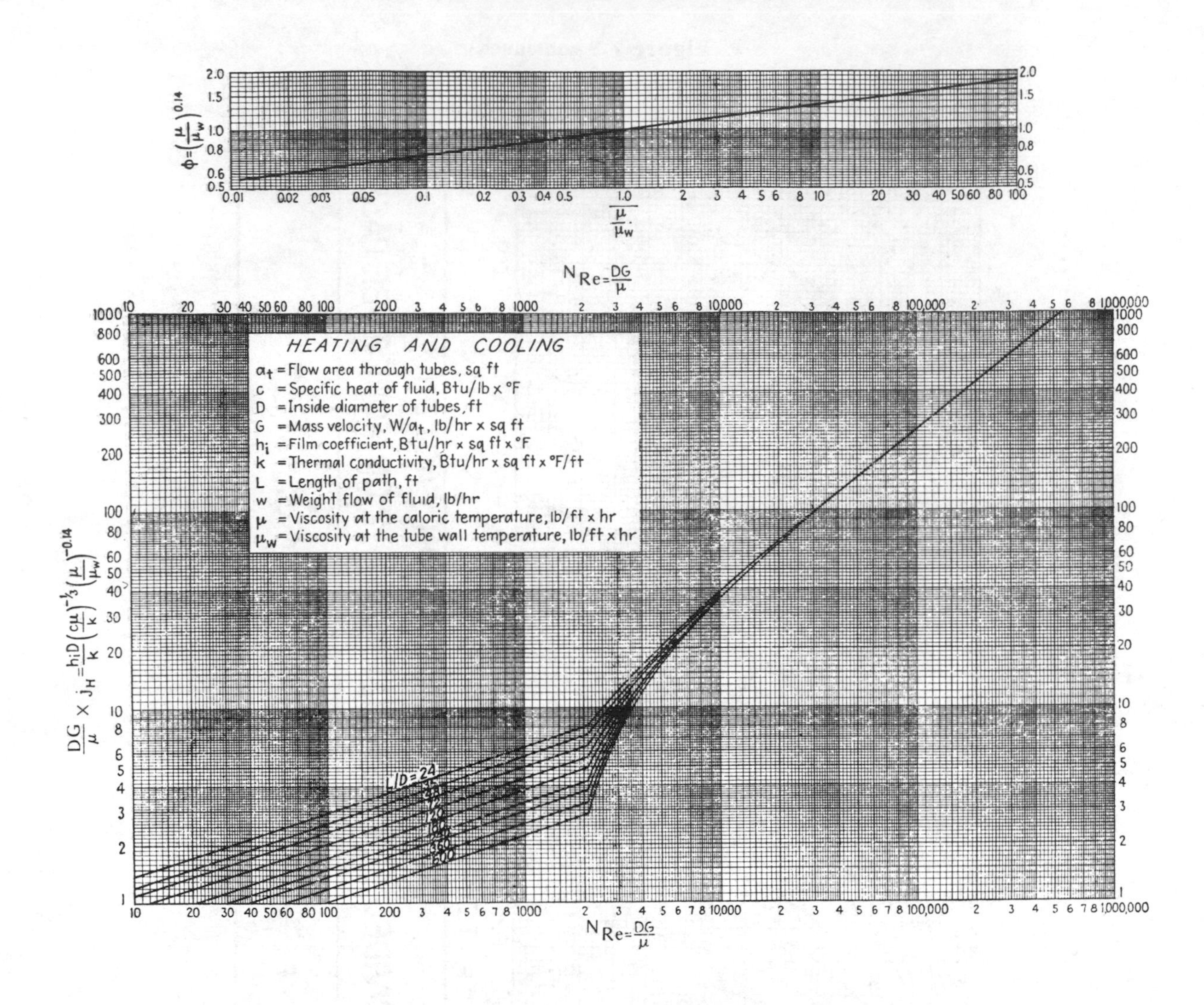

Figure 7.9 Tube-Side Heat Transfer Curve

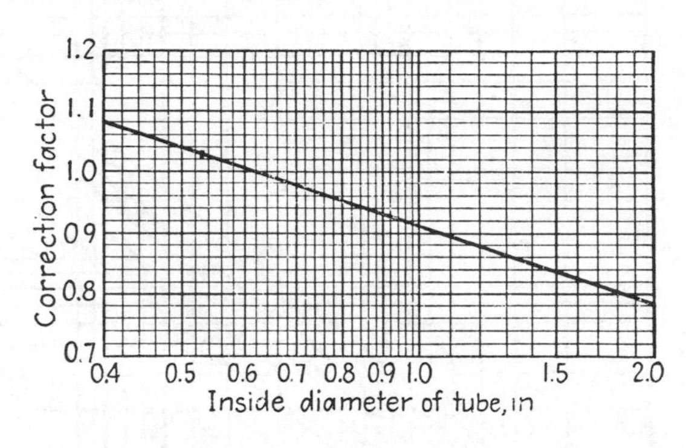

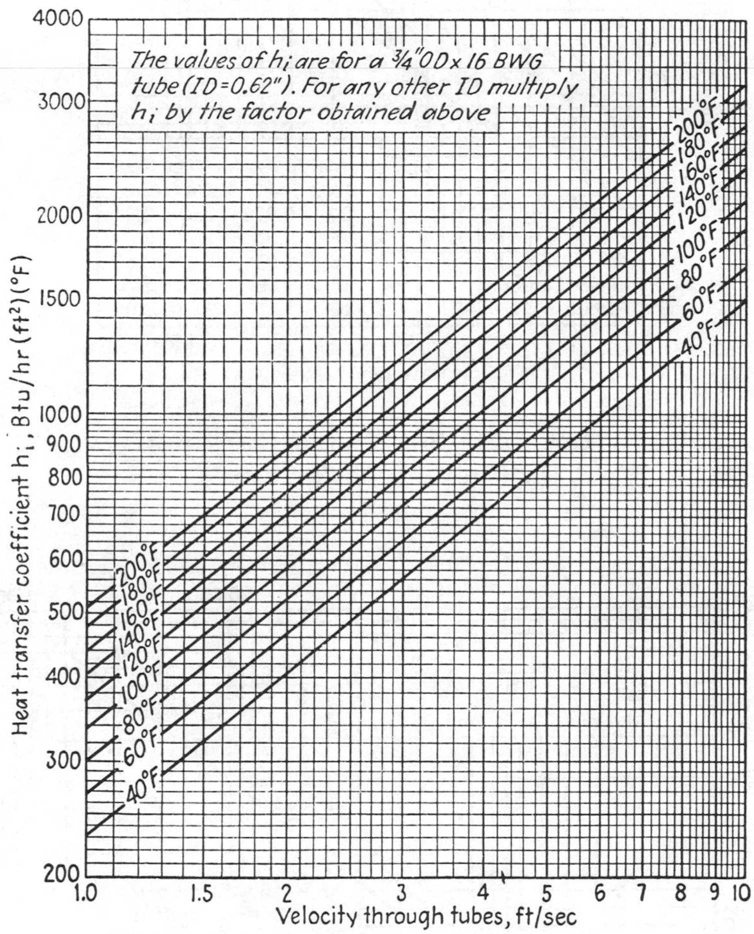

Figure 7.10 Tube-Side Water-Heat Transfer Curve

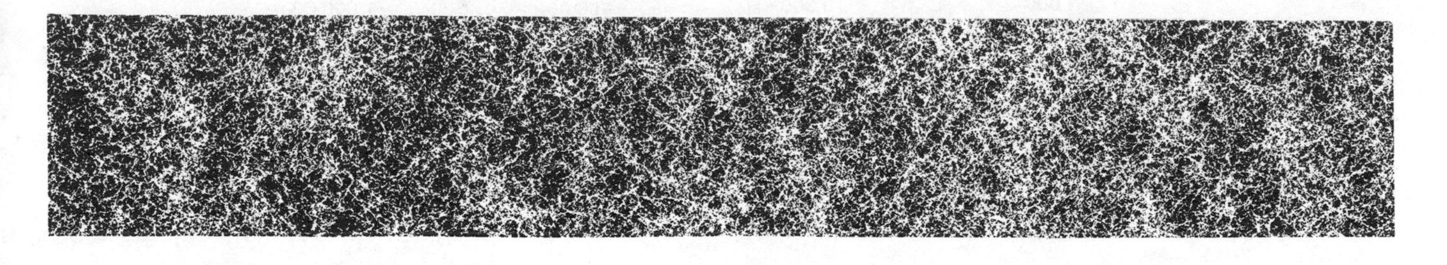

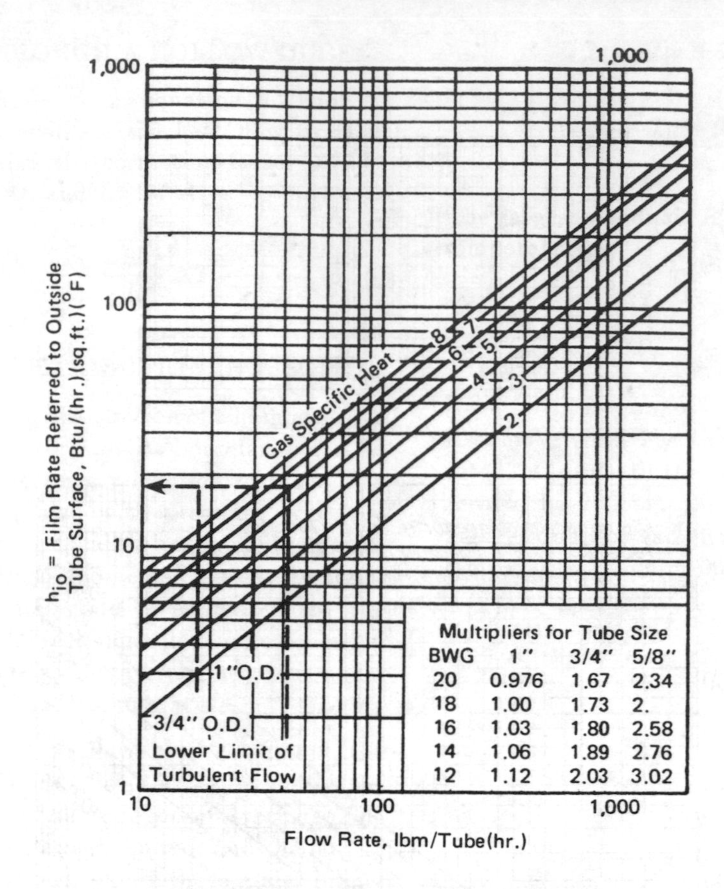

Figure 7.11 Heat Transfer Coefficients for Gases in Turbulent Flow

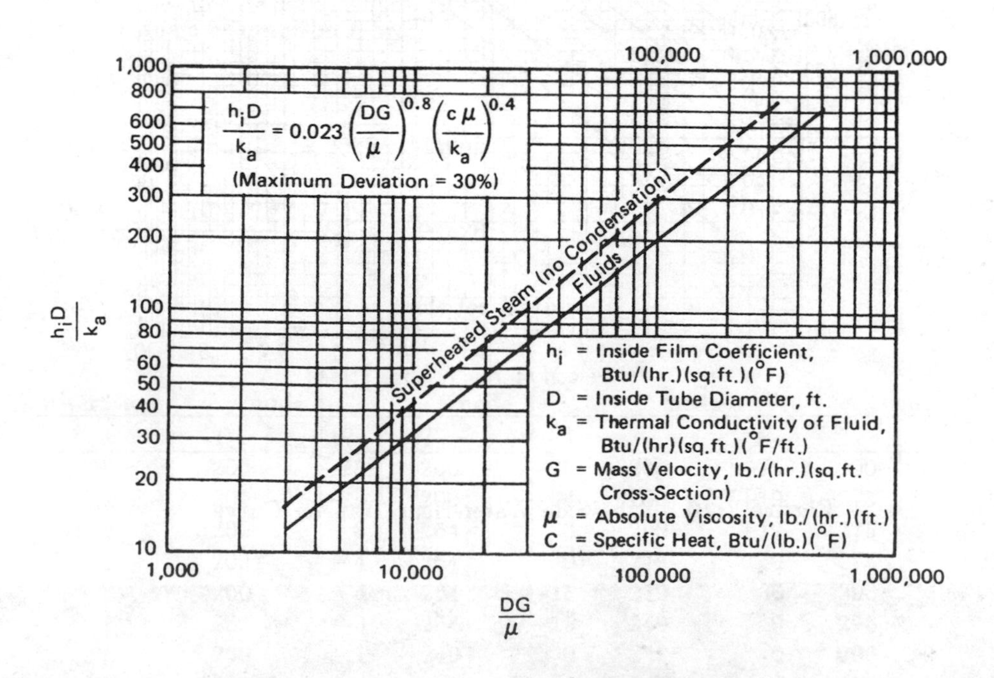

Figure 7.12 Nusselt Number – Reynolds Number Plot for Superheated Steam and Fluids

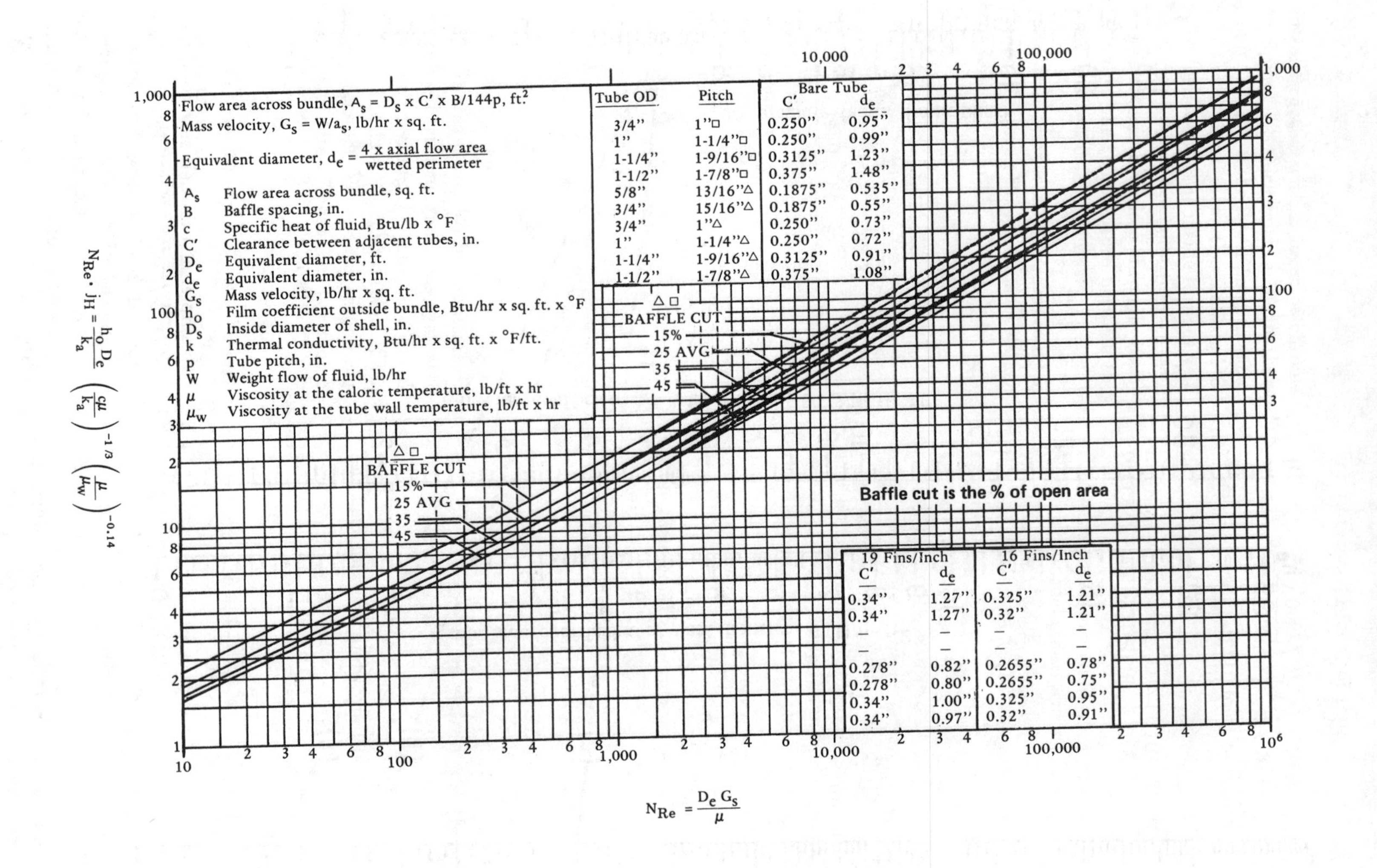

Tube OD	Pitch	Bare Tube C'	Bare Tube d_e
3/4"	1"□	0.250"	0.95"
1"	1-1/4"□	0.250"	0.99"
1-1/4"	1-9/16"□	0.3125"	1.23"
1-1/2"	1-7/8"□	0.375"	1.48"
5/8"	13/16"Δ	0.1875"	0.535"
3/4"	15/16"Δ	0.1875"	0.55"
3/4"	1"Δ	0.250"	0.73"
1"	1-1/4"Δ	0.250"	0.72"
1-1/4"	1-9/16"Δ	0.3125"	0.91"
1-1/2"	1-7/8"Δ	0.375"	1.08"

19 Fins/Inch C'	19 Fins/Inch d_e	16 Fins/Inch C'	16 Fins/Inch d_e
0.34"	1.27"	0.325"	1.21"
0.34"	1.27"	0.32"	1.21"
—	—	—	—
—	—	—	—
0.278"	0.82"	0.2655"	0.78"
0.278"	0.80"	0.2655"	0.75"
0.34"	1.00"	0.325"	0.95"
0.34"	0.97"	0.32"	0.91"

Figure 7.13 Colburn Plot for Multiple and Finned Tube Exchangers

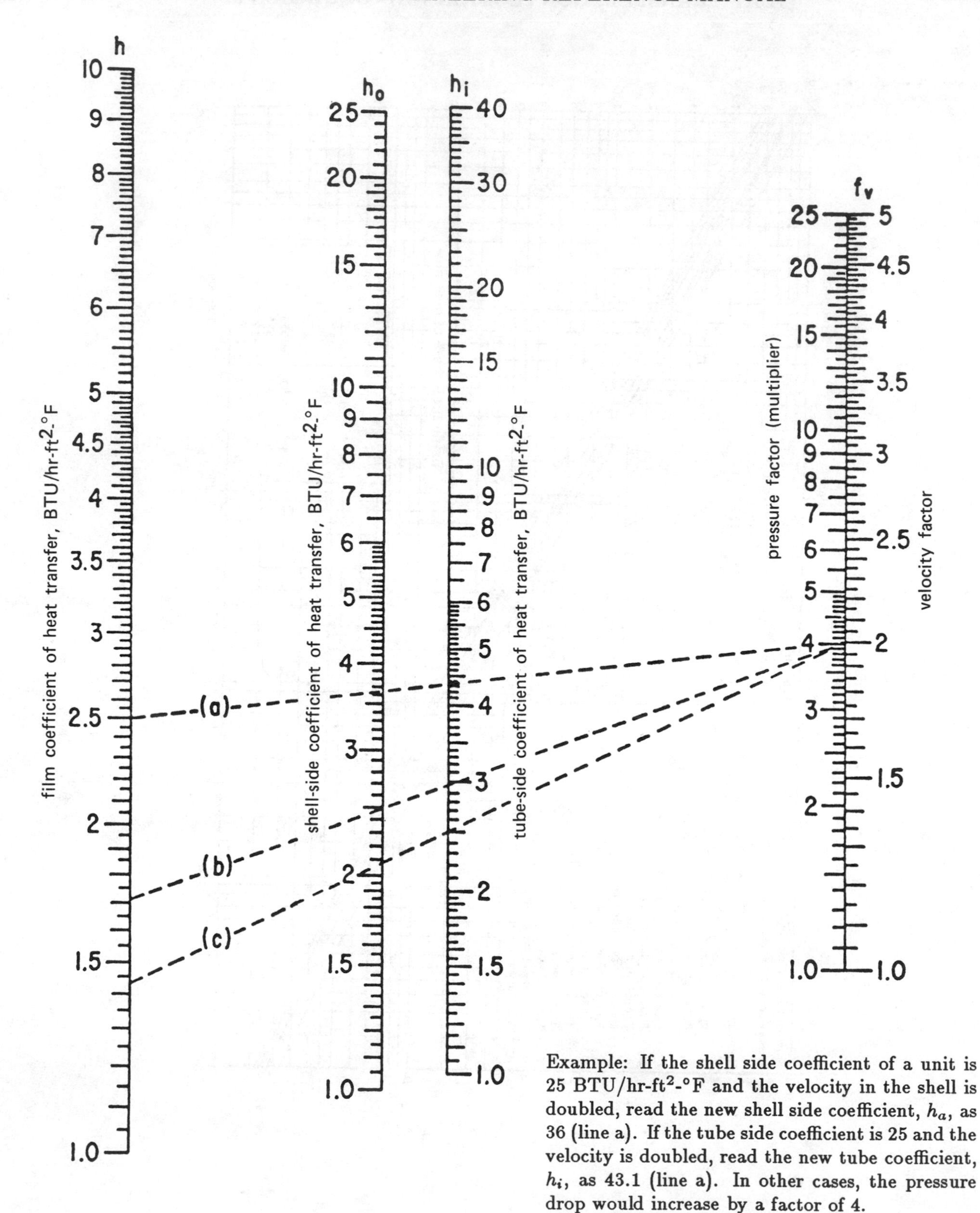

Example: If the shell side coefficient of a unit is 25 BTU/hr-ft^2-°F and the velocity in the shell is doubled, read the new shell side coefficient, h_a, as 36 (line a). If the tube side coefficient is 25 and the velocity is doubled, read the new tube coefficient, h_i, as 43.1 (line a). In other cases, the pressure drop would increase by a factor of 4.

Note: This may be used in reverse for reduced flow.

Figure 7.14 Nomograph of Effect of Fluid Velocity On Heat Transfer Coefficient

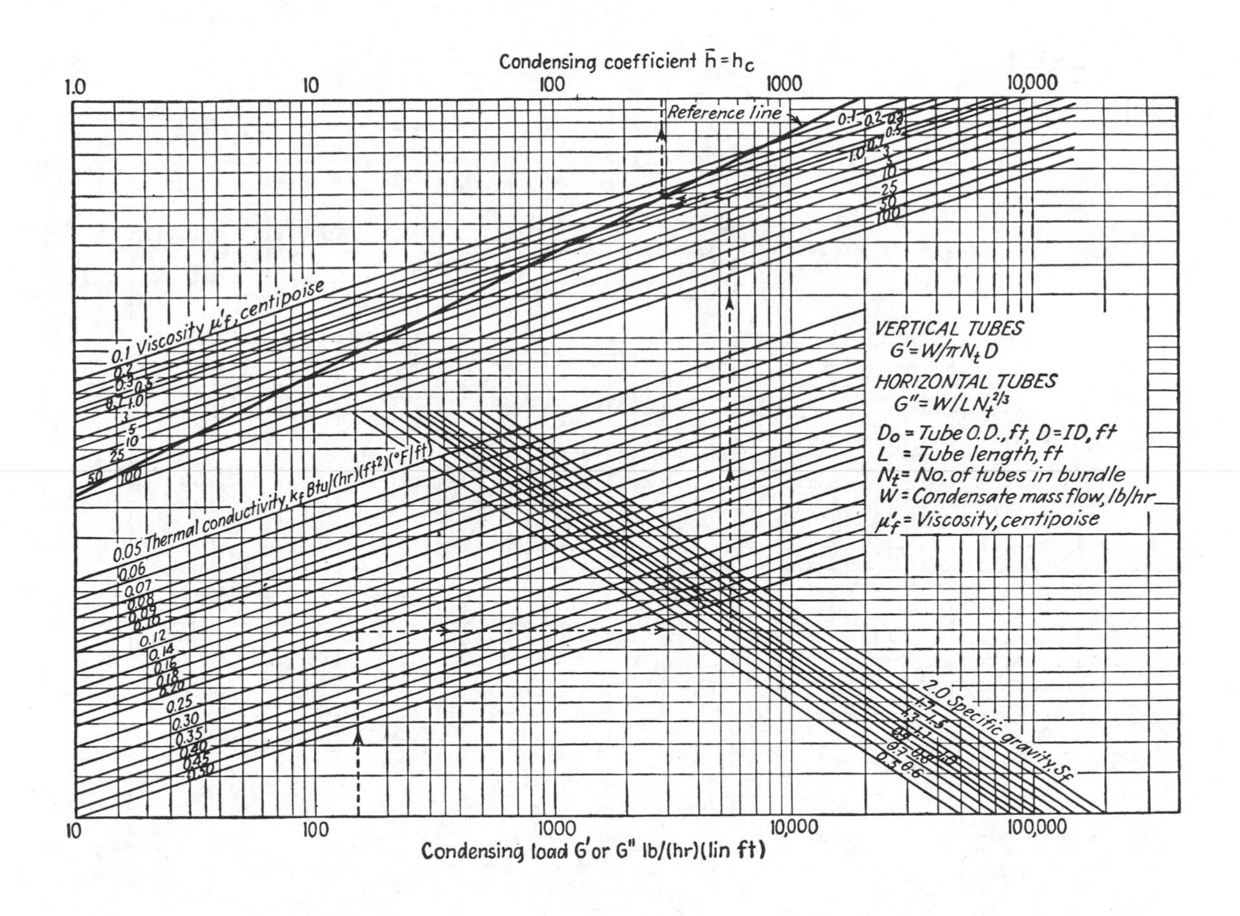

Figure 7.15 Condensing Coefficients

CONVECT

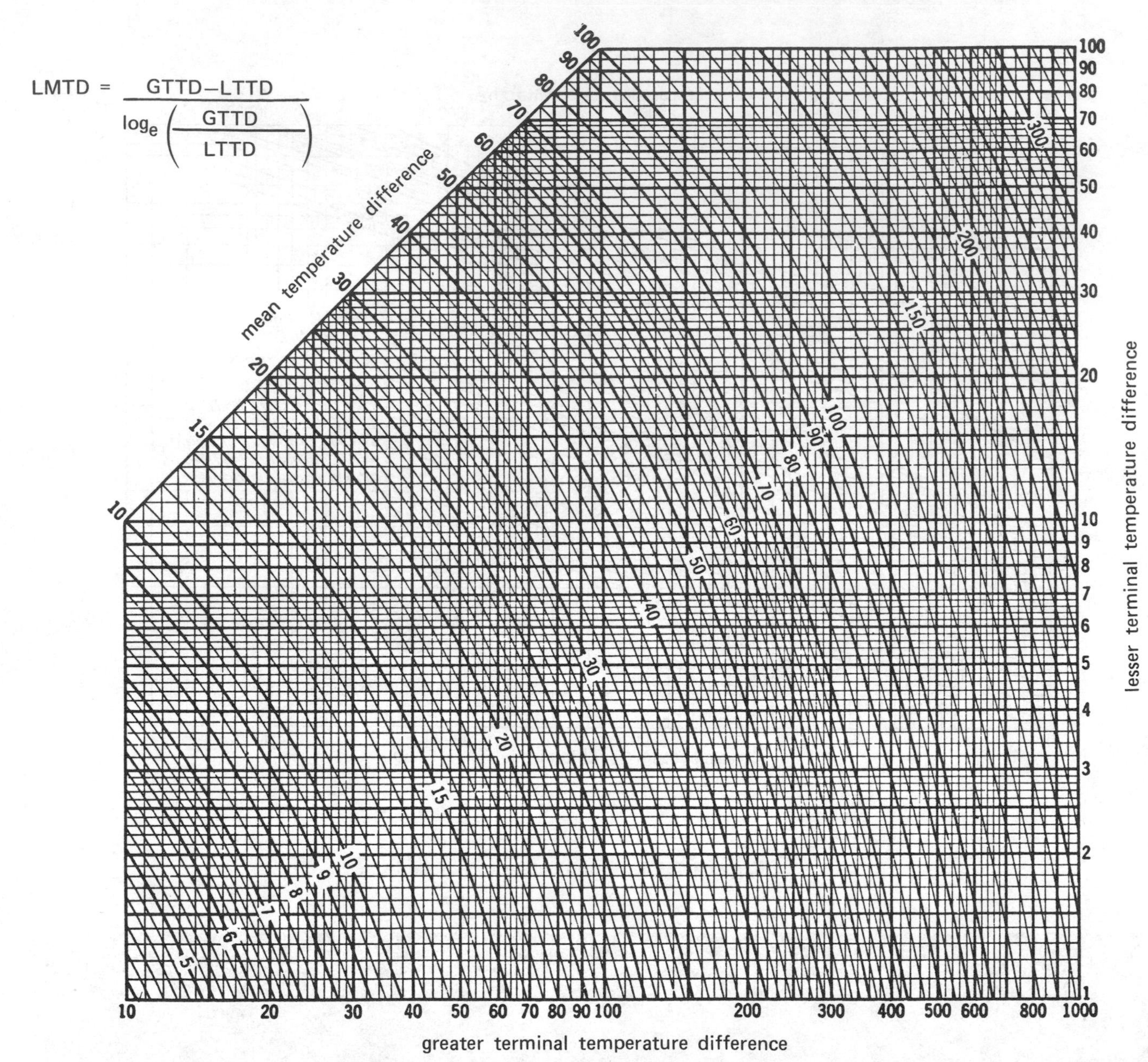

NOTE—For points not included on this sheet multiply Greater Terminal Temperature Difference and Lesser Terminal Temperature Difference by any multiple of 10 and divide resulting value of curved lines by same multiple.

CONVECT

Figure 7.16 Logarithmic Mean Temperature Differences

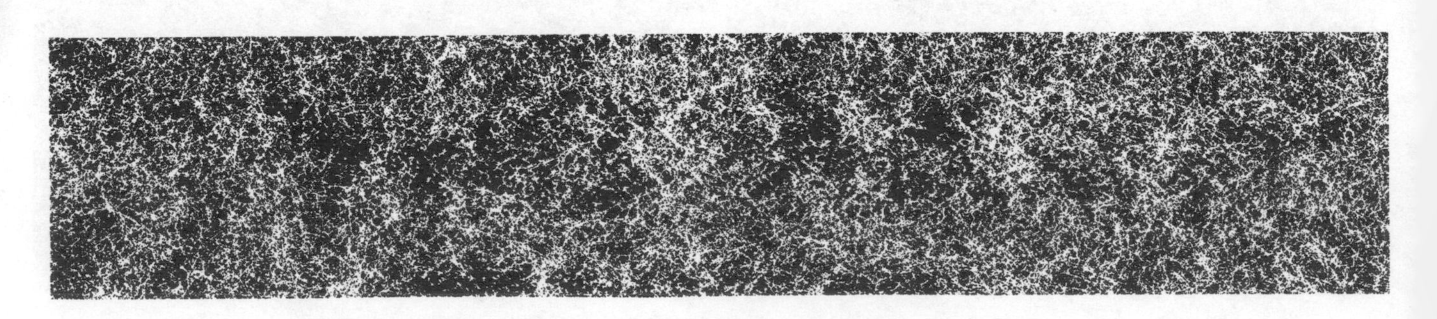

PRACTICE PROBLEMS

1. The heat from 10,000 pounds per hour of hot pressurized (350 °F) water is used to preheat 10,000 pounds per hour of distilled water from 70 °F. The distilled water is ultimately to be heated to 400 °F. What distilled water outlet temperature could be reasonably obtained, and what heat could be recovered in a concurrent exchanger? (Assume that "reasonable" means a 5 °F approach.)

2. Repeat problem 1 for a countercurrent exchanger. (Assume that "reasonable" for a countercurrent exchanger means a LMTD of 10 °F.)

3. Repeat problem 1 for a one-two exchanger with distilled water in the tubes. (Use MTD factor at maximum efficiency.)

4. Water flows through a $\frac{3}{4}''$ 16 BWG exchanger tube at 40 pounds per minute. The entering temperature is 50 °F, and the leaving temperature is 150 °F. Heat is supplied to the water by steam condensation at 250 °F on the outside of the tube. The outside heat transfer coefficient, h_o, is 1000 BTU/hr-ft^2-°F. Neglecting viscosity changes, calculate the inside heat transfer film coefficient.

5. For problem 4, calculate the overall heat transfer coefficient based on the inside area, U_i.

6. For problem 4, calculate the length of tube required.

7. For problem 4, calculate the value of the inside scale coefficient if the original overall coefficient decreases to one-half of the value calculated in problem 5.

8. During the summer, a black asphalt street ($\varepsilon = 1$) absorbs radiant energy at the rate of 350 BTU/hr-ft^2. If a mild breeze is blowing (so that the heat transfer coefficient is 2.8 BTU/hr-ft^2-°F when the air is 90 °F), what is the equilibrium temperature of the street?

9. 10,000 pounds per hour of methanol vapor condense in a single-pass condenser. Methanol at the saturation temperature, 160 °F, is introduced and is withdrawn as a saturated liquid at 160 °F. Cooling water is introduced at 70 °F and removed at 115 °F. (a) Calculate the water flow required. (b) Calculate the heat transfer area required. (c) List the assumptions used by the method of calculation in (b).

methanol heat of vaporization = 600 BTU/lbm
heat capacity water = 1.0 BTU/lbm-°F
heat capacity methanol = 0.8 BTU/lbm-°F
overall heat transfer coefficient, based on outside area = 400 BTU/hr-ft^2-°F

10. A saturated vapor passes through a 4″ schedule-40 pipeline at a rate of 15,000 pounds per hour. The line is covered with 2″ of 85% magnesia insulation, and the outer surface of the insulation is at a constant temperature of 100 °F along the entire 75′ of the line. The combined radiation-convection coefficient at the insulation surface is 1.75 BTU/hr-ft^2-°F, and the ambient temperature is 80 °F. If the latent heat of vaporization of the vapor is 55 BTU/lbm, calculate the quality of the vapor at the end of the line.

11. A solution of organic colloids is concentrated from 20% to 65% solids (by weight) in a vertical tube evaporator. The solution has negligible elevation in boiling point, and the specific heat of the feed is 0.93. Saturated steam at 10 psia is used, and the pressure of the condenser is 4″ mercury absolute. The feed enters at 60 °F. The overall heat transfer coefficient is 250 BTU/hr-ft^2-°F. The evaporator must evaporate 40,000 pounds per hour of water. (a) How many square feet of surface are required? (b) What is the steam consumption in pounds per hour?

12. A single-pass, double-pipe heat exchanger is used to cool 8000 pounds per hour of light oil (specific heat 0.4 BTU/lbm-°F) from 150 °C to 40 °C. Water at 20 °C is used for cooling, and the exchanger is operated in countercurrent fashion. The exit water temperature is 40 °C. The heat exchanger is constructed as a 1″ pipe passing through a 2″ pipe, both schedule-40. The film coefficients for water are 2000 BTU/hr-ft^2-°F in the center of the pipe and 840 BTU/hr-ft^2-°F in the annulus. Values for oil are 90 BTU/hr-ft^2-°F in the center of the pipe and 41 BTU/hr-ft^2-°F in the annulus. (a) Should the water or the oil be put inside the smaller pipe? Why? (b) What length exchanger is required?

13. If a countercurrent concentric pipe heat exchanger can heat air from 80 °F to 180 °F using condensing steam at 220 °F, estimate how many times as much air could be heated within the same range by steam condensing at 250 °F.

14. A shell and tube heat exchanger must be designed to heat 650,000 pounds per hour of a solution (specific heat of 0.78, specific gravity of 1.30) from 148 °F to 198 °F. Steam is available at 20 psig. At 6 feet per second, the average liquor film coefficient is 600 BTU/hr-ft^2-°F. Assume the steam side film coefficient to be 1300 BTU/hr-ft^2-°F. Other considerations have dictated that type 316 stainless steel tubes $1\frac{1}{2}''$ outside diameter (16-gauge BWG, 12′ long) will be used. The thermal conductivity of 316 stainless steel is 105 BTU-in/hr-ft^2-°F. Calculate the number of tubes and passes required, neglecting all fouling factors.

15. Crude oil flows at the rate of 2000 pounds per hour through the inside pipe of a double-pipe heat exchanger, and is heated from 90 °F to 200 °F. The heat is supplied by kerosene, initially at 450 °F, flowing through the annular space. The overall coefficient, U, is 80 BTU/hr- ft^2-°F. The specific heat of the oil is 0.56 BTU/lbm-°F, and the specific heat of the kerosene is 0.60 BTU/lbm-°F. For safety reasons, the minimum temperature difference between the fluids must be 20 °F. (a) Would the flow rate of the kerosene be greater for countercurrent or concurrent flow? (b) Would the heat transfer area required be greater for the countercurrent or the concurrent flow? Prove your answers, showing calculations.

16. A heat exchanger has $1\frac{1}{2}''$ outside diameter by 0.120″ wall thickness tubes of material A, which has a thermal conductivity of 460 BTU-in/hr-ft^2-°F. The exchanger has an overall heat transfer coefficient based on an outside area of 250 BTU/hr-ft^2-°F. On the shell side, the condensate film coefficient is 1200, and scaling is not a factor. The liquor film coefficient varies with the square root of the velocity. Tubes with $1\frac{1}{2}''$ outside diameter and 0.065 wall of material B, which has a thermal conductivity of 105 BTU-in/hr-ft^2-°F, are installed in place of material A. (a) What revised overall heat transfer is to be expected? (b) What is the difference between a BTU and a PCU?

17. Saturated benzene at 5000 pounds per hour is condensed at atmospheric pressure on the shell side of a shell and tube heat exchanger. The condenser contains 36 tubes, 1″ outside diameter by 16 BWG, 18′ long, made of type 304 stainless steel. Measurements show the temperature of the cooling water to be 35 °C at the inlet and 65 °C at the outlet. The condenser currently operates with a single pass on both shell and tube sides. The normal boiling point of benzene is 80.1 °C, and its latent heat of vaporization is 170 BTU/lbm. (a) What is the overall heat transfer coefficient based on the outside area? (b) If the film coefficient for condensing benzene vapor is 300 BTU/hr-ft^2-°F, what is the water film coefficient? Include any surface fouling in the water coefficient. (c) It is suspected that there is some fouling on the water side of the tubes. Does the data validate the suspicion?

18. 7000 pounds per hour of aniline are heated from 100 °F to 150 °F by cooling 10,000 pounds per hour of toluene (with an initial temperature of 185 °F) in 15′ long 2″ × 1″ double pipe hairpin exchangers. The total dirt factor is 0.005 ft^2-hr-°F/BTU. The aniline is in the inner pipe. How many hairpin sections are required?

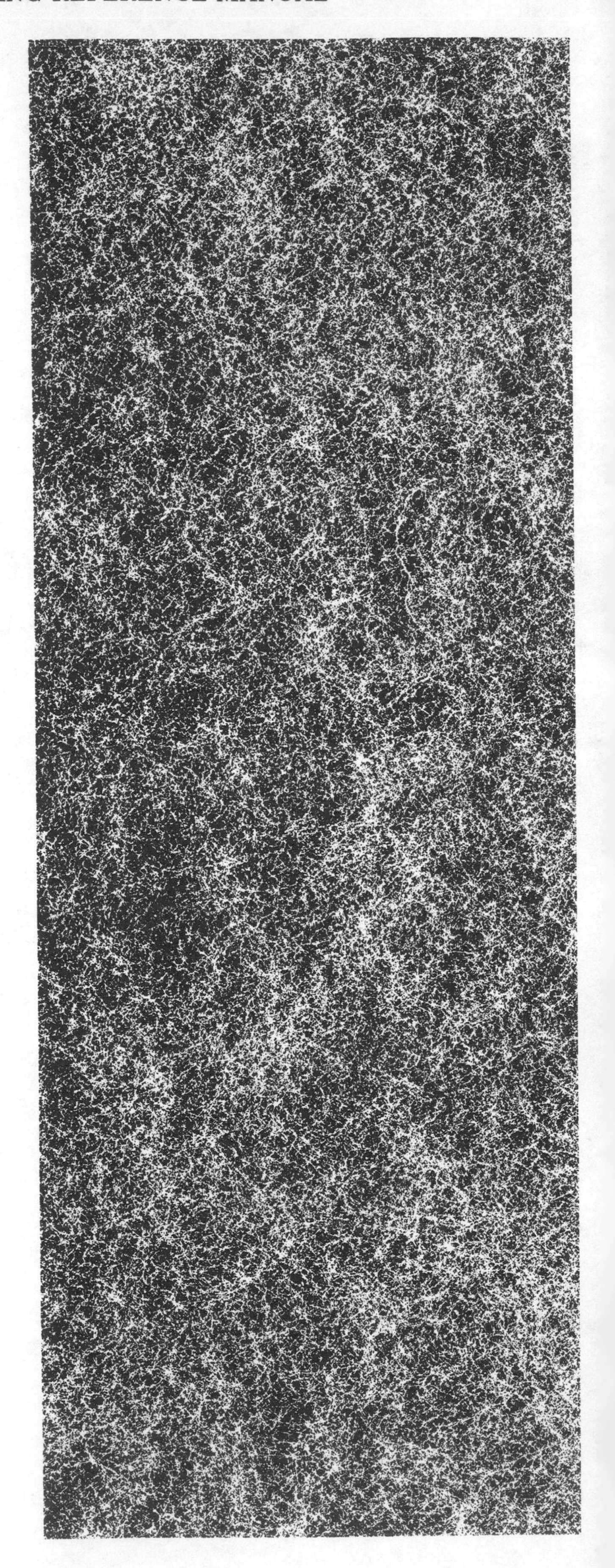

CONVECT

8 VAPOR-LIQUID PROCESSES

Nomenclature

a	interfacial area/unit volume	ft^2/ft^3
A	cross-sectional area, or absorption factor	ft^2, –
B	film thickness	cm
c	molar concentration	$lbmole/ft^3$
C	number of components in a system	–
D_{AB}	mass diffusivity of component A into B	cm^2/sec
E	absorption, or stripping efficiency	–
f_i	molar flow of component i in feed	gmole/hr
F	overall feed molar flow	gmole/hr
G	molar gas flow	$lbmole/hr\text{-}ft^2$
G_s	solute-free inert gas flow	$lbmole/hr\text{-}ft^2$
h	packing height	ft
H	Henry's Law constant	$atm\text{-}cm^3/gmole$, or atm/mole fraction
H	height of transfer unit	ft
HETP	height equivalent to a theoretical plate	ft
HTU	height of a transfer unit	ft
k	Boltzmann constant (1.380×10^{-16})	erg/molecule-°K
k_c	gas mass transfer coefficient	cm/sec
k'_c	gas equimolal mass transfer coefficient	cm/sec
k_G	gas mass transfer coefficient	$gmole/sec\text{-}cm^2\text{-}atm$
k'_G	gas equimolal mass transfer coefficient	$gmole/sec\text{-}cm^2\text{-}atm$
k_L	liquid mass transfer coefficient	$gmole/sec\text{-}cm^2$
k'_L	liquid equimolal mass transfer coefficient	$gmole/sec\text{-}cm^2$
k_x	liquid mass transfer coefficient	$gmole/sec\text{-}cm^2$
k'_x	liquid equimolal mass transfer coefficient	$gmole/sec\text{-}cm^2$
k_y	gas mass transfer coefficient	$gmole/sec\text{-}cm^2$
k'_y	gas equimolal mass transfer coefficient	$gmole/sec\text{-}cm^2$
k_Y	gas mass transfer coefficient	$gmole/sec\text{-}cm^2$
K	vapor-liquid equilibrium constant	–
K^0	ideal vapor-liquid equilibrium constant	–
K_G	overall gas mass transfer coefficient	$gmole/sec\text{-}cm^2$
K_L	overall gas mass transfer coefficient	$gmole/sec\text{-}cm^2$
l_i	molar flow of component i in liquid	gmole/hr
L	overall liquid molar flow	$lbmole/hr\text{-}ft^2$
L_M	molar liquid flux	$lbmole/hr\text{-}ft^2$
L_s	solute-free inert liquid flow	$lbmole/hr\text{-}ft^2$
m	liquid-vapor equilibrium constant	–
m_c	equilibrium distribution coefficient	$gmole/cm^3\text{-}atm$
M	molecular weight	g/gmole, or lbm/lbmole
n	number of components in a system	–
N	mass transfer rate, or number of theoretical plates	$gmole/sec\text{-}cm^3$, –
N_{OG}	number of transfer units (gas)	–
NTU	number of transfer units	–
p_i	partial pressure of component i	atm
p^0	vapor pressure	atm
P	total pressure	atm
R	universal gas constant (82.1)	$atm\text{-}cm^3/gmole°K$

VAP/LIQ

S	stripping factor	–
T	absolute temperature	°K
v_i	molar flow of component i in vapor	gmole/hr
V	overall vapor molar flow	gmole/hr
W	mass transfer rate	g/cm^3-sec
x	mole fraction in liquid	–
X	mole ratio in liquid	–
y	mole fraction in vapor	–
Y	mass ratio	–
z	mole fraction in feed, or length or direction of diffusion	–, or ft or cm
Z	height of tower packing	ft

Symbols

α	volatility ratio	–
ϕ	number of phases in a system	–
ψ	number of intensive variables in a system	–
$\frac{\varepsilon}{\kappa}$	Lennard-Jones temperature	°K
ρ	mass density	grams/cm^3
σ	Lennard-Jones intermolecular distance	angstroms
ω	mass fraction	–
Ω	Lennard-Jones mass transfer function	–

Superscripts

*	equilibrium composition
0	pure component

Subscripts

A	component A
B	component B
G	gas phase
i	component i in an n-component system, interface
lm	log mean
L	liquid phase
M	molar
o	overall
s	solute-free
t	total

1 THE PHASE RULE

A problem confronting early thermodynamicists was determining the number of extensive variables needed to completely define a system consisting of a known number of phases and components. As an example, was knowing the temperature and index of refraction of a pure liquid enough to completely determine the state of the liquid? Gibbs determined that knowing two variables was enough. For a pure single component liquid, temperature and index of refraction are all that are needed to determine all the other extensive variables of the system. Extensive variables such as density, molar concentration, vapor pressure, etc., can each be expressed as functions of temperature and index of refraction. Gibbs derived the phase rule which relates the number of intensive variables to the number of components and phases:

$$\psi = C + 2 - \phi \qquad 8.1$$

A *phase* is a physically distinct and homogeneous substance, and may be a solid, liquid, gas, or plasma. Only one plasma phase or gas phase can exist in a system. Several liquid phases can exist simultaneously. Many solid phases can exist simultaneously, most notably as diverse crystalline structures. A *component* is typically an element or compound.

The number of *intensive variables* (sometimes called *degrees of freedom*) is the number of variables that can be varied independently to change the state of the system. An intensive variable is independent of the total quantity of the phase. Temperature and pressure are the most common intensive variables of a system. For a system composed of vapor and liquid phases, the number of intensive variables is $\psi = C + 2 - 2 = C$. Thus, for vapor and liquid to be in equilibrium, the number of intensive variables needed to define such a system is equal to the number of components in the system.

The most common system encountered is the binary (two component) liquid-vapor system. Therefore, for a binary liquid-vapor system, temperature and pressure alone will completely define the system. For an n-component two-phase system, the number of intensive variables is n.

Example 8.1

A solvent extraction process is used to extract uranium from an aqueous mixture consisting of uranyl nitrate and nitric acid in an immiscible organic phase of tributyl phosphate (TBP). Can the equilibrium between the two phases be calculated by knowing only the aqueous phase uranium and nitric acid concentrations and the organic phase TBP concentration?

Since

$C = 4 \quad (H_2O, UO_2(NO_3)_2, HNO_3, \text{and TBP})$

and

$$\phi = 2,$$

then,

$$\psi = 4 + 2 - 2 = 4$$

The four independent concentrations are enough to define the equilibrium in the system.

2 VAPOR-LIQUID EQUILIBRIUM

The simplest expression describing the equilibrium between vapor and liquid assumes that the vapor and liquid both are ideal. An *ideal liquid* is one whose properties can be related linearly to the mole fraction of its components (which usually only applies for dilute solutions). *Raoult's Law* relates the ideal vapor and liquid phase, and states that the partial pressure of a component in the vapor phase is directly proportional to the product of the component liquid phase mole fraction and the pure component vapor pressure. The equilibrium of component A between the vapor and liquid phase is

$$p_A = p^0 x_A \qquad 8.2$$

Dalton's Law for ideal gases (where each component is usually dilute) can be used to find the composition of the gas phase.

$$p_A = y_A P \qquad 8.3$$

If a system obeys Raoult's and Dalton's laws, then the vapor evolved from a liquid mixture will be a mixture of the same components (not usually in the same proportions) as the liquid. The vapor will normally be richer in the component having the highest vapor pressure at the temperature of the system. The component with the higher vapor pressure is called the *light* component. Vapor pressures of common hydrocarbons and water are found in figure 8.1, known as a *Cox chart.*

In systems where Raoult's Law does not apply, the phase composition can be predicted using vapor liquid equilibrium constants defined by equation 8.4.

$$y_A = K_A x_A \qquad 8.4$$

The vapor-liquid equilibrium constant is an experimentally determined constant, and is a function of temperature, pressure, and composition. In general terms, the equilibrium constant is

$$K_A = f(T, P, x_A) \qquad 8.5$$

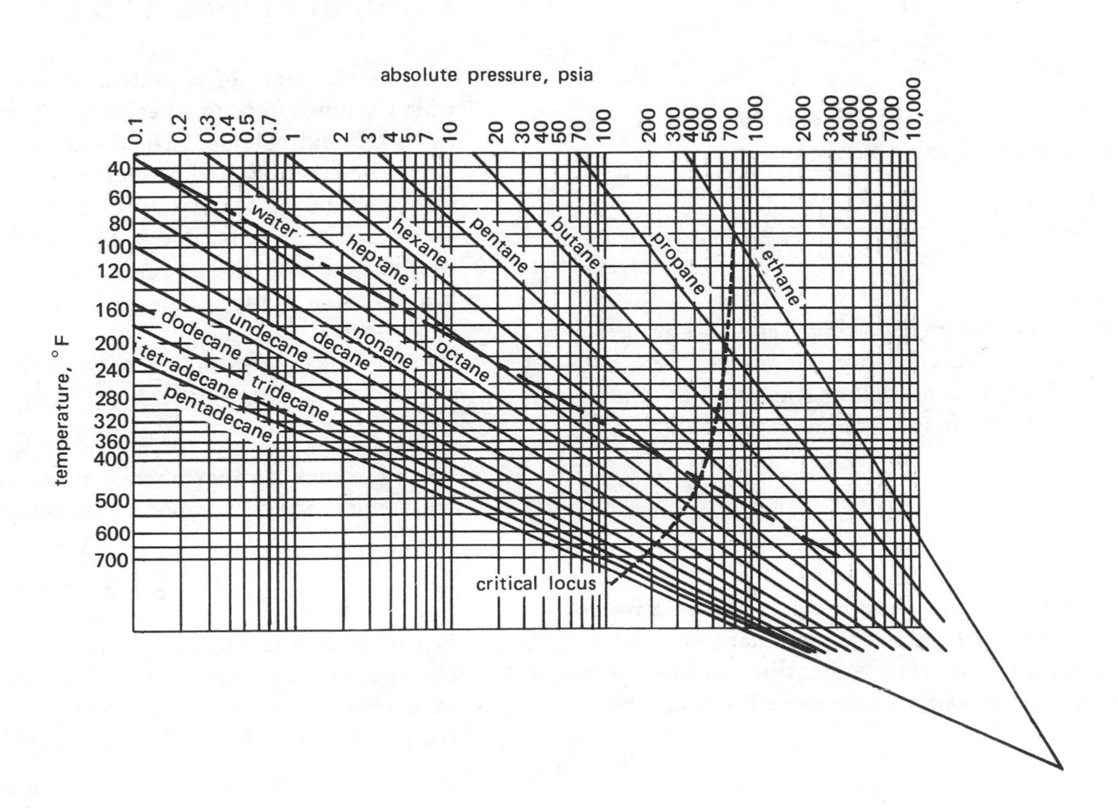

Figure 8.1 Vapor Pressure of Normal Paraffin Hydrocarbons and Water (Cox Chart)

Vapor-liquid equilibrium constants for common hydrocarbons are found in figures 8.2 and 8.3. If K_A is a function of temperature only, then the constant is ideal and is related to the vapor and liquid phase compositions by equation 8.6.

$$y_A = K_A^0 x_A \tag{8.6}$$

Henry's Law is very similar to the ideal equilibrium relationship, except that it relates the vapor phase partial pressure to the liquid phase concentration:

$$p_A = H_A x_A \tag{8.7}$$

For *binary systems* where Raoult's Law applies, and where component A is the more volatile component, the relative volatility can be determined using the equation

$$\frac{y_A x_B}{y_B x_A} = \frac{p_A^0}{p_B^0} = \alpha_{AB} \tag{8.8}$$

Binary systems are particularly interesting because only one component concentration is necessary to define the other in the same phase. The gas phase and liquid phase concentrations satisfy equations 8.9 and 8.10.

$$y_A + y_B = 1 \tag{8.9}$$

$$x_A + x_B = 1 \tag{8.10}$$

Therefore, for a binary system, the *relative volatility* can be computed using the equation

$$\frac{y_A (1 - x_A)}{x_A (1 - y_A)} = \alpha_{AB} \tag{8.11}$$

For a binary system following Raoult's Law or Henry's Law, the relative volatility is constant. For a binary system which does not follow Raoult's Law, the relative volatility can be found from equation 8.12.

$$\frac{K_A}{K_B} = \alpha_{AB} \tag{8.12}$$

For systems obeying equation 8.12, the relative volatility decreases with increasing pressure or temperature. The Raoult's Law relative volatility can be rearranged to calculate the vapor phase mole fraction:

$$y_A = \frac{\alpha_{AB} x_A}{1 + x_A (\alpha_{AB} - 1)} \tag{8.13}$$

The values of y_A as functions of α_{AB} and x are tabulated in table 9.1.

Equation 8.13 can be used to construct an equilibrium line on an x-y diagram when α_{AB} is given. If Raoult's Law does not apply, then the vapor phase mole fraction can be determined from equation 8.14.

$$y_A = \frac{K_A x_A}{K_B + (K_A - K_B) x_A} \tag{8.14}$$

Example 8.2

What is the vapor phase composition in equilibrium with an equimolar liquid solution of n-hexane and n-heptane at 50 °F and 30 psia?

From figure 8.2, the equilibrium constants at 50 °F and 30 psia are 0.064 and 0.019 for hexane and heptane, respectively. Therefore, using equation 8.14,

$$y_{hexane} = \frac{(0.064)(0.5)}{0.019 + (0.064 - 0.019)(0.5)}$$
$$= 0.7711$$
$$y_{heptane} = 1 - 0.7711 = 0.2289$$

3 BUBBLE POINT AND DEW POINT

The *bubble point* of a system is the temperature at which a liquid mixture begins to vaporize. If there is enough liquid volume, then the composition of the liquid will essentially be unchanged when the first bubble is formed during vaporization. The composition of the vapor in the first bubble is the vapor composition in equilibrium with the liquid in which it was formed. If Raoult's Law applies, then for each component,

$$y_i = \frac{p_i^0 x_i}{P} \tag{8.15}$$

Since the sum of vapor phase component mole fractions must equal one, the vapor phase composition is calculated:

$$\sum y_i = \left(\frac{1}{P}\right) \sum p_i^0 x_i = 1 \tag{8.16}$$

Equation 8.16 is solved by trial and error by varying the temperature (and therefore the vapor pressures of each component) until the equation is satisfied. The temperature at which this occurs is the bubble point.

If Raoult's Law does not apply, then the bubble point is calculated using equation 8.17.

$$\sum y_i = \sum K_i x_i = 1 \tag{8.17}$$

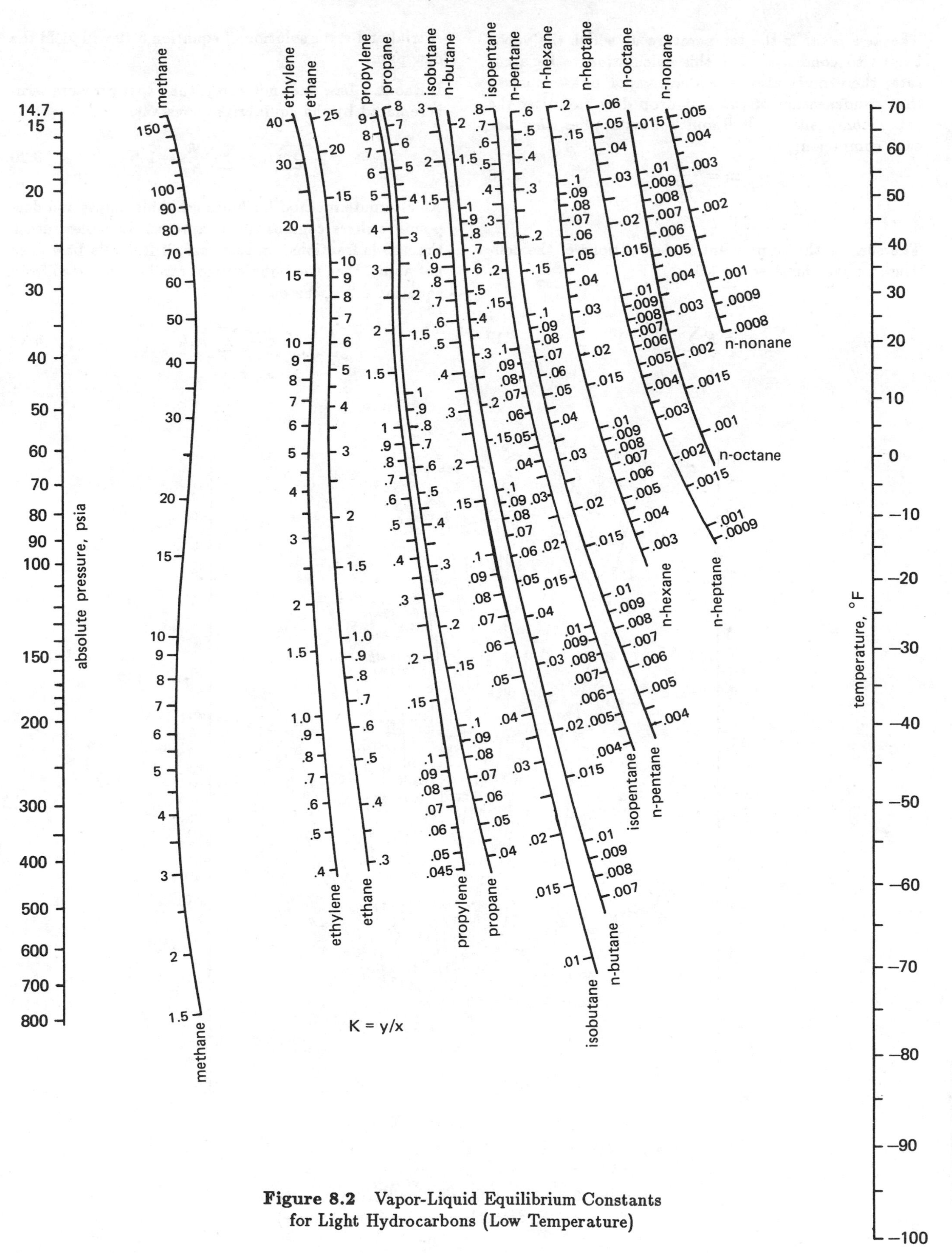

Figure 8.2 Vapor-Liquid Equilibrium Constants for Light Hydrocarbons (Low Temperature)

The *dew point* is the temperature at which the vapor begins to condense. For this calculation to be accurate, there must also be enough vapor phase mass so that condensation of the first drop does not alter the vapor composition. If Raoult's Law applies, then for each component,

$$x_i = \frac{P y_i}{p_i^0} \qquad 8.18$$

The sum of the component mole fractions of the condensed phase must equal 1.0.

$$\sum x_i = P \sum \frac{y_i}{p_i^0} = 1 \qquad 8.19$$

A trial-and-error solution of equation 8.19 will yield the dew point.

If Raoult's Law does not apply, the vapor pressure term is replaced by the equilibrium constant.

$$\sum x_i = \sum \frac{y_i}{K_i} = 1 \qquad 8.20$$

An alternate method for finding bubble points and dew points utilizes compositions expressed in moles rather than mole fractions. In this case, if Raoult's Law does not apply, then the bubble point can be calculated from equation 8.21, where $l_i = f_i$.

$$F = \sum K_i l_i \qquad 8.21$$

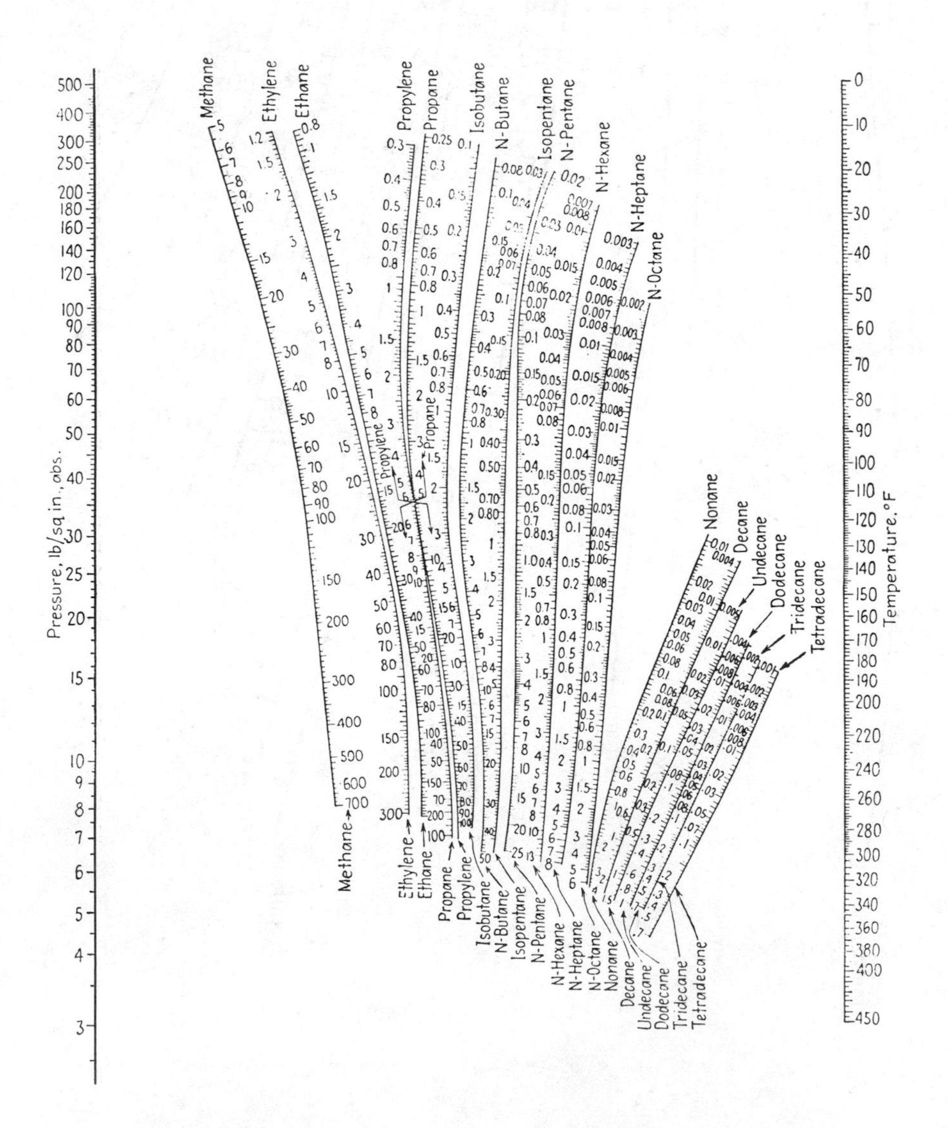

Figure 8.3 Vapor-Liquid Equilibrium Constants for Light Hydrocarbons (High Temperature)

The dew point can be calculated from equation 8.22.

$$F = \sum \frac{v_i}{K_i} \quad 8.22$$

$$V = \sum v_i \quad 8.23$$

$$L = \sum l_i \quad 8.24$$

$$F = \sum f_i \quad 8.25$$

$$y_i = \frac{v_i}{V} \quad 8.26$$

$$x_i = \frac{l_i}{L} \quad 8.27$$

Trial-and-error determinations of bubble or dew points require familiarity with the properties of the equilibrium constants. The following generalizations will help during calculations:

- K usually increases with increasing temperature.
- K usually decreases with increasing pressure.
- Calculations may be simplified if a reference component is chosen whose K is nearest 1.0 at the temperature in question. This reference component is used as a basis for the second trial according to the following rules:

For bubble point,

$$(K_{ref})_{trial\,2} = (K_{ref})_{trial\,1} \frac{F}{F_{calc}} \quad 8.28$$

For dew point,

$$(K_{ref})_{trial\,2} = (K_{ref})_{trial\,1} \frac{F_{calc}}{F} \quad 8.29$$

The use of equations 8.28 and 8.29 is illustrated in example 8.3.

4 FLASH VAPORIZATION

Flash vaporization occurs when a saturated liquid is pumped into a chamber held at a pressure lower than the saturation pressure. The liquid stream partially vaporizes, and two phases exit the chamber, one vapor and the other liquid. The net result is a liquid stream richer in the *heavier* components and a vapor richer in the *lighter* components. Vapor-liquid equilibrium calculations are used to determine the exit stream concentrations. With the following material balance and equilibrium equations, flash vaporization calculations can be made.

$$F = V + L \quad 8.30$$

$$f_i = v_i + l_i \quad 8.31$$

$$V = \sum v_i \quad 8.32$$

$$L = \sum l_i \quad 8.33$$

$$y_i = K_i x_i = \frac{v_i}{V} \quad 8.34$$

$$x_i = \frac{l_i}{L} = \frac{f_i - v_i}{L} \quad 8.35$$

An equation for the vapor phase composition can be obtained by combining the above equations, multiplying by $\frac{L}{K_i}$, and rearranging. The result is the equation

$$v_i = \frac{f_i}{1 + \frac{L}{K_i V}} \quad 8.36$$

Equation 8.36 gives the vapor flow rate of a *flashing feed*. A similar equation of the liquid flow rate of a flashing feed is

$$l_i = \frac{f_i}{1 + \frac{K_i V}{L}} \quad 8.37$$

Flash vaporization is evaluated by trial and error calculations to find vapor flow. A more direct method can be used to calculate flash vaporization with quick convergence of the trial and error results. This method normalizes the feed molar flow by letting $F = 1$.

$$z_i = \frac{f_i}{F} \quad 8.38$$

$$F = L + V = 1 \quad 8.39$$

$$y_i = \frac{z_i K_i}{K_i V + 1 - V} \quad 8.40$$

It is convenient to utilize a graphical method to converge to the result quickly. A plot of $\sum_i y_i$ versus V on linear coordinates is made. Start with $V = 0.5$ for convenience. The correct value of V is where the straight line (of negative slope) crosses $\sum_i y_i = 1.0$.

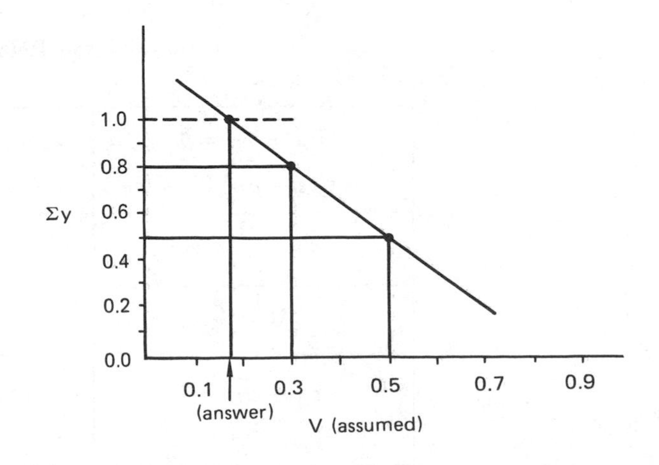

Figure 8.4 Typical Plot Used for the Solution of Flash Vaporization

VAP/LIQ

Example 8.3

Calculate the vapor and liquid phase compositions in a flash vaporization (140 °F and 200 psia) chamber for the starting composition shown.

i	component	f_i	K_i
1	ethane	15	3.70
2	propane	15	1.38
3	butane	15	0.57
4	pentane	15	0.21

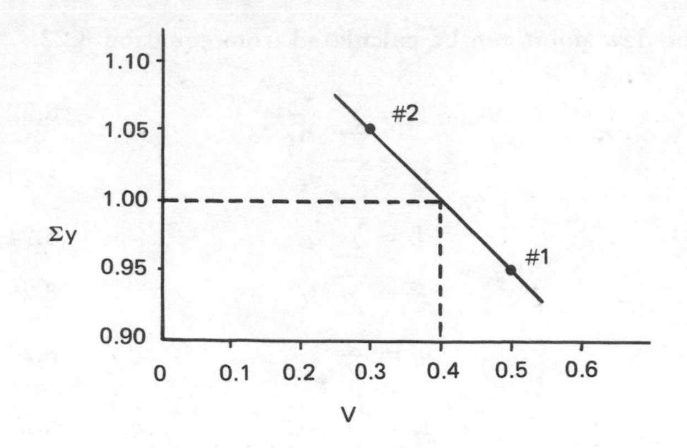

trial 1 $V = 0.5$

i	z_i	$K_i z_i$	$K_i V$	$K_i V + 1 - V$	y
1	0.25	0.926	1.85	2.35	0.384
2	0.25	0.348	0.69	1.19	0.292
3	0.25	0.143	0.285	0.785	0.182
4	0.25	0.053	0.105	0.605	0.088
	1.00				0.946

trial 2 $V = 0.3$

i	z_i	$K_i z_i$	$K_i V$	$K_i V + 1 - V$	y
1	0.25	0.926	1.11	1.81	0.511
2	0.25	0.348	0.414	1.114	0.313
3	0.25	0.143	0.171	0.871	0.164
4	0.25	0.053	0.063	0.763	0.069
	1.00				1.057

$V = 0.4$ (from plot)

i	z_i	$K_i z_i$	$K_i V$	$K_i V + 1 - V$	y
1	0.25	0.926	1.48	2.08	0.445
2	0.25	0.348	0.552	1.152	0.302
3	0.25	0.143	0.228	0.828	0.173
4	0.25	0.053	0.084	0.684	0.077
	1.00				0.997

$$V = 0.4(60) = 24$$
$$L = 60 - 24 = 36$$

Applying equation 8.35,

$$x_1 = \frac{15 - 24\,(0.445)}{36} = 0.12$$
$$x_2 = \frac{15 - 24\,(0.302)}{36} = 0.2153$$
$$x_3 = \frac{15 - 24\,(0.173)}{36} = 0.3013$$
$$x_4 = \frac{15 - 24\,(0.077)}{36} = 0.3653$$

$$1.0019$$

Table 8.1
Concentration Relationships for Binary Systems

$V_A + V_B = 1$ $\rho = (\rho_A - \rho_B) V_A + \rho_B$	$\rho_A = c_A M_A$	$\omega_A = \frac{\rho_A}{\rho}$
$c = c_A + c_B$	$c_A = \frac{\rho_A}{M_A}$	$x_A = \frac{c_A}{c}$
$M = \frac{\rho}{c}$		
$x_A + x_B = 1$	$x_A M_A + x_B M_B = M$	$x_A = \frac{W_A/M_A}{\frac{W_A}{M_A} + \frac{W_B}{M_B}}$
$\omega_A + \omega_B = 1$	$\frac{\omega_A}{M_A} + \frac{\omega_B}{M_B} = \frac{1}{M}$	$\omega_A = \frac{x_A M_A}{x_A M_A + x_B M_B}$

5 MASS TRANSFER

Mass transfer is the movement or exchange of components between two phases from a higher concentration to a lower concentration. Part of the flow is due to *molecular motion* (diffusion). Concentration is measured several different ways in mass transfer equations. Table 8.1 is a collection of equations which relate the various concentrations in a binary system.

6 DIFFUSION

The net rate of diffusion of A, N_A, at a point in a stationary fluid mixture of A and B is given by *Fick's Law of Diffusion*:

$$N_A = -D_{AB}\frac{dc_A}{dz} \tag{8.41}$$

Equation 8.41 states that substance A diffuses in the direction of decreasing concentration of A (just as heat flows by conduction in the direction of decreasing temperature). The rate of diffusion, measured by the diffusion coefficient, D_{AB}, is rapid in gases, and slower in liquids. The units of mass diffusivity are the same as the units of kinematic viscosity: cm^2/sec or ft^2/hr. For a one-dimensional system, Fick's Law is analogous to Newton's Law for mass transfer and Fourier's Law for heat transfer.

For a binary gas mixture at low pressure, the diffusivity, D_{AB}, is inversely proportional to pressure, increases with increasing temperature, and is almost independent of composition. The *Chapman-Enskog equation* for diffusivity is the best estimate for unmeasured binary pairs.

$$D_{AB} = 1.8583\times 10^{-3}\,\frac{\sqrt{T^3\left(\frac{1}{M_A}+\frac{1}{M_B}\right)}}{P\sigma_{AB}^2\Omega} \tag{8.42}$$

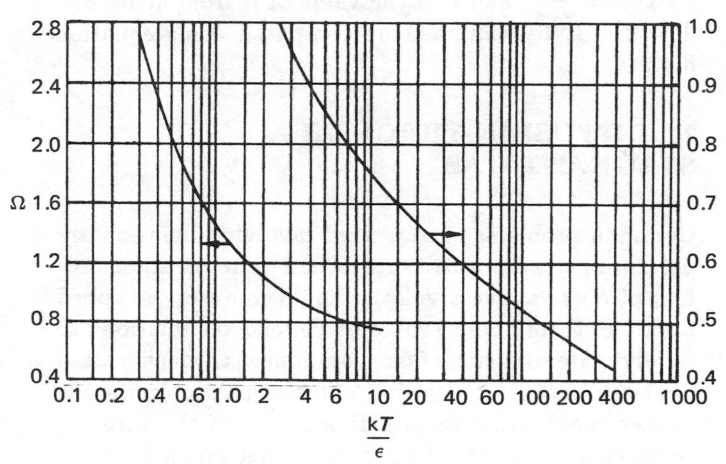

Figure 8.5 Plot of Ω as a Function of kT/ε

Table 8.2
Force Constants ($\frac{\varepsilon}{k}$) and Collision Diameters (σ) for Various Gases

gas	ε/k, °K	σ, angstroms	gas	ε/k, °K	σ, angstroms
air	97.0	3.617	hydrogen	33.3	2.968
ammonia	315	2.624	HCl	360	3.305
argon	124.0	3.418	iodine	550	4.982
benzene	440	5.270	methane	136.5	3.882
CO_2	190	3.996	neon	35.7	2.80
CO	110.8	3.590	NO	119	3.47
CCl_4	327	5.881	nitrogen	91.5	3.681
diphenyl	600	6.223	N_2O	220	3.879
ethane	230	4.418	n-octadecane	820	7.963
ethanol	391	4.455	n-octane	320	7.451
ethyl ether	350	5.424	oxygen	113.2	3.433
fluorocarbon F-12	288	5.110	propane	254	5.061
helium	6.03	2.70	SO_2	252	4.290
n-heptadecane	800	7.923	water	363	2.655

The function Ω is a dimensionless function of $\frac{kT}{\varepsilon_{AB}}$. Its value is determined using figure 8.5. The *Lennard-Jones parameters*, σ_{AB} and ε_{AB} are calculated using equations 8.43 and 8.44.

$$\sigma_{AB} = \frac{\sigma_A + \sigma_B}{2} \qquad 8.43$$

$$\varepsilon_{AB} = \sqrt{\varepsilon_A \varepsilon_B} \qquad 8.44$$

Table 8.2 lists some of the Lennard-Jones parameters for common gases. To calculate the diffusivity of a *gas pair*, find the values of $\frac{\varepsilon_A}{k}$, $\frac{\varepsilon_B}{k}$, σ_A, and σ_B from table 8.2. Calculate $\frac{\varepsilon_{AB}}{k}$ using equation 8.44 to find ε_{AB}. Calculate $\frac{kT}{\varepsilon_{AB}}$, and find the value of Ω from figure 8.5. The diffusivity can then be computed using equation 8.42.

7 DIFFUSION THROUGH A STAGNANT GAS

Common problems encountered involving diffusion are determining evaporation rates and concentration profiles of a gas above a volatile liquid in an open-topped tank. Both of these problems depend on solutions to steady state diffusion of one component through a stagnant component. Figure 8.6 represents a problem of the former type. The concentration profile of the diffusing component A in the open tank is determined by the solution of Fick's Law. Equation 8.45 is that solution.

$$ln\left(\frac{1 - y_A}{1 - y_{A1}}\right) = \frac{z - z_1}{z_2 - z_1}\, ln\left(\frac{1 - y_{A2}}{1 - y_{A1}}\right) \qquad 8.45$$

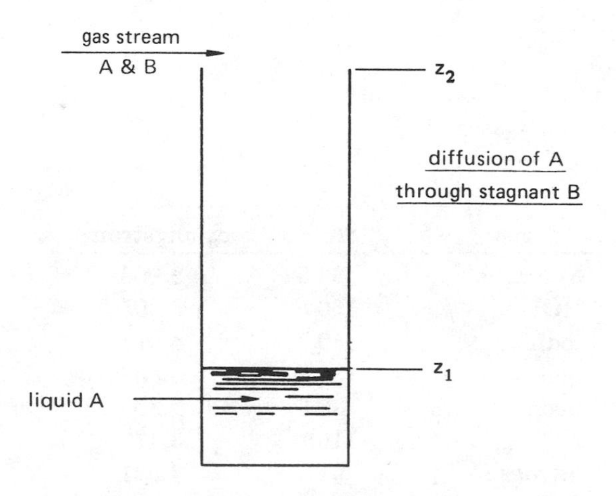

Figure 8.6 Diffusion of A Through Stagnant B

The *rate of evaporation* is given by another solution of Fick's Law. Referring to figure 8.6, the rate of evaporation of A at the surface of the liquid is given by equation 8.46.

$$N_A = \frac{P\, D_{AB}\,(p_{A1} - p_{A2})}{(z_2 - z_1)\, RT\, (p_B)_{lm}} \qquad 8.46$$

The *log-mean partial pressure* is

$$(p_B)_{lm} = \frac{p_{B2} - p_{B1}}{ln\,\frac{p_{B2}}{p_{B1}}} \qquad 8.47$$

Equation 8.46 can be written in terms of mole fractions:

$$N_A = \frac{c\, D_{AB}\,(y_1 - y_2)}{(z_2 - z_1)(y_B)_{lm}} \qquad 8.48$$

8 DIFFUSION THROUGH TWO MOVING PHASES

The *dual film theory* for the transfer of substance A through nontransferring substance B, illustrated in figure 8.7, is a situation analogous to convective heat transfer. In mass transfer, the two films are in contact with each other instead of being separated by a surface. The mass transfer through both phases is given by equations 8.49 through 8.53.

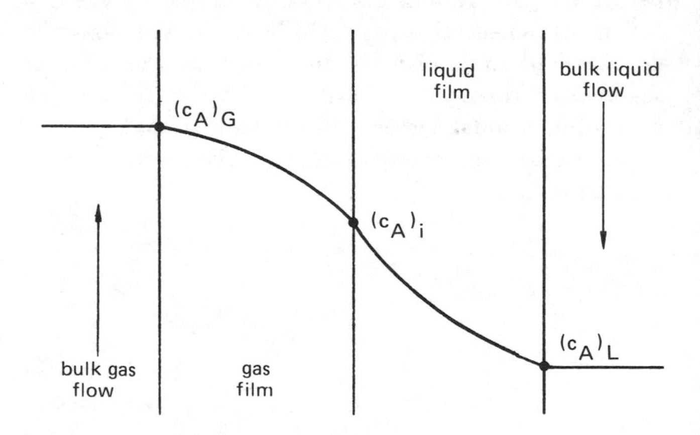

Figure 8.7 Transfer of Substance A Through Nontransferring B

For gases:

$$N_A = k_G\,[(p_A)_G - (p_A)_i] \qquad 8.49$$

$$= k_y\,[(y_A)_G - (y_A)_i] \qquad 8.50$$

$$= k_c\,[(c_A)_G - (c_A)_i] \qquad 8.51$$

For liquids:

$$N_A = k_x\,[(x_A)_L - (x_A)_i] \qquad 8.52$$

$$= k_L\,[(c_A)_L - (c_A)_i] \qquad 8.53$$

VAP/LIQ

Table 8.3
Definition of Mass Transfer Coefficients

(a) gases

counter diffusion	through stagnant B	units of coefficients
$N_A = k'_G \, \Delta p_A$	$N_A = k_G \, \Delta p_A$	$\frac{\text{moles transferred}}{\text{(time)(area)(pressure)}}$
$N_A = k'_y \, \Delta y_A$	$N_A = k_c \, \Delta y_A$	$\frac{\text{moles transferred}}{\text{(time)(area)(mole fraction)}}$
$N_A = k'_c \, \Delta c_A$	$N_A = k_c \, \Delta c_A$	$\frac{\text{moles transferred}}{\text{(time)(area)(moles/volume)}}$
	$W_A = k_Y \, \Delta y_A$	$\frac{\text{mass transferred}}{\text{(time)(area)(mass A/mass B)}}$

$$\text{conversions: } k_G (p_B)_{lm} = k_y \frac{(p_B)_{lm}}{P_t} = k_c \frac{(p_B)_{lm}}{RT} = \frac{k_Y}{M_B} = k'_y = k'_c \frac{P}{RT}$$

(b) liquids

counter diffusion	through stagnant B	units of coefficients
$N_A = k'_L \, \Delta c_A$	$N_A = k_L \, \Delta c_A$	$\frac{\text{moles transferred}}{\text{(time)(area)(moles/volume)}}$
$N_A = k'_x \, \Delta x_A$	$N_A = k_x \, \Delta x_A$	$\frac{\text{moles transferred}}{\text{(time)(area)(mole fraction)}}$

$$\text{conversions: } k_x (x_B)_{lm} = k_L (x_B)_{lm} \, c = k'_L c = k'_L \frac{\rho}{M} = k'_x$$

9 RATE OF MASS TRANSFER

Mass transfer rate coefficients can be defined in terms of the rate of transfer across the interface, N_A, and the driving force as shown in table 8.3.

As an example, the equation for a stripper can be written:

$$N_A = k_y (y_i - y) \tag{8.54}$$

$$= k_x (x - x_i) \tag{8.55}$$

$$\frac{y_i - y}{x - x_i} = \frac{k_x}{k_y} = \frac{L_M H_G}{G_M H_L} \tag{8.56}$$

The *rate of transfer* can be solved graphically from a plot of the equilibrium vapor and the liquid compositions (equilibrium line) and the operating line (from material balances), as shown in figure 8.8.

The rate of transfer, N_A, is proportional to the difference of bulk concentration in one phase and the concentration in the same phase that would be in equilibrium with the bulk concentration of the other phase (i.e., $y^* - y$ or $x - x^*$).

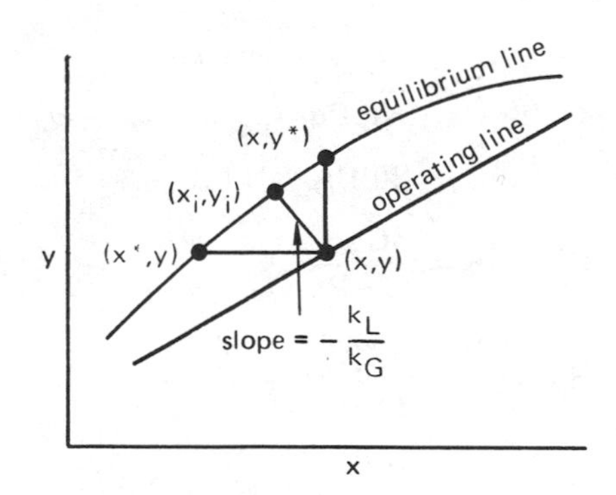

Figure 8.8 Operating Diagram of a Stripper

An alternate method for finding the rate of transfer is to solve equation 8.56 for $\frac{k_x}{k_y}$. Then, the equilibrium relation, y_i versus x_i can be determined. If overall coefficients are used, the rate of transfer is

$$N_A = K_y (y^* - y) \tag{8.57}$$

$$= K_x (x - x^*) \tag{8.58}$$

It can be shown that, where $y = mx$, the overall gas coefficient is

$$\frac{1}{K_G} = \frac{1}{k_G} + \frac{m}{k_L} \tag{8.59}$$

VAP/LIQ

The overall liquid coefficient is

$$\frac{1}{K_L} = \frac{1}{mk_G} + \frac{1}{k_L} \qquad 8.60$$

When m has a high value (low solubility in liquid), the mass transfer is *liquid phase controlled* ($K_L \sim k_L$). When m has a low value (high solubility in the liquid), the mass transfer is *gas phase controlled* ($K_G \sim k_G$).

For design usage, the interfacial area between phases is not easily determined. To compensate for this, the product $K_G a$ or $K_L a$ is used. The previous equations become

$$\frac{1}{K_G a} = \frac{1}{k_G a} + \frac{m}{k_L a} \qquad 8.61$$

$$\frac{1}{K_L a} = \frac{1}{mk_G a} + \frac{1}{k_L a} \qquad 8.62$$

10 NUMBER OF TRANSFER UNITS AND HEIGHT EQUIVALENT TO A UNIT

The *transfer unit* concept has evolved as a way of representing mass transfer data that is nearly independent of flow rate. The transfer units concept finds application in designing countercurrent contacting equipment, such as packed absorption towers. The height of packing required can be calculated:

$$h = \frac{G_M}{K_G P a} \int_{y_{A\,in}}^{y_{A\,out}} \frac{dy_A}{y^* - y_A} \qquad 8.63$$

$$= \mathrm{HTU}_{OG}\ \mathrm{NTU}_{OG} \qquad 8.64$$

$$\mathrm{HTU}_{OG} = \frac{G_M}{K_G P a} \qquad 8.65$$

$$\mathrm{NTU}_{OG} = \int_{y_{A\,in}}^{y_{A\,out}} \frac{dy_A}{y^* - y_A} \qquad 8.66$$

If y^* is constant or linear in y_A (i.e., there is a straight equilibrium line), the expression is

$$\mathrm{NTU}_{OG} = \frac{(y_A)_{out} - (y_A)_{in}}{(y_A^* - y_A)_{lm}} \qquad 8.67$$

If the mass transfer is liquid controlled (one which has a large Henry's constant), similar equations for HTU_{OL} and NTU_{OL} can be derived for a straight equilibrium line:

$$\mathrm{NTU}_{OL} = \frac{(x_A)_{in} - (x_A)_{out}}{(x_A - x_A^*)_{lm}} \qquad 8.68$$

$$\mathrm{HTU}_{OL} = \frac{L_M}{K_L \rho_M a} \qquad 8.69$$

$$L_M = \frac{L}{A} \qquad 8.70$$

In general, HTU_{OG} and HTU_{OL} are different, because the driving forces are different. For the same reason, NTU_{OG} and NTU_{OL} are different. When y^* is not constant or linear in y_A (or x^* is not linear in x_A), then the integrals must be solved graphically. Graphical integrations can be avoided by using short cuts which assume linear x^* and y^*. A *Colburn chart*, such as figure 8.9, can be used to determine NTU_{OG} if $y^* = mx$. Figure 8.9 plots NTU_{OG} (sometimes called N_{OG}) as a function of the ratio

$$\frac{(y_A)_{in} - m(x_A)_{out}}{(y_A)_{out} - m(x_A)_{out}} = \frac{y_1 - mx_2}{y_2 - mx_2} \qquad 8.71$$

for various values of $\frac{mG_M}{L_M}$.

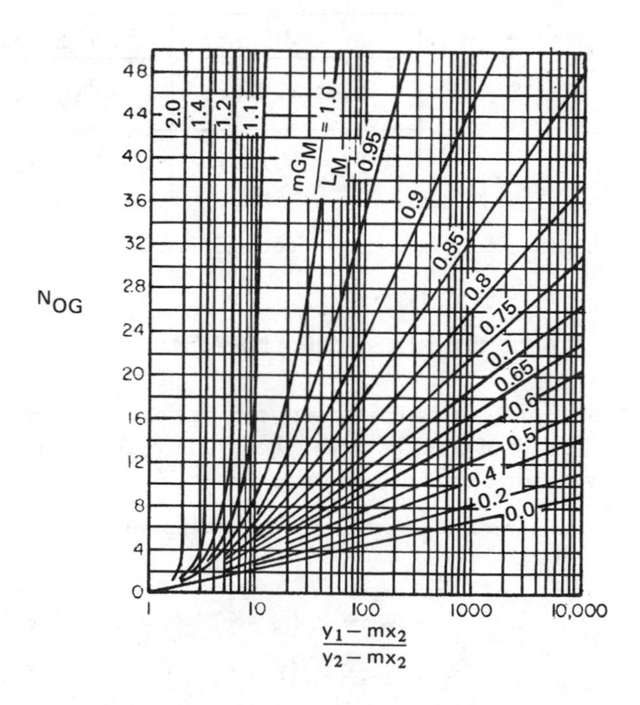

Figure 8.9 Number of Units in an Absorption Column for Constant $\frac{mG_M}{L_M}$ (Colburn Chart)

11 MASS TRANSFER OPERATIONS

Mass transfer operations are involved in making component separations. Table 8.4 classifies the major transfer operations.

Table 8.4
Major Transfer Operations

process	light phase	heavy phase	transfer direction
gas absorption	gas	liquid	gas to liquid
stripping	gas	liquid	liquid to gas
extraction	liquid	liquid	either
leaching	liquid	solid	solid to liquid
humidification	air	water	either
drying	gas	solid	solid to gas
adsorption	gas or liquid	solid	gas or liquid to solid
ion exchange	liquid	solid	liquid to solid
distillation	vapor	liquid	either

12 ABSORPTION AND STRIPPING

When a gas mixture is contacted with a liquid to preferentially remove one or more of the components of the gas, thereby forming a solution of gases in the liquid, the operation is called *gas absorption*. Such an operation requires mass transfer of a component from the gas stream to the liquid. When mass transfer occurs from the liquid to the gas, the operation is called *desorption* or *stripping*.

Problems involving gas absorption include the calculation of tower height, number of equilibrium stages, efficiency, and vapor-liquid handling capacity of the equipment involved (e.g., calculation of tower diameter). Since solutions to problems of the last type usually are empirical in nature, they will not be dealt with here. Material balances and vapor-liquid equilibrium are required to solve height, number of stages, or efficiency problems.

13 MATERIAL BALANCES: ABSORPTION AND STRIPPING

A material balance around an absorber or stripper tower is made without accumulation. The material balance (see figure 8.10) is

$$G_1 - G_2 = L_1 - L_2 \qquad 8.72$$

A component balance at any height, z, from the bottom of the tower gives the material balance:

$$Lx - L_1x_1 = Gy - G_1y_1 \qquad 8.73$$

If a small amount of substance is transferring between phases, the gas phase flow changes little, and $G_1 \sim G$ and $L_1 \sim L$. The operating line equation results:

$$\frac{G}{L} = \frac{x - x_1}{y - y_1} \qquad 8.74$$

The operating line equation can also be written in the more common slope form:

$$y = \left(\frac{L}{G}\right) x + y_1 - \left(\frac{L}{G}\right) x_1 \qquad 8.75$$

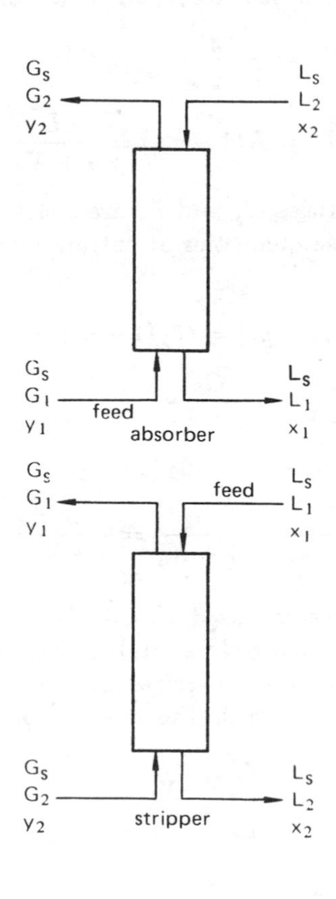

Figure 8.10 Absorption and Stripping Line Definitions

VAP/LIQ

This equation is only approximate since $\frac{L}{G}$ is not constant throughout the tower. If the material balance is performed around the top of the tower, the operating line equation becomes

$$y = \left(\frac{L}{G}\right) x + y_2 - \left(\frac{L}{G}\right) x_2 \quad 8.76$$

Figure 8.10 shows a countercurrent tower, which may be a packed tower, a spray tower, a tower filled with bubble-cap trays, or a tower having any internal construction to bring about liquid-gas contact. The gas stream at any point in the tower consists of G total moles/hr-ft^2 (where the area basis is the cross section of the tower), and is made up of diffusing solute A of mole fraction y, partial pressure p_A, or mole ratio Y, and a non-diffusing, insoluble gas G_s moles/hr-ft^2. The relationships between these quantities are:

$$Y_A = \frac{y_A}{1 - y_A} = \frac{p_A}{P - p_A} \quad 8.77$$

$$G_s = G(1 - y_A) = \frac{G}{1 + Y_A} \quad 8.78$$

Similarly, the liquid stream at any point in the tower consists of L total moles/hr-ft^2, containing x mole fraction solubilized gas A, or mole ratio X, and an essentially non-volatile solvent, L_s moles/hr-ft^2.

$$X_A = \frac{x_A}{1 - x_A} \quad 8.79$$

$$L_s = L(1 - x_A) = \frac{L}{1 + X_A} \quad 8.80$$

The flow quantities G_s and L_s are constant. The relationship of these quantities at entrance and exit conditions are:

$$G_s = G_1(1 - y_1) = G_2(1 - y_2) = G(1 - y) \quad 8.81$$

$$G_s = \frac{G_1}{1 + Y_1} = \frac{G_2}{1 + Y_2} = \frac{G}{1 + Y} \quad 8.82$$

$$L_s = L_1(1 - x_1) = L_2(1 - x_2) = L(1 - x) \quad 8.83$$

$$L_s = \frac{L_1}{1 + X_1} = \frac{L_2}{1 + X_2} = \frac{L}{1 + X} \quad 8.84$$

These quantities are used to calculate G and L with the assumption that only a small amount of component transfer occurs, which requires $L \sim L_1$ and $G \sim G_1$. If mole ratios are used, the exact expression for the operating line is

$$\frac{G_s}{L_s} = \frac{X_1 - X}{Y_1 - Y} \quad 8.85$$

In slope form, equation 8.85 is

$$Y = \frac{L_s}{G_s} X + Y_1 - \frac{L_s}{G_s} X_1 \quad 8.86$$

If mass transfer is substantial enough to alter the gas or liquid flow quantity, mole ratios are the preferred concentration units.

14 *X-Y* PLOTS FOR ABSORBERS AND STRIPPERS

The number of *theoretical plates* can be determined in absorbers and strippers in a manner similar to distillation processes using McCabe-Thiele diagrams. First, the *X-Y* diagram is set up by plotting the equilibrium line. The operating line is then determined by the material balance around the tower. The lines in figure 8.11 are generalized plots of the equilibrium and operating lines for an absorber and stripper.

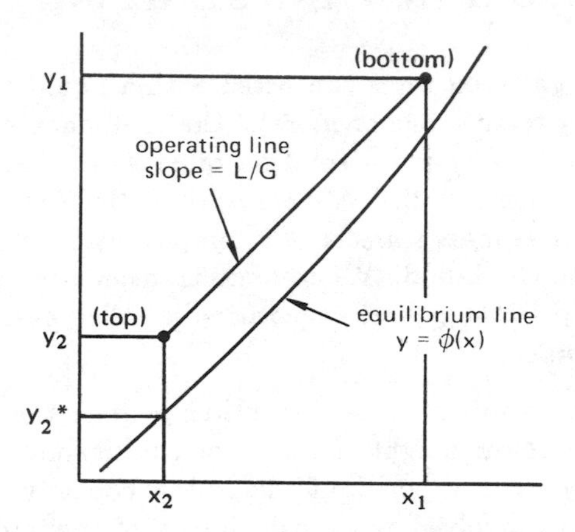

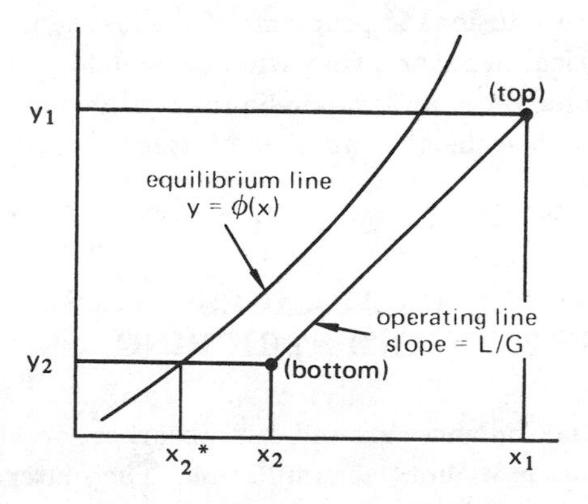

Figure 8.11 *X-Y* Plot for Absorber and Stripper

In the absorber plot, the operating line is above the equilibrium curve, and has a slope $\frac{L_s}{G_s}$. The stripper has an operating line below the equilibrium curve, with a slope $\frac{L_s}{G_s}$. The operating line in these plots is not straight, but since the endpoints of the line are all that are known, a straight line is assumed. The operating line in an *X-Y* plot (mole ratio) is straight. Care must be taken when using *X-Y* plots, since the equilibrium curve usually is given as a straight line in terms of mole fraction, (e.g., $y = mx$), which when converted to mole ratios, is not a straight line. If the mole fraction equilibrium curve is a straight line, then the mole ratio equilibrium curve becomes

$$Y = \frac{mX}{1 + X(1-m)} \qquad 8.87$$

15 MINIMUM $\frac{L}{G}$

An X-Y plot of absorber performance is shown in figure 8.12.

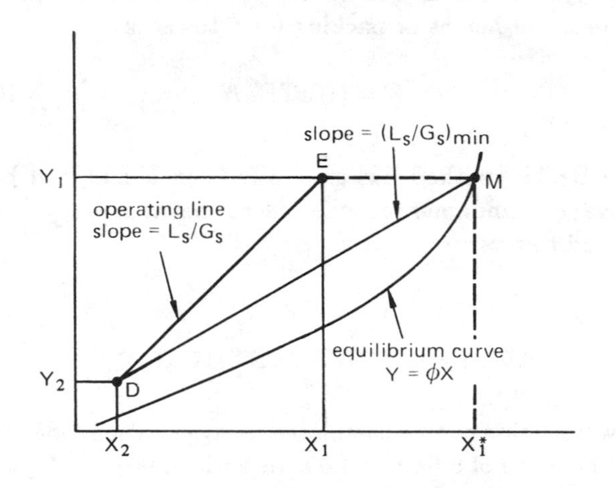

Figure 8.12 Absorber Performance

The terminal concentrations Y_1, Y_2, and the composition of entering liquid X_2 are ordinarily fixed by process requirements. The operating line must pass through point D, and must end at the ordinate Y_1. The quantity of liquid L or gas G usually can be varied. If G or L is varied so as to give operating line DE, the exit liquid composition will be X_1. If less liquid (or more gas) is used (and hence decreasing the slope of the operating line), the exit liquid composition will be greater, but since the diffusional driving forces will be smaller, the absorption will be slower. The time of contact between liquid and gas must be greater (and correspondingly the tower must be taller).

The minimum liquid rate corresponds to operating line DM having the greatest slope touching the equilibrium line. At point M, the driving force is zero, the time of contact is infinite, and an infinitely tall tower results. The minimum liquid-gas ratio represents a limiting value. The minimum $\frac{L}{G}$ can be calculated if $y = mx$ (mole fractions). In terms of mole ratios,

$$Y = mX\left(\frac{Y+1}{X+1}\right) \qquad 8.88$$

$$\left(\frac{L_s}{G_s}\right)_{min} = \frac{Y_1 - Y_2}{\dfrac{1}{m - 1 + \frac{m}{Y_1}} - X_2} \qquad 8.89$$

16 ABSORPTION FACTORS

If the equilibrium relationship is $y = mx$, then the *absorption factor* can be calculated using equation 8.90.

$$A = \frac{L}{mG} \qquad 8.90$$

When A is large, the component i tends to be absorbed in the liquid phase. The *stripping factor* is defined as

$$S = \frac{mG}{L} \qquad 8.91$$

When S is large, the component i tends to be absorbed in the vapor phase and is thus stripped from the liquid. As a guide, typical values are 1.0 to 1.5 for A, and 1.5 to 2.0 for S. Table 8.5 gives the different forms of absorption and stripping factors for various equilibrium relationships.

Table 8.5
Absorption and Stripping Factors

equilibrium expression	A	S
$y = Kx$ (vaporization constant)	$\frac{L}{KG}$	$\frac{KG}{L}$
$y = \frac{Hx}{P}$ (Henry's Law)	$\frac{PL}{GH}$	$\frac{HG}{PL}$
$y = \frac{p^0}{P}x$ (Raoult's Law)	$\frac{PL}{p^0G}$	$\frac{p^0G}{PL}$

17 KREMSER-BROWN ABSORPTION FACTORS

The *Kremser-Brown technique* is a method which rapidly determines the number of theoretical stages needed in absorbers or strippers. The method assumes that the operating line is straight (in an x-y plot), and that $L_1 \sim L$ and $G_1 \sim G$. With these assumptions, the Kremser-Brown equation can be derived:

$$E = \frac{y_1 - y_2}{y_1 - y_2^*} = \frac{A - A^{N+1}}{1 - A^{N+1}} \qquad 8.92$$

In equation 8.92, y_2^* is the value of y_2 which would be in equilibrium with x_2, and E is the *effectiveness* or fraction absorbed. For strippers, the effectiveness is

$$E = \frac{S - S^{N+1}}{1 - S^{N+1}} \qquad 8.93$$

The Kremser-Brown factors are plotted in figure 8.13 as A or S versus E, and as number of theoretical plates versus E, respectively. These graphs are useful for determining the number of theoretical stages for an absorber or stripper having straight equilibrium and operating lines (x-y plot). Equations 8.92 and 8.93 can be solved for N, which results in equation 8.94.

$$N = \frac{\ln\left(\frac{A - E}{A - AE}\right)}{\ln A} \qquad 8.94$$

In the special case where $A = 1$, the efficiency and number of plates can be calculated using equations 8.95 and 8.96.

$$E = \frac{N}{N + 1} \qquad 8.95$$

$$N = \frac{E}{1 - E} \qquad 8.96$$

The equations for strippers are the same as equations 8.94 to 8.96, when S is substituted for A.

18 COLBURN FACTOR

An alternate method of determining the number of theoretical absorber or stripper stages is to use the *Colburn method.* This method is based on a straight equilibrium line only. The operating line can be curved, so that $G_1 \neq G$ and $L_1 \neq L$. The absorption factors are calculated at the feed end (bottom for an absorber) and at the effluent end (top for an absorber).

$$A_1 = \frac{L_1}{mG_1} \qquad 8.97$$

$$A_2 = \frac{L_2}{mG_2} \qquad 8.98$$

Where $A_2 > A_1 > 1$, the Colburn equation results

$$E = 1 - \frac{A_1 (A_2 - 1)^2}{A_2 (A_1 - 1)(A_2^{N+1} + 1)} \qquad 8.99$$

The number of plates can be computed by rearranging equation 8.99:

$$N = \frac{\ln\left[\frac{A_1}{(1 - E)(A_1 - 1)}\left(\frac{A_2 - 1}{A_2}\right)^2 - \frac{1}{A_2}\right]}{\ln A_2} \qquad 8.100$$

19 HEIGHT EQUIVALENT OF A THEORETICAL PLATE

A simplistic method for determining the height of an absorber tower ignores the differences between stagewise and continuous contact. This method utilizes the number of theoretical trays determined by the Kremser-Brown or Colburn method, and multiplies that quantity by the *height of a theoretical plate* (HETP) to give the required *height of packing.* The HETP must be experimentally determined for each type of packing or plates used in the tower. The HETP varies with type and size of packing, with flow rate of each fluid, and with the type of system and concentration. If the HETP is known, the height of packing for a tower is

$$Z = (\text{HETP})\, N \qquad 8.101$$

The HETP method has generally been abandoned because of fundamental differences between tray and packed towers.

20 TRANSFER UNIT METHOD

Tower heights can be calculated using equation 8.63 or a combination of 8.68 and 8.69. In both equation sets, the integral term is equal to the total composition change divided by the driving force.

From the discussion on mass transfer, the tower height can be determined as the product of the number of transfer units and the transfer unit height:

$$Z = \text{HTU}_{OG}\,\text{NTU}_{OG} \qquad 8.102$$

$$\text{HTU}_{OG} = \frac{G}{K_G a} = \frac{G}{K_y a\,(1 - y)_{lm}}$$

$$= \frac{G}{K_G a P\,(1 - y)_{lm}} \qquad 8.103$$

$$\text{NTU}_{OG} = \int_{y_1}^{y_2} \frac{1}{\ln\left(\frac{1-y}{1-y^*}\right)}\left(\frac{dy}{y - y^*}\right) \qquad 8.104$$

Equation 8.104 defines the transfer unit used with absorbers. If the absorption factor is constant throughout the tower, and the equilibrium line is linear, the solution to equation 8.104 is

$$\text{NTU}_{OG} = \frac{A}{A - 1} \ln\left(\frac{A - E}{A - AE}\right) \qquad 8.105$$

VAP/LIQ

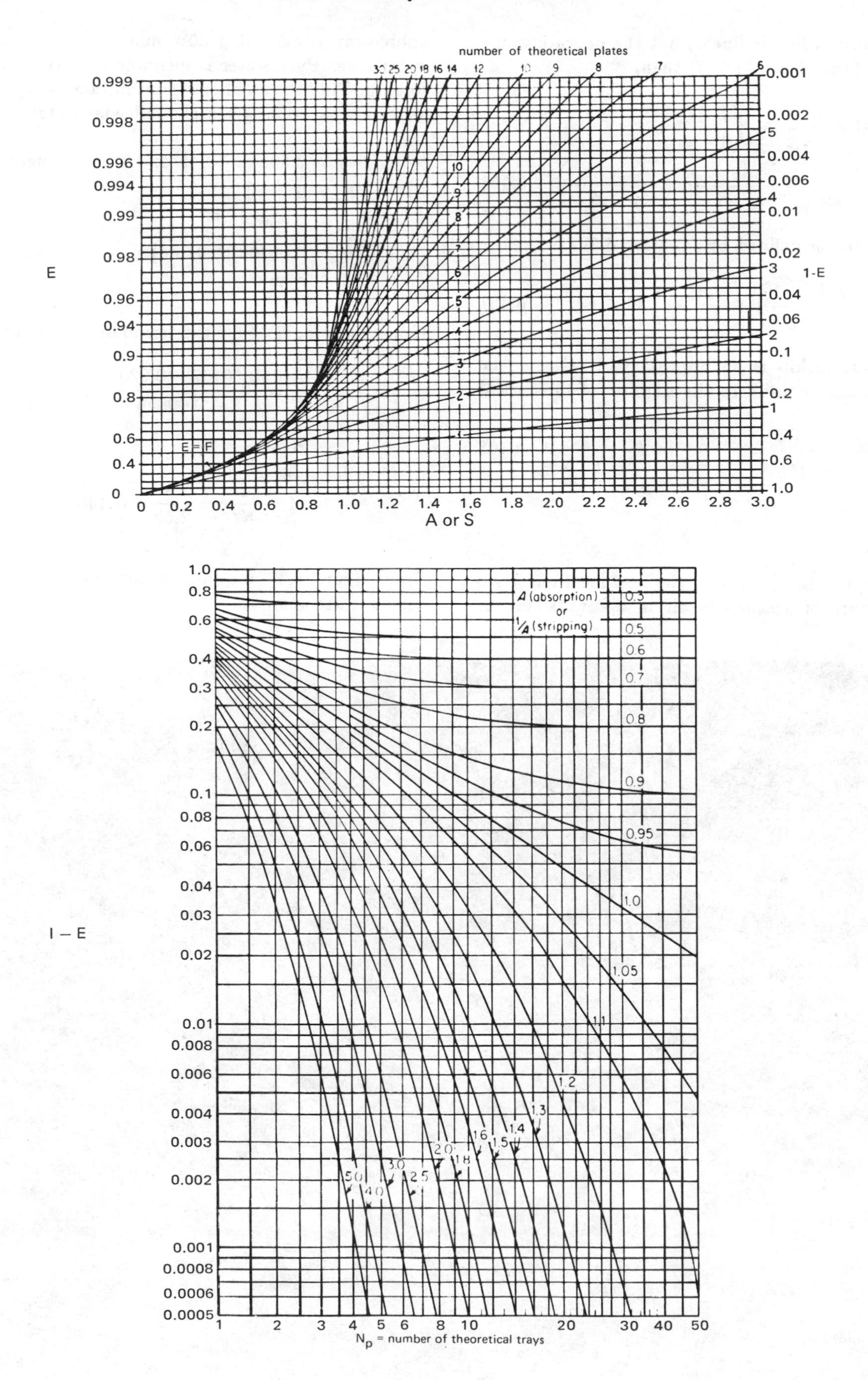

Figure 8.13 Kremser-Brown Absorption Factors

If the equilibrium line is linear, but the operating line is not linear (and $A_2 > A_1 > 1$) then,

$$\text{NTU}_{OG} = \frac{1}{2}\, ln \left(\frac{1-y_2}{1-y_1}\right) + \frac{A_2}{A_2-1}\, ln \left(\frac{A_2-\varepsilon}{A_2-A_2\varepsilon}\right) \qquad 8.106$$

where ε is a factor defined by equation 8.107.

$$1-\varepsilon = \left(\frac{A_1-1}{A_1}\right)\left(\frac{A_2}{A_2-1}\right)(1-E) \qquad 8.107$$

A useful relationship is an approximate ratio of the number of transfer units to the number of theoretical plates:

$$\frac{\text{NTU}}{N} = \frac{A_2}{A_2-1}\, ln\, A_2 \qquad 8.108$$

Example 8.4

It is desired to absorb 95% of the acetone from a 2.0 mole% mixture of acetone in air in a countercurrent bubble-cap tower using 20% more than the minimum water rate. Pure water is introduced to the top of the column. Find the number of equilibrium stages required for this separation, given the following data:

$$y = 2.53x \text{ at } 80^\circ \text{ F and 1 atmosphere}$$
$$x_2 = 0 \text{ (top)}$$
$$x_1 = 0.00658$$
$$y_2 = 0.00102 \text{ (bottom)}$$
$$y_1 = 0.02$$

Using equation 8.74 for the terminal conditions,

$$\frac{L}{G} = \frac{0.00102 - 0.020}{0 - 0.00658} = 2.90$$

$$K = \frac{y}{x} = 2.53$$

$$A = \frac{L}{KG} = \frac{2.90}{2.53} = 1.146$$

From figure 8.13 for 95% absorption: 9 theoretical plates is the answer.

PRACTICE PROBLEMS

1. A mixture of ethane, propane, and n-butane exists at −10 °C and a total pressure of 3000 mm mercury. Determine the compositions of the vapor and liquid phases if the propane and n-butane compositions are equal in the liquid phase. Vapor pressures at −10 °C are (in mm mercury): ethane, 14,000; propane, 2700; and n-butane, 500.

2. A 1000 lbm/hr mixture of acetone in air (2.0 mole%) is treated in a countercurrent gas absorber, 1.4′ in diameter and packed with Raschig rings. The equilibrium equation is $y = 2.53x$. Assume that the mass rate of lean gas mixture remains unchanged. What is the minimum rate (in lbm/hr) of acetone-free water flow to the absorber for 95% recovery of the acetone?

3. A mixture of two high boiling point organic acids is separated from a small amount of non-volatile, carbonaceous material by continuous steam distillation in a small still operating at 212 °F and 200 mm mercury. The organic acid mixture is equimolar with respect to the acids, whose vapor pressures are 32 and 14 mm mercury at 212 °F, respectively. Assume that the acid mixture obeys Raoult's Law, the acid mixture is immiscible in water, the carbonaceous material has no effect on the equilibrium, and the vapor leaves in equilibrium with the liquid in the still. How many pounds of steam are required per mole of acid recovered?

4. A vent pipe (2″ I.D.) contains n-octane at its bottom. The pipe outlet is 5′ from the liquid surface. The n-octane temperature is 31.5 °C, and the total pressure is one atmosphere. The vapor pressure of n-octane is 20 mm mercury at 31.5 °C. Wind blows across the top of the pipe, where the concentration of octane is negligible. The lower explosive limit of n-octane is 1.0% by volume. The total diffusivity of octane is 0.577×10^{-3} lbmoles/ft-hr in air. (a) What is the rate of evaporation for the n-octane? (b) How far below the top of the pipe is the explosive limit reached?

5. A mixture of hydrocarbons (by mole, 10% methane, 20% ethane, 30% propane, 15% isobutane, 20% n-butane, and 5% n-pentane) is flashed into a separator at 80 °F and 150 psia. 50% leaves as a liquid. What is the composition of the liquid? Equilibrium constants are: CH_4, 17.0; C_2H_6, 3.1; C_3H_8, 1.00; i-C_4H_{10}, 0.44; C_4H_{10}, 0.32; C_5H_{12}, 0.0986.

6. A 20 mole% mixture of hexane in octane at one atmosphere is fed to the top plate of a stripping tower. 90% of the hexane is to be removed in the distillate, and the waste is not to contain more than 3 mole% hexane. The vapor-liquid equilibrium for the mixture (expressed in mole%) is determined by the equation $y = 6x - 5xy$. The feed enters the column at its bubble point. The latent heat of vaporization of the still waste and product are 14,500 BTU/lbmole. The temperature in the reboiler is 250 °F, maintained by heating with 50 psia saturated steam. The cost of the steam is $0.50 per lbm, and the cost of cooling water is $0.12/1000 gallons. Condenser water enters at 70 °F and leaves at 170 °F. The feed (top of the column) is 60 lbmoles/hr at 225 °F. Overall plate efficiency is 40%. The maximum allowable vapor velocity of the column is 3.0 feet per second. The cost of the bubble cap column and auxiliaries is:

diameter, feet:	2	3	4
cost, $/tray:	300	350	420

The life of the column is 15 years. Decide (on the basis of total cost) whether the column should be run on an external reflux ratio, $\frac{L_R}{D}$, of two or five.

7. An air stream containing 0.015 mole fraction acetone is currently being exhausted to the atmosphere. Recent pollution laws require that the acceptable limit of acetone in the air cannot exceed 0.00015 mole fraction. It has been suggested that a countercurrent packed tower water scrubber might be effective. Data is as follows:

$$\text{gas flow} = 800 \text{ ft}^3/\text{hr-ft}^2 \text{ at STP}$$
$$\text{water flow} = 1250 \text{ lbm/hr-ft}^2$$
$$K_y a = 1.82 \text{ lbmoles/ft}^3\text{-hr-mole fraction}$$
$$y = 1.75x \text{ (mole fraction equilibrium)}$$

Henry's Law applies

(a) What is the number of overall gas phase transfer units required to meet the specified conditions? (b) What is the height of a transfer unit? (c) What is the required packing height?

8. A mixture of 60 mole% n-hexane, 10 mole% n-heptane, and 30 mole% steam is cooled at constant pressure of 1 atmosphere from an initial temperature of 400 °F. Assume that the hydrocarbons and water are immiscible. The vapor pressure of the hydrocarbons is given by the equation $ln\, p^0 = A + \frac{B}{T}$ where p^0 is in mm mercury and T is in °R. The values of A and B are:

	A	B
n-hexane	17.7109	−6816.4
n-heptane	17.9184	−7547.4

(a) What is the temperature at which the first liquid appears? (b) What is the composition of the first liquid phase as it appears? (c) What is the temperature at which the second liquid phase appears? (d) What is the composition of the second liquid phase as it appears?

9 DISTILLATION, EVAPORATION, AND HUMIDIFICATION

Nomenclature

a	contact area	ft^2/ft^3 tower volume
A	drying surface	ft^2
b	intercept for operating line equation	–
B	bottoms flow	lbmole/hr
c_p	heat capacity	BTU/lbm-°F
c_s	heat capacity of moist air	BTU/lbm-°F
D	distillate flow	lbmole/hr
f	feed condition factor	–
F	feed flow, or feed batch	lbmole/hr, or lbmole
G	gas flow rate	lbmole/hr
h	liquid enthalpy	BTU/lbmole
h_c	heat transfer coefficient (convection and radiation)	BTU/hr-ft²-°F
H	absolute humidity, or enthalpy	lbm water/lbm air, or BTU/lbmole
k	thermal conductivity	BTU/hr-ft-°F
k_1	mass transfer coefficient	lbm/hr-ft²
k_G	mass transfer coefficient	lbm/hr-ft²-atm
K	mass transfer coefficient, or Underwood parameter	lbm/hr-ft², or –
l	number of moles of liquid	lbmoles
L	liquid flow rate	lbmole/hr
Le	Lewis number	–
L_n	liquid flow from plate n	lbmole/hr
m	equilibrium constant $\left(\frac{y}{x}\right)$	–
M	molecular weight	lbm/lbmole
M_R	Colburn equation rectifying parameter	–
M_S	Colburn equation stripping parameter	–
n_m	minimum number of plates	–
n_R	number of plates in rectifying section	–
n_S	number of plates in stripping section	–
N	number of moles	–
p	partial pressure	atm
p^o	vapor pressure	atm
P	total pressure	atm
q	mole fraction	–
Q	heat flow	BTU/hr
R	reflux ratio $\left(\frac{L_o}{D}\right)$	–
T	temperature	°F
v	volume	ft³/lbm dry air
V	cooling volume, or vapor flow rate	ft³/ft² plan area or lbmole/hr
V_n	vapor flow from plate n	lbmole/hr
w	moisture content	lbm water/lbm dry solid
W	amount remaining after distillation	lbmoles
x	liquid mole fraction	–
y	vapor mole fraction	–

Symbols

α	volatility ratio $\left(\frac{y_A x_B}{y_B x_A}\right)$	–
λ	distillation factor $\left(\frac{mV}{L}\right)$, or heat of vaporization	–, or BTU/lbm
θ	time	hr
μ	viscosity	centipoise
ρ	density	lbm/ft³
Φ	drying factor $\left(\frac{-L_s}{A\left(\frac{dw}{d\theta}\right)_{CR}}\right)$	–
$*$	equilibrium	–

Subscripts

0	initial
1	entering
2	leaving
a	air
avg	average
A	water

b	bottom plate, or saturated liquid
B	bottom, or air
c	critical, or condenser
CR	constant rate
d	saturated vapor
D	distillate
eq	equilibrium
f	final
F	feed
FR	falling rate
i	initial, or intersection
m	minimum, or plate m
n	plate n
o	lowest stage in rectifying section
p	highest stage in rectifying section
r	reboiler
R	rectifying, or relative
s	saturated, steam, or humid
S	stripping
w	water

1 DISTILLATION

Distillation is a method of separating components of a solution by vaporization and condensation. In distillation, all components are present in both liquid and gas phases. Distillation is mainly used where all components are volatile to large extents. By repeated vaporizations and condensations, it is possible to make as complete a separation as is desired.

Distillation requires the introduction or removal of heat, which introduces certain limitations. If a component is sensitive to the addition of heat, that component might be changed undesirably. Also, the vapor that is formed from the application of heat to a liquid consists only of the components comprising the liquid. Since the vapor is very similar chemically to the liquid, the change in composition resulting from the distribution of the components between the two phases may not be very great. In some cases, the change in composition is so insignificant that the use of distillation as a separation process is impractical when $\alpha < 1.05$ (as a rule of thumb).

In practice, distillation is achieved by two principal methods. The first method is based on boiling the liquid mixture to be separated, and condensing the vapor without letting any liquid return to the still in contact with the vapor. A *still* is the vessel in which distillation takes place. The second method is based on the return of part of the condensate to the still, where the liquid is brought into contact with the vapors on their way to the condenser.

Flash distillation and *simplified batch distillation* are processes that use the first method. *Steam, continuous,* and *batch distillation with reflux* are processes that use the second method. *Reflux* is the part of the overhead condensate that is returned to the column as a liquid stream. In steam distillation, the heat of vaporization is obtained by injecting steam into the liquid phase, which is not usually miscible with water.

A. BATCH DISTILLATION

Batch distillation is where the separation of components of an initial volume of liquid mixture (i.e., a *batch charge*) is carried out in a heated vessel (*pot*). The vapors may pass through a rectifying column on their way to the condenser. The condensed stream is collected and/or refluxed. In the simplified form of batch distillation, no rectification or reflux is provided, and the batch material is boiled and condensed. The batch charge is constantly changing in composition and volume. Therefore, batch distillation is not a steady state process. The *Rayleigh equation* can be used to solve for the pot composition in a batch distillation process.

$$ln\,\frac{F}{W} = \int_{x_f}^{x_i} \frac{dx}{y^* - x} \qquad 9.1$$

Relative volatility is the ratio of two components in one phase to that in the other phase. It is a measure of separability. For binary systems,

$$\alpha = \frac{\frac{y}{x}}{\frac{1-y}{1-x}} \qquad 9.2$$

Table 9.1 lists binary vapor-liquid equilibrium compositions for various relative volatilities. x is the liquid mole fraction, and y is the vapor mole fraction. The body of table 9.1 contains values of y.

If the relative volatility is constant throughout the process such that α is the same at the original pot concentration as at the final pot concentration, then the Rayleigh batch distillation equation can easily be integrated.

$$ln\,\frac{F}{W} = \frac{1}{\alpha - 1}\left(ln\,\frac{x_i}{x_f} - \alpha\,ln\,\frac{1-x_i}{1-x_f}\right) \qquad 9.3$$

Table 9.1
y as a Function of x and Relative Volatility for Binary Systems

	relative volatility									
x	1.2	1.4	1.6	1.8	2.0	2.2	2.4	2.6	2.8	3.0
0.050	0.059	0.069	0.078	0.087	0.095	0.104	0.112	0.120	0.128	0.136
0.100	0.118	0.135	0.151	0.167	0.182	0.196	0.211	0.224	0.237	0.250
0.150	0.175	0.198	0.220	0.241	0.261	0.280	0.298	0.315	0.331	0.346
0.200	0.231	0.259	0.286	0.310	0.333	0.355	0.375	0.394	0.412	0.429
0.250	0.286	0.318	0.348	0.375	0.400	0.423	0.444	0.464	0.483	0.500
0.300	0.340	0.375	0.407	0.435	0.462	0.485	0.507	0.527	0.545	0.563
0.350	0.393	0.430	0.463	0.492	0.519	0.542	0.564	0.583	0.601	0.618
0.400	0.444	0.483	0.516	0.545	0.571	0.595	0.615	0.634	0.651	0.667
0.450	0.495	0.534	0.567	0.596	0.621	0.643	0.663	0.680	0.696	0.711
0.500	0.545	0.583	0.615	0.643	0.667	0.688	0.706	0.722	0.737	0.750
0.550	0.595	0.631	0.662	0.688	0.710	0.729	0.746	0.761	0.774	0.786
0.600	0.643	0.677	0.706	0.730	0.750	0.767	0.783	0.796	0.808	0.818
0.650	0.690	0.722	0.748	0.770	0.788	0.803	0.817	0.828	0.839	0.848
0.700	0.737	0.766	0.789	0.808	0.824	0.837	0.848	0.858	0.867	0.875
0.750	0.783	0.808	0.828	0.844	0.857	0.868	0.878	0.886	0.894	0.900
0.800	0.828	0.848	0.865	0.878	0.889	0.898	0.906	0.912	0.918	0.923
0.850	0.872	0.888	0.901	0.911	0.919	0.926	0.932	0.936	0.941	0.944
0.900	0.915	0.926	0.935	0.942	0.947	0.952	0.956	0.959	0.962	0.964
0.950	0.958	0.964	0.968	0.972	0.974	0.977	0.979	0.980	0.982	0.983
1.000	1.000	1.000	1.000	1.000	1.000	1.000	1.000	1.000	1.000	1.000

Simple batch distillation is also called *differential distillation*. Equations 9.1 and 9.3 can be used to determine F, W, x_f, or x_i. If an algebraic equilibrium relationship (such as the relative volatility) is not known, then equation 9.1 is usually solved by graphic integration. $\frac{1}{y^*-x}$ is plotted as the ordinate, and x as the abscissa. The area under the curve between the indicated limits can then be found. The composite distillate composition, y_{avg}, can be determined by the material balance around the process.

$$Fx_F = Dy_{avg} + Wx_f \qquad 9.4$$

and

$$F = D + W \qquad 9.5$$

B. STEAM (INERT) DISTILLATION

A component with a high boiling point can be distilled at a lower overall temperature by introducing steam into the still pot. However, the steam must be immiscible with the mixture being distilled. Assuming the heat of vaporization comes from the steam, the total pressure in the still pot is the sum of the vapor pressure of the high boiler (A) and the steam.

$$P = p_A^o + p_s \qquad 9.6$$

The ratio of the moles of component A to the moles of steam is

$$\frac{N_A}{N_s} = \frac{p_A^o}{p_s} \qquad 9.7$$

If the steam is superheated, then it will function as an inert *stripping gas*. The total pressure of the still pot is

$$P = p_A^o + p_s^o \qquad 9.8$$

C. CONTINUOUS DISTILLATION-BINARY SYSTEMS

Continuous rectification or *fractionation* is a multistage, countercurrent distillation operation. It is possible (in a binary system) to recover each component into any desired state of purity. Feed is introduced approximately in the center of a vertical cascade of stages. Vapor rising above the *feed stage* (in the section called the *rectifying* or *enriching section*) is washed with liquid to remove the less volatile component. The washing liquid is produced by condensing the vapor leaving the top of the rectifying section. This wash liquid is richer in the more volatile component.

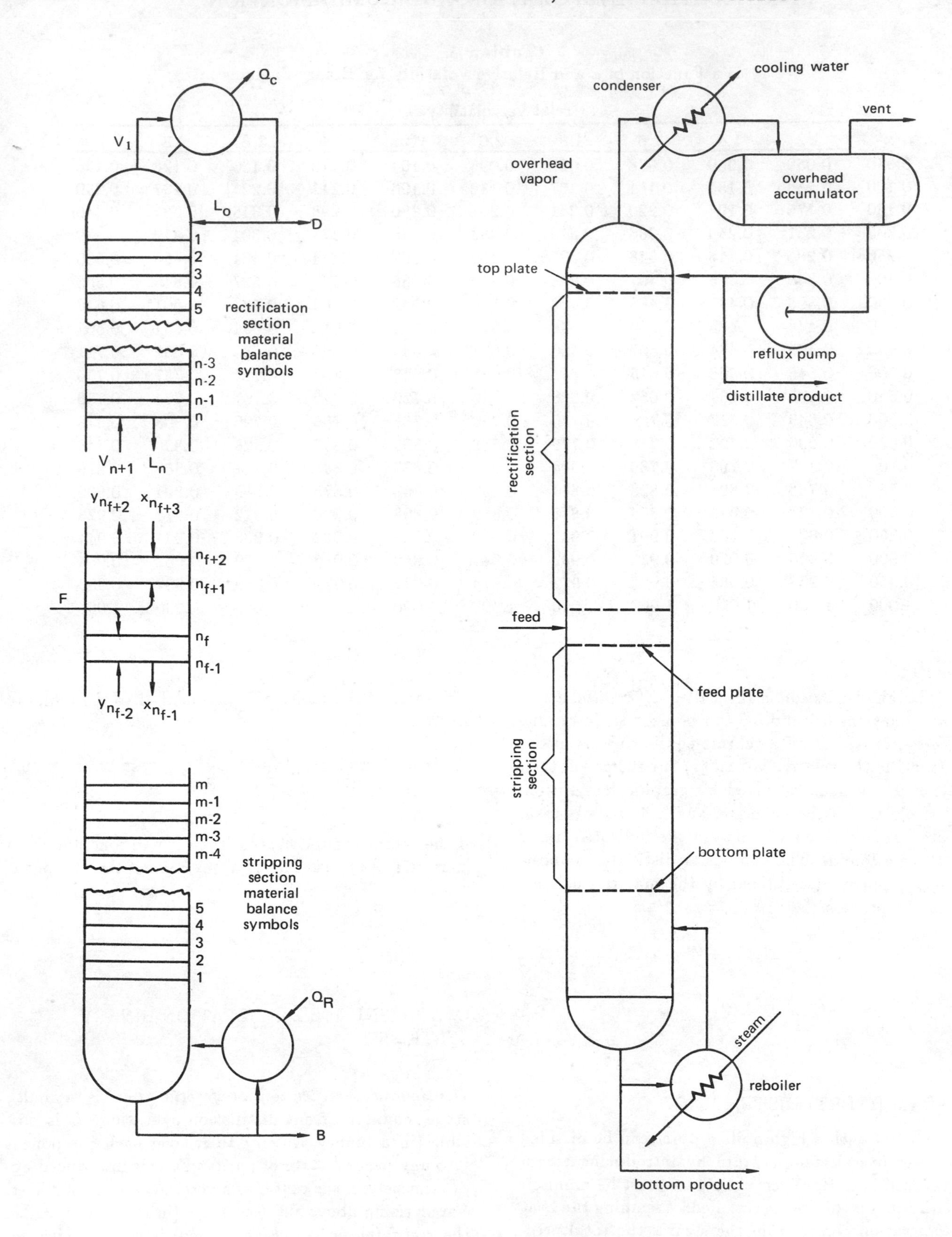

Figure 9.1 A Fractionator

The liquid returned from the top of the rectifying section is called the *reflux*, and the material permanently removed is called the *distillate* (which may be a vapor or a liquid). In the section below the feed stage (the *stripping* or *exhausting* section), the liquid is stripped of the more volatile component with vapor produced by a partial vaporization of the liquid using a reboiler at the bottom of the column. The liquid removed from the bottom of the tower is called the *residue* or *bottoms*, and is richer in less volatile components.

Inside the tower, the liquids and vapors are at their *bubble points* and *dew points*, respectively. The highest temperature is at the bottom, and the lowest at the top. The entire device is called a *fractionator*. The purities obtained from such a device depends on the liquid-gas ratio, equilibrium, and the number of stages provided in the two sections of the tower. Figure 9.1 illustrates a typical fractionator.

Total and component material balances around the distillation column results in two equations. (Mole fractions refer to the more volatile component.)

$$F = D + B \qquad 9.9$$

$$Fx_F = Dx_D + Bx_B \qquad 9.10$$

Since there are two sections in the column, there are also two operating lines—one for the rectifying section and one for the stripping section. Material, component (more volatile), and energy balances around the rectifying section produce the following equations:

$$V_{n+1} = L_n + D \qquad 9.11$$

$$V_{n+1}\,y_{n+1} = L_n x_n + Dx_D \qquad 9.12$$

$$V_{n+1}\,H_{n+1} = L_n h_n + Dh_D + Q_c \qquad 9.13$$

Combining the rectifying section material and component balances results in an equation for the *operating line* in the rectifying section.

$$y_{n+1} = \frac{L_n}{V_{n+1}} x_n + \frac{D}{V_{n+1}} x_D \qquad 9.14$$

The rectifying section energy balance becomes

$$(L_n + D)\,H_{n+1} = L_n h_n + Dh_D + Q_c \qquad 9.15$$

In a similar manner, the equations for the operating line and energy balance in the stripping section become:

$$y_m = \frac{L_{m+1}}{V_m} x_{m+1} - \frac{B}{V_m} x_B \qquad 9.16$$

$$(V_m + B)\,h_{m+1} = V_m H_m + Bh_B - Q_R \qquad 9.17$$

The slope of the operating line in each section is the ratio of the liquid to vapor flow in that section.

The analysis of fractionating columns is facilitated by the use of a quantity called the *reflux ratio*. The reflux ratio can be interpreted in two ways. One way is the ratio of the reflux to the overhead product, called the *external reflux ratio*. The other, rarely used, is the ratio of the reflux to vapor, called the *internal reflux ratio*. The former is defined:

$$R = \frac{L}{D} \qquad 9.18$$

The operating line for the rectifying section in terms of reflux ratio is

$$y_{n+1} = \frac{R}{R+1} x_n + \frac{1}{R+1} x_D \qquad 9.19$$

The equation of the operating line has a y-intercept of $\frac{x_D}{R+1}$. The concentration x_D is set by the conditions of the design, and R is an operating variable that can be controlled by varying the flow rate of the reflux for a given flow rate of the overhead product.

D. McCABE-THIELE METHOD

When the operating lines represented by equations 9.14 and 9.16 are plotted with the equilibrium curve on an x-y diagram, the McCabe-Thiele construction can be used to compute the number of ideal stages needed to accomplish a desired concentration difference in either the rectifying or stripping sections. Unless L_n and L_m are constant, the operating lines are curved. In real operations, they are nearly constant, so only a small error is introduced if it is assumed that the operating line is linear. A rigorous energy balance can be performed to show that the vapor and liquid flows are relatively constant, provided that the following assumptions are valid:

- the heat losses from the column are small
- the molal heats of vaporization of the components are substantially equal
- the temperature change across the column is not excessive

When L and V are constant, the column has *constant molal vaporization* and *constant molal overflow*. When the vaporization and overflow are constant in any section, the operating lines are straight and can be plotted if two points, or one point and the slope, are known.

Condenser and Top Plate

At constant molal overflow, the operating line for the rectification section (dropping the plate identification subscripts) becomes:

$$y = x\left(\frac{L}{L+D}\right) + \frac{Dx_D}{L+D} \quad 9.20$$

The intersection of the operating line represented by equation 9.20 and the diagonal represented by the equation $y = x$ occurs at the point (x_D, x_D). The McCabe-Thiele construction for the top plate does not depend on the action of the condenser. If the condenser is a partial condenser, then the top plate vapor composition is y_1, and the composition for the reflux to the top plate is x_c.

The upper terminus of the operating line is at the point (y_1, x_c). Figure 9.2 illustrates the graphical construction for the top plate using a total condenser or a partial condenser.

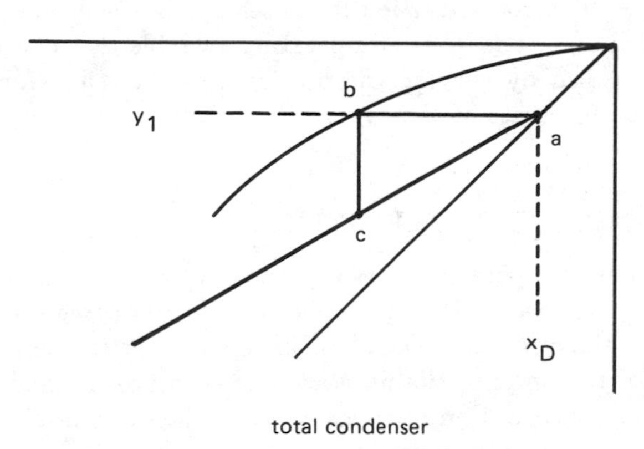

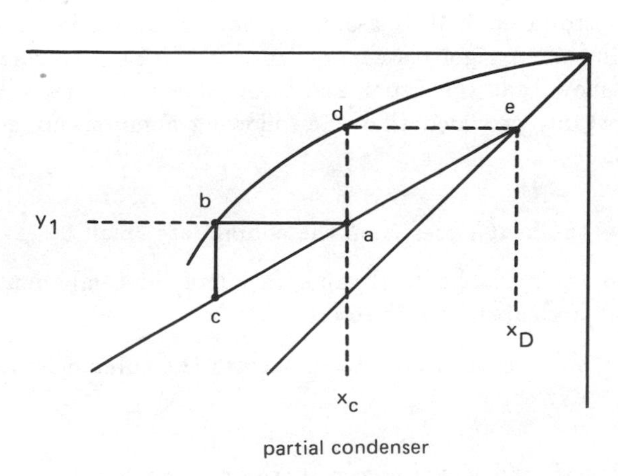

Figure 9.2 McCabe-Thiele Construction for the Top Plate

The simplest arrangement for obtaining reflux and liquid products is the *single total condenser*, which condenses all of the vapor from the column, and supplies both reflux and product. When a *total condenser* is used, the top plate vapor composition and the reflux are at the same concentration, and each can be denoted as x_D. The operating line terminus becomes point (x_D, x_D). Triangle *abc* in figure 9.2 represents the top plate.

When a *partial condenser* or *dephlegmator* is used, the liquid composition does not have the same composition as the vapor from the top plate. The vapor leaving the partial condenser has composition $y = x_D$, and the liquid refluxing back to the column has liquid composition x_c. As far as the column is concerned, the operating line ends at point (y_1, x_c). Triangle *abc* represents the top plate, and triangle *ade* represents the additional theoretical stage of the partial condenser.

Bottom Plate and Reboiler

The bottom plate and reboiler performance is similar to the top plate. The operating line in the stripping section is

$$y = x\left(\frac{L}{L-B}\right) - \frac{Bx_B}{L-B} \quad 9.21$$

The operating line crosses the diagonal at point (x_B, x_B). The terminus of the operating line is point (x_b, y_r), where x_b and y_r are the concentrations of liquid from the bottom plate and of vapor from the reboiler, respectively. In common types of reboilers, the liquid leaving as bottoms is in equilibrium with the vapor leaving the reboiler. The reboiler then acts as an ideal plate if all of the bottoms liquid is not vaporized. Figure 9.3 is the graphical construction for a reboiler. Triangle *cde* represents the reboiler, and the bottom plate is represented by triangle *abc*.

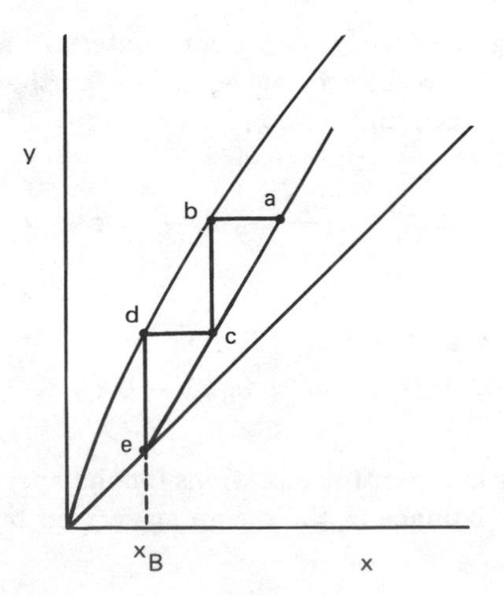

Figure 9.3 McCabe-Thiele Construction for the Bottom Plate

Feed Plate

The feed plate is located near the intersection of the rectifying and stripping operating lines. The addition of feed increases reflux in the stripping section, or increases vapor in the rectifying section, or both. In the rectifying section, the vapor flow rate is greater than the vapor flow rate in the stripping section. It follows that the slope of the operating line in the rectifying section is less than one, and the slope of the operating line in the stripping section is greater than one.

The feed may be introduced as a subcooled liquid, liquid at its bubble point, liquid-vapor mixture, saturated vapor, or superheated vapor. All five feed types can be correlated by the use of a single feed condition factor, denoted by the variable[1] f and defined as the moles of vapor flow in the rectifying section that result by the introduction of each mole of feed. Table 9.2 tabulates the values of f for the various feed conditions.

Table 9.2
Feed Condition and f

feed	f
cold feed	< 0
saturated liquid	0
liquid-vapor mixture	$0 < f < 1$
saturated vapor	1
superheated vapor	> 1

The feed line is the line on which all intersections of the operating lines must fall. The equation for the feed line is

$$y = x\,\frac{f-1}{f} + \frac{x_F}{f} \qquad 9.22$$

The value for f is found from equations 9.23 and 9.24, in which the heat capacity is for the feed stream (either liquid or gas).

$$f = -c_p\,\frac{T_b - T_F}{\lambda} \quad \text{(subcooled liquids)} \qquad 9.23$$

$$f = 1 + c_p\,\frac{T_F - T_d}{\lambda} \quad \text{(superheated vapor)} \qquad 9.24$$

Construction of Operating Lines

A typical problem involving a McCabe-Thiele construction is where F, α, D, R (or some percentage over the minimum reflux ratio), x_F, and x_D are known, and the number of stages and feed plate location are needed. The reflux ratio for any separation decreases as the number of theoretical stages increases. There is a limit below which the reflux ratio cannot be lowered. Any practical separation process requires that the operating lines intersect below the equilibrium line.

The minimum reflux ratio occurs when the intersection lies on the equilibrium line. This requires an infinite number of stages to accomplish the given separation. Generally, the minimum reflux ratio, R_m, is found graphically. The rectifying operating line is drawn at the intersection of the feed line and the equilibrium line. The slope of this line is $\frac{R_m}{R_m+1}$. The actual operating line is found by factoring up the minimum reflux (i.e., 30 percent above minimum reflux is $R = 1.3\,(R_m)$).

The following steps will aid in the construction of a McCabe-Thiele diagram:

step 1: Use a material balance to find B and x_B.

step 2: Using table 9.1 and α, construct an equilibrium curve.

step 3: Draw the 45-degree line.

step 4: Locate x_F on the 45-degree line. (The feed line must intersect this point.) Locate and draw the feed line between the equilibrium curve and point x_F on the 45-degree line.

step 5: Draw the rectifying operating line for minimum reflux through point (x_D, x_D) and the intersection of the feed line with the equilibrium line. Compute the *minimum reflux ratio* from the slope of this line. ($R_m = \frac{\text{slope}}{1-\text{slope}}$)

step 6: Calculate the slope of the rectifying operating line (if R is known), or calculate R and the slope of the rectifying operating line (if the percent over R_m is known).

step 7: Draw the rectifying operating line by calculating the y-axis intercept $\frac{x_D}{R+1}$, drawing it through that point and point (x_D, x_D).

step 8: Draw the *stripping line* through the intersection of the rectifying line with the feed line and the point (x_B, x_B).

step 9: Step off the stages between the operating lines and the equilibrium curve, starting from the end of the operating line intersection with the diagonal where the composition is given.

step 10: The feed plate is located at the stage at the feed line. The feed tray is the top stripping tray in the column. Since operating lines

[1] Sometimes q is used, where $q = 1-f$. Equation 9.22 then becomes $y = x\,\frac{q}{q-1} - \frac{x_F}{q-1}$.

represent liquid and vapor compositions passing one another between trays, the highest value of y that can be read on the stripping operating line is the vapor composition of the feed tray. This highest value of y is found by moving horizontally to the left of the intersection of the operating lines to the equilibrium line.

Figure 9.5 illustrates a general McCabe-Thiele diagram for a distillation column. The feed line, equilibrium line, operating lines, and the ideal stages are shown.

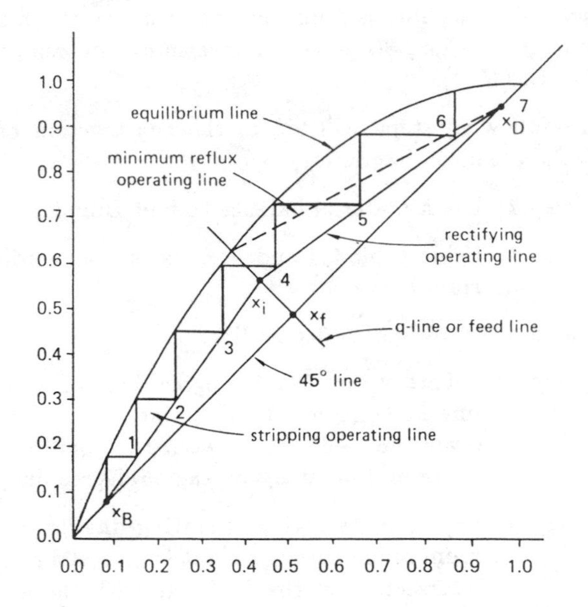

Figure 9.5 McCabe-Thiele Diagram for Distillation

Feed Plate Location

After the operating lines are plotted and the number of ideal plates has been established by counting the number of stages between the distillate and bottoms compositions, the optimum location of the feed plate can be established. If the stages are stepped off from the top of the diagram, a decision has to be made when the steps should be transferred to the stripping line.

The decision should be made to minimize the number of plates and maximize the enrichment per plate. This criterion is met if the transfer is made where x becomes less than the x coordinate of the intersection of the two operating lines. The feed plate is always represented by the triangle that has one corner on the rectifying line and the other on the stripping line. The optimum feed plate location is where the feed plate triangle straddles the operating lines intersection.

Minimum Stages

Since the slope of the rectifying line is $\frac{R}{R+1}$, the slope increases as the reflux increases. When R is infinite, the slope is one, and $V = L$. The operating lines then both coincide with the diagonal. At this point, the column is operating at total reflux. The product, feed, and bottoms flows are zero. The minimum number of theoretical stages in a column is found by constructing equilibrium steps on an x-y diagram between compositions x_D and x_B, using the diagonal as the operating lines for both the rectifying section and the stripping sections.

E. FENSKE METHOD — MINIMUM STAGES

Since stage requirements for any separation are reduced as the reflux ratio is increased, the minimum number of stages occurs at infinite reflux ratio or total reflux, a condition realized when no product is withdrawn from an operating column. On an x-y diagram, the infinite reflux case occurs when the operating lines coincide with the diagonal. When the relative volatility is constant, the minimum number of stages can be found from equation 9.25. The number of stages computed from equation 9.25 assumes use of a total condenser. The reboiler (partial) adds one stage. The use of a partial condenser also adds one stage.

$$n_m = \frac{ln\left(\frac{x_D(1-x_B)}{(1-x_D)x_B}\right)}{ln\,\alpha} - 1 \qquad 9.25$$

F. OPTIMUM DESIGN

For a binary distillation, there is an infinite number of design possibilities ranging from minimum reflux and infinite stages to total reflux and minimum stages. From an economic standpoint, the optimum reflux will do the job at the lowest cost (operating and investment). As the reflux ratio increases, the operating cost increases, but the number of stages decreases (decreasing investment costs). For estimation purposes, optimum performance usually occurs at 1.1 to 1.5 times minimum reflux. When energy costs are high, the optimum will be nearer the lower end of this range, or approximately 1.15 to 1.25 times minimum reflux.

G. ANALYTICAL SOLUTION — UNDERWOOD METHOD

Analytical solutions for tray requirements can be made when the equilibrium curve is relatively straight or when α is constant. The Underwood method can be used when there is:

- total reflux and constant relative volatility, or
- partial reflux, constant relative volatility, and constant molal overflow, or
- partial reflux, linear equilibrium, and constant molal overflow

The Underwood equation is

$$\frac{L(\alpha-1)}{V}K^2+\left(\frac{L}{V}+b(\alpha-1)-\alpha\right)K+b=0 \quad 9.26$$

The operating line equation is

$$y=\left(\frac{L}{V}\right)x+b \quad 9.27$$

Since equation 9.26 is a quadratic, it has two roots: K_1 and K_2. The value of K_1 represents the lower intersection point of the rectifying operating line and the equilibrium curve. The value of K_2 represents the fictitious upper intersection of the operating line and the equilibrium curve lying outside the x-y diagram. For the case of the stripping section, K_1 is the lower intersection of the operating line and equilibrium line lying outside the x-y diagram, and K_2 is the upper intersection of the operating line and the equilibrium line. For the rectifying section, the value of K_1 lies between zero and one, while K_2 is greater than one. The number of rectifying stages, n_R, is

$$n_R\, ln\,\frac{\alpha\frac{V}{L}}{[1+(\alpha-1)K_1]^2}=ln\,\frac{(x_D-K_1)(K_2-x_i)}{(x_i-K_1)(K_2-x_D)} \quad 9.28$$

The number of stages in the stripping section, n_S, is

$$n_S\, ln\,\frac{\alpha\frac{V}{L}}{[1+(\alpha-1)K_1]^2}=ln\,\frac{(x_i-K_1)(K_2-x_B)}{(x_B-K_1)(K_2-x_i)} \quad 9.29$$

H. ANALYTICAL SOLUTION — COLBURN METHOD

If the equilibrium relationship is a straight line having slope m, then the factor λ can be used.

$$\lambda=\frac{mV}{L} \quad 9.30$$

For the rectifying section,

$$n_R\, ln\,\frac{1}{\lambda}=ln\,[(1-\lambda)\,M_R+\lambda] \quad 9.31$$

The Colburn rectifying parameter, M_R, can be computed from equation 9.32.

$$M_R=\frac{(1-y_o)-m(1-x_p)}{(1-y_p)-m(1-x_p)} \quad 9.32$$

In equation 9.32, y_o is the vapor concentration at the lowest stage in the rectifying section, y_p is the vapor concentration at the highest point in the rectifying section, and x_p is the operating line point in the rectifying section corresponding to y_p.

If a total condenser is used, then $y_p=x_D$. If a stripping section meets the criterion, the number of stripping stages can be computed:

$$n_S\, ln\,\lambda=ln\left(M_s-\frac{(M_s-1)}{\lambda}\right) \quad 9.33$$

$$M_S=\frac{x_p-\frac{x_o}{m}}{x_o-\frac{x_o}{m}} \quad 9.34$$

In equation 9.34, x_o is the liquid concentration at the lowest plate in the stripping section (usually $x_o=x_B$), and x_p is the liquid concentration at the highest plate in the stripping section (usually x_F).

If the stripping section uses live steam,

$$M_S=\frac{x_p}{x_o} \quad 9.35$$

I. SPECIAL PROBLEM: NUMBER OF STAGES KNOWN

In some cases, a column of known design may be specified. With the number of stages known, the reflux ratio will be the unknown. Although this problem can be solved graphically by adjusting the operating line slopes until the desired separation is obtained, the result is inaccurate. This type of problem must be solved stage by stage. Four equations must be set up for a stage-by-stage computation.

First, solve the equilibrium relationship for x (e.g., solve the equilibrium relationship for x in terms of relative volatility and y). Next, solve the feed line in terms of x and f. Finally, solve both the operating lines in terms of R (rectifying line) or x_D (stripping line).

Start at the top of the column. For the first step, let $y=x_D$. Assume a reflux ratio R. Proceed with the following steps until the actual number of stages has been computed. Plot x_B as a function of the reflux ratio. When x_B is straddled by the reflux ratio, linearly interpolate to find the actual reflux ratio. Repeat the calculation steps one additional time to verify the answer. The following procedure formalizes the method.

step 1: With y known, compute x from the equilibrium relationship. If the stripping operating line is used, go to step 6.

step 2: Compute y using x from step 1 in the rectifying operating line equation.

step 3: Compute y for the feed line using x from step 1. If the feed line y is greater than y computed in step 2, then the stripping operating line should be used for the rest of the steps, and go to step 6.

step 4: One more stage in the rectifying section has been completed if you are at this step.

step 5: Let y equal the value computed in step 2. Go to step 1.

step 6: Compute y using x from step 1 in the stripping operating line. One more stage in the stripping section has been completed. If the total number of stages computed is equal to the given stages, go to step 7. Otherwise, let y be the value computed in this step and go to step 1.

step 7: If x_n is less than x_B, then the reflux ratio is too large. Assume a smaller reflux ratio, and start all calculations from the beginning. If x is greater than x_B, then assume a larger reflux ratio and start all calculations from the beginning. If $x = x_B$, then the correct reflux ratio has been chosen. The problem is solved.

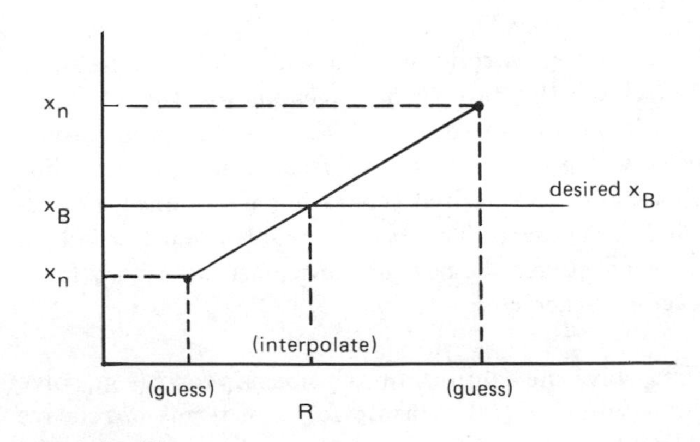

Figure 9.6 Method with Known Number of Stages

2 HUMIDIFICATION

Humidification operations involve the transfer of material between a pure liquid phase and a fixed insoluble gas phase. In a sense, this process is simpler than other vapor-liquid operations because the liquid phase is pure and, therefore, there are no concentration gradients (and no resistance to transfer in the liquid phase).

Special definitions are applied in humidification operations, especially in the common air-water system. When the term "vapor" is used, it refers to the gaseous form of the component that is also present as a liquid. "Gas" refers to the component that is only present in gaseous form. In the gas phase, the vapor will be referred to as component A, and the fixed phase will be component B. For air-water systems, A will be water and B will be air. It is also assumed that most humidification operations occur at a pressure of one atmosphere, and that the gas-vapor mixture is an ideal gas.

Humidity is the ratio of mass of vapor in one unit mass (usually one pound) of vapor-free gas. When the total pressure is fixed, the humidity depends on the partial pressure of the vapor in the mixture. Therefore, for any gas-vapor system,

$$H = \frac{M_A p_A}{M_B (1 - p_A)} \qquad 9.36$$

For problems involving air-water, humidity is given by equation 9.37 (p is in atmospheres).

$$H = \frac{18 p_A}{29 (1 - p_A)} \qquad 9.37$$

The mole fraction of the vapor is

$$y = \frac{\frac{H}{M_A}}{\frac{1}{M_B} + \frac{H}{M_A}} \qquad 9.38$$

The *saturation humidity* (where the gas has vapor in equilibrium with the liquid at the gas temperature) is

$$H_s = \frac{M_A p_A^o}{M_B (1 - p_A^o)} \qquad 9.39$$

The ratio of the actual humidity to the saturation humidity is called the *percent humidity.*

$$H_A = 100 \frac{H}{H_s} = H_R \frac{1 - p_A^o}{1 - p_A} \qquad 9.40$$

Humid heat is the heat necessary to increase the temperature of one pound of gas and whatever vapor it contains by 1 °F.

$$c_s = (c_p)_A + H(c_p)_B \qquad 9.41$$

For air-water,

$$c_s \approx 0.24 + 0.45\,H \qquad 9.42$$

Humid volume is the total volume, in cubic feet, of one pound of vapor-free gas plus whatever vapor it contains at one atmosphere and at the gas temperature. For air-water systems,

$$v_s = \frac{\left(\frac{1}{29} + \frac{H}{18}\right) 359 T_{°R}}{492} \quad 9.43$$

The *total enthalpy* of the humid system is the enthalpy of one pound of gas plus the enthalpy of whatever vapor it contains. This is the sum of the sensible heat of the vapor, the latent heat of the liquid, and the sensible heat of the vapor-free gas. In equation 9.44, T_o is the *reference state temperature* for the vapor.

$$h_a = c_s(T - T_o) + H\lambda_o \quad 9.44$$

A. WET BULB AND ADIABATIC SATURATION TEMPERATURE

Wet bulb temperature is the dynamic equilibrium temperature attained by a liquid surface when the rate of heat transfer to the surface by convection equals the rate of heat required for evaporation away from the surface. Neglecting changes in dry bulb temperature, a heat balance for the surface of the liquid at equilibrium is

$$k_G \lambda (p_w^o - p) = h_c (T - T_w) \quad 9.45$$

Equation 9.45 implies an adiabatic process in which there is negligible heat exchange with the surroundings. The amount of gas is assumed to be very large compared to the amount of liquid, so that the humidity and temperature of the air mass remain constant. Under ordinary conditions, the partial pressure and the vapor pressure are small relative to the total pressure, and the wet bulb equation can be written in terms of humidity differences:

$$H_s - H = h_c \frac{T - T_w}{\lambda k_1} \quad 9.46$$

For an air-water system, the wet bulb equation becomes

$$H_s - H = \frac{18 h_c (T - T_w)}{29\, k_G \lambda} \quad 9.47$$

The *adiabatic saturation temperature* is reached when a stream of air is intimately mixed with a quantity of water at a temperature, T_s, in an adiabatic system. The stream of water is recirculated to keep its temperature uniform. The enthalpy of the moist air stream will remain essentially constant because as the temperature of the gas decreases, the heat is transferred into vaporization of liquid which increases the humidity of the gas. If T_s is such that the air leaving the system is in equilibrium with the water, T_s will be the adiabatic saturation temperature. The equation for the adiabatic saturation temperature is

$$H_s - H = \frac{c_s (T - T_s)}{\lambda} \quad 9.48$$

Combining the equations for the wet bulb temperature and the adiabatic saturation temperature,

$$\frac{h_c}{c_s k_1} = \frac{T - T_s}{T - T_w} \quad 9.49$$

The term $\frac{h_c}{c_s k_1}$ is related to the *Lewis number*, *Le*:

$$\frac{h_c}{c_s k_1} = Le^{\frac{2}{3}} \quad 9.50$$

The term $Le^{\frac{2}{3}}$ is called the *psychrometric ratio*. For air-water systems, the psychrometric ratio is approximatedly equal to one. It is for this reason that, for air-water systems, T_w and T_s can be used interchangeably. In general, however, T_s and T_w are not the same.

B. PSYCHROMETRIC (HUMIDITY) CHARTS

A psychrometric (humidity) chart provides a convenient method for determining the properties of mixtures of a gas and a condensable vapor. Figure 9.4 is for a mixture of air and water at one atmosphere. Any point on the chart represents a specific mixture of air and water. The *lever rule* can be used with this chart when mixing two air masses together to find the resultant mixture properties. The curved line marked 100% is the humidity of saturated air as a function of air temperature. (This is a plot of equation 9.39.) Any point above and to the left of the saturation line represents a mixture of saturated air and liquid water (fog). Any point below the saturation line represents unsaturated air.

The curved lines between the saturation line and the temperature axis represent mixtures of air and water at specific percentage humidities. Linear interpolation between the lines may be used for other humidities. The nearly straight, slanting lines running downward to the right of the saturation line are the *adiabatic cooling lines*. Saturated volume and dry volume are also plotted against temperature; the volumes are read on the scale on the right. The heat of evaporation is plotted against temperature; the heat of evaporation is read on the scale on the right. Humid heat is plotted against humidity; humid heat is read on the scale at the top. Saturation pressure is read on the scale on the right.

C. HUMIDIFICATION EQUIPMENT

Air can be brought to a desired temperature and humidity by bringing it in contact with a water spray. A humidifier uses heating coils and a warm water spray. Figure 9.7 shows two processes which can take place in a humidifying device. Point A represents the entering air with dry bulb temperature T_1 and humidity H_1. A dry bulb temperature of T_2 and humidity of H_2 (point B) are desired. Path 1 consists of going from point A to point C by water spray (to reach the desired humidity) and then heating to reach point B. Path 2 consists of heating the air from A to D, cooling by water along an adiabatic cooling line to reach point C, then heating to reach point B.

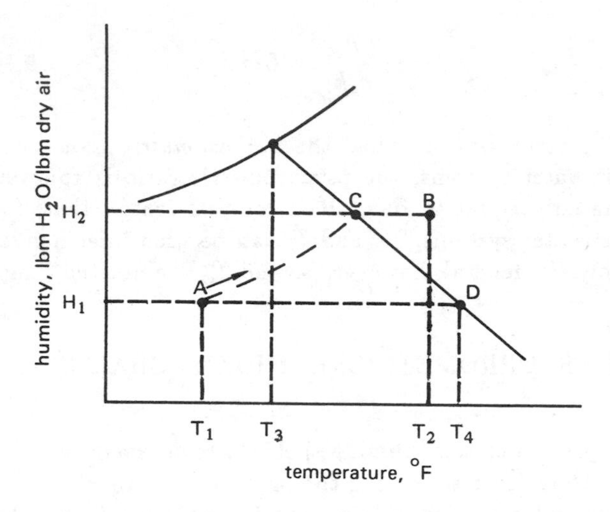

Figure 9.7 Humidification Process

D. EVAPORATIVE COOLING IN COOLING TOWERS

A recirculating water spray will be cooled when it gives up latent and sensible heat. Approximately 80% of this heat transfer is due to latent heat, while the rest is due to sensible heat. The theoretical heat transfer is dependent on the air temperature and humidity. The air's wet bulb temperature is the limiting temperature to which the water can be cooled, but in actual practice, cooling towers are seldom designed for approaches closer than 5 °F to the wet bulb temperature.

The *Merkel equation* is generally used for cooling tower design:

$$\frac{KaV}{c_p L} = \int_{T_2}^{T_1} \frac{dT}{h_1 - h_a} \qquad 9.51$$

An enthalpy-temperature diagram of a cooling tower is shown in figure 9.8, which illustrates the air and water operating lines. Points A and B represent the entering and leaving water properties. Point C represents the properties of the air entering the cooling tower. The slope of the air operating line is $\frac{L}{G}$, the water-to-air flow ratio.

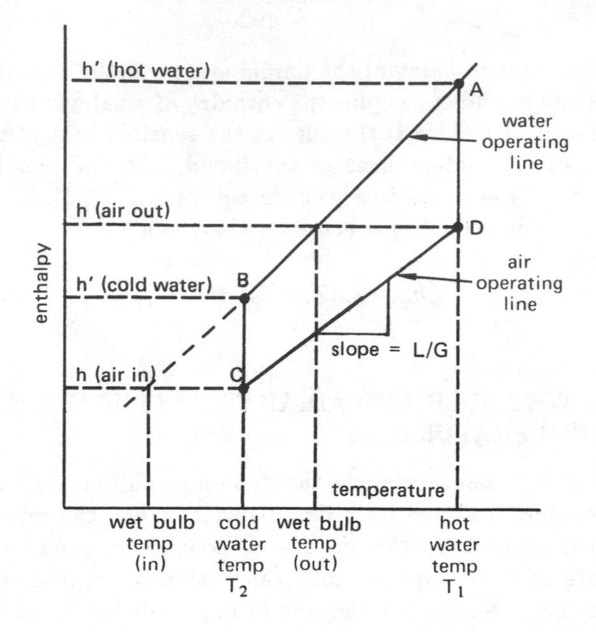

Figure 9.8 Cooling Tower Operation

Example 9.1

Given moist air at a dry bulb temperature of 85 °F and wet bulb temperature of 68 °F, find the (a) absolute humidity, (b) saturation humidity, (c) dew point, (d) enthalpy, (e) humid heat, (f) humid volume, (g) percent absolute humidity, (h) relative humidity, (i) heat of vaporization, (j) vapor pressure, and (k) partial pressure of water.

(a) 0.0105 lbm water/lbm dry air

(b) 0.0265 lbm water/lbm dry air

(c) 59 °F

(d) 25 BTU/lbm dry air

(e) 0.245 BTU/lbm-°F dry air

(f) $13.7 + (14.3 - 13.7)(\frac{0.0105}{0.0265}) = 13.95$ ft^3/lbm dry air

(g) 40% ($\frac{0.0105}{0.0265} \times 100 = 40\%$, or from chart)

(h) 42% ($H_r = \frac{p}{p_s} \times 100 = \frac{0.25}{0.60} \times 100 = 42\%$, or from chart)

(i) 1046 BTU/lbm water

(j) 0.6 psia

(k) 0.25 psia

3 DRYING

Drying is the removal of some or all liquid from solid materials. Drying reduces the residual liquid content of the solid to some acceptable level, and is usually the final step in the production of powdered materials. The majority of liquid removed from a solid is usually removed mechanically (i.e., is *dewatered*). Thermal moisture removal is generally more expensive than dewatering. The moisture content of the finished product varies. If the product contains no water when dried, it is referred to as *bone dry*. It is common for dried products to contain some residual water.

Dryers can be classified as *direct dryers* (where hot gases directly contact the solid), and *indirect dryers* (where heat for drying is transferred through a surface). Both types can be operated in batch or continuous modes.

In direct drying calculations, the first step is to write a material balance for the system. A direct drying process consists of preheating dry ambient air and passing it over the wet product to remove moisture. The evaporated moisture joins the air stream leaving the dryer. The entire process is considered adiabatic. Therefore, the sensible heat in the entering air equals the latent heat of evaporation of the water from the feed solid.

Since this is the definition of an *adiabatic saturation* temperature of the air, the drying process can be represented by the constant enthalpy (or, approximately, the wet bulb temperature) line on a psychrometric chart.

A. RATE OF DRYING

Drying is a combination of heat transfer and mass transfer. A temperature gradient between the air and the solid is required for heat transfer, and a pressure gradient between the liquid and the gas is required for mass transfer. Initially, the surface of the solid is wet, and the liquid can exert its vapor pressure at the temperature of the solid. With heating, the evaporating surface will move inward, towards the interior of the solid. As soon as the surface is no longer completely wet (and dry spots appear), the rate of drying will decrease. Pressure and temperature gradients will increase in the solid, adding to the drying resistance.

When the surface is wet, the drying rate is constant. When the surface starts to dry, the drying rate decreases. If the drying rate is plotted as a function of time or moisture content, figure 9.9 will be typical of the results.

The curve for drying rate versus free moisture content has noteworthy characteristics. The initial free moisture content is $w_o - w_{eq}$. From the initial moisture content to the critical moisture content the drying occurs at a constant rate, $(\frac{dw}{d\theta})_{CR}$. At the *critical moisture content*, the drying rate starts to decrease. The drying rate continually decreases thereafter. For simplification, the falling rate period has constant slope, and K can be calculated:

$$K = \frac{\left(\frac{-dw}{d\theta}\right)_{CR}}{w_c - w_{eq}} \quad 9.52$$

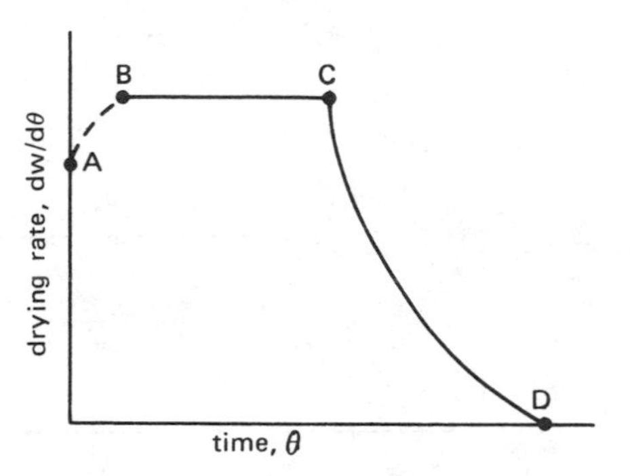

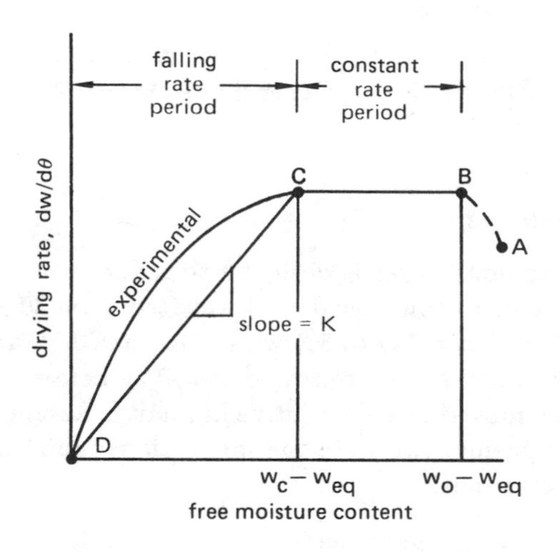

Figure 9.9 Drying Rates

If K is constant, the equation for the falling rate period is

$$\frac{dw}{d\theta} = -Kw_F \quad 9.53$$

If equation 9.53 is integrated,

$$ln\,\frac{w_{F1}}{w_{F2}} = K\,\Delta\theta_{FR} \quad 9.54$$

From figure 9.10, the drying time between moisture contents w_1 and w_c is

$$\theta_c = \Phi\,(w_1 - w_c) \quad 9.55$$

The drying time from w_c to w_2 (where w_2 is greater than w_{eq}) is

$$\theta_f = \Phi(w_c - w_{eq}) \ln \frac{w_c - w_{eq}}{w_2 - w_{eq}} \qquad 9.56$$

Φ is known as the *drying factor.*

$$\Phi = \frac{-L_s}{A\left(\frac{dw}{d\theta}\right)_{CR}} \qquad 9.57$$

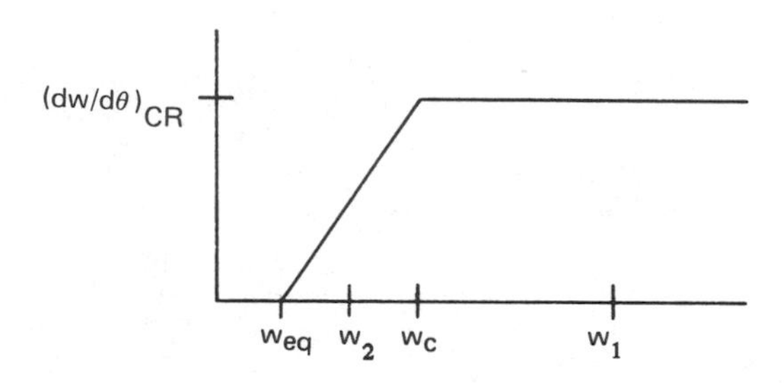

Figure 9.10 Generalized Drying Plot

Example 9.2

A continuous dryer is designed to reduce the moisture content in 10 tons per day of wet feed from 20% water (wet basis) per day to 5% water. Air at 70 °F and 40% relative humidity is preheated to 250 °F before entering. Air is removed at 65% relative humidity. Assuming adiabatic drying, calculate the input air required in cubic feet per minute.

Basis: 1-hour operation

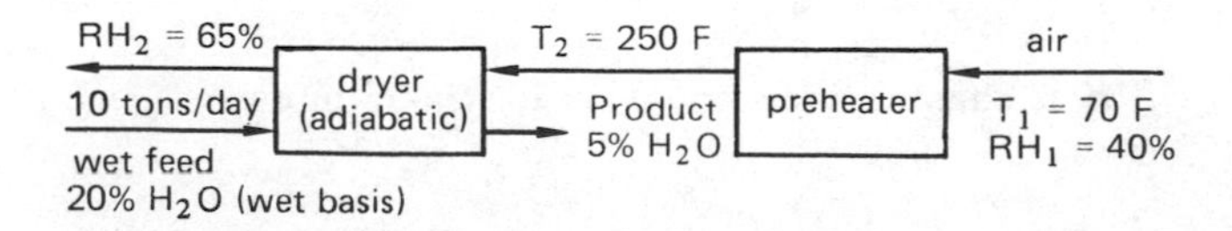

- material balance:

$$\text{feed in: } \frac{10\,(2000)}{24} = 833.3 \text{ lbm/hr wet feed}$$

$$\text{water: } 833.3(0.20) = 166.7 \text{ lbm/hr water in feed}$$

$$\text{dry solid: } 833.3 - 166.7 = 666.6 \text{ lbm/hr dry solid}$$

$$\text{lbm water/lbm dry solid} = \frac{166.7}{666.6} = 0.25$$

- water evaporated:

$$\text{lbm water/lbm dry solid} = \frac{0.05}{0.95} = 0.0527$$

water: 0.0527 (666.6) = 35.1 lbm water/hr in product

166.7 − 35.1 = 132.6 lbm water/hr evaporated

- air humidity:

entering:

H = 0.006 lbm water/lbm dry air

wet bulb temperature = 99 °F

leaving:

H = 0.039 lbm water/lbm dry air

wet bulb temperature = 110 °F

(found by moving horizontally from 70 °F, 40% humidity to 250 °F horizontally to 65% humidity on wet bulb line)

- humidity difference:

$$0.039 - 0.006 = 0.033 \text{ lbm water/lbm dry air}$$

- air requirements:

$$\text{dry air: } \frac{132.6}{0.033} = 4050 \text{ lbm/hr dry air}$$

$$\text{moist air: } 13.32 + (13.69 - 13.32)(0.4) = 13.47 \text{ ft}^3\text{/lbm dry air}$$

- total volume: $\frac{4050\,(13.47)}{60} = 910 \text{ ft}^3\text{/minute}$ at 70 °F and 40% relative humidity

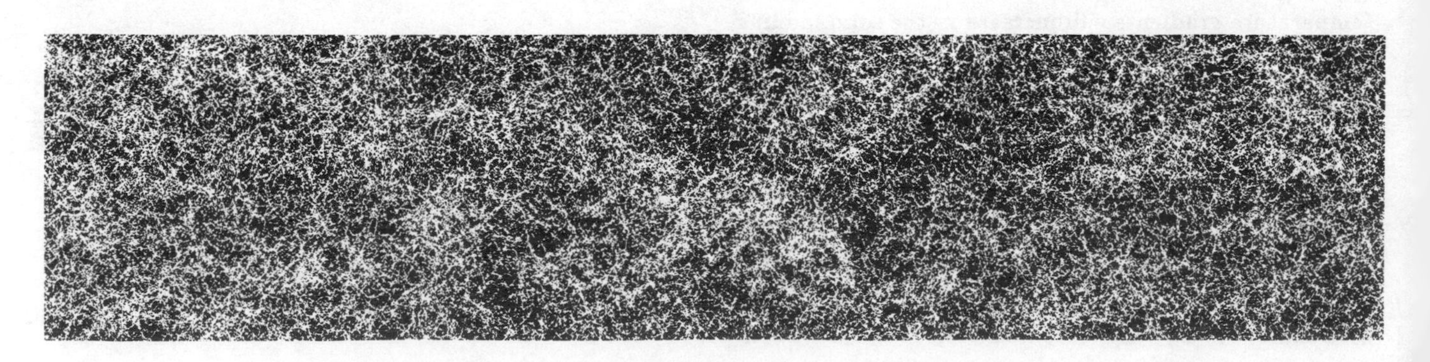

Note: The enthalpy and wet bulb temperature scales have been drawn parallel for convenience. They actually diverge slightly. Use the enthalpy deviation curves to correct.

Note: To obtain water vapor pressure, use the steam saturation table to get the saturation pressure at the air's dew point. Then, multiply by the relative humidity.

$$P_V = \phi P_{sat}$$

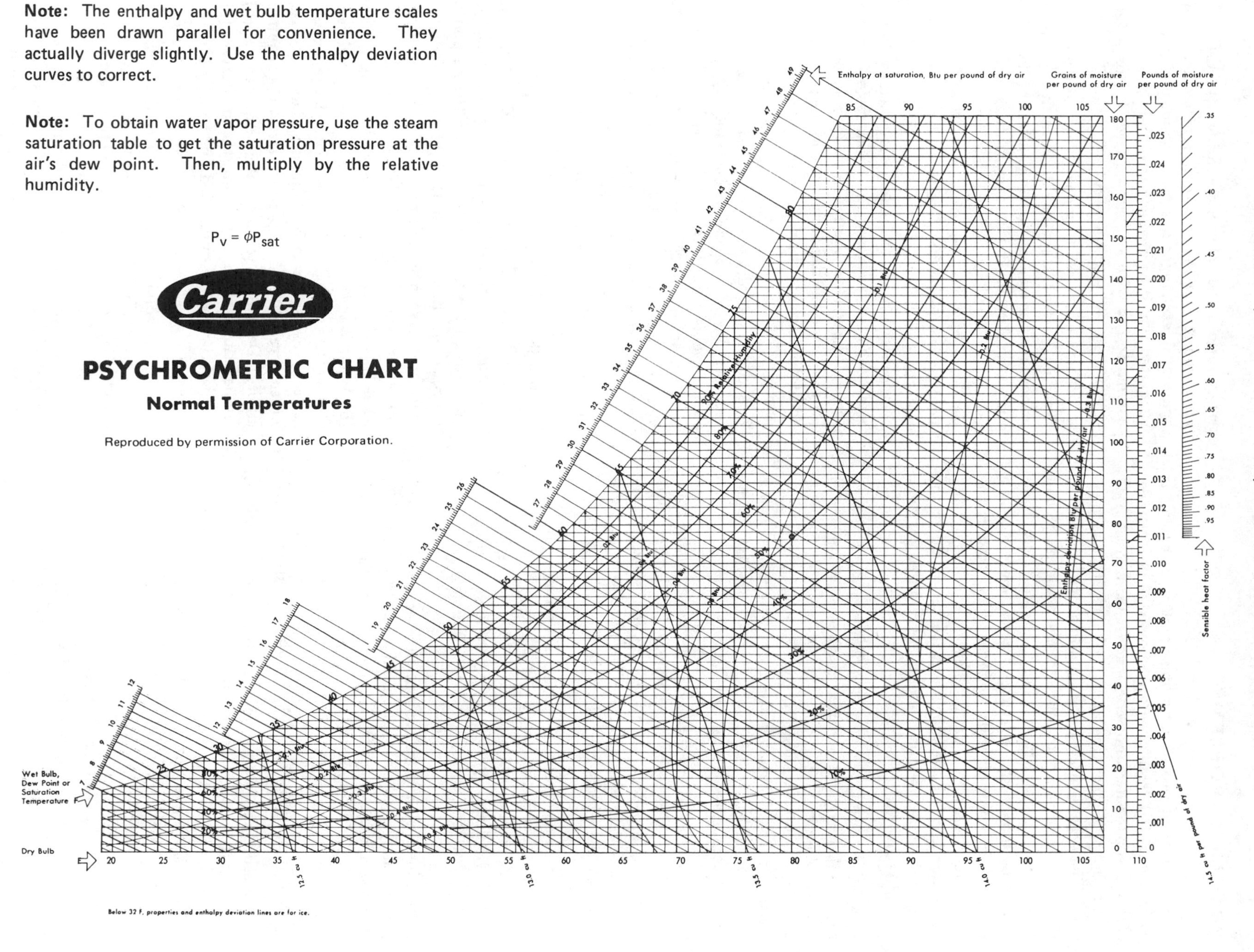

Figure 9.11 Psychrometric Chart

PRACTICE PROBLEMS

1. A mixture of two high boiling point organic acids is to be separated from a small amount of non-volatile, carbonaceous material by continuous steam distillation in a small still operating at 212 °F and 200 mm mercury. The organic mixture is equimolar with respect to the acids. The vapor pressures of the acids are 32 and 14 mm mercury at 212 °F. The following assumptions should be made: the mixture of acids obeys Raoult's Law; the mixture of acids is immiscible in water; the carbonaceous material has no effect on equilibrium; and the vapor leaves in equilibrium with the liquid in the still. How many pounds of steam are required per mole of acid recovered?

2. A 537 kg solution containing 50% heptane and 50% (by weight) octane is batch distilled until only 4.74 lbmoles remain in the pot. (a) Assuming a relative volatility of 2.0, what will be the pot composition after distillation? (b) What is the composition of the first vapor? (c) If the relative volatility is 1.0, what will be the pot composition after distillation?

3. A binary liquid mixture of A (the more volatile component) and B is to be continuously distilled in a plate column. Components A and B form ideal solutions over the entire composition range. The relative volatility is constant and equals 2.0. The design conditions are:

feed condition	saturated liquid
feed composition x_F	50 mole% A
feed rate, F	100 lbmoles/hr
distillate, x_D	90 mole% A
bottoms, x_B	10 mole%

Determine (a) the minimum number of theoretical plates (analytically), (b) the minimum reflux ratio $\left(\frac{L}{D}\right)$ (analytically), and (c) the number of plates required if the column is operated at a reflux ratio of 1.2 times the minimum.

4. Repeat parts (a) and (b) of problem 3 graphically.

5. A distillation column composed of theoretical plates processes a benzene-toluene mixture. The feed is 1000 lbmoles/hr as a saturated liquid and contains 30 mole% C_6H_6. During a run, liquid and vapor samples of streams passing each other at a point in the rectifying section are taken, and the analyses are:

$$y = 0.61 \text{ mole fraction } C_6H_6$$

$$x = 0.52 \text{ mole fraction } C_6H_6$$

The total condenser produces a stream containing 95% C_6H_6, and the thermosyphon reboiler produces a stream with 6 mole% C_6H_6. The equilibrium curve for benzene follows the equation $y = x\,(2-x)$. (a) At what reflux ratio is the column operating? (b) Estimate the area required for the countercurrent condenser. (c) A spare heat exchanger has been found, and it has been suggested that it be used as a feed preheater to allow the column to run with a saturated feed. Is this suggestion reasonable? Why or why not?

6. A 3′ diameter, 25-tray distillation column has been designed to separate n-hexane from n-heptane under atmospheric pressure. Upon installation, the column is run at total reflux until steady state is reached. Analysis of the distillate and bottoms gives mole fractions (in terms of n-hexane) as $x_D = 0.98$ and $x_B = 0.03$. Feed $x_F = 0.45$ (at 10 °F below its boiling point) is put into the column. Estimate the effect on the product composition of each of the following actions, considered separately: (a) reducing the reflux ratio from infinity to 10, (b) reducing the reflux ratio from infinity to 1, (c) increasing the steam pressure in the partial reboiler, and (d) increasing the vapor and liquid rates in the column.

7. A continuous dryer is being designed to produce 15 tons per day of a product containing 3% water. The feed contains 33% H_2O. The air used for drying has a temperature of 68 °F and a relative humidity of 45%, and will be preheated to 300 °F. Calculate (a) the volume of input air (in ft^3/min) required if the exhaust is 68% relative humidity, and (b) the heat required by the preheater in BTU/hr.

8. A certain process requires air with a moisture content of 0.014 lbm water/lbm dry air. (The air temperature is unimportant.) The air is to be produced in a system consisting of a preheater followed by an adiabatic humidification tower. The flow rates of the air and water are constant. In a test run, the system was found to operate properly when the air was preheated to 100 °F and the water temperature was 70 °F. (a) In the test run, what was the humidity of the air entering the tower? (b) What was the temperature of the humidified air? On another day, the air drawn into the preheater was at 75 °F and 20% relative humidity. The preheater and water were adjusted to produce air at the required humidity. Determine (c) the water temperature, (d) the temperature of the air leaving the preheater, (e) the temperature of the humidified air, and (f) the amount of heat added per pound of dry air in the heater.

9. A porous solid is dried in a batch dryer under constant drying conditions. Six hours are required to reduce the moisture content from 30 to 10 lbm water/lbm dry solid. The critical moisture content is 16 lbm water/lbm dry solid and the equilibrium moisture is 2 lbm water/lbm dry solid. (All moisture contents are on a dry basis.) Assume that the drying rate during the falling rate period is a straight line through the origin. Calculate the time required to dry the solid from 30 to 6 lbm water/lbm dry solid.

10. Air at 81 °F and 80% relative humidity is dehumidified in a silica gel absorber. Air leaving the absorber is required to be at 75 °F and humidity of 0.005 lbm water/ lbm dry air. The silica gel reduces the moisture to 0.001 lbm water/lbm dry air, so part of the incoming air is bypassed around the absorber and mixed with air leaving the absorber to produce the final required air conditions. What weight fraction of the incoming air is bypassed?

11. The operator of a chemical plant wants to install a cooling tower. Water would enter the tower at 2000 gpm at 115 °F. Water must be returned to process coolers at a rate of 2000 gpm and 85 °F. Atmospheric pressure is 760 mm, and air enters the tower at 75 °F and 45% relative humidity. (a) Calculate the air volume in ft^3/min supplied to the tower. (b) If windage loss is 90% of the evaporation loss in the tower, calculate the makeup water required in gallons per minute. (c) Define azeotropic boiling.

12. Air enters a dryer at 70 °F and 20% humidity, and leaves at 180 °F and 50% relative humidity. If the dryer operates at a pressure of 14.3 psia, how many cubic feet of entering air would be needed if 12.0 pounds per hour of water are evaporated in the dryer?

13. The equilibrium data for a binary mixture of organic compounds A and B at one atmosphere total pressure are given below. A is the more volatile component. A mixture of 55 mole% A is fed to a fractionating column equipped with a total condenser and reboiler. The entire system operates at one atmosphere. One-half of the feed vaporizes as it enters the column. The overhead product must be 95 mole% A. The bottoms should not contain more than 5 mole% A.

mole fraction A in liquid	mole fraction A in vapor
0.0	0.0
0.100	0.225
0.200	0.460
0.300	0.600
0.400	0.695
0.500	0.775
0.600	0.825
0.700	0.875
0.800	0.915
0.900	0.955
1.000	1.000

Determine (a) the minimum reflux ratio, (b) the number of actual plates required under total reflux conditions, (c) the number of actual plates required at a reflux ratio of twice the minimum and column efficiency of 60%, and (d) the theoretical plate on which the feed is admitted.

14. It is desired to separate isobutanol and 1,1,2,2 tetrachloroethane in a continuous distillation column. The column will operate with the following isobutanol mole fractions: feed, 0.60; distillate, 0.97; and bottoms, 0.04. The reflux ratio will be 3.0, and the feed will be at its bubble point. The heater in the bottom is a still pot, and it should be considered as a plate. The equilibrium data are given below. (a) If the plate efficiency is 65%, how many plates are required? (b) On which plate should the feed be introduced?

liquid	vapor
0.04	0.209
0.072	0.270
0.097	0.394
0.145	0.480
0.230	0.611
0.365	0.690
0.504	0.785
0.710	0.855
0.860	0.925
0.964	0.988

10 LIQUID-LIQUID AND SOLID-LIQUID PROCESSES

PROCESSES

Nomenclature

A	area	ft^2
B	intercept of constant pressure filtration line	sec/ft^3
c	mass fraction of solids slurry	lbm solid/lbm slurry
C	intercept of constant rate filtration line	ft^3/sec
C	mass of crystals in final magma	lbm
c_p	specific heat	BTU/lbm-°F
D	diameter	ft
E	extract stream rate, or evaporation during the process	lbm or lbm/hr
f	fraction of cycle time for filtration	–
F	feed stream rate	lbm, or lbm/hr
F_c	centrifugal force	gravities
h	enthalpy	BTU/lbm
H_o	total mass of solvent in original batch	lbm
K_p	slope of constant pressure filtration	sec/ft^6
K_r	inverse of slope of constant rate filtration	$lbm\text{-}sec/ft^5$
L	cake thickness	ft
L	liquid flow rate	lbm/hr
L_V	latent heat of evaporation from solution	BTU/lbm
M	total flow rate	lbm, or lbm/hr
M_1	ideal mixer stream in extraction	lbm or lbm/hr
n	rotational speed	rpm
N	mass ratio of component B, or total number of stages	–
Δp	pressure drop	lbf/ft^2
q	heat	BTU
Q	quantity flow rate	ft^3/sec
r	resistance of unit area of medium	ft^{-1}
R	raffinate flow rate	lbm, or lbm/hr
R	ratio of molecular weights	–
R_c	resistance of filter cake	ft^{-1}
s	cake compressibility	–
S	solvent flow rate	lbm, or lbm/hr
S	steam flow rate	lbm/hr
S	anhydrous solubility	parts/100 parts solvent
t	time	seconds
T	temperature	°F
U	heat transfer coefficient	BTU/hr-ft²-°F
v	velocity	ft/sec
V	filtrate volume	ft^3
V	vapor flow rate	lbm/hr
w	mass of dry cake (solids)	lbm/ft^3 filtrate
w_o	mass of anhydrous solute in original batch	lbm
W	mass of dry cake (solids)	lbm/hr
x	liquid mole fraction, or mass or weight fraction of solvent-poor phase	–
X	mass fraction (B-free basis)	–
y	mass fraction of solvent-rich phase	–
Y	weight of fraction (B-free phase)	–

Symbols

α	specific cake resistance	ft/lbf
Δ_R	diffraction point, or net flow outward at last stage	–
λ	heat of vaporization	BTU/lbm

μ	dynamic viscosity	lbf-sec/ft^2
ε	porosity	–
ρ	density	lbm/ft^3

Subscripts

b	bowl
c	cake, condensate, or crystallization
f	feed
i	initial
L	liquor
m	medium
M_1	at mixed point
n	intermediate nth stage
N	last stage
o	original
p	constant pressure, particle
r	constant rate
s	solvent phase, or solid, or steam
t	terminal
v	vapor
π	total cycle

1 EXTRACTION

A. EXTRACTION TERMS

Liquid-liquid extraction, sometimes called *solvent extraction*, is the separation of components dissolved in a liquid through contact with another insoluble liquid. If the components attain different concentrations in the two liquids, a certain degree of separation will be achieved.

In extraction operations, the solution to be extracted is called the *feed*. Dissolved in the feed is the component to be extracted. The insoluble liquid brought into contact with the feed is called the *solvent*. The solvent may be totally insoluble or only slightly soluble in the feed medium. The solvent-rich product is called the *extract*, and the residual liquid from which most of the extractable components have been removed is called the *raffinate*.

B. EXTRACTION SYSTEMS

A single-stage extraction system consists of a *mixer* (which brings the insoluble liquids into contact with each other), and a *settler* (where the product of the mixer, usually an *emulsion*, is allowed to separate into two distinct phases).

A *single-stage contact scheme* is shown in figure 10.1. This simple system approaches one equilibrium step. The amount of solute extracted is fixed by equilibrium conditions and by the amount of solvent used.

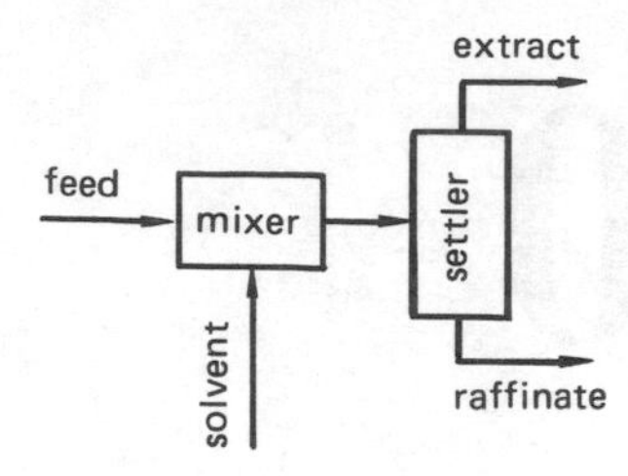

Figure 10.1 Single-stage Contact Scheme

Figure 10.2 illustrates a *simple multistage contact* (also called *co-current*) or *cross-current extraction process*. The solvent is divided into several portions. The feed is, essentially, treated with each of these fresh solvent portions, while extract is withdrawn from each stage. The extract quantity is optimized when equal amounts of solvent are used in each stage.

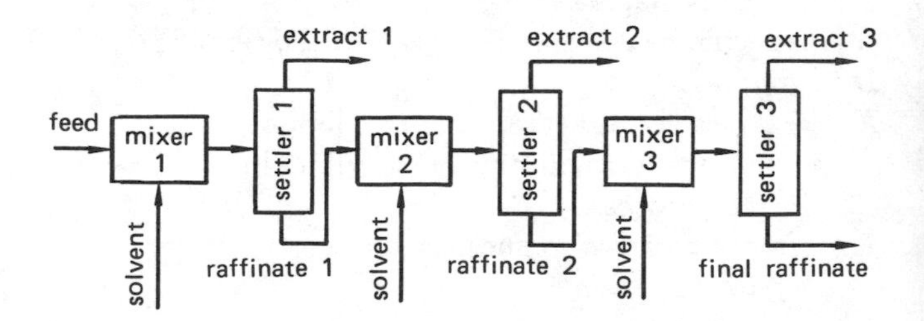

Figure 10.2 Simple Multistage Contact

Figure 10.3 illustrates a *countercurrent multistage contact extraction process*. Fresh solvent and feed enter the series extraction stages at opposite ends. Extract and raffinate phases pass continously and countercurrently from stage to stage through the system. Countercurrent multistage contact is analogous to distillation in a plate column.

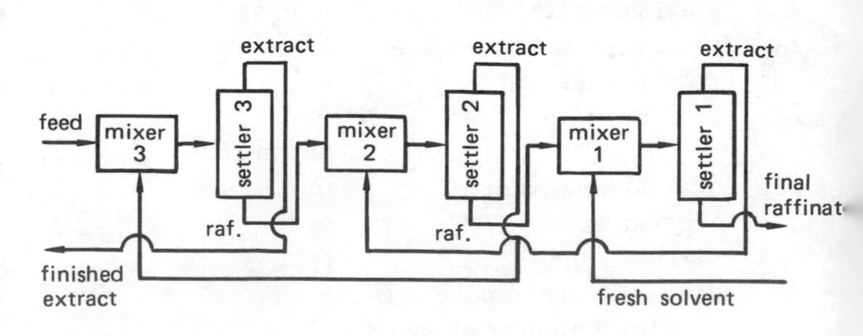

Figure 10.3 Countercurrent Multistage Contact Extraction

In general, extraction is analogous to the stripping operation in a distillation column. The solvent is analogous to reboiler vapor, the raffinate is analogous to reboiler bottoms, and the extract is analogous to the vapor leaving the feed stage.

C. MUTUALLY INSOLUBLE SYSTEMS

If the component to be extracted is the only soluble component in each phase, and if the solvent and the feed medium are not soluble in one another, then the system can be designed or analyzed graphically with techniques similar to the McCabe-Thiele diagrams for distillation.

1. Single-Stage Contact

With single-stage contact, a material balance should be made around the contactor.

$$Fx_f + Sy_s = Ey* + Rx* \qquad 10.1$$

Since the mass of solvent and feed medium remain unchanged within the contactor, the concentration term is slightly altered to be on a *solute-free basis.* Since the solvents are insoluble in one another, the operating line equation can be derived from equation 10.1:

$$\frac{-F}{S} = \frac{y_s - y*}{x_f - x^*} \qquad 10.2$$

Equation 10.2 is the equation for the operating line of a single-stage contactor with entering concentrations x_f and y_s, and leaving equilibrium concentrations $x*$ and $y*$. The slope of this line is the ratio $\frac{-F}{S}$. A diagram should be constructed showing the equilibrium curve and entering conditions. The leaving conditions can be graphically determined by plotting the slope of the operating line through the entering conditions to the equilibrium line. Figure 10.4 illustrates the graphical solution for single-stage extraction using insoluble solvents. The solute concentrations can also be conveniently expressed in units of mass per unit volume (i.e., lbm/ft^3).

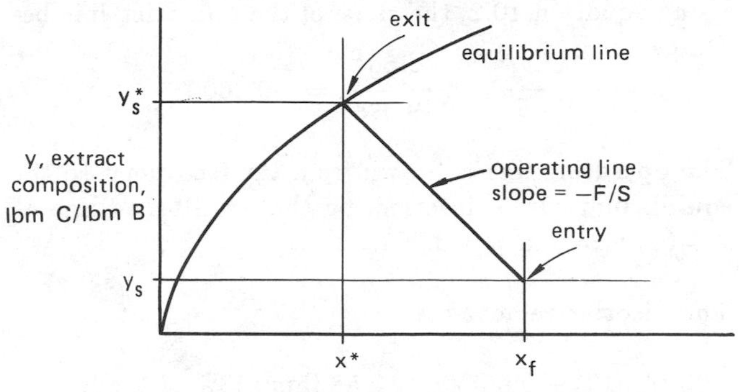

Figure 10.4 Single-Stage Extraction with Insoluble Solvents

Example 10.1

The nicotine equilibrium relationship between water, F, and kerosene, S, can be expressed by the equilibrium relationship: $y = 0.9167x$, where x the mass ratio of nicotine to water, and y is the mass ratio of nicotine to kerosene. Water and kerosene are essentially insoluble. Determine the percentage extraction of nicotine if 100 lbm of feed solution containing 1% nicotine is processed with 150 lbm of kerosene containing 0.001% nicotine.

Since $x_f = 0.01$ weight fraction, and the equilibrium is a weight ratio,

$$x_f = \frac{0.01}{1 - 0.01} = 0.0101 \text{ lbm nicotine/lbm water}$$

$$y_s = \frac{0.00001}{1 - 0.00001} = 0.00001 \text{ lbm nicotine/lbm kerosene}$$

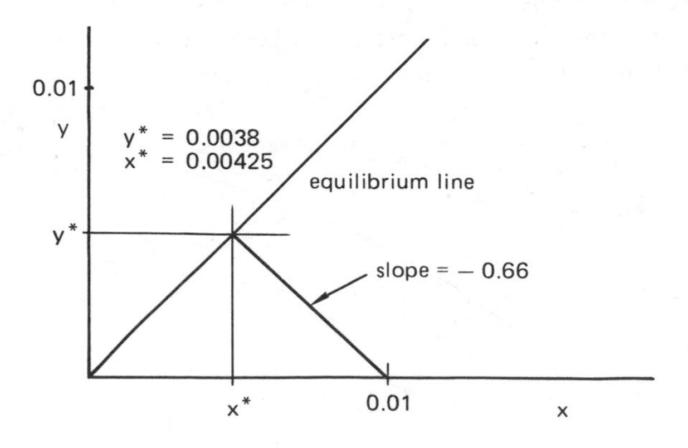

For a solute free basis,

$$F = 100 - 0.01(100) = 99$$
$$S = 150 - 0.00001(150) = 149.9985$$

Using equation 10.2, the slope of the operating line becomes

$$\frac{-F}{S} = \frac{-99}{149.9985} = -0.6600$$

The operating line is drawn from the feed point to the equilibrium curve, intersecting the equilibrium line at $x = 0.00425$ and $y = 0.0038$.

The nicotine removed is

$$99\ (0.0101 - 0.00425) = 0.58 \text{ lbm } (58\% \text{ of the feed})$$

2. Cross-Current Extraction

For cross-current extraction, the equilibrium concentrations are plotted as in figure 10.4, again using lbm/lbm solvent as the concentration unit. Since the liquids A and B are insoluble, there are A pounds of this substance in all raffinates. Similarly, the extract from each stage contains all solvent B fed to that stage. The balance around any stage n is

$$Fx_{n-1} + S_n y_s = E_n y_n + R_n x_n \qquad 10.3$$

Since $F = R$, and $S = E$, the operating line for each stage in a cross-current extraction process is

$$\frac{-F}{S_n} = \frac{y_s - y_n}{x_{n-1} - x_n} \qquad 10.4$$

The operating line for each stage n, has slope $\frac{-F}{S_n}$, and passes through the points (x_{n-1}, y_s), and (x_n, y_n) (where the second point also lies on the equilibrium curve). Each operating line intersects the equilibrium curve at the raffinate and extract compositions. No raffinate of concentration smaller than that in equilibrium with the entering solvent is possible. Figure 10.5 illustrates the construction of operating lines in cross-current extraction with insoluble solvents.

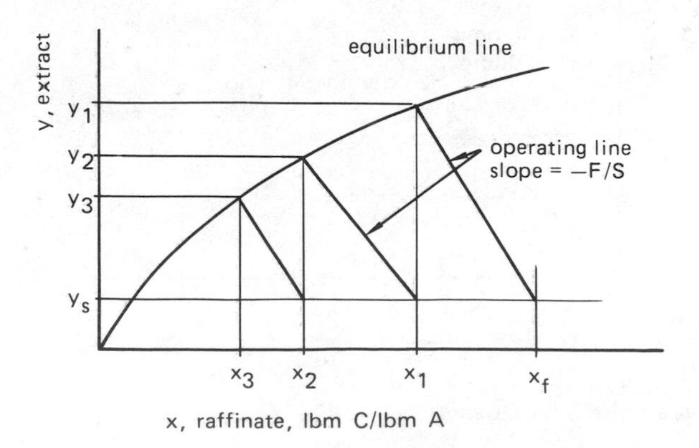

Figure 10.5 Cross-Current Solvent Extraction with Insoluble Solvents

Example 10.2

Repeat example 10.1 using three ideal, cross-current extractions of 50 lbm solvent for each stage.

For each stage,

$$\frac{-F}{S_n} = \frac{-99}{50} = -1.98$$

The construction of the operating line starts at 0.0101, 0.00001, with slope of −1.98. The final raffinate composition obtained from the plot is $x_3 = 0.0034$, and the nicotine extracted is calculated as

$$99\,(0.0101 - 0.0034) = 0.633 \text{ } (63\% \text{ of that in the feed})$$

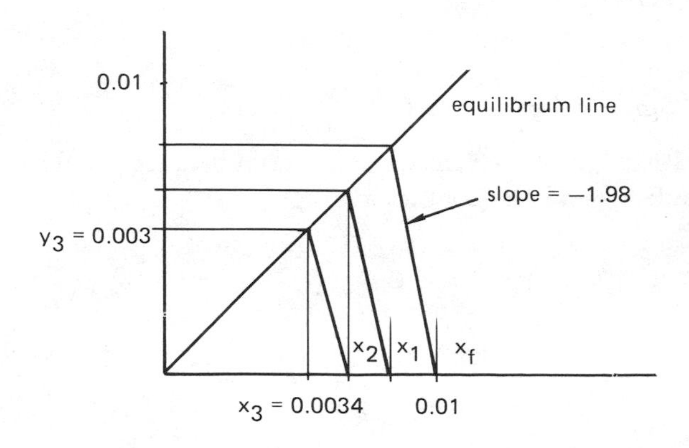

3. Continuous Counter-Current Multistage Extraction

When the liquids A and B are insoluble, the stage calculations are usually made on a solute-free basis. Thus, if substance C is to be extracted, the concentrations in the two phases would be represented by the concentration units of lbm C/lbm A or lbm C/lbm B. The overall balance for component C for N stages becomes

$$Fx_f + Sy_s = Rx_N + Ey_1 \qquad 10.5$$

For insoluble solvents, equation 10.5 is

$$\frac{F}{S} = \frac{y_1 - y_s}{x_f - x_N} \qquad 10.6$$

Therefore, the operating line is a straight line with slope $\frac{F}{S}$ passing through points (x_f, y_1) and (x_N, y_s). Figure 10.6 illustrates the various stages for counter-current extraction.

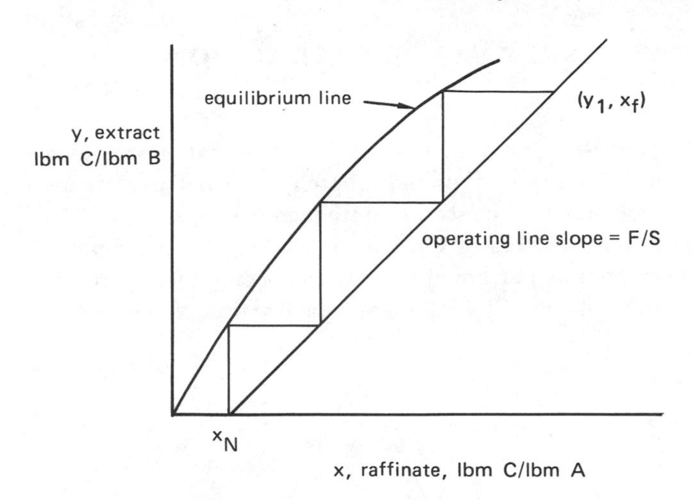

Figure 10.6 Counter-Current Extraction with Insoluble Solvents

For the special case where the equilibrium curve is straight and has slope $m = \frac{y*}{x*}$, the equation for the leaving and entering concentrations as a function of the number of equilibrium stages can be related in a manner similar to how the Kremser equation relates stages for absorption or stripping. The equation relating exit and entering concentrations is

$$\frac{x_f - x_N}{x_f - \frac{y_s}{m}} = \frac{\left(\frac{mS}{F}\right)^{N+1} - \frac{mS}{F}}{\left(\frac{mS}{F}\right)^{N+1} - 1} \qquad 10.7$$

$\frac{mS}{F}$ is the *extraction factor.* This factor can be used with a Kremser chart by substituting the extraction factor for the absorption factor.

Example 10.3

1000 lbm/hr of a 1% nicotine water solution is to be counter-currently extracted with 1150 lbm/hr of pure kerosene, reducing the nicotine content to 0.1%. What is the theoretical number of stages required?

F = 990 lbm/hr (water rate), and S = 1150 lbm/hr. m from example 10.1 is 0.9167. The extraction factor is

$$\frac{(0.9167)(1150)}{990} = 1.0649$$

The efficiency of the overall extraction is $1 - \frac{0.001001}{0.0101} = 0.901$. Using the Kremser-Brown Absorption Factor chart in chapter 8, the number of theoretical stages is 8.4.

D. SYSTEMS WITH PARTIALLY SOLUBLE SOLVENTS

1. *Triangular Diagrams*

When the mutual solubility of the dilutent and solvent cannot be neglected, the solubility and equilibrium relationships can be shown on triangular coordinates. The composition of any three component mixtures can be shown as a point inside the triangle, as shown in figure 10.7.

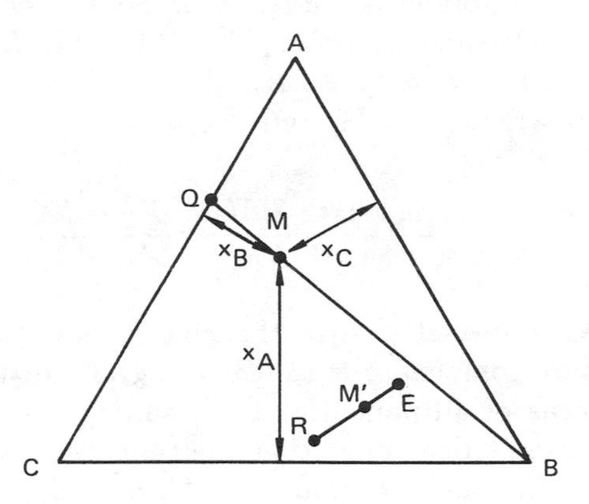

Figure 10.7 Triangular Solubility Diagram

The triangular diagram has several important characteristics:

- Concentrations are based on the entire mixture, not on one or two components, so that

$$x_A + x_B + x_C = 1$$
$$y_A + y_B + y_C = 1 \qquad 10.8$$

- Either mole fractions or mass fractions can be used. Mass fractions are more common.
- Since the sum of the concentrations is unity, the sum of the perpendicular distances from a point in a triangle to the sides of an equilateral triangle is equal to the altitude of the triangle.
- An apex represents a pure component.
- A sideline represents binary mixtures of the components represented by the apexes.
- The straight lines connecting two points on the solubility curve are known as *tie lines.* The ends of a tie line represent two phases in equilibrium with each other.
- When the length of a tie line is zero, the curve is said to be at the *plait point.* At the plait point the phases are identical. Point E in figure 10.8, type I, is a plait point.

PROCESSES

- Outside of the insoluble envelope lies the area where all components are soluble in one phase. This area is called the *homogeneous liquid area.* Within the envelope the area is called *heterogeneous.*
- The *center of gravity principle* (also known as *lever rule* or *mixing rule*) applies to mixtures on the tie lines: the amount of phase present is proportional to the fraction of the tie line from the mixture to the solubility curve. If R pounds of mixture (represented by point R) is added to E pounds of mixture (at point E), the resultant mixture is shown at point M, and

$$\frac{R}{E} = \frac{\text{line length } EM}{\text{line length } RM'} = \frac{X_E - X_M}{X_M - X_R} \qquad 10.9$$

As a special case of the mixing rule, when one pure component is added to a given mixture, the locus of all points for the resulting solution is a straight line connecting the point with the apex for the added component. Thus, if B is stripped from mixture M shown in figure 10.7, the result would be the two-component mixture represented by point Q.

Two common generalized systems are shown in figure 10.8. Type I is the more common system.

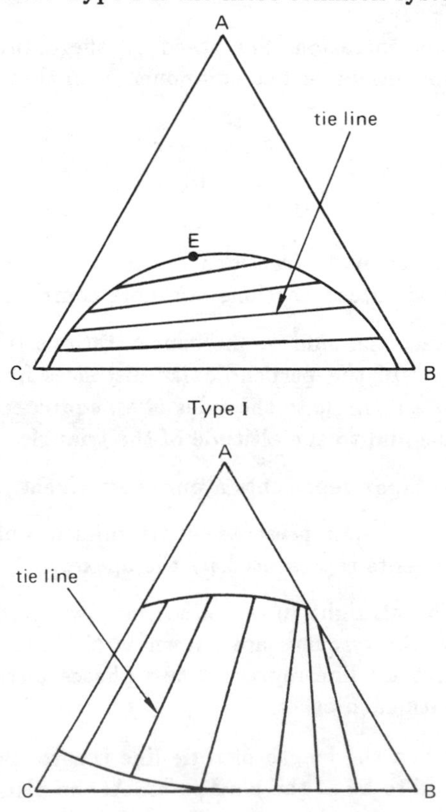

Figure 10.8 Tie Lines in a Solubility Diagram

2 SINGLE-STAGE EXTRACTION

Single-stage extraction can be either batch or continuous. The feed, F pounds (batch) or F pounds per hour (continuous), contains component A and C (with XE weight fraction of C). This is contacted with S pounds (S pounds per hour) solvent containing mostly B (with some A and C) with y_S weight fraction of solute C.

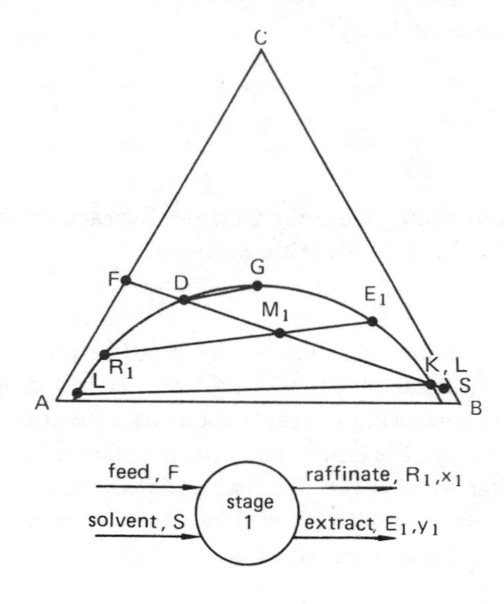

Figure 10.9 Single Stage Extraction

If it is desired to find E and R compositions given F and S compositions, then adding F to S by the mixing rule produces a mixture at M. The equilibrium phases E_1 and R_1 are joined by the tie line through M_1. The total material balance is

$$F + S = M_1 = E_1 + R_1 \qquad 10.10$$

The component C balance can be used to compute x_{M_1}.

$$Fx_F + Sy_S = M_1 x_{M_1} \qquad 10.11$$

If it is desired to find the composition of S (the amount of solvent), F, and M_1 (or F, the composition of S, E_1, and R_1), the amount of solvent can be calculated by the mixing rule.

$$\frac{S_1}{F} = \frac{x_F - x_{M_1}}{x_{M_1} - y_S} \qquad 10.12$$

If the quantities of raffinate or extract are unknown, equation 10.13 can be used.

$$E_1 = M_1 \frac{x_{M_1} - x_1}{y_1 - x_1} \qquad 10.13$$

The point M must lie in the heterogeneous liquid area for extraction to occur. The *minimum solvent* is found by locating M_1 at D, which provides an infinitesimal amount of extract at G. The *maximum solvent* is found by locating M_1 at K, producing an infinitesimal amount of raffinate at L.

3 SIMPLE MULTISTAGE EXTRACTION

Simple multistage extraction (also known as *cross-current extraction*) is an extension of single-stage extraction. The raffinate is contacted with fresh solvent (which may be done continuously or in batch). A single final raffinate results. The extracts may be kept separate or may be composited as shown in figure 10.10.

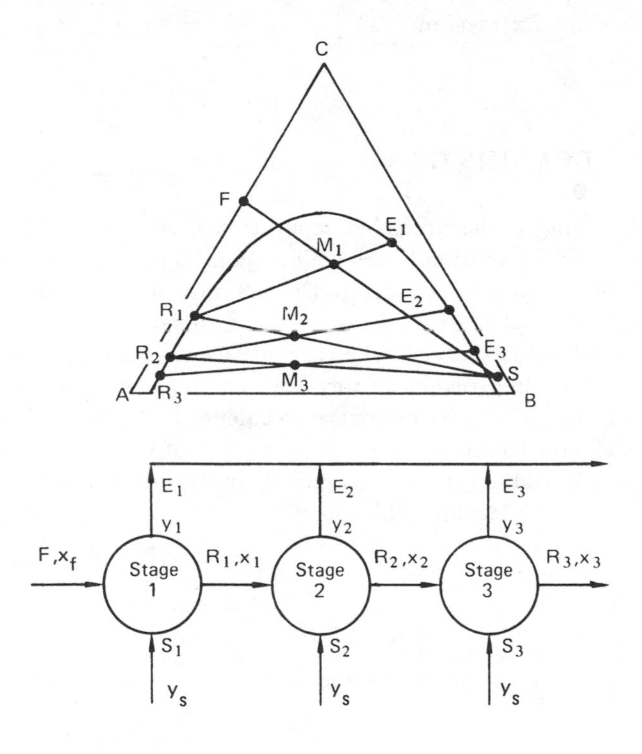

Figure 10.10 Multistage Extraction

The solution method is similar to that for a single-stage extraction. The material balances for the single-stage apply to the first stage. Material balances for subsequent stages are calculated with the feed to any stage being the raffinate of the preceding stage.

For any stage, n,

$$R_{n-1} + S_n = M_n = E_n + R_n \qquad 10.14$$

The material balance for component C is

$$\begin{aligned} R_{n-1}x_{n-1} + S_n y_s &= M_n x_{Mn} \\ &= E_n y_n + R_n x_n \end{aligned} \qquad 10.15$$

4 CONTINUOUS COUNTER-CURRENT MULTISTAGE EXTRACTION

The flow sheet for a continuous countercurrent multistage extraction operation is shown in figure 10.11. Extract and raffinate streams flow countercurrently from stage to stage, resulting in two products: raffinate, R_N, and extract, E_1. For a given degree of separation and amount of solvent, this type of operation requires fewer stages for a given amount of solvent than the cross-current method.

The total material balance around the system is

$$F + S = E_1 + R_N = M \qquad 10.16$$

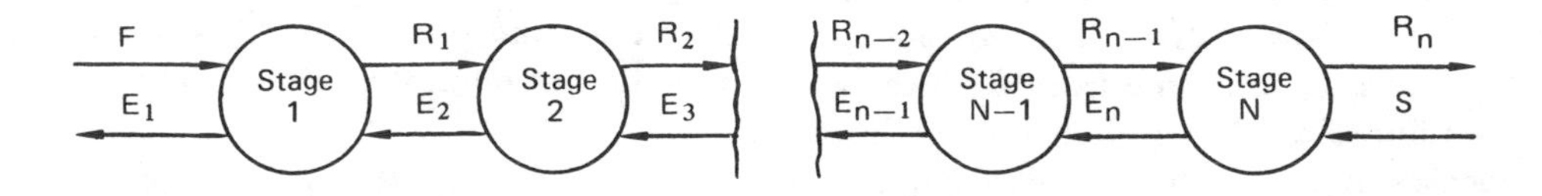

Figure 10.11 Flow Diagram of Continuous Countercurrent Multistage Extraction

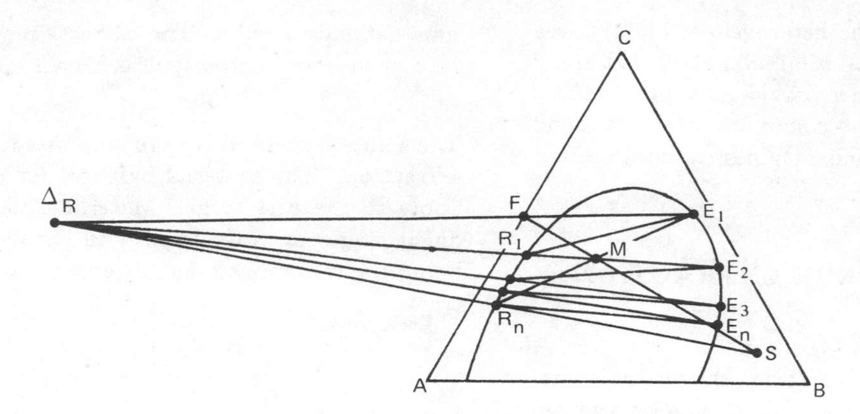

Figure 10.12 Triangular Diagram Construction of Stages for Continuous Countercurrent Multistage Extraction

Point M is located on line FS by a material balance of ingredient C.

$$Fx_F + Sy_s = E_1y_1 + R_Nx_N$$

$$= Mx_M \qquad 10.17$$

$$x_M = \frac{Fx_F + Sy_s}{F+S} \qquad 10.18$$

The overall material balance suggests that M must also lie on the line R_NE_1.

$$R_N - S = F - E_1 = \Delta_R \qquad 10.19$$

Δ_R is the *difference point*, the net flow outward at the last stage, N. Equation 10.19 shows that the lines extended from E_1F and SR_N must intersect at point Δ_R. A material balance at any stage S through stage N is

$$R_{S-1} + S = R_N + E_S \qquad 10.20$$

$$R_N - S = R_{S-1} - E_S = \Delta_R \qquad 10.21$$

The difference in flow rates at the location of any two adjacent stages is constant Δ_R. Line E_SR_{S-1} must also pass through Δ_R. These properties can be used with a graphical solution as follows: after locating points F, S, M, E_1, R_N, and Δ_R, a tie line from E_1 locates R_1. (Extract and raffinate are in equilibrium.) A line from Δ_R to R_1 locates E_2, a tie line locates R_2, etc. If the amount of solvent is increased, point M moves toward S, and Δ_R moves further left.

If a line from Δ_R is parallel with a tie line, an infinite number of stages will be required to separate the components.

5 LEACHING

Leaching is the preferred separation process when one or more constituents are solid. Quantities unknown in leaching problems may be the amount of soluble substance leached from a solid, the number of washings with leaching solvent, the concentration of the solute in the leaching solvent, if any, and the method of leaching. Or, it may be necessary to calculate the number of washings to reduce the solute content of the solid to a specific value (given the amount and solute concentration of the leaching solvent).

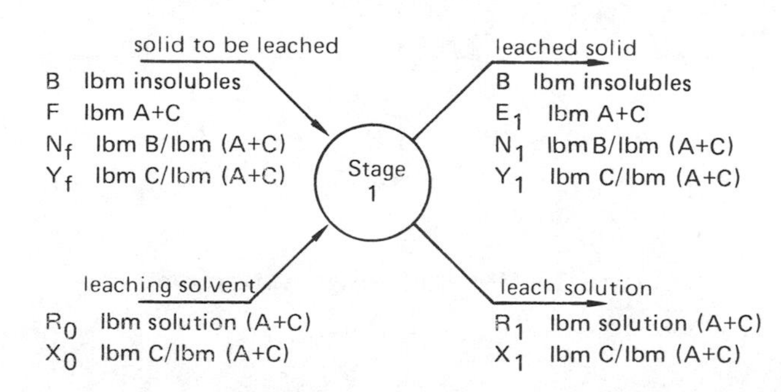

Figure 10.13 Generalized Leaching Process

Single-stage leaching is shown in figure 10.13. Assume B pounds of insolubles, F pounds of $A + C$ in solid, solute C, and solvent A. The concentration of B is N pounds per pound of $A + C$. Then,

$$B = N_FF = E_1N_1 \qquad 10.22$$

If the weight fractions X_A and X_B are on B free basis, the solute balance is

$$FY_F + R_oX_o = E_1Y_1 + R_1X_1 \qquad 10.23$$

The solvent A balance is

$$F(1-Y_F)+R_o(1-X_o) = E_1(1-Y_1)+R_1(1-X_1) \quad 10.24$$

The solution balance (solute and solvent) is

$$F + R_o = E_1 + R_1 = M_1 \quad 10.25$$

The mixture of the solids (to be leached) and the leaching solvent produces a mixture M_1, defined (on a B-free basis) by equations 10.26 and 10.27.

$$N_{M_1} = \frac{B}{F + R_o} = \frac{B}{M_1} \quad 10.26$$

$$Y_{M_1} = \frac{Y_F F + R_o X_o}{F + R_o} \quad 10.27$$

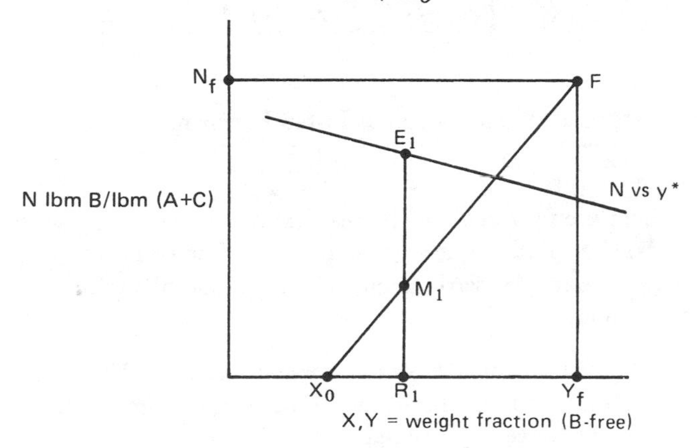

Figure 10.14 Leaching Process Operating Lines

6 FILTRATION

A. DESCRIPTION

Filtration is the separation of a fluid from solid (where the solid is suspended as particles in the fluid), by passing the mixture through a porous medium onto which the solids are deposited. The fluid can be a liquid or gas. A pressure gradient is needed to force the fluid through the medium (filter surface). The pressure gradient needed to force the liquid through the cake and filter medium can be supplied by gravity, pump pressure, or vacuum. The solids are retained on the filter surface as a *cake*. When the amount of solids present is small, the process is called *clarification, polishing*, or *finishing*. As the percentage of solid in suspension becomes higher, the process is called *filtration* or *pressing*.

Either the liquid,solid, or both phases may be the valuable phase.

B. TYPES OF FILTERS

Filters can be generally categorized into four groups: strainers, clarifiers, cake filters, and filter thickeners. A *strainer* is little more than a screen or cloth. In the case of the strainers, the solid is usually the discarded material. *Clarifiers*, on the other hand, remove small quantities of solids to produce clear liquids. An example of a clarifier is the *sand filter*. Sand filters are vessels filled with graded sand providing the porous medium through which the solid-liquid mixture is passed. The solids are trapped on the inlet side of the sand filter. When the pressure drop gets too high, a back washing operation clears the collected solids. A clarifier can also be cloth, paper, or a cartridge that is simply discarded when plugged.

Cake filters separate large amounts of solids from a liquid or slurry, forming a cake of crystals or sludge. The cake can be washed with clear liquid to remove residual impurities before discharge. Cake filters are usually used when the solids are the more valuable phase.

A *thickener* will only partially separate a thin slurry, discharging some clear liquid and a thickened but still flowable suspension of solids.

Filters are also classified into categories depending on how the pressure drop across the filter is generated. *Pressure filters* are those which operate with the upstream pressure higher than atmospheric. *Vacuum filters* operate with atmospheric pressure on the upstream side and vacuum on the downstream side. Pressure filters can obtain their high pressure from gravity acting on a column of liquid above the filter, by pump pressure, or by centrifugal force (rotation).

Most industrial filters use pressure or vacuum. They are either continuous or discontinuous, depending on whether the solids are discharged at a steady rate or intermittently. A common type of *continuous filter* is the *rotary vacuum drum filter*.

C. FILTER SELECTION

Several characteristics of the system and feed affect the selection of a filtration system. The percentage of solids in the feed has an influence on the selected filter medium as well as the distinction of the desired product: cake, filtrate, or both. The value and quantity of the product influences the investment that should be made for the process. If production rate is important, a decision between batch versus continuous operation must be made.

The characteristics of the feed (granular or open versus colloidal and dense) have the greatest impact on filtration selection. The characteristics of the cake, (i.e., dryness, compressibility, washing properties, etc.) are also important.

The most important caveat about selection of a filtration system is that type selection cannot be done without prior experimental measurements of the actual filtrate in the actual type of filter. There is no way to predict, beforehand, the filtration characteristics of a slurry.

D. THE EFFECT OF PRESSURE

During cake filtration, the cake itself acts as a filtering medium, rather than the filter cloth holding the cake. The formation of the initial layer of cake on the cloth is of prime importance. For a given slurry in a specific filter, the major variable under control is the overall pressure drop. The outlet pressure is usually constant, so pressure drop is controlled by varying inlet pressure.

- *Constant pressure filtration* – If the pressure drop is constant, the flow rate will be maximum at the start, and will decrease continuously. High pressure will tend to blind the media initially; low pressure will have a lower filtration rate.
- *Constant rate filtration* – If the pressure drop is varied (kept small at the start and increased continuously or in steps) to keep the flow rate constant, the filtration rate will be constant.
- *Combination* – The combination method starts at low pressure to build up the cake without blinding, and then goes to maximum pressure for maximum rate.

A *filter press* or *pressure leaf filter* may be operated in one of the above ways. A continuous *rotary filter* operates under constant vacuum. This amounts to constant pressure filtration, but with the constant pressure operating on the filtrate side rather than on the cake side. The resistance of the cake reduces the pressure gradient on the slurry side.

E. FILTER AIDS

Very fine solids form dense, impermeable cakes which can quickly plug a filter medium that is fine enough to retain them. Filtration requires that the porosity of the cake be increased to permit passage of the liquid at a reasonable rate. This is done by adding a filter aid (e.g. diatomaceous earth, purified wood cellulose, or other internally porous solid), to the slurry before filtration. The filter aid can be subsequently separated from the cake by burning or dissolution.

Filter aids can also be used as *precoats* (deposits of a thin layer on the filter medium before filtration begins). In batch filtration, the precoat is usually thin. In continuous filtration, the precoat is thick and the top layer is continually cut off by an advancing blade to expose fresh surface to the slurry.

F. FILTRATION THEORY

Figure 10.15 shows flow rate through a filter measured versus time. Volume for a clean liquid will plot as a straight line. If the liquid has suspended solids, the resistance of the porous layer increases as particles accumulate on it. The amount of liquid collected per unit time will gradually decrease. The line is always curved for a liquid with suspended solids.

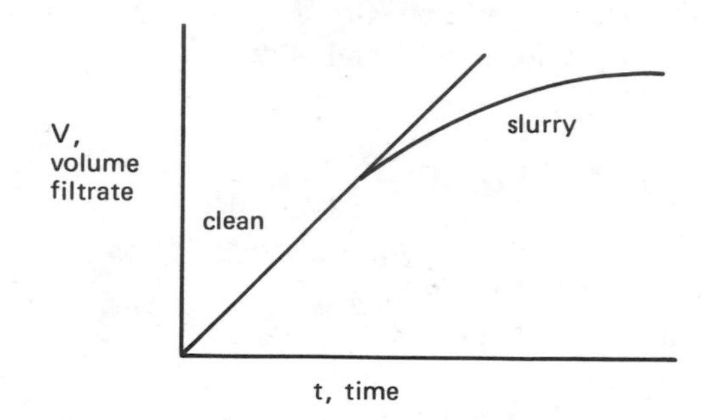

Figure 10.15 Typical Filter Performance Curve

There are four general shapes that are encountered with filtration plots. Three (and possibly four) mechanisms can account for various curvatures of the filtration volume line:

- *Cake formation* occurs when cake continuously forms on the filter medium, leaving the medium resistance unchanged.
- *Complete blocking* occurs when the particles plug the pores or capillaries of the medium.
- *Standard blocking* occurs when the solids adhere to the walls of the capillaries, progressively reducing flow path diameter.
- *Intermediate blocking* is between complete and standard blocking.

With these mechanisms in mind, various types of questions about filter performance can be asked:

- Given the filtration conditions and design of a filter, what volume of filtrate or cake will be obtained in a given length of time?
- What volume of wash water can be passed through a cake in a definite length of time?
- What will be the relationship between the concentration of recovered material in the wash water and the amount of wash water used?

Complete answers cannot be given to the above questions because filtration remains an art instead of a science. The main variable which determines filtration characteristics is the nature of the solids. Laboratory leaf filter tests serve as a starting point for application of theory. Generally, pilot tests are required under actual filtration conditions to accurately preduct filter operation. Once the characteristics of a filtering operation are found from pilot tests, theory is useful in predicting changes in interpreting filtration data.

Traditionally, filtration data are plotted as straight line graphs. Table 10.1 lists the filtration expressions used to make straight line plots for the four different types of mechanisms. The only data taken are the filtrate volume, V, and time, t. Using Q as the rate of flow, $\frac{dV}{dt}$, and Q_o as the rate of flow at the beginning ($t = 0$) of filtration, k is a constant for the type of filtration. Figure 10.16 shows how k and Q can be determined graphically. The units of k are dependent upon which mechanism is used.

Table 10.1
Expressions for Various Filtration Mechanisms

mechanism	filtration expression	units of k
cake	$\frac{kV}{2} + \frac{1}{Q_o} = \frac{t}{V}$	sec/ft^6
intermediate	$kt = \frac{1}{Q} - \frac{1}{Q_o}$	ft^{-3}
standard	$\frac{kt}{2} - \frac{1}{Q_o} = \frac{t}{V}$	ft^{-3}
complete	$V = Q_o(1 - e^{-kt})$	sec^{-1}

Filtration rate is directly proportional to the applied pressure. Because the flow of liquor through the capillaries is always laminar, flow can be predicted by Poiseuille's equation:

$$\text{v} = \frac{dV}{Adt} = \frac{\Delta p}{\mu\left[\alpha\left(\frac{W}{A}\right) + r\right]} = \frac{\Delta p}{\mu(R_c + r)} \qquad 10.28$$

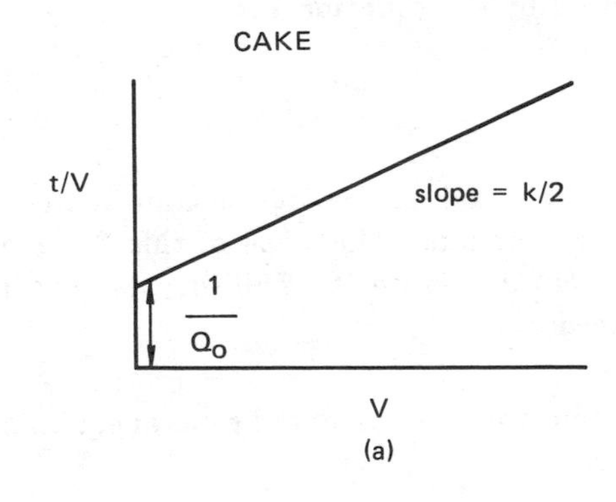

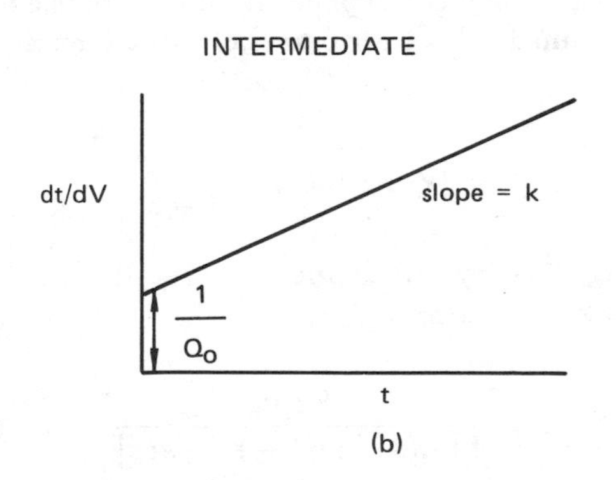

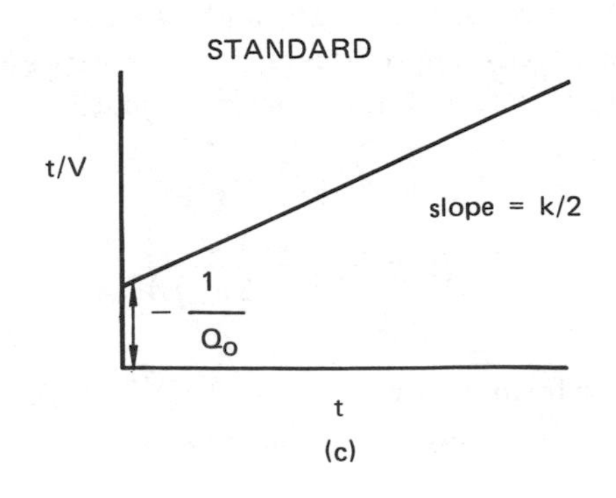

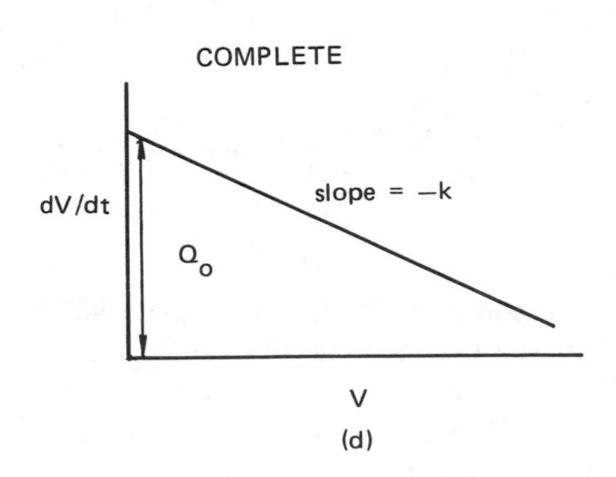

Figure 10.16 Volume Time Plots for Various Filtration Mechanisms

The average specific *cake resistance*, α, is related to the cake compressibility by the empirical relationship

$$\alpha = \alpha_o(\Delta p)^s \quad 10.29$$

The *filter cake resistance* is

$$R_c = \frac{\alpha W}{A} \quad 10.30$$

G. FILTRATION CALCULATIONS

The weight of solids in the cake equals the weight of the solids in the slurry fed to the filter, provided no solid passes into the filtrate. The weight of the solids in the cake is then,

$$W = LA(1-\varepsilon)\rho_s \quad 10.31$$

The weight of the solids in the slurry feed equals the weight fraction of the solids in the slurry, times the weight of the slurry corresponding to the volume of the filtrate obtained, plus the volume of filtrate retained in the cake.

$$W = (V + \varepsilon LA)\,\rho \times \frac{c}{1-c} \quad 10.32$$

Combining equations 10.31 and 10.32 results in an equation for cake thickness:

$$L = \frac{Vc\rho}{A\left[\rho_s(1-c)(1-\varepsilon) - \varepsilon c\rho\right]} \quad 10.33$$

For constant rate filtration, the inverse of Poiseuille's equation is used, and is integrated for constant pressure. The only variables are V and t. The result is equation 10.34.

$$t = \frac{\mu}{\Delta p}\left[\frac{\alpha w}{2}\left(\frac{V}{A}\right)^2 + r\left(\frac{V}{A}\right)\right] \quad 10.34$$

When using equation 10.35, it is assumed that time starts when the first filtrate is obtained, so that $V = 0$ at $t = 0$. To evaluate α and r for a specific pressure drop, data of V versus t is needed. To facilitate the calculations, equation 10.28 can be written as equation 10.35.

$$\frac{dt}{dV} = K_pV + B \quad 10.35$$

The slope and intercept of this line are

$$K_p = \frac{w\alpha\mu}{A^2\Delta p} \quad 10.36$$

$$B = \frac{r\mu}{A\Delta p} \quad 10.37$$

Assume a number of observations have been made of V versus t. For any two successive observations, $\frac{dt}{dV}\left(\frac{\Delta t}{\Delta V}\right)$ can be calculated, where Δt is the time between observations, and ΔV is the increment of filtrate collected over time Δt. Since $\frac{dt}{dV}$ is linear with V, the value of $\frac{\Delta t}{\Delta V}$ is the slope of the line at a point halfway between the values of V that define ΔV (i.e., $\frac{V_1+V_2}{2}$).

Thus, $\frac{\Delta t}{\Delta V}$ is plotted against the arithmetic mean of two successive V readings. The slope of this plot is K_p; the ordinate intercept is B. The points near the early stages of filtration may not fall accurately on the line. Less weight should be given to these points.

The values of α for each Δp can be obtained in the following manner. The specific cake resistance can be represented by the equation

$$\alpha = \alpha_o(\Delta p)^s \quad 10.38$$

Plotting α for each Δp on log-log coordinates will produce a straight line. The slope of this line is the *cake compressibility*. Figure 10.17 illustrates how it can be determined.

For constant rate filtration, the general equation can be integrated:

$$t = \frac{\mu w\alpha}{2\Delta p_c}\left(\frac{V}{A}\right)^2 \quad 10.39$$

Δp_c is the pressure drop through the cake only. This equation is only approximate, since α varies with Δp. However, this is not the working equation. If $\Delta p = \Delta p_c + \Delta p_m$, then,

$$\frac{dt}{dV} = \frac{\mu wV\alpha}{(\Delta p - \Delta p_m)A^2} \quad 10.40$$

The linear form of equation 10.40 can be used to determine the approximate working equation.

$$\Delta p = K_rV + C \quad 10.41$$

The slope and intercept of equation 10.41 are related to the constant flow rate, Q_c.

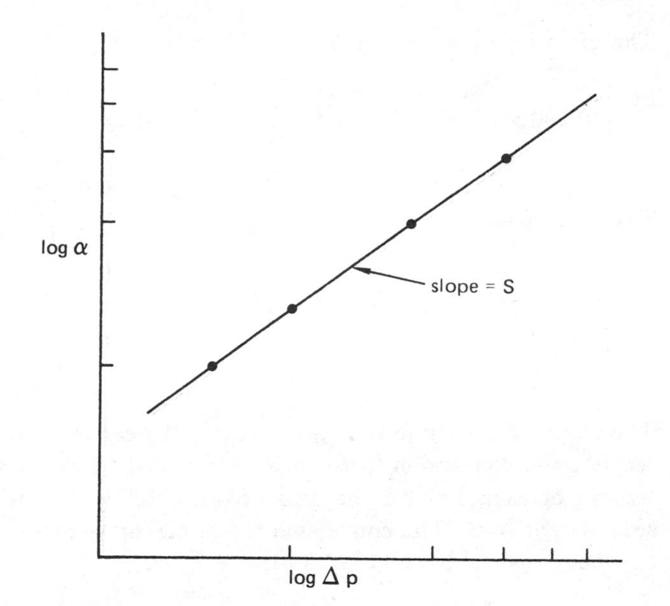

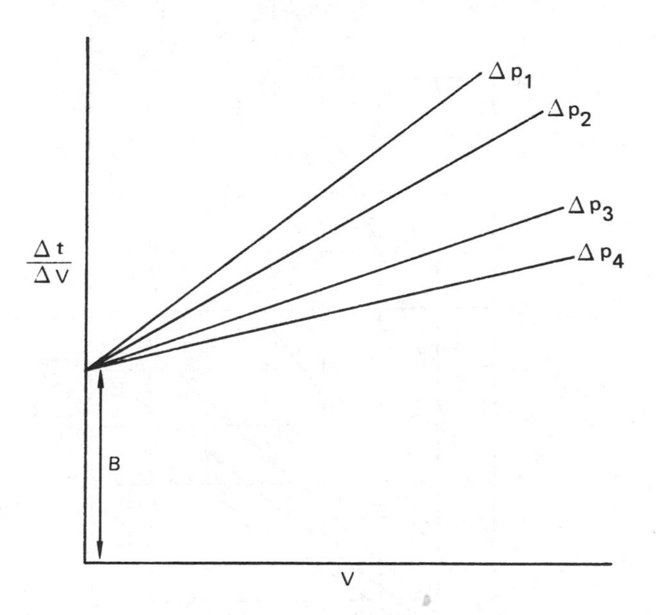

Figure 10.17 Plots Used to Determine Cake Compressibility

$$K_r = \mu \frac{wQ_c}{A^2} \qquad 10.42$$

$$C = \mu r \frac{Q_c}{A} \qquad 10.43$$

Equations 10.43 and 10.44 are only valid if the cake is incompressible. If it is not, the pressure drop through the medium is negligible, and s is known, then the constant rate equation can be rewritten:

$$V = \left(\frac{1}{K_s}\right) \Delta p^{1-s} - \left(\frac{1}{C_s}\right) \Delta p^{-s} \qquad 10.44$$

$$K_s = \mu \frac{wQ_c}{A^2} \qquad 10.45$$

$$C_s = \alpha_0 \frac{w}{rA} \qquad 10.46$$

Figure 10.18 illustrates how the constants can be graphically determined from experimental data.

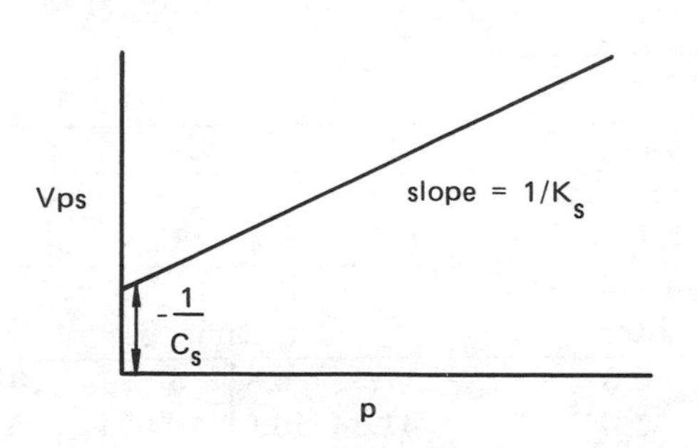

Figure 10.18 Constant Rate Filtration Plots

In a continuous filter, (i.e., typical of a rotary drum), the feed, filtrate, and cake move at steady rates. However, for any particular section of the filter surface, conditions are not steady, but are cyclic. For example, a cycle consists of cake formation, filtrate removal, washing, drying, and cake removal. The conditions are continuously changing. If we consider only the cake formation step (which essentially occurs at constant pressure), the equations for discontinuous constant pressure filtration can be used with modification.

$$\frac{t}{\frac{V}{A}} = \frac{\mu \alpha w}{2\Delta p}\left(\frac{V}{A}\right) + \frac{\mu r}{\Delta p} \qquad 10.47$$

In continuous filtration, the filter medium resistance is negligible in comparison with the cake resistance. Therefore, the $\frac{\mu r}{\Delta p}$ term drops out.

$$t = \frac{\mu \alpha w}{2\Delta p}\left(\frac{V}{A}\right)^2 = ft_\pi \qquad 10.48$$

Solving for the filtrate flow per unit area,

$$\frac{V}{A} = \sqrt{\frac{2\Delta p f t_\pi}{\mu \alpha w}} \qquad 10.49$$

The average filtrate flow per unit area is

$$\frac{Q}{A} = \frac{V}{At_\pi} = \sqrt{\frac{2\Delta p f}{\mu \alpha w t_\pi}} \qquad 10.50$$

Example 10.4

Laboratory filtration tests are conducted with a constant pressure drop using a slurry of $CaCO_3$ in H_2O. The filter area is 440 cm^2, the mass of the solid per unit volume of filtrate is 23.5 g/liter, and the temperature was 25 °C. (a) Determine (in foot, pound force, and second units) the quantities α and R_m as functions of the pressure drop. (b) Obtain an empirical relationship between α and Δp.

test number	I	II	III	IV	V
pressure drop $-\Delta p$ lbf/in^2	6.7	16.2	28.2	36.3	49.1
filtrate volume V, liters	time, sec				
0.5	17.3	6.8	6.3	5.0	4.4
1.0	41.3	19.0	14.0	11.5	9.5
1.5	72.0	34.6	24.2	19.8	16.3
2.0	108.3	53.4	37.0	30.1	24.6
2.5	152.1	76.0	51.7	42.5	34.7
3.0	201.7	102.0	69.0	56.8	46.1
3.5		131.2	88.8	73.0	59.0
4.0		163.0	110.0	91.2	73.6
4.5			134.0	111.0	89.4
5.0			160.0	133.0	107.3
5.5				156.8	
6.0			182.5		

The first step is to prepare plots for each of the five constant-pressure experiments of $\frac{\Delta t}{\Delta V}$ versus V, which is $\frac{(V_1+V_2)}{2}$ of each increment of filtrate volume.

filtrate volume, V (liters)	time t (sec)	Δt	ΔV	$\frac{\Delta t}{\Delta V}$	$\overline{V}$
0	0				
0.5	17.3	17.3	0.5	34.6	0.25
1.0	41.3	24.0	0.5	48.0	0.75
1.5	72.0	30.7	0.5	61.4	1.25
2.0	108.3	36.3	0.5	72.6	1.75
2.5	152.1	43.8	0.5	87.6	2.25
3.0	201.7	49.6	0.5	99.2	2.75

The viscosity of water is 0.886 centipoise, or

$$0.886 \times 6.72 \times 10^{-4} = 5.95 \times 10^{-4} \text{ lb/ft-sec}$$

The filter area is

$$\frac{440}{30.48^2} = 0.474 \text{ ft}^2$$

The slope of each line is K_p, in sec/l^2. To convert to sec/ft^6, the conversion factor is $28.31^2 = 801$. The intercept of each line on the axis of ordinates is B, in seconds per liter. The conversion factor to convert this to seconds per cubic foot is 28.31.

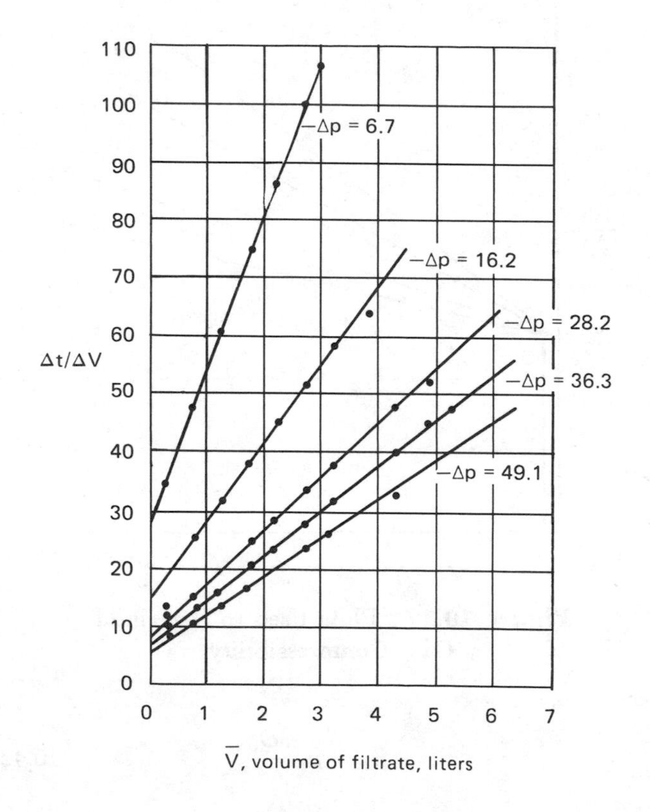

test	pressure drop $-\Delta p$ lbf/in^2	pressure drop $-\Delta p$ lbf/ft^2	slope K_p sec/liter^2	slope K_p sec/ft^6	intercept B sec/liter	intercept B sec/ft^3	R_m ft^3 $\times 10^{10}$	α, ft/lb $\times 10^{11}$
I	6.7	965	25.8	20,700	28.5	807	1.99	1.65
II	16.2	2330	13.9	11,150	13.2	374	2.23	2.15
III	28.2	4060	9.3	7450	8.5	241	2.50	2.50
IV	36.3	5230	7.7	6170	7.0	198	2.65	2.67
V	49.1	7070	6.4	5130	5.5	156	2.82	3.00

The concentration c is

$$\frac{(23.5)(28.31)}{454} = 1.47\ \text{lb/ft}^3$$

From the values of K_p and B, corresponding quantities for α and R_m are found.

$$\begin{aligned}\alpha &= \frac{A^2(-\Delta p)g_cK_p}{c\mu} = \frac{(0.474^2)(32.17)(-\Delta p)K_p}{(5.95\times10^{-4})(1.47)}\\ &= 8.28\times10^3(-\Delta p)K_p\\ R_m &= \frac{A(-\Delta p)g_cB}{\mu} = \frac{(0.474)(32.17)(-\Delta p)B}{5.95\times10^{-4}}\\ &= 2.56\times10^{-4}(-\Delta p)B\end{aligned}$$

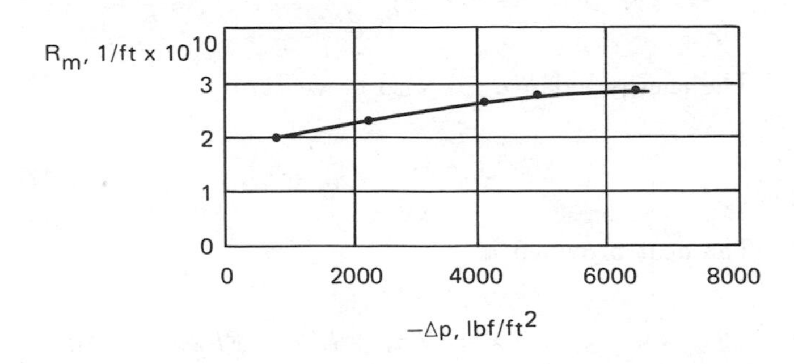

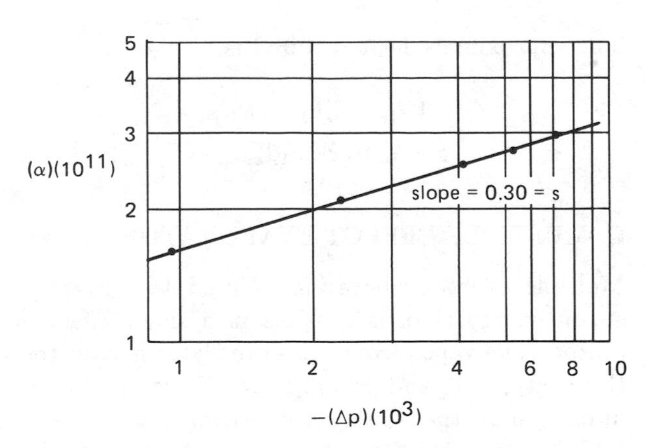

$$\alpha' = 2.08\times10^{10}$$

$$\alpha = 2.08\times10^{10}(-\Delta p)^{0.3}$$

7 EVAPORATION

A. EVAPORATORS

An *evaporator* is a vessel in which liquid is boiled to concentrate any solids that are in solution. Generally, the term *evaporation* implies water is the evaporated phase, but it could apply to any liquid (e.g., a hydrocarbon). In contrast to distillation, no attempt is made to separate components that are in the vapor. The heating medium for evaporation is usually steam at 5 to 50 psig, utilizing heat exchanger tubing as the surface area. An evaporator may be *single* or *multiple effect*.

Table 10.2 lists the types of evaporators that are generally used, and their typical configurations.

Table 10.2
Types of Evaporators

type and tube design	heat exchange bundles	tubes inside	tubes outside
horizontal tube	inside evaporator	steam	solution
standard vertical	inside evaporator	solution	steam
vertical basket	inside evaporator	solution	steam
vertical long tube	inside evaporator	solution	steam
vertical forced	inside evaporator	solution	steam
horizontal forced	outside evaporator	solution	steam

Figure 10.19 shows a standard vertical tube evaporator and its temperature profile along the length of the tube.

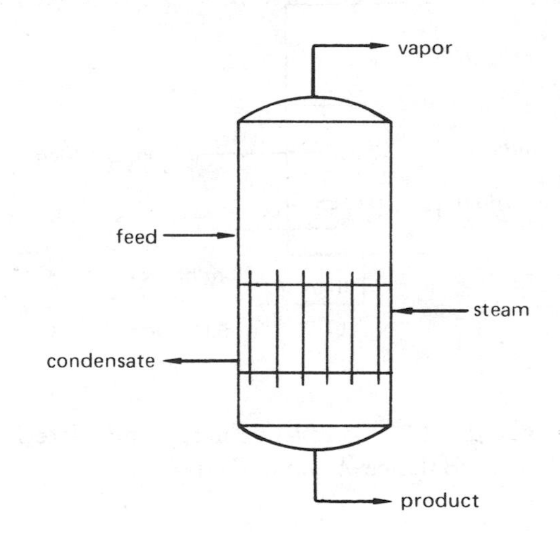

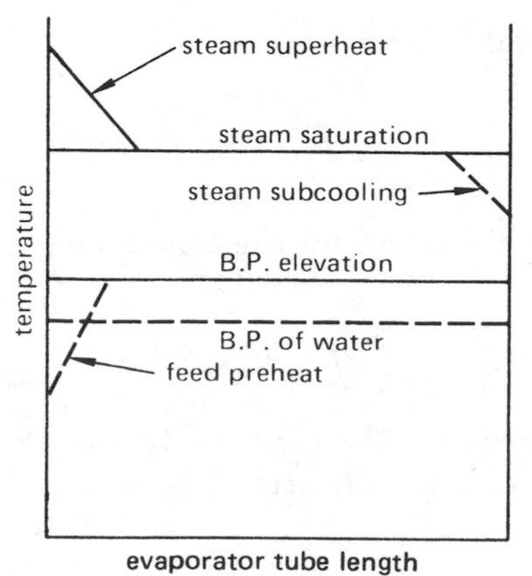

Figure 10.19 Vertical Tube Evaporator and Temperature Profile

B. EVAPORATOR CALCULATIONS

If the heat exchange in an evaporator takes place in the heat exchanger tubing, and there are no heat losses from the evaporator vessel, the generalized heat exchanger equation can be used to calculate the energy balance.

$$Q = UA\Delta T \qquad 10.51$$

ΔT is the temperature difference between the saturated steam outside the tube and the elevated boiling point of the solution. For first approximations, it is allowable to ignore the effect of heating the feed to the final solution temperature. The complete energy and material balance around the evaporator is shown in figure 10.20

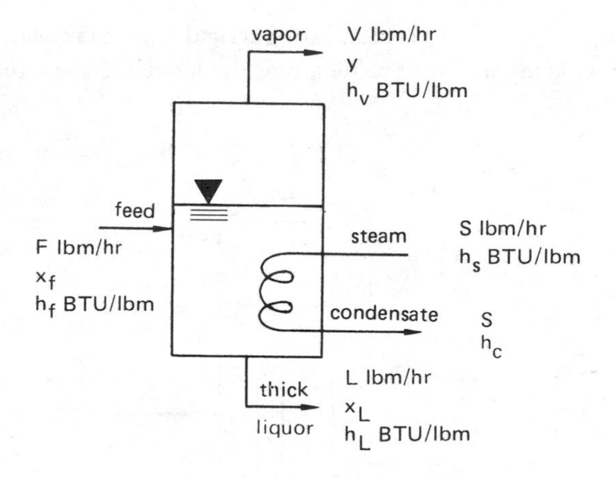

Figure 10.20 Complete Energy and Material Balance Around Evaporator

The material balance is

$$F = V + L \qquad 10.52$$

The material balance for the solute alone is

$$Fx_f = Lx_L + Vy \qquad 10.53$$

In steam systems, the mole fraction of the product in the vapor, y, is usually zero. The material balance becomes

$$V = F\left(1 - \frac{x_f}{x_L}\right) \qquad 10.54$$

When writing the heat balance, some simplifying assumptions can be made:

- superheat in steam is negligible
- subcooling of steam condensate is negligible
- heat loss by radiation is negligible
- heat of concentration is negligible
- heat of crystallation is not negligible

The heat balance becomes

$$Fh_f + Sh_s = Vh_v + Lh_L + Sh_c \qquad 10.55$$

The latent heat of vaporization of the steam is

$$\lambda_s = h_s - h_c \qquad 10.56$$

The energy balance can also be written

$$Fh_f + S\lambda_s = Vh_v + Lh_L \qquad 10.57$$

The heat provided is

$$Q = S\lambda_s = Vh_v + Lh_L - Fh_f \qquad 10.58$$

The approximate heat required is

$$\begin{aligned} Q &= Vh_v + Lh_L - Fh_f \\ &= Vh_v + Fc_p\left(T_{evap} - T_{feed}\right) \end{aligned} \qquad 10.59$$

C. MULTIPLE EFFECT EVAPORATORS

Multiple effect evaporation is used to extract more steam energy than is possible in a single effect evaporator. The vapor from one evaporator is used to heat the next effect, and so on. The effects are commonly numbered in the direction of steam flow. The liquid feed flow can be either forward or backward. Figure 10.21 shows a triple effect evaporator.

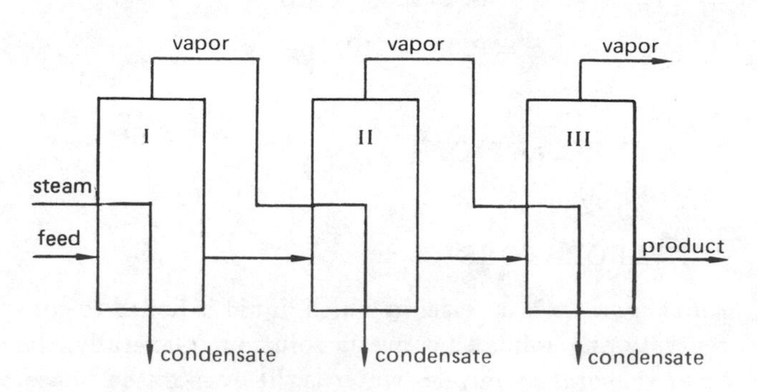

Figure 10.21 Schematic of a Triple Effect Evaporator

The performance of a triple effect evaporator is given by equations 10.60 through 10.65. Performance equations for other numbers of effects are similar.

$$Q_{total} = Q_1 + Q_2 + Q_3 \quad 10.60$$

$$U_{avg}(A)(\Delta T) = U_1 A_1 \Delta T_1 + U_2 A_2 \Delta T_2 + U_3 A_3 \Delta T_3 \quad 10.61$$

$$Q_1 = Q_2 = Q_3 \quad 10.62$$

$$A_1 = A_2 = A_3 \quad 10.63$$

$$S_o = V_1 = V_2 = V_3 \quad 10.64$$

The total water evaporated from the solution is approximately

$$W = S_o \text{ (number of effects)} \quad 10.65$$

D. EFFECT OF BOILING POINT ELEVATION ON EVAPORATORS

The effect of dissolved salt is to increase the boiling point above the boiling point of pure water. This is called *boiling point elevation* (BPE). The boiling point of strong solutions can be found using an empirical rule known as *Duhring's rule*: the boiling point of a given solution is a linear function of the boiling point of water.

If the boiling point of a given solution is plotted against that of water at the same pressure, the result will be a straight line. Different lines are obtained for different concentrations. Over wide pressure ranges (encountered with multiple effect evaporators), the lines are nearly straight, but not necessarily parallel. Figure 10.22 illustrates a generalized plot according to Duhring's rule.

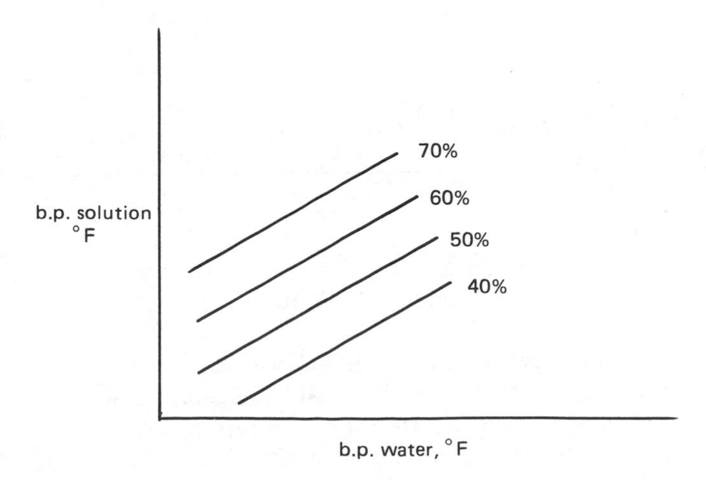

Figure 10.22 Generalized Plot of Duhring's Rule

In an evaporator, the vapor is at the solution temperature, and is superheated by the amount of the BPE. This superheat represents a very small fraction of the total applied heat, since the heat capacity of the superheated steam is about 0.5 BTU/lbm. Since this energy essentially dissipates in the steam chest, the chest will be at the temperature corresponding to the pressure of the previous effect, rather than the temperature corresponding to its own pressure.

The temperature drop across any effect is calculated on the basis of saturated steam at the pressure of the chest. This means that the BPE in any effect is not part of the total available temperature drop. The loss occurs in every effect, and results in capacity decrease of the system. Thus, if T_i is the steam temperature entering effect i, and T_f is the temperature of the vapor leaving the last effect, the overall temperature drop for n effects

$$\Delta T_{overall} = T_i - T_f - [(BPE)_1 + (BPE)_2 \ldots + (BPE)_n] \quad 10.66$$

For estimation, if the overall heat transfer coefficient is given for each effect, the temperature drop can be estimated by assuming that the areas of each effect are equal.

$$\frac{\Delta T_{n-1}}{\Delta T_n} = \frac{U_n}{U_{n-1}} \quad 10.67$$

$$\Delta T_1 + \Delta T_2 + \Delta T_3 + \ldots + \Delta T_n = \Delta T_{overall} \quad 10.68$$

The temperature drop across effect 1 is

$$\Delta T_1 = \frac{\Delta T_{overall}}{1 + \frac{U_1}{U_2} + \frac{U_1}{U_3} + \frac{U_1}{U_4} + \ldots + \frac{U_1}{U_n}} \quad 10.69$$

In general, the temperature drop across the jth effect is

$$\Delta T_j = \frac{\Delta T_{overall}}{U_j \sum_{i=1}^{n} \left(\frac{1}{U_i}\right)} \quad 10.70$$

Example 10.5

A continuous, horizontal tube evaporator is fed with a 5% caustic soda solution at 65 °F. The product is continuously discharged at 230 °F. The product contains 30 pounds of caustic soda per 100 pounds of water. The heat capacity of caustic soda is

$$c_p = 1 - 0.32\sqrt{\%\text{ caustic soda}}\ \text{(BTU/lbm)}$$

The overall heat transfer coefficient is 300 BTU/hr-ft^2-°F. Steam is available at 30 psig. How many square feet of surface area are needed per 100 pounds of dilute liquor?

The basis is 100 lbm feed/hr $= F$

The water evaporated is $V = F\left(1 - \frac{x_f}{x_L}\right)$

$$x_L = \frac{30}{(100+30)} = 0.231$$

$$V = 100\left(1 - \frac{0.05}{0.231}\right) = 78.3\ \text{lbm/hr}$$

The heat balance is

$$Q = V\lambda_v + Fc_p(T_{evap} - T_f)$$

From steam tables

$$\lambda_v = 958.8\ \text{BTU/lbm at } 230°\text{F}$$

$$c_p = 1 - 0.32\sqrt{5} = 0.929$$

$$Q = 78.3(958.8) + 100(0.929)(230 - 65) = 90{,}300\ \text{BTU/hr}$$

The area calculation is based on

$$A = \frac{Q}{U\Delta T}$$

From steam tables,

$$T_{sat}\ 44.7\ \text{psia} = 274\ °\text{F}$$

$$T = 274 - 230 = 44\ °\text{F}$$

$$A = \frac{90{,}300}{(300)(44)} = 6.84\ \text{ft}^2$$

8 CRYSTALLIZATION

A. THEORY

Crystallization of a solute from a solution can be accomplished by any of the following processes:

- cooling
- evaporation of the solvent
- adiabatic evaporation
- "salting out" (adding a substance that reduces the solubility of the solute)

Crystallization is the formation of solid particles from a homogeneous phase. It can occur from a gas phase (as snow), or from a liquid phase (freezing or precipitation). Although a crystal will be relatively pure, it can form from a relatively impure mother liquor. Therefore, good separations can be made with a crystallization process. When the crystals are removed in the final magma, mother liquor may be included in the solid mass which may contaminate the pure crystals. Mother liquor contamination can be removed by washing, followed by filtration, or centrifuging.

When crystallization is used to purify a component, the process is called *fractional crystallization.* This is accomplished by progressively cooling a saturated solution until crystallization takes place. In most cases, solute and solvent are insoluble in the solid phase.

For binary solutions a solid-liquid equilibrium diagram can be constructed as in figure 10.23. In figure 10.23, M is the melting point of component B; and P is the melting point of A. The curves MFU and PDU are the solubilities of these components in the liquid solution. A liquid solution at C, if cooled, precipitates pure A at the temperature corresponding to D. As the temperature is lowered, additional pure A crystallizes, while the liquid composition moves along the the curve DU towards U.

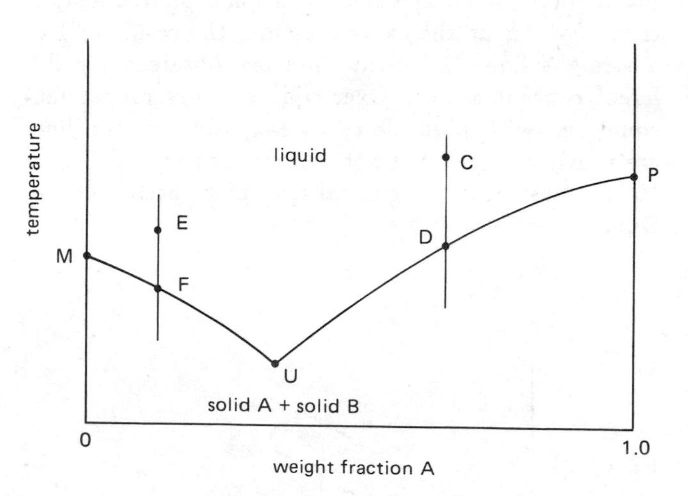

Figure 10.23 Generalized Binary Liquid-Solid Equilibrium

At a temperature, T_u, the liquid remaining is a *eutectic mixture*, and further cooling results in complete solidification of a mixed crystal called the *eutectic solid.* If the liquid solution was originally at E and cooled to F, the solid precipitating would be pure B. Further cooling at this condition will result in pure B until the eutectic is reached again.

Crystallization from a solution is the creation of a new phase within a homogenous mixture. Crystallization usually occurs in two steps: *nucleation* and *crystal growth*. The driving force for these two steps is supersaturation. Neither nucleation nor crystal growth can take place in saturated or undersaturated solutions.

A *supersaturated solution* is one in which the solution contains more solute than it can hold at equilibrium solubility for a particular temperature. The system readjusts by forming minute crystalline nucleii (which serve as the basis for crystal growth) until the system readjusts to equilibrium solubility. The degree of supersaturation of the solution will be greater for large crystals than for small crystals, since the small crystal nucleii have a higher solubility.

Supersaturation can be obtained in three ways. (1) If the solubility of the solute increases with temperature, as is the case with many common inorganic salts, supersaturation can occur when the solution is cooled. (2) If the solubility is relatively independent of temperature, such as with NaCl, supersaturation can be produced by evaporating a portion of the solvent. (3) If neither cooling nor evaporation is desirable, supersaturation may be generated by adding a third component and causing precipitation. The third component may act physically with the original solvent, forming a mixed solvent in which the solute solubility is greatly reduced. This process is called *salting*. If complete precipitation is required, the third substance may act chemically with the solute to form an insoluble substance. In either case, the addition of a third component can rapidly cause high supersaturations.

B. YIELD CALCULATIONS

Yields from a crystallizer are calculated from a material balance based on the solute content in the original and final solutions. The material balances are simple, with few complications arising from evaporation of the solvent and *solvated crystals* (water of hydration).

The yield and saturation are calculated as:

$$\% \text{ yield} = \frac{\text{weight of crystals} \times 100}{\text{total weight of solute actually present}} \qquad 10.71$$

$$\% \text{ saturation} = \frac{\text{actual weight of solute} \times 100}{\text{weight of solute at saturation temperature}} \qquad 10.72$$

The theoretical yield from any crystallizer (valid for both hydrated or anhydrous crystals) is

$$C = R\left[\frac{100W_o - S(H_o - E)}{100 - S(R-1)}\right] \qquad 10.73$$

The yield from a vacuum crystallizer is more difficult to determine since it operates adiabatically. The heat released by the solution on cooling to the equilibrium temperature, and the heat of crystallization, are available for vaporizing the water in solution. The thermal effects must balance.

$$E = \frac{(W_o + H_o)(c_p)\Delta T[100 - S(R-1)] + q_c R(100W_o - SH_o)}{L_v[100 - S(R-1)] - q_c RS} \qquad 10.74$$

If E is known, equation 10.71 can be used to calculate the yield.

Example 10.6

100 pounds of water are saturated at 100 °C with 42.5 pounds of anhydrous sodium sulfate. The solution is cooled to 0 °C, with an evaporation loss of 5 pounds, where the solubility for the decahydrate is 12.2 g/100 g of water. How much decahydrate is formed?

Molecular weights:

Na_2SO_4 142.1 lbm/mole
$Na_2SO \cdot 10\ H_2O$ 322.2 lbm/lbmole

lbm decahydrate = lbm anhydrous (decahydrate/anhydrous)

$96.4 = 42.5\left(\frac{322.5}{142.1}\right)$

water needed for hydration = 96.4 − 42.5 = 53.9 lbm

water as solvent = (100 − 5) − 53.9 = 41.1 lbm

decahydrate left in mother liquor = 0.122(41.1) = 5 lbm

yield = 96.4 − 5.0 = 91.4 lbm

% yield - 91.4 $\left(\frac{100}{96.4}\right)$ = 94%

Example 10.7

A 30% solution of Na_2CO_3 weighing 10,000 pounds is slowly cooled to 20 °C. The crystals formed are sal-soda ($Na_2CO_3 \cdot 10\ H_2O$). The solubility of Na_2CO_3 at 20 °C is 21.5 parts of anhydrous salt per 100 parts water. During cooling, 3% of the weight of the original solution is lost by evaporation. (a) What is the weight of the original solution lost by evaporation? (b) What is the weight of $Na_2CO_3 \cdot 10H_2O$ formed?

Molecular weights:

$$Na_2CO_3 = 106$$

$$Na_2CO_3 \cdot 10H_2O = 286.2$$

$$R = \frac{286.2}{106} = 2.7$$

$$w_o = 0.3(10,000) = 3000 \text{ lbm}$$
$$H_o = 10,000 - 3000 = 7000 \text{ lbm}$$
$$E = 0.03(10,000) = 300$$
$$C = 2.7\frac{[100(3000) - 21.5(7000 - 300)]}{[100 - 21.5(2.7 - 1)]}$$
$$= 6636 \text{ lbm}$$

Example 10.8

How much solution containing 620 lbm $CaCl_2$/1000 lbm H_2O is needed to dissolve 250 lbm of $CaCl_2 \cdot 6H_2O$? The solubility at 25 °C is 7.38 lbmoles of $CaCl_2$/1000 lbm H_2O.

Molecular weights:

$$CaCl_2 = 111$$
$$CaCl_2 \cdot 6H_2O = 219$$

At saturation:

$CaCl_2 \cdot H_2O$ requires 7.38(6)(18)

$$= \frac{797 \text{ lbm } H_2O}{1000 \text{ lbm } H_2O \text{ solvent}}$$

The rest of the water acts as solvent.

Original solution:

$$\frac{620}{111} = 5.59 \text{ moles } CaCl_2/1000 \text{ lbm } H_2O$$

$$\begin{aligned}\text{lbm solution} &= \text{moles hexahydrate} \\ &\times \text{lbm solvent/mole hexahydrate} \\ &\times \text{lbm solution/lbm solvent}\end{aligned}$$

$$210 \text{ lbm} = \frac{250}{219}\left(\frac{1000 - 797}{7.38 - 5.59}\right)\left(\frac{1000 + 620}{1000}\right)$$

9 CENTRIFUGATION

Centrifugation is similar to filtration, except that centrifugal force is used instead of a pressure gradient (or gravity) to separate solid from a liquid. Centrifuges are also used to separate emulsions of two immiscible liquids. The centrifugal acceleration used is many times that of gravity.

Centrifugal force, commonly expressed as multiples of gravity force, varies with rotational speed and with the radial distance from the center of rotation. At the wall of a bowl of diameter d''_B, rotating at n revolutions per minute, the centrifugal force F_c, in multiples of gravity is

$$F_c = 1.42 \times 10^{-5} n^2 d_B \qquad 10.75$$

Solid particles settling through a liquid in a centrifugal force field are subjected to an increasing force as they travel away from the axis of rotation. The particles never reach terminal velocity, and they continue to accelerate toward the bowl wall. At any given radial distance, r feet, the *settling velocity*, v_t ft/sec, is given by the *Stokes equation*.

$$v_t = 1.09 \times 10^{-2} n^2 r(\rho_p - \rho)\frac{Dp^2}{18\mu} \qquad 10.76$$

If Stokes settling within a dilute suspension of uniform particles occurs in a tubular bowl of radius r feet, containing a thin layer of liquid of thickness s feet, the flow rate through the centrifuge at which half the solid particles will be removed is

$$Q_c = 1.097 \times 10^{-2} Dp^2 V r n^2 \frac{\rho_p - \rho}{9\mu s} \qquad 10.77$$

With a given flow rate, Q, the *critical diameter*, or *cut point*, D_{pc}, is found by rearranging equation 10.77.

$$D_{pc} = \sqrt{\frac{9\mu s}{1.097 \times 10^{-2} V r n^2(\rho_p - \rho)}} \qquad 10.78$$

Most particles larger than D_{pc} will be eliminated by the centrifuge, while most of the particles smaller than D_{pc} will appear in the effluent. Particles with diameter D_{pc} will be divided equally between the effluent and settled solids phase. When the space for sedimentation in the centrifuge bowl is not cylindrical, or the liquid layer is thick, the appropriate averaged values for r and s are required for equations 10.75 and 10.76. The variables involved in centrifugation are illustrated in figure 10.24.

A centrifuge should be selected over a filter when continuous operation is desired, and the filter feed is at a temperature near its boiling point. This condition precludes the use of a rotary vacuum drum filter. A centrifuge also produces a drier cake than a filter.

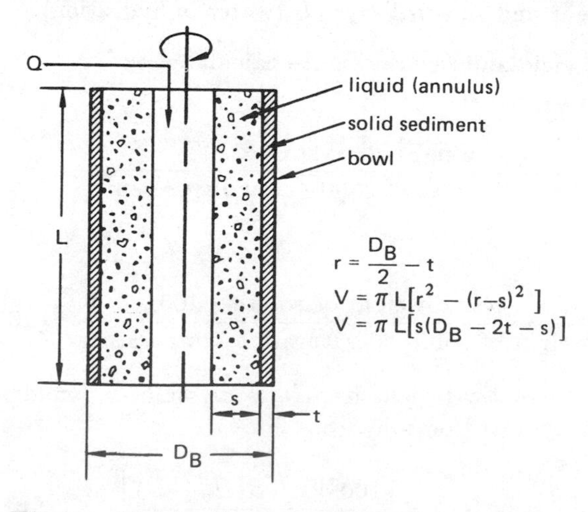

Figure 10.24 Common Centrifuge Variables

10 SEDIMENTATION AND SCREENING

Sedimentation is the removal of suspended solid particles from a liquid stream by gravitational settling. Sedimentation operations may be divided into *thickening* and *clarification*. The primary purpose of thickening is to increase the concentration of the feed stream. The purpose of clarification is to remove the solids from a relatively dilute stream.

Sedimentation can also be considered to perform the function of separation, because particle size affects settling velocity. *Stokes law* of settling is derived for laminar flow region of a spherical particle moving through a liquid. Under such conditions, the entire resistance to movement is caused by the internal friction of the fluid and particle. Inertial forces are negligible. It is generally useful to use Stokes law for first approximations in all settling problems. However, corrections have to be made because of other geometric shapes, settling in transition or turbulent regions, and *hindered settling* (rather than *free settling*).

The *terminal velocity* according to Stokes law is

$$v_t = \frac{D^2(\rho_s - \rho)g}{18\mu} \qquad 10.79$$

The above equation is valid only for Reynold's numbers less than 0.3.

Example 10.9

Spheres having diameters of 0.005 cm and densities of 1.2 gm/cc are settling in water ($\mu = 0.01$ poise) under free settling conditions. The particles do not interfere with each other. What will be the free settling velocity?

Stokes law:

$$v_t = \frac{(0.005)^2(1.2 - 1)(981)}{(18)(0.01)} = 0.027 \text{ cm/sec}$$

Reynolds number:

$$\frac{Dv_t\rho}{\mu} = \frac{(0.005)(0.027)(1)}{0.01} = 0.135 < 0.3 \text{ (o.k.)}$$

Screening is the simplest method of obtaining size separation if the particles are of relatively large diameter. In ore processing plants, screens may be used for particles down to the 2 or 3 mesh size.

PRACTICE PROBLEMS

1. A 1000 pound mixture of 20% sand and 80% KCl is to be heap leached at constant temperature with 1000 pounds of a 5% solution of KCl in water. The pile is drained to a moisture content of 1 pound solution per pound dry solids. A second leach is made with an amount of pure water equal to the weight of drained solids. The remaining solids are drained and dried. The solubility at the temperature of the leaching is 32.0% KCl. What is the weight and composition of the drained, dried solids after the second leach?

2. In problem 1, what is the weight and composition of the collective drained solutions?

3. Water containing nicotine (1%) is extracted with kerosene. (Water and kerosene are essentially insoluble.) Determine the percent extraction of nicotine if 100 pounds of feed solution are extracted with 150 pounds of kerosene. The equilibrium equation is $y = 0.91x$, where y = pound nicotine/pound kerosene, and x = pound of nicotine/pound water.

4. A solid B contains a soluble component, A, (mass fractions $x_A = 0.3$, and $x_B = 0.7$) and is to be treated to recover A by solvent extraction with C. Solid B and solvent C are mutually and totally insoluble. The extracted solid is to be screw pressed to a 1 pound solution/pound B in the under flow. The entrainment of B in the overflow can be ignored. Calculate the pounds of solvent C (A-free basis) that must be fed, per pound of $A + B$ feed solid, to obtain 90% of the A in the extract overflow. Calculate the concentration of A in the extract overflow solution.

5. A countercurrent extraction battery is used to extract NaOH from a feed of NaOH and $CaCO_3$ produced by the causticization of lime. The reaction is carried out not with stoichiometric amount of water, but with 4 pounds of H_2O/lbm $CaCO_3$. The final extract solution is to contain 10% NaOH on recovery of 98% of the NaOH. How much wash water is required, and how many theoretical stages will be required? The underflow can be assumed constant at 2 pounds solution/pound $CaCO_3$. If the operation of the extraction units could be improved so that the underflow contained only 1 pound solution/pound $CaCO_3$, how many stages will be required?

6. The triple effect evaporator shown is operated with steam at 260 °F, quality 93%, and at a rate of 20,000 lbm/hr. The pressure in effect III is 0.8153 psia. The boiling point elevations are 12 °F, 6 °F, and 3 °F for effects I, II, III, respectively. The corrected heat transfer coefficients for each effect are $U_1 = 500$, $U_2 = 400$, $U_3 = 300$ BTU/hr-ft^2-°F, and for the preheater, $U_p = 200$. The feed contains 6% dissolved solids, has a heat capacity of $c_p = 0.8$ BTU/lbm-°F, and enters the preheater at 50 °F. The feed enters effect III at 80 °F. Using reasonable assumptions, (a) estimate the heating area of each effect in ft^2, (b) the temperature in the chest and body of each effect, and (c) the feed rate in lbm/hr.

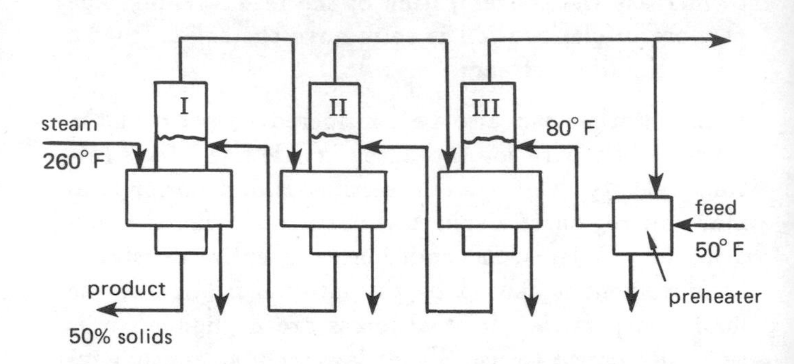

7. A double effect evaporator is to be designed to produce 100 tons of anhydrous NaOH per 22-hour day as a 50% solution containing 1% NaCl. The cell liquor feed contains 10.5% NaOH and 15.5% NaCl. A centrifuge associated with the evaporator will discharge crystalline NaCl with a moisture content of 5% when it receives the crystals from the evaporator in a mother liquor containing 20% NaOH. Assume there is no washing of salt crystals and no miscellaneous losses. All percentages are on weight basis. Prepare a material balance for the system on an hourly basis, showing the total cell liquor feed, the evaporation, and the crystalline NaCl discharge, all in pounds per hour.

8. It is desired to use a thermal recompressor for the first effect of a multiple effect evaporator. High pressure steam is available at 120 psia. The thermal recompression equipment has the facility to entrain 1.0 pounds of low pressure steam per pound of motive steam at a pressure of 120 psia. The first effect evaporator operates at atmospheric pressure, and receives a feed of 100 gpm of an aqueous solution having a specific gravity of 1.05 and a temperature of 185 °F. Evaporation is 14,000 lbm/hr of water. Assume a specific heat of 1.00 and zero boiling point elevation. There is no subcooling of the condensate. (a) Determine the steam required in lbm/hr. (b) Describe the other type of recompression required.

9. How long will it take for the spherical particles listed to settle under free settling conditions through 5′ of water at 70°F?

substance	specific gravity	diameter, inches
galena	7.5	0.01
galena	7.5	0.0001
quartz	2.65	0.01
quartz	2.65	0.001

10. What is the capacity (in gpm) of a cylindrical clarifying centrifuge operating at 1000 rpm, having a bowl diameter of 24″, liquid thickness of 3″, bowl depth of 16″, and with no sedimentation thickness? The liquid has a viscosity of 3 centipoise, and a specific gravity of 1.3. The solid has a specific gravity of 1.6. The required cut size is 30 microns.

11. A small filter press is being designed to remove suspended solids from human plasma. One of these designs is an experimental unit having a cross section of 3 square inches. The unit is tested by passing the plasma through the press with a constant pressure head of 2 atmospheres. Data for the tests are shown. Clinical applications require that a constant plasma filtration rate be achieved. The extreme clinical case is a rate of 0.1 ml/sec. What is the required pressure for clinical applications?

t, minutes	V, ml
2.0	27
7.5	66
11.7	78
14.6	100

12. A crystalline product is to be filtered, washed, and dewatered in a rotary drum vacuum filter. The slurry contains 0.01 lbm of crystals/lbm slurry. The filter and cake have the following characteristics:

drum: 4′ diameter × 6′ long, 40° submergence
rotation: 1 rpm
pressure: 2.4″ Hg vacuum
cake: thickness 2″
incompressible
density to 30 lbm/ft^3 (dry basis)

It is proposed to increase production by increasing the rotation to 2 rpm. Provided that the vacuum can be maintained and the filter medium has negligible resistance, what percent increase in production can be realized?

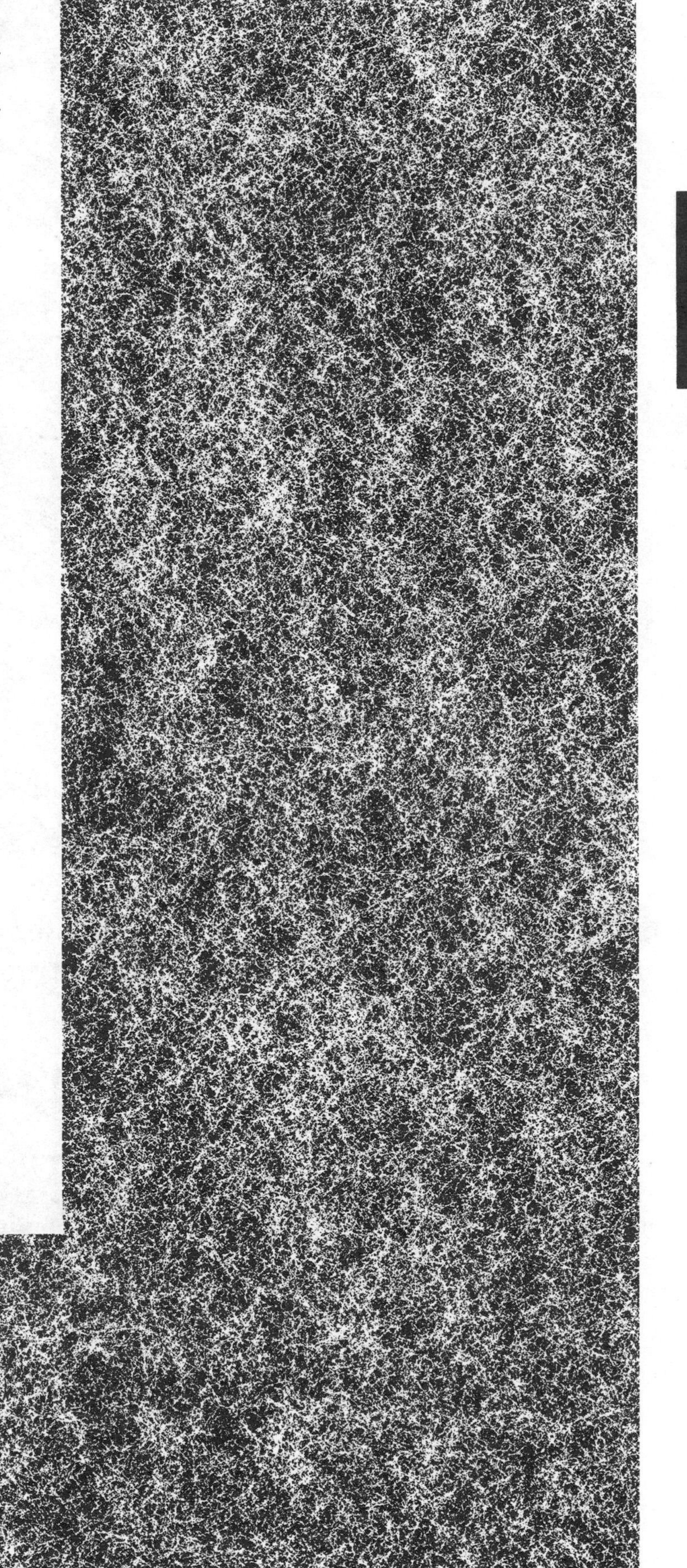

PROCESSES

11 KINETICS

KINETICS

Nomenclature

C	concentration	moles/cm^3, or moles/liter
E	activation energy	cal/moles
F	$F_o(1-x)$, molar flow rate	moles/sec
k	reaction rate constant	(moles/cm^3)$^{1-n}$/sec, or (moles/l)$^{1-n}$/sec
K	equilibrium constant	–
m	mass	grams
n	number of moles, or order of reaction	– –
N	number of equal-size mixed reactors in series, or number of moles	–
r	component reaction rate	moles/cm^3-sec, or moles/l-sec
r'	rate of reaction based on unit mass of catalyst	moles/g-sec
p	partial pressure	atm
R	universal gas law constant $= 1.987$ cal/g-moles-°K $= 1.987$ BTU/lbm-moles-°R $= 82.06$ cm^3-atm/g-moles-°K	
s	space velocity	sec^{-1}
S	surface area	cm^2
t	time	sec
$\bar{t}$	$\frac{V}{v}$, reactor holding time or mean residence time of fluid in a flow reactor	sec
T	temperature	°K
v	volumetric flow rate	cm^3/sec, or l/sec
V	volume	cm^3, or liter
X	fraction of reactant converted into a product	–
y	$\frac{p}{\pi}$, mole fraction in gas	–
z	$\frac{1}{L}$, fractional distance through reactor	

Symbols

ε	fractional volume change on complete conversion	
π	total pressure	atm
ρ	molar density	moles/cm^3, or moles/liter
τ	$\frac{C_{Ao}V}{F_{Ao}} = \frac{V}{v_o}$, space-time	sec

Subscripts

$\frac{1}{2}$	half life
A	component A
b	batch
e	equilibrium
f	leaving, or final
g	gas phase of main gas stream
i	entering, or component i, or inert
l	liquid phase
p	plug flow
o	entering, or reference

1 DEFINITIONS

Catalyst: a material entering into a reaction which is neither a reactant or a product. The presence of a catalyst can either accelerate or hinder the reaction process without being modified itself.

Heterogeneous reaction: a reaction that requires the presence of at least two phases for the reaction to proceed as it does.

Homogeneous reaction: a reaction that takes place in one phase alone.

Kinetics: the study of factors that influence the rate of reaction.

Rate-determining step: the slowest step of a series of reaction steps which comprise the whole reaction.

2 INTRODUCTION TO CHEMICAL REACTORS

Reactors are used in many industries. There are catalytic crackers for oil refining; blast furnaces for iron making; activated sludge ponds for sewage treatment; polymerization tanks for plastics, paints, and fibers; pharmaceutical vats for producing drugs; and fermentation jugs for wine making. To find out what a reactor is able to do, three things must be known: the kinetics, the contacting pattern, and the performance equation.

Kinetics is a study of how fast things happen in the reactor. If a reaction is fast, equilibrium calculations will predict the amounts of each component. If the reaction is not fast, then the rate of chemical reaction, the rate of heat transfer, or the rate of mass transfer will determine the amounts of the components present.

The *contacting pattern* refers to how the materials flow through the contactor, when they mix, and the clumpiness or segregation.

Finally, the most important quantitative relationship is the *performance equation.* The performance equation relates the input to the output for various kinetics and contacting patterns. A generalized form of a performance equation is

$$\text{output} = \mathbf{f}\,(\text{input, kinetics, contacting pattern}) \qquad 11.1$$

The performance equation predicts the effects of changing various parameters, and enables the best design to be selected.

Single-phase systems will usually be a single gas or single liquid flowing through a reactor. Only three idealized contacting patterns are considered simple enough for manual solutions: batch contacting, plug flow, and mixed flow. These three contacting patterns are useful because (a) they have simple performance equations, and (b) they are the most efficient contacting patterns. Other contacting patterns are available (e.g., the recycle reactor or the staged reactor).

3 REACTION RATE

The most useful prediction of reaction velocity (reaction rate) in a single-phase system is the differential equation of equation 11.2[1].

$$-r = \frac{-dN}{V\,dt} = \frac{\text{moles reacting}}{(\text{volume of reactor})(\text{time})} \qquad 11.2$$

Although reaction rates are normally defined in terms of unit volume of fluid, other bases can also be used, such as volume of reactor or mass of the catalyst.

For the common homogeneous reaction, the volume in equation 11.2 is the volume of the fluid. The reaction rates for liquids are usually expressed in terms of concentrations. Since $C = \frac{N}{V}$, the reaction rate expression can be simplified.

$$-r = \frac{-dC}{dt} \qquad 11.3$$

The reaction rate is a function of the properties of the system (i.e., pressure, temperature, composition, etc.). The rate of reaction is influenced by the composition and the energy of the fluid. The energy available to the fluid can be from heat, light, magnetic fields, etc. Ordinarily, only heat energy is considered.

$$-r = \mathbf{f}\,(\text{temperature terms, concentration terms}) \qquad 11.4$$

The rate expression is a function of a temperature-dependent term and a composition-dependent term. Reaction kinetics tries to determine the nature of this rate function. The rate term usually rises exponentially with temperature and often follows the *Arrhenius Law* expression:

$$k = k_o e^{\frac{-E}{RT}} \qquad 11.5$$

[1] Equation 11.2 and similar reaction rate equations which follow are implicitly written for a single component. For example, using equation 11.2, the reaction rate for component A would be $-r_A = \frac{-dN_A}{V\,dt}$

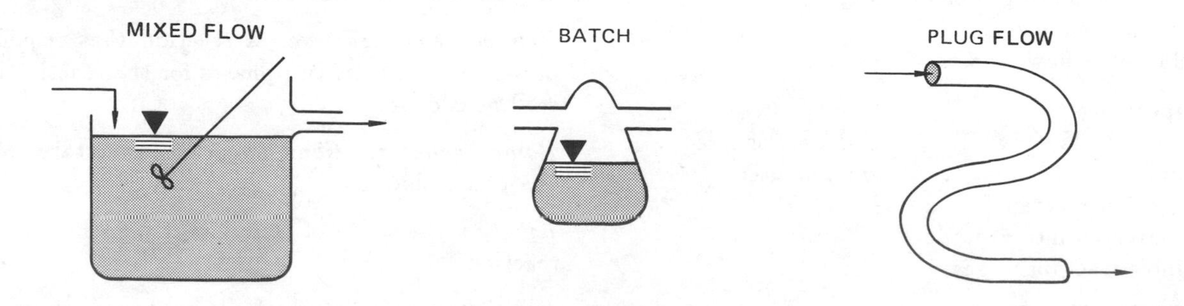

Figure 11.1 Reactor Flow Patterns

KINETICS

The concentration term often fits a simple expression. However, it depends on the type of reaction and the type of reactor. An example of a simple expression would be an elementary reaction of order a:

$$-r = kC^a = k_o e^{\frac{-E}{RT}} C^a \qquad 11.6$$

The product of the temperature-dependent and composition-dependent terms constitutes the rate equation.

4 THE TEMPERATURE-DEPENDENT TERM

The temperature-dependent term in the rate expression will ordinarily be the *reaction rate constant, k*. The rate constant, k, is a constant only at a given temperature, and varies with temperature by Arrhenius' Law. The constant k_o is the *frequency factor*; E is the *activation energy*. If equation 11.5 is made linear, a plot of $ln\ k$ versus $\frac{1}{T}$ yields a straight line. The slope is $\frac{-E}{R}$, and the intercept is $ln\ k_o$.

$$ln\ k = \frac{-E}{RT} + ln\ k_o \qquad 11.7$$

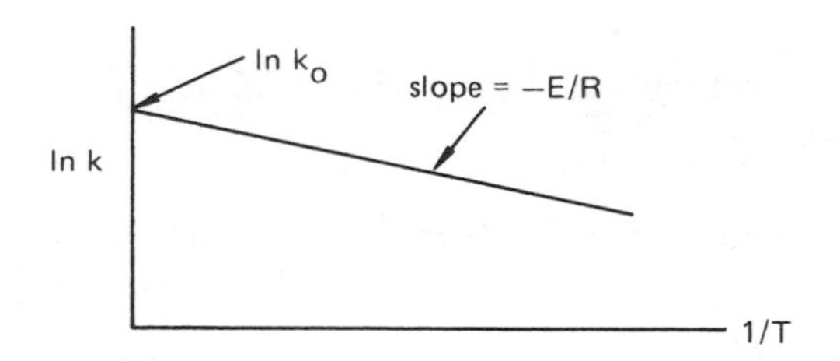

Figure 11.2 Reaction Rate Constant Graph

The Arrhenius' equation can be used to calculate the reaction rate constant at a third temperature when it is given for two different temperatures.

$$ln\ \frac{k_2}{k_1} = \frac{E}{R}\left(\frac{1}{T_1} - \frac{1}{T_2}\right) \qquad 11.8$$

The differential form of the temperature-dependent term is

$$\frac{d\ ln\ k}{dT} = \frac{E}{RT^2} \qquad 11.9$$

A steep slope on the logarithm of k versus $\frac{1}{T}$ plot indicates a high value for E. This implies a temperature sensitive reaction. High E values (50,000 to 100,000 cal/mole) are typical for gas phase reactions which proceed significantly only at high temperatures. Low E values (5000 to 20,000 cal/mole) are typical of enzyme, biochemical, and biological reactions (which occur at room temperatures). Table 11.1 illustrates the temperature dependency of the rate constant as a function of the activation energy and temperature level.

Table 11.1
Relative Reaction Rates as a Function of E and T

temp	E (cal/mole)		
°C	10,000	40,000	70,000
0	10^{48}	10^{24}	1
400	7×10^{52}	10^{43}	2×10^{33}
1000	2×10^{54}	10^{49}	10^{44}

The important characteristics of the temperature-dependent term can be summarized as follows:

- A plot of $ln\ k$ versus $\frac{1}{T}$ is a straight line, with large negative slopes for large values of E and small negative slopes for small values of E.
- Reactions with large E are temperature-sensitive, and reactions with small E are temperature-insensitive.
- A given reaction is more temperature-sensitive at low temperatures than high temperatures.
- The frequency factor, k_o, does not affect the temperature sensitivity of a reaction.

5 KINETIC TERMS

Consider an ideal gas reaction taking place in a constant volume reactor:

$$aA + bB \rightarrow rR + sS \qquad 11.10$$

The change in the number of moles is

$$\Delta n = r + s - (a + b) \qquad 11.11$$

For reactant A, the partial pressure can be computed:

$$p_A = C_A RT = p_{Ao} + \frac{a}{\Delta n}(\pi_o - \pi) \qquad 11.12$$

KINETICS

For product R, the partial pressure can also be computed:

$$p_R = C_R RT = p_{Ro} - \frac{r}{\Delta n}(\pi_o - \pi) \quad 11.13$$

For an ideal gas in a constant pressure reaction (where the reactor volume changes with time), the *fractional volume change*, ε_A, is defined as

$$V = V_o\ (1 + \varepsilon_A X_A) \quad 11.14$$

$$\varepsilon_A = \frac{V_{all\ A\ reacted} - V_{no\ reaction}}{V_{no\ reaction}}$$

$$= \frac{V_{X=1} - V_{X=0}}{V_{X=0}} \quad 11.15$$

For constant density ($\varepsilon_A = 0$), X_A and C_A are related:

$$X_A = 1 - \frac{C_A}{C_{Ao}} \quad 11.16$$

$$dX_A = \frac{-dC_A}{C_{Ao}} \quad 11.17$$

For systems of changing density where the volume of an element changes linearly with conversion ($V = V_o\ (1 + \varepsilon_A X_A)$), C_A and X_A are related:

$$X_A = \frac{C_{Ao} - C_A}{C_{Ao} + \varepsilon_A C_A} \quad 11.18$$

$$\frac{C_A}{C_{Ao}} = \frac{1 - X_A}{1 + \varepsilon_A X_A} \quad 11.19$$

$$dX_A = \frac{-C_{Ao}(1 + \varepsilon_A)}{(C_{Ao} + \varepsilon_A C_A)^2}\ dC_A \quad 11.20$$

$$dC_A = \frac{-C_{Ao}(1 + \varepsilon_A)}{(1 + \varepsilon_A X_A)^2}\ dX_A \quad 11.21$$

Given the general reaction of equation 11.10, for varying density systems,

$$\frac{C_{Ao} X_A}{a} = \frac{C_{Bo} X_B}{b} \quad 11.22$$

For the special case of constant densities,

$$\frac{C_{Ao} - C_A}{a} = \frac{C_{Bo} - C_B}{b} = \frac{C_{Ro} - C_R}{-r} \quad 11.23$$

In general,

$$\frac{a\varepsilon_A}{C_{Ao}} = \frac{b\varepsilon_B}{C_{Bo}} \quad 11.24$$

Example 11.1

Consider a gaseous feed to a reactor with the following initial concentrations:

$$C_{Ao} = 100$$
$$C_{Bo} = 200$$
$$C_{Ro} = 50$$
$$C_{io} = 250$$

(i stands for an inert material)

In the reactor, materials A and B react according to

$$A + 3B \rightarrow R$$

In the exit stream, $C_A = 80$. Find C_B, X_A, and X_B in the exit stream. Assume ideal gas behavior, constant temperature, and constant pressure.

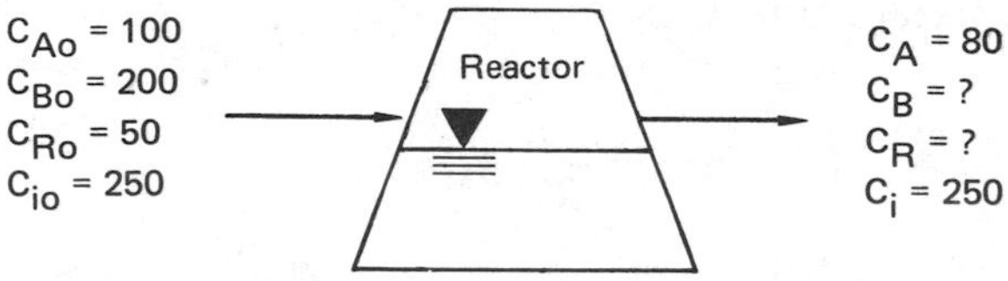

Take 600 volumes of the entering fluid as a basis. At $X_A = 0$,

$$V_o = 100A + 200B + 50R + 250i = 600\ \text{volumes}$$

At $X_A = 1$,

$$V = 0A + (-100)B + (50 + 100)R + 250i$$
$$= 300\ \text{volumes}$$

$$\varepsilon_A = \frac{300 - 600}{600} = \frac{-1}{2}$$

Using equation 11.18,

$$X_A = \frac{100 - 80}{100 - \frac{1}{2}(80)} = \frac{1}{3}$$

Using equation 11.24,

$$\varepsilon_B = \frac{200}{100}\left(\frac{1}{3}\right)\left(\frac{-1}{2}\right) = \frac{-1}{3}$$

Using equation 11.22,

$$X_B = \frac{(3)100\left(\frac{1}{3}\right)}{200} = \frac{1}{2}$$

$$C_B = C_{Bo}\frac{1 - X_B}{1 + \varepsilon_B X_B} = \frac{1 - \frac{1}{2}}{1 - \frac{1}{6}}(200) = 120$$

6 CONCENTRATION-DEPENDENT TERM

The factors which affect the performance of a reactor include the reaction type, molecularity and order of the reaction, reaction reversibility, type of reactor, reaction rate constant,[2] k.

A. TYPES OF REACTIONS

The distinction between reaction types is based on the form and number of kinetic equations used to describe the progress of the reaction. From the stoichiometry or other information in the problem, it is easy to decide whether single or multiple reactions are occurring. Multiple reactions can be classified into four types:

- series reactions: $A \rightarrow B \rightarrow S$
- parallel reactions:

$$A \nearrow S, \quad A \searrow R \qquad \text{or} \qquad \begin{array}{c} A \rightarrow R \\ A \rightarrow S \end{array}$$

- complex reactions:

$$\begin{array}{c} A + B \rightarrow S \\ S + B \rightarrow R \end{array}$$

- autocatalytic reactions: $A + R \rightarrow R + R$

An *elementary reaction* occurs in a single step. The stoichiometric equation generates the rate equation. For example, consider the reaction

$$A + B \rightarrow R \qquad 11.25$$

The rate of reaction in an elementary reaction will be controlled by collisions of single molecules of A and B. The frequency of collisions is proportional to the rate of reaction. At a given temperature, the frequency of collisions is proportional to the concentration of reactants. The rate of disappearance of component A is

$$-r_A = kC_A C_B \qquad 11.26$$

Similar rate equations can be written for other stoichiometric equations:

[2] It is assumed that the temperature of the system is kept constant so that k is constant.

Table 11.2
Rate Equations for Elementary Reactions

stoichiometric equations	rate
$A \rightarrow R$	$-r_A = kC_A$
$2A \rightarrow R$	$-r_A = kC_A^2$
$3A \rightarrow R$	$-r_A = kC_A^3$
$A + B \rightarrow R$	$-r_A = kC_A C_B$
$2A + B \rightarrow R$	$-r_A = kC_A^2 C_B$

Because of complex mechanisms, there is no relationship between stoichiometric and rate equations for *nonelementary reactions*. Information about the rate equation must be determined experimentally. A classic example is the ammonia synthesis reaction which has the form

$$A + 3B \rightarrow 2S \qquad 11.27$$

The rate equation for this complex mechanism is

$$r_S = k_1 \frac{C_A C_B^{1.5}}{C_S^2} - k_2 \frac{C_S}{C_B^{1.5}} \qquad 11.28$$

B. MOLECULARITY AND ORDER OF A REACTION

The *molecularity* of an elementary reaction is the number of molecules involved in the rate-determining step of a reaction. It has values of one, two, or in very rare cases, three.

The *order of a reaction* refers to the power to which the concentration is raised in an empirical rate equation. Consider the rate equation

$$-r_A = kC_A^a C_B^b \ldots C_D^d$$
$$a + b + \ldots + d = n \qquad 11.29$$

$a, b, \ldots, d$ are not necessarily related to the stoichiometric coefficients (except in elementary reactions). This reaction is ath order with respect to A, bth order with respect to B, etc., and nth order overall. The order is found experimentally, and need not be an integer. When elementary reactions are involved, the order follows from the stoichiometry. If it were elementary, the above reaction would have the stoichiometric reaction

$$aA + bB + \ldots + dD \rightarrow \text{products} \qquad 11.30$$

For gases, partial pressures rather than concentrations may be used.

KINETICS

C. REVERSIBLE AND IRREVERSIBLE REACTIONS

Every homogeneous reaction is theroretically reversible. In many cases, however, it can be considered irreversible in the forward direction to simplify problem calculation. This occurs when the equilibrium constant, K, is large (because $k_1 > k_2$). Consider the elementary reversible reaction

$$A + B \longleftrightarrow R + S \qquad 11.31$$

The rate of formation of R is

$$r_R = k_1 C_A C_B \qquad 11.32$$

The rate of disappearance of R is

$$-r_R = k_2 C_R C_S \qquad 11.33$$

At equilibrium there is no net formation of R, so that

$$k_1 C_A C_B = k_2 C_R C_S \qquad 11.34$$

$$K = \frac{k_1}{k_2} = \frac{C_{R_e} C_{S_e}}{C_{A_e} C_{B_e}} \qquad 11.35$$

D. TYPE OF REACTOR

The type of reactor influences how the compositions vary within the region of volume V. This shows up in the integration of the basic differential performance equations.

E. REACTION RATE CONSTANT

The *reaction rate constant* is a function of temperature only. Its units depend on the concentration term. The units of k for an nth order reaction are

$$(\text{time})^{-1}(\text{concentration})^{1-n} \qquad 11.36$$

F. REPRESENTATION OF REACTION RATE

For brevity, elementary reactions are often represented by an equation showing the molecularity and the rate constant. For example, a bimolecular irreversible reaction with second-order rate constant k should be represented as[3]

$$2A \xrightarrow{k} 2B \qquad 11.37$$

The rate equation is

$$-r_A = kC_A^2 \qquad 11.38$$

The condensed rate expression form can be ambiguous. To eliminate any possible confusion, write the stoichiometric equation followed by the complete rate equation. Check to make sure that the units of the rate constant are consistent.

7 DETERMINATION OF RATE EQUATION FROM EXPERIMENTAL DATA

Reaction rate equations are determined experimentally in batch or flow type reactors. Using a batch reactor isothermally is a simple way to obtain homogeneous data. Flow reactors are used primarily for heterogeneous reactions, although sometimes they are used for special homogeneous reactions that are difficult to follow in batch reactors. (A *batch reactor* is simply a container to hold the contents while they react.) The extent of reaction can be determined at various time to give kinetic data.

There are two procedures for analyzing kinetic data: the *integral method* and the *differential method.* The integral method is easy to use and is recommended for testing relatively simple mechanisms. In the integral method of analysis, the form of the rate equation is assumed, and the concentration term (of the assumed form) is plotted against time. If you have assumed the right form, the plot will be a straight line with a slope of k. If the plot is curved, the assumed form is wrong.

In the differential method, the fit of the data to the rate expression is tested (without integration). The concentration of the limiting reactant is plotted against time, and its slope is determined at many points to determine $\frac{dC_A}{dt} = r_A$. A search for a relationship between C_A and r_A is made.

---

[3] It would not be proper to represent the equation as $A \xrightarrow{k} B$. This would imply that the rate equation is $-r_A = kC_A$.

8 PERFORMANCE EQUATIONS FOR BATCH REACTORS

There are three general types of reactors (batch, plug flow, and mixed flow), four reaction orders (zero, first, second, and third), constant and variable volume systems, and reversible and irreversible type reactions, giving a total of 48 possible integrated performance equation forms. All these forms can be generated from six basic performance equations.

To develop the performance equations for the batch reactors, a material balance for the reactor is required. If the fluid is well mixed and uniform in composition at all times, but concentration is constantly changing with time since $C_A = C_{Ao}$ at $t = 0$, and $C_A = C_A$ at $t > 0$, then the material balance is

$$\text{input} - \text{output} = \text{accumulation} + \text{disappearance}$$

$$0 - 0 = \frac{dN_A}{dt} + (-r_A)V$$

$$\frac{-dN_A}{dt} = (-r_A)V \qquad 11.39$$

KINETICS

Table 11.3
Integrated Forms of the Constant Volume, Irreversible Batch Reactor

order	reaction	rate equation	integrated form
0	$A \rightarrow R$	$-r_A = k$	$kt = C_{Ao} - C_A = C_{Ao}X_A \quad (t \leq \frac{C_{Ao}}{k})$
			$C_A = 0 \quad (t \geq \frac{C_{Ao}}{k})$
1	$A \rightarrow R$	$-r_A = kC_A$	$kt = ln\frac{C_{Ao}}{C_A} = ln\frac{1}{1-X_A}$
2	$2A \rightarrow R$	$-r_A = kC_A^2$	$kt = \frac{1}{C_A} - \frac{1}{C_{Ao}} = \frac{X_A}{C_{Ao}(1-X_A)}; \; \frac{C_A}{C_{Ao}} = \frac{1}{1+ktC_{Ao}}$
	$A + bB \rightarrow R$	$-r_A = kC_AC_B$	$ktbC_{Ao}(M-1) = ln\frac{C_B}{MC_A} = ln\frac{M-X_A}{M(1-X_A)} \quad (M = \frac{C_{Bo}}{bC_{Ao}} \neq 1)$
			$ktC_{Bo} = \frac{C_{Ao}-C_A}{C_A} = \frac{X_A}{1-X_A} \quad (M = 1)$
3	$A + B + C \rightarrow R$	$-r_A = kC_AC_BC_C$	$kt = \frac{1}{\Delta_{AB}\Delta_{AC}} ln\frac{C_{Ao}}{C_A} + \frac{1}{\Delta_{BA}\Delta_{BC}} ln\frac{C_{Bo}}{C_B} + \frac{1}{\Delta_{CA}\Delta_{CB}} ln\frac{C_{Co}}{C_C}$ where $C_{Ao} \neq C_{Bo} \neq C_{Co}$ and $\Delta_{YZ} = C_{Yo} - C_{Zo}$
			$kt(2C_{Bo} - C_{Ao})^2 = \frac{(2C_{Bo}-C_{Ao})(C_{Ao}-C_A)}{C_{Ao}C_A} + ln\frac{C_{Bo}C_A}{C_{Ao}C_B}$ where $C_{Ao} = C_{Co} \neq C_{Bo}$
3	$A + B + C \rightarrow R$	$-r_A = kC_AC_BC_C$	$2kt = \frac{1}{C_A^2} - \frac{1}{C_{Ao}^2}$ where $C_{Ao} = C_{Bo} = C_{Co}$
3	$2A + B \rightarrow R$	$-r_A = kC_A^2C_B$	$kt(2C_{Bo} - C_{Ao})^2 = \frac{(2C_{Bo}-C_{Ao})(C_{Ao}-C_A)}{C_{Ao}C_A} + ln\frac{C_{Bo}C_A}{C_{Ao}C_B}$ where $C_{Ao} \neq 2C_{Bo}$
			$kt = \frac{1}{C_A^2} - \frac{1}{C_{Ao}^2}$ where $C_{Ao} = 2C_{Bo}$
3	$3A \rightarrow R$	$-r_A = kC_A^3$	$2kt = \frac{1}{C_A^2} - \frac{1}{C_{Ao}^2}$

If V is constant (or $\varepsilon = 0$), the result is the performance equation for the constant density batch reactor:

$$t = -\int_{C_{Ao}}^{C_A} \frac{dC_A}{-r_A} = C_{Ao}\int_0^{X_A} \frac{dX_A}{-r_A} \qquad 11.40$$

If $V = V_0\,(1 + \varepsilon_A X_A)$, the result is the performance equation for the variable volume batch reactor:

$$t = -\int_{C_{Ao}}^{C_A} \frac{dC_A}{\left(1+\varepsilon_A \frac{C_A}{C_{Ao}}\right)(-r_A)}$$

$$= C_{Ao}\int_0^{X_A} \frac{dX_A}{(1+\varepsilon_A X_A)(-r_A)} \qquad 11.41$$

A list of the integrated forms of the constant volume irreversible batch, variable volume irreversible batch, and the constant volume reversible batch reactors is given in tables 11.3 through 11.5.

Example 11.2

What is the integrated form of a first-order, irreversible, constant volume batch reactor?

The rate expression for the reaction $A \rightarrow$ products is

$$-r_A = kC_A$$

Substituting this rate expression into equation 11.40 results in

$$t = -\int_{C_{A_o}}^{C_A} \frac{dC_A}{kC_A}$$

Integrating,

$$kt = ln\frac{C_{Ao}}{C_A} = ln\left(\frac{1}{1-X_A}\right)$$

Example 11.3

What is the integrated form of a first-order, irreversible, variable volume batch reactor?

The rate expression is

$$-r_A = kC_A$$

Combining the rate expression with equation 11.19 gives the variable volume rate equation:

$$-r_A = kC_{Ao}\frac{1-X_A}{1+\varepsilon_A X_A}$$

Substituting this into equation 11.41,

$$t = \frac{1}{k}\int_0^{X_A} \frac{dX_A}{1-X_A}$$

Table 11.4
Integrated Forms of the Variable Volume, Irreversible Batch Reactor

$$V = V_o(1+\varepsilon_A X_A)$$
$$\Delta V = V_o\varepsilon_A X_A$$

order	reaction	rate equation	integrated form
0	$A \rightarrow R$	$-r_A = k$	$kt = \frac{C_{Ao}}{\varepsilon_A} ln\,(1+\varepsilon_A X_A) = \frac{C_{Ao}}{\varepsilon_A} ln\,\frac{V}{V_o}$
1	$A \rightarrow R$	$-r_A = kC_A$	$kt = -ln(1-X_A) = -ln\left(1-\frac{\Delta V}{\varepsilon_A V_o}\right)$
2	$2A \rightarrow R$	$-r_A = kC_A^2$	$C_{Ao}kt = \frac{(1+\varepsilon_A)X_A}{1-X_A} + \varepsilon_A\, ln\,(1-X_A)$
	$A+B \rightarrow R$	$-r_A = kC_AC_B$	$(C_{Bo}-C_{Ao})kt = \varepsilon_A M\, ln\,(1-\frac{X_A}{M}) - \varepsilon_A\, ln\,(1-X_A)$ $+\ ln\,\frac{M-X_A}{M-MX_A}$ where $M = \frac{C_{Bo}}{C_{Ao}} \neq 1$

KINETICS

Table 11.5
Integrated Forms of the Constant Volume, Reversible Batch Reactor

order	reaction	rate equation	integrated form
1	$A \leftrightarrow rR$	$-r_A = k_1C_A - k_2C_R$	$k_1t = \frac{M+rX_{Ae}}{M+r}\ ln\ \frac{C_{Ao}-C_{Ae}}{C_A-C_{Ae}}$ where $M = \frac{C_{Ro}}{C_{Ao}}$; C_{Ae} = equilibrium
2	$A+B \leftrightarrow R+S$	$-r_A = k_1C_AC_B$ $-k_2C_RC_S$	$k_1t = \frac{X_{Ae}}{2C_{Ao}(1-X_{Ae})}\ ln\ \frac{X_{Ae}-X_A(2X_{Ae}-1)}{X_{Ae}-X_A}$ where $C_{Ao} = C_{Bo}$ and $C_{Ro} = C_{So} = 0$

Integrating,

$$kt = ln\left(\frac{1}{1-X_A}\right)$$

Comparing this result with the constant volume system, the fractional conversion is the same at any time whether or not it is constant volume or not for the case of a first order reaction. However, the concentration of materials is not the same because of the volume difference.

Example 11.4

What is the integrated form of a first-order, reversible, constant volume batch reactor?

The rate expression for the reaction $A \underset{k_2}{\overset{k_1}{\longleftrightarrow}} R$ is

$$-r_A = k_1C_A - k_2C_R = C_{Ao}\frac{dX_A}{dt}$$

At equilibrium $-r_A = 0$, the equilibrium constant is

$$K = \frac{C_{Re}}{C_{Ae}} = \frac{M+X_{Ae}}{1-X_{Ae}}$$

$$M = \frac{C_{Ro}}{C_{Ao}}$$

Combining equations,

$$\frac{dX_A}{dt} = \frac{k_1(M+1)}{M+X_{Ae}}(X_{Ae}-X_A)$$

Substituting into equation 11.40, and integrating,

$$\left[\frac{M+1}{M+X_{Ae}}\right]k_1t = ln\left(\frac{X_{Ae}}{X_{Ae}-X_A}\right)$$
$$= ln\left(\frac{C_{Ao}-C_{Ae}}{C_A-C_{Ae}}\right)$$

Example 11.5

The pyrolysis of ethane proceeds with an activation energy of approximately 75,000 cal/gmole. How much faster is the decomposition 650 °C than at 500 °C?

From equation 11.8,

$$650\,^\circ\text{C} = 650 + 273 = 923\,^\circ\text{K}$$
$$500\,^\circ\text{C} = 773\,^\circ\text{K}$$
$$\text{R} = 1.99\ \text{cal/gmole-}^\circ\text{K}$$

$$ln\ \frac{k_2}{k_1} = \frac{E}{R}\left(\frac{1}{T_1}-\frac{1}{T_2}\right) = \frac{E}{R}\left(\frac{T_2-T_1}{T_1T_2}\right)$$

$$ln\ \frac{k_{650}}{k_{500}} = \frac{75,000}{1.99}\left(\frac{923-773}{923\times 773}\right) = 7.924$$

$$\frac{k_{650}}{k_{500}} = 2761$$

The rate at 650 °C is 2761 times faster than at 500 °C.

Example 11.6

Determine the order of the reaction and the rate constant for the reaction

$$A + B \rightarrow R + S$$

The initial concentrations are

$$C_{Ao} = 0.311\ \text{moles/liter}$$
$$C_{Bo} = 0.564\ \text{moles/liter}$$

Conversion data are:

time, t(minutes)	C_R, moles/liter
0	0
6.55	0.077
11.15	0.117
16.82	0.153
21.5	0.176

Data for Example 11.6

t	C_R	X_A	$M-X_A$	$M(1-X_A)$	$\frac{M-X_A}{M(1-X_A)}$	$ln\frac{M-X_A}{M(1-X_A)}$
0	0	0	1.814	1.814	1.000	0
6.55	0.077	0.248	1.566	1.364	1.148	0.138
11.15	0.117	0.376	1.438	1.132	1.270	0.239
16.82	0.153	0.492	1.322	0.922	1.434	0.361
21.5	0.176	0.566	1.248	0.787	1.586	0.46

Since the stoichiometric equation is of the form of second-order reaction, assume a second-order reaction.

From table 11.3,

$$ln\frac{M-X_A}{M(1-X_A)} = (C_{Bo}-C_{Ao})kt$$

$$M = \frac{C_{Bo}}{C_{Ao}} = \frac{0.564}{0.311} = 1.814$$

$$X_A = \frac{C_{Ao}-C_A}{C_{Ao}}$$

The straight line confirms that the reaction is second order.

$$\text{slope} = 0.0215 = (C_{Bo}-C_{Ao})k$$

$$= (0.564-0.311)k$$

$$k = \frac{0.0215}{0.253} = 0.085 \text{ liters/mole-min}$$

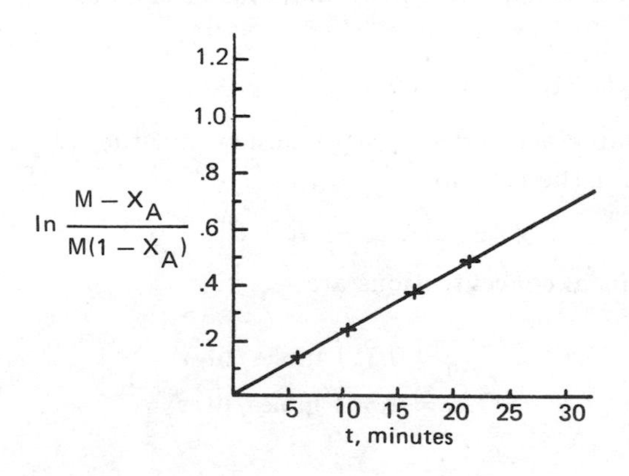

Example 11.7

In a homogeneous, isothermal liquid polymerization, 20% of the monomer disappears in 34 minutes regardless of whether the initial monomer concentration is 0.04 or 0.8 moles/liter. What is the rate of disappearance of the monomer?

Since the fractional disappearance is independent of initial concentration, the reaction is first-order. Let C be the monomer concentrations.

$$\frac{-dC}{dt} = kC$$

$$ln\frac{C_o}{C} = kt$$

$$ln\frac{C_o}{0.8\,C_o} = k(34)$$

$$k = \frac{-ln\,0.8}{34} = 0.00656 \text{ min}^{-1}$$

The rate equation is

$$-r = \frac{-dC}{dt} = 0.00656\ C$$

Example 11.8

The first-order reversible liquid reaction that takes place in a batch reactor is

$$A \longleftrightarrow R$$

$$C_{Ao} = 0.5 \text{ moles/liter}$$

$$C_{Ro} = 0$$

After eight minutes, the conversion of A is 33.3%. At equilibrium, the conversion is 66.7%. Find the rate equation.

From table 11.5,

$$-ln\left(1-\frac{X_A}{X_{Ae}}\right) = (k_1+k_2)t$$

$$-ln\left(1-\frac{\frac{1}{3}}{\frac{2}{3}}\right) = (k_1+k_2)\ 8 \text{ minutes}$$

$$k_1+k_2 = \frac{ln\,2}{8} = 0.086625 \text{ min}^{-1}$$

$$K = \frac{C_{Re}}{C_{Ae}} = \frac{k_1}{k_2}$$

$$= \frac{0.5\times\frac{2}{3}}{0.5\times\frac{1}{3}} = 2$$

$$k_1 = 2k_2$$

Solving,

$$k_2 = \frac{0.086625}{3} = 0.028875$$
$$k_1 = 0.057750$$

The rate expression for disappearance of A is

$$-r_A = 0.05775\ C_A - 0.028875\ C_R$$

Example 11.9

A small reaction bomb fitted with a sensitive pressure-measuring device is filled with a mixture of 76.94% reactant A and 23.06% inert at one atmosphere pressure and 14 °C, (a temperature low enough that the reaction does not proceed to any appreciable extent). The temperature is raised rapidly to 100 °C, and the total pressure monitored. The stoichiometry of the reaction is $A \rightarrow 2R$, and after sufficient time the reaction proceeds to completion. Find a rate equation which will satisfactorily fit the data.

t, minutes	π, atmospheres
0.5	1.5
1	1.65
1.5	1.76
2	1.84
2.5	1.90
3	1.95
3.5	1.99
4	2.025
5	2.08
6	2.12
7	2.15
8	2.175

Take 100 moles as the initial basis. At $t = 0$, the numbers of moles present are

$$A \rightarrow 2R + \text{inerts}$$

$$\text{moles } 76.94 + 0 + 23.06$$

The raised temperature is

$$100 + 273 = 373\ ^\circ\text{K}$$

The total pressure after raising the temperature is

$$\pi_o = 1 \text{ atm} \times \frac{373\ ^\circ\text{K}}{287\ ^\circ\text{K}} = 1.30 \text{ atm}$$

Since $a = 1$, the change in number of moles is

$$\Delta n = r - a = 2 - 1 = 1$$

From equation 11.12 (rearranged),

$$p_A = p_{Ao} - \frac{a}{\Delta n}(\pi - \pi_o)$$
$$p_{Ao} = 0.7694 \times 1.30 = 1.00 \text{ atm}$$
$$p_A = 1.00 - \frac{1}{1}(\pi - 1.30) = 2.30 - \pi$$

t	π	p_A	p_A^{1-n} $n = 2$	p_A^{1-n} $n = 1.5$
0	1			
0.5	1.5	0.80		
1	1.65	0.65	1.538	1.24
1.5	1.76	0.54		
2	1.84	0.46	2.174	1.47
2.5	1.90	0.40		
3	1.95	0.35	2.857	1.69
3.5	1.99	0.31		
4	2.025	0.275	3.636	1.91
5	2.08	0.22	4.545	2.13
6	2.12	0.18	5.556	2.36
7	2.15	0.15	6.667	2.58
8	2.175	0.125	8.000	2.83

From the tabulation of t versus p_A, the order of reaction must be decided.

$$p_A = 0.8 \text{ @ } t = 0.5$$
$$p_A = 0.4 \text{ @ } t = 2.5$$
$$\Delta t = 2.5 - 0.5 = 2 \text{ min}$$

For the three to five minute period, the concentration does not halve. So, the order is not first order (half-life test).

Application of integral tests for zero order and second order (by plotting appropriate forms) also does not work out.

Therefore, we have to try a general form for the nth order $aA \rightarrow R$. The integrated form is

$$C_A^{1-n} - C_{Ao}^{1-n} = (n-1)kt$$
$$n \neq 1$$
$$p_A^{1-n} - p_{Ao}^{1-n} = (n-1)kt$$

Plot p_A^{1-n} versus t.

A trial-and-error approach must be used to determine n. The plotted data is a straight line when n is assumed to be 1.5.

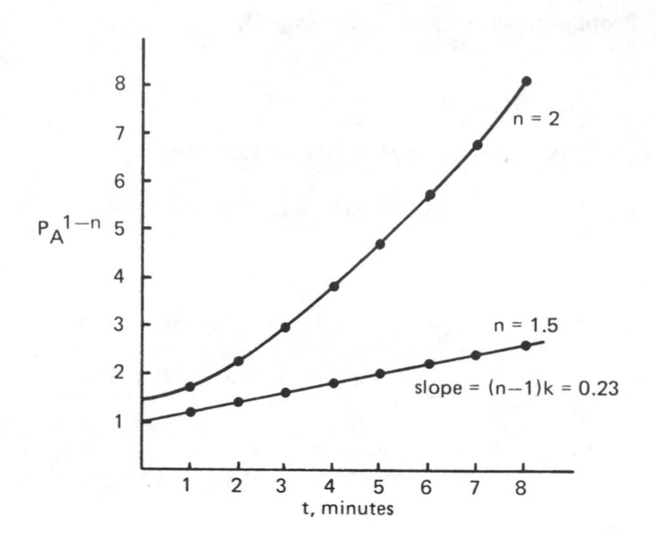

$$k = \frac{\text{slope}}{n-1} = \frac{0.23}{0.5}$$
$$= 0.46 \text{ min}^{-1} \text{ atm}^{-0.5}$$

Then, the equation relating in terms of partial pressure as a function of time is

$$p_A^{-0.5} - p_{Ao}^{-0.5} = 0.23\,t$$

Since $C^{0.5} = \left(\frac{N}{V}\right)^{0.5} = \left(\frac{P}{RT}\right)^{0.5}$, then multiply the equation relating partial pressure and time by $\sqrt{RT}$ to express it in terms of C:

$$C_A^{-0.5} - C_{Ao}^{-0.5} = 0.23t\sqrt{RT}$$
$$= 0.23t\sqrt{T}\sqrt{.08206 \text{ liter-atm/mole-°K}}$$
$$C_A^{-0.5} - C_{Ao}^{-0.5} = 0.0659t\sqrt{T}$$

Example 11.10

Consider the irreversible chemical reaction $A + B \rightarrow C$.

The reaction rate constant at 760 °R is 3.1 ft³/lbmole-sec. The reaction occurs at a constant pressure of five atmospheres and a constant temperature of 760 °R. Initially, there are five moles of A, six moles of B, one mole of C, and four moles of an inert D. All species are gaseous and follow the ideal gas law. Compute the time required for 95% of A to react.

The reaction can be shown graphically as follows:

component	no. of moles		component	no. of moles
A	5	batch reaction	A	0
B	6	→ 5 atm →	B	1
C	1	760 °R	C	6
D (inert)	4	constant pressure	D	4
ϵ	16			11
$X_A = 0$			$X_A = 1$	

The units of reaction rate imply a second-order reaction.

$$\varepsilon_A = \frac{V_{X_A=1} - V_{X_A=0}}{V_{X_A} = 0} = \frac{11-16}{16}$$
$$\varepsilon_A = -0.313$$
$$C_{Ao} = \frac{N_{Ao}}{V} = \frac{N_{Ao}}{\frac{NRT}{p}} = \frac{5}{\frac{16 \times 10.73 \times 760}{5 \times 14.7}}$$
$$C_{Ao} = 0.0028 \text{ moles/ft}^3$$

Use second order integrated equation when $C_{A_o} \neq C_{B_o}$,

$$t = \frac{1}{K}\,\frac{1}{C_{B_o} - C_{A_o}}\left\{\varepsilon_A\left[M\ln\left(1 - \frac{X_A}{M}\right) - \ln(1 - X_A)\right] + \ln\left(\frac{1 - \frac{X_A}{M}}{1 - X_A}\right)\right\}$$

$$M = \frac{C_{B_o}}{C_{A_o}} = \frac{6}{5} = 1.2$$
$$C_{B_o} = 1.2C_{A_o} = 1.2(0.0028) = 0.0034$$

When $X_A = 0.95$,

$$t = \frac{1}{3.1}\,\frac{1}{0.0034 - 0.0028}\left\{-0.313\left[1.5\ln\left(1 - \frac{0.95}{1.2}\right) - \ln(1 - 0.95)\right] + \ln\left(\frac{1 - \frac{0.95}{1.2}}{1 - 0.95}\right)\right\}$$

$t = 580$ sec or 9.67 minutes

9 FLOW REACTORS – GENERAL

In contrast to batch reactors, concentration in steady state flow reactors is not a function of time but of position. Flow reactors also experience a pressure drop which, for gas reactions, affects concentrations.

The time of reaction in a flow reactor cannot be measured directly and is, therefore, not a convenient variable for correlation of rate data. The basic relationship for a flow reactor is between the reactor volume, the kinetics, the feed rate, and the conversion.

There are two ideal types of flow reactors:

- The *continuous stirred tank reactor* (CSTR), also known as a *constant flow stirred tank reactor* (CFSTR), *back mixed,* or *perfectly mixed reactor* (PMR). The contents of the CSTR are perfectly mixed so that they have the same concentration as the effluent. Material is held in the reactor for varying periods of time based on statistical probability theory.
- The *plug flow reactor* (PFR), also called the *tubular, piston flow,* or *unmixed flow reactor.* All material leaving the PFR will have been in the reactor for the same length of time. It is assumed that there is no longitudinal mixing or diffusion.

10 SPACE-TIME AND SPACE-VELOCITY CONCEPTS

While reaction time is a natural measure of the processing rate for batch reactors, space-time and space-velocity are the typical performance measures for flow reactors.

Space-time is the time required to process one reactor volume of feed, measured at specified conditions.

$$\tau = \frac{1}{s} \qquad 11.42$$

Space-velocity is the number of reactor volumes of feed at specified conditions which can be treated in unit time.

$$s = \frac{1}{\tau} \qquad 11.43$$

The choice of temperature, pressure, and phase (gas, liquid, or solid) at which the volume of material fed to the reactor is measured is arbitrary. The value of the space-velocity or space-time depends upon the conditions selected. For most applications, the space-time is based on feed conditions. If the conditions of the feed stream entering the reactor are chosen, the relationship between s, τ, and the other pertinent variables becomes

$$\tau = \frac{1}{s} = \frac{C_{Ao}V}{F_{Ao}}$$

$$= \frac{\dfrac{\text{moles } A \text{ entering}}{\text{volume of feed}} \times \text{volume of reactor}}{\dfrac{\text{moles } A \text{ entering}}{\text{time}}} \qquad 11.44$$

$$\tau = \frac{V}{v_o} = \frac{\text{reactor volume}}{\text{volumetric feed rate}} \qquad 11.45$$

It may be more convenient to measure the volumetric feed rate at some standard conditions, especially when the reactor is to operate at a number of different temperatures. The relationship between space-time at actual entering conditions and at standard conditions (designated by primes) is

$$\tau' = \frac{C'_{Ao}V}{F_{Ao}} = \tau\frac{C'_{Ao}}{C_{Ao}} \qquad 11.46$$

An example of the use of a standard feed state for space-velocity is found in the petroleum refining industry which uses *liquid hourly space-velocity* (LHSV). A liquid hydrocarbon is measured at 60 °F and one atmosphere even though it may be heated and vaporized before entering the reactor. (In equation 11.47, BPD stands for barrels per day.)

$$\text{LHSV} = \frac{\text{ft}^3 \text{ feed/hour}}{\text{ft}^3 \text{ reactor}} = \frac{\text{BPD } (0.234)}{\text{ft}^3 \text{ reactor}} \qquad 11.47$$

Table 11.6
Performance Equations

	$\varepsilon = 0$	$\varepsilon \neq 0$
batch reactor	$t = -\int_{C_{Ao}}^{C_A} \frac{dC_A}{-r_A}$	$t = C_{A_o}\int_0^{X_A} \frac{dX_A}{(1+\varepsilon_A X_A)(-r_A)}$
CSTR	$\tau = \frac{C_{Ao} - C_A}{-r_A}$	$\tau = \frac{C_{Ao}X_A}{-r_A}$
PFR	$\tau = -\int_{C_{Ao}}^{C_A} \frac{dC_A}{-r_A}$	$\tau = C_{A_o}\int_0^{X_A} \frac{dX_A}{-r_A}$

11 THE CSTR

The performance equation for a CSTR is developed from a steady state material balance.

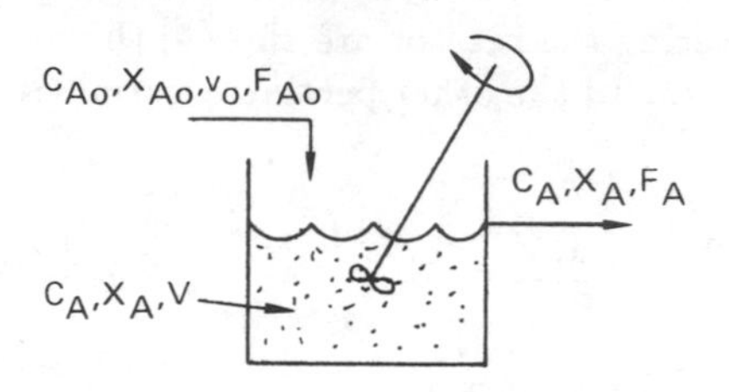

Figure 11.3 The CSTR

The basic material balance for a component is

input − output = disappearance by reaction

The feed rate of a component to the reactor, considering the reactor as a whole, is

$$F_o = v_o C_o \qquad 11.48$$

The input term of A is

$$\text{input} = F_{A_o}(1 - X_{A_o}) = F_{A_o}(1 - 0) = F_{A_o} \qquad 11.49$$

$$\text{output} = F_A = F_{A_o}(1 - X_A) \qquad 11.50$$

The disappearance by reaction can be written as

$$-r_A V = \frac{\text{moles reacting}}{(\text{time})\ (\text{volume of fluid})}(\text{volume of reactor}) \qquad 11.51$$

Substituting,

$$F_{Ao} X_A = -r_A V \qquad 11.52$$

Substituting into equation 11.44 results in the performance equation for a CSTR:

$$\tau = \frac{C_{Ao} X_A}{-r_A} \qquad 11.53$$

The values of X_A and $-r_A$ are evaluated at the exit stream conditions, which are the same as the conditions within the reactor. The power and simplicity of equation 11.53 will be quickly realized in solving specific problems. The procedure is to merely insert the rate expression into equation 11.53 to obtain the performance equation. No integration will be required.

For the special case of constant density ($\varepsilon = 0$) so that $C_A = C_{A_o}(1 - X_A)$, the performance equation for the CSTR is

$$\tau = \frac{C_{A_o} X_A}{-r_A} = \frac{C_{A_o} - C_A}{-r_A} \qquad 11.54$$

For any specific kinetic form, the performance equation can be written directly. As an example, for a constant density, first-order reaction, the performance equation for a CSTR is

$$k\tau = \frac{C_{A_o} - C_A}{C_A} = \frac{X_A}{1 - X_A} \qquad 11.55$$

If the density is changing within the CSTR, then

$$V = V_o(1 + \varepsilon_A X_A) \qquad 11.56$$

$$C_A = C_{A_o}\left(\frac{1 - X_A}{1 + \varepsilon_A X_A}\right) \qquad 11.57$$

The CSTR performance equation for a first-order reaction with a changing density is

$$k\tau = X_A \frac{1 + \varepsilon_A X_A}{1 - X_A} \qquad 11.58$$

Similar expressions can be written for any form of rate equation. The expressions can be written in terms of concentrations or conversions. The latter form is simpler for systems of changing density, while either form can be used for systems of constant density.

To illustrate the distinction between the *holding time* (or *mean residence time*) and the space-time for a flow reactor, consider the following case. Suppose one liter per second of gaseous reactant A is introduced into a CSTR. The stoichiometry is $A \rightarrow 3R$, the conversion is 50%, and under these conditions, the leaving flow rate is two liter per second. By definition, the space-time for this operation is

$$\tau = \frac{V}{v_o} = \frac{1 \text{ liter}}{1 \text{ liter/sec}} = 1 \text{ sec}$$

However, since the fluid has expanded to twice its initial volume upon entering the reactor, the holding time is

$$\bar{t} = \frac{V}{v_o(1 + \varepsilon_A X_A)} = \frac{V}{v_f} = \frac{1 \text{ liter}}{2 \text{ liter/sec}} = \frac{1}{2} \text{ sec}$$

If the reaction had been in a liquid phase instead of gas phase, expansion would have been negligible, and the holding time and space time would have been identical. Holding time, by itself, has no practical significance in reaction kinetics. Space-time is important because it represents the ratio $\frac{V}{v_o}$, which always appears in the performance equations for both types of flow reactors.

A CSTR is inherently less efficient than a batch reactor because the "driving force" to carry out the reaction is much lower. The reactant concentrations correspond to the final values, rather than decreasing from initial to final values as the reaction progresses. Even though more total reactor volume may be required for a CSTR as compared to a batch or PFR, its use is frequently more economically attractive because:

- A CSTR permits continuous, rather than batch, operation.
- At atmospheric pressure, inexpensive tanks can be used (as opposed to obtaining volume with a small diameter pipe).
- A CSTR allows heat addition or removal in jacketed vessels *(Pfaudler kettles)*.

The staging of CSTR vessels in a series is a method of improving the efficiency of this type of reactor. Consider a system of N mixed reactors connected in series. Though the concentration is uniform in each reactor, there is a change in concentration as the fluid moves from reactor to reactor. With an infinite number of vessels, performance would approach that of a PFR. The performance of a series of N equal-sized CSTR's operating with constant density can be evaluated by equations 11.59 and 11.60.

First-order (τ is identical in all reactors):

$$\frac{C_N}{C_o} = 1 - X_N = \frac{1}{(1+k\tau)^N} \qquad 11.59$$

Second-order:

$$C_N = \frac{1}{4k\tau}\left(-2 + 2\sqrt{-1 \ldots + 2\sqrt{-1 + 2\sqrt{1 + 4C_o k\tau}}}\ \begin{matrix} N=N \\ N=2 \\ N=1 \end{matrix}\right) \qquad 11.60$$

The greatest improvement in adding series reactors to a system occurs with the addition of a second vessel to a one-vessel system. Generally, it will be optimal economically to use equal-sized reactors rather than different-sized reactors in series.

Example 11.11

One liter/min of liquid containing A and B($C_{Ao} = 0.10$ mole/liter, $C_{Bo} = 0.01$ mole liter) flows into a mixed reactor of one liter volume. The materials react in a complex manner for which the stoichiometry is unknown. The outlet stream from the reactor contains A, B, and C ($C_{Af} = 0.02$ mole/liter, $C_{Bf} = 0.03$ mole/liter, $C_{Cf} = 0.04$ mole liter). Find the rate of A, B, and C for the conditions within the reactor.

For a liquid in a mixed reactor, $\varepsilon_A = 0$. Equation 11.54 applies to each of the reacting components, predicting the rate of disappearance.

$$-r_A = \frac{C_{Ao} - C_A}{\tau} = \frac{C_{Ao} - C_A}{\frac{V}{v_o}} = \frac{0.10 - 0.02}{\frac{1}{1}}$$
$$= 0.08 \text{ mole/liter-min}$$
$$-r_B = \frac{C_{Bo} - C_B}{\tau} = \frac{0.01 - 0.03}{1}$$
$$= -0.02 \text{ mole/liter-min}$$
$$-r_C = \frac{C_{Co} - C_C}{\tau} = \frac{0 - 0.04}{1} = -0.04 \text{ mole/liter-min}$$

A is disappearing while B and C are being formed.

Example 11.12

Consider the following liquid-phase reaction

$$A + B \underset{k_2}{\overset{k_1}{\longleftrightarrow}} R + S$$

$$k_1 = 7 \text{ liter/mole-min}$$
$$k_2 = 3 \text{ liter/mole-min}$$

This reaction is to take place in a 120-liter, steady-state mixed reactor. Two feed streams, one containing 2.8 mole/liter of A and the other containing 1.6 mole/liter of B, are to be introduced in equal volumes into the reactor, and 75% conversion of limiting component is desired. What should be the flow rate of each stream? Assume a constant density throughout.

The concentrations of components in the mixed feed stream are

$$C_{Ao} = 1.4 \text{ mole/liter}$$
$$C_{Bo} = 0.8 \text{ mole/liter}$$
$$C_{Ro} = 0$$
$$C_{So} = 0$$

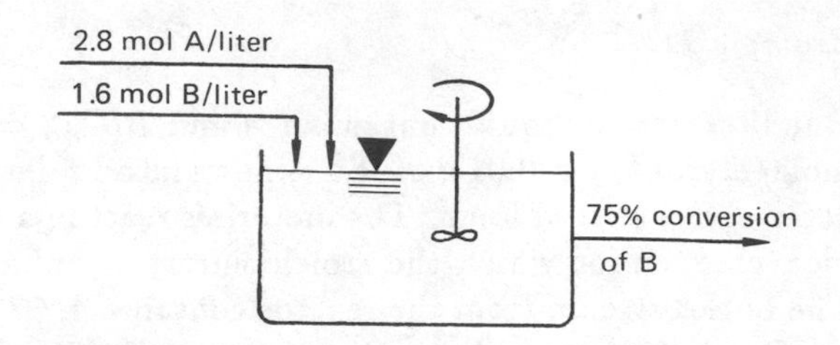

For 75% conversion of B (and $\varepsilon = 0$), the compositions within the reactor and in the exit stream are

$$C_A = 1.4 - 0.6 = 0.8 \text{ mole/liter}$$
$$C_B = 0.8 - 0.6 = 0.2 \text{ mole/liter}$$
$$C_R = 0.6 \text{ mole/liter}$$
$$C_S = 0.6 \text{ mole/liter}$$

The rate of reaction within the reactor is

$$\begin{aligned} -r_A = -r_B &= k_1 C_A C_B - k_2 C_R C_S \\ &= \left(7 \frac{\text{liter}}{\text{mole-min}}\right)\left(0.8 \frac{\text{mole}}{\text{liter}}\right)\left(0.2 \frac{\text{mole}}{\text{liter}}\right) \\ &\quad - \left(3 \frac{\text{liter}}{\text{mole-min}}\right)\left(0.6 \frac{\text{mole}}{\text{liter}}\right)\left(0.6 \frac{\text{mole}}{\text{liter}}\right) \\ &= (1.12 - 1.08) \text{ mole/liter-min} \\ &= 0.04 \text{ mole/liter-min} \end{aligned}$$

For no change in density, $\varepsilon = 0$,

$$\tau = \frac{V}{\text{v}} = \frac{C_{Ao} - C_A}{-r_A} = \frac{C_{Bo} - C_B}{-r_B}$$

The volumetric flow rate into and out of the reactor is

$$\begin{aligned} \text{v}_o &= \frac{V(-r_A)}{C_{Ao} - C_A} = \frac{V(-r_B)}{C_{Bo} - C_B} \\ &= \frac{(120 \text{ liter})(0.04 \text{ mole/liter-min})}{0.6 \text{ mole/liter}} = 8 \text{ liter/min} \end{aligned}$$

Each of the two input streams should feed at four liters per minute.

12 THE PLUG FLOW REACTOR

Figure 11.4 illustrates the definitions used to derive the performance equation for a plug flow reactor (PFR).

The material balance for an element of reactor volume, dV, for a component is

$$\begin{aligned} \text{input} - \text{output} &= \text{disappearance by reaction} \\ \text{input} &= F_A \\ \text{output} &= F_A + dF \\ \text{disappearance} &= -r_A dV \end{aligned}$$

$$F_A = F_A + dF_A + (-r_A)dV \qquad 11.61$$

However,

$$dF_A = d[F_{A_o}(1 - X_A)] = -F_{A_o} dX_A \qquad 11.62$$

The resulting differential equation is

$$F_{A_o} dX_A = (-r_A)dV \qquad 11.63$$

Equation 11.63 accounts for the component in the differential element of the reactor of volume dV. For the reactor as a whole, the expression must be integrated. Note that F_{A_o} is constant, but that $-r_A$ is dependent on the conversion or concentration of materials. Separating variables, the equation 11.63 becomes

$$\frac{dV}{F_{A_o}} = \frac{dX_A}{-r_A} \qquad 11.64$$

This can be integrated to give the performance equation for a plug flow reactor.

$$\frac{C_{A_o} V}{F_{Ao}} = \tau = C_{A_o} \int_0^{X_A} \frac{dX_A}{-r_A} \qquad 11.65$$

In terms of concentrations,

$$\tau = -\int_{C_{A_o}}^{C_A} \frac{dC_A}{-r_A} \qquad 11.66$$

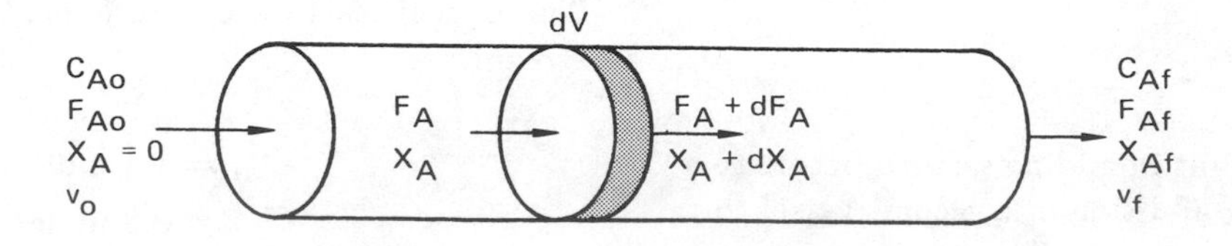

Figure 11.4 The PFR

Table 11.7
Performance Equations of the Constant Volume, Irreversible CSTR

order	reaction	rate equation	performance equation
0	$A \rightarrow R$	$-r_A = k$	$k\tau = C_{Ao} - C_A = C_{Ao}X_A$
1	$A \rightarrow R$	$-r_A = kC_A$	$k\tau = \frac{C_{Ao}-C_A}{C_A} = \frac{X_A}{1-X_A}$
2	$2A \rightarrow R$	$-r_A = kC_A^2$	$k\tau = \frac{C_{Ao}-C_A}{C_A^2} = \frac{X_A}{C_{Ao}(1-X_A)^2}$
	$A + bB \rightarrow R$	$-r_A = kC_AC_B$	$k\tau = \frac{C_{Ao}-C_A}{bC_A[C_{Ao}(M-1)+C_A]} = \frac{X_A}{bC_{Ao}(1-X_A)(M-X_A)}$ where $M = \frac{C_{Bo}}{bC_{Ao}} \neq 1$
			$k\tau = \frac{C_{Ao}-C_A}{bC_A^2} = \frac{X_A}{bC_{Ao}(1-X_A)^2}$ where $M = 1$
3	$3A \rightarrow R$	$-r_A = kC_A^3$	$k\tau = \frac{C_{Ao}-C_A}{C_A^3} = \frac{X_A}{C_{Ao}^2(1-X_A)^3}$
	$2A + bB \rightarrow R$	$-r_A = kC_A^2C_B$	$k\tau = \frac{2(C_{Ao}-C_A)}{bC_A^2[C_{Ao}(M-1)+C_A]} = \frac{2X_A}{bC_{Ao}^2(1-X_A)^2(M-X_A)}$ where $M = \frac{2C_{Bo}}{bC_{Ao}} \neq 1$
3	$2A + bB \rightarrow R$	$-r_A = kC_A^2C_B$	$k\tau = \frac{2(C_{Ao}-C_A)}{bC_A^3} = \frac{2X_A}{bC_{Ao}^2(1-X_A)^3}$ where $M = \frac{2C_{Bo}}{bC_{Ao}} = 1$
	$A + bB + cC \rightarrow R$	$-r_A = kC_AC_BC_C$	$k\tau = \frac{X_A}{bcC_{Ao}^2(1-X_A)(M_1-X_A)(M_2-X_A)}$ where $M_1 = \frac{C_{Bo}}{bC_{Ao}} \neq 1;\ M_2 = \frac{C_{Co}}{cC_{Ao}} \neq 1$
			$k\tau = \frac{X_A}{bcC_{Ao}^2(1-X_A)^2(M_1-X_A)}$ $M_1 \neq 1;\ M_2 = 1$
			$k\tau \frac{X_A}{bcC_{Ao}^2(1-X_A)^2(M_2-X_A)}$ $M_1 = 1;\ M_2 \neq 1$
			$k\tau = \frac{X_A}{bcC_{Ao}^2(1-X_A)^3}$ $M_1 = 1;\ M_2 = 1$

The PFR performance equations are the same for the constant density and variable density cases. (For systems of changing density, it is more convenient to use conversions.) Whatever its form, the performance equation correlates the rate of reaction, the extent of reaction, the reactor volume, and feed rate.

The space-time ratio of reactor volume and feed rate is independent of reactor size, shape, and value of feed rate as long as longitudinal diffusion is negligible. This permits easy scale-up of PFR's even though the reaction kinetics theory may not be well-understood. The pilot plant data, in effect does the necessary integration to predict conversion as a function of space-velocity (space-time).

The following observations are relevant to tables 11.3 and 11.4 and PFR performance:

- For systems with constant density, the performance equations are identical for batch reactions and PFR. τ for plug flow is equivalent to t in the batch reactor. The equations can be used interchangeably.

- For systems with changing density, there is no direct correspondence between the batch and the plug flow reactor equations. The correct equation must be used for each particular situation.

Table 11.8
Performance Equations of the Variable Volume, Irreversible CSTR

order	reaction	rate equation	performance equation
0	$A \rightarrow R$	$-r_A = k$	$k\tau = C_{Ao}X_A$
1	$A \rightarrow R$	$-r_A = kC_A$	$k\tau \frac{X_A(1+\varepsilon_A X_A)}{1-X_A}$
2	$2A \rightarrow R$	$-r_A = kC_A^2$	$k\tau = \frac{X_A(1+\varepsilon_A X_A)^2}{C_{Ao}(1-X_A)^2}$
	$A + bB \rightarrow R$	$-r_A = kC_A C_B$	$k\tau \frac{X_A(1+\varepsilon_A X_A)\left(1+\frac{\varepsilon_A X_A}{M}\right)}{bC_{Ao}(1-X_A)(M-X_A)}$ where $M = \frac{C_{Bo}}{bC_{Ao}} \neq 1$
			$k\tau = \frac{X_A(1+\varepsilon_A X_A)^2}{bC_{Ao}(1-X_A)^2}$ $M = 1$
3	$3A \rightarrow R$	$-r_A = kC_A^3$	$k\tau \frac{X_A(1+\varepsilon_A X_A)^3}{C_{Ao}^2(1-X_A)^3}$
	$2A + bB \rightarrow R$	$-r_A = kC_A^2 C_B$	$k\tau = \frac{2X_A(1+\varepsilon_A X_A)^2\left(1+\frac{\varepsilon_A X_A}{M}\right)}{bC_{Ao}^2(1-X_A)^2(M-X_A)}$ where $M = \frac{2C_{Bo}}{bC_{Ao}} \neq 1$
3	$2A + bB \rightarrow R$	$-r_A = kC_A^2 C_B$	$k\tau = \frac{2X_A(1+\varepsilon_A X_A)^3}{bC_{Ao}^2(1-X_A)^3}$ where $M = \frac{2C_{Bo}}{bC_{Ao}} = 1$
	$A + bB + cC \rightarrow R$	$-r_A = kC_B C_A C_C$	$k\tau = \frac{X_A(1+\varepsilon_A X_A)\left(1+\frac{\varepsilon_A X_A}{M_1}\right)\left(1+\frac{\varepsilon_A X_A}{M_2}\right)}{bcC_{Ao}^2(1-X_A)(M_1-X_A)(M_2-X_A)}$ where $M_1 = \frac{C_{Bo}}{bC_{Ao}} \neq 1$; $M_2 = \frac{C_{Co}}{cC_{Ao}} \neq 1$
			$k\tau = \frac{X_A(1+\varepsilon_A X_A)^2\left(1+\frac{\varepsilon_A X_A}{M_2}\right)}{bcC_{Ao}^2(1-X_A)^2(M_2-X_A)}$ $M_1 = 1$; $M_2 \neq 1$
			$k\tau = \frac{X_A(1+\varepsilon_A X_A)^2\left(1+\frac{\varepsilon_A X_A}{M_1}\right)}{bcC_{Ao}^2(1-X_A)^2(M_1-X_A)}$ $M_1 \neq 1$; $M_2 = 1$
			$k\tau = \frac{X_A(1-\varepsilon_A X_A)^3}{bcC_{Ao}^2(1-X_A)^3}$ $M_1 = M_2 = 1$

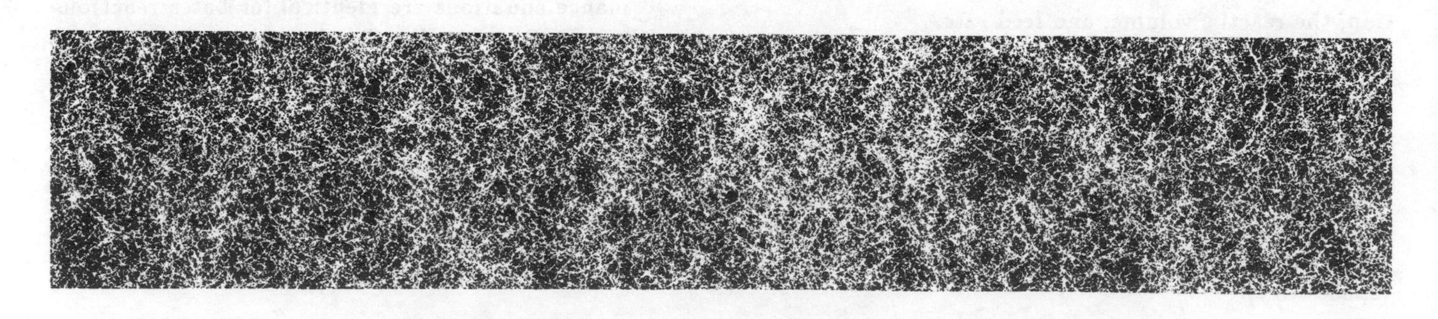

Table 11.9
Integrated Forms of the Constant Volume, Irreversible PFR

order	reaction	rate equation	integrated form
0	$A \rightarrow R$	$-r_A = k$	$k\tau = C_{Ao} - C_A = C_{Ao}X_A \quad (\tau \leq \frac{C_{Ao}}{k})$
			$C_A = 0 \quad (\tau \geq \frac{C_{Ao}}{k})$
1	$A \rightarrow R$	$-r_A = kC_A$	$k\tau = ln \frac{C_{Ao}}{C_A} = ln \frac{1}{1-X_A}$
2	$2A \rightarrow R$	$-r_A = kC_A^2$	$k\tau = \frac{1}{C_A} - \frac{1}{C_{Ao}} = \frac{X_A}{C_{Ao}(1-X_A)}; \quad \frac{C_A}{C_{Ao}} = \frac{1}{1+k\tau C_{Ao}}$
	$A + bB \rightarrow R$	$-r_A = kC_AC_B$	$k\tau bC_{Ao}(M-1) = ln \frac{C_B}{MC_A} = ln \frac{M-X_A}{M(1-X_A)} \quad \left(M = \frac{C_{Bo}}{bC_{Ao}} \neq 1\right)$
			$k\tau C_{Bo} = \frac{C_{Ao}-C_A}{C_A} = \frac{X_A}{1-X_A} \quad (M = 1)$
3	$A + B + C \rightarrow R$	$-r_A = kC_AC_BC_C$	$k\tau = \frac{1}{\Delta_{AB}\Delta_{AC}} ln \frac{C_{Ao}}{C_A} + \frac{1}{\Delta_{BA}\Delta_{BC}} ln \frac{C_{Bo}}{C_B} + \frac{1}{\Delta_{CA}\Delta_{CB}} ln \frac{C_{Co}}{C_C}$ where $C_{Ao} \neq C_{Bo} \neq C_{Co}$ and $\Delta_{YZ} = C_{Yo} - C_{Zo}$
			$k\tau(2C_{Bo} - C_{Ao})^2 = \frac{(2C_{Bo}-C_{Ao})(C_{Ao}-C_A)}{C_{Ao}C_A} + ln \frac{C_{Bo}C_A}{C_{Ao}C_B}$ where $C_{Ao} = C_{Co} \neq C_{Bo}$
3	$A + B + C \rightarrow R$	$-r_A = kC_AC_BC_C$	$2k\tau = \frac{1}{C_A^2} - \frac{1}{C_{Ao}^2}$ where $C_{Ao} = C_{Bo} = C_{Co}$
3	$2A + B \rightarrow R$	$-r_A = kC_A^2C_B$	$k\tau(2C_{Bo} - C_{Ao})^2 = \frac{(2C_{Bo}-C_{Ao})(C_{Ao}-C_A)}{C_{Ao}C_A} + ln \frac{C_{Bo}C_A}{C_{Ao}C_B}$ where $C_{Ao} \neq 2C_{Bo}$
			$k\tau = \frac{1}{C_A^2} - \frac{1}{C_{Ao}^2}$ where $C_{Ao} = 2C_{Bo}$
3	$3A \rightarrow R$	$-r_A = kC_A^3$	$2k\tau = \frac{1}{C_A^2} - \frac{1}{C_{Ao}^2}$

Table 11.10
Integrated Forms of the Variable Volume, Irreversible PFR

order	reaction	rate equation	integrated form
0	$A \rightarrow R$	$-r_A = k$	$k\tau = C_{Ao}X_A$
1	$A \rightarrow R$	$-r_A = kC_A$	$k\tau = -(1+\varepsilon_A)\, ln\,(1 - X_A) - \varepsilon_A X_A$
2	$2A \rightarrow R$	$-r_A = kC_A^2$	$C_{Ao}k\tau = 2\varepsilon_A(1+\varepsilon_A)\, ln\,(1-X_A) + \varepsilon_A^2 X_A + (\varepsilon_A + 1)^2 \frac{X_A}{1-X_A}$

KINETICS

PRACTICE PROBLEMS

1. At 500 °K, the rate of a bimolecular reaction is 10 times greater than at 400 °K. (a) What is the activation energy of this reaction? (b) How much faster is the reaction at 450 °K?

2. A 10 minute reaction shows that 75% of a liquid reactant is converted to a product at a half-order rate. What would be the amount converted in a half-hour test?

3. Liquid A decomposes by first order kinetics. In a batch reactor, 50% of A is converted in a five minute run. How much longer will it take to reach 75% conversion?

4. A small reaction bomb fitted with a pressure sensor is purged of inerts, and it is filled with pure reactant A at one atmosphere pressure at 25 °C. At 25 °C, the reaction rate is so slow that the reaction does not proceed to any extent. The bomb is plunged into a 100 °C bath, and additional readings are taken. The reaction proceeds stoichiometrically as $2A \rightarrow B$. After leaving the bomb in the bath for two days, the contents are analyzed and no reactant A is found. Find the rate equation.

t, (min)	π, (atm)	t, (min)	π, (atm)
1	1.14	7	0.850
2	1.04	8	0.832
3	0.982	9	0.815
4	0.940	10	0.800
5	0.905	15	0.754
6	0.870	20	0.728

5. The reaction of ethylene chlorohydrin with sodium bicarbonate to produce ethylene glycol is

$$\underset{80.52}{ClCH_2CH_2OH} + \underset{84.02}{NaHCO_3} \rightarrow \underset{62.07}{HOCH_2CH_2OH} + \underset{58.46}{NaCl} + \underset{44.01}{CO_2}$$

MW: 80.52, 84.02, 62.07, 58.46, 44.01

This is a simple, second-order, irreversible reaction. At 200 °F, $k_c = 1250$ gal/lbmole-hr. It is desired to design a pilot plant stirred continuous overflow reactor to study the economic feasibility of producing ethylene glycol by this reaction. The feed streams available are a 25% aqueous solution of ethylene chlorohydrin and pure dry sodium bicarbonate. The design production rate is 10.0 lbm/hr of ethylene glycol (assume 100% recovery), with a reaction temperature of 200 °F and 99% conversion. A 10% excess bicarbonate solution is used.

Carbon dioxide will evolve from the reacting solution as soon as it is formed. The density of the reacting solution at 200 °F is 1.10 g/cm^3. What reactor volume is required?

6. The liquid phase hydrolysis of dilute aqueous acetic anhydride solutions is second order and irreversible as indicated by the reaction

$$(CH_3CO)_2O + H_2O \rightarrow 2CH_3OOH$$

A batch reactor for carrying out the hydrolysis is charged with 200 liters of anhydride solution at 15 °C. The specific gravity and specific heat of the mixture are constant and equal to 1.05 and 0.9 cal/g-°C, respectively. The heat of reaction has been investigated over a range of temperatures. The rate is found to be a linear function of the acetic anhydride concentration (C) in gmoles/cc.

T °C	rate, gmoles/cm^3-min
10	$0.0567 \times C$
15	$0.0806 \times C$
25	$0.1580 \times C$
40	$0.380 \times C$

(a) Explain why the rate expressions can be written as shown, even though the reaction is second order. (b) If the reactor is cooled so that its operation is isothermal at 15 °C, how much time would be required to obtain a conversion of 70% on the anhydride? (c) Determine an analytical expression for the rate of reaction in terms of temperature and concentration.

7. The oxidation of nitric oxide is a third-order reaction according to the equation

$$2NO + O_2 \rightarrow 2NO_2$$

The rate constant is $k_c = 26,500$ liter^2/gmole^2-sec at 30 °C. Assume the ideal gas law applies. A reactor is charged with a mixture containing 18.42 vol% NO, 10.53 vol% O_2, and 71.05 vol% nitrogen. The initial pressure is 760 mm Hg. Assume an isothermal batch reaction at 30 °C. (a) How much time is required to reach 90% conversion? (b) What is the total pressure and composition at that point?

8. An irreversible reaction between A and B is represented by the stoichiometric equation

$$A + B \rightarrow R + S$$

The reactants and products are liquids, and the reaction is carried out isothermally in a backmix reactor.

The inlet flow rates are 5 gpm for A, and 4 gpm for B. The concentrations of A and B in their respective feed streams are 3 lbmole/gal and 1.5 lbmole/gal. The reaction rate constant is given as $k = 1.0$ gal/lbmole-min.

Assume the design equation will be

$$\frac{V_{reactor}}{\text{v}} = \frac{C_{Bo} - C_B}{kC_A C_B}$$

Determine the volume of the reactor required if the conversion of B is 80% complete. Assume ideal conditions.

9. The statement is commonly made that the rate of a chemical reaction is approximately doubled for each 10 °C rise in temperature. Referring to the differential form of the Arrhenius equation, derive a general relation between absolute temperature and activation energy which must hold if this statement is to be true. Complete the following table.

T, °K	300	400	600	800	1000
$E \frac{\text{kcal}}{\text{gmole}}$					

10. Data taken for the decomposition of compound A in a batch reactor at constant volume is given. The reaction is irreversible. In each case, the initial concentration of A was $C_{Ao} = 15$ gram moles/liter. (a) Determine the order of the reaction. (b) Develop an equation relating the reaction velocity constant to the temperature.

	time, seconds		
C_A gmole/liter	100 °C	120 °C	150 °C
15.0	0.0	0.0	0.0
12.0	111.0	55.3	22.7
10.5	178.0	–	–
7.5	346.0	172.0	70.5
6.0	457.0	–	–
4.5	600.0	300.0	122.4
3.0	805.0	402.2	164.1
1.5	1145.0	573.1	232.0

11. The research department in your company has given you information on the following reaction

$$E \rightarrow 2G$$

Two trial reactions were run for 10 minutes. In both of the cases, the initial concentration of E was 0.175 moles/liter, and the concentration of G was zero. The data given is

at $T = 70$ °F, $(G) = 0.022$ moles/liter after 10 min
at $T = 100$ °F, $(G) = 0.059$ moles/liter after 10 min

Assuming first-order kinetics, how long must a batch reactor be run at 170 °F to produce a conversion of 80% E with initial composition of 0.065 moles/liter?

12. In the isothermal aqueous-phase reaction $A \rightarrow B$, the half-life of A is 1.3 hours. A feed containing A is to be reduced in concentration by 70%. A 200 liter batch reactor and a 400 liter continuous stirred tank reactor are both available. Which one should be chosen in order to get the maximum throughput? Neglect batch charging and discharging times.

13. A compound Q is produced according to the reaction $2P \rightarrow Q + R$. The reaction is second-order and irreversible. The rate constant for the formation of Q is 0.2 liter/mole-hr. (a) What volume batch reactor is required to produce 100 moles of Q in one hour if the concentration of P in the feed is 1.5 moles/liter? (b) Can the batch reactor be run as a continuous stirred tank reactor and produce the required 100 moles of Q per hour? If so, what must the volumetric feed rate be?

14. An aqueous-phase, isothermal, irreversible chemical reaction is to be used to prepare a product D as $A+B \rightarrow D$. The reaction at 100 °F is found to have the kinetic equation

$$r_D = kC_A C_B$$

r_D = rate of formation of D, lbmole/ft^3-min
k = kinetic rate constant, ft^3/lbmole-min
C_A = concentration of A, lbmole/ft^3
C_B = concentration of B, lbmole/ft^3

At 100 °F, $k = 17.1$. An aqueous stream of 6000 ft^3/hr ($C_A = C_B = 0.2$ and $C_D = 0$) is to be processed. A 2000 ft^3 stirred tank jacketed reactor is to be used. Assuming this reactor behaves as an ideal continuous stirred tank reactor, calculate C_D in the outlet stream.

15. The decomposition of a certain gas proceeds according to a second-order reaction

$$2A(g) \rightarrow 2R(g) + S(g)$$

The reverse reaction is negligible. At 950 °C, the reaction velocity constant is 1200 cm^3/gmole-sec. The initial reaction mixture consists of pure A. Calculate the time required to decompose 90% of A at 950 °C in a batch reactor when the pressure is kept constant at one atmosphere. The molecular weight of A is 50.

12 MANAGEMENT THEORIES

1 INTRODUCTION

Effective management techniques are based on behavioral science studies. Behavioral science is an outgrowth of the *human relations theories* of the 1930's, in which the happiness of employees was the goal (i.e., a happy employee is a productive employee ...). Current behavioral science theories emphasize minimizing tensions that inhibit productivity.

There is no evidence yet that employees want a social aspect to their jobs. Nor is there evidence that employees desire job enlargement or autonomy. Behavioral science makes these assumptions anyway.[1]

2 BEHAVIORAL SCIENCE KEY WORDS

Cognitive system: method we use to interpret our environment.

Collaboration: influence through a mutual agreement, relationship, respect, or understanding, without a formal or contractual authority relationship.

Equilibrium: maintaining the status quo of a group or an individual.

Job enrichment: letting employees have more control over their activities and working conditions.

Manipulation: influencing others by recognizing and building upon their needs.

MBO: management by objectives—setting job responsibilities and standards for each group and employee.

Normative judgment: judging others according to our own values.

Paternalism: corporate subsidy—showering employees with benefits and expecting submission in return.

[1] In an exhaustive literature survey up through 1955, researchers found no conclusive relationships between satisfaction and productivity. There was, however, a relationship between lack of satisfaction and absenteeism and turnover.

Personal map: a person's expectations of his environment.

Selective perception: seeing what we want to see—a form of defense mechanism, since things first must be perceived to be ignored.

Superordinate goals: goals which are outside of the individual, such as corporate goals.

3 HISTORY OF BEHAVIORAL SCIENCE STUDIES AND THEORIES

A. HAWTHORNE EXPERIMENTS

From 1927 to 1932, the Hawthorne (Chicago) Works of the Western Electric Company experimented with working conditions in an attempt to determine what factors affected output.[2] Six average employees were chosen to assemble and inspect phone relays.

Many factors were investigated in this exhaustive test. After weeks of observation without making any changes (to establish a baseline), Western Electric varied the number and length of breaks, the length of the work day, the length of the work week, and the illumination level of the lighting. Group incentive plans were tried, and in several tests, the company even provided food during breaks and lunch periods.

The employees reacted in the ways they thought they should. Output (as measured in relay production) increased after every change was implemented. In effect, the employees reacted to the attention they received, regardless of the working conditions.

Western Electric concluded that there was no relationship between illumination and other conditions to productivity. The increase in productivity during the testing procedure was attributed to the sense of value each employee felt in being part of an important test.

[2] Experiments were conducted by Elton Mayo from the Harvard Business School.

MGMT

The employees also became a social group, even after hours. Leadership and common purpose developed, and even though the employees were watched more than ever, they felt no supervision anxiety since they were, in effect, free to react in any way they wanted.

One employee summed up the test when she said, "It was fun."

B. BANK WIRING OBSERVATION ROOM EXPERIMENTS

In an attempt to devise an experiment which would not suffer from the problems associated with the Hawthorne studies, Western Electric conducted experiments in 1931 and 1932 on the effects of wage incentives.

The group of nine wiremen, three soldermen, and two inspectors was interdependent. This was supposed to prevent any individual from slacking off. However, wage incentives failed to improve productivity. In fact, fast employees slowed down to protect their slower friends. Illicit activities, such as job trading and helping, also occurred.

The group was reacting to the notion of a *proper day's work*. When the day's work (or what the group considered to be a day's work) was assured, the whole group slacked off. The group also varied what it reported as having been accomplished and claimed more unavoidable delays than actually occurred. The output was essentially constant.

Western Electric concluded that social groups form as protection against arbitrary management decisions, even when such decisions have never been experienced. The effort to form the social groups, to protect slow workers, and to develop the notion of a proper day's work is not conscious. It develops automatically when the company fails to communicate to the contrary.

C. NEED HIERARCHY THEORY

During World War II, Dr. Abraham Maslow's *need hierarchy theory* was implemented into leadership training for the U.S. Air Force. This theory claims that certain needs become dominant when lesser needs are satisfied. Although some needs can be sublimated and others overlap, the need hierarchy theory generally requires the lower-level needs to be satisfied before the higher-level needs are realized. (The ego and self-fulfillment needs rarely are satisfied.)

Table 12.1
The Need Hierarchy

(In order of lower to higher needs)

1. Physiological needs: air, food, water.
2. Safety needs: protection against danger, threat, deprivation, arbitrary management decisions. Need for security in a dependent relationship.
3. Social needs: belonging, association, acceptance, giving and receiving of love and friendship.
4. Ego needs: self-respect and confidence, achievement, self-image. Group image and reputation, status, recognition, appreciation.
5. Self-fulfillment needs: realizing self potential, self development, creativity.

The need hierarchy theory explains why money is a poor motivator of an affluent individual. The theory does not explain how management should apply the need hierarchy to improve productivity.

D. THEORY OF INFLUENCE

In 1948, the Human Relations Program (under the direction of Donald C. Pelz) at Detroit Edison studied the effectiveness of its supervisors. The most effective supervisors were those who helped their employees benefit. Supervisors who were close to their employees (and sided with them in disputes) were effective only if they were influential enough to help the employees. The study results were formulated into the *theory of influence*.

- Employees think well of supervisors who help them reach their goals and meet their needs.
- An influential supervisor will be able to help employees.
- An influential supervisor who is also a disciplinarian will breed dissatisfaction.
- A supervisor with no influence will not be able to affect worker satisfaction in any way.

The implication of the theory of influence is that whether or not a supervisor is effective depends on his influence. Training of supervisors is useless unless they have the power to implement what they have learned. Also, increases in supervisor influence are necessary to increase employee satisfaction.

E. HERZBERG MOTIVATION STUDIES

Frederick W. Herzberg interviewed 200 technical personnel in 11 firms during the late 1950's. Herzberg was especially interested in exceptional occurrences resulting in increases in job satisfaction and performance. From those interviews, Herzberg formulated his *motivation-maintenance theory.*

According to this theory, there are satisfiers and dissatisfiers which influence employee behavior. The *dissatisfiers* (also called *maintenance/motivation factors*) do not motivate employees; they can only dissatisfy them. However, the dissatisfiers must be eliminated before the satisfiers work. Dissatisfiers include company policy, administration, supervision, salary, working conditions (environment), and interpersonal relations.

Satisfiers (also known as *motivators*) determine job satisfaction. Common satisfiers are achievement, recognition, the type of work itself, responsibility, and advancement.

An interesting conclusion based on the motivation-maintenance theory is that fringe benefits and company paternalism do not motivate employees since they are related to dissatisfiers only.

F. THEORY X AND THEORY Y

During the 1950's Douglas McGregor (Sloan School of Industrial Management at MIT) introduced the concept that management had two ways of thinking about its employees. One way of thinking, which was largely pessimistic, was theory X. The other theory, theory Y, was largely optimistic.

Theory X is based on the assumption that the average employee inherently dislikes and avoids work. Therefore, employees must be coerced into working by threats of punishment. Rewards are not sufficient. The average employee wants to be directed, avoids responsibility, and seeks the security of an employer-employee relationship.

This assumption is supported by much evidence. Employees exist in a continuum of wants, needs, and desires. Many of the need satisfiers (salary, fringe benefits, etc.) are effective only off the job. Therefore, work is considered a punishment or a price paid for off-the-job satisfaction.

Theory X is pessimistic about the effectiveness of employers to satisfy or motivate their employees. By satisfying the physiological and safety (lower level) needs, employers have shifted the emphasis to higher level needs which they cannot satisfy. Employees, unable to derive satisfaction from their work, behave according to theory X.

Theory Y, on the other hand, assumes that the expenditure of effort is natural and is not inherently disliked. It assumes that the average employee can learn to accept and enjoy responsibility. Creativity is widely distributed among employees, and the potentials of average employees are only partially realized.

Theory Y places the blame for worker laziness, indifference, and lack of cooperation in the lap of management, since the integration of individual and organization needs is required. This theory is not fully validated, nor is its full use ever likely to be implemented.[3]

4 JOB ENRICHMENT

In an effort to make their employees happier, companies have tried to enrich the jobs performed by employees. Enrichment is a subjective result felt by employees when their jobs are made more flexible or are enlarged. Adding flexibility to a job allows an employee to move from one task to another, rather than doing the same thing continually. Horizontal job enlargement adds new production activities to a job. Vertical job enlargement adds planning, inspection, and other non-production tasks to the job.

There are advantages to keeping a job small in scope. Learning time is low, employee mental effort is reduced, and the pay rate can be lower for untrained labor. Supervision is reduced. Such simple jobs, however, also result in high turnover, absenteeism, and lower pride in job (and subsequent low quality rates).

Job enlargement generally results in better quality products, reduces inspection and material handling, and counteracts the disadvantages previously mentioned. However, training time is greater, tooling costs are higher, and inventory records are more complex.

5 QUALITY IMPROVEMENT PROGRAMS

A. ZERO DEFECTS PROGRAM

Employees have been conditioned to believe that they are not perfect and that errors are natural. However, we demand zero defects from some professions (e.g., doctors, lawyers, engineers). The philosophy of a zero defect program is to expect zero defects from everybody.

[3] Theory Y is not synonymous with soft management. Rather than emphasize tough management (as does theory X), theory Y depends on commitment of employees to achieve mutual goals.

Zero defects programs develop a constant, conscious desire to do the job right the first time. This is accomplished by giving employees constant awareness that their jobs are important, that the product is important, and that management thinks their efforts are important.

Zero defects programs try to correct the faults of other types of employee programs.[4] Programs are based on what the employee has for his own: pride and desire. The programs present the challenge of perfection and explain the importance of that perfection. Management sets an example by expecting zero defects of itself. Standards of performance are set and are related to each employee. Employees are checked against these performance requirements periodically, and recognition is given when goals are met.

[4] Motivational programs are not honest, according to the zero-defects theory, since management tries to convince employees to do what management wants. Wage incentive programs encourage employee dishonesty and errors by emphasizing quantity, not quality. Theory X management, with its implied punitive action if goals are not achieved, never has been effective.

B. QUALITY CIRCLES/TEAM PROGRAMS

Quality circle programs are voluntary or required programs in which employees within a department actively participate in measuring and improving quality and performance. It involves periodic meetings on a weekly or a monthly basis. Workers are encouraged to participate in volunteering ideas for improvement.

MGMT

13 MISCELLANEOUS TOPICS

PART 1: Accuracy and Precision Experiments

1 ACCURACY

An experiment is said to be *accurate* if it is unaffected by experimental error. In this case, *error* is not synonymous with *mistake*, but rather includes all variations not within the experimenter's control.

For example, suppose a gun is aimed at a point on a target and five shots are fired. The mean distance from the point of impact to the sight-in point is a measure of the alignment accuracy between the barrel and sights. The difference between the actual value and the experimental value is known as *bias*.

2 PRECISION

Precision is not synonymous with accuracy. Precision is concerned with the repeatability of the experimental results. If an experiment is repeated with identical results, the experiment is said to be precise.

In the previous example, the average distance of each impact from the centroid of the impact group is a measure of the precision of the experiment. Thus, it is possible to have a highly precise experiment with a large bias.

Most techniques applied to experiments to improve the accuracy of the experimental results (e.g., repeating the experiment, refining the experimental methods, or reducing variability) actually increase the precision.

Sometimes the word *reliability* is used with regards to the precision of an experiment. A reliable estimate is used in the same sense as a precise estimate.

3 STABILITY

Stability and *insensitivity* are synonymous terms. A stable experiment is insensitive to minor changes in the experiment parameters. Suppose the centroid of a bullet group is 2.1 inches away from the sight-in point at 65 °F and 2.3 inches away at 80 °F. The experiment's sensitivity to temperature changes would be $\frac{2.3 - 2.1}{80 - 65} = 0.0133$ inches/°F.

PART 2: Dimensional Analysis

Nomenclature

c_p	specific heat	BTU/lbm-°F
C_i	a constant	–
D	diameter	ft
F	force	lbf
g_c	gravitational constant (32.2)	lbm-ft/sec²-lbf
$\overline{h}$	average film coefficient	BTU/hr-ft²-°F
J	Joule's constant (778)	ft-lbf/BTU
k	number of pi-groups $(m - n)$	–
L	length	ft
m	number of relevant independent variables	–
M	mass	lbm
n	number of independent dimensional quantities	–
N_{Nu}	Nusselt number	–
N_{Pe}	Peclet number	–
N_{Re}	Reynolds number	–
v	velocity	ft/sec
x_i	the ith independent variable	various
y	dependent variable	various

MISC

Symbols

ρ	density	lbm/ft^3
θ	time	sec
π_i	ith dimensionless group	–
μ	viscosity	lbm/ft-sec

Dimensional analysis is a means of obtaining an equation for some phenomenon without understanding the inner mechanism of the phenomenon. The most serious limitation to this method is the need to know beforehand which variables influence the phenomenon. Once these variables are known or are assumed, dimensional analysis can be applied by a routine procedure.

The first step is to select a system of primary dimensions. Usually the MLθT system (mass, length, time, and temperature) is used, although this choice may require the use of g_c and J in the final results. The dimensional formulas and symbols for variables most frequently encountered are given in table 13.1.

The second step is to write a functional relationship between the dependent variable and the independent variables, x_i.

$$y = \mathbf{f}(x_1, x_2, \ldots, x_m) \qquad 13.1$$

This function can be expressed as an exponentiated series.

$$y = C_1 x_1^{a_1} x_2^{b_1} x_3^{c_1} \cdots x_m^{z_1} + C_2 x_1^{a_2} x_2^{b_2} x_3^{c_2} \cdots x_m^{z_2} + \cdots \qquad 13.2$$

The C_i, a_i, b_i, $\cdots z_i$ in equation 13.2 are unknown constants.

The key to solving the above equation is that each term on the right-hand side must have the same dimensions as y. Simultaneous equations are used to determine some of the a_i, b_i, c_i, and z_i. Experimental data is required to determine the C_i and the remaining exponents. In most analyses, it is assumed that the $C_i = 0$ for $i = 2$ and up.

Table 13.1
Units and Dimensions of Typical Variables

quantity	symbol	MLθT system	MLθTFQ system	units in engineering system
length	L or x	L	L	ft
time	θ	θ	θ	sec or hour
mass	M	M	M	lbm
force	F	ML/θ^2	F	lbf
temperature	T	T	T	°F
heat	Q	ML^2/θ^2	F	BTU
velocity	V	L/θ	L/θ	ft/sec
acceleration	a or g	L/θ^2	L/θ^2	ft/sec^2
dimensional conversion factor	g_c	none	ML/θ^2F	32.2 lbm-ft/sec^2-lbf
energy conversion factor	J	none	FL/Q	778 ft-lbf/BTU
work	W	ML^2/θ^2	FL	ft-lbf
pressure	p	M/θ^2L	F/L^2	lbf/ft^2
density	ρ	M/L^3	M/L^3	lbm/ft^3
internal energy and enthalpy	u, h	L^2/θ^2	Q/M	BTU/lbm
specific heat	c	L^2/θ^2T	Q/MT	BTU/lbm-°F
dynamic viscosity	μ_f	M/Lθ	Fθ/L^2	lbf-sec/ft^2
absolute viscosity	μ	M/Lθ	M/Lθ	lbm/ft-sec
kinematic viscosity	$\nu = \mu\rho$	L^2/θ	L^2/θ	ft^2/sec
thermal conductivity	k	ML/θ^3T	Q/LTθ	BTU/hr-ft-°F
coefficient of expansion	β	1/T	1/T	1/°F
surface tension	σ	M/θ^2	F/L	lbf/ft
stress	σ or τ	M/Lθ^2	F/L^2	lbf/ft^2
film coefficient	h	M/θ^3T	Q/θL^{2}T	BTU/hr-ft^2-°F
mass flow rate	m	M/θ	M/θ	lbm/sec

MISC

Example 13.1

A sphere submerged in a fluid rolls down an incline. Find an equation for the velocity, v.

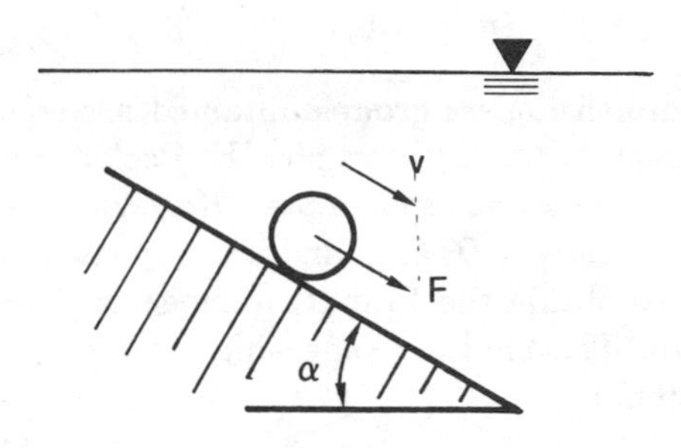

It is assumed that the velocity depends on the force, F, due to the inclination, the diameter of the sphere, D, the density of the fluid, ρ, and the viscosity of the fluid, μ.

$$\mathrm{v} = \mathbf{f}(F, D, \rho, \mu)$$

This equation can be written in terms of the dimensions of the variables.

$$\frac{L}{\theta} = C\left(\frac{ML}{\theta^2}\right)^a (L)^b \left(\frac{M}{L^3}\right)^c \left(\frac{M}{L\theta}\right)^d$$

Since L on the left-hand side has an implied exponent of one, the necessary equation is

$$1 = a + b - 3c - d \qquad (L)$$

Similarly, the other necessary equations are

$$-1 = -2a - d \qquad (\theta)$$
$$0 = a + c + d \qquad (M)$$

Solving simultaneously yields

$$b = -1$$
$$c = a - 1$$
$$d = 1 - 2a$$

or

$$\mathrm{v} = C\left(\frac{\mu}{D\rho}\right)\left(\frac{F\rho}{\mu^2}\right)^a$$

C and a would have to be determined experimentally.

Since the above method requires working with m different variables and n different independent dimensional quantities (such as M, L, T, and θ), an easier method is desirable. One simplification is to combine the m variables into dimensionless groups, called *pi-groups*.

If these dimensionless groups are represented by π_1, π_2, π_3, $\cdots \pi_k$, the equation expressing the relationship between the variables is given by the *Buckingham π-theorem.*

$$\mathbf{f}(\pi_1, \pi_2, \pi_3, \ldots, \pi_k) = 0 \qquad 13.3$$
$$k = m - n \qquad 13.4$$

The dimensionless pi-groups usually are found from the m variables according to an intuitive process. A formalized method is possible as long as the following conditions are met.

- The dependent variable and independent variables chosen contain all of the variables affecting the phenomenon. Extraneous variables can be included at the expense of obtaining extra pi-groups.
- The pi-groups must include all of the original x_i at least once.
- The dimensions all must be independent.

The formal procedure is to select n variables (x_i) out of the total m as repeating variables to appear in all k pi-groups. These variables are used in turn with the remaining variables in each successive pi-group. Each of the repeating variables must have different dimensions, and the repeating variables collectively must contain all of the dimensions. This procedure is illustrated in example 13.2.

Example 13.2

It is desired to determine a relationship giving the heat transfer to air flowing across a heated tube. The following variables affect the heat flow.

variable	symbol	dimensional equation
tube diameter	D	L
fluid conductivity	k	ML/θ^3T
fluid velocity	v	L/θ
fluid density	ρ	M/L^3
fluid viscosity	μ	$M/L\theta$
fluid specific heat	c_p	L^2/θ^2T
film coefficient	$\bar{h}$	M/θ^3T

There are $m = 7$ variables and $n = 4$ primary dimensions (L, M, θ, and T). Accordingly, there are $k = 7 - 4 = 3$ dimensionless groups that are required to correlate the data. The four repeating variables are chosen such that all dimensions are represented. Then the π_i are written as functions of these repeating variables in turn with the remaining variables.

The repeating variables should not include any of the unknown quantities. For example, $\bar{h}$ should not be chosen as a repeating variable since it is directly related to the unknown heat flow. In addition, important material properties, such as c_p and k, often are omitted. Trial and error is required to include all four primary dimensions.

Using trial and error, omitting $\overline{h}$ as a repeating variable, and representing all four primary dimensions, arbitrarily choose the variables as D, k, v, and ρ.

The pi-groups are

$$\pi_1 = D^{a_1} k^{a_2} v^{a_3} \rho^{a_4} \mu$$
$$\pi_2 = D^{a_5} k^{a_6} v^{a_7} \rho^{a_8} c_p$$
$$\pi_3 = D^{a_9} k^{a_{10}} v^{a_{11}} \rho^{a_{12}} \overline{h}$$

Since the π_i are dimensionless, we write for π_1

$$\begin{aligned} 0 &= a_1 + a_2 + a_3 - 3a_4 - 1 && \text{(L)} \\ 0 &= a_2 + a_4 + 1 && \text{(M)} \\ 0 &= -3a_2 - a_3 - 1 && (\theta) \\ 0 &= -a_2 && \text{(T)} \end{aligned}$$

Therefore,

$$a_2 = 0 \quad a_3 = -1 \quad a_4 = -1 \quad a_1 = -1$$
$$\pi_1 = \frac{\mu}{Dv\rho}$$

π_1 is the reciprocal of the Reynolds number. Proceeding similarly with π_2,

$$\begin{aligned} 0 &= a_5 + a_6 + a_7 = 3a_8 + 2 && \text{(L)} \\ 0 &= a_6 + a_8 && \text{(M)} \\ 0 &= -3a_6 - a_7 - 2 && (\theta) \\ 0 &= -a_6 - 1 && \text{(T)} \end{aligned}$$

Therefore,

$$a_6 = -1 \quad a_7 = 1 \quad a_8 = 1 \quad a_5 = 1$$
$$\pi_2 = \frac{D v \rho c_p}{k}$$

π_2 is the *Peclet number* (product of the Reynolds number and the Prandtl number).

π_3 is found to be $\frac{D\overline{h}}{k}$, which is the *Nusselt number*.

The seven original variables have been combined into three dimensionless groups, making data correlation much easier. The implicit equation for heat transfer is

$$\mathbf{f}_1(\pi_1, \pi_2, \pi_3) = \mathbf{f}_1(N_{Re}, N_{Nu}, N_{Pe}) = 0$$

Rearrangement of the pi-groups is needed to isolate the dependent variable (in this case, $\overline{h}$).

$$N_{Nu} = \mathbf{f}_2(N_{Re}, N_{Pe}) = C(N_{Re})^{e_1}(N_{Pe})^{e_2}$$

C, e_1, and e_2 are found experimentally.

The selection of the repeating and non-repeating variables is the key step. The choice of repeating variables determines which dimensionless groups are obtained. The theoretical maximum number of valid dimensionless groups is

$$\frac{m!}{(n+1)!(m-n-1)!} \qquad 13.5$$

Not all dimensionless groups obtained are equally useful to researchers. For example, the Peclet number was obtained in the above example. However, researchers would have chosen D, k, ρ, and μ as repeating variables in order to obtain the Prandtl number as a dimensionless group. This choice of repeating variables is a matter of intuition.

PART 3: Reliability

Nomenclature

$\mathbf{f}(t)$	probability density function	–
$\mathbf{F}(t)$	cumulative density function	–
k	minimum number for operation	–
MTBF	mean time before failure	time
n	number of items in the system	–
R*	system reliability	–
$\mathbf{R}_i(t)$	ith item reliability	–
t	time	time
x	number of failures	–
X	binary ith item performance variable	–
Y	arbitrary event	–
$\mathbf{z}(t)$	hazard function	1/time

Symbols

λ	constant failure or hazard rate	1/time
ϕ	binary system performance variable	–

1 ITEM RELIABILITY

Reliability as a function of time, $\mathbf{R}(t)$, is the probability that an item will continue to operate satisfactorily up to time t. Although other distributions are possible, reliability often is described by the *negative exponential distribution.* Specifically, it is assumed that an item's reliability is

$$\mathbf{R}(t) = 1 - \mathbf{F}(t) = e^{-\lambda t} = e^{-t/\text{MTBF}} \qquad 13.6$$

MISC

This infers that the probability of x failures in a period of time is given by the Poisson distribution.

$$p\{x\} = \frac{e^{-\lambda}\lambda^x}{x!} \qquad 13.7$$

The negative exponential distribution is appropriate whenever an item fails only by random causes but never experiences deterioration during its life. This implies that the *expected future life* of an item is independent of the previous duration of operation.

Example 13.3

An item exhibits an exponential time to failure distribution with MTBF of 1000 hours. What is the maximum operating time such that the reliability does not drop below 0.99?

$$0.99 = e^{-t/1000}$$
$$t = 10.05 \text{ hours}$$

The *hazard function* is defined as the conditional probability of failure in the next time interval given that no failure has occurred thus far. For the exponential distribution, the hazard function is

$$\mathbf{z}(t) = \lambda \qquad 13.8$$

Since this is not a function of t, exponential failure rates are not dependent on the length of time previously in operation.

In general,

$$\mathbf{z}(t) = \frac{\mathbf{f}(t)}{\mathbf{R}(t)} = \frac{\frac{d\mathbf{F}(t)}{dt}}{1 - \mathbf{F}(t)} \qquad 13.9$$

The exponential distribution is summarized by equations 13.10 through 13.13.

$$\mathbf{f}(t) = \lambda e^{-\lambda t} \qquad 13.10$$
$$\mathbf{F}(t) = 1 - e^{-\lambda t} \qquad 13.11$$
$$\mathbf{R}(t) = 1 - \mathbf{F}(t) = e^{-\lambda t} \qquad 13.12$$
$$\mathbf{z}(t) = \frac{\lambda e^{-\lambda t}}{e^{-\lambda t}} = \lambda \qquad 13.13$$

2 SYSTEM RELIABILITY

The binary variable, X_i, is defined as 1 if item i operates satisfactorily and 0 otherwise. Similarly, the binary variable, ϕ, is 1 only if the system operates satisfactorily. ϕ will be a function of the X_i.

A. SERIAL SYSTEMS

The *performance function* for a system of n serial items is

$$\phi = X_1X_2X_3\ldots X_n = \min\{X_i\} \qquad 13.14$$

Equation 13.18 implies that the system will fail if any of the individual items fail. The system reliability is

$$R^* = R_1R_2R_3\ldots R_n \qquad 13.15$$

Example 13.4

A block diagram of a system with item reliabilities is shown. What is the performance function and the system reliability?

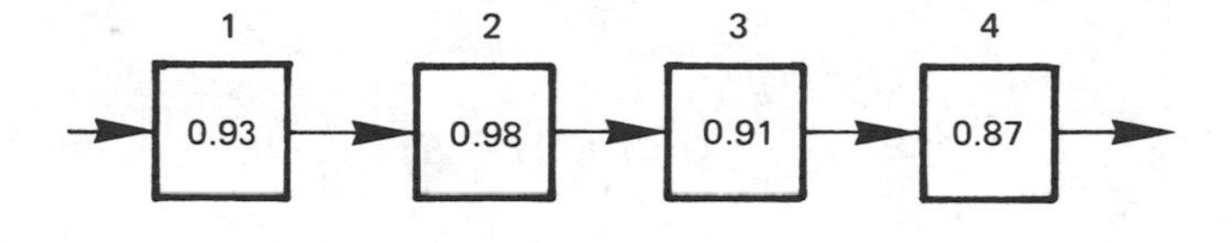

$$\phi = X_1X_2X_3X_4$$
$$R^* = (0.93)(0.98)(0.91)(0.87) = 0.72$$

B. PARALLEL SYSTEMS

A parallel system with n items will fail only if all n items fail. This property is called *redundancy*, and such a system is said to be redundant. Using redundancy, a highly reliable system can be produced from components with relatively low individual reliabilities.

The performance function of a redundant system is

$$\phi = 1 - (1 - X_1)(1 - X_2)(1 - X_3)\cdots(1 - X_n)$$
$$= \max\{X_i\} \qquad 13.16$$

The reliability is

$$R^* = 1 - (1 - R_1)(1 - R_2)(1 - R_3)\cdots(1 - R_n) \qquad 13.17$$

Example 13.5

What is the reliability of the system shown?

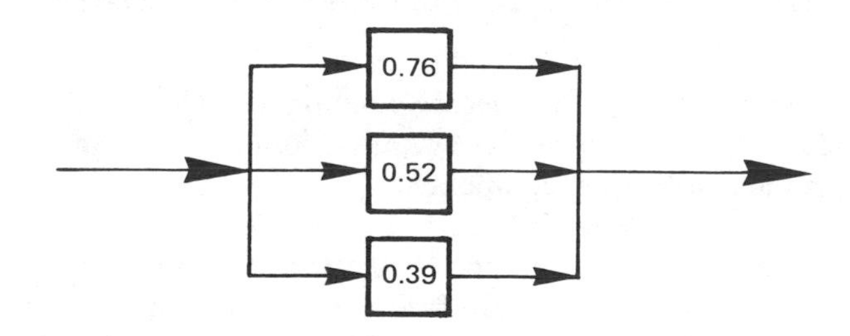

$$R^* = 1 - (1 - 0.76)(1 - 0.52)(1 - 0.39) = 0.93$$

C. k-out-of-n SYSTEMS

If the system operates with an k of its elements operational, it is said to be a k-out-of-n system. The performance function is

$$\phi = \begin{cases} 1 \text{ if } \Sigma X_i \geq k \\ 0 \text{ if } \Sigma X_i < k \end{cases} \qquad 13.18$$

The evaluation of the system reliability is quite difficult unless all elements are identical and have identical reliabilities, **R**. In that case, the system reliability follows the binomial distribution.

$$R^* = \sum_{j=k}^{n} \binom{n}{j} R^j (1-R)^{n-j} \qquad 13.19$$

D. GENERAL SYSTEM RELIABILITY

A general system can be represented by a graphical network. Each path through the network from the starting node to the finishing node represents a possible operating path. For the 5-path network in figure 13.1, even if BD and AC are cut, the system will operate by way of path ABCD.

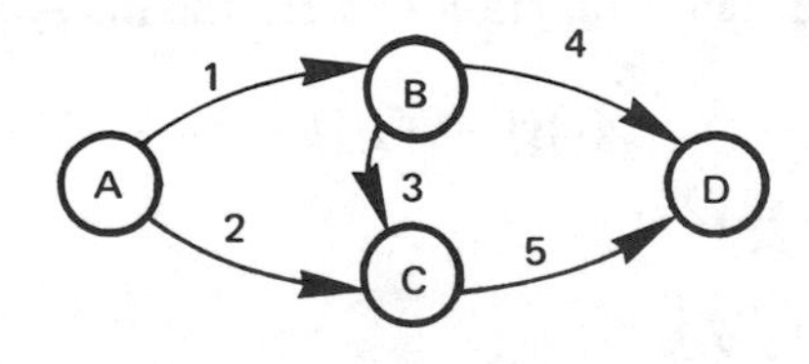

Figure 13.1 A 5-Path Network

The reliability of the system will be the sum of the serial reliabilities, summed over all possible paths in the system. However, the concepts of minimal paths and minimal cuts are required to facilitate the evaluation of the system reliability.

A *minimal path* is a set of components that, if operational, will ensure the system's functioning. In the previous example, components [1 with 4] are a minimal path, as are [2 with 5] and [1 with 3 with 5]. A *minimal cut* is a set of components that, if non-functional, inhibits the system from functioning. Minimal cuts in the previous example are [1 with 2], [4 with 5], [1 with 5], and [2 with 3 with 4].

Since it usually is easier to determine all minimal paths, a method of finding the exact system reliability from the set of minimal paths is needed. In general, the probability of a union of n events contains $(2^n - 1)$ terms and is given by

$$\begin{aligned} &p\{Y_1 \text{ or } Y_2 \text{ or } \cdots Y_n\} = p\{Y_1\} + p\{Y_2\} + p\{Y_3\} \\ &+ \cdots + p\{Y_n\} - p\{Y_1 \text{ and } Y_2\} - p\{Y_1 \text{ and } Y_3\} \\ &- \cdots - p\{Y_1 \text{ and } Y_n\} - p\{Y_1 \text{ and } Y_2 \text{ and } Y_3\} \\ &- p\{Y_1 \text{ and } Y_2 \text{ and } Y_4\} - \cdots - p\{Y_i \text{ and } Y_j \text{ and } Y_k\} \\ &\text{all } i \neq j \neq k \\ &+ \{-1\}^{n-1} p\{Y_1 \text{ and } Y_2 \text{ and } Y_3 \text{ and } \ldots \text{ and } Y_n\} \end{aligned} \qquad 13.20$$

Returning to the 5-path example,

$$\begin{aligned} Y_1 &= [1 \text{ with } 4] \\ Y_2 &= [2 \text{ with } 5] \\ Y_3 &= [1 \text{ with } 3 \text{ with } 5] \end{aligned}$$

Then,

$$\begin{aligned} p\{\phi = 1\} &= p\{Y_1 \text{ or } Y_2 \text{ or } Y_3\} \\ &= p\{X_1X_4 = 1\} + p\{X_2X_5 = 1\} \\ &\quad + p\{X_1X_3X_5 = 1\} - p\{X_1X_2X_4X_5 = 1\} \\ &\quad - p\{X_1X_3X_4X_5 = 1\} \\ &\quad - p\{X_1X_2X_3X_5 = 1\} \\ &\quad + p\{X_1X_2X_3X_4X_5 = 1\} \end{aligned}$$

In terms of the individual item reliabilities, this is

$$\begin{aligned} R^* = {} & R_1R_4 + R_2R_5 + R_1R_3R_5 \\ & - R_1R_2R_4R_5 - R_1R_3R_4R_5 - R_1R_2R_3R_5 \\ & + R_1R_2R_3R_4R_5 \end{aligned}$$

In the 5-path example given,

$$R^* \leq p\{X_1X_4 = 1\} + p\{X_2X_5 = 1\} + p\{X_1X_3X_5 = 1\}$$

This method requires considerable computation, and an upper bound on R^* would be sufficient. Such an upper bound is close to R^* since the product of individual reliabilities is small. The upper bound is given by

$$p\{\phi = 1\} \leq p\{Y_1\} + p\{Y_2\} + \cdots + p\{Y_n\} \qquad 13.21$$

MISC

PART 4: Replacement

Nomenclature

C_1	item replacement cost with group replacement
C_2	item replacement cost after individual failure
$\mathbf{F}(t)$	number of units failing in the interval ending at t
$\mathbf{K}(t)$	total cost of operating from $t = 0$ to $t = T$
MTBF	mean time before failure
n	number of units in original system
$p\{t\}$	probability of failing in the interval ending at t
$\mathbf{S}(t)$	number of survivors at the end of time t
t	time
T	time to replacement (life)
$\mathrm{v}\{t\}$	conditional probability of failure in the interval $(t-1)$ to t given non-failure before $(t-1)$

1 INTRODUCTION

Replacement and renewal models determine the most economical time to replace existing equipment. Replacement processes fall into two categories, depending on the life pattern of the equipment, which either deteriorates gradually (becomes obsolete or less efficient) or fails suddenly.

In the case of gradual deterioration, the solution consists of balancing the cost of new equipment against the cost of maintenance or decreased efficiency of the old equipment. Several models are available for cases with specialized assumptions, but no general solution methods exist.

In the case of sudden failure, of which light bulbs are examples, the solution consists of finding a replacement frequency which minimizes the costs of the required new items, the labor for replacement, and the expected cost of failure. The solution is made difficult by the probabilistic nature of the life spans.

2 DETERIORATION MODELS

The replacement criterion with deterioration models is the present worth of all future costs associated with each policy. Solution is by trial and error, calculating the present worth of each policy and incrementing the replacement period by one time period for each iteration.

Example 13.6

Item A currently is in use. Its maintenance cost is \$400 this year, increasing each year by \$30. Item A can be replaced by item B at a current cost of \$3500. However, the cost of B is increasing by \$50 each year. Item B has no maintenance costs. Disregarding income taxes, find the optimum replacement year. Use 10% as the interest rate.

Calculate the present worth of the various policies.

policy 1: Replacement at $t = 5$ (starting the 6th year)

$$PW(A) = -400\left(\frac{P}{A}, 10\%, 5\right) - 30\left(\frac{P}{G}, 10\%, 5\right) = -1722$$

$$PW(B) = -[3500 + 5(50)]\left(\frac{P}{F}, 10\%, 5\right) = -2328$$

policy 2: Replacement at $t = 6$

$$PW(A) = -400\left(\frac{P}{A}, 10\%, 6\right) - 30\left(\frac{P}{G}, 10\%, 6\right) = -2033$$

$$PW(B) = -[3500 + 6(50)]\left(\frac{P}{F}, 10\%, 6\right) = -2145$$

policy 3: Replacement at $t = 7$

$$PW(A) = -400\left(\frac{P}{A}, 10\%, 7\right) - 30\left(\frac{P}{G}, 10\%, 7\right) = -2330$$

$$PW(B) = -[3500 + 7(50)]\left(\frac{P}{F}, 10\%, 7\right) = -1975$$

The present worth of B drops below the present worth of A at $t = 6$. Replacement should take place at that time.

3 FAILURE MODELS

The time between installation and failure is not constant for members in the general equipment population. Therefore, in order to solve a failure model, it is necessary to have the distribution of individual item lives (*mortality curve*). The *conditional probability of failure* in a small time interval, say from t to $(t + \delta t)$, is calculated from the mortality curve. This probability is *conditional* since it is conditioned on non-failure up to time t.

The conditional probability can decrease with time (e.g., *infant mortality*), remain constant (as with an exponential reliability distribution and failure from random causes), or increase with time (as with items that deteriorate with use). If the conditional probability decreases or remains constant over time, operating items should never be replaced prior to failure.

It usually is assumed that all failures occur at the end of a period. The problem is to find the period which minimizes the total cost.

Example 13.7

100 items are tested to failure. Two failed at $t = 1$, five at $t = 2$, seven at $t = 3$, 20 at $t = 4$, 35 at $t = 5$, and 31 at $t = 6$. Find the probability of failure in any period, the conditional probability of failure, and the mean time before failure.

The MTBF is

$$\frac{(2)(1) + (5)(2) + (7)(3) + (20)(4) + (35)(5) + (31)(6)}{100}$$

$$= 4.74$$

elapsed time t	failures $\mathbf{F}(t)$	survivors $\mathbf{S}(t)$	probability of failure $p\{t\} = 0.01\mathbf{F}(t)$	conditional probability of failure $v\{t\} = \frac{\mathbf{F}(t)}{\mathbf{S}(t-1)}$
0	0	100	—	—
1	2	98	0.02	0.02
2	5	93	0.05	0.051
3	7	86	0.07	0.075
4	20	66	0.20	0.233
5	35	31	0.35	0.530
6	31	0	0.31	1.00

4 REPLACEMENT POLICY

The expression for the number of units failing in time t is

$$\mathbf{F}(t) = n\left[p\{t\} + \sum_{i=1}^{t-1} p\{i\}p\{t-i\} + \sum_{j=2}^{t-1}\left[\sum_{i=1}^{j-1} p\{i\}p\{j-i\}\right]p\{t-j\} + \cdots\right] \quad 13.22$$

The term $np\{t\}$ gives the number of failures in time t from the original group.

The term $n\sum p\{i\}p\{t-i\}$ gives the number of failures in time t from the set of items which replaced the original items.

The third probability term times n gives the number of failures in time t from the set of items which replaced the first replacement set.

It can be shown that $\mathbf{F}(t)$ with replacement will converge to a steady state limiting rate of

$$\overline{\mathbf{F}(t)} = \frac{n}{\text{MTBF}} \quad 13.23$$

The optimum policy is to replace all items in the group, including items just installed, when the total cost per period is minimized. That is, we want to find T such that $\mathbf{K}(T)/T$ is minimized.

$$\mathbf{K}(T) = nC_1 + C_2\sum_{t=0}^{T-1}\mathbf{F}(t) \quad 13.24$$

Discounting usually is not included in the total cost formula since the time periods are considered short. If the equipment has an unusually long life, discounting is required.

There are some cases where group replacement always is more expensive than replacing just the failures as they occur. Group replacement will be the most economical policy if equation 13.25 holds.

$$C_2[\overline{\mathbf{F}(t)}] > \left.\frac{\mathbf{K}(T)}{T}\right|_{\text{minimum}} \quad 13.25$$

If the opposite inequality holds, group replacement still may be the optimum policy. Further analysis is required.

PART 5: FORTRAN Programming

The FORTRAN language currently exists in several versions. Although differences exist between compilers, these are relatively minor. However, some of the instructions listed in this chapter may not be compatible with all compilers.

This section is not intended as instruction in FORTRAN programming, but rather serves as a documentation of the language.

1 STRUCTURAL ELEMENTS

Symbols are limited to the upper-case alphabet, digits 0 through 9, the blank, and the following special characters.

+ = - * / ( ) , . $

Statements written with these characters generally are prepared in an 80-column format. Statements are executed sequentially regardless of the statement numbers.

position	use
1	The letter C is used for a *comment.* Comments are not executed.
2–5	The statement number, if used, is placed in positions 2 through 5. Statement numbers can be any integers from 1 through 9999.
6	Any character except *zero* can be placed in position 6 to indicate a continuation from the previous statement.
7–72	The FORTRAN statement is placed here.
73–80	These positions are available for any use and are ignored by the compiler. Usually, the final debugged program is numbered sequentially in these positions.

FORTRAN compilers pack the characters. Therefore, blanks can be inserted at any place in most statements. For example, the following statements are compiled the same way.

```
IF (AGE.LT.YEARS) GO TO 10
IF(AGE.LT.YEARS)GOTO10
```

2 DATA

Numerical data can be either real or integer. *Integers* usually are limited to nine digits. Unsigned integers and integers preceded by a plus sign are the same. Commas are not allowed in integer constants. For example, ninety thousand would be written as 90000, not 90,000.

Real numbers are distinguished from integers by a decimal point and may contain a fractional part. Scientific notation is indicated by the single letter E. Real numbers are limited to one decimal point and usually seven digits.

value	FORTRAN notation
2 million	2. E6
0.00074	7.4 E−4
2.	2.

3 VARIABLES

Variable names can be formed from up to six alphanumeric characters. The first character must be a letter. Variable names starting with the letters I, J, K, L, M, or N are assumed by the compiler to be integers unless defined otherwise by an *explicit typing statement.* All other variable names represent real variables, unless explicitly typed.

The type convention can be overridden in an explicit typing statement. This is done by defining the desired variable type in the first part of the program with an INTEGER or REAL statement. For example, the statements

```
INTEGER TIME, CLOCK
REAL INSTANT
```

would establish TIME and CLOCK as integer variables and INSTANT as a real variable. The order of such declarations is unimportant. Variables following the standard type convention (implicit typing) do not have to be declared.

Subscripted variables with up to seven dimensions are allowed. They always must be defined in size by the DIMENSION statement. For example, the statements

```
DIMENSION SAMPLE(5)
REAL DIMENSION INCOME(2,7)
```

would establish a 1×5 real *array* called SAMPLE and a 2×7 real array called INCOME. INCOME would have been an integer array without the REAL declaration.

Elements of arrays are addressed by placing the subscripts in parentheses.

```
SAMPLE(2)
INCOME(1,6)
```

The subscripts also can be variables. SAMPLE(K) would be permitted as long as K was defined, was between 1 and 5, and was an integer.

Variables and arrays once defined and declared are not initialized automatically. If it is necessary to initialize a storage location prior to use, the DATA statement can be used. Consider the following statements.

```
REAL X,Y,Z
DIMENSION ONEDIM(5)
DIMENSION TWODIM(2,3)
DATA X,Y,Z/3*0.0/(ONEDIM(I),I = 1,5)
1/5*0.0/ ((TWODIM(I,J),I = 1,2),J = 1,3)
2/1.,2.,3.,4.,5.,6.
```

MISC

Variables X, Y, and Z will be initialized to 0.0. The entries in ONEDIM will have the values (0,0,0,0,0). The TWODIM array will be initialized to

$$\begin{pmatrix} 1.0 & 2.0 & 3.0 \\ 4.0 & 5.0 & 6.0 \end{pmatrix}$$

After being initialized with a DATA statement, variables can have their values changed by arithmetic operations.

4 ARITHMETIC OPERATIONS

FORTRAN provides for the usual arithmetic operations. These are listed in table 13.2.

Table 13.2
FORTRAN Operators

symbol	meaning
=	replacement
+	addition
−	subtraction
*	multiplication
/	division
**	exponentiation
( )	preferred operation

The *equals* symbol is used to replace one quantity with another. For example, the following statement is algebraically incorrect. However, it is a valid FORTRAN statement.

$$Z = Z + 1$$

Each statement is scanned from left-to-right (except that a right-to-left scan is made for exponentiation).

Operations are performed in the following order.

exponentiation first

multiplication and division second

addition and subtraction last

Parentheses can be used to modify this hierarchy.

Each operation must be stated explicitly and unambiguously. Thus, AB is not a substitute for A^*B. Two operations in a row, as in $(A + -B)$ also are unacceptable. Some FORTRAN compilers allow mixed-mode arithmetic. Others, ANSI FORTRAN among them, require all variables in an expression to be either integer or real. Where mixed-mode arithmetic is permitted, care must be taken in the conversion of real data to the integer mode.

Integer variables used to hold the results of a mixed-mode calculation will have their values truncated. This is illustrated in the following example.

Example 13.8

Evaluate J in the following expression.

$$J = (6.0 + 3.0)^*3.0/6.0 + 5.0 - 6.0^{**}2.0$$

The expressions within parentheses are evaluated in the first pass.

$$J = 9.0^*3.0/6.0 + 5.0 - 6.0^{**}2.0$$

The exponentiation is performed in the second pass.

$$J = 9.0^*3.0/6.0 + 5.0 - 36.0$$

The multiplication and division are performed in the third pass.

$$J = 4.5 + 5.0 - 36.0$$

The addition and subtraction are performed in the fourth pass.

$$J = -26.5$$

However, J is an integer variable, so the real number -26.5 is truncated and converted to integer. The final result is -26.

5 PROGRAM LOOPS

Loops can be constructed from IF and GO TO statements. However, the DO statement is a convenient method of creating loops. The general form of the DO statement is

$$\text{DO } s\ i\ = j, k, l$$

where s is a statement number.

i is the integer loop variable.

j is the initial value assigned to i.

k, which must exceed j, is an inclusive upper bound on i.

l is the increment for i, with a default value of 1 if omitted.

The DO statement causes the execution of the statements immediately following it through statement s until i equals k or greater.[1] A loop can be *nested* by

[1] A peculiarity of FORTRAN DO statements is that they are executed at least once, regardless of the values of i and j.

MISC

placing it within another loop. The loop variable may be used to index arrays.

When i equals or exceeds k, the statement following s is executed. However, the loop may be exited at any time before i reaches k if the logic of the loop provides for it.

6 INPUT/OUTPUT STATEMENTS

The READ, WRITE, and FORMAT statements are FORTRAN's main I/O statements. Forms of the READ and WRITE statements are

READ (u_1, s) [list]
WRITE (u_2, s) [list]

where

u_1 is the unit number designation for the desired input device, usually 5 for the card reader.

u_2 is the unit designation for the desired output device, usually 6 for the line printer.

s is the statement number of an associated FORMAT statement.

[list] is a list of variables separated by commas whose values are being read or written.

The [list] also can include an implicit DO loop. The following example reads six values, the first five into the array PLACE and the last into SHOW.

```
READ (5,85) (PLACE(J),J = 1,5) SHOW
```

The purpose of the FORMAT statement is to define the location, size, and type of the data being read. The form of the FORMAT statement is

s FORMAT [field list]

As before, s is the statement number. [field list] consists of specifications, set apart by commas, defining the I/O fields. [list] can be shorter than [field list].

The format code for integer values is nIw.

n is an optional repeat counter which indicates the number of consecutive variables with the same format. w is the number of character positions. The format codes for real values are

$$nFw.d \text{ or } nEw.d$$

Again, w is the number of character positions allocated, including the space required for the decimal point. d is the implied number of spaces to the right of the decimal point. In the case of input data, decimal points in any position take precedence over the value of d.

The F format will print a total of $(w-1)$ digits or blanks representing the number. The E format will print a total of $(w-2)$ digits or blanks and give the data in a standard scientific notation with an exponent.

Other formats which can be used in the FORMAT statement are

X	horizontal blanks
/	skipping lines
H	alphanumeric data
D	double precision real
T	position (column) indicator
Z	hexadecimal
P	decimal point modification
L	logical data
' '	literal data

The usual output device is a line printer with 133 print positions. The first print position is used for *carriage control*. The data (control character) in the first output position will control the printer advance according to the rules in table 13.3.

Table 13.3
FORTRAN Printer Control Characters

control character	meaning
blank	advance one line
0	advance two lines
1	skip to line one on next page
+	do not advance (overprint)

Carriage control usually is accomplished by the use of literal data. Consider the following statements.

```
INTEGER K
K = 193
WRITE (6,100) K
WRITE (6,101) K
100 FORMAT (' ',I3)
101 FORMAT (I3)
```

The above program would print the number 193 on the next line of the current page and the number 93 on the first line of the next page.

Data can be written to or read from an array by including the array subscripts in the I/O statement.

```
    DIMENSION CLASS (2,5)
    READ (5,15) ((CLASS(I,J),I = 1,2),J = 1,5)
15  FORMAT (10F3.0)
```

MISC

7 CONTROL STATEMENTS

The STOP statement is used to indicate the logical end of the program. The format is *s* STOP.

STOP should not be the last statement. When it is reached, program execution is terminated. The value of *s* is printed out or made available to the next program step. The use of STOP rarely is recommended.

The END statement is required as the last statement. It tells the compiler that there are no more lines in the program to be compiled. A program cannot be compiled or executed without an END statement.

The PAUSE statement will cause execution to stop temporarily. Its format is *s* PAUSE.

When the PAUSE statement is reached, the number *s* is transmitted to the computer operator. This gives the operator a chance to set various control switches on the console (the choice of switches being dependent on the value of *s* and the program logic), prior to pushing the START button. The PAUSE statement is used only if the programmer is operating the computer.

The CALL statement is used to transfer execution to a *subroutine.* CALL EXIT will terminate execution and turn control over to the operating system. The CALL EXIT and STOP statements have similar effects. The RETURN statement ends execution of a program called subroutine and passes execution to the main program. The CONTINUE statement does nothing. It can be used with a statement number as the last line of a DO loop.

The GO TO [s] statement transfers control to statement *s*.

The arithmetic IF statement is written

$$\text{IF[e]}s_1, s_2, s_3$$

[e] is any numerical variable or arithmetic expression, and the s_i are statement numbers. The transfer occurs according to the following table.

[e]	statement executed
[e] < 0	s_1
[e] = 0	s_2
[e] > 0	s_3

The logical IF statement has the form

IF[le][**statement**]

[le] is a logical expression, and [**statement**] is any executable statement except DO and IF. Only if [le] is true will [**statement**] be executed. Otherwise, the next instruction will be executed.

The logical expression [le] is a relational expression using one of several operators.

Logical expressions also can incorporate the connectors .AND., .OR., and .NOT..

Table 13.4
FORTRAN Logical Operations

operator	meaning
.LT.	less than
.LE.	less than or equal to
.EQ.	equal to
.NE.	not equal to
.GT.	greater than
.GE.	greater than or equal to

Example 13.9

$$\text{IF } (A.GT.25.6) \ A = 27.0$$

Meaning: If A is greater than 25.6, set A equal to 27.0.

IF (Z.EQ.(T−4.0).OR.Z.EQ.0.) GO TO 17

Meaning: If Z is equal to (T−4.0) or if Z is equal to *zero*, go to statement 17.

8 LIBRARY FUNCTIONS

The following single-precision library functions are available. Most are accessed by placing the argument in parentheses after the function name. Placing the letter *D* before the function name will cause the calculation to be performed in double precision. Arguments for trigonometric functions are expressed in radians.

Table 13.5
Some FORTRAN Library Functions

function	use
EXP	e^x
ALOG	natural logarithm
ALOG10	common logarithm
SIN	sine
COS	cosine
TAN	tangent
SINH	hyperbolic sine
SQRT	square root
ASIN	arcsine
MOD	remaindering modulus (integer)
AMOD	remaindering modulus (real)
ABS	absolute value (real)
IABS	absolute value (integer)
FLOAT	convert integer to real
FIX	convert real to integer

9 USER FUNCTIONS

A user-defined function can be created with the FUNCTION statement. Such functions are governed by the following rules.

- The function is defined as a variable in the main program even though it is a function.
- When used in the main program, the function is followed by its arguments in parentheses.
- In the function itself, the function name is type-declared and defined by the word FUNCTION.
- The arguments (parameters) need not have the same names in the main program and function.
- Only the function has a RETURN statement.
- Both the main program and the function have END statements.
- The arguments (parameters) must agree in number, order, type, and length.

These construction rules are illustrated by example 13.10.

Example 13.10

```
REAL HEIGHT, WIDTH, AREA, MULT
HEIGHT = 2.5
WIDTH = 7.5
AREA = MULT(HEIGHT,WIDTH)
END

REAL FUNCTION MULT(HEIGHT,WIDTH)
REAL HEIGHT, WIDTH
MULT = HEIGHT*WIDTH
RETURN
END
```

10 SUBROUTINES

A subroutine is a user-defined subprogram. It is more versatile than a user-defined function as it is not limited to mathematical calculations. Subroutines are governed by the following rules.

- The subroutine is activated by the CALL statement.
- The subroutine has no type.
- The subroutine does not take on a value. It performs operations on the arguments (parameters) which are passed back to the main program.
- A subroutine has a RETURN statement.
- Both the main program and the subroutine have END statements.
- The arguments (parameters) need not have the same names in the main program and the subroutine.
- The arguments (parameters) must agree in number, order, type, and length.
- It is possible to return to any part of the main program. It is not necessary to return to the statement immediately below the CALL statement.

These rules are illustrated by example 13.11.

Example 13.11

```
REAL HEIGHT, WIDTH, AREA
CALL GET(HEIGHT, WIDTH)
AREA = HEIGHT*WIDTH
END

SUBROUTINE GET(A,B)
REAL A,B
READ (5,100) A,B
100 FORMAT(2F3.1)
    RETURN
    END
```

MISC

Variables in functions and subroutines are completely independent of the main program. Subroutine and main program variables which have the same names will not have the same values. A link between the main program and the subroutine can be established, however, with the COMMON statement.

The COMMON statement assigns storage locations in memory to be shared by the main program and all of its subroutines. Even the COMMON statement, however, allows different names. It is the order of the common variables which fixes their position in upper memory.

Example 13.12

What are the values of X and Y in the subroutine?

```
COMMON X, Y          main program
     X  =  2.0
     Y  =  10.0
COMMON Y, X          subroutine
```

Since Y is the first common subroutine variable which corresponds to X in the main program, $Y = 2.0$. Similarly, $X = 10.0$.

If variables are to be shared with only some of the subroutines, the *named* COMMON statement is required. Whereas there can be only one regular COMMON statement, there can be multiple-named COMMON statements.

```
COMMON/PLACE/CAT, COW, DOG  main program

COMMON/PLACE/HORSE, PIG, EXPENSE
                            subroutine
```

PART 6: Fire Safety Systems

1 INTRODUCTION

In many cases, the design of fire detection, fire alarm, and sprinkler systems is governed by state or local codes. It is necessary to review all applicable codes and to meet the most stringent of them. Generally, insurance carrier requirements are more stringent than code minimums. Although codes mandate minimum standards, common sense and professional prudence should be used in specifying the level of fire protection in a building.

In buildings that are partially or wholly sprinklered, provisions for alarm and evacuation as well as provisions for sprinkler supervision must be made.

The type of occupancy greatly affects the degree of protection. Nursing homes, schools, hospitals, and office buildings all have greatly different needs. Furthermore, multi-story buildings with limited escape routes affect the degree of protection.

2 DETECTION DEVICES

• *Manual Fire Alarm Stations*

Mandatory in any system, large or small. Locate in the natural path of exit with the maximum traveling distance to any manual station of 200 feet. Identification of an activated manual station which when opened should be readily visible from the side, down a corridor for at least 200 feet.

• *Heat Detectors—Fixed Temperature*

135 °F is the typical temperature threshold in open spaces. 190 ° to 200 °F is generally used in enclosed or confined spaces such as boiler rooms, closets, etc., where the heat build-up will be fast and confined.

• *Heat Detectors—Rate of Rise*

Combination rate of rise and fixed temperature of 135 ° or 200 °F. Rate of rise portion operates when the temperature rises in excess of 15 degrees per minute. More sensitive than the fixed temperature detector.

• *Heat Detector—Rate Compensated*

Considered to be the most responsive of all thermal detectors. Operates at 135 ° or 200 °F. Detects both slow and fast developing fires by anticipating the temperature increase and moving towards the alarm point as the temperature gradually increases.

• *Photoelectric Smoke Detectors*

Operates on a photo beam or light scattering principle. The photoelectric detector responds best to products of combustion or smoke with a particle size from approximately 10.0 microns down to 0.1 micron and of the proper concentration. Proper concentration is defined by Underwriters Laboratories as the ability to sense smoke in the 0.2 to 4.0 percent obscuration per foot range. Photoelectric detectors generally are considered to be the best for cold smoke fires.

• *Ionization Detectors*

Ionization detectors detect products of combustion by sensing the disruption of conductivity in an ionized chamber due to the presence of smoke. The ionization detector responds best to fast burning fires where particle sizes range from approximately 1.0 micron down to 0.01 micron and of the proper concentration. Proper concentration is defined by Underwriters Laboratories as the ability to sense smoke from 0.2 to 4.0 percent obscuration per foot range. **Note:** Ionization and photoelectric detectors can be intermixed within a system to provide the best form of detection suitable to the environment.

• *Infrared Flame Detectors*

Generally, infrared detectors respond to radiation in the 6500 to 8500 angstrom range. Good detectors will filter out solar interference and respond to radiation in the 4000 to 5500 angstrom range. It is preferrable to use a detector with a dual sensing circuit in order to discriminate unwanted or false alarms.

• *Ultraviolet Flame Detectors*

Ultraviolet flame detectors respond to radiation in the spectral range of 1700 to 2900 angstroms. It is not sensitive to solar interference. Built-in time delays prevent false alarms.

• *Waterflow Detectors*

Used in wet sprinkler systems to indicate a flow of water. Use on the main sprinkler risers and throughout the building to indicate sub-sections or floors of the building to locate sprinkler discharges quickly. These detectors employ a retard mechanism to prevent false alarms from water surges.

• *Pressure Switches*

Used in dry or pre-action sprinkler systems to provide an alarm when water is discharged.

• *Valve Monitor Switches*

Closed water supply valves are a major weakness of sprinkler systems. This is the most frequently neglected and forgotten item in the sprinkler system. Closed valves also have accounted for countless millions of dollars worth of damage because the system would not operate.

• *Low Temperature Monitor Switches*

Low air temperature (under 40 °F) or low water temperature is detrimental to a sprinkler system and creates a great nuisance by freezing and bursting the sprinkler system pipes.

3 ALARM DEVICES

• *Bells*

This is the most commonly accepted form of alarm. However, bells should not be used in schools where bells are used for other signaling purposes. Bells should not be used in any area where the same sound is used for any other function.

• *Chimes*

Single stroke devices in coded systems and used in certain types of applications such as quiet areas of hospitals or nursing homes.

• *Horns*

Horns generally are capable of producing a higher sound level than either a bell or a chime.

• *Alarm Lights*

Used individually as a fire alarm visual indicator or used in conjunction with a horn. The light can be either a flashing incandescent bulb or a flashing strobe. It sometimes is required by certain codes or types of occupancy in order to provide an alarm for handicapped persons.

• *Remote Annunciators*

Generally, these duplicate the main control panel and have a light for every fire alarm zone. They are used at the second entrance or at the main entrance to assist the fire department in locating the fire zone. Also used in nursing homes, at nursing stations in hospitals or in the engineering room of a factory or a building.

• *Speakers*

Used in emergency voice evacuation systems, generally in high-rise buildings. Specific quality, construction, and performance have been established by the NFPA code for speakers used in voice evacuation systems.

4 SIGNAL TRANSMISSION

• *Reverse Polarity*

Uses a dedicated leased telephone directly between the protected premises and the municipal fire department or a commercial central station.

• *Central Station*

These are central monitoring facilities which monitor fire and security alarms from protected premises for a monthly fee.

• *Telephone Dialers*

Tape dialers have had a very bad reputation due to high failure rate and susceptibility to false alarms. Solid state digital dialers with compatible solid state digital receivers have increased the reliability and dependability of this product very dramatically and are becoming more acceptable as an alternate form of signal transmission.

• *Radio Transmitters*

Radio transmitter boxes are used in certain applications to transmit alarms between the protected premises and some monitoring point. Before using, it is advisable to check that the line of transmission is clear from obstruction and interference.

5 AUXILIARY CONTROL

• *Smoke Doors*

Generally held open with floor or wall mounted electromagnets. Generally all doors close in all parts of the building on the first alarm.

• *Fire Doors*

Generally treated the same way as smoke doors.

• *Stairwell Exit Doors*

Sometimes for security reasons, these doors are held latched with a door strike. The door also must employ a panic exit bar to override and manually open the door.

• *Elevator Capture*

Generally accepted by all codes that the elevator immediately return to the first floor or to some designated alternate floor. The specification should designate that the elevator manufacturer is responsible for accepting low voltage signal or a dry contact from the fire alarm panel and programming the elevator to return to the designated floor.

• *Fire Dampers*

Either motorized or fusible link type are closed to prevent the spread of smoke to other areas. Care should be exercised in connecting these to a fire alarm system because a fire alarm system generally operates with small amounts of D.C. power. The responsibility should be stated for coordinating the voltages and contact ratings necessary to do the job.

• *Fans—Supply and Exhaust*

Sometimes, all fans are shut down on the first alarm. However, the design of some buildings and other considerations make it desirable to exhaust the smoke from the building and shut down the input supply fans.

• *Pressurization*

Pressurization creates positive air pressure to inhibit the influx of smoke from adjacent areas or floors above or below. Exit stairwell escape routes out of the high-rise building should be pressurized to provide a smoke-free exit path.

6 SPECIAL CONSIDERATIONS

• *Handicapped Persons*

There may be requirements from Federal health officials (HHS) or OSHA to provide consideration for handicapped persons.

• *Weather Protection of Devices*

Check to make sure that none of the devices will be exposed to adverse conditions such as excessive moisture.

• *Open Plenums*

Drop ceilings that use the space above as an open air plenum present special problems and are treated as a potential hazard in most codes. Smoke detectors located in open air plenums need to be located properly, and it is advisable to locate them so that they are accessible for maintenance. A remote alarm lamp should be brought down and mounted below the ceiling level to indicate which detector is an alarm.

• *Duct Detectors*

Duct heat detectors are considered to be of negligible value. Duct smoke detectors can provide warning of smoke in an air duct and can prevent costly damage to expensive HVAC equipment. Duct detectors are not a substitute for open area smoke detectors. Problem areas can develop with duct detectors from excessive humidity, poor or no maintenance, and improper location.

• *Sound Pressure Level of Alarm Devices*

The sound pressure level required to provide an adequate alarm in the environment in which it is intended to operate generally is considered to have been accomplished if the alarm sound is 12 decibels above the normal ambient level.

• *Emergency Generators*

If the fire alarm system does not use its own standby battery pack, it is advisable to coordinate the details of how the emergency generator feeds power back into the building distribution system and to insure that the fire

MISC

alarm system will be provided with emergency power. In some cases, it is advisable to provide the fire alarm system with a small amount of standby battery power in order to keep the fire alarm system on line until the generator gets to full power.

• *Fire Pump Supervision*

Insure that fire pumps, if used, are supervised adequately for all critical functions and that the building fire alarm system is provided with one or more zones to interface with fire pump signals.

7 SPECIAL SUPPRESSION SYSTEMS

• *Deluge Systems*

The control panel for the water deluge system is specialized and generally is mounted in a hostile environment. The control equipment is mounted inside an enclosure which is watertight and dust tight. Deluge systems are actuated manually and also by thermal detectors or flame detectors. The entire system, including the electronic control panel, is provided by the sprinkler contractor.

• *Foam Systems*

High expansion foam or water suppression systems are specialized systems requiring special handling. They generally are activated by infrared or ultraviolet flame detectors.

• *Halon Systems*

Halon systems generally are used in clean environments such as computer rooms or any room that contains high value equipment. Examples of this are tapes, microfilms, computer records, laboratories for research and design, and medical laboratories. Generally, the criterion is the high value of the equipment or the data stored, and it is desirable that water not touch the equipment. Halon systems, because of their great expense, usually are engineered specially for the particular room in which they are to be located. The control panel uses cross-zoning techniques to prevent unnecessary discharges.

• *High/Low Pressure CO_2 Systems*

CO_2 systems are used in industrial applications to suppress fires where the use of water, extinguishing powders, or Halon would be unsuitable because of expense, hazard, or the size of the equipment to be protected. Because CO_2 is hazardous to life, it has to be an engineered system designed for the particular application.

PART 7: Nondestructive Testing

1 MAGNETIC PARTICLE TESTING

This procedure is based on the attraction of magnetic particles to leakage flux at surface flaws. The particles accumulate and become visible at the flaw. This method works for magnetic materials in locating cracks, laps, seams, and in some cases, subsurface flaws. The test is fast and simple and is easy to interpret. However, parts must be relatively clear and demagnetized. A high current (power) source must be available.

2 EDDY CURRENT TESTING

Alternating currents from a source coil induce eddy currents in metallic objects. Flaws and other material properties affect the current flow. The change in current flow is observed on a meter or a screen. This method can be used to locate defects of many types, including changes in composition, structure, and hardness, as well as locating cracks, voids, inclusions, weld defects, and changes in porosity.

Intimate contact between the material and the test coil is not required. Operation can be continuous, automatic, and monitored electronically. Therefore, this method is ideal for unattended continuous processing. Sensitivity is easily adjustable. Many variables, however, can affect the current flow.

3 LIQUID PENETRANT TESTING

Liquid penetrant (dye) is drawn into surface defects by capillary action. A developer substance then is used to develop the penetrant to aid in visual inspection. This method can be used with any nonporous material, including metals, plastics, and glazed ceramics. It locates cracks, porosities, pits, seams, and laps.

Liquid penetrant tests are simple to perform, can be used with complex shapes, and can be performed on site. Parts must be clean, and only small surface defects are detectable.

4 ULTRASONIC TESTING

Mechanical vibrations in the 0.1 to 25 MHz range are induced in an object. The transmitted energy is reflected and scattered by interior defects. The results are interpreted from a screen or a meter. The method can be used for metals, plastics, glass, rubber, graphite, and concrete. It detects inclusions, cracks, porosity, laminations, structure, and other interior defects.

This test is extremely flexible. It can be automated and is very fast. Results can be recorded or interpreted electronically. Penetration of up to 60 feet of steel is possible. Only one surface needs to be accessed. However, rough surfaces or complex shapes may cause difficulties.

5 INFRARED TESTING

Infrared testing emitted from objects can be detected and correlated with quality. The detection can be recorded electronically. Any discontinuity that interrupts heat flow, such as flaws, voids, and inclusions, can be detected.

Infrared testing requires access to only one side, and it is highly sensitive. It is applicable to complex shapes and assemblies of dissimilar components, but it is relatively slow. Results are affected by material variations, coatings, and colors, and hot spots can be hidden by cool surface layers.

6 RADIOGRAPHY

X-ray and gamma-ray sources can be used to penetrate objects. The intensity is reduced in passing through, and the intensity changes are recorded on film or screen. This method can be used with most materials to detect internal defects, material structure, and thickness. It also can be used to detect the absence of internal parts.

Up to 30 inches of steel can be penetrated by x-ray sources. Gamma sources, which are more portable and lower in cost than x-ray sources, are applicable to 10-inch thickness of steel.

There are health and government standards associated with these tests. Electric power and cooling water may be required in large installations. Shielding and film processing also is required, making this the most expensive nondestructive test.

PART 8: Mathematical Programming

1 INTRODUCTION

Mathematical programming is a modeling procedure applicable to problems for which the goal and resource limitations can be described mathematically. If the *goal function* and all *resource constraints* are linear (polynomials of degree 1 only), the procedure is known as *linear programming.*

If the variables can take on only integer values, a procedure known as *integer programming* is required. If the polynomials are of any degree or contain other functions, a procedure known as *dynamic programming* is required.

2 FORMULATION OF A LINEAR PROGRAMMING PROBLEM

All linear programming problems have a similar format. Each has an *objective function* which is to be optimized. Usually, the objective function is to be maximized, as in the case of a *profit function.* If the objective is to minimize some function, such as cost, the problem may be turned into a maximization problem by maximizing the negative of the original function.

Example 13.13

A cattle rancher buys three types of cattle food. The rancher wants to minimize the cost of feeding his cattle. Write the objective function for this problem on a per animal basis.

food type	cost per pound
1	1.5
2	2.5
3	3.5

Let x_i be the number of pounds of food i purchase per animal. Then, the objective function to be minimized is

$$Z = 1.5x_1 + 2.5x_2 + 3.5x_3$$

Each linear programming problem also has a set of limitation functions called *constraints.* Constraints are used to set the bounds for the objective function.

Example 13.14

The rancher is concerned with meeting published nutritional information on minimum daily requirements (MDR) given in milligrams per animal. The composition of each food type is known and the contributions for each vitamin in mg/pound are

		food type		
vitamin	MDR (mg)	1	2	3
A	100	1	7	13
B	200	3	9	15
C	300	5	11	17

MISC

It is also physically impossible for an animal to eat more than the following amounts per day.

food type	maximum feeding
1	50 lbm
2	40 lbm
3	30 lbm

The constraints on this problem are

$$
\begin{aligned}
x_1 + 7x_2 + 13x_3 &\geq 100 \\
3x_1 + 9x_2 + 15x_3 &\geq 200 \\
5x_1 + 11x_2 + 17x_3 &\geq 300 \\
x_1 &\leq 50 \\
x_2 &\leq 40 \\
x_3 &\leq 30 \\
x_1 &\geq 0 \\
x_2 &\geq 0 \\
x_3 &\geq 0
\end{aligned}
$$

Linear programming problems are generally solved by computer. Some simple problems can be solved by hand with a procedure known as the *simplex method*. Specialized methods allowing easy manual solutions are available for certain classes of problems, primarily the *transportation problem* and *assignment problem*.

Once a solution is found, it is possible to determine the effect on the objective function of changing one of the program parameters. This is known as *sensitivity analysis* and is very important in instances where the accuracy of collected data is unknown.

3 SOLUTION TO 2-DIMENSIONAL PROBLEMS

If a linear programming problem can be formulated in terms of only two variables, x_1 and x_2, it can be solved graphically by the following procedure:

step 1: Graph all of the constraints and determine the *feasible region*. Usually this will result in a *convex hull*.

step 2: Evaluate the objective function, Z, at each corner of the hull.

step 3: The values of x_1 and x_2 which optimize Z are the coordinates of the corner at which Z is optimized.

Example 13.15

Solve the following linear programming problem graphically.

$$
\begin{aligned}
\max Z &= x_1 + 2x_2 \\
\text{such that } 4x_1 + x_2 &\leq 24 \\
2x_1 + x_2 &\leq 14 \\
-x_1 + 2x_2 &\leq 8 \\
-x_1 + x_2 &\leq 3 \\
x_1 &\geq 0 \\
x_2 &\geq 0
\end{aligned}
$$

The region enclosed by the constraints is shown.

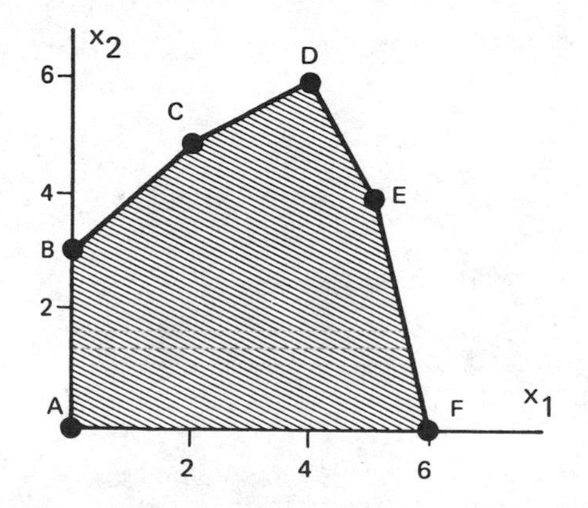

The coordinates and objective function values for each corner are

corner	coordinates (x_1, x_2)	Z
A	(0,0)	0
B	(0,3)	6
C	(2,5)	12
D	(4,6)	16
E	(5,4)	13
F	(6,0)	6

Z is maximized when $x_1 = 4$ and $x_2 = 6$.

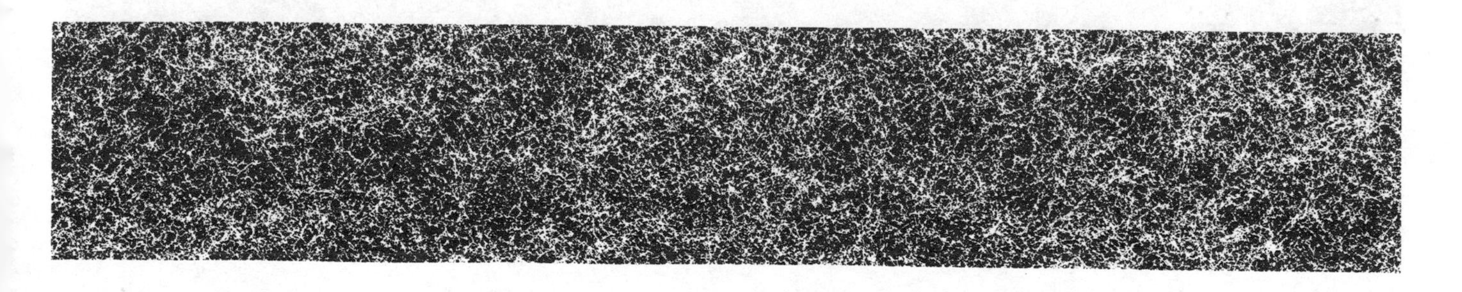

14 SYSTEMS OF UNITS

1 CONSISTENT SYSTEMS OF UNITS

A set of units used in a problem is said to be *consistent*[1] if no conversion factors are needed. For example, a moment with units of foot-pounds cannot be obtained directly from a moment arm with units of inches. In this illustration, a conversion factor of $\frac{1}{12}$ feet/inch is needed, and the set of units used is said to be *inconsistent*.

On a larger scale, a system of units is said to be consistent if Newton's second law of motion can be written without conversion factors. Newton's law states that the force required to accelerate an object is proportional to the amount of matter in the item.

$$F = ma \tag{14.1}$$

The definitions of the symbols, F, m, and a, are familiar to every engineer. However, the use of Newton's second law is complicated by the multiplicity of available unit systems. For example, m may be in kilograms, pounds, or slugs. All three of these are units of mass. However, as figure 14.1 illustrates, these three units do not represent the same amount of mass.

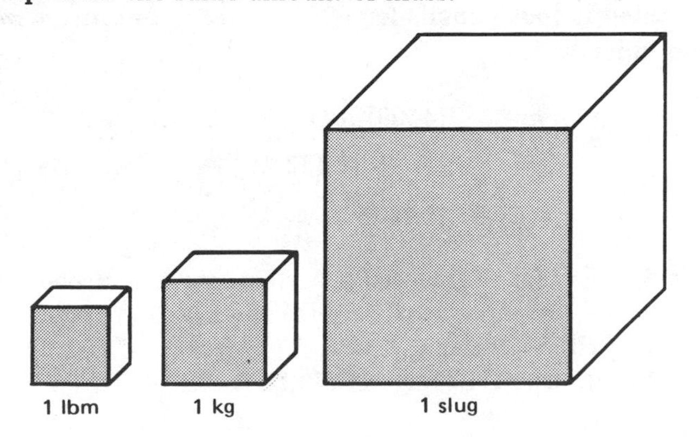

Figure 14.1 Common Units of Mass

[1] The terms *homogeneous* and *coherent* also are used to describe a consistent set of units.

It should be mentioned that the decision to work with a consistent set of units is arbitrary and unnecessary. Problems in fluid flow and thermodynamics commonly are solved with inconsistent units. This causes no more of a problem than working with inches and feet in the calculation of a moment.

2 THE ABSOLUTE ENGLISH SYSTEM

Engineers are accustomed to using pounds as a unit of mass. For example, density typically is given in pounds per cubic foot. The abbreviation *pcf* tends to obscure the fact that the true units are pounds of *mass* per cubic foot.

If pounds are the units for mass, and feet per second squared are the units of acceleration, the units of force for a consistent system can be found from Newton's second law.

$$\begin{aligned} \text{units of } F &= (\text{units of } m)(\text{units of } a) \\ &= (\text{lbm})(\text{ft/sec}^2) = \frac{\text{lbm-ft}}{\text{sec}^2} \end{aligned} \tag{14.2}$$

The units for F cannot be simplified any more than they are in equation 14.2. This particular combination of units is known as a *poundal*.[2]

The absolute English system, which requires the poundal as a unit of force, seldom is used, but it does exist. This existence is a direct outgrowth of the requirement to have a consistent system of units.

3 THE ENGLISH GRAVITATIONAL SYSTEM

Force frequently is measured in pounds. When the thrust on an accelerating rocket is given as so many

[2] A poundal is equal to 0.03108 pounds force.

UNITS

pounds, it is understood that the pound is being used as a unit of force.

If acceleration is given in feet per second squared, the units of mass for a consistent system of units can be determined from Newton's second law.

$$\text{units of } m = \frac{\text{units of } F}{\text{units of } a} = \frac{\text{lbf}}{\text{ft/sec}^2} = \frac{\text{lbf-sec}^2}{\text{ft}} \qquad 14.3$$

The combination of units in equation 14.3 is known as a *slug*.[3] Slugs and pounds-mass are not the same, as illustrated in figure 14.1. However, units of mass can be converted, using equation 14.4.

$$\text{number of slugs} = \frac{\text{number of lbm}}{g_c} \qquad 14.4$$

g_c[4] is a dimensional conversion factor having the following value.

$$g_c = 32.1740 \frac{\text{lbm-ft}}{\text{lbf-sec}^2} \qquad 14.5$$

32.1740 commonly is rounded to 32.2 when six significant digits are unjustified. That practice is followed in this book.

Notice that the number of slugs cannot be determined from the number of pounds-mass by dividing by the local gravity. g_c is used regardless of the local gravity. However, the local gravity can be used to find the weight of an object. *Weight* is defined as the force exerted on a mass by the local gravitational field.

$$\text{weight in lbf} = (m \text{ in slugs})(g \text{ in ft/sec}^2) \qquad 14.6$$

If the effects of large land and water masses are neglected, the following formula can be used to estimate the local acceleration of gravity in ft/sec^2 at the earth's surface. ϕ is the latitude in degrees.

$$g_{surface} = 32.088[1 + (5.305 \text{ EE} - 3)\sin^2\phi - (5.9 \text{ EE} - 6)\sin^2 2\phi] \qquad 14.7$$

If the effects of the earth's rotation are neglected, the gravitational acceleration at an altitude, h, in miles is given by equation 14.8. R is the earth's radius (approximately 3960 miles).

$$g_h = g_{surface}\left(\frac{R}{R+h}\right)^2 \qquad 14.8$$

[3] A slug is equal to 32.1740 lbm.

[4] Three different meanings of the symbol g commonly are used. g_c is the dimensional conversion factor given in equation 14.5. g_o is the standard acceleration due to gravity with a value of 32.1740 ft/sec^2. g is the local acceleration due to gravity in ft/sec^2.

4 THE ENGLISH ENGINEERING SYSTEM

Many thermodynamics and fluid flow problems freely combine variables containing pound-mass and pound-force terms. For example, the steady-flow energy equation mixes enthalpy terms in BTU/lbm with pressure terms in lbf/ft^2. This requires the use of g_c as a mass conversion factor.

Newton's second law becomes

$$F \text{ in lbf} = \frac{(m \text{ in lbm})(a \text{ in ft/sec}^2)}{\left(g_c \text{ in} \frac{\text{lbm-ft}}{\text{lbf-sec}^2}\right)} \qquad 14.9$$

Since g_c is required, the English Engineering System is inconsistent. However, that is not particularly troublesome, and the use of g_c does not overly complicate the solution procedure.

Example 14.1

Calculate the weight of a 1.0 lbm object in a gravitational field of 27.5 ft/sec^2.

Since weight commonly is given in pounds-force, the mass of the object must be converted from pounds-mass to slugs.

$$F = \frac{ma}{g_c} = \frac{(1) \text{ lbm } (27.5) \text{ ft/sec}^2}{(32.2)\frac{\text{lbm-ft}}{\text{lbf-sec}^2}} = 0.854 \text{ lbf}$$

Example 14.2

A rocket with a mass of 4000 lbm is traveling at 27,000 ft/sec. What is its kinetic energy in ft-lbf?

The usual kinetic energy equation is $E_k = \frac{1}{2}mv^2$. However, this assumes consistent units. Since energy is wanted in foot-pounds-force, g_c is needed to convert m to units of slugs.

$$E_k = \frac{mv^2}{2g_c} = \frac{(4000) \text{ lbm } (27{,}000)^2 \text{ ft}^2/\text{sec}^2}{(2)(32.2)\frac{\text{lbm-ft}}{\text{lbf-sec}^2}} = 4.53 \text{ EE10 ft-lbf}$$

In the English Engineering System, work and energy typically are measured in ft-lbf (mechanical systems) or in British Thermal Units, BTU (thermal and fluid systems). One BTU equals 778.26 ft-lbf.

5 THE cgs SYSTEM

The cgs system has been used widely by chemists and physicists. It is named for the three primary units used to construct its derived variables. The *c*entimeter, the *g*ram, and the *s*econd form the basis of this system.

The cgs system avoids the lbm versus lbf type of ambiguity in two ways. First, the concept of weight is not used at all. All quantities of matter are specified in grams, a mass unit. Second, force and mass units do not share a common name.

When Newton's second law is written in the cgs system, the following combination of units results.

$$\begin{aligned}\text{units of force} &= (m \text{ in } g)\left(a \text{ in } \frac{\text{cm}}{\text{sec}^2}\right)\\ &= \frac{g\text{–cm}}{\text{sec}^2}\end{aligned} \qquad 14.10$$

This combination of units for force is known as a *dyne*.

Energy variables in the cgs system have units of dyne-cm or, equivalently, of

$$\frac{g\text{–cm}^2}{\text{sec}^2}$$

These combinations are known as an *erg*. There is no uniformly accepted unit of power in the cgs system, although calories per second frequently is used. Ergs can be converted to calories by multiplying by 2.389 EE–8.

The fundamental volume unit in the cgs system is the cubic centimeter (cc). Since this is the same volume as one thousandth of a liter, units of millimeters (ml) are used freely in this system.

6 THE mks SYSTEM

The mks system is appropriate when variables take on values larger than can be accomodated by the cgs system. This system uses the *m*eter, the *k*ilogram, and the *s*econd as its primary units. The mks system avoids the lbm versus lbf ambiguity in the same ways as does the cgs system.

The units of force can be derived from Newton's second law.

$$\begin{aligned}\text{units of force} &= (m \text{ in kg})\left(a \text{ in } \frac{\text{m}}{\text{sec}^2}\right)\\ &= \frac{\text{kg–m}}{\text{sec}^2}\end{aligned} \qquad 14.11$$

This combination of units for force is known as a *newton*.

Energy variables in the mks system have units of N-m or, equivalently, $\frac{\text{kg–m}^2}{\text{sec}^2}$. Both of these combinations are known as a *joule*. The units of power are joules per second, equivalent to a *watt*. The common volume unit is the liter, equivalent to one-thousandth of a cubic meter.

Example 14.3

A 10 kg block hangs from a cable. What is the tension in the cable?

$$\begin{aligned}F = ma &= (10)\text{ kg }(9.8)\frac{\text{m}}{\text{sec}^2}\\ &= 98\,\frac{\text{kg–m}}{\text{sec}^2} = 98\ N\end{aligned}$$

Example 14.4

A 10 kg block is raised vertically 3 meters. What is the change in potential energy?

$$\begin{aligned}\Delta E_p = mg\Delta h &= (10)\text{ kg }(9.8)\ \frac{\text{m}}{\text{sec}^2}\ (3)\text{ m}\\ &= 294\,\frac{\text{kg–m}^2}{\text{sec}^2} = 294\ J\end{aligned}$$

7 THE SI SYSTEM

Strictly speaking, both the cgs and the mks systems are *metric* systems. Although the metric units simplify solutions to problems, the multiplicity of possible units for each variable sometimes is confusing.

The SI system (International System of Units) was established in 1960 by the General Conference of Weights and Measures, an international treaty organization. The SI system is derived from the earlier metric systems, but it is intended to supersede them all.

The SI system has the following features.

- There is only one recognized unit for each variable.
- The system is fully consistent.
- Scaling of units is done in multiples of 1000.
- Prefixes, abbreviations, and symbol-syntax are rigidly defined.

Table 14.1
SI Prefixes

prefix	symbol	value
exa	E	EE18
peta	P	EE15
tera	T	EE12
giga	G	EE9
mega	M	EE6
kilo	k	EE3
hecto	h	EE2
deca	da	EE1
deci	d	EE−1
centi	c	EE−2
milli	m	EE−3
micro	μ	EE−6
nano	n	EE−9
pico	p	EE−12
femto	f	EE−15
atto	a	EE−18

Three types of units are used: base units, supplementary units, and derived units. The base units (table 14.2) are dependent on only accepted standards or reproducible phenomena. The supplementary units (table 14.3) have not yet been classified as being base units or derived units. The derived units (tables 14.4 and 14.5) are made up of combinations of base and supplementary units.

The expressions for the derived units in symbolic form are obtained by using the mathematical signs of multiplication and division. For example, units of velocity are m/s. Units of torque are $N \cdot m$ (not N-m or Nm).

Table 14.2
SI Base Units

quantity	name	symbol
length	meter	m
mass	kilogram	kg
time	second	s
electric current	ampere	A
temperature	kelvin	K
amount of substance	mole	mol
luminous intensity	candela	cd

Table 14.3
SI Supplementary Units

quantity	name	symbol
plane angle	radian	rad
solid angle	steradian	sr

In addition, there is a set of non-SI units which can be used. This temporary concession is due primarily to the significance and widespread acceptance of these units. Use of the non-SI units listed in table 14.6 usually will create an inconsistent expression requiring conversion factors.

In addition to having standardized units, the SI system also specifies syntax rules for writing the units and combinations of units. Each unit is abbreviated with a specific *symbol*. The rules for writing these symbols should be followed.

- The symbols are always printed in roman type, irrespective of the type used in the rest of the text. The only exception to this is in the use of the symbol for *liter*, where the lower case *l* (ell) may be confused with the number 1 (one). In this case, *liter* should be written out in full or the script *l* used. There is no problem with such symbols as cl (centiliter) or ml (milliliter).
- Symbols are never pluralized: 1 kg, 45 kg (not 45 kgs).
- A period is not used after a symbol, except when the symbol occurs at the end of a sentence.
- When symbols consist of letters, there always is a full space between the quantity and the symbols; e.g., 45 kg (not 45kg). However, when the first character of a symbol is not a letter, no space is left, e.g., 32°C (not 32° C or 32 °C) or 42° 12′ 45″ (not 42 ° 12 ′ 45″).
- All symbols are written in lower case, except when the unit is derived from a proper name. For example, m for meter, s for second, but A for ampere, Wb for weber, N for newton, W for watt. Prefixes are printed roman type without spacing between the prefix and the unit symbol, e.g., km for kilometer.
- In text, symbols should be used when associated with a number. When no number is involved, the unit should be spelled out. For example, the area of a carpet is 16 m^2, not 16 square meters, and carpet is sold by the square meter, not by the m^2.
- A practice in some countries is to use a comma as a decimal marker, while the practice in North America, the United Kingdom, and some other countries is to use a period (or a dot) as the decimal marker. Further, in some countries using the decimal comma, a dot frequently is used to divide long numbers into groups of three. Because of these differing practices, spaces must be used instead of commas to separate long lines of digits into easily-readable blocks of three digits with respect to the

UNITS

decimal marker, e.g., 32 453.246 072 5. A space (a half space is preferred) is optional with a four-digit number, e.g., 1 234, 1 234, or 1234.

- Where a decimal fraction of a unit is used, a zero should be placed before the decimal marker; e.g., 0.45 kg (not .45 kg). This practice draws attention to the decimal marker and helps avoid errors of scale.
- Some confusion may arise with the word *tonne* (1 000 kg). When this word occurs in French text of Canadian origin, the meaning may be a ton or 2 000 pounds.

Table 14.4
Some SI Derived Units with Special Names

quantity	name	symbol	expressed in terms of other units
frequency	hertz	Hz	
force	newton	N	
pressure, stress	pascal	Pa	N/m^2
energy, work, quantity of heat	joule	J	N·m
power, radiant flux	watt	W	J/s
quantity of electricity, electric charge	coulomb	C	
electric potential, potential difference, electromotive force	volt	V	W/A
electric capacitance	farad	F	C/V
electric resistance	ohm	Ω	V/A
electric conductance	siemen	S	A/V
magnetic flux	weber	Wb	V·s
magnetic flux density	tesla	T	Wb/m^2
inductance	henry	H	Wb/A
luminous flux	lumen	lm	
illuminance	lux	lx	lm/m^2

UNITS

Table 14.5
Some SI Derived Units

quantity	description	expressed in terms of other units
area	square meter	m^2
volume	cubic meter	m^3
speed—linear	meter per second	m/s
angular	radian per second	rad/s
acceleration—linear	meter per second squared	m/s^2
angular	radian per second squared	rad/s^2
density, mass density	kilogram per cubic meter	kg/m^3
concentration (of amount of substance)	mole per cubic meter	mol/m^3
specific volume	cubic meter per kilogram	m^3/kg
luminance	candela per square meter	cd/m^2
dynamic viscosity	pascal second	Pa·s
moment of force	newton meter	N·m
surface tension	newton per meter	N/m
heat flux density, irradiance	watt per square meter	W/m^2
heat capacity, entropy	joule per kelvin	J/K
specific heat capacity, specific entropy	joule per kilogram kelvin	J/(kg·K)
specific energy	joule per kilogram	J/kg
thermal conductivity	watts per meter kelvin	W/(m·K)
energy density	joule per cubic meter	J/m^3
electric field strength	volt per meter	V/m
electric charge density	coulomb per cubic meter	C/m^3
surface density of charge, flux density	coulomb per square meter	C/m^2
permittivity	farad per meter	F/m
current density	ampere per square meter	A/m^2
magnetic field strength	ampere per meter	A/m
permeability	henry per meter	H/m
molar energy	joule per mole	J/mole
molar entropy, molar heat capacity	joule per mole kelvin	J/(mole·K)
radiant intensity	watt per steradian	W/sr

Table 14.6
Acceptable Non-SI Units

quantity	unit name	symbol	relationship to SI unit
area	hectare	ha	1 ha = 10,000 m^2
energy	kilowatt-hour	kWh	1 kWh = 3.6 MJ
mass	metric ton[5]	t	1 t = 1000 kg
plane angle	degree (of arc)	°	1° =0.017 453 rad
speed of rotation	revolution per minute	r/min	1 r/min = $\frac{2\pi}{60}$ rad/s
temperature interval	degree Celsius	°C	1°C = 1 K
time	minute	min	1 min = 60 s
	hour	h	1 h = 3600 s
	day (mean solar)	d	1 d = 86 400 s
	year (calendar)	a	1 a = 31 536 000 s
velocity	kilometer per hour	km/h	1 km/h = 0.278 m/s
volume	liter[6]	l	1 l = 0.001 m^3

Numbers in parentheses are the number of ESU or EMU units, per single SI unit, except for the permittivity and the permeability of free space, where each actual values of ϵ_o and μ_o are given.

[5] The international name for metric ton is *tonne*. The metric ton is equal to the *megagram*, Mg.

[6] The international symbol for liter is the lowercase "l," which can be confused easily with the numeral "1." Several English speaking countries have adopted the script l as the symbol for liter in order to avoid any misinterpretation.

Appendix A
Selected Conversion Factors to SI Units

	SI Symbol	Multiplier to Convert From Existing Unit to SI Unit	Multiplier to Convert From SI Unit to Existing Unit
Area			
Circular Mil	µm²	506.7	0.001 974
Foot Squared	m²	0.092 9	10.764
Mile Squared	km²	2.590	0.386 1
Yard Squared	m²	0.836 1	1.196
Energy			
Btu (International)	kJ	1.055 1	0.947 8
Erg	µJ	0.1	10.0
Foot Pound-Force	J	1.355 8	0.737 6
Horsepower Hour	MJ	2.684 5	0.372 5
Kilowatt Hour	MJ	3.6	0.277 8
Meter Kilogram-Force	J	9.806 7	0.101 97
Therm	MJ	105.506	0.009 478
Kilogram Calorie (International)	kJ	4.186 8	0.238 8
Force			
Dyne	µN	10.	0.1
Kilogram-Force	N	9.806 7	0.101 97
Ounce-Force	N	0.278 0	3.597
Pound-Force	N	4.448 2	0.224 8
KIP	N	4 448.2	0.000 224 8
Heat			
Btu Per Hour	W	0.293 1	3.412 1
Btu Per (Square Foot Hour)	W/m²	3.154 6	0.317 0
Btu Per (Square Foot Hour °F)	W/(m²·°C)	5.678 3	0.176 1
Btu Inch Per (Square Foot Hour °F)	W/(m·°C)	0.144 2	6.933
Btu Per (Cubic Foot °F)	MJ/(m³·°C)	0.067 1	14.911
Btu Per (Pound °F)	J/(kg·°C)	4 186.8	0.000 238 8
Btu Per Cubic Foot	MJ/m³	0.037 3	26.839
Btu Per Pound	J/kg	2 326.	0.000 430
Length			
Angstrom	nm	0.1	10.0
Foot	m	0.304 8	3.280 8
Inch	mm	25.4	0.039 4
Mil	mm	0.025 4	39.370
Mile	km	1.609 3	0.621 4
Mile (International Nautical)	km	1.852	0.540
Micron	µm	1.0	1.0
Yard	m	0.914 4	1.093 6
Mass (weight)			
Grain	mg	64.799	0.015 4
Ounce (Avoirdupois)	g	28.350	0.035 3
Ounce (Troy)	g	31.103 5	0.032 15
Ton (short 2000 lb.)	kg	907.185	0.001 102
Ton (long 2240 lb.)	kg	1 016.047	0.000 984 2
Slug	kg	14.593 9	0.068 522
Pressure			
Bar	kPa	100.0	0.01
Inch of Water Column (20°C)	kPa	0.248 6	4.021 9
Inch of Mercury (20°C)	kPa	3.374 1	0.296 4
Kilogram-force per Centimeter Squared	kPa	98.067	0.010 2
Millimeters of Mercury (mm·Hg) (20°C)	kPa	0.132 84	7.528
Pounds Per Square Inch (P.S.I.)	kPa	6.894 8	0.145 0
Standard Atmosphere (760 torr)	kPa	101.325	0.009 869
Torr	kPa	0.133 32	7.500 6

Appendix A (continued)
Selected Conversion Factors to SI Units

	SI Symbol	Multiplier to Convert From Existing Unit to SI Unit	Multiplier to Convert From SI Unit to Existing Unit
Power			
Btu (International) Per Hour	W	0.293 1	3.412 2
Foot Pound-Force Per Second	W	1.355 8	0.737 6
Horsepower	kW	0.745 7	1.341
Meter Kilogram-Force Per Second	W	9.806 7	0.101 97
Tons of Refrigeration	kW	3.517	0.284 3
Torque			
Kilogram-Force Meter (kg·m)	N·m	9.806 7	0.101 97
Pound-Force Foot	N·m	1.355 8	0.737 6
Pound-Force Inch	N·m	0.113 0	8.849 5
Gram-Force Centimeter	mN·m	0.098 067	10.197
Temperature			
Fahrenheit	°C	$\frac{5}{9}$ (°F − 32)	($\frac{9}{5}$°C) + 32
Rankine	K	(°F + 459.67)$\frac{5}{9}$	(°C + 273.16)$\frac{9}{5}$
Velocity			
Foot Per Second	m/s	0.304 8	3.280 8
Mile Per Hour	m/s	0.447 04	2.236 9
	or	*or*	*or*
	km/h	1.609 34	0.621 4
Viscosity			
Centipoise	mPa·s	1.0	1.0
Centistoke	μm²/s	1.0	1.0
Volume (Capacity)			
Cubic Foot	*l* (dm³)	28.316 8	0.035 31
Cubic Inch	cm³	16.387 1	0.061 02
Cubic Yard	m³	0.764 6	1.308
Gallon (U.S.)	*l*	3.785	0.264 2
Ounce (U.S. Fluid)	ml	29.574	0.033 8
Pint (U.S. Fluid)	*l*	0.473 2	2.113
Quart (U.S. Fluid)	*l*	0.946 4	1.056 7
Volume Flow (Gas-Air)			
Standard Cubic Foot Per Minute	m³/s	0.000 471 9	2119.
	or	*or*	*or*
	l/s	0.471 9	2.119
	or	*or*	*or*
	ml/s	471.947	0.002 119
Standard Cubic Foot Per Hour	ml/s	7.865 8	0.127 133
	or	*or*	*or*
	μ*l*/s	7 866.	0.000 127
Volume Liquid Flow			
Gallons Per Hour (U.S.)	*l*/s	0.001 052	951.02
Gallons Per Minute (U.S.)	*l*/s	0.063 09	15.850

UNITS

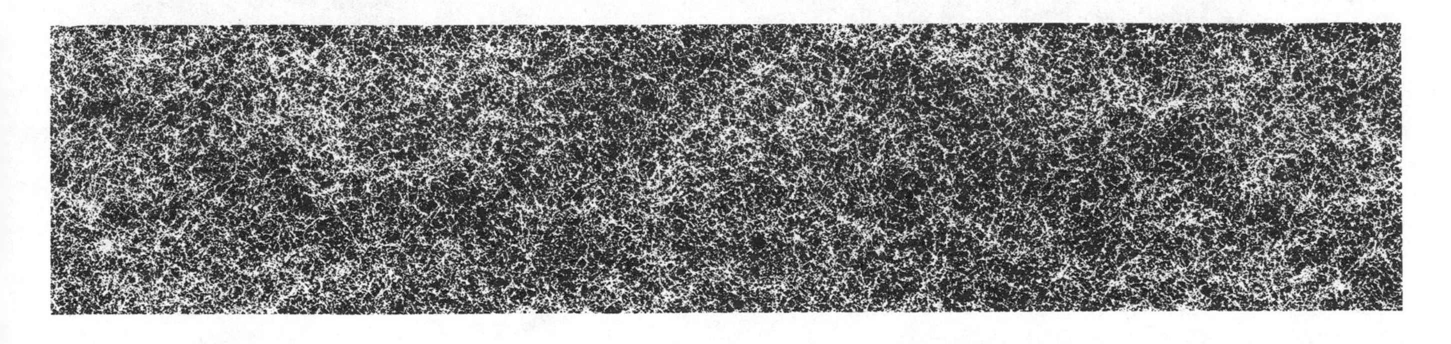

15 POSTSCRIPTS

This chapter collects comments, revisions, and commentary that cannot be incorporated into the body of the text until the next edition. New postscript sections are added as needed when the *Chemical Engineering Reference Manual* is reprinted. Subjects in this chapter are not necessarily represented by entries in the index. **It is suggested that you make a note in the appropriate text pages to refer to this chapter.**

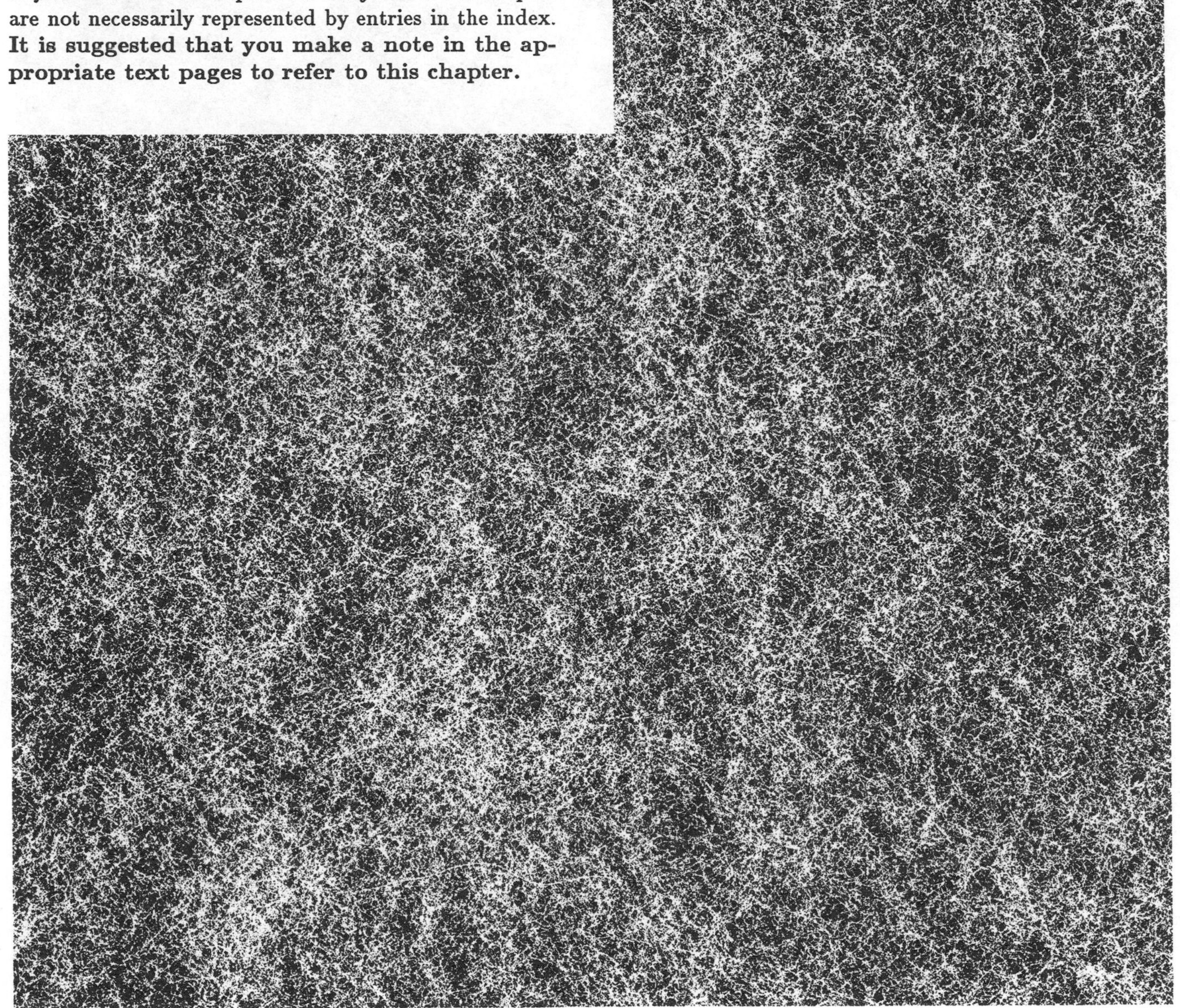

INDEX

J

K

L

M

N

INDEX OF FIGURES AND TABLES

P

R

S

T

U

V

Concentrate Your Studies . . .

Studying for the exam requires a significant amount of time and effort on your part. Make every minute count towards your success with these additional study materials in the Engineering Reference Manual Series:

Solutions Manual for the Chemical Engineering Reference Manual

64 pages, $8\frac{1}{2} \times 11$, paper 0-912045-26-4

Don't forget that there is a companion **Solutions Manual** that provides step-by-step solutions to the practice problems given at the end of each chapter in this reference manual. This important study aid will provide immediate feedback on your progress. Without the **Solutions Manual**, you may never know if your methods are correct.

Chemical Engineering Quick Reference Cards

56 pages, $8\frac{1}{2} \times 11$, paper 0-912045-36-1

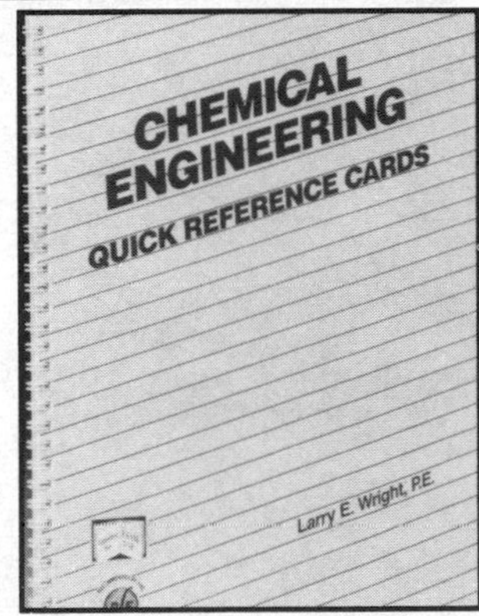

Because speed is important during the exam, you will welcome the advantage provided by the **Chemical Engineering Quick Reference Cards**. This handy resource is divided into eight basic subject areas and summarizes all pertinent formulas, tables, and other data, providing quick access to the key concepts needed during the exam.

Chemical Engineering Practice Exam Set

136 pages, $8\frac{1}{2} \times 11$, paper 0-932276-93-8

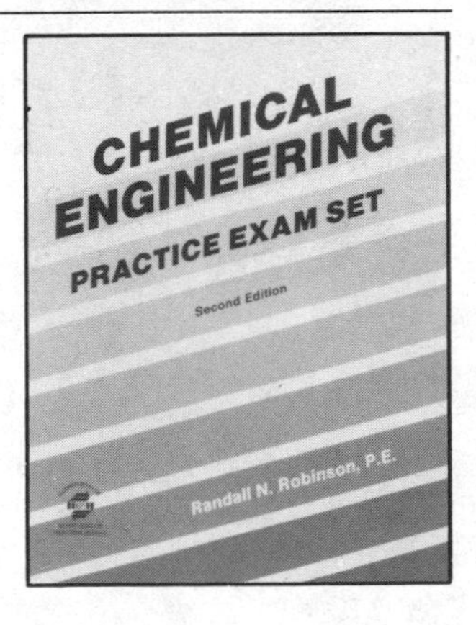

Reinforce your knowledge and increase your speed in solving the types of questions that may appear on the exam with the **Chemical Engineering Practice Exam Set**. This collection of seven different eight-hour exams will familiarize you with the format, range of topics, and the degree of difficulty involved in the actual exam. Complete step-by-step solutions to each problem are provided so that you can check your own work for completeness and accuracy.

Professional Publications, Inc.
1250 Fifth Avenue
Department 77
Belmont, CA 94002
(415) 593-9119

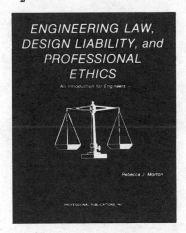

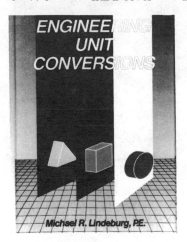